CW01455620

VOLUME SIXTY EIGHT

ADVANCES IN MICROBIAL PHYSIOLOGY

Advances in Bacterial Electron Transport Systems and Their Regulation

VOLUME SIXTY EIGHT

ADVANCES IN MICROBIAL PHYSIOLOGY

Advances in Bacterial Electron Transport Systems and Their Regulation

Edited by

ROBERT K. POOLE

West Riding Professor of Microbiology
Department of Molecular Biology and Biotechnology
The University of Sheffield
Firth Court, Western Bank
Sheffield, UK

AMSTERDAM • BOSTON • HEIDELBERG • LONDON
NEW YORK • OXFORD • PARIS • SAN DIEGO
SAN FRANCISCO • SINGAPORE • SYDNEY • TOKYO
Academic Press is an imprint of Elsevier

Academic Press is an imprint of Elsevier
125 London Wall, London, EC2Y 5AS, UK
The Boulevard, Langford Lane, Kidlington, Oxford OX5 1GB, UK
525 B Street, Suite 1800, San Diego, CA 92101-4495, USA
50 Hampshire Street, 5th Floor, Cambridge, MA 02139, USA

First edition 2016

Notices
Knowledge and best practice in this field are constantly changing. As new research and experience broaden our understanding, changes in research methods, professional practices, or medical treatment may become necessary.

Practitioners and researchers must always rely on their own experience and knowledge in evaluating and using any information, methods, compounds, or experiments described herein. In using such information or methods they should be mindful of their own safety and the safety of others, including parties for whom they have a professional responsibility.

To the fullest extent of the law, neither the Publisher nor the authors, contributors, or editors, assume any liability for any injury and/or damage to persons or property as a matter of products liability, negligence or otherwise, or from any use or operation of any methods, products, instructions, or ideas contained in the material herein.

ISBN: 978-0-12-804823-8
ISSN: 0065-2911

For information on all Academic Press publications visit our website at https://www.elsevier.com

Publisher: Zoe Kruze
Acquisition Editor: Mary Ann Zimmerman
Editorial Project Manager: Helene Kabes
Production Project Manager: Vignesh Tamil
Designer: Greg Harris

Typeset by SPi Global, India

CONTENTS

CONTRIBUTORS

V. Bautista
Universidad de Alicante, Alicante, Spain

E.J. Bedmar
Estación Experimental del Zaidín, CSIC, Granada, Spain

M.T. Bes
University of Zaragoza, Zaragoza, Spain

M.J. Bonete
Universidad de Alicante, Alicante, Spain

J.N. Butt
School of Biological Sciences and School of Chemistry, University of East Anglia, Norwich, United Kingdom

M. Camacho
Universidad de Alicante, Alicante, Spain

T.A. Clarke
School of Biological Sciences and School of Chemistry, University of East Anglia, Norwich, United Kingdom

N.M. de Almeida
Institute of Water and Wetland Research, Radboud University Nijmegen, Nijmegen, The Netherlands

M.J. Delgado
Estación Experimental del Zaidín, CSIC, Granada, Spain

M.J. Edwards
School of Biological Sciences and School of Chemistry, University of East Anglia, Norwich, United Kingdom

J. Esclapez
Universidad de Alicante, Alicante, Spain

D. Falke
Institute for Biology/Microbiology, Martin-Luther University Halle-Wittenberg, Halle (Saale), Germany

M.F. Fillat
University of Zaragoza, Zaragoza, Spain

M. Fischer
Institute for Biology/Microbiology, Martin-Luther University Halle-Wittenberg, Halle (Saale), Germany

A.J. Gates
School of Biological Sciences, University of East Anglia, Norwich Research Park; Centre for Molecular and Structural Biochemistry, University of East Anglia, Norwich, United Kingdom

L. Gomez-Perez
School of Biological Sciences and School of Chemistry, University of East Anglia, Norwich, United Kingdom

A. González
University of Zaragoza, Zaragoza, Spain

B. Kartal
Institute of Water and Wetland Research, Radboud University Nijmegen, Nijmegen, The Netherlands; Laboratory of Microbiology, Ghent University, Ghent, Belgium

J.T. Keltjens
Institute of Water and Wetland Research, Radboud University Nijmegen, Nijmegen, The Netherlands

R.M. Martínez-Espinosa
Universidad de Alicante, Alicante, Spain

C. Monzel
Institute for Microbiology and Wine Research, University of Mainz, Mainz, Germany

M.L. Peleato
University of Zaragoza, Zaragoza, Spain

C. Pire
Universidad de Alicante, Alicante, Spain

D.J. Richardson
School of Biological Sciences and School of Chemistry; University of East Anglia, Norwich Research Park; Centre for Molecular and Structural Biochemistry, Norwich, United Kingdom

G. Rowley
School of Biological Sciences, University of East Anglia, Norwich Research Park, Norwich, United Kingdom

F. Sargent
School of Life Sciences, University of Dundee, Dundee, Scotland, United Kingdom

R.G. Sawers
Institute for Biology/Microbiology, Martin-Luther University Halle-Wittenberg, Halle (Saale), Germany

E. Sevilla
University of Zaragoza, Zaragoza, Spain

J. Simon
Technische Universität Darmstadt, Darmstadt, Germany

J. Torregrosa-Crespo
Universidad de Alicante, Alicante, Spain

M.J. Torres
Estación Experimental del Zaidín, CSIC, Granada, Spain

G. Unden
Institute for Microbiology and Wine Research, University of Mainz, Mainz, Germany

H.J.C.T. Wessels
Nijmegen Center for Mitochondrial Disorders, Radboud Proteomics Centre, Translational Metabolic Laboratory, Radboud University Medical Center, Nijmegen, The Netherlands

G.F. White
School of Biological Sciences and School of Chemistry, University of East Anglia, Norwich, United Kingdom

S. Wörner
Institute for Microbiology and Wine Research, University of Mainz, Mainz, Germany

PREFACE

A dominant theme in microbiology for many decades has been the diversity of respiratory and energy-generating pathways. These stand in stark contrast to the 'simpler', more focused pathways that exist in higher organisms, where the general rule is that respiration is aerobic, utilising oxygen as sole terminal electron acceptor; moreover, the oxygen-reducing chemistry is generally contained in one terminal oxidase.

The area of diverse energy-yielding pathways in microbes was covered in *Advances in Microbial Physiology* Volume 61, where some of the work presented at an international symposium in Sweden was published. In the intervening years, many other reviews have appeared in this series on related topics; now Volume 68 publishes eight reviews based on talks presented at the meeting on Bacterial Electron Transfer and Its Regulation held in Vimeiro, Portugal, March 2015, and organised by Jeff Cole and Ligia Saraiva. Topics covered are aspects of anaerobic respiration, roles for iron, proteomic approaches to respiratory complexes, transmembrane signalling and interactions with extracellular redox species.

I am grateful to the many authors who contributed authoritative and fascinating accounts of their specialist research areas so promptly and efficiently and to my Elsevier colleagues for their tireless efforts.

Robert K. Poole
March 2016

CHAPTER ONE

Oxygen and Nitrate Respiration in *Streptomyces coelicolor* A3(2)

R.G. Sawers[1], D. Falke, M. Fischer
Institute for Biology/Microbiology, Martin-Luther University Halle-Wittenberg, Halle (Saale), Germany
[1]Corresponding author: e-mail address: gary.sawers@mikrobiologie.uni-halle.de

Contents

Abstract

Streptomyces species belong to the phylum Actinobacteria and can only grow with oxygen as a terminal electron acceptor. Like other members of this phylum, such as corynebacteria and mycobacteria, the aerobic respiratory chain lacks a soluble cytochrome *c*. It is therefore implicit that direct electron transfer between the cytochrome bc_1 and the cytochrome aa_3 oxidase complexes occurs. The complex developmental cycle of streptomycetes manifests itself in the production of spores, which germinate in the presence of oxygen into a substrate mycelium that greatly facilitates acquisition of nutrients necessary to support their saprophytic lifestyle in soils. Due to the highly

Advances in Microbial Physiology, Volume 68
ISSN 0065-2911
http://dx.doi.org/10.1016/bs.ampbs.2016.02.004

variable oxygen levels in soils, streptomycetes have developed means of surviving long periods of hypoxia or even anaerobiosis but they fail to grow under these conditions. Little to nothing is understood about how they maintain viability under conditions of oxygen limitation. It is assumed that they can utilise a number of different electron acceptors to help them maintain a membrane potential, one of which is nitrate. The model streptomycete remains *Streptomyces coelicolor* A3(2), and it synthesises three nonredundant respiratory nitrate reductases (Nar). These Nar enzymes are synthesised during different phases of the developmental cycle and they are functional only under oxygen-limiting (<5% oxygen in air) conditions. Nevertheless, the regulation of their synthesis does not appear to be responsive to nitrate and in the case of Nar1, it appears to be developmentally regulated. This review highlights some of the novel aspects of our current, but somewhat limited, knowledge of respiration in these fascinating bacteria.

1. INTRODUCTION

Members of the genus *Streptomyces* belong to the order *Streptomycetales* in the phylum Actinobacteria (Kämpfer, 2006, 2012; Waksman & Henrici, 1943). They are high G-C gram-positive bacteria that are characterised by requiring oxygen for growth (Chater, Biró, Lee, Palmer, & Schrempf, 2010; Hodgson, 2000; Hopwood, 2007). *Streptomyces* species are widespread in nature but are generally found in soil, which is the ideal environment for their saprophytic lifestyle (Kämpfer, 2012). They are classical chemoorganotrophs that use a wide range of organic compounds as sole carbon and energy sources. In particular, they produce numerous exoenzymes (Peczynska-Czoch & Mordarski, 1988) allowing them to hydrolyse polymers such as lignocellulose, hemicelluloses, pectin, keratin and chitin (Goodfellow & Williams, 1983). Being aerobes, they also synthesise catalase to protect themselves from the deleterious effects of reactive oxygen species, generated during their strongly oxidative metabolism.

Streptomycetes form extensively branched substrate (or vegetative) mycelium, as well as an aerial mycelium. Growth occurs at the tips of the nonfragmenting vegetative hyphae, which are between 0.5 and 2 μm in diameter. This unusual morphology for a bacterium means it is ideally suited to its saprophytic lifestyle, greatly increasing its surface area and allowing it to compete very effectively for growth substrates by enveloping soil particles containing nutrients (Korn-Wendisch & Kutzner, 1992). At maturity *Streptomyces* spp. produce aerial hyphae that develop into chains of spores. The hydrophobic spores function as a means of dispersal (Ruddick & Williams, 1972), and although they do not show the robust characteristics typical of firmicute endospores, they can nevertheless, if necessary, survive for decades before germinating (Elliot & Flärdh, 2012; Kämpfer, 2006). *Streptomyces* species account for 10^4–10^7 colony-forming units in every gram of soil (Korn-Wendisch & Kutzner, 1992). Consequently, they make a major contribution to the ecology of aerated soils.

A further important influence of streptomycetes on soil ecology is their production of secondary metabolites. Amongst these compounds are antibiotics, fungicides as well as compounds with cytostatic activity. Synthesis of secondary metabolites requires considerable amounts of ATP and metabolic precursors. Thus, the onset of secondary metabolism has major implications for the energy metabolism of the bacterium and has perhaps selected for maintenance of an aerobic respiratory metabolism to fulfil these enegetic demands.

The model species of the genus *Streptomyces* is *S. coelicolor* A3(2) and this review will focus on the respiration of this bacterium. The publication of the complete genome sequence of *S. coelicolor* has provided much insight into the respiratory capabilities of the bacterium. The linear genome is 8.7 Mbp in length and encompasses 7825 genes (Bentley et al., 2002). This large gene set reflects the diverse and complex lifestyle of the bacterium. An analysis of the genome has revealed some surprising features, both with respect to aerobic respiratory capabilities but also with regard to the discovery of three operons encoding respiratory nitrate reductases (Nar) (van Keulen, Alderson, White, & Sawers, 2005). Despite extensive research efforts, it has never been possible to demonstrate growth of the bacterium by nitrate respiration (van Keulen, Alderson, White, & Sawers, 2007). Nevertheless, as will be described below, these respiratory Nar enzymes are not redundant and presumably help the bacterium to maintain a proton gradient in the absence of growth and when oxygen is limited (Fischer, Alderson, van Keulen, White, & Sawers, 2010). Nitrate is an excellent electron acceptor and is present in aerated soils due to the action of bacterial nitrification, whereby ammonium is oxidised via nitrite to nitrate by ammonia- and nitrite-oxidising bacteria, respectively (Prosser, 2007).

Moreover, coping with hypoxia also appears to be important during the life-cycle of *Streptomyces* and the genome has revealed that the bacterium encodes a high-affinity respiratory cytochrome *bd* respiratory quinol: O_2 oxidoreductase (Bentley et al., 2002). In this review, we will focus on respiration in *S. coelicolor* and how it is likely employed by the bacterium to maintain its complex developmental cycle, as well as how it helps drive secondary metabolism. Surprisingly, however, there is comparatively little known about the aerobic respiration of *S. coelicolor*, or for that matter, other species of the genus. Consequently, we will, where appropriate, draw on parallels between respiratory metabolism of *S. coelicolor* and that of its actinobacterial relatives *Mycobacterium* and *Corynebacterium* species.

2. GENERAL ASPECTS OF RESPIRATION

2.1 Electron Transport and Proton-Motive Force Generation

A proton-motive force (*pmf*) is required by all microorganisms to grow and remain viable under replicating, as well as under nonreplicating, conditions, including in spores. During respiration, protein complexes are arranged asymmetrically across the membrane such that when electrons flow through these complexes they cause net consumption of protons in the cytoplasmic compartment and net release or accumulation of protons on the outside of the cytoplasmic membrane. Proton movement across the cytoplasmic membrane can be achieved by directly translocating protons across the membrane, whereby the protein complexes directly pump protons through the respective complex, often involving conformational pumping mechanisms (Nicholls & Ferguson, 2013), induced by electron flow through these complexes. The electron transport chains of mitochondria or obligately aerobic, oxidase-positive bacteria function in this manner (Fig. 1). These complexes include NADH:quinone oxidoreductase (also known as NADH dehydrogenase I; complex I), quinol: cytochrome *c* oxidoreductase (cytochrome bc_1 complex, or complex III) and cytochrome *c* oxidase (cytochrome aa_3 oxidase, or complex IV), which has a covalently bound cytochrome *c*. Current evidence suggests that formally 4 H^+ are pumped per electron pair through complex I ($H^+/2e^- = 2$), a total of $4H^+$ are translocated per $2e^-$ by the combined action of the Q-cycle ($2H^+$) and complex III ($2H^+$), and $2H^+$ are pumped by complex IV (Nicholls & Ferguson, 2013). Electron transfer between complex I and complex III is mediated by quinones, which simultaneously receive two electrons from complex I and two protons at the cytoplasmic side of the membrane. Electron delivery to, and flow through, complex III involves electron bifurcation and the protons are released on the opposite side of the cytoplasmic membrane (Fig. 1). The process forms the basis of the Q-cycle. Electron transfer between complex III and complex IV is performed by soluble cytochrome *c*, which shuttles between the complexes and is located on the external leaflet of the cytoplasmic membrane.

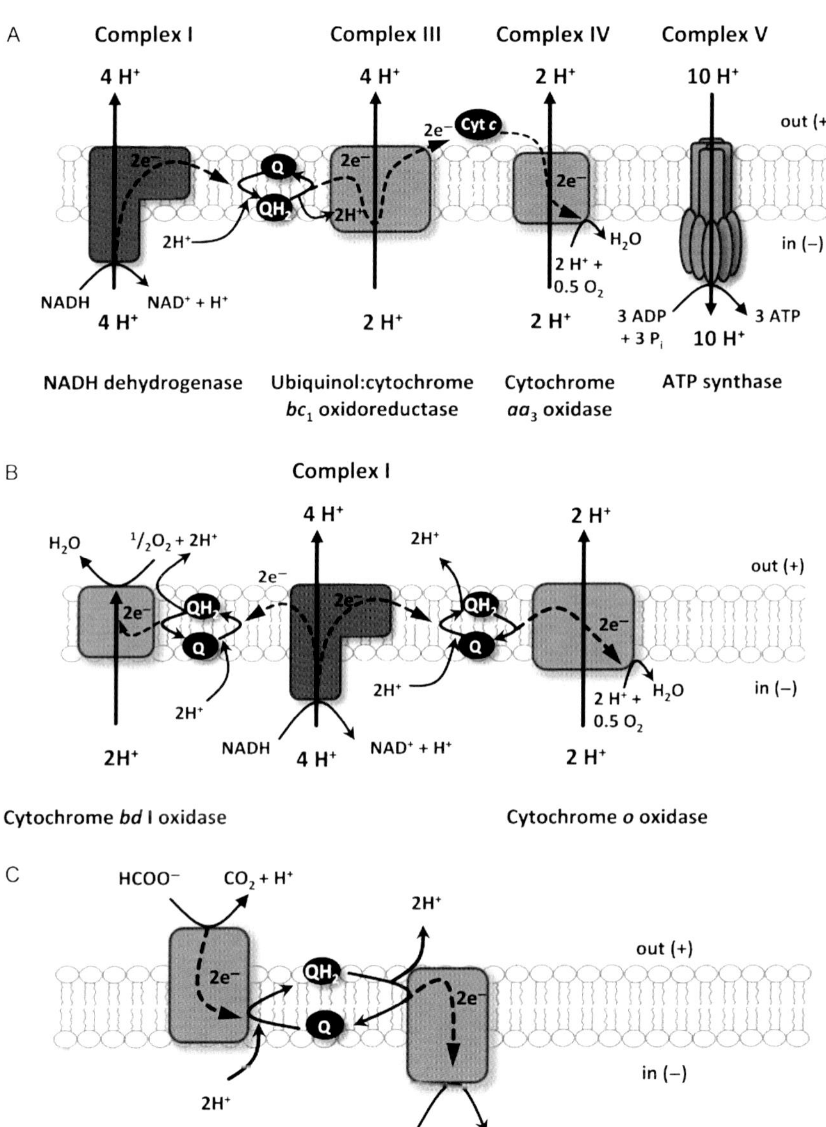
A
Complex I
Complex III
Complex IV
Complex V
4 H+
4 H+
2 H+
10 H+
out (+)
2e−
Cyt c
2e−
Q
QH2
2e−
2H+
2e−
2H+
H2O
2 H+ +
0.5 O2
in (−)
NADH
NAD+ + H+
4 H+
2 H+
2 H+
3 ADP
+ 3 Pi
10 H+
3 ATP
NADH dehydrogenase
Ubiquinol:cytochrome bc1 oxidoreductase
Cytochrome aa3 oxidase
ATP synthase
B
Complex I
4 H+
2 H+
H2O
1/2O2 + 2H+
2H+
out (+)
2e−
2e−
QH2
Q
2e−
QH2
Q
2e−
2H+
2H+
H2O
2 H+ +
0.5 O2
in (−)
2H+
NADH
4 H+
NAD+ + H+
2 H+
Cytochrome bd I oxidase
Cytochrome o oxidase
C
HCOO−
CO2 + H+
2H+
out (+)
2e−
QH2
Q
2e−
in (−)
2H+
2H+ + NO3−
NO2− + H2O
Formate dehydrogenase
Nitrate reductase

Fig. 1—Cont'd

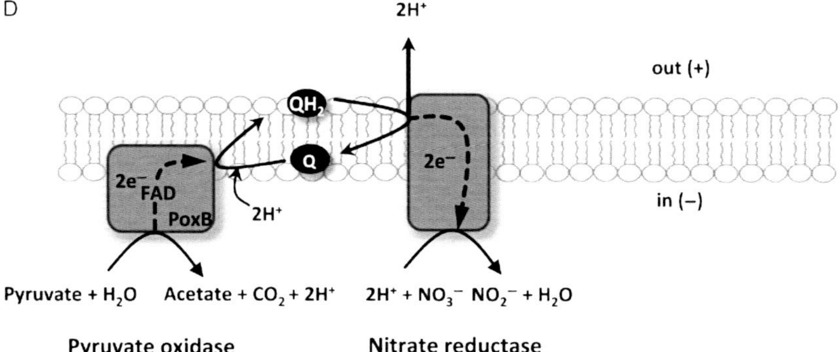

Fig. 1 Schematic representation of different modes of proton-motive force generation. (A) The organisation of the respiratory complexes in mitochondria or oxidase-positive bacteria is shown. The electron flow is indicated, as is the number of protons translocated from the negative (cytoplasm in bacteria) to the positive (periplasm in Gram-negative bacteria) of the cytoplasmic membrane. (B) The electron flow from complex I to either cytochrome *bd* oxidase (*left*) or cytochrome *o* oxidase (*right*) is shown, for example, as occurs in *E. coli* depending on the external oxygen concentration (Poole & Cook, 2000). (C) An example of a redox loop is shown in which electrons flow from an outwardly facing formate dehydrogenase to an inwardly facing nitrate reductase via quinones. (D) Putative electron flow and proton translocation via an inwardly oriented membrane-associated flavoprotein dehydrogenase (eg, pyruvate oxidase) and an inwardly facing terminal nitrate reductase. *Modified from Bott M. and Niebisch A., The respiratory chain of* Corynebacterium glutamicum, Journal of Biotechnology ***104****, 2003, 129–153 with permission.*

The *pmf* can be used to drive flagellar rotation or the transport of subtrates into or out of the cell. Crucially, the *pmf* also drives the ATP synthase (complex V), a proton turbine converting the proton gradient into chemical energy in the form of ATP (Dimroth & Cook, 2004). The ATP synthase is thus a reversible ATP-driven proton pump. It is generally accepted that one ATP is synthesised by ATP synthase for every 3–4 H^+ that pass through the complex and for oxidase-positive bacteria roughly three ATP (equivalent to roughly 10 translocated H^+; ATP/O or ATP/2e$^-$ = 3) are synthesised for every NADH (2e$^-$) oxidised by the combined actions of the respiratory complexes.

2.2 Oxygen Respiration Under Hypoxic Conditions

Many bacteria can respire with oxygen, but they do not use complex III, IV or soluble cytochrome *c* in this type of respiration. Instead, this form of respiration, which is typically found in oxidase-negative bacteria, use different terminal oxidases that accept electrons directly from reduced quinones (Fig. 1B). For example, *Escherichia coli* synthesises the heme-copper cytochrome *o* oxidase (proton-translocating H^+/e^- ratio of 2) and a cytochrome *bd* oxidase (nonproton translocating), which shows no sequence homology to the heme-copper oxidase family that includes cytochrome aa_3 oxidase (Borisov, Gennis, Hemp, & Verkhovsky, 2011). These oxidases have higher affinities for oxygen than complex IV and confer upon the bacterium the ability to grow at variable oxygen concentrations. In particular, cytochrome *bd* oxidase synthesis is induced under oxygen-limiting conditions (Aung, Berney, & Cook, 2014) and allows aerobic respiration to proceed under microaerobic or hypoxic conditions where oxygen is in the low micromolar range. The compromise for the organism of producing an enzyme with increased affinity for oxygen is a reduction in the H^+/e^- ratio of the complex, which means they cannot translocate as many protons as the heme-copper oxidase. This is particularly relevant for cytochrome *bd* oxidase, which does not translocate protons but nevertheless achieves a H^+/e^- ratio of 1 (Bekker, de Vries, Ter Beek, Hellingwerf, & de Mattos, 2009; Poole, 1994; Fig. 1B).

2.3 Anaerobic Respiration

A *pmf* can also be generated indirectly, by a scalar or redox loop mechanism, orignially proposed by Mitchell (Mitchell, 1979; Mitchell & Moyle, 1967), whereby the net 'consumption' of protons by reductases localised on the cytoplasmic side of the membrane, which reduce oxidised electron acceptors, is coupled to the net 'release' of protons on the external leaflet of the membrane by the action of membrane-associated dehydrogenases, which oxidise electron donors, concomitantly yielding protons (Fig. 1C). These various circuits are frequently found in respiratory processes employing alternative electron acceptors in various anaerobic bacteria (see also later).

Under oxygen-limiting or anaerobic conditions, many respiratory microbes can use alternative electron acceptors, for example nitrate or fumarate, and in these cases, although the principle of *pmf* generation is essentially the same, the electron-transfering complexes do not pump protons but rather they employ a redox loop mechanism (Jormakka, Byrne, & Iwata, 2003; Jormakka, Törnroth, Byrne, & Iwata, 2002; Richardson & Sawers, 2002; Fig. 1C). Respiratory chains comprising both proton pumps and redox loops occur, for example, when denitrifying bacteria couple NADH oxidation via complex I with nitrate reduction, using nitrate reductase

(Nicholls & Ferguson, 2013). These 'simple' respiratory chains confer upon bacteria great flexibility in energy conservation mechanisms.

2.4 Electron Donation to the Respiratory Chain

Respiratory flexibility can also be achieved through the electron donor and substrate oxidation is frequently catalysed by flavin-based dehydrogenases that associate with the inner leaflet of the cytoplasmic membrane; these enzymes bypass complex I and they are not proton translocating, for example, pyruvate oxidase, PoxB, which could pass electrons via the quinone pool to either cytochrome oxidase or nitrate reductase (Fig. 1D). The classic example of this is complex II, or succinate dehydrogenase, of the respiratory chain, which channels electrons to complex III via quinol (Hartman et al., 2014). There are, however, other reduced primary metabolic intermediates that can serve as effective electron donors for oxidative phosphorylation and which use membrane-associated flavin-based dehydrogenases to deliver electrons to the quinone pool and thence to complexes III and IV. In the actinobacterium *Corynebacterium glutamicum*, L-malate, pyruvate and glyceraldehyde 3-phosphate all can deliver electrons to the respiratory chain using flavoprotein dehydrogenases (Bott & Niebisch, 2003).

3. THE AEROBIC RESPIRATORY CHAIN OF *S. COELICOLOR*

Early studies focussing on the requirements of *S. griseus* for streptomycin production revealed that aeration was important for both antibiotic production and growth (Gottlieb & Anderson, 1948; Rake & Donovick, 1946; Waksman, Schatz, & Reilly, 1946), suggesting oxygen was required for both processes. These findings were also in accord with an earlier discovery of cytochrome in streptomycetes (Sato, 1940). Heim et al. then provided spectroscopic verification for cytochrome *b* (Heim, Silver, & Birk, 1957), and shortly thereafter, Inoue (1958) and Niederpruem and Hackett (1961) identified spectral features of *a*-, *b*- and *c*-type cytochromes and it was also demonstrated that cyanide inhibited O_2 respiration. This latter observation is commensurate with *Streptomyces* species lacking cyanide-resistant CIO oxidase. That NADH could act as electron donor for O_2 respiration was demonstrated for *S. antibioticus* (Rehácek, Ramankutty, & Kozová, 1968) and these authors confirmed the cyanide sensitivity of oxygen reduction.

Like both *Corynebacterium* and *Mycobacterium* species (Collins, Pirouz, Goodfellow, & Minnikin, 1977), streptomycetes lack ubiquinone and are solely reliant on menaquinones (mainly MKH_6-9 and MKH_8-9) as lipid-soluble electron and proton carriers (Collins & Jones, 1981). Moreover, the genome sequence of *S. coelicolor* (Bentley et al., 2002) revealed that no soluble cytochrome *c* is present in the respiratory chain and this is a characteristic of other members of the actinobacteria (Cook et al., 2009; Niebisch & Bott, 2001; Sone et al., 2003). Unusually, *C. glutamicum*, *M. tuberculosis*, *Rhodococcus rhodochrous* and, based on the genome sequence (Fig. 2; Bentley et al., 2002), *S. coelicolor* all have a cytochrome *c* component, encoded by *qcrC*, of ubiquinol–cytochrome *c* oxidoreductase (complex III) that is a diheme cytochrome *c* (Niebisch & Bott, 2003; Sone et al., 2001), in contrast to the more typical monoheme cytochrome *c* of the mitochondrial respiratory chain and that of other aerobic bacteria (Xia et al., 1997). The hemes are bound by classical CXXCH heme-binding motifs. Moreover, QcrC is anchored in the cytoplasmic membrane (Fig. 3). These findings invoke a mechanism of electron transfer between the cytochrome bc_1 and cytochrome aa_3 oxidase complexes involving direct protein–protein

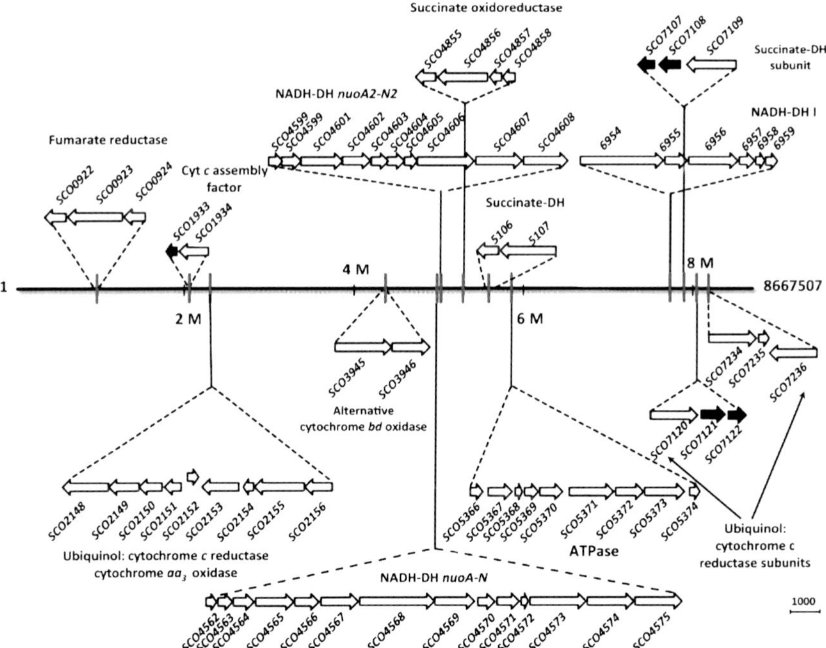

Fig. 2 Genes and operons on the genome of *Streptomyces coelicolor* A3(2) whose products hypothetically encode components of the aerobic respiratory chain. The linear 8.7 Mbp *S. coelicolor* genome (Bentley et al., 2002) is represented and the locations of the listed genes, including their SCO numbers (http://strepdb.streptomyces.org.uk), with their corresponding encoded gene products are indicated.

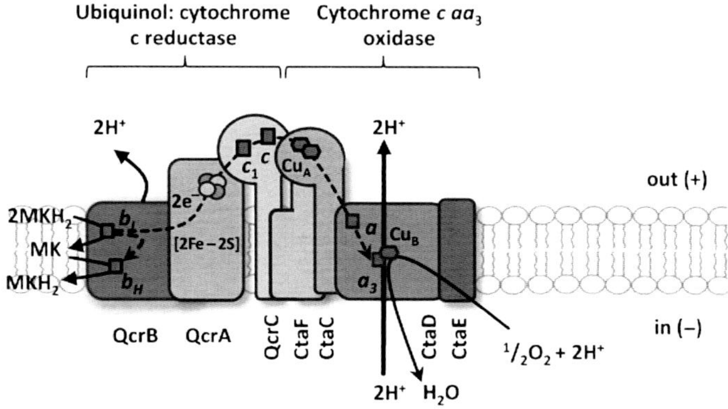

Fig. 3 Schematic representation of a putative respiratory 'super-complex' between the cytochrome bc_1 complex and cytochrome aa_3 oxidase in *S. coelicolor*. The respective subunits are indicated in the membrane and the electron flow is indicated by *dashed arrows*. The cofactors present in the complexes include hemes (*squares*), copper ions (*black balls*) and iron–sulphur clusters (groups of *light* and *dark* grey balls). The schematic is based on findings originally reported for *C. glutamicum* (Niebisch & Bott, 2001) and is published with permission.

interaction and this has indeed been demonstrated for the *C. glutamicum* respiratory chain (Niebisch & Bott, 2003).

The cytochrome bc_1 complex is encoded by the *qcrCAB* genes and these three genes appear to be cotranscribed with the *ctaE* gene encoding subunit III of cytochrome aa_3 oxidase (Fig. 2). This gene organisation is also conserved in several actinobacterial species (Boshoff & Barry, 2005; Matsoso et al., 2005; Niebisch & Bott, 2001; Sone et al., 2003). Another unusual feature of the cytochrome bc_1 complex in the actinobacteria is that the cytochrome *b* component, QcrB, has nine transmembrane helices rather than the typical eight, with a further approximately 100 amino acid C-terminal extension predicted to be exposed on the outer leaflet of the cytoplasmic membrane (Bott & Niebisch, 2003; Hopkins, Buchanan, & Palmer, 2014; Niebisch & Bott, 2001). Otherwise, QcrB has two noncovalently bound hemes *b* (*b*-566, *b*-560), each coordinated by bis-histidine residues.

A further distinction of the cytochrome bc_1 complex in the actinobacteria is that the Rieske iron–sulphur protein (QcrA) has three transmembrane helices (Bott & Niebisch, 2003; Hopkins et al., 2014) instead of the typical single-membrane-spanning helix (Xia et al., 1997) making it considerably more hydrophobic. The soluble iron–sulphur domain is located on the outside of the cytoplasmic membrane and requires (*t*win *a*rginine *t*ranslocation) Tat-dependent transport; however, the transmembrane helices do not show

Tat dependence indicating that this membrane protein component of complex III exhibits dual targeting (Hopkins et al., 2014).

3.1 The Terminal Oxidases

S. coelicolor has a branched respiratory chain whereby the electrons from menaquinol can be transferred either to the heme-copper *aa*$_3$ oxidase (complex IV in Fig. 1A) via complex III or to the high-affinity cytochrome *bd* oxidase (Bentley et al., 2002; Borisov et al., 2011). The fact that O_2 reduction in *S. coelicolor* and other *Streptomyces* species is sensitive towards cyanide rules out that they have alternative oxidases (Borisov et al., 2011; Rogov, Sukhanova, Uralskaya, Aliverdieva, & Zvyagilskaya, 2014), which is borne out by analysis of the genome sequences. As mentioned above, due to the lack of a soluble cytochrome *c*, electron transfer between the cytochrome *bc*$_1$ oxidoreductase and cytochrome *aa*$_3$ oxidase requires formation of a supercomplex between complexes III and IV. Notably, in actinobacteria, the *ctaE* gene encoding subunit III of the cytochrome *aa*$_3$ oxidase of *S. coelicolor* is located adjacent to the *qcrCAB* genes (Fig. 2). The *ctaCDF* genes, encoding subunits II, I and IV, respectively, of complex IV are in the vicinity and separated from *ctaEqcrCAB* by genes SCO2153, encoding a hypothetical secreted protein and SCO2152, encoding a two-component system response regulator of unknown function (Fig. 2). The CtaF protein (subunit IV) is unique to the actinobacteria and might be important for interaction with complex III. A second copy of *ctaF* (SCO7235) is located on the genome of *S. coelicolor* in a putative transcriptional unit together with a further copy of *ctaD* (SCO7234) (Fig. 2). While only *S. albulus* and *S. hygroscopicus* have an additional copy of *ctaF* in their respective genome, it is more common for *Streptomyces* species to harbour more than one copy of the *ctaD* encoding subunit I. Indeed, many species have between 4 and 7 copies of the *ctaD* gene (NCBI-Microbial-Genomes-database); however, what the functions of their gene products might be and whether they can interact with the other components of complex IV remains to be established.

The additional copies of the *ctaD* and *ctaF* genes on the genome of *S. coelicolor* lie adjacent to a further copy of *qcrB* (*qcrB3*; SCO7236), which is convergently transcribed with respect to them (Fig. 2). Transcriptional studies reveal that all three of these genes are mainly transcribed in spores (D. Falke, M. Fischer, & R. G. Sawers, unpublished). A third copy of the *qcrB* gene (*qcrB2*; SCO7120) is located at a distinct site on the genome and also appears to be expressed exclusively in spores. Comparison of the deduced amino acid sequences of all three cytochromes *b* indicates that they are highly hydrophobic proteins with nine transmembrane helices and they share roughly 50% identity at the amino acid sequence level. While *C. glutamicum*, mycobacteria and *Rhodococcus* species each have a single copy of *qcrB*, several, but importantly not all, *Streptomyces* species have more than one copy. Of those species with a completely sequenced genome, nine have two or more *qcrB* genes. *S. lividans* strain TK24 has the same complement of three *qcrB* genes as *S. coelicolor*, while *S. bingchenggensis* has four copies of the gene, two of whose gene products share more that 60% amino acid sequence identity with QcrB (SCO2148), while the other two are more similar to QcrB2 (SCO7120). Those species with two *qcrB* genes include *S. hygroscopicus*, *S. lydicus*, *S. violaceusniger*, *S. collinus*, *S.* sp. CNQ-509 and *S.* sp. SirexAA-E.

A single copy of each of the *cydAB* genes, encoding cytochrome *bd* oxidase, is located next to a gene (SCO3947) that encodes a product exhibiting amino acid sequence similarity to both CydC and CydD, suggesting a fusion of both proteins (Fig. 2); CydCD forms an ABC transporter (ATP-binding cassette) for cysteine and glutathione and they play a key role in assembly of the cytochrome *bd* oxidase, presumably by tuning the redox status of the outer leaflet of the cytoplasmic membrane (Cruz-Ramos, Cook, Wu, Cleeter, & Poole, 2004; Holyoake, Poole, & Shepherd, 2015). Early transcriptional studies in *S. coelicolor* (Brekasis & Paget, 2003) indicated that expression of *cydAB* is induced by hypoxia and is dependent on the Rex transcriptional regulator. Rex proteins sense the NAD^+/NADH ratio, which reflects the redox status of the cell.

C. glutamicum and *M. smegmatis* mutants lacking either complex III or complex IV are severely compromised for aerobic growth (Matsoso et al., 2005; Niebisch & Bott, 2001; Niebisch & Bott, 2003), while in *M. tuberculosis* this pathway of oxygen reduction appears to be essential (Cook et al., 2009). Analogous *S. coelicolor* mutants lacking complexes III and IV also grow considerably more slowly than the wild type (Falke et al., unpublished). While *S. coelicolor* mutants devoid of both cytochrome aa_3 oxidase and cytochrome *bd* oxidase cannot be stably generated, similar mutants of *C. glutamicum* ferment glucose to lactate (Koch-Koerfges, Pfelzer, Platzen, Oldiges, & Bott, 2013). These findings confirm that *S. coelicolor* has an absolute requirement for oxygen to grow, while *C. glutamicum* is not an 'obligate' aerobe. As observed for *S. coelicolor*, synthesis of the CydAB oxidase in both *C. glutamicum* and *M. smegmatis* is induced by oxygen limitation (Kabus, Niebisch, & Bott, 2007; Kana et al., 2001). In *C. glutamicum* CydAB is important for stationary phase growth and mutants lacking *cydAB* yield 40% less biomass when cultured in minimal medium (Kabus et al., 2007). Moreover, constitutive overproduction of CydAB also negatively impacts biomass yields, suggesting that optimal balanced growth requires fine-tuning of terminal oxidase levels and activities. Future studies will be required to determine what role CydAB plays in streptomycete biology.

3.2 NADH Dehydrogenase 1 and 2

Two classes of NADH dehydrogenase exist in bacteria: the proton- or sodium-pumping multisubunit NADH-1 enzyme complex, usually comprising up to 14 Nuo (NuoA-N) subunits (Schneider et al., 2008); or NADH-2, which is a nonproton-translocating, single subunit enzyme encoded by the *ndh* gene. Actinobacterial species that have been characterised so far appear to use preferentially the NADH-2 type enzyme (Weinstein et al., 2005). Indeed, *C. glutamicum* lacks the *nuo* operon and has a single *ndh* gene (Bott & Niebisch, 2003; Molenaar, van der Rest, Drysch, & Yücel, 2000), as does *M. leprae* (Cole et al., 2001), while the *nuo* operon is dispensible in both *M. tuberculosis* and *M. smegmatis* (Miesel, Weisbrod, Marcinkeviciene, Bittman, & Jacobs, 1998; Sassetti, Boyd, & Rubin, 2003; Vilcheze et al., 2005). The *S. coelicolor* genome includes a complete putative 14-gene *nuoA-N* operon (SCO4562–4575) as well as a second, incomplete, *nuo* operon (SCO4599–4608; Fig. 2), but which lacks *nuoDEFG*. Moreover, a further copy of the *nuoL* and *nuoN* genes (SCO6954 and SCO6956, respectively) is colocalised on the genome in an apparent six-gene operon (Fig. 2). The other four genes in this putative operon are predicted to encode a cation/proton antiporter similar to proteins involved in drug resistance or pH regulation (Riedel, Cohn, Stabler, Wren, & Brøndsted, 2012). To date, limited information is available on the synthesis of the *nuo* gene products in *Streptomyces* species; however, some polypeptides have been detected by proteome analysis (Thomas et al., 2012).

S. coelicolor has several copies of putative *ndh* genes (Fig. 2). These include SCO6496, which exhibits 30% amino acid identity with Ndh from *M. tuberculosis* (Weinstein et al., 2005); SCO4119, SCO3092 and SCO7101 with respective amino acid identities of 38%, 30% and 32% to Ndh from *E. coli*. To date, no expression studies have been performed on any of these genes.

3.3 Flavin-Based Electron-Donating Complexes

Succinate dehydrogenase (Sdh) (complex II) is a nonproton-translocating complex that donates electrons to the quinone pool (Fig. 4). In analogy to *M. tuberculosis* (Hartman et al., 2014), and based on genome analyses, *S. coelicolor* has two putative operons encoding Sdh and a further operon encoding a putative fumarate reductase (FR) (SCO0922–0924). FR in *M. tuberculosis* has been shown to be upregulated during hypoxia and the enzyme is thought to help maintain a membrane potential under oxygen-limiting growth (Watanabe et al., 2011) but the putative FR in *S. coelicolor* has yet to be characterised.

Notably, Sdh-1 of *M. tuberculosis* has been shown to be essential for long-term survival of the bacterium in the stationary phase and its absence leads to loss of respiratory control with concomitant high O_2 consumption (Hartman et al., 2014). *M. smegmatis* also synthesises two succinate: menaquinone oxidoreductases, and while Sdh-2 proved to be essential in the bacterium, mutants devoid of Sdh-1 lacked any obvious growth phenotype (Pecsi et al., 2014). Sdh-1 from *M. tuberculosis* is highly similar to the combined SCO5106–5109 gene products of *S. coelicolor* (Bentley et al., 2002). Also in analogy to *M. tuberculosis*, in *S. coelicolor* the C and D membrane anchor components appear to be fused in a single gene product. The flavin-containing catalytic subunit SdhA and the iron–sulphur cluster-containing SdhB small subunit each share around 70% amino acid identity with their *M. tuberculosis* counterparts.

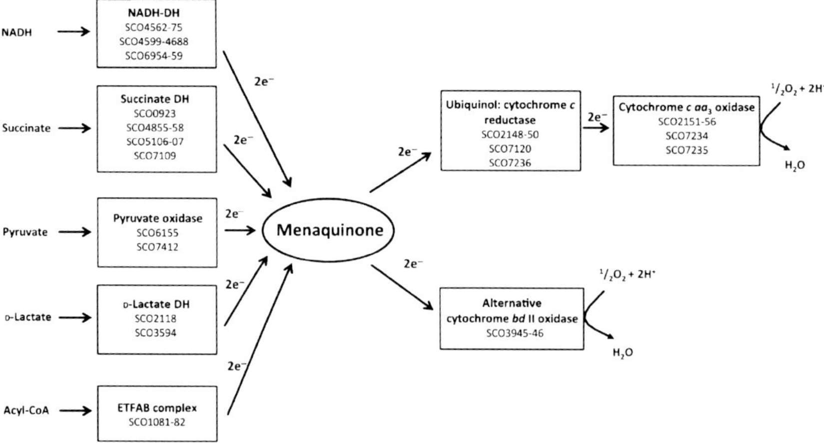

Fig. 4 The possible electron donors and cytochrome oxidases encoded in the genome of *S. coelicolor* A3(2).

Sdh-2 in *M. tuberculosis* is similar to the SCO4855–SCO4858 gene products with the catalytic subunit exhibiting 48% amino acid identity with SdhA of *E. coli*. In the predicted Sdh-2 enzyme, the membrane anchor comprises two distinct proteins. A further putative operon exists in *S. coelicolor* (SCO0922–SCO0924) that encodes proteins with similarity to FR from *M. tuberculosis* (Hartman et al., 2014). The polypeptides of both Sdh-2 (SCO4855–4858) and FR were identified in a proteomic study of a *S. coelicolor phoP* mutant (Thomas et al., 2012). Finally, an apparent monocistronic gene (SCO7109) encodes a product with 41% amino acid identity to the catalytic subunit of Sdh from *Methanobacterium thermoautotrophicum*. However, in none of these cases is any further information available regarding functional studies on these gene products in *Streptomyces* species.

The Sdh of *C. glutamicum* has been purified to homogeneity and is comprised of three polypeptides of 67, 29 and 23 kDa (Kurokawa & Sakamoto, 2005) and the enzyme has been proposed to have a role in the regulation of sulphur metabolism and general redox homeostasis (Lee, Park, Kim, & Lee, 2014).

Although to our knowledge no functional characterisation studies have yet been carried out on SCO6155 and SCO7412 in *S. coelicolor*, the products of both genes are predicted to be PoxB-like pyruvate oxidase flavoenzymes (Bentley et al., 2002; Fig. 4). PoxB is membrane-associated and catalyses the oxidative decarboxylation of pyruvate to acetate and CO_2 with the reducing equivalents being delivered to menaquinone rather than NAD^+ (Neumann, Weidner, Pech, Stubbs, & Tittmann, 2008), as is the case with the pyruvate dehydrogenase complex. SCO6155 encodes a 580 amino acid protein with 53% amino acid identity to PoxB from *E. coli*; SCO7412 encodes a 600 amino acid-long polypeptide with only 34% amino acid similarity to the *E. coli* PoxB enzyme.

Further gene products likely to be involved in delivering electrons to the menaquinone pool via flavin-based electron transfer and membrane association (Fig. 4) are encoded by SCO1081 and SCO1082, which are two subunits of electron transfer flavoprotein (Etf) (Thomas et al., 2012). EtfA and EtfB exhibit high identity (59% and 52%, respectively) with the corresponding subunits of the *M. tuberculosis* enzyme but in contrast to *C. glutamicum* (Bott & Niebisch, 2003; Molenaar, van der Rest, & Petrovic, 1998), *S. coelicolor* apparently lacks a malate: quinone oxidoreductase.

A single *atp* operon, organised as in most other bacteria *atpIBEFHAGDC* (SCO5366–SCO5374), encoding ATP synthase (complex V in Fig. 1A) is present on the *S. coelicolor* genome (Bentley et al., 2002). The complex has been purified from both *S. lividans* (Hensel, Deckers-Hebestreit, & Altendorf, 1991) and *S. fradiae*, whereby phosphorylation studies suggested that the b and β subunits might undergo phosphorylation (Alekseeva et al., 2015). A proteomic study carried out with *S. coelicolor* identified the alpha chain of the ATP synthase as one of the most abundant proteins in the bacterium (Thomas et al., 2012).

4. RESPIRATION WITH NITRATE

Several *Streptomyces* species encode one or more respiratory Nar; however, in no case reported to date has it been plausibly shown that the bacterium can grow by nitrate respiration (Fischer et al., 2010), contrasting what has been observed for many other bacterial species such as *E. coli*, *Bacillus subtilis*, *Paracoccus denitrificans*, *Pseudomonas aeruginosa* and *Bradyrhizobium japonicum* (Fernández-López, Olivares, & Bedmar, 1994; Nakano & Zuber, 1998; van Spanning, Richardson, & Ferguson, 2007; Zumft, 1997). Rather, it appears that in *Streptomyces* species Nar has the function of contributing to maintenance of a membrane potential when oxygen is limiting or absent (Fischer, Falke, Pawlik, & Sawers, 2014; van Keulen et al., 2005, 2007). In the absence of O_2, nitrate is an excellent electron acceptor ($E^{0\prime} = +433$ mV) and it is generated by nitrification (Prosser, 2007) in environments where streptomycetes are also abundant. In nitrification (Fig. 5A), ammonia is oxidised to nitrite by ammonia-oxidising bacterial genera, such as *Nitrosomonas*, *Nitrosococcus* and *Nitrosospira*, as well as by the recently discovered ammonia-oxidising archaea (Stahl & de la Torre, 2012); nitrite is subsequently further oxidised to nitrate by members of the genera *Nitrobacter*, *Nitrococcus* and *Nitrospira*, which act in syntrophy with the ammonia oxidisers. While many soil bacteria (eg, *Bacillus*, *Pseudomonas* and *Paracoccus* species) reduce nitrate all the way to nitrogen gas in the process of denitrification, others, such as *E. coli*, can perform a six-electron reduction of nitrite to ammonia catalysed by a soluble siroheme-containing nitrite reductase (Cole & Richardson, 2008). Despite early reports suggesting that some streptomycetes can perform partial denitrification (Albrecht, Ottow, Benckiser, Sich, & Russow, 1997; Chèneby, Philippot, Hartmann, Henault, & Germon, 2000; Kumon et al., 2002; Shoun, Kano, Baba, Takaya, & Matsuo, 1998) the mechanisms underlying this process in these bacteria remain unclarified. In the studies that focused on enzyme activity, the release of nitrous oxide was tested and for

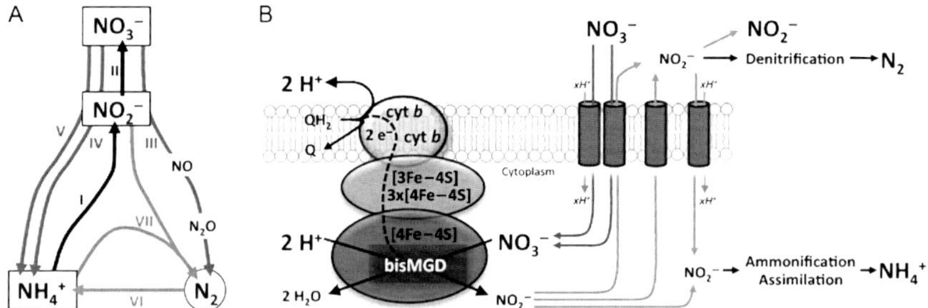

Fig. 5 Nitrate reduction and the nitrogen cycle. (A) A simplified scheme of the bacterial nitrogen cycle shows nitrification (I and II), denitrification (III), nitrate/nitrite ammonification (IV), assimilatory nitrate/nitrite reduction (V), nitrogen fixation (VI) and anaerobic ammonia oxidation (anammox; VII). Note that scheme I also includes ammonia-oxidising archaea (AOA—see text for details). (B) Schematic representation of respiratory nitrate reductase including the bis-MGD (bis-molybdenum guanine dinucleotide) cofactor in the catalytic subunit. Nitrate reduction requires nitrate uptake into the cytoplasm, which can occur by nitrite-driven antiport (Zheng, Wisedchaisri, & Gonen, 2013) or by proton-symport (Moir & Wood, 2001). Nitrite extrusion can theoretically be achieved without being coupled to nitrate import. Toxic nitrite can be reduced by ammonification in the cytoplasm or on the periplasmic side of the membrane by denitrification. (See the colour plate.)

S. nitrosporeus, *S. antibioticus* and *S. thioluteus* reduction to the level of N_2O could be determined (Albrecht et al., 1997; Kumon et al., 2002; Shoun et al., 1998). In the case of *S. antibioticus* activities of nitrate reductase, nitrite reductase and nitrous oxide reductase could be verified (Kumon et al., 2002) and indeed it was claimed for *S. antibioticus* and *S. thioluteus* that anaerobic growth was detected (Kumon et al., 2002; Shoun et al., 1998). However, upon close examination of the data anaerobic growth amounted to only 2.5% of the dry weight attained during the same period of aerobic growth (Kumon et al., 2002), calling into question whether this was really growth. Nevertheless, a bioinformatic analysis of the genome sequences of these *Streptomyces* species will prove to be very informative.

On the basis of published genome sequences, the majority of streptomycetes do not perform denitrification, and despite the presence of siroheme-dependent nitrite reductase, nitrate ammonification, at least in the case of *S. coelicolor*, does not occur (Bentley et al., 2002; Borodina, Krabben, & Nielsen, 2005; Fischer, Schmidt, Falke, & Sawers, 2012), which contrasts what is observed for *E. coli* (Simon, 2002) and *B. subtilis* (Hoffmann, Frankenberg, Marino, & Jahn, 1998). Rather, the nitrite reductase is induced under nitrogen-limiting conditions and is required for the organism to use nitrate as a nitrogen source (Fischer et al., 2012; Tiffert et al., 2008).

4.1 Respiratory Nitrate Reductases

Nars form a membrane-associated enzyme complex (Fig. 5B), which has the site of nitrate reduction on the cytoplasmic face of the membrane (Richardson, 2000; Richardson, Berks, Russell, Spiro, & Taylor, 2001). The Nar complex comprises three different protein subunits, the catalytic subunit NarG (135–138 kDa), an electron-transferring subunit NarH (55–57 kDa) and a membrane anchor subunit NarI (26–29 kDa) and structural studies have provided molecular details of the organisation of the cofactors within the complex (Bertero et al., 2003; Jormakka, Richardson, Byrne, & Iwata, 2004). The NarG subunit harbours the bis-MGD-containing catalytic site, while the NarH subunit has an electron-transferring function and binds four iron–sulphur clusters that facilitate electron delivery to the active site. The whole complex is anchored to the membrane via the di-*b*-heme integral membrane NarI subunit, which has a quinol dehydrogenase function. The genes encoding Nar enzymes have a conserved organisation in most organisms synthesising these enzymes and they form an operon comprising *narGHJI*. The product of the *narJ* gene is not part of the enzymatically active complex but rather is a system-specific chaperone that interacts with the NarG subunit and is involved in bis-MGD cofactor insertion and folding of the protein (Chan et al., 2014; Lorenzi et al., 2012).

Due to the inherent toxicity of the reduced oxyanion nitrite, it is necessary to remove it efficiently and quickly from the cytoplasm. Moir and Wood proposed two classes of NarK protein, both of which belong to the major facilitator family of secondary transporters (Goddard, Moir, Richardson, & Ferguson, 2008; Moir & Wood, 2001). NarK-type secondary transporters appear to be primarily associated with respiratory nitrate respiration while ABC transporters can also function in some bacteria during nitrate assimilation (Moir & Wood, 2001).

The NarK type I class is proposed to catalyse nitrate/proton (cation) symport, and the type II class probably functions in nitrate–nitrite antiport (Fig. 5B). Recent structural studies of NarK and NarU from *E. coli*, both of which belong to the NarK type II class, indicate that NarK is a nitrate–nitrite antiporter, while NarU is a nitrate–sodium ion symporter (Fukuda et al., 2015; Yan et al., 2013; Zheng et al., 2013). Some bacteria also have NirC proteins that appear to be channels specific for bidirectional nitrite transport (Jia & Cole, 2005; Jia, Tovell, Clegg, Trimmer, & Cole, 2009; Lü et al., 2013, 2012). Although the genome of *S. coelicolor* does not encode a NirC, it does encode two possible NarK orthologues (Fischer et al., 2014; van Keulen et al., 2005). Both orthologues have 12-transmembrane helices, characteristic of NarK proteins (Goddard et al., 2008) and NarK1 belongs to the type II class along with NarK1, NarK3 and NarU from *M. tuberculosis* (Cole et al., 1998; Moir & Wood, 2001), while NarK2 from *S. coelicolor* belongs to the type I class of NarK proteins, along with NarK2 from *M. tuberculosis* (Giffin, Raab, Morganstern, & Sohaskey, 2012; Sohaskey & Wayne, 2003) and NarK from *B. subtilis* (Nakano, Zuber, Glaser, Danchin, & Hulett, 1996). NarK in *C. glutamicum* is encoded as part of the *narKGHJI* operon (Bott & Niebisch, 2003) and like *S. coelicolor, C. glutamicum* fails to grow anaerobically by nitrate respiration (Bott & Niebisch, 2003). In contrast, nitrate reduction was induced in the transition to the so-called nonreplicating state and a mutant of *M. bovis* lacking Nar exhibited decreased persistence in some tissues of immunocompetent mice, suggesting Nar and nitrate respiration are important for survival of mycobacteria in the lung (Fritz, Maass, Kreft, & Bange, 2002; Wayne & Hayes, 1998). While Nar synthesis in *M. tuberculosis* is constitutive, NarK2 synthesis is induced by oxygen limitation (Sohaskey & Wayne, 2003), resulting in an increased level of nitrate respiration, which is commensurate with nitrate reduction being a major factor for sustaining and perhaps inducing nonreplicating persistence (Wayne & Sohaskey, 2001). NarK synthesis is also induced by oxygen limitation in *B. subtilis* (Reents et al., 2006).

4.2 Genes Whose Products Are Involved in Nitrate Reduction in *S. coelicolor*

Being a Gram-positive bacterium lacking a periplasm, actinobacteria do not synthesise Nap periplasmic nitrate reductases (Jepson et al., 2007; Moreno-Vivián, Cabello, Martínez-Luque, Blasco, & Castillo, 1999). *S. coelicolor*, however, synthesises three Nar enzymes and it also synthesises a functional assimilatory nitrate reductase (NasA: SCO2473) (Fischer et al., 2010; van Keulen et al., 2005). To date, *S. coelicolor* remains the only example within the streptomycetes that synthesises three Nar enzymes. The Nar and Nas enzymes are molybdoenzymes and a mutant lacking the molybdenum cofactor biosynthetic gene *moaA* fails to synthesise either Nar or Nas enzymes and concomittantly loses the ability to reduce nitrate and to use nitrate as a nitrogen source (Fischer et al., 2010, 2012). The gene *nasA* (SCO2473) encodes the assimilatory Nas enzyme (Fischer et al., 2012; Wang & Zhao, 2009) and the demonstration that a *nasA* mutant fails to utilise nitrate as a nitrogen source on solid medium indicates that the Nar enzymes are not able to channel intracellular nitrite to the NirBD assimilatory nitrite reductase (Fischer et al., 2012); however, nitrite added exogenously restores growth to the *nasA* mutant, which also implies that nitrite produced by Nar enzymes in *S. coelicolor* is immediately exported from the cell. Although the *nasA* and *nirBD* genes are not colinear on the genome, they are in proximity to each other (Fig. 6). This contrasts the situation in *Amycolatopsis mediterranei* U32, a rifamycin-producing actinobacterium where the *nasACKBDEF* operon encodes assimilatory Nas (*nasA* and *nasC*), nitrite reductase (*nasBD*) and a putative nitrate–nitrite transporter (*nasK*) (Shao et al., 2011). Notably, induction of expression of the *nasACKBDEF* operon during growth on nitrate as a nitrogen source also stimulates rifamycin production, indicating a physiological link between nitrate and antibiotic metabolism.

The genes so far identified to be most relevant to respiratory nitrate reduction, as well as *nasA* and *nirB* for nitrate assimilation, in *S. coelicolor* are shown in Fig. 6. All of the genes necessary to encode the key enzymes for molybdenum cofactor biosynthesis (Schwarz, Mendel, & Ribbe, 2009) are distributed throughout the linear genome. *S. coelicolor* lacks *moaD* and

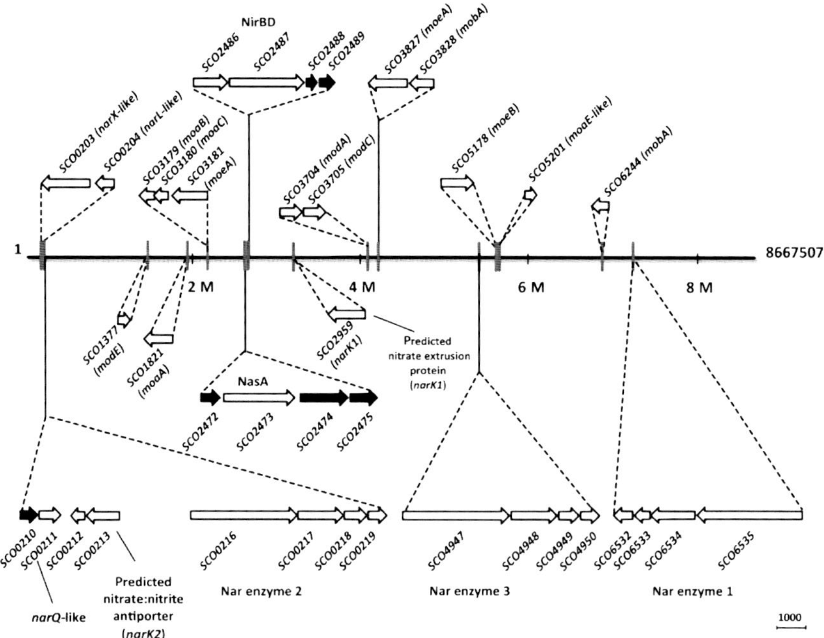

Fig. 6 Genes and operons whose products encode components involved in nitrate respiration in *Streptomyces coelicolor* A3(2). The locations of the listed genes, including their SCO numbers, with their corresponding encoded enzymes are highlighted on the linear representation of the 8.7 Mbp genome of *S. coelicolor*.

moaB genes and it is unclear whether other proteins within streptomycetes take over their respective functions. Like the tungsten-dependent archaeon *Pyrocccus furiosus* (Bevers et al., 2008), *S. coelicolor* does not have a *mogA* gene but instead encodes a MoaB-like protein that probably takes over the function of MogA in adenylating molybdopterin. Furthermore, and in analogy to *P. furiosus*, *S. coelicolor* has two *moeA* genes (Fig. 6). In *P. furiosus*, these have been suggested to be needed for metal selectivity (Bevers et al., 2008); however, there is no evidence to date that *S. coelicolor* synthesises tungsten-dependent enzymes.

4.3 Phylogeny of Nar Enzymes in *S. coelicolor*

The ability of streptomycetes to reduce nitrate has been recognised for a long time and used in taxonomy to allocate *Streptomyces* species into taxonomy-based clades (Williams, Goodfellow, Alderson, et al., 1983; Williams, Goodfellow, Wellingtion, et al., 1983). Subsequent implementation of 16S rRNA-based comparisons (Woese, 1987) allowed a more accurate assessment of the phylogenetic relationship within the genus *Streptomyces* and between other members of the actinobacteria (van Keulen et al., 2005). That study also demonstrated the close relationship between the Nar enzymes of *S. coelicolor* and those of the mycobacteria, while the rhodococci were on a separate clade. Significantly, the catalytic subunits of the three *S. coelicolor* Nar enzymes show less than 75% amino acid sequence identity, which is in a similar range to that between the *S. coelicolor* enzymes and those of the mycobacteria and rhodococci (van Keulen et al., 2005). The three *narGHJI* operons of *S. coelicolor* are located separately on the genome (Fig. 6). The *narGHJI* (*nar1*) operon encoding Nar1 is located on the right arm, the *narG2H2J2I2* (*nar2*) operon on the extreme end of the left arm and the *narG3H3J3I3* (*nar3*) operon is in the core region of the genome. Together, these findings suggest distinct phylogeny of these enzymes and therefore, possibly also point to differences in either synthesis or physiological roles.

The accumulation of numerous genome sequences in recent years has allowed a reassessment of the distribution and phylogeny of the NarG catalytic subunits within the streptomycetes in particular, and the actinobacteria in general (Fig. 7). It is important to note that only the catalytic NarG subunits were considered in this comparison and only sequences from completely sequenced genomes of bacteria with a dedicated species name. For example, in the current NCBI database (November 2015) there are

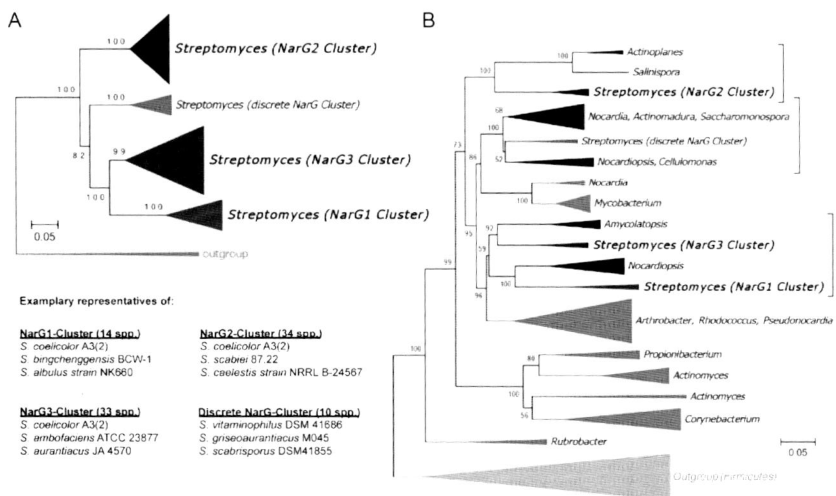

Fig. 7 Phylogenetic relationships between NarG polypeptides in the actinobacteria. Phylogenetic analyses were conducted in MEGA6 (Tamura, Stecher, Peterson, Filipski, & Kumar, 2013) by the neighbour-joining method (Saitou & Nei, 1987). The phylogenetic distances were computed using the Poisson correction method (Zuckerkandl & Pauling, 1965) and are presented in units of the number of amino acid substitutions per site. The percentage of replicate trees in which the associated taxa clustered together in the bootstrap test (1000 replicates) is shown next to the branches (Felsenstein, 1985). The analysis involved 91 complete NarG amino acid sequences for the tree in part A and 141 complete amino acid sequences for the tree shown in part B. All sequences containing gaps and missing data were eliminated. (A) A total of 91 amino acid sequences of translated full-length *narG* gene products were used, which were found in the NCBI-Microbial-Genomes-database for complete genomes and draft genomes of *Streptomyces*. The outgroup is represented by NarG sequences of *Desulfitobacterium hafniense* Y51 and *Thermodesulfobacterium narugense* DSM 14796. (B) For the comparison of actinobacterial NarGs, 141 amino acid sequences of translated and full-length *narG* gene products were used. These were representatively selected from complete genomes (NCBI-Microbial-Genomes-database). The selection includes 137 species, which were as follows: *Streptomyces* (12), *Actinomyces* (9), *Actinoplanes* (2), *Actinomadura* (2), *Amycolatopsis* (5), *Arthrobacter* (3), *Cellulomonas* (2), *Corynebacterium* (12), *Saccharomonospora* (3), *Mycobacterium* (24), *Nocardia* (12), *Nocardiopsis* (12), *Propionibacterium* (5), *Pseudonocardia* (4), *Rhodococcus* (5), *Rubrobacter* (3), *Salinispora* (2). The outgroup comprised: *Lactobacillus* (3), *Geobacillus* (4), *Bacillus* (7), *Paenibacillus* (3), *Staphylococcus* (3), *Macrococcus* (1), *Anoxybacillus* (1), *Desulfitobacterium* (1), *Thermodesulfobium* (1).

more than 500 listed 'draft' genome sequences of *Streptomyces*; however, only 156 of these have species names associated with them and only these were considered in this analysis.

Some general conclusions are perhaps relevant at the outset of this analysis. When the complete sequences of bacteria are considered there are a number of bacteria with more than one Nar enzyme and these include the enterobacteria *E. coli*, *Salmonella enterica*, *Klebsiella pneumoniae* and *Enterobacter cloacae*, each of which has two Nars encoded in their genomes. The Betaproteobacterium *Burkholderia thailandensis*, the Deltaproteobacterium *Geobacter metallidurans* and the Firmicutes species *Bacillus licheniformis*, *Geobacillus thermodenitrificans* and *Paenibacillus terrae* also each have two Nars. To date only the actinobacteria *S. coelicolor* and *Nocardia cyriacigeorgica* have three Nar enzymes.

The Alphaproteobacteria, and the phyla *Deferribacteres*, *Deinococcus* and the halophilic archaeal genus *Haloarcula* never have more than one respiratory Nar enzyme and there are only two species belonging to the clostridia (*Desulfitobacterium hafniense* and *Thermodesulfobacterium narugense*) that have a respiratory Nar. The following classes or phyla have no recorded example of a respiratory nitrate reductase encoded in the sequenced genomes: *Cyanobacteria*, *Chlamydiae*, *Chlorobi*, *Chloroflexi*, *Cytophagia*, *Epsilonproteobacteria*, *Flavobacteriia*, *Planctomycetes*, *Tenericutes* (including the *Mollicutes*) *Thermotogae*, *Thermodesulfobacteria*, *Verrucomicrobia* and the genus *Aquifex*.

Three genera of actinobacteria lack Nar and these include the plant-associated *Frankia* species, *Bifidobacterium* and *Kineococcus*, while the following genera have never more than one Nar enzyme: *Arthrobacter*, *Cellulomonas*, *Corynebacterium*, *Micromonospora*, *Mycobacterium*, *Nocardiopsis*, *Propionibacteriu*, *Rhodococcus* and *Rubrobacter*.

With regard to Nar enzymes in *Streptomyces* species, it is important to stress that there seems to be no evidence for a correlation between the presence of a particular Nar enzyme and the existence of a particular aerobic respiratory enzyme complex. Moreover, there appears to be no specific taxonomic clustering of nitrate-reducing vs non-nitrate-reducing streptomycetes. Of the 27 complete genome sequences of *Streptomyces* species in the NCBI-Microbial-Genomes-database, 11 (41%) species have one or more NarG orthologues (Fig. 7). There is a similar distribution in the other 156 available genome sequences (drafts) of *Streptomyces* species whereby 63 (40%) encode one or more NarG polypeptides. Importantly, in the *Streptomyces* species with more than one Nar enzyme, these Nars belong to Nar1, Nar2 or the Nar3 classes.

When the deduced amino acid sequences of NarG within the *Streptomyces* and across the actinobaceria are compared they exhibit between 60% and 70% amino acid similarity (Fig. 7). This suggests that the three different Nar enzymes from *S. coelicolor* A3(2) form distinct clusters and are functionally distinct with regard to their role in metabolism or when they are synthesised. The Nar enzymes in other streptomycetes fall within one of these classes of enzyme. An alignment of the NarG amino acid sequences from complete genome sequences within the actinobacteria with subsequent phylogenetic analysis using either the neighbour-joining or the maximum-likelihood methods supports the hypothesis that minimally three NarG clusters are present within *Streptomyces* species (Fig. 7A). The phylogram also indicates that the NarG1 and NarG3 classes are more closely related than the more distant NarG2 class. Moreover, there is an indication that a fourth cluster or class of NarG might exist (see Fig. 7A), which is distinct from NarG1, NarG2 or NarG3; however, caution is advised here because these sequences are incomplete.

The clustering of the *S. coelicolor* NarG polypeptides into three classes is also in accord with the lack of redundancy in their function during growth of the bacterium (Fischer et al., 2010) and with the apparent differential regulation of the operons encoding them (see Sections 4.4 and 4.6). A comparison of 141 NarGs from sequenced actinobacterial genomes (NCBI-Microbial-Genomes-database) shows how the NarG sequences are related to the NarG classes found in *S. coelicolor* (Fig. 7B). While this phylogram gives a first indication of which Nar enzyme a particular actinobacterial might have, the allocation to a particular class of Nar enzyme must await experimental verification.

4.4 Tissue-Specific Synthesis and Functionality of Nars in *S. coelicolor*

Each of the three Nar enzymes in *S. coelicolor* is functional in the reduction of nitrate to nitrite and one of the first detailed studies performed revealed a stoichiometric uptake of nitrate from the growth medium and concomitant release of nitrite into the medium by mycelial cultures (Fischer et al., 2010). Construction of a triple *nar* operon deletion mutant and a *moaA* gene mutant clearly distinguished respiratory from assimilatory nitrate reduction and Nar activity could be detected in spores as well as in mycelium (Fischer et al., 2010). The use of the nitrate analogue chlorate, which is reduced by Nar enzymes to toxic chlorite (Azoulay, Puig, & Pichinoty, 1967), suggested that different enzymes were synthesised in spores and mycelium and this was later confirmed (Fischer et al., 2014; Fischer, Falke, & Sawers, 2013). Being an obligate aerobe, initial results suggested that Nar-dependent nitrate reduction might occur in the presence of oxygen (Fischer et al., 2010). However, these experiments were performed in rich medium and in shaking cultures, whereby *S. coelicolor* grows in aggregates that rapidly become oxygen limited in their centre (Fischer & Sawers, 2013).

Subsequent studies have revealed that for all three Nar enzymes nitrate reduction only occurs under strongly oxygen-limiting or anaerobic conditions and oxygen above a concentration of 5 μM inhibits Nar-dependent nitrate reduction in spore suspensions or in intact mycelium (Fischer et al., 2014, 2013, unpublished data).

Through the use of defined *nar* operon mutants it was possible to demonstrate that Nar1 is exclusively functional in spores (Fischer et al., 2013), Nar2 is induced during germination of spores, as well as during hypoxic stress of spores, and is the main Nar enzyme active in exponentially growing mycelium (Fischer et al., 2014) and that Nar3 synthesis is induced upon substrate limitation, in particular by phosphate limitation, and entry into the stationary phase (Fischer et al., 2014; Falke et al., unpublished data). The link between Nar3 synthesis and secondary metabolism has been recently verified (Li et al., 2015).

4.5 Coupling of Nar Activity to Nitrate-Nitrite Transport

When Nar enzyme activity is determined in cell extracts an artificial electron donor must be provided and this is usually dithionite-reduced methyl viologen ($E^{o\prime} = -446$ mV) (Enoch & Lester, 1975). Under these conditions it is not possible to couple nitrate reduction with nitrate transport. However, when *S. coelicolor* spores are suspended anaerobically in buffer or water, addition of mM concentrations of nitrate results in the quantitative reduction to nitrite by Nar1 (Fischer et al., 2013). Nitrate reduction does not require the supply of an exogenous electron donor, indicating that spores supply the electrons for nitrate reduction. The electron donor is likely to be NADH derived from the degradation of trehalose, which comprises up to 20–25% of the dry weight of spores (McBride & Ensign, 1990; Quirós, Hardisson, & Salas, 1986).

Addition of oxygen leads to the immediate cessation of nitrate-reducing activity in spore suspensions and this indicates that nitrate reduction and nitrate–nitrite transport are tightly coupled. This is analogous to oxygen inhibition of nitrate transport activity in *E. coli* (Noji & Taniguchi, 1987) and in *M. tuberculosis* (Sohaskey & Wayne, 2003). Oxygen inhibition most likely results from a re-routing of electron transport to the thermodynamically more efficient electron acceptor oxygen. It is perhaps noteworthy that a functional complex III is required for Nar1-dependent nitrate reduction in spores (Falke et al., unpublished data). Complex III is not required, however, for nitrate reduction in intact mycelium, suggesting that electrons can take an alternative route to Nar2 and Nar3 in mycelium. Presumably, this route is directly from menaquinol to Nar2 and Nar3, while complex III might transfer the electrons directly to Nar1 in spores, because actinobacteria do not have a soluble cytochrome *c* (Bott & Niebisch, 2003; Cook et al., 2009).

The nitrate–nitrite transporter in spores has not yet been identified. Spore suspensions derived from knock-out mutants lacking either NarK1 or NarK2 (Fig. 6) are unimpaired in nitrate reduction (Fischer et al., 2014). This suggests that the NarK enzymes are functionally redundant or another new transporter is responsible for nitrate uptake and nitrite export in spores of *S. coelicolor*.

A different scenario is observed in exponentially growing mycelium, where NarK2 is solely responsible for nitrate delivery to the Nar2 enzyme (Fischer et al., 2014). The *narK2* gene is located in the immediate neighbourhood of the *nar2* operon (Fig. 6).

As shown for Nar1, nitrate reduction by Nar3 also appears to be unaffected in mutants devoid of NarK1 or NarK2 (Fischer et al., 2014).

4.6 Regulation of Nar Enzyme Synthesis

The regulation of respiratory nitrate reduction has been studied in detail for *E. coli* (Stewart, 2003) as well as for *B. subtilis* (Tielen, Schobert, Härtig, & Jahn, 2012). Generally, these *nar* genes underlie a dual regulation primarily through a global effect caused by oxygen limitation and subsequently by nitrate and in *E. coli* this involves the oxygen-sensing Fnr protein working together with the nitrate-responsive NarXL two-component system (Stewart, 2003). In *S. coelicolor* no nitrate-dependent synthesis of any Nar enzyme has so far been observed (Fischer et al., 2014; Thomas et al., 2012; our unpublished data). Circumstantial evidence for a lack of nitrate- or oxygen-dependent regulation of *nar* operon expression was suggested based on Nar activity measurements in *C. glutamicum* (Takeno et al., 2007), although this has been contested in another study (Nishimura, Vertès, Shinoda, Inui, & Yukawa, 2007). In the case of *Mycobacterium* species *narGHJI* operon expression does not respond either to anaerobiosis or to nitrate (Cook et al., 2009; Sohaskey & Wayne, 2003). Together, these data suggest that actinobacterial *nar* operon expression is regulated differently compared with that observed in other bacteria and the current data strongly suggests that nitrate does not regulate Nar synthesis; for *S. coelicolor* Nar synthesis is clearly not regulated in response to nitrate (Fischer et al., 2014, 2013; Thomas et al., 2012).

Regulation of Nar1 synthesis appears not to be affected by oxygen. We could observe that Nar1 enzyme activity is always detected in spores and this activity is unaffected by the protein synthesis inhibitor chloramphenicol, strongly suggesting that Nar1 synthesis occurs prior to completion of spore formation (Fischer et al., 2013). Precisely when expression of the *narGHJI* operon is induced and in response to what signal remains to be elucidated; currently, it cannot be excluded that operon expression might be developmentally regulated. It is perhaps noteworthy that *narG* transcripts can be readily detected in spores and they are present at low levels in mycelium, suggesting that post-transcriptional regulation might also be involved in controlling enzyme synthesis.

Nar2 synthesis appears to initiate upon germination of spores and can also be induced during hypoxic stress of mycelial filaments. Both of these processes can be inhibited by chloramphenicol (Fischer et al., 2014). Surprisingly, Nar2 synthesis can also be induced to a low level in spore suspensions that have been incubated for several days in the absence of oxygen; however, these spores do not germinate in the absence of oxygen (Fischer et al., 2014). In mycelial cultures Nar2 is synthesised only at a low level if the cultures are abruptly switched to anaerobic conditions or retained under high aeration. However, if the cultures are shifted to hypoxic conditions (<5% oxygen) then high-level Nar2 synthesis is observed (Fischer et al., 2014). This could be due either to increased translation of *nar2* operon transcripts and/or to increased transcription. This is reminiscent of hypoxic control of Nar activity in *M. tuberculosis*, where hypoxic upregulation of *narK2* gene expression results in improved supply of nitrate to the constitutively synthesised Nar enzyme and thereby to increased levels of nitrate reduction (Sohaskey & Wayne, 2003). In *S. coelicolor* expression of both the *nar2* operon and the *narK2* gene, encoding the Nar2-specific nitrate–nitrite antiporter NarK2 (Fischer et al., 2014), appears to be dependent on the SCO0203 gene product (Fig. 6). SCO0203 encodes a predicted membrane-associated histidine kinase similar to the NarX nitrate-sensor from *E. coli* (our unpublished data). The neighbouring SCO0204 gene encodes a response regulator with amino acid similarity to NarL (Stewart, 2003). As the expression of the *nar2* operon is not influenced by nitrate, it is conceivable that SCO0203 responds instead to hypoxia.

Synthesis of Nar3 is restricted to nutrient-limited mycelium and correlates strongly with the onset of secondary metabolism (Fischer et al., 2010). Our recent data indicate that Nar3 synthesis occurs only during phosphate limitation and inclusion of phosphate in the growth medium inhibits enzyme synthesis (Falke et al., unpublished data). Whether *narG3H3J3I3* operon expression also responds to phosphate levels remains to be established.

Our current understanding of the regulation of Nar enzyme synthesis is summarised in Fig. 8. To what extent Nar enzyme synthesis is regulated by putative transcriptional regulators (van Keulen et al., 2007) such as DosRS (Sivaramakrishnan & Ortiz de Montellano, 2013; Unden, Müllner, & Reinhart, 2009), Rex (Brekasis & Paget, 2003), ResDE-type regulators (Hoffmann et al., 1998; Nakano & Zuber, 1998), in response to oxygen or the redox status or to phosphate (Rodríguez-García, Barreiro, Santos-Beneit, Sola-Landa, & Martín, 2007; Sola-Landa et al., 2003) and other predicted two-component regulatory systems remains to be established. The sequence of *S. coelicolor* genome (Bentley et al., 2002; van Keulen et al., 2007) indicates that there is no classical Fnr-type regulator (Zumft, 1997)

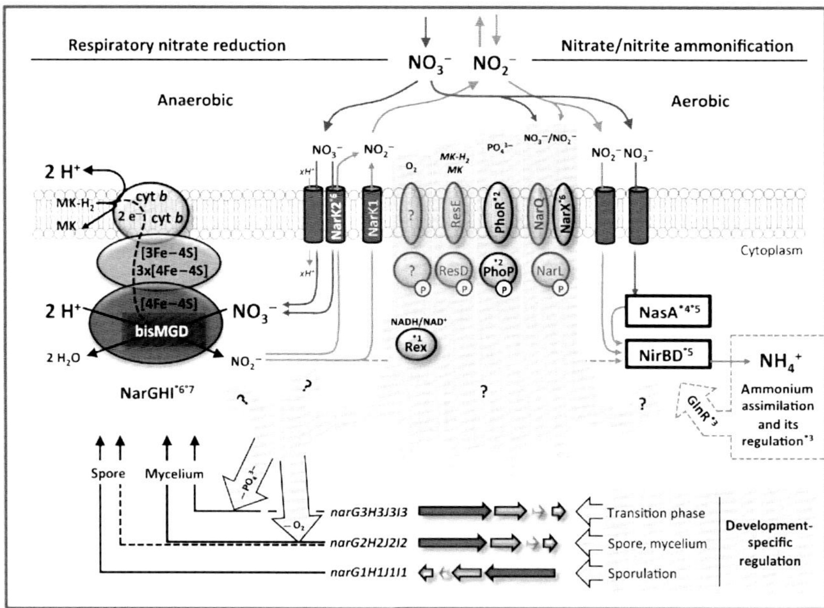

Fig. 8 Model of the regulation and function of respiratory and assimilatory nitrate reduction in *S. coelicolor* A3(2). The regulatory features or functions of particular proteins are based on the following papers, which are included within the diagram: *1, Brekasis & Paget, 2003; *2, Sola-Landa, Moura, & Martín, 2003; *3, Fink, Weissschuh, Reuther, Wohlleben, & Engels, 2002; *4, Wang & Zhao, 2009. Those papers, which are based on our work are as follows: °5, Fischer et al., 2012; °6, Fischer et al., 2014; °7, Fischer et al., 2010 and Fischer et al., 2013. Hypothetical regulatory control nodes are indicated by *question marks*. (See the colour plate.)

encoded on the genome. The marginal overlap of Nar enzyme activities in *S. coelicolor* suggests a highly coordinated regulation of gene expression and enzyme synthesis.

4.7 Physiological Consequences of Nitrate Reduction for *Streptomyces*

With the exception of evidence for increased survival of *M. tuberculosis* in the long-term persistent state upon a sudden shift to anaerobiosis (Sohaskey, 2008), or potentially some very slow growth under nitrate respiratory conditions for *C. glutamicum* (Nishimura et al., 2007; Takeno et al., 2007), it appears clear the primary function of actinobacterial Nar enzymes is in contributing to the maintenance of a membrane potential in the absence of oxygen. Why does nitrate respiration not support growth in these bacteria?

One simple answer to this question might be that Nar is not an abundant enzyme in actinobacterial species. Comparison of the specific enzyme activity of Nar measured in *S. coliecolor* extracts of spores (Fischer et al., 2013) and mycelium (Fischer et al., 2014) with Nar activity measured in extracts of *C. glutamicum* (Takeno et al., 2007) and *M. tuberculosis* (Sohaskey, 2008), reveals that for each bacterium values are around 10–25 nmol min^{-1} mg^{-1} protein. In contrast, extracts of *E. coli* growing by nitrate respiration have Nar specific activities that are roughly 400- to 500-fold higher (7–10 μmol min^{-1} mg^{-1} protein) (Enoch & Lester, 1975). Thus, while this mode of respiratory nitrate reduction is sufficient to contribute to maintaining a proton gradient, the activity is insufficient to support growth of the bacterium. An analogous observation has been made recently in *M. smegmatis*, as well as members of the acidobacteria phylum of soil bacteria, which use high-affinity NiFe-hydrogenases to scavenge atmospheric hydrogen gas to help maintain a membrane potential and give the bacteria a competitive edge during hypoxia and anoxia (Berney, Greening, Conrad, Jacobs, & Cook, 2014; Greening et al., 2015). Based on the genome sequence *S. coelicolor* is predicted to encode over 200 different oxidoreductases of unknown function (Bentley et al., 2002) and it is conceivable that some of these might also be able to contribute to survival of the bacterium under anoxic conditions by using alternative electron donors or acceptors to help generate a proton gradient.

Streptomycetes grow only at the tips of their mycelium and our findings suggest that the Nar enzymes are distributed throughout the mycelium (M. Fischer & R. G. Sawers, unpublished data). This is also commensurate with the Nar enzyme contributing to 'fitness' of the organism by maintenance of a membrane potential.

Because *S. coelicolor* reduces nitrate to nitrite and no further, the bacterium has to deal with the toxic nitrite oxyanion, which can accumulate and is known to have antimicrobial properties, for example, it can attack iron–sulphur clusters and other cofactors in enzymes (Cammack et al., 1999; Reddy, Lancaster, & Cornforth, 1983). Moreover, it can be oxidised chemically to NO under acidic conditions or by a side reaction of Nar (Davidson, Juneja, & Branen, 2002; Dykhuizen et al., 1998; Platzen, Koch-Koerfges, Weil, Brocker, & Bott, 2014). It has been demonstrated recently for both *S. coelicolor* (Fischer et al., 2014) and *C. glutamicum* (Platzen et al., 2014) that nitrite inhibits growth of the bacteria, presumably through generation of reactive nitrogen species (Platzen et al., 2014). Nitrite might, therefore, also be beneficial for the saprophytic lifestyle of *S. coelicolor* by inhibiting growth of competitors.

5. RESPIRATORY ENZYME COMPLEXES—AN OUTLOOK AND PERSPECTIVES

The fact that actinobacteria lack a soluble cytochrome *c* in their respiratory chain (Bott & Niebisch, 2003) immediately raises the question regarding how electron transfer between complex III, the cytochrome bc_1 complex and cytochrome aa_3 oxidase occurs. This question was answered by the discovery of respiratory super-complexes in *C. glutamicum* (Niebisch & Bott, 2003), which suggests direct electron transfer between hemes within each complex. This might also explain the need for the unusual diheme cytochrome *c* subunit C in complex III (Cook et al., 2009; Niebisch & Bott, 2001). Although not yet categorically demonstrated, it is anticipated that in other actinobacteria, including streptomycetes, a similar respiratory super-complex exists. Meanwhile, respiratory super-complexes have also been discovered in other bacteria (Kim et al., 2015; Sousa et al., 2012; Stroh et al., 2004), suggesting that this occurrence is perhaps more widespread than previously realised. The fact that it has been possible to demonstrate in spores of *S. coelicolor* that Nar1 activity is dependent on an assembled, active complex III suggests that this respiratory Nar enzyme might also interact directly with a modified form of complex III (Falke et al., unpublished data).

In this context, the question remains: what is the physiological function of respiratory nitrate reduction in spores and mycelium of *Streptomyces* species, and in actinobacteria generally? The likely explanation is a contribution to long-term survival and general 'fitness' of the spores and the organism by contributing to generation and maintenance of a proton gradient, as postulated earlier (van Keulen et al., 2007). It is clear, however, that nitrate reduction does not function as the sole means of allowing the bacterium to retain a proton gradient and it makes sense that the organism would not rely on a single electron acceptor or donor to achieve this goal. Indeed, the recent suggestion that hydrogenases also contribute to the overall 'fitness' of mycobacteria under hypoxic or anaerobic conditions (Berney et al., 2014; Greening et al., 2015) supports this hypothesis. The lifestyle of streptomycetes living in a soil environment where the oxygen status is continually changing affords a back-up system to ensure short-, medium- or even long-term survival during extended periods of anaerobiosis. The fact that both spores and mycelium accrue energy-rich reserves, such as trehalose (McBride & Ensign, 1987), during aerobic growth affords credence to the hypothesis that they can employ a variety of electron acceptors to maintain survival. Indeed, the fact that streptomycetes grow only at their tips, yet the main hyphal 'tissue' remains viable and metabolically active attests to a system being in place to ensure energy conservation for vital cellular processes (Hodgson, 2000; Hopwood, 2006). These considerations have relevance to how organisms enter and survive the *viable-but-non-culturable* state (Oliver, 2005) as well how streptomycete spores, and for that matter endospores, are able to survive for decades on meagre energy reserves and an absolutely limited metabolism. The challenges will be to determine how they shut down metabolism to an 'idling' status, and to identify what alternative electron acceptors can be used by actinobacteria to facilitate this survival mechanism. A further important challenge will be to determine how widespread this phenomenon is amongst other microorganisms.

Future research should also focus on the biochemistry, bioenergetics and physiology underlying metabolic restriction because this has relevance to ageing processes in higher organisms. Streptomycete spores provide an excellent system to study respiratory homeostasis, and in particular, how the dramatic effects of hypoxia and the sudden onset of anaerobiosis for an aerobe are offset by appropriate regulatory control of respiration and electron flow through respiratory complexes. The biochemical analysis of the postulated actinobacterial respiratory super-complexes will be a first step towards providing new insight into how membrane potential is maintained efficiently with limited metabolic resources.

ACKNOWLEDGEMENTS

Research in the authors' laboratory was supported by the Deutsche Forschungsgemeinschaft (SA 494-4).

Competing interests: The authors declare that they have no competing interest.

REFERENCES

Albrecht, A., Ottow, J. C. G., Benckiser, G., Sich, I., & Russow, R. (1997). Incomplete denitrification (NO and N_2O) from nitrate by *Streptomyces violaceoruber* and *S. nitrosporeus* revealed by acetylene inhibition and ^{15}N gas chromatography-quadrupole mass spectrometry analyses. *Naturwissenschaften*, *84*, 145–147.

Alekseeva, M. G., Mironcheva, T. A., Mavletova, D. A., Elizarov, S. M., Zakharevich, N. V., & Danilenko, V. N. (2015). F_oF_1-ATP synthase of *Streptomyces fradiae* ATCC 19609: Structural, biochemical, and functional characterization. *Biochemistry (Moscow)*, *80*, 296–309.

Aung, H. L., Berney, M., & Cook, G. M. (2014). Hypoxia-activated cytochrome *bd* expression in *Mycobacterium smegmatis* is cyclic AMP receptor protein dependent. *Journal of Bacteriology*, *196*, 3091–3097.

Azoulay, E., Puig, J., & Pichinoty, F. (1967). Alteration of respiratory particles by mutation in *Escherichia coli* K 12. *Biochemical and Biophysical Research Communications*, *27*, 270–274.

Bekker, M., de Vries, S., Ter Beek, A., Hellingwerf, K. J., & de Mattos, M. J. (2009). Respiration of *Escherichia coli* can be fully uncoupled via the nonelectrogenic terminal cytochrome *bd*-II oxidase. *Journal of Bacteriology*, *191*, 5510–5517.

Bentley, S. D., Chater, K. F., Cerdeño-Tárraga, A.-M., Challis, G. L., Thomson, N. R., James, K. D., et al. (2002). Complete genome sequence of the model actinomycete *Streptomyces coelicolor* A3(2). *Nature*, *417*, 141–147.

Berney, M., Greening, C., Conrad, R., Jacobs, W. R., Jr., & Cook, G. M. (2014). An obligately aerobic soil bacterium activates fermentative hydrogen production to survive reductive stress during hypoxia. *Proceedings of the National Academy of Sciences of the United States of America*, *111*, 11479–11484.

Bertero, M. G., Rothery, R. A., Palak, M., Hou, C., Lim, D., Blasco, F., et al. (2003). Insights into the respiratory electron transfer pathway from the structure of nitrate reductase A. *Nature Structural Biology*, *10*, 681–687.

Bevers, L. E., Hagedoorn, P. L., Santamaria-Araujo, J. A., Magalon, A., Hagen, W. R., & Schwarz, G. (2008). Function of MoaB proteins in the biosynthesis of the molybdenum and tungsten cofactors. *Biochemistry*, *47*, 949–956.

Borisov, V. B., Gennis, R. B., Hemp, J., & Verkhovsky, M. I. (2011). The cytochrome *bd* respiratory oxygen reductases. *Biochimica et Biophysica Acta*, *1807*, 1398–1413.

Borodina, I., Krabben, P., & Nielsen, J. (2005). Genome-scale analysis of *Streptomyces coelicolor* A3(2) metabolism. *Genome Research*, *15*, 820–829.

Boshoff, H. I. M., & Barry, C. E. (2005). Tuberculosis—Metabolism and respiration in the absence of growth. *Nature Reviews. Microbiology*, *3*, 70–80.

Bott, M., & Niebisch, A. (2003). The respiratory chain of *Corynebacterium glutamicum*. *Journal of Biotechnology*, *104*, 129–153.

Brekasis, D., & Paget, M. S. (2003). A novel sensor of NADH/NAD^+ redox poise in *Streptomyces coelicolor* A3(2). *EMBO Journal*, *60*, 687–696.

Cammack, R., Joannou, C. L., Cui, X. Y., Torres Martinez, C., Maraj, S. R., & Hughes, M. N. (1999). Nitrite and nitrosyl compounds in food preservation. *Biochimica et Biophysica Acta*, *1411*, 475–488.

Chan, C. S., Bay, D. C., Leach, T. G., Winstone, T. M., Kuzniatsova, L., Tran, V. A., et al. (2014). "Come into the fold": A comparative analysis of bacterial redox enzyme maturation protein members of the NarJ subfamily. *Biochimica et Biophysica Acta*, *1838*, 2971–2984.

Chater, K. F., Biró, S., Lee, K. J., Palmer, T., & Schrempf, H. (2010). The complex extracellular biology of *Streptomyces*. *FEMS Microbiology Reviews*, *34*, 171–198.

Chèneby, D., Philippot, L., Hartmann, A., Henault, C., & Germon, J.-C. (2000). 16S rDNA analysis for characterization of denitrifying bacteria isolated from three agricultural soils. *FEMS Microbiology Ecology*, *34*, 121–128.

Cole, S. T., Brosch, R., Parkhil, L. J., Garnier, T., Churcher, C., Harris, D., et al. (1998). Deciphering the biology of *Mycobacterium tuberculosis* from the complete genome sequence. *Nature*, *393*, 537–544.

Cole, S. T., Eiglmeier, K., Parkhill, J., James, K. D., Thomson, N. R., Wheeler, P. R., et al. (2001). Massive gene decay in the leprosy bacillus. *Nature*, *409*, 1007–1011.

Cole, J. A., & Richardson, D. J. (2008). Respiration of nitrate and nitrite. *EcoSal Plus*, *3*(1). http://dx.doi.org/10.1128/ecosal.3.2.5.

Collins, M. D., & Jones, D. (1981). Distribution of isoprenoid quinone structural types in bacteria and their taxonomic implication. *Microbiological Reviews*, *45*, 316–354.

Collins, M. D., Pirouz, T., Goodfellow, M., & Minnikin, D. E. (1977). Distribution of menaquinones in actinomycetes and corynebacteria. *Journal of General Microbiology*, *100*, 221–230.

Cook, G. M., Berney, M., Gebhard, S., Heinemann, M., Cox, R. A., Danilchanka, O., et al. (2009). Physiology of mycobacteria. *Advances in Microbial Physiology*, *55*, 81–319.

Cruz-Ramos, H., Cook, G. M., Wu, G., Cleeter, M. W., & Poole, R. K. (2004). Membrane topology and mutational analysis of *Escherichia coli* CydDC, an ABC-type cysteine exporter required for cytochrome assembly. *Microbiology*, *150*, 3415–3427.

Davidson, P. M., Juneja, V. K., & Branen, J. K. (2002). Antimicrobial agents. In L. Branen, P. M. Davidson, S. Salminen, & J. H. Thorngate (Eds.), *Food additives* (2nd ed., pp. 97–100). New York, NY: Marcel Dekker.

Dimroth, P., & Cook, G. M. (2004). Bacterial Na^+- or H^+-coupled ATP synthases operating at low electrochemical potential. *Advances in Microbial Physiology*, *49*, 175–218.

Dykhuizen, R. S., Fraser, A., McKenzie, H., Golden, M., Leifert, C., & Benjamin, N. (1998). *Helicobacter pylori* is killed by nitrite under acidic conditions. *Gut*, *42*, 334–337.

Elliot, M. A., & Flärdh, K. (2012). Streptomycete spores. In A. Finazzi-Agrò (Ed.), *Encyclopedia of life sciences (eLS)*. Chichester: John Wiley & Sons. http://www.els.net.

Enoch, H. G., & Lester, R. L. (1975). The purification and properties of formate dehydrogenase and nitrate reductase from *Escherichia coli*. *The Journal of Biological Chemisry*, *250*, 6693–6705.

Felsenstein, J. (1985). Confidence limits on phylogenies: An approach using the bootstrap. *Evolution*, *39*, 783–791.

Fernández-López, M., Olivares, J., & Bedmar, E. J. (1994). Two differentially regulated nitrate reductases required for nitrate-dependent, microaerobic growth of *Bradyrhizobium japonicum*. *Archives of Microbiology*, *162*, 310–315.

Fink, D., Weissschuh, N., Reuther, J., Wohlleben, W., & Engels, A. (2002). Two transcriptional regulators GlnR and GlnRII are involved in regulation of nitrogen metabolism in *Streptomyces coelicolor* A3(2). *Molecular Microbiology*, *46*, 331–347.

Fischer, M., Alderson, J., van Keulen, G., White, J., & Sawers, R. G. (2010). The obligate aerobe *Streptomyces coelicolor* A3(2) synthesizes three active respiratory nitrate reductases. *Microbiology*, *156*, 3166–3179.

Fischer, M., Falke, D., Pawlik, T., & Sawers, R. G. (2014). Oxygen-dependent control of respiratory nitrate reduction in mycelium of *Streptomyces coelicolor* A3(2). *Journal of Bacteriology*, *196*, 4152–4162.

Fischer, M., Falke, D., & Sawers, R. G. (2013). A respiratory nitrate reductase active exclusively in resting spores of the obligate aerobe *Streptomyces coelicolor* A3(2). *Molecular Microbiology*, *89*, 1259–1273.

Fischer, M., & Sawers, R. G. (2013). A universally applicable and rapid method for measuring the growth of *Streptomyces* and other filamentous microorganisms by methylene blue adsorption-desorption. *Applied and Environmental Microbiology*, *79*, 4499–4502.

Fischer, M., Schmidt, C., Falke, D., & Sawers, R. G. (2012). Terminal reduction reactions of nitrate and sulfate assimilation in *Streptomyces coelicolor* A3(2): Identification of genes encoding nitrite and sulfite reductases. *Research in Microbiology*, *163*, 340–348.

Fritz, C., Maass, S., Kreft, A., & Bange, F. C. (2002). Dependence of *Mycobacterium bovis* BCG on anaerobic nitrate reductase for persistence is tissue specific. *Infection and Immunity*, *70*, 286–291.

Fukuda, M., Takeda, H., Kato, H. E., Doki, S., Ito, K., Maturana, A. D., et al. (2015). Structural basis for dynamic mechanism of nitrate/nitrite antiport by NarK. *Nature Communications*, *6*, 7097.

Giffin, M. M., Raab, R. W., Morganstern, M., & Sohaskey, C. D. (2012). Mutational analysis of the respiratory nitrate transporter NarK2 of *Mycobacterium tuberculosis*. *PLoS One*, *7*, e45459.

Goddard, A. D., Moir, J. W., Richardson, D. J., & Ferguson, S. J. (2008). Interdependence of two NarK domains in a fused nitrate/nitrite transporter. *Molecular Microbiology*, *70*, 667–681.

Goodfellow, M., & Williams, S. T. (1983). Ecology of actinomycetes. *Annual Review of Microbiology*, *37*, 189–216.

Gottlieb, D., & Anderson, H. W. (1948). The respiration of *Streptomyces griseus*. *Nature*, *107*, 172–173.

Greening, C., Constant, P., Hards, K., Morales, S. E., Oakeshott, J. G., Russell, R. J., et al. (2015). Atmospheric hydrogen scavenging: From enzymes to ecosystems. *Applied and Environmental Microbiology*, *81*, 1190–1199.

Hartman, T., Weinrick, B., Vilchèze, C., Berney, M., Tufariello, J., Cook, G. M., et al. (2014). Succinate dehydrogenase is the regulator of respiration in *Mycobacterium tuberculosis*. *PLoS Pathogens*, *10*, e1004510.

Heim, A. H., Silver, W. S., & Birk, Y. (1957). Cytochrome composition of some strains of *Streptomyces*. *Nature*, *180*, 608–609.

Hensel, M., Deckers-Hebestreit, G., & Altendorf, K. (1991). Purification and characterization of the F_1 portion of the ATP synthase (F_1F_o) of *Streptomyces lividans*. *European Journal of Biochemistry*, *202*, 1313–1319.

Hodgson, D. A. (2000). Primary metabolism and its control in streptomycetes: A most unusual group of bacteria. *Advances in Microbial Physiology*, *42*, 47–238.

Hoffmann, T., Frankenberg, N., Marino, M., & Jahn, D. (1998). Ammonification in *Bacillus subtilis* utilizing dissimilatory nitrite reductase is dependent on *resDE*. *Journal of Bacteriology*, *180*, 186–189.

Holyoake, L. V., Poole, R. K., & Shepherd, M. (2015). The CydDC family of transporters and their roles in oxidase assembly and homeostasis. *Advances in Microbial Physiology*, *66*, 1–53.

Hopkins, A., Buchanan, G., & Palmer, T. (2014). Role of the twin arginine protein transport pathway in the assembly of the *Streptomyces coelicolor* cytochrome bc_1 complex. *Journal of Bacteriology*, *196*, 50–59.

Hopwood, D. A. (2006). Soil to genomics: The *Streptomyces* chromosome. *Annual Review of Genetics*, *40*, 1–23.

Hopwood, D. A. (2007). Actinomycetes and antibiotics. In *Streptomyces in nature and medicine—The antibiotic makers* (pp. 1–27). Oxford, UK: Oxford University Press.

Inoue, Y. (1958). The metabolism of *Streptomyces griseus*. IV. The terminal pathway of the respiration of Streptomyces griseus. *The Journal of Antibiotics (Tokyo)*, *11*, 109–115.

Jepson, B. J., Mohan, S., Clarke, T. A., Gates, A. J., Cole, J. A., Butler, C. S., et al. (2007). Spectropotentiometric and structural analysis of the periplasmic nitrate reductase from *Escherichia coli*. *Journal of Biological Chemistry*, *282*, 6425–6437.

Jia, W., & Cole, J. A. (2005). Nitrate and nitrite transport in *Escherichia coli*. *Biochemical Society Transactions*, *33*, 159–161.

Jia, W., Tovell, N., Clegg, S., Trimmer, M., & Cole, J. (2009). A single channel for nitrate uptake, nitrite export and nitrite uptake by *Escherichia coli* NarU and a role for NirC in nitrite export and uptake. *Biochemical Journal*, *417*, 297–304.

Jormakka, M., Byrne, B., & Iwata, S. (2003). Protonmotive force generation by a redox loop mechanism. *FEBS Letters*, *545*, 25–30.

Jormakka, M., Richardson, D., Byrne, B., & Iwata, S. (2004). Architecture of NarGH reveals a structural classification of Mo-bis-MGD enzymes. *Structure*, *12*, 95–104.

Jormakka, M., Törnroth, S., Byrne, B., & Iwata, S. (2002). Molecular basis of proton motive force generation: Structure of formate dehydrogenase-N. *Science*, *295*, 1863–1868.

Kabus, A., Niebisch, A., & Bott, M. (2007). Role of cytochrome *bd* oxidase from *Corynebacterium glutamicum* in growth and lysine production. *Applied and Environmental Microbiology*, *73*, 861–868.

Kämpfer, P. (2006). The family *Streptomycetaceae*. Part I: Taxonomy. In M. Dworkin, S. Falkow, E. Rosenberg, K. H. Schleifer, & E. Stackebrandt (Eds.), *The prokaryotes: A handbook on the biology of bacteria. Archaea. Bacteria: Firmicutes, actinomycetes: Vol. 3* (3rd ed., pp. 538–604). New York, NY: Springer.

Kämpfer, P. (2012). Genus I. Streptomyces. In M. Goodfellow, P. Kämpfer, H. J. Busse, M. E. Trujillo, K. I. Suzuki, W. Ludwig, & W. B. Whitman (Eds.), *Bergey's manual of systematic bacteriology, part B: Vol. 5* (2nd ed., pp. 1455–1767). New York, NY: Springer.

Kana, B. D., Weinstein, E. A., Avarbock, D., Dawes, S. S., Rubin, H., & Mizrahi, V. (2001). Characterization of the *cydAB*-encoded cytochrome *bd* oxidase from *Mycobacterium smegmatis*. *Journal of Bacteriology*, *183*, 7076–7086.

Kim, M. S., Jang, J., Ab Rahman, N. B., Pethe, K., Berry, E. A., & Huang, L. S. (2015). Isolation and characterization of a hybrid respiratory supercomplex consisting of *Mycobacterium tuberculosis* cytochrome *bcc* and *Mycobacterium smegmatis* cytochrome aa_3. *Journal of Biological Chemistry*, *290*, 14350–14360.

Koch-Koerfges, A., Pfelzer, N., Platzen, L., Oldiges, M., & Bott, M. (2013). Conversion of *Corynebacterium glutamicum* from an aerobic respiring to an aerobic fermenting bacterium by inactivation of the respiratory chain. *Biochimica et Biophysica Acta*, *1827*, 699–708.

Korn-Wendisch, F., & Kutzner, H. J. (1992). The family *Streptomycetaceae*. In A. Balows, H. G. Trüper, M. Dworkin, W. Harder, & K. H. Schleifer (Eds.), *The prokaryotes: A handbook on the biology of bacteria: Ecophysiology, isolation, identification, applications* (2nd ed., pp. 921–995). New York, NY: Springer.

Kumon, Y., Sasaki, Y., Kato, I., Takaya, N., Shoun, H., & Beppu, T. (2002). Co-denitrification and denitrification are dual metabolic pathways through which dinitrogen evolves from nitrate in *Streptomyces antibioticus*. *Journal of Bacteriology*, *184*, 2963–2968.

Kurokawa, T., & Sakamoto, J. (2005). Purification and characterization of succinate: menaquinone oxidoreductase from *Corynebacterium glutamicum*. *Archives of Microbiology*, *183*, 317–324.

Lee, D.-S., Park, J.-S., Kim, Y., & Lee, H.-S. (2014). *Corynebacterium glutamicum sdhA* encoding succinate dehydrogenase subunit A plays a role in cysR-mediated sulfur metabolism. *Applied Microbiology and Biotechnology*, *98*, 6751–6759.

Li, X., Wang, J., Li, S., Ji, J., Wang, W., & Yang, K. (2015). ScbR- and ScbR2-mediated signal transduction networks coordinate complex physiological responses in *Streptomyces coelicolor*. *Scientific Reports*, *5*, 14831.

Lorenzi, M., Sylvi, L., Gerbaud, G., Mileo, E., Halgand, F., Walburger, A., et al. (2012). Conformational selection underlies recognition of a molybdoenzyme by its dedicated chaperone. *PLoS One*, 7, e49523.

Lü, W., Du, J., Schwarzer, N. J., Wacker, T., Andrade, S. L., & Einsle, O. (2013). The formate/nitrite transporter family of anion channels. *Biological Chemistry*, *394*, 715–727.

Lü, W., Schwarzer, N. J., Du, J., Gerbig-Smentek, E., Andrade, S. L., & Einsle, O. (2012). Structural and functional characterization of the nitrite channel NirC from *Salmonella typhimurium*. *Proceedings of the National Academy of Sciences of the United States of America*, *109*, 18395–18400.

Matsoso, L. G., Kana, B. D., Crellin, P. K., Lea-Smith, D. J., Pelosi, A., Powell, D., et al. (2005). Function of the cytochrome bc_1-aa_3 branch of the respiratory network in mycobacteria and network adaptation occurring in response to its disruption. *Journal of Bacteriology, 187*, 6300–6308.
McBride, M. J., & Ensign, J. C. (1987). Effects of intracellular trehalose content on *Streptomyces griseus* spores. *Journal of Bacteriology, 169*, 4995–5001.
McBride, M. J., & Ensign, J. C. (1990). Regulation of trehalose metabolism by *Streptomyces* griseus spores. *Journal of Bacteriology, 172*, 3637–3643.
Miesel, L., Weisbrod, T. R., Marcinkeviciene, J. A., Bittman, R., & Jacobs, W. R., Jr. (1998). NADH dehydrogenase defects confer isoniazid resistance and conditional lethality in *Mycobacterium smegmatis*. *Journal of Bacteriology, 180*, 2459–2467.
Mitchell, P. (1979). Keilin's respiratory chain concept and its chemiosmotic consequences. *Science, 206*, 1148–1159.
Mitchell, P., & Moyle, J. (1967). Chemiosmotic hypothesis of oxidative phosphorylation. *Nature, 213*, 137–139.
Moir, J. W., & Wood, N. J. (2001). Nitrate and nitrite transport in bacteria. *Cellular and Molecular Life Sciences, 58*, 215–224.
Molenaar, D., van der Rest, M. E., Drysch, A., & Yücel, R. (2000). Functions of the membrane-associated and cytoplasmic malate dehydrogenases in the citric acid cycle of *Corynebacterium glutamicum*. *Journal of Bacteriology, 182*, 6884–6891.
Molenaar, D., van der Rest, M. E., & Petrovic, S. (1998). Biochemical and genetic characterization of the membrane-associated malate dehydrogenase (acceptor) from *Corynebacterium glutamicum*. *European Journal of Biochemistry, 254*, 395–403.
Moreno-Vivián, C., Cabello, P., Martínez-Luque, M., Blasco, R., & Castillo, F. (1999). Prokaryotic nitrate reduction: Molecular properties and functional distinction among bacterial nitrate reductases. *Journal of Bacteriology, 181*, 6573–6584.
Nakano, M. M., & Zuber, P. (1998). Anaerobic growth of a "strict aerobe" (*Bacillus subtilis*). *Annual Review of Microbiology, 52*, 165–190.
Nakano, M. M., Zuber, P., Glaser, P., Danchin, A., & Hulett, F. M. (1996). Two-component regulatory proteins ResD-ResE are required for transcriptional activation of *fnr* upon oxygen limitation in *Bacillus subtilis*. *Journal of Bacteriology, 178*, 3796–3802.
Neumann, P., Weidner, A., Pech, A., Stubbs, M. T., & Tittmann, K. (2008). Structural basis for membrane binding and catalytic activation of the peripheral membrane enzyme pyruvate oxidase from *Escherichia coli*. *Proceedings of the National Academy of Sciences of the United States of America, 105*, 17390–17395.
Nicholls, D. G., & Ferguson, S. J. (2013). *Bioenergetics* (4th ed.). Waltham, MA: Academic Press.
Niebisch, A., & Bott, M. (2001). Molecular analysis of the cytochrome bc_1–aa_3 branch of the *Corynebacterium glutamicum* respiratory chain containing an unusual diheme cytochrome c_1. *Archives of Microbiology, 175*, 282–294.
Niebisch, A., & Bott, M. (2003). Purification of a cytochrome *bc-aa3* supercomplex with quinol oxidase activity from *Corynebacterium glutamicum*. Identification of a fourth subunit of cytochrome *aa3* oxidase and mutational analysis of diheme cytochrome *c1*. *Journal of Biological Chemistry, 278*, 4339–4346.
Niederpruem, D. J., & Hackett, D. P. (1961). Respiratory chain of *Streptomyces*. *Journal of Bacteriology, 81*, 557–563.
Nishimura, T., Vertès, A. A., Shinoda, Y., Inui, M., & Yukawa, H. (2007). Anaerobic growth of *Corynebacterium glutamicum* using nitrate as a terminal electron acceptor. *Applied Microbiology and Biotechnology, 75*, 889–897.
Noji, S., & Taniguchi, S. (1987). Molecular oxygen controls nitrate transport of *Escherichia coli* nitrate-respiring cells. *Journal of Biological Chemistry, 262*, 9441–9443.

Oliver, J. D. (2005). The viable but non-culturable state in bacteria. *The Journal of Microbiology*, *43*, 93–100.
Pecsi, I., Hards, K., Ekanayaka, N., Berney, M., Hartman, T., Jacobs, W. R., Jr., et al. (2014). Essentiality of succinate dehydrogenase in *Mycobacterium smegmatis* and its role in the generation of the membrane potential under hypoxia. *mBio*, *5*. e01093-14.
Peczynska-Czoch, W., & Mordarski, M. (1988). Actinomycete enzymes. In M. Goodfellow, S. T. Williams, & M. Motdarksi (Eds.), *Actinomycetes in biotechnology* (pp. 219–283). San Diego, CA: Academic Press.
Platzen, L., Koch-Koerfges, A., Weil, B., Brocker, M., & Bott, B. (2014). Role of flavohaemoprotein Hmp and nitrate reductase NarGHJI of *Corynebacterium glutamicum* for coping with nitrite and nitrosative stress. *FEMS Microbiology Letters*, *350*, 239–248.
Poole, R. K. (1994). Oxygen reactions with bacterial oxidases and globins: Binding, reduction and regulation. *Antonie Van Leeuwenhoek*, *65*, 289–310.
Poole, R. K., & Cook, G. M. (2000). Redundancy of aerobic respiratory chains in bacteria? Routes, reasons and regulation. *Advances in Microbial Physiology*, *43*, 165–224.
Prosser, J. I. (2007). The ecology of nitrifying bacteria. In H. Bothe, S. J. Ferguson, & W. E. Newton (Eds.), *Biology of the nitrogen cycle* (1st ed., pp. 223–243). Amsterdam, The Netherlands: Elsevier.
Quirós, L. M., Hardisson, C., & Salas, J. A. (1986). Isolation and properties of *Streptomyces* spore membranes. *Journal of Bacteriology*, *165*, 923–928.
Rake, G., & Donovick, R. (1946). Studies on the nutritional requirements of *Streptomyces griseus* for the formation of streptomycin. *Journal of Bacteriology*, *52*, 223–226.
Reddy, D., Lancaster, J. R., Jr., & Cornforth, D. P. (1983). Nitrite inhibition of *Clostridium botulinum*: Electron spin resonance detection of iron-nitric oxide complexes. *Science*, *221*, 769–770.
Reents, H., Gruner, I., Harmening, U., Böttger, L. H., Layer, G., Heathcote, P., et al. (2006). *Bacillus subtilis* Fnr senses oxygen via a [4Fe-4S] cluster coordinated by three cysteine residues without change in the oligomeric state. *Molecular Microbiology*, *60*, 1432–1445.
Rehácek, Z., Ramankutty, M., & Kozová, J. (1968). Respiratory chain of antimycin A-producing *Streptomyces antibioticus*. *Applied Microbiology*, *16*, 29–32.
Richardson, D. J. (2000). Bacterial respiration: A flexible process for a changing environment. *Microbiology*, *146*, 551–571.
Richardson, D. J., Berks, B. C., Russell, D. A., Spiro, S., & Taylor, C. J. (2001). Functional, biochemical and genetic diversity of prokaryotic nitrate reductases. *Cellular and Molecular Life Sciences*, *58*, 165–178.
Richardson, D., & Sawers, G. (2002). Structural biology: Through the redox-loop. *Science*, *295*, 1842–1843.
Riedel, C. T., Cohn, M. T., Stabler, R. A., Wren, B., & Brøndsted, L. (2012). Cellular response of Campylobacter jejuni to trisodium phosphate. *Applied and Environmental Microbiology*, *78*, 1411–1415.
Rodríguez-García, A., Barreiro, C., Santos-Beneit, F., Sola-Landa, A., & Martín, J. F. (2007). Genome-wide transcriptomic and proteomic analysis of the primary response to phosphate limitation in *Streptomyces coelicolor* M145 and in a Δ*phoP* mutant. *Proteomics*, *7*, 2410–2429.
Rogov, A. G., Sukhanova, E. I., Uralskaya, L. A., Aliverdieva, D. A., & Zvyagilskaya, R. A. (2014). Alternative oxidase: Distribution, induction, properties, structure, regulation, and functions. *Biochemistry (Moscow)*, *79*, 1615–1634.
Ruddick, S. M., & Williams, S. T. (1972). Studies on the ecology of actinomycetes in soil V. Some factors influencing the dispersal and adsorption of spores in soil. *Soil Biology and Biochemistry*, *4*, 93–103.
Saitou, N., & Nei, M. (1987). The neighbor-joining method: A new method for reconstructing phylogenetic trees. *Molecular Biology and Evolution*, *4*, 406–425.

Sassetti, C. M., Boyd, D. H., & Rubin, E. J. (2003). Genes required for mycobacterial growth defined by high density mutagenesis. *Molecular Microbiology, 48*, 77–84.

Sato, S. (1940). Cytochromes in bacteria, especially *Actinomyces*. *The Kitasato Archives of Experimental Medicine, 17*, 2.

Schneider, D., Pohl, T., Walter, J., Dörner, K., Kohlstädt, M., Berger, A., et al. (2008). Assembly of the *Escherichia coli* NADH:Ubiquinone oxidoreductase (complex I). *Biochimica et Biophysica Acta, 1777*, 735–739.

Schwarz, G., Mendel, R. R., & Ribbe, M. W. (2009). Molybdenum cofactors, enzymes and pathways. *Nature, 460*, 839–847.

Shao, Z., Gao, J., Ding, X., Wang, J., Chiao, J., & Zhao, G. (2011). Identification and functional analysis of a nitrate assimilation operon *nasACKBDEF* from *Amycolatopsis mediterranei* U32. *Archives of Microbiology, 193*, 463–477.

Shoun, H., Kano, M., Baba, I., Takaya, N., & Matsuo, M. (1998). Denitrification by actinomycetes and purification of dissimilatory nitrite reductase and azurin from *Streptomyces thioluteus*. *Journal of Bacteriology, 180*, 4413–4415.

Simon, J. (2002). Enzymology and bioenergetics of respiratory nitrite ammonification. *FEMS Microbiology Reviews, 26*, 285–309.

Sivaramakrishnan, S., & Ortiz de Montellano, P. R. (2013). The DosS-DosT/DosR mycobacterial sensor system. *Biosensors, 3*, 259–282.

Sohaskey, C. D. (2008). Nitrate enhances the survival of *Mycobacterium tuberculosis* during inhibition of respiration. *Journal of Bacteriology, 190*, 2981–2986.

Sohaskey, C. D., & Wayne, L. G. (2003). Role of narK2X and narGHJI in hypoxic upregulation of nitrate reduction by *Mycobacterium tuberculosis*. *Journal of Bacteriology, 185*, 7247–7256.

Sola-Landa, A., Moura, R. S., & Martín, J. F. (2003). The two-component PhoR-PhoP system controls both primary metabolism and secondary metabolite biosynthesis in *Streptomyces lividans*. *Proceedings of the National Academy of Sciences of the United States of America, 100*, 6133–6138.

Sone, N., Fukuda, M., Katayama, S., Jyoudai, A., Syugyou, M., Noguchi, S., et al. (2003). *qcrCAB* operon of a nocardia-form actinomycete *Rhodococcus rhodochrous* encodes cytochrome reductase complex with diheme cytochrome *cc* subunit. *Biochimica et Biophysica Acta, 1557*, 125–131.

Sone, N., Nagata, K., Kojima, H., Tajima, J., Kodera, Y., Kanamaru, T., et al. (2001). A novel hydrophobic diheme *c*-type cytochrome. Purification from Corynebacterium glutamicum and analysis of the qcrCBA operon encoding three subunit proteins of a putative cytochrome reductase complex. *Biochimica et Biophysica Acta, 1503*, 279–290.

Sousa, P. M. F., Videira, M. M., Bohn, A., Hood, B. L., Conrads, T. P., Goulao, L. F., et al. (2012). The aerobic respiratory chain of *Escherichia coli*: From genes to supercomplexes. *Microbiology, 158*, 2408–2418.

Stahl, D. A., & de la Torre, J. R. (2012). Physiology and diversity of ammonia-oxidizing archaea. *Annual Review of Microbiology, 66*, 83–101.

Stewart, V. (2003). Nitrate- and nitrite-responsive sensors NarX and NarQ of proteobacteria. *Biochemical Society Transactions, 31*, 1–10.

Stroh, A., Anderka, O., Pfeiffer, K., Yagi, T., Finel, M., Ludwig, B., et al. (2004). Assembly of respiratory complexes I, III, and IV into NADH oxidase supercomplex stabilizes complex I in *Paracoccus denitrificans*. *Journal of Biological Chemistry, 279*, 5000–5007.

Takeno, S., Ohnishi, J., Komatsu, T., Masaki, T., Sen, K., & Ikeda, M. (2007). Anaerobic growth and potential for amino acid production by nitrate respiration in *Corynebacterium glutamicum*. *Applied Microbiology and Biotechnology, 75*, 1173–1182.

Tamura, K., Stecher, G., Peterson, D., Filipski, A., & Kumar, S. (2013). MEGA6: Molecular evolutionary genetics analysis version 6.0. *Molecular Biology and Evolution, 30*, 2725–2729.

Thomas, L., Hodgson, D. A., Wentzel, A., Nieselt, K., Ellingsen, T. E., Moore, J., et al. (2012). Metabolic switches and adaptations deduced from the proteomes of *Streptomyces coelicolor* wild type and *phoP* mutant grown in batch culture. *Molecular and Cellular Proteomics*, *11*. M111.013797.

Tielen, P., Schobert, M., Härtig, E., & Jahn, D. (2012). Anaerobic regulatory networks in bacteria. In A. A. M. Filloux (Ed.), *Bacterial regulatory networks* (pp. 273–305). Norfolk, UK: Caister Academic Press.

Tiffert, Y., Supra, P., Wurm, R., Wohlleben, W., Wagner, R., & Reuther, J. (2008). The *Streptomyces coelicolor* GlnR regulon: Identification of new GlnR targets and evidence for a central role of GlnR in nitrogen metabolism in actinomycetes. *Molecular Microbiology*, *67*, 861–880.

Unden, G., Müllner, M., & Reinhart, F. (2009). Sensing of oxygen by bacteria. In R. Krämer & K. Jung (Eds.), *Bacterial signaling* (pp. 289–305). Weinheim: Wiley-VCH.

van Keulen, G., Alderson, J., White, J., & Sawers, R. G. (2005). Nitrate respiration in the actinomycete *Streptomyces coelicolor*. *Biochemical Society Transactions*, *33*, 210–212.

van Keulen, G., Alderson, J., White, J., & Sawers, R. G. (2007). The obligate aerobic actinomycete *Streptomyces coelicolor* A3(2) survives extended periods of anaerobic stress. *Environmental Microbiology*, *9*, 3143–3149.

van Spanning, R. J. M., Richardson, D. J., & Ferguson, S. J. (2007). Introduction to the biochemistry and molecular biology of denitrification. In H. Bothe, S. J. Ferguson, & W. E. Newton (Eds.), *Biology of the nitrogen cycle* (1st ed., pp. 3–20). Amsterdam, The Netherlands: Elsevier.

Vilcheze, C., Weisbrod, T. R., Chen, B., Kremer, L., Hazbon, M. H., Wang, F., et al. (2005). Altered NADH/NAD^+ ratio mediates coresistance to isoniazid and ethionamide in mycobacteria. *Antimicrobial Agents and Chemotherapy*, *49*, 708–720.

Waksman, S. A., & Henrici, A. T. (1943). The nomenclature and classification of the actinomycetes. *Journal of Bacteriology*, *46*, 337–341.

Waksman, S. A., Schatz, A., & Reilly, H. C. (1946). Metabolism and the chemical nature of *Streptomyces griseus*. *Journal of Bacteriology*, *51*, 753–759.

Wang, J., & Zhao, G.-P. (2009). GlnR positively regulates *nasA* transcription in *Streptomyces coelicolor*. *Biochemical and Biophysical Research Communications*, *386*, 77–81.

Watanabe, S., Zimmermann, M., Goodwin, M. B., Sauer, U., Barry, C. E., & Boshoff, H. I. (2011). Fumarate reductase activity maintains an energized membrane in anaerobic *Mycobacterium tuberculosis*. *PLoS Pathogens*, 7, e1002287.

Wayne, L. G., & Hayes, L. G. (1998). Nitrate reduction as a marker for hypoxic shiftdown of *Mycobacterium tuberculosis*. *Tubercle Lung Disease*, *79*, 127–132.

Wayne, L. G., & Sohaskey, C. D. (2001). Nonreplicating persistence of *Mycobacterium tuberculosis*. *Annual Review of Microbiology*, *55*, 139–163.

Weinstein, E. A., Yano, T., Li, L. S., Avarbock, D., Avarbock, A., Helm, D., et al. (2005). Inhibitors of type II NADH:menaquinone oxidoreductase represent a class of antitubercular drugs. *Proceedings of the National Academy of Sciences of the United States of A*, *102*, 4548–4553.

Williams, S. T., Goodfellow, M., Alderson, G., Wellington, E. M., Sneath, P. H., & Sackin, M. J. (1983a). Numerical classification of *Streptomyces* and related genera. *Journal of General Microbiology*, *129*, 1743–1813.

Williams, S. T., Goodfellow, M., Wellington, E. M., Vickers, J. C., Alderson, G., Sneath, P. H., et al. (1983b). A probability matrix for identification of some streptomycetes. *Journal of General Microbiology*, *129*, 1815–1830.

Woese, C. R. (1987). Bacterial evolution. *Microbiological Reviews*, *51*, 221–271.

Xia, D., Yu, C. A., Kim, H., Xia, J. Z., Kachurin, A. M., Zhang, L., et al. (1997). Crystal structure of the cytochrome bc_1 complex from bovine heart mitochondria. *Science*, *277*, 60–66.

Yan, H., Huang, W., Yan, C., Gong, X., Jiang, S., Zhao, Y., et al. (2013). Structure and mechanism of a nitrate transporter. *Cell Reports*, *3*, 716–723.
Zheng, H., Wisedchaisri, G., & Gonen, T. (2013). Crystal structure of a nitrate/nitrite exchanger. *Nature*, *497*, 647–651.
Zuckerkandl, E., & Pauling, L. (1965). Evolutionary divergence and convergence in proteins. In V. Bryson & H. J. Vogel (Eds.), *Evolving genes and proteins* (pp. 97–166). New York, NY: Academic Press.
Zumft, W. G. (1997). Cell biology and molecular basis of denitrification. *Microbiology and Molecular Biology Reviews*, *61*, 533–616.

CHAPTER TWO

Anaerobic Metabolism in *Haloferax* Genus: Denitrification as Case of Study

J. Torregrosa-Crespo*, R.M. Martínez-Espinosa*[,1], J. Esclapez*, V. Bautista*, C. Pire*, M. Camacho*, D.J. Richardson[†], M.J. Bonete*

*Universidad de Alicante, Alicante, Spain
[†]University of East Anglia, Norwich, United Kingdom
[1]Corresponding author: e-mail address: rosa.martinez@ua.es

Contents

Abstract

A number of species of *Haloferax* genus (halophilic archaea) are able to grow microaerobically or even anaerobically using different alternative electron acceptors such as fumarate, nitrate, chlorate, dimethyl sulphoxide, sulphide and/or trimethylamine. This metabolic capability is also shown by other species of the Halobacteriaceae and Haloferacaceae families (Archaea domain) and it has been mainly tested by physiological

Advances in Microbial Physiology, Volume 68
ISSN 0065-2911
http://dx.doi.org/10.1016/bs.ampbs.2016.02.001

studies where cell growth is observed under anaerobic conditions in the presence of the mentioned compounds. This work summarises the main reported features on anaerobic metabolism in the *Haloferax*, one of the better described haloarchaeal genus with significant potential uses in biotechnology and bioremediation. Special attention has been paid to denitrification, also called nitrate respiration. This pathway has been studied so far from *Haloferax mediterranei* and *Haloferax denitrificans* mainly from biochemical point of view (purification and characterisation of the enzymes catalysing the two first reactions). However, gene expression and gene regulation is far from known at the time of writing this chapter.

ABBREVIATIONS

DDC diethyldithiocarbamate
DMSO dimethyl sulphoxide
DTE dithioerythritol
EDTA ethylenediaminetetraacetic acid
PHA polyhydroxyalcanoates
PHB polyhydroxybutyrates
TMAO trimethylamine *N*-oxide

1. INTRODUCTION

In general, microorganisms are metabolically versatile growing in a range of environments. In a basic aerobic microbial model, cells use glucose as a carbon and energy source and oxygen acts as terminal electron acceptor. Thus, glucose becomes oxidised resulting in an electrons flux force and a proton-motive force, which are really important to produce ATP and reduced coenzymes such as NADH, NADPH or $FADH_2$. At produced is the vital high-energy molecule that supports growth and synthesis of all the major cellular compounds. Many bacteria and archaea, for instance, can grow in environments without oxygen using anaerobic respiration and fermentation. In the anaerobic respiration, a compound is oxidised using something besides oxygen as the terminal electron acceptor and resulting in a proton-motive force. Both, aerobic and anaerobic respiration, share the same end goal: the generation of a proton-motive force that can be used to synthesise ATP using the ATP synthase.

Archaea, one of the three life domains, make up a significant fraction of the microbial biomass on Earth. This domain is one of the three phylogenetic domains established for the first time by Woese and coworkers (Woese & Fox, 1977; Woese, Kandler, & Wheelis, 1990). This three-domain model for the deepest branches in evolution is now well grounded by considerable further sequence information and biochemical correlations (Graham, Overbeek, Olsen, & Woese, 2000; Woese, 2004).

Archaea were for a long time thought to be restricted to extreme environments, such as those with elevated temperatures, low or really high pH, high salinity or strict anoxia (Valentine, 2007). Thus, the species grouped in this domain were initially viewed as extremophiles inhabiting hostile environments, such as hot springs and salted lakes. In all of these extreme situations, archaea are found together with bacterial and eukaryal organisms also showing extreme phenotypes adapted to these restrictive conditions.

However, environmental sampling analysis based on rRNA sequences has revealed that archaea are ubiquitous in 'normal' ecosystems, including soils, oceans, marshlands, human colon, human oral cavity and even in human skin. Archaea are particularly numerous in the oceans; thus, archaea in plankton may be one of the most abundant groups of organisms on the planet. From a metabolic point of view, archaea have evolved a variety of energy metabolisms using organic and/or inorganic electron donors and acceptors. Because of that reason, microorganisms of the Archaea domain play important roles in the Earth's global geochemical cycles and greenhouse gas emissions (Offre, Spang, & Schleper, 2013). In general, organisms of the Archaea domain are difficult to culture, which impairs experimental manipulation for many of them. Genetic systems exist for all of them (Leigh, Albers, Atomi, & Allers, 2011), but physiological, biochemical and genetic tools have not been developed to an extent similar to *Escherichia coli* (Kletzin, 2007).

Salty environments are dominated by halotolerant and halophilic organisms. Halotolerant organisms do not require salt (mainly NaCl) but their growth is not impaired under saline conditions either; on the contrary, halophiles must have NaCl for growth. In fact, the name 'halophile' comes from the Greek word for 'salt-loving'. Halophiles can be classified into three groups according to their NaCl requirements: slight halophiles (2–5% or 0.34–0.85 M), moderate halophiles (5–20% or 0.85–3.4 M) and extreme halophiles (20–30% or 3.4–5.1 M) (Larsen, 1962).

Archaebacterial halophiles, also called haloarchaea, are extreme or moderated halophilic species inhabiting neutral saline environments such as salt lakes, marine salterns, marshes, saltern crystalliser ponds and the Dead Sea (Grant, Kamekura, McGenity, & Ventosa, 2001; Oren, 2002), for instance. In those environments, salt concentrations are around 1.5–4 M, which corresponds to 9–30% of salts (*p/v*). NaCl is the predominant salt in these ecosystems, and ionic proportions are quite similar to those dissolved salts in seawater. These salted waters/habitats are termed 'thalassohalines', which arise from seawater evaporation and therefore are dominated by NaCl, such as the crystalliser ponds of coastal solar salterns. In contrast, 'athalassohalines' waters/habitats are not of marine origin but from evaporation of freshwater in a system usually dominated by calcium, magnesium and sulphate (Remane & Schleper, 1971).

Hypersaline environments can show neutral (Dead Sea, for instance) or alkaline pH (Big Soda Lake in Nevada). Those ecosystems harbour a large diversity of microorganisms of all three domains: primary producers as the green algae *Dunaliella* (Oren, 2005), aerobic heterotrophic bacteria (mainly belonging to the family Halomonadaceae), anaerobic fermentative bacteria (families Halanaerobiaceae and Halobacteroidaceae) and archaeal microorganisms of the families Halobacteriaceae and Haloferacaceae. In fact, while cell counts of bacterial and eukaryal species decrease with increasing salt concentrations, haloarchaea become the dominant populations (Kletzin, 2007).

Haloarchaea have also been isolated from fossil halite deposits (Stan-Lotter, 2004), and it has also been reported that those microorganisms can be trapped in salt crystals remaining viable for a long time (Grant, 2004). Haloarchaea are also of interest for astrobiological studies and the search for life on Mars, due to their apparent longevity in dry salty environments (Fendrihan et al., 2006) and their ability to cope with extreme temperatures, pH and radiation. Regarding these extreme capabilities, *Haloferax mediterranei*, for instance, was successfully subjected to simulate microgravity (Dornmayr-Pfaffenhuemer, Legat, Schwimbersky, Fendrihan, & Stan-Lotter, 2011). On the other hand, during the last 20 years special attention has being paid on potential uses of haloarchaea in biotechnology and biomedicine due to the capacity of some species producing secondary metabolites of high interest (carotenoids, enzymes showing catalytic properties useful for some industrial processes, etc.).

This chapter offers a summary about anaerobic metabolism in *Haloferax*, one of the better known haloarchaeal genus. Special attention is paid in nitrate respiration (also called denitrification) as model of anaerobic respiration in haloarchaea.

2. GENERAL CHARACTERISTICS OF THE *HALOFERAX* GENUS

Haloferax is the name used to identify a genus of the Haloferacaceae family (Gupta, Naushad, & Baker, 2015), one of the two families grouping haloarchaea. This genus was first described by Torreblanca et al. (1986) and currently comprises several well described species and few strains partially characterised. Probably the better known species are those with the following validly published names: *Haloferax volcanii* (Mullakhanbhai & Larsen, 1975), *Haloferax denitrificans* (Tomlinson, Jahnke, & Hochstein, 1986), *Haloferax gibbonsii* (Juez, Rodriguez-Valera, Ventosa, & Kushner, 1986), *Hfx. mediterranei* (Rodriguez-Valera, Juez, & Kushner, 1983), *Haloferax alexandrinus* (Asker & Ohta, 2002) and *Haloferax sulfurifontis* (Elshahed et al., 2004). Other species included in this genus are *Haloferax lucentense* (formerly *Haloferax alicantei*) (Gutierrez, Kamekura, Holmes, Dyall-Smith, & Ventosa, 2002), *Haloferax prahovense* (Enache, Itoh, Kamekura, Teodosiu, & Dumitru, 2007), *Haloferax larsenii* (Xu et al., 2007), *Haloferax elongans* (Allen et al., 2008), *Haloferax mucosum* (Allen et al., 2008) and *Haloferax chudinovii* (Saralov, Baslerov, & Kuznetsov, 2013).

Members of the genus *Haloferax* are characterised by extreme pleomorphism and a relatively low salt requirement compared with other haloarchaea. Thus, species such as *Hfx. mediterranei* are able to grow even at low salt concentration (D'Souza, Altekar, & D'Souza, 1997).

Looking cell structure in detail, it is interesting to highlight that most of the outer surfaces of the *Haloferax* species are covered with a hexagonally packed surface called S-layer (surface-layer), which is mainly constituted by glycoproteins forming a regularly structured array. These glycoprotein subunits join via both N- and O-glycosidic bonds and are held together by divalent cations (probably magnesium) (Mengele & Sumper, 1992; Sumper, Berg, Mengele, & Strobel, 1990). In fact, cell shape and cell wall structure in haloarchaea, in general, and in *Haloferax* in particular, are unusual due to the S-layer. This layer is plenty of pores and can be removed by treating cells with chelating agents, such as ethylenediaminetetraacetic acid (EDTA). The S-layer is a common feature of many genera of archaea.

The main characteristics of species belonging to this genus are (i) cell shape includes irregular rods, cups or disks ($1.0–3.0 \times 2.0–3.0$ μm); (ii) stain Gram (−); (iii) in general, they show aerobic metabolism but few species are denitrifiers (they use nitrate as terminal electron acceptor under anaerobic conditions); (iv) they are chemoheterotrophic microorganisms able to use carbohydrates, alcohols, carboxylic acids, amino acids and nitrogen compounds such as nitrate, nitrite and ammonium as carbon and nitrogen sources; (v) acidic compounds are produced from sugars; (vi) polyhydroxyalcanoates (PHA) and polyhydroxybutyrates (PHB) are accumulated under certain growth conditions by some *Haloferax* species

(Antón, Meseguer, & Rodríguez-Valera, 1988; Lillo & Rodriguez-Valera, 1990); (vii) polar lipids are characterised by C_{20},C_{20} derivatives of S-DGD-1; (viii) some species are bacteriorhodopsin producers; (ix) carotenoids such as β-carotene, canthaxanthin, astaxanthin and bacterioruberin are produced at high concentrations by some species under certain conditions (Asker & Ohta, 2002; Rodrigo-Baños, Garbayo, Vílchez, Bonete, & Martínez-Espinosa, 2015) and (x) at least one of the species (*Hfx. mediterranei*) produces gas vesicles (Englert, Horne, & Pfeifer, 1990).

Salty environments are highly hostile in terms of life because oxygen is limited and nutrients are scarce. However, haloarchaea have adopted several strategies to sustain metabolism and life under such restricted conditions. Some of the main adaptations are summarised as follows:

- Cells accumulate molar KCl concentrations to maintain osmotic balance instead of accumulating compatible solutes. Thus, cells are isotonic with their surroundings (salt-in strategy). Haloarchaea contain potent transport systems to expel sodium ions, which are predominant in the medium, from the interior of the cell (Madigan & Oren, 1999). This haloadaptation implies that the whole cellular machinery of the haloarchaea is used to K^+ concentrations around 3–5 M, which requires far-reaching alterations of proteins to enable intracellular enzymatic systems to be active.
- Proteins are rich in acidic amino acids as a consequence of the adaptation mentioned before (Madern, Ebel, & Zaccai, 2000). This adaptation allows the proteins to maintain their proper conformation and activity at near-saturating salt concentrations (Oren, 2008). In fact, aspartic and glutamic acid could constitute up to 10% of the overall amino acidic composition. Haloarchaeal proteins have, therefore, become strictly dependent on salt presence (Madern et al., 2000) and most of them denature in solutions containing less than 1–2 M salt (Eisenberg, 1995).
- Modulation of the N-linked glycans decorating the S-layer glycoprotein exits as an adaptive response to salinity changes (Guan, Naparstek, Calo, & Eichler, 2012).
- Synthesis of archaeocins (Besse, Peduzzi, Rebuffat, & Carré-Mlouka, 2015) to be more competitive in the environment. Halocin, which is a type or archaeocin, is produced by *Hfx. mediterranei* as a molecule to inhibit the growth of other halophilic archaea (Cheung, Danna, O'Connor, Price, & Shand, 1997; Naor, Yair, & Gophna, 2013). This strategy reports competitive advantages when *Hfx. mediterranei* is colonising one specific environment.

– Some species such as *Hfx. mediterranei* shows DNA restriction pattern modifications under different salt concentrations (Juez, Rodriguez-Valera, Herrero, & Mojica, 1990).

Genetically, the members of *Haloferax* genus usually have one main chromosome and a variable number of plasmids (Soppa et al., 2008). All of them are characterised by high G+C content (around 65%) (Soppa et al., 2008). This feature increases the stability of the genome within a cytoplasm with high ionic strength. Another possible advantage would be that a G+C-rich genome decreases the possible targets of insertion sequences (IS) elements, which recognise A+T-rich regions (Leigh et al., 2011). IS sequences were studied in *Hfx. volcanii*, where they were located in nonessential regions of the megaplasmids (López-García, St Jean, Amils, & Charlebois, 1995).

The model for genetic research in *Haloferax* has always been *Hfx. volcanii*: it was the first specie discovered in 1936 by Benjamin Elazari-Volcanii and the first genome fully sequenced. It has managed to develop robust transformation protocols and selection markers for the construction of knockout cells (Allers, Ngo, Mevarech, & Lloyd, 2004). Currently, progress has been made in understanding the genomes of other species such as *Hfx. mediterranei* ATCC 33500 (Han et al., 2012), *Hfx. gibbonsii* strain ARA 6 (Pinto, D'Alincourt Carvalho-Assef, Vieira, Clementino, & Albano, 2015) or *Hfx. denitrificans* ATCC 35960, *Hfx. mucosum* ATCC BAA-1512, *Hfx. sulfurifontis* ATCC BAA-897 (Lynch et al., 2012).

Lateral gene transfer (LGT) is a process closely associated with the dynamics of the genomes in *Haloferax* species. Although there are not so many studies of this mechanism as in the Bacteria domain, there are evidences demonstrating that it is present in such microorganisms. For example, the UvrABC complex, whose origin is bacterial, is in haloarchaeas such as *Hfx. volcanii*, where it is fully active (Lestini, Duan, & Allers, 2010). The LGT process could also explain why there are multiple isoforms of genes in a lot of *Haloferax* species, while in other microorganisms exist only in a single form. In fact, it has been proposed that haloarchaea descended from methanogens that acquire the genes for aerobic respiration from bacteria (Leigh et al., 2011).

At the time of writing this work, around 650 papers have been published (http://www.ncbi.nlm.nih.gov/pubmed/?term=haloferax) about items related to *Haloferax*'s physiology, molecular metabolism or molecular biology, which is quite scarce information if we compared with the knowledge reported from other microbial groups. In fact, only 13 of the mentioned papers are focused on anaerobiosis in *Haloferax* (http://www.ncbi.nlm.nih.gov/pubmed/?term=haloferax+%26+anaerobic). However, from the details reported for now, it is possible to conclude that haloarchaea exhibit some characteristics close to eukarya and many others close to prokarya, resulting in interesting phenotypes able to be adapted to very restrictive environmental conditions.

3. ANAEROBIC METABOLISM IN THE *HALOFERAX* GENUS

Anaerobic metabolism, in general, and in particular anaerobic respiration, plays a major role in the global nitrogen, sulphur and carbon cycles through the reduction of the oxyanions of nitrogen, sulphur and carbon to more reduced compounds. Climate change, anthropogenic activities as well as seasonal features cause oxygen availability changes and consequently, sequential changes in redox conditions. Environmental redox cycling often has strong effects on natural biogeochemical cycling as well as biodegradation of anthropogenic organic pollutants. In that context, microorganisms, mainly those showing anaerobic metabolism, play an important role.

As mentioned before, haloarchaea are oxygen-respiring heterotrophs that derive from methanogens-strictly anaerobic, hydrogen-dependent autotrophs (Leigh et al., 2011; Nelson-Sathi et al., 2012). Haloarchaeal genomes are known to have acquired, via LGT, several genes from eubacteria.

Extreme or moderate marine origin environments are inhabited by halophiles, and haloarchaea constitute the major populations as mentioned before. The vast majority of these populations include members of the following haloarchaeal genera: *Haloarcula*, *Haloquadratum*, *Halobacterium*, *Haloferax* or even *Natronomonas* when the pH is alkaline. Most of these species generally grow chemoorganotrophically thanks to a respiratory chain that enables them to use oxygen as electron acceptor (Oren, 1991). Nevertheless, some species have facultative anaerobic capabilities (DasSarma & Arora, 2002). Due to the low solubility of gases and other nutrients in salt-saturated brines, oxygen may easily become a limiting factor for reproducing (flourishing) haloarchaea in these environments. In fact, in saline and hypersaline environments inhabited by species of *Haloferax* genus, oxygen is a limiting factor for cell growth. The often high temperatures of their natural habitats, the high salt concentrations as well as the presence of communities of other halophilic archaea and bacteria that consume oxygen are also responsible for the low availability of it in these environments (Müller & DasSarma, 2005).

Related to this aspect, it is interesting to highlight that a few representative species of the *Haloferax* genus are able to float to the air–water interface, thanks to the production of gas vesicles (Oren, 2002, 2012). Gas vesicles increase the buoyancy of cells and allow them to migrate vertically in the water body from low oxygen conditions to regions with optimal conditions to sustain microaerobic or aerobic growth (Pfeifer, 2015). Haloarchaeal gas vesicles consist of mainly one protein called GvpA, but their formation occurs along a complex pathway involving 14 different *gvp* genes, some of which regulate the process (Zimmermann & Pfeifer, 2003). Gas vesicles synthesis depends on environmental factors, such as light, temperature, salt concentration and oxygen supply. Thus, the production of these vesicles is inhibited under anaerobic conditions at least in *Hfx. mediterranei* and *Hfx. volcanii* (Hechler & Pfeifer, 2009), and glucose also inhibits the formation of gas vesicles in *Hfx. volcanii* transformants (Hechler, Frech, & Pfeifer, 2008).

Apart from gas vesicle strategy, as a mechanisms to optimise cell location in an oxic or microaerobic environment, there are many other strategies supporting haloarchaeal growth under anaerobic conditions: denitrification (Bonete, Martínez-Espinosa, Pire, Zafrilla, & Richardson, 2008; Mancinelli & Hochstein, 1986; Martinez-Espinosa et al., 2007), arginine fermentation (Ruepp & Soppa, 1996), and use of the bacteriorhodopsin (a light-driven proton pump) (Dassarma et al., 2001; Papke, Douady, Doolittle, & Rodríguez-Valera, 2003; Sharma et al., 2007). However, the last strategy has not been developed by *Hfx. mediterranei*, for instance, which lacks energy-generating retinal-based, light-driven ion pumps such as bacteriorhodopsin and halorhodopsin (Oren & Hallsworth, 2014).

In general terms, *Haloferax* growth under those microaerobic or even strict anaerobic conditions is possible because oxygen is replaced by other final electron acceptors such as nitrate, nitrite (Bonete et al., 2008; Esclapez, Zafrilla, Martínez-Espinosa, & Bonete, 2013; Lledó, Martínez-Espinosa, Marhuenda-Egea, & Bonete, 2004; Nájera-Fernández, Zafrilla, Bonete, & Martínez-Espinosa, 2012), (per)chlorate (Martínez-Espinosa, Richardson, & Bonete, 2015; Oren, Elevi Bardavid, & Mana, 2014), sulphur or sulphide (Elshahed et al., 2004), arsenate (Rascovan, Maldonado, Vazquez, & Eugenia Farías, 2015), dimethyl sulphoxide (DMSO), trimethylamine *N*-oxide (TMAO) and fumarate (Müller & DasSarma, 2005; Oren, 1991, 1999; Oren & Trüper, 1990).

Biochemical characterisation of these pathways as well as the enzymes involved in is still scarce, with the exception of denitrification. What is clear is that Rieske-like proteins and cytochromes play an important role to sustain bioenergetics not only under aerobic but also under anaerobic conditions in haloarchaea (Baymann, Schoepp-Cothenet, Lebrun, van Lis, & Nitschke, 2012). More details about these processes are summarised below.

3.1 Denitrification

Denitrification is probably the most studied anaerobic metabolic pathway in *Haloferax* genus (mainly in *Hfx. mediterranei* and *Hfx. denitrificans*). It is based on the use of nitrate (NO_3^-) as final electron acceptor, which is further reduced to gaseous products: nitric oxide (NO), nitrous oxide (N_2O) and dinitrogen (N_2). Generally, organisms able to perform denitrification are classified as complete or incomplete denitrifiers: in the first case, NO_3^- (nitrate) is completely reduced to N_2 (dinitrogen); in the second case, NO_3^- is reduced partially to NO (nitric oxide) or N_2O (nitrous oxide) (NO_x gases). The release of these gases (NO_x) to the atmosphere is harmful since they are responsible of the destruction of the ozone layer and contribute to the greenhouse effect (Ravishankara, Daniel, & Portmann, 2009; Thomson, Giannopoulos, Pretty, Baggs, & Richardson, 2012).

Some enzymes involved in denitrification have been purified and characterised from *Haloferax* species (see Section 4). However, there are no detailed physiological studies on denitrification and NO_x production in species of *Haloferax* genus, thus revealing the depth process. With the information available today, it appears that some organisms such as *Hfx. mediterranei* are complete denitrifiers (Bonete et al., 2008), other such as *Hfx. volcanii* are incomplete and in some cases (*Hfx. denitrificans*) are complete or incomplete depending on the initial amount of NO_3^- (Tindall, Tomlinson, & Hochstein, 1989). Nevertheless, it is clear that denitrification is a form of anaerobic respiration really significant in haloarchaea, and probably the best characterised anaerobic pathway for now from this kind of microorganisms.

Denitrification occurs in many environments including soils, oceans and freshwaters, and it is usually carried out by facultative anaerobes growing under microaerophilic or anoxic conditions (Zumft, 1997). Denitrifying organisms include various bacteria, some archaea, and even eukaryotes (Cabello, Roldán, & Moreno-Vivián, 2004). Only a few cultured archaea are capable of denitrification; *Haloferax* genus plays an important role in that sense grouping several species able to perform partial or even complete denitrification as mentioned before (*Hfx. mediterranei*, *Hfx. volcanii* and *Hfx. denitrificans*, for instance).

The predominance of anaerobic metabolism in archaea and the biogeochemical significance of archaeal denitrification have been little investigated (for review Offre et al., 2013). There are few studies focused on the metabolism of halophilic archaea; however, most of them analyse this subject through systematic metabolic reconstruction and comparative analysis of available genomes (Falb et al., 2008). At the time of writing this chapter, when systems biology approaches have been used to construct predictive models of gene expression and metabolism in bacteria and eukarya, only few studies summarise details about the status of genomics, functional genomics and molecular genetics of haloarchaea (Soppa et al., 2008). This situation could be due to the fact that not too many haloarchaeal genomes are completely sequenced and assembled.

It also remains unclear the role of denitrification in haloarchaea in terms of its ecological relevance. In saline and hypersaline habitats, nitrate is rarely found at high concentrations mainly because of the lack of autotrophic nitrification. So, why these microorganisms have been maintained over time those denitrification genes? Are they an adaptive advantage? These are open questions that should be addressed in the next future.

3.2 Perchlorate and Chlorate Reduction

Microbial reduction of chlorine oxyanions can be found in diverse habitats and different environmental conditions (temperature, salinities, pH) (Nilsson, Rova, & Smedja Bäcklund, 2013). This metabolic process commonly involves the enzymes perchlorate reductase (Pcr), chlorate reductase (Clr) and chlorite dismutase (Cld). The final products are oxygen and chloride (Cl^-). Horizontal gene transfer seems to play an important role for the acquisition of functional genes. Novel and efficient Clds were isolated from microorganisms incapable of growing on chlorine oxyanions (Liebensteiner, Oosterkamp, & Stams, 2015).

One of the latest anaerobic pathways studied in haloarchaea is the anaerobic respiration of perchlorate and chlorate. It has been found that *Hfx. mediterranei* can grow in anaerobic environment using (per)chlorate as final electron acceptors (Martínez-Espinosa et al., 2015). *Hfx. mediterranei* genome analysis revealed that there are not genes coding for the enzymes involved in the (per)chlorate reduction. So, compounds such as chlorate or perchlorate may be reduced through the respiratory nitrate reductase enzyme (Martínez-Espinosa et al., 2015). This is the reason why cells can grow with (per)chlorate as terminal electron acceptor, but only if they have previously been exposed to nitrate. Consequently, it is necessary a preinduction of respiratory nitrate reductase to support the anaerobic growth with (per)chlorate. What it has been suggested for now is that haloarchaea is able to reduce (per)chlorate using the respiratory nitrate reductase located at the positive side of the membrane (pNar) for perchlorate reduction and lack a functional Cld. Chlorite is possibly eliminated by alternative (abiotic) reactions (Martínez-Espinosa et al., 2015).

3.3 Dimethyl Sulphoxide, Trimethylamine *N*-Oxide and Fumarate as Final Electron Acceptors

Some members of the *Haloferax* genus can reduce DMSO, TMAO (Oren & Trüper, 1990) and fumarate using them as final electron acceptors (Oren, 1991). These results come from physiological experiments where cells are grown in the presence of TMAO or DMSO, but there is no information about the regulation of those metabolic pathways or the enzymes catalysing those reactions. On the one hand, the reduction of the first two compounds is usually coupled, producing dimethylsulfide and trimethylamine as final products. *Hfx. mediterranei* can reduce both while *Hfx. volcanii* only can use DMSO. Other members of the *Haloferax* group like *Hfx. gibbonsii* do not grow in anaerobic conditions with addition of DMSO or TMAO (Oren & Trüper, 1990). Although the bases of the DMSO and TMAO respiratory systems have not been described for any member or the domain Archaea, some studies revealed that the genetic machineries involved in two pathways are closely related in haloarchaea (Müller & DasSarma, 2005). On the other hand, the use of fumarate as terminal electron acceptor producing succinate has been described for *Hfx. volcanii* and *Hfx. denitrificans*, but not for *Hfx. mediterranei* and *Hfx. gibbonsii* (Oren, 1991). The ability to reduce fumarate is not correlated with the ability to reduce nitrate, (per)chlorate, DMSO or TMAO.

The presence of the named compounds is minority in environments with high salt concentrations. Their ecological role is still uncertain in the context of the habitat where *Haloferax* species live. In this field, not only physiological but also genetic and biochemical studies are needed.

4. ENZYMES INVOLVED IN ANAEROBIC METABOLISM IN *HALOFERAX* GENUS: DENITRIFICATION AS STUDY OF CASE

Denitrification pathway, carried out under anoxic conditions, is a key process involved in the nitrogen cycle of the Earth. In the complete pathway, the nitrate is reduced to N_2 by the action of four metalloenzymes: respiratory nitrate reductase, respiratory nitrite reductase, nitric oxide reductase and nitrous oxide reductase. Physiological, biochemical and genetic data have provided a detailed process for this pathway in the Bacteria domain (Zumft, 1997). However, the biochemical and genomic data related to denitrification process in extremophiles, and specifically in *Haloferax* genus, are still scarce. Although during the last years the number of available genomes of *Haloferax* genus has increased allowing the identification of denitrification genes, the biochemical studies related to this pathway are basically restricted to the purification and characterisation of respiratory nitrate and nitrite reductases from *Hfx. mediterranei*, *Hfx. denitrificans* and *Hfx. volcanii.*

In view of this, the present section describes the biochemical characteristic of the denitrifying enzymes from *Haloferax* microorganisms.

4.1 Respiratory Nitrate Reductases in *Haloferax* Genus

Denitrifying microorganisms contain nitrate reductase as the terminal enzyme of the nitrate respiration (Zumft, 1997). According to the structural and catalytic characteristics, dissimilatory nitrate reductases can be classified into two groups: periplasmic nitrate reductase (Nap) and membrane-bound nitrate reductase (Nar). The Nap enzymes are mainly found in Gram-negative bacteria, and they are involved in different processes depending on the organism in which are found (Ellington, 2003; Gavira, Roldan, Castillo, & Moreno-Vivian, 2002). Generally, Nap enzymes are heterodimers composed of a catalytic subunit (NapA) and a cytochrome *c* (NapB) which receives electrons from NapC, a membrane cytochrome *c* (Richardson, Berks, Russell, Spiro, & Taylor, 2001). On the other hand, Nar enzymes are distributed more widely in the nitrate-respiring microorganisms, and they are the responsible for the generation of metabolic energy using nitrate as a terminal electron acceptor. Not surprisingly, they are negatively regulated by oxygen, induced by the presence of nitrate and unaffected by ammonium. In general, Nar complex is a heterotrimer composed of a catalytic subunit (NarG) that binds a bis-molybdopterin guanine dinucleotide (bis-MGD) cofactor for nitrate reduction, an electron transfer subunit with four iron-sulphur centres (NarH) as well as a di-*b*-haem integral membrane quinol dehydrogenase subunit (NarI). The NarG and NarH are membrane-extrinsic domain, whereas the NarI is a hydrophobic membrane protein which connects the NarGH complex to the membrane (Cabello et al., 2004; Martínez-Espinosa, Richardson, Butt, & Bonete, 2006; Richardson et al., 2001).

At the time of writing, all the purified and characterised nitrate reductases from haloarchaea, belonging to *Haloferax* genus, are membrane-bound Nar enzymes (Table 1). In general, the characteristics of these enzymes showed marked resemblance with the bacterial NarGH complex, underscoring the fact that there was a relevant difference related to the subcellular localization between the halophilic and bacterial enzymes (Martinez-Espinosa et al., 2007; Yoshimatsu, Iwasaki, & Fujiwara, 2002).

In *Haloferax* genus, purification of respiratory Nar enzymes has been reported from three halophilic microorganisms (Table 1), being the best studied Nar enzyme that one which belongs to *Hfx. mediterranei* specie.

The first respiratory nitrate reductase purified and characterised from the *Haloferax* genus was the *Hfx. denitrificans* membrane-bound Nar (Table 1). This enzyme is a heterodimer with a K_m for nitrate of 0.2 mM. The enzyme is able to reduce not only nitrate but also chlorate, the electron donor is methyl viologen (MV) and is inhibited by azide and cyanide. Azide and cyanide are inhibitors with respect to nitrate. The first one acts directly in the molybdenum containing site or the Nar, probably by metal chelation. The second one blocks electron transfer in oxygen respiration and acts as a noncompetitive inhibitor of nitrate reduction. Curiously, unlike other

Table 1 Characteristics of Respiratory Nitrate Reductases and Cu-Nitrite Reductases from *Haloferax* Genus

Microorganism	Enzyme	Structure Features	Optimal Activity Conditions	Inhibitors
Haloferax denitrificans	Nitrate reductase	Heterodimer: 116 and 60 kDa	Absence of salt	Azide Cyanide
	Cu-nitrite reductase	Homotrimer	4 M NaCl pH 4.8–5.0	DDC
Haloferax volcanii	Nitrate reductase	Heterotrimer: 100, 61 and 31 kDa	Absence of salt Temperature 80°C pH 7.5	Azide Cyanide Thiocyanate
Haloferax mediterranei	Nitrate reductase	Heterodimer: 112 and 61.5 kDa	Absence of salt pH 7.9 at 40°C pH 8.2 at 60°C	Dithiothreitol Azide Cyanide EDTA
	Cu-nitrite reductase	Homotrimer	2 M NaCl pH 5.5	Nondetermined

halophilic enzymes, this nitrate reductase is stable in the absence of salt and its activity decreases with increasing salt concentration. Besides, it was suggested that the enzyme contains molybdenum because tungstate represses nitrate reductase synthesis (Hochstein & Lang, 1991).

Bickel-Sandkotter and Ufer described the properties of respiratory Nar from *Hfx. volcanii* in 1995 (Table 1), whose activity was induced by the addition of nitrate as nitrogen source to the culture media and anaerobic conditions. This enzyme was also located on the surface of the cytoplasmic membrane. It was described as a trimeric protein, whose putative subunits showed molecular masses of approximately 100, 61 and 31 kDa. The kinetic constant of respiratory Nar from *Hfx. volcanii* was determined using MV and dithionite in saturating conditions and high NaCl concentration (1.75 M). The K_m calculated for nitrate was 0.36 mM, being similar to the date calculated for other dissimilatory Nar. Like the *Hfx. denitrificans* Nar, this enzyme showed optimal activity in the absence of NaCl. Moreover, *Hfx. volcanii* Nar reached its optimum activity in a buffer with medium pH of 7.5 and high temperatures up to 80°C, and it was inhibited in the presence of cyanide, azide and a relatively high concentrations of thiocyanate (Bickel-Sandkotter & Ufer, 1995).

In *Hfx. mediterranei* two different nonassimilatory nitrate reductases have been purified and characterised: a dissimilatory nitrate reductase described by Alvarez-Ossorio, Muriana, de la Rosa, and Relimpio (1992) and Nar characterised by Lledó et al. (2004). The expression of the first one was induced not only by means of the presence of nitrate as nitrogen source but also by the switch to anaerobic conditions. This enzyme was purified in five steps (ammonium sulphate precipitation followed by Sepharose DL-4B, calcium phosphate, DEAE-Sephacel and Sephacryl S-200 chromatographies), being its estimated molecular weight 170,000 Da. Unlike previously described dissimilatory nitrate reductases, the activity of this enzyme was salt dependent, showing its optimal activity at 89°C in 3.2 M NaCl. Its kinetic parameters depend also on salt concentration in the assay. In fact, the K_m for nitrate changed from 2.5 to 6.7 mM when the salt concentration increased from 0.8 to 3.4 M. This halophilic enzyme was strongly inhibited in the presence of *p*-hydroxymercuribenzoate, dithioerythritol (DTE), azide and cyanide while the cyanate, potassium chlorate or EDTA produced a partial inhibition. The electron donor studies revealed that methyl and benzyl viologen are the best for this enzyme, while FMA or FAD was ineffective electron donors (Alvarez-Ossorio et al., 1992). According to its molecular mass and enzymatic properties, Lledó et al. (2004) proposed that the enzyme purified by Alvarez-Ossorio allows the dissipation of reducing power for redox balancing. The *Hfx. mediterranei* Nar was purified by means of three chromatographic steps: two DEAE-Sepharose CL-6B and Sephacryl S-300 chromatographies. The enzyme was described as a heterodimer with a K_m for nitrate of 0.82 mM, which is in the range of the values determined from other nitrate reductases (Zumft, 1997). Like other nitrate reductases, cyanide and azide were strong inhibitors of this enzyme. Other compounds as dithiothreitol (DTT) and EDTA were also tested, but they were not effective inhibitors since only decreased partially the activity. The *Hfx. mediterranei* Nar did not exhibit a strong dependence on temperature at the different NaCl concentrations assayed (0–3.8 M NaCl), showing the maximum activity at 70°C for all NaCl concentrations. Hence, this halophilic enzyme also presented a remarkable thermophilicity although the Nar activity did not show a dependence on salt concentration, as was described for *Hfx. denitrificans* Nar (Hochstein & Lang, 1991) and *Hfx. volcanii* Nar (Bickel-Sandkotter & Ufer, 1995). Not all nitrate reductases activities found in halophilic archaea exhibit similar dependence (Alvarez-Ossorio et al., 1992; Yoshimatsu, Sakurai, & Fujiwara, 2000). Even though most proteins for haloarchaea are stable and active at high ionic strength, there are some that are either active or stable in the absence of salt. The origin of haloarchaeal enzymes which does not require salt is unclear, but it has been proposed that Nar could be acquired by the extreme halophiles from a eubacterial source (Hochstein & Lang, 1991). The absorption spectrum of the *Hfx. mediterranei* Nar showed a broad band around 400–415 nm indicating that this enzyme has Fe–S clusters as other Nar purified from denitrifying microorganisms (Lledó et al., 2004).

Classically, it has been considered that the subunits NarG and NarH are located in the cytoplasm and associate with NarI at the membrane potential-negative cytoplasmic face of the cytoplasmic membrane. These data indicate that the nitrate reduction must produce it on the inside of this membrane. This arrangement is conserved in Gram-negative bacteria and indeed, for many years, it was assumed that this orientation would be conserved among prokaryotes in general. However, the presence of a typical twin-arginine signal in *Hfx. mediterranei* NarG and another halophilic microorganisms suggests that nitrate reductases from archaea could be translocated across the membrane by Tat export pathway. Later, the analysis of N-terminal region of the archaeal nitrate reductases revealed the conservation of a twin-arginine motif (Martinez-Espinosa et al., 2007). These data underline the fact that NarG protein could be strongly attached to the membrane fraction and requires detergent solubilisation to release it (Lledó et al., 2004). In order to study the location of the NarG, amino acid sequence analysis and bioinformatic studies were carried out with *Hfx. mediterranei* (Martinez-Espinosa et al., 2007). The results obtained revealed that the electron donation to the active site of an enzyme is on the outside, rather than inside, of the cytoplasmic membrane. These experiments have not yet been reported for the other archaeal Nars with Tat sequences thus far identified. Nonetheless, the available data support the fact that the active site of these archaeal Nar systems is indeed on the outside of the cytoplasmic membrane (Martinez-Espinosa et al., 2007).

Hence, according to the subunit composition and subcellular location in *Hfx. mediterranei*, it can suggest that archaeal Nars are a new type of enzymes with the active site facing the outside and connected to the membrane by cytochrome *b*. The location of archaeal NarG (catalytic site) has important implications because to be energy-conserving require the coupling of this process to a proton-motive complex, instead of the typical redox loop mechanism, the NarI subunit described in bacteria. On the other hand, it appears that an active nitrate-uptake system would not be required for respiratory nitrate reduction in archaea, consequently the energetic yield of the nitrate reduction process increases (Bonete et al., 2008; Martinez-Espinosa et al., 2007).

The last advances related to the knowledge of respiratory Nar has been carried out in *Hfx. mediterranei* (Martínez-Espinosa et al., 2015), where it has been tested the capacity of the whole cells and pure NarGH to reduce different substrates as chlorate, perchlorate, bromate, iodate and selenate. The results demonstrated that not only the whole *Hfx. mediterranei* cells but also pure NarGH were able to reduce chlorate, bromate and perchlorate, but no reduction activity was observed with iodate or selenate. Therefore, it is clear that the same microorganism is able to reduce nitrate and chlorate thanks to the nitrate reductase under microaerobic or anaerobic conditions. Undoubtedly, due to most of the wastewater samples containing nitrate also include chlorate and other oxyanions, these results are crucial for wastewater bioremediation aims. Although the removal procedure is not really fast (4.8 mM chlorate after 150 h incubation) the removed concentration using microorganisms is one of the highest described up to now (Bardiya & Bae, 2005; van Ginkel, van Haperen, & van der Togt, 2005). In addition, one of the advantages of using *Hfx. mediterranei* cells as well as its NarGH is that nitrate reduction is not inhibited by the presence of chlorate or perchlorate at high ionic strength. These results are of great interest for bioremediation processes based on the use of haloarchaea, as it has been explained earlier, or even to improve the knowledge of biological chlorate reduction in early Earth or Martian environments (Martínez-Espinosa et al., 2015).

4.2 Respiratory Nitrite Reductases in *Haloferax* Genus

One of the most important steps in denitrification pathway involves the reduction of nitrite to nitric oxide by the respiratory nitrite reductase (NiR), a key enzyme used to distinguish between nitrate reducers and denitrifiers. This reaction represents the return of nitrite to the gaseous state leading to a significant loss of fixed nitrogen from the terrestrial environment. According to structural features and prosthetic metal, the respiratory nitrite reductases have been classified into two types: cytochrome cd_1-nitrite reductase (encoded by *nirS*) and Cu-containing dissimilatory nitrite reductase (encoded by *nirK*). The cd_1-nitrite reductase is homodimeric and contains haem *c* and d_1 as prosthetic cofactors, whereas Cu-nitrite reductase (Cu-NiR) is homotrimeric and contains two Cu atoms per subunit molecule. Cu-NiR enzymes can be easily distinguished according to their spectra and its sensitivity to diethyldithiocarbamate (DDC) (Shapleigh & Payne, 1985). The two NiR types are functionally and physiologically equivalent, but while cd_1-nitrite reductase predominates in denitrifying bacteria, Cu-nitrite reductase is present in a greater variety of physiological groups and bacteria from various habitats (Heylen et al., 2006; Zumft, 1997).

The first evidence related to the activity of respiratory nitrite reductase in *Haloferax* genus was reported in *Hfx. denitrificans*, where it was described that the reduction of nitrite to nitric oxide by their membranes was inhibited by DDC. These results suggested that the Cu-NiR was involved in that reaction (Hochstein & Tomlinson, 1988). It was in 1996 when the first extremophilic respiratory nitrite reductase from *Hfx. denitrificans* was purified and characterised (Table 1) from soluble and membrane fractions (Inatomi & Hochstein, 1996). Electrophoretic analysis of the purified protein revealed the presence of two peptides of 64 and 51 kDa. The molecular mass of this protein was solved by gel filtration chromatography, suggesting that the enzyme was a dimer with 127 kDa. The authors proposed that the small band present in polyacrylamide gel was the result of a degradation of the larger subunit, although nowadays it is known that these data are inaccurate. Although the protein showed its maximum activity in the presence of 4 M NaCl (Table 1), in the absence of salt the enzyme did not loss activity. Its absorption spectrum was characterised by maxima located at 462, 594 and

682 nm, which disappeared after the addition of dithionite. These data indicated that the halophilic NiR belongs to the green Cu-NiR. The inhibition of Cu-NiR in the presence of low concentrations of DDC supported that this enzyme was a Cu-NiR. Even though the membrane-bound Cu-NiR was not totally purified, its characteristics were similar to those of the enzyme purified from the soluble fraction (Inatomi & Hochstein, 1996).

The last advances in the study of respiratory nitrite reductases in extremophilic microorganisms, in general, and in *Haloferax* genus, specifically, have been carried out in *Hfx. mediterranei* (Table 1) (Esclapez et al., 2013). This halophilic respiratory nitrite reductase was expressed in the halophilic host *Hfx. volcanii*. The enzymatic activity of the recombinant protein was detected in cytoplasmic fraction and membranes as well as in the culture media. The enzymes isolated from cytoplasmic fraction and culture media were purified and characterised. The cytoplasmic NiR was described as a trimeric protein, which presented its maximum activity in the presence of 2 M of salt (NaCl or KCl), and around 70°C. The presence of four significant regions in its structure were established from bioinformatics analysis, which are

- Probable Tat motif. Consequently, that region could act as the Tat motif for the protein to be exported via Tat system.
- Possible cutting targets recognised for proteases in positions 27 and 34 from the N-terminal end. The presence of this sequence is associated with the Tat signals because of the mature protein exportation through the cytoplasmic membrane requires the removal of the signal peptide.
- Type 1 copper centre constituted by His129, Cys170, His178 and Met183.
- Type 2 copper centre constituted by Asp132, His134 and His169.

On the other hand, two different maxima absorption at 453 and 587 nm were identified in the UV–vis spectrum suggesting that the enzyme belongs to the green Cu-NiR group. In order to elucidate the composition of the native enzyme, a native PAGE of pure enzyme followed by activity NiR staining showed that the intracellular Cu-NiR is composed of at least five different isoforms of the enzyme. The SDS-PAGE of each of the five bands revealed that each one presents a different combination of two isoforms with 44.3 and 39.8 kDa. The smaller form was the predominant isoform protein in this cellular fraction (Fig. 1A). According to the two cleavage sites present in *Hfx. mediterranei* Cu-NiR sequence, it is logical to think that the expression of recombinant protein could conclude with the maturation of the initial polypeptide through a cut in one of the two targets present at its

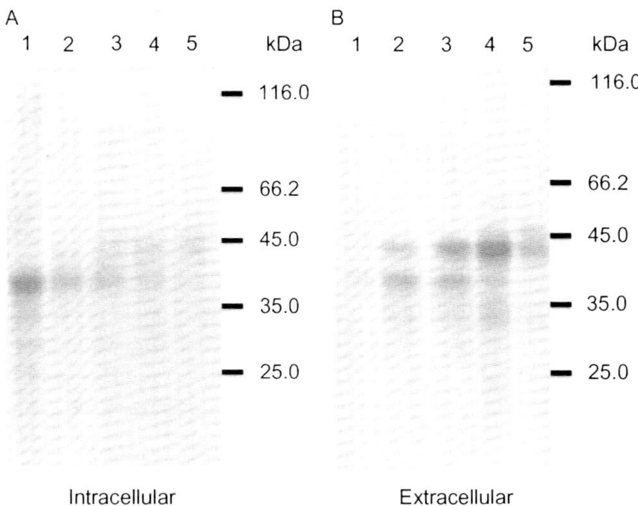

Fig. 1 SDS-PAGE for each of the five isoforms of the intracellular (A) and extracellular (B) Cu-NiR obtained from a native PAGE. *Lanes 1–5*: isoforms detected in the native PAGE for the intracellular (A) and extracellular (B) Cu-NiR; *Lane 6*: molecular weight standards (Thermo Scientific).

N-terminal end. Then, the two possible isoforms could combine to form a pool of active trimers. This maturation mechanism could also explain why it is possible to observe two bands with different masses to NiR purification carried out in *Hfx. denitrificans*. On the other hand, the extracellular pool of recombinant NiR was also purified and characterised. The results obtained with this fraction were similar to those obtained with the intracellular Cu-NiR fraction. However, the comparison of the isoform expression pattern of both samples in the SDS-PAGE revealed a significant difference. In the intracellular fraction, the 39.8 kDa isoform was predominant and the 44.3 kDa isoform appeared slightly, whereas in the extracellular fractions the 44.3 kDa isoform was the predominant or even the only one (Fig. 1B). According to these electrophoretic analyses, the halophilic Cu-NiR could be involved in a maturation process and exportation via the Tat system. To elucidate the nature of the two isoforms the first eight amino acids of each one were sequenced. The results showed that the 44.3 kDa isoform is obtained due to the cleavage between the 33rd and 34th residues. Therefore, this isoform could be exported via the Tat system, being cleaved by the twin-arginine signal sequence after its translocation to extracellular medium. The sequence of the small isoform started in the 52nd position, but no cutting target was predicted around this location. Consequently, it seems more likely that this isoform could be obtained as a result of an alternative translation mechanism (Hering, Brenneis, Beer, Suess, & Soppa, 2009) or mRNA processing rather than as a cleavage process. Once the two possible transcripts are translated, a combination of the two isoforms to form the trimer occurs between them. This process originates the pool of possible isoforms found both in the cytoplasmic and extracellular fractions. Finally, the Tat system of *Hfx. volcanii* could promote the exportation of recombinant Cu-NiR active trimers whenever any of the three contain the signal peptide. In the process of exportation through the membrane, the signal peptides of the large isoform are cleaved. Therefore, outside the cell it can find a mixture of the cleaved and signal-avoided NiR, prevailing over the large isoform. Otherwise only the trimers remain inside the cell exclusively composed by untargeted peptides that not are able to cross the membrane and go outside the cell.

This difference between targeted and nontargeted peptides could be a mechanism for regulating the system and final Cu-NiR location. The location of recombinant Cu-NiR outside the cell agrees with the results related with the extracellular location of membrane-associated NarGH from *Hfx. mediterranei* detailed earlier (Martinez-Espinosa et al., 2007). For this reason, there are considerable evidences to propose that the complete reduction of nitrate could take place through an extracellular enzymatic complex which is part of the machinery associated with the outer face of the cytoplasmic membrane, whereas the rest of soluble enzymes and metabolites are embedded in the porous S-layer. This unusual respiratory structure offers advantages to these microorganisms in oxygen-poor environments such as hypersaline ecosystems. With this modification, the presence of NO_3^- transporters is not needed and the electron acceptor can be reduced directly in the growth media improving the efficiency of the process. Finally, the mobilisation of the proteins involved in NO_3^- respiration appears to be regulated by the Tat system so that they are folded and loaded with metallic cofactors inside the cell before being exported out of the cell where they will take part in their physiological role.

4.3 Nitric Oxide Reductases in *Haloferax* Genus

Nitric oxide is the product of the reaction catalysed by respiratory nitrite reductase. This compound is toxic for the cells and for that reason it is immediately reduced to N_2O by nitric oxide reductases (Nor). The toxicity of NO is due to its reactivity with transition metal proteins and oxygen and its capacity to produce adducts with amines and thiols of fluctuating stability. In fact, knockout mutation for Nor enzymes results lethal for the microorganisms (Zumft, 1997). There are various difficulties that complicate the biochemical analysis of this type of enzyme as, for example:

- Nor enzymes are membrane-bound proteins which require a detergent for solubilisation.
- NO reactivity with cellular components complicates the enzymatic purification.
- Nor enzymes must be isolated in anaerobic conditions.
- Nor enzymes could be part of other protein complexes.

Nonetheless, different enzymes with Nor activities have been described up to now, among which are the following (Bonete et al., 2008; Cabello et al., 2004; Nakara, Tanimoto, Hatano, Usuda, & Shoun, 1993):

- Denitrifying fungi: Nor enzymes are soluble and monomeric. They belong to the cytochrome P-450 family. Its expression is induced by the presence of nitrate or nitrite under anaerobic conditions.
- Denitrifying bacteria: Nor enzymes are heterodimer and constitute membrane complex of a cytochrome *c* (encoded by *norC*) and a cytochrome *b* with 12 transmembrane regions (encoded by *norB*). These enzymes are known as cNor.
- Other bacteria: Nor enzymes are monomeric with 14 transmembrane regions. These enzymes are called qNor due to its quinol-oxidising activity. qNor enzyme is similar to NorB subunit, although it contains an N-terminal extension, with a quinone-binding site, absent in NorB.

Despite the fact that there is only one study related to the characterisation of Nor in extremophilic microorganisms thus far, gas formation from nitrite has been reported for two microorganisms belonging to *Haloferax* genus, which are *Hfx. denitrificans* and *Hfx. mediterranei* (Zumft & Kroneck, 2006). Regarding to the genetic analysis, *Hfx. volcanii* and *Hfx. mediterranei* contain in their genomes a copy of a *norB* gene (see Section 5). However, at the time of writing this review, no Nor enzyme has been characterised and purified neither in microorganisms belonging to *Haloferax* genus nor in extreme halophilic microorganisms.

4.4 Nitrous Oxide Reductases in *Haloferax* Genus

Conversion of N_2O to N_2 is the last step of denitrification pathway and represents a respiratory process in its own right. This reaction is of high environmental importance because it closes the N-cycle. For that reason, no wonder that this pathway is found in a broad spectrum of microorganisms, ranging from psychrophiles to hyperthermophiles or from halophiles to barophiles. Therefore, extreme habitats and N_2O use are compatible, being its reduction catalysed by nitrous oxide reductases (Nos). However, N_2O is less toxic that NO or nitrite and the vast majority of microorganisms could manage without converting N_2O into N_2, performing a partial denitrification.

The nitrous oxide reductases are structurally complex enzymes with different copper centres. Two different novel Cu centres have been described in bacteria: the mixed-valent dinuclear Cu_A species at the electron entry site of the enzyme, and the tetranuclear Cu_z centre as the first catalytically active Cu–S complex described. Furthermore, the synthesis of this type of enzymes is very complicated, being accessory proteins (Cu chaperone and ABC transporters) involved in the biogenesis of the catalytic centre. This is the reason why there are few studies related with this enzyme. The presence of other important bioelements as Mo, Mn, Co, Ni or Zn appears not to be required for Nos activity. Nevertheless, calcium could play an important function in the stabilisation of the protein structure, which is critical for catalysis. Related to Nos inhibition studies, it is known that the acetylene acts as non-competitive inhibitor, although its mechanism of action is still unknown (Zumft & Kroneck, 2006).

Total denitrification pathway has been studied in a high number of bacteria, which contain nitrous oxide reductases encoded by the *nosZ* gene. These bacterial enzymes are located in the periplasm and they are multi-copper homodimers whose electron donor is cytochrome *c* or pesudoazurin (Zumft, 1997). Putative *nosZ* gene has been identified in *Hfx. mediterranei* and *Hfx. denitrificans*, but no Nos enzyme has been purified and characterised from these microorganisms yet, due to the difficulty of working with this type of enzymes and the scarce information available related to the denitrification enzymes in extreme halophilic microorganisms. Despite this, preliminary assays carried out with *Hfx. mediterranei* have revealed that this enzyme is expressed in anaerobic conditions and it is located in its membranes.

5. GENES CODING FOR THE ENZYMES SUSTAINING DENITRIFICATION

From the previous section it is possible to conclude that some studies about biochemical characterisation of denitrifying enzymes have been reported from *Haloferax* species. However, no details have been published so far from genes coding these enzymes (with the exception of *nar* operon from *Hfx. mediterranei*; Lledó et al., 2004) or their regulation.

A general analysis of the *Haloferax* genome sequences already published has been designed and performed in our research group. The aim of this analysis was to look for genes coding for enzymes and proteins involved in denitrification. This analysis has been tedious due to two main reasons: (i) most of the sequences are in contig format (only few genomes are completely sequenced, assembled and annotated), (ii) the nomenclature used to identify the genes is confuse and not conserved. To perform this study the following tools have been used: DNA and amino acid sequences were analysed using the available database on the NCBI (National Center for Biotechnology Information) server (http://www.ncbi.nlm.nih.gov/protein; http://www.ncbi.nlm.nih.gov/bioproject/; http://www.ncbi.nlm.nih.gov/gene) and the Ensembl genome annotation system (http://ensemblgenomes.org/).

Figs. 2–4 summarised the organisation of the respiratory nitrate reductase, respiratory nitrite reductase, nitric oxide reductase and nitrous oxide reductase gene clusters in several species of the *Haloferax* genus: *Hfx. mediterranei* (which is the model organisms in our group to study nitrogen metabolism), *Hfx. volcanii*, *Hfx. denitrificans*, *Hfx. larsenii*, *Hfx. elongans*, *Hfx. lucentense*, *Hfx. alexandrinus*, *Hfx. gibbonsii*, *Hfx. prahovense* and *Hfx. sulfurifontis*. The gene cluster organisation in *Hfx. larsenii*, *Hfx. elongans*, *Hfx. lucentense*, *Hfx. alexandrinus*, *Hfx. prahovense* and *Hfx. sulfurifontis* was obtained from genome sequencing projects in which the DNA sequences are in different contigs. As the ORFs have been automatically annotated, in some of them the nomenclature used to identify the genes is confusing. The homology of these ORFs was compared using the Blast tool (http://blast.ncbi.nlm.nih.gov/Blast.cgi). Because of the previous reason, the bioinformatic analysis is difficult and in some cases could contain gaps due to the information lacking in the sequenced genomes.

Regarding the respiratory nitrate reductase gene cluster (Fig. 2), the organisation is almost the same in the seven compared species. The three ORFs encoding NarJ and the two adjacent hypothetical proteins cannot be found in *Hfx. alexandrinus*. It is notorious that in the genome of *Hfx. prahovense* nitrate reductase operon was not identified, although the nitrite reductase and the nitric oxide reductase gene clusters were present. The nitrate reductase cluster is encoded in a plasmid in *Hfx. mediterranei* and in *Hfx volcanii*, but it is impossible to know their localization in the other species. It is possible to think that the ability of some species for nitrate respiration has been acquired by lateral transference of the plasmid encoded genes. This cluster contains genes coding for the catalytic and the electron transfer subunit (*narG* and *NarH*, respectively), a gene (*mobA*) involved

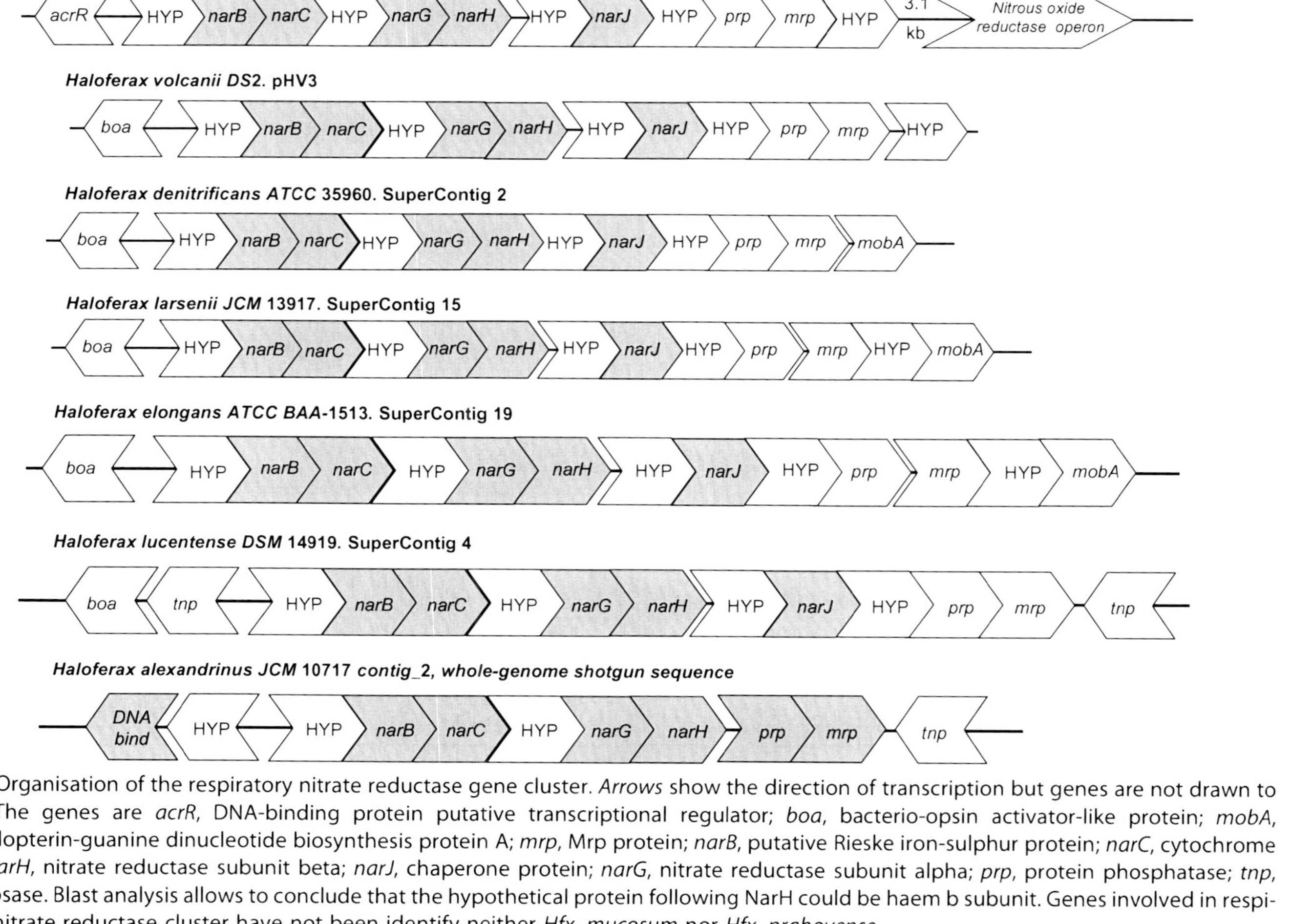

Fig. 2 Organisation of the respiratory nitrate reductase gene cluster. *Arrows* show the direction of transcription but genes are not drawn to scale. The genes are *acrR*, DNA-binding protein putative transcriptional regulator; *boa*, bacterio-opsin activator-like protein; *mobA*, molybdopterin-guanine dinucleotide biosynthesis protein A; *mrp*, Mrp protein; *narB*, putative Rieske iron-sulphur protein; *narC*, cytochrome *b*/*b6*; *narH*, nitrate reductase subunit beta; *narJ*, chaperone protein; *narG*, nitrate reductase subunit alpha; *prp*, protein phosphatase; *tnp*, transposase. Blast analysis allows to conclude that the hypothetical protein following NarH could be haem b subunit. Genes involved in respiratory nitrate reductase cluster have not been identify neither *Hfx. mucosum* nor *Hfx. prahovense.*

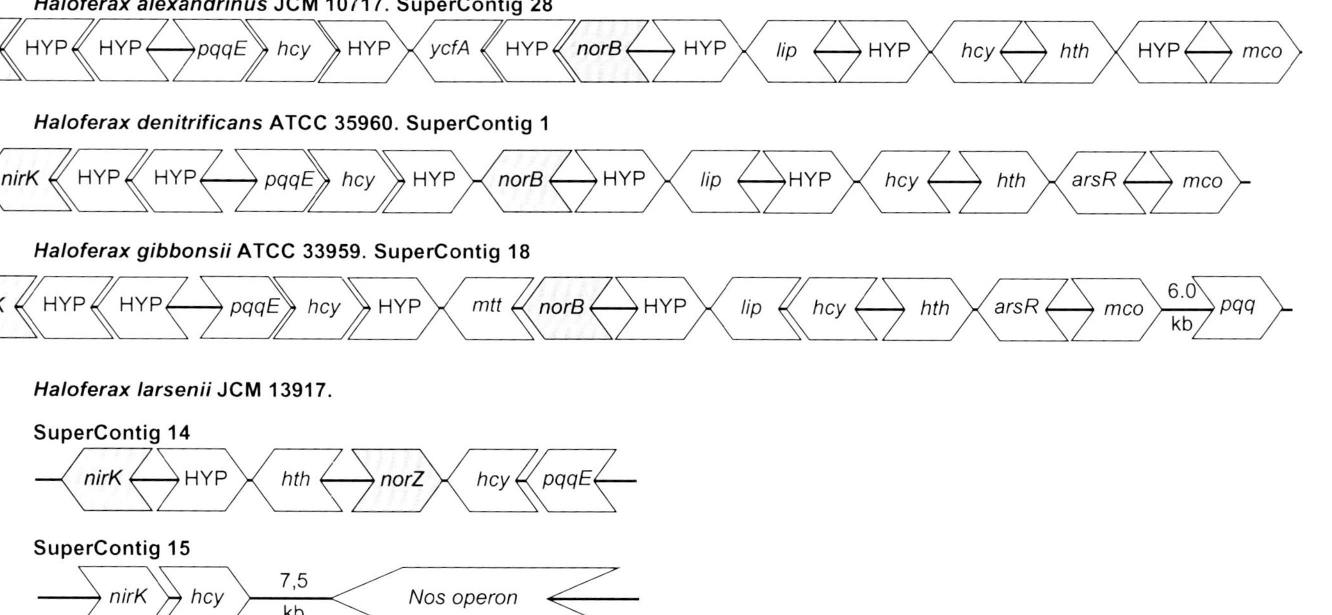

Fig. 3 Organisation of the nitrite reductase copper containing (NirK) and cytochrome *b* subunit of nitric oxide reductase (NorB) clusters. *Arrows* show the direction of transcription but genes are not drawn to scale. The genes are *arsR*, putative transcriptional regulator, ArsR family; *cox*, cytochrome *c* oxidase subunit I; *gdhA1*, glutamate dehydrogenase (NAD(P)+); *hcy*, halocyanin precursor-like protein; HYP, hypothetical protein; *hth*, HTH DNA-binding domain, family protein; *lip*, lipoprotein; *mco*, multicopper oxidase; *mtt*, methyltransferase type 12; *nirK*, nitrite reductase copper containing protein; n*orB*, cytochrome *b* subunit of nitric oxide reductase; *norZ*, nitric oxide reductase, NorZ apoprotein; *pqqE*, coenzyme PQQ synthesis protein; *ycfA*, YcfA family protein.

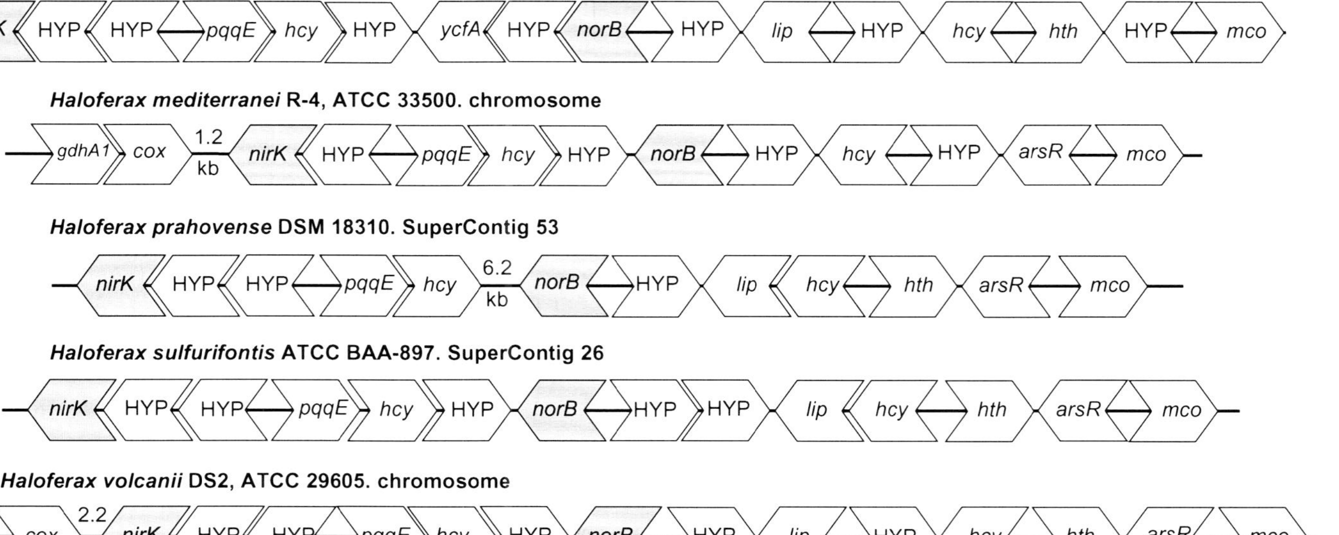
Haloferax lucentense DSM 14919. SuperContig 10
nirK
HYP
HYP
pqqE
hcy
HYP
ycfA
HYP
norB
HYP
lip
HYP
hcy
hth
HYP
mco
Haloferax mediterranei R-4, ATCC 33500. chromosome
gdhA1
cox
1.2 kb
nirK
HYP
pqqE
hcy
HYP
norB
HYP
hcy
HYP
arsR
mco
Haloferax prahovense DSM 18310. SuperContig 53
nirK
HYP
HYP
pqqE
hcy
6.2 kb
norB
HYP
lip
hcy
hth
arsR
mco
Haloferax sulfurifontis ATCC BAA-897. SuperContig 26
nirK
HYP
HYP
pqqE
hcy
HYP
norB
HYP
HYP
lip
hcy
hth
arsR
mco
Haloferax volcanii DS2, ATCC 29605. chromosome
cox
2.2 kb
nirK
HYP
HYP
pqqE
hcy
HYP
norB
HYP
lip
HYP
hcy
hth
arsR
mco

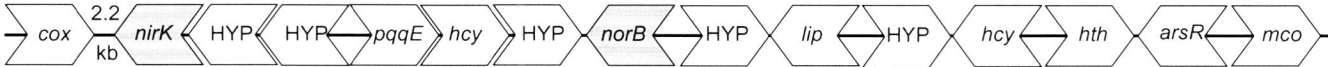

Fig. 3—Cont'd

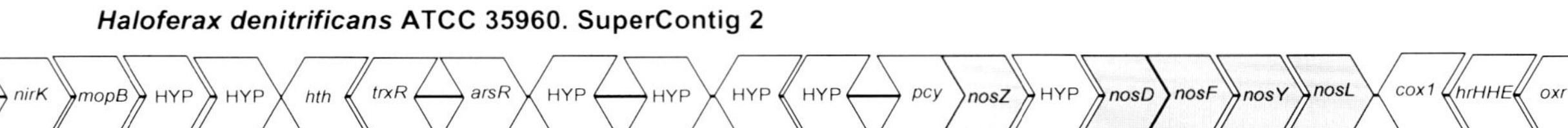

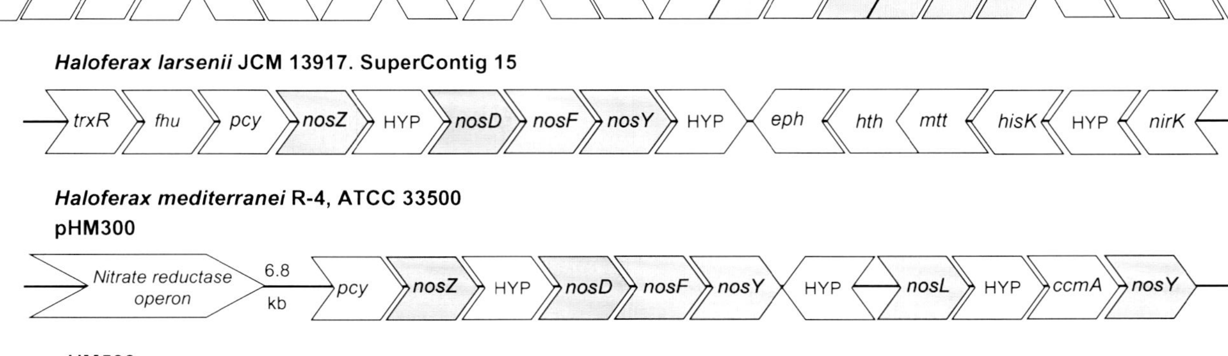

Fig. 4 Organisation of the nitrous oxide reductase cluster. *Arrows* show the direction of transcription but genes are not drawn to scale. The genes are *arsR*, ArsR family transcriptional regulator; *ccmA*, ABC transporter ATP-binding protein; *cox1*, cytochrome *c* oxidase subunit I; *eph*, epoxide hydrolase-related protein; *fhu*, ferrichrome-binding protein; *hisK*, signal-transducing histidine kinase; *hrHHE*, hemerythrin HHE cation-binding region; *hth*, DNA-binding protein; *mopB*, oxidoreductase molybdopterin-binding protein; *mtt*, type 11 methyltransferase; *nifU*, nifU C-terminal domain-containing protein; *nirK*, nitrite reductase; *nosD*, copper-binding protein; *nosF*, ABC transporter ATP-binding protein; *nosL*, Cu(I) protein of the nitrous oxide reductase (nos) gene cluster; *nosY*, ABC-type transport system involved in multicopper enzyme maturation permease component; *nosZ*, nitrous oxide reductase; *oxr*, Fe-S oxidoreductase; *paaB*, phenylacetic acid degradation protein B; *pcy*, copper-binding plastocyanin-like protein; *trxR*, thioredoxin reductase (trxR). Genes involved in this cluster have been identified only in three of the nine analysed species.

in the MOCO cofactor synthesis, genes coding for proteins involved in the electron transfer (*narB* and *narC*), and a gene coding for chaperone-like protein (*narJ*), which is presumably involved in maturation and assembly of the αβγ complex (Blasco et al., 1998; Liu & DeMoss, 1997; Palmer et al., 1996). The hypothetical protein codified by the gene following *narH* belongs to the DMSO reductase family type II enzyme, haem b subunit, which can be identified as the homologue to the *narI* gene coding for the gamma subunit in bacteria. *mrp* codes for a Mrp protein; this protein belongs to the conserved protein domain family MRP-like (multiple resistance and pH adaptation), a homologue of the Fer4 NifH superfamily, which is found in bacteria as a membrane-spanning protein and functions as a Na^+/H^+ antiporter. Like the other members of the superfamily, it contains an ATP-binding domain. The N-terminal presents also homology with a FeS assembly SUF system protein, and thus, could be involved in cofactor biosynthesis or electron transport. The gene *acrR* encodes a DNA-binding protein putative transcriptional regulator, *boa* codes for bacterio-opsin activator-like protein and *prp* encodes a protein phosphatase. These three elements could be implied in a regulation mechanism of nitrate respiration but physiological and biochemical studies should be done properly to elucidate this role. Finally, *tnp* codes for a transposase. The Nar operon has been very well described from *Hfx. mediterranei* (Lledó et al., 2004) and other haloarchaea such as *Haloarcula marismortui* (Yoshimatsu et al., 2002); however, its regulation has not been explored yet.

Genes coding for respiratory nitrite reductase and nitric oxide reductase are closely located and constitute a gene cluster summarised in Fig. 3 for several species. These genes are encoded in the chromosome of *Hfx. mediterranei* and in *Hfx. volcanii*. The first important feature to be highlighted when comparing between species is that the gene coding for respiratory nitrite reductase copper containing protein (*nirK*) is close to *hcy*, a gene coding for a halocyanin precursor-like protein. Halocyanin is small blue copper protein with a molecular mass of about 15 kDa which serves as a mobile electron carrier.

Although more work must be done to elucidate the role of this protein in the context of the haloarchaeal denitrification, it is possible to conclude from the bibliography that it may be involved in electron transfer during nitrite and/or nitric oxide reduction (Brischwein, Scharf, Engelhard, & Mäntele, 1993; Mattar et al., 1994; Scharf & Engelhard, 1993). There is also a gene coding for multicopper oxidase (*mco*) which oxidises their substrate by accepting electrons at a mononuclear copper centre and transferring them to a trinuclear copper centre. This protein as well as halocyanin could be involved in the reaction catalysed in vivo by NirK. This cluster also contains several genes related to nitric oxide reductase: *norB*, coding for cytochrome *b* subunit of nitric oxide reductase; *norZ*, which encodes nitric oxide reductase, also called NorZ apoprotein. Other genes code for proteins related to electron transfer or cofactor biosynthesis. Thus, *cox* codes for cytochrome *c* oxidase subunit I; while, *pqqE* encodes Coenzyme PQQ synthesis protein, which is a protein involved in the pathway pyrroloquinoline quinone biosynthesis (cofactor biosynthesis). Other genes included in these clusters are *ycfA*, which codes for YcfA family protein (most of these proteins are hypothetical proteins of unknown function); *lip* which encodes a lipoprotein; *arsR*: putative transcriptional regulator, ArsR family; *gdhA1*, coding for glutamate dehydrogenase (NAD(P)+); *hth*, encoding HTH DNA-binding domain and *mtt*, coding for methyltransferase type 12.

Fig. 4 summarised the organisation of nitrous oxide reductase cluster. In this case, only the information from three species has been included because it was impossible to identify genes coding for neither the enzyme nor their accessory proteins from other species. In this cluster, the following genes coding for nitrous oxide reductase and its accessory proteins are identified: *nosL*, a gene coding for Cu(I) protein of the nitrous oxide; *nosY*, which codes for an ABC-type transport system involved in multicopper enzyme maturation permease component; and *nosZ*, a gene coding for nitrous oxide reductase. The rest of the gees included in this cluster code for proteins involved in electron transfer (ie, *pcy*: copper-binding plastocyanin-like protein) as well as proteins involved in protein maturation can be found. *Hfx. larsenii* and *Hfx. denitrificans* clusters also contain the gene *trxR* which codes for a thioredoxin reductase. This protein is ubiquitous and it is involved in the defence against oxidative damage due to oxygen metabolism, and redox signalling using molecules like hydrogen peroxide and nitric oxide.

In *Hfx. denitrificans* as well as in *Hfx. larsenii* genes coding for NirK are closely located to genes encoding Nos. The genomes of the mentioned species are not completely sequenced, assembly and annotated so it is impossible to identify genes localization (chromosome, plasmids) or even operons. *Hfx. mediterranei* genome offers a more detailed analysis. In that case, genes coding for NarGH and Nos are located in one of the three plasmids (pHM300), whilst genes encoding Nirk and Nor are located in the main chromosome. This fact supports the theory of LGT as an important tool in haloarchaea resulting in several phenotypes better adapted to the environment they are inhabiting. Nar genes constitute an operon. However, *nir* and *nor* genes are close and probably constitute an operon (molecular biology studies must be done to address this hypothesis). On the other hand, *nos* genes in *Hfx. mediterranei* could constitute another operon which should be analysed in terms of regulation in the next future.

6. POTENTIAL USES OF THE DENITRIFICATION CARRIED OUT BY *HALOFERAX* IN BIOTECHNOLOGY

Nowadays many industrial processes such as food processing, oil production or handling of pharmaceuticals generate compounds containing nitrogen in different forms (Lefebvre & Moletta, 2006). Organic and inorganic nitrogen are harmful to the environment and lead to many health problems (Li et al., 2013). High concentrations of nitrates and nitrites in water are very toxic to humans, fauna and flora (Philips, Laanbroek, & Verstraete, 2002). Furthermore, the presence of this anion in the bloodstream transforms the haemoglobin to methaemoglobin irreversibly, hindering the release of oxygen to tissues (van Leeuwen, 2000) and causing respiratory problems in aquatic and terrestrial animals, including humans (Nájera-Fernández et al., 2012; Philips et al., 2002).

Although most biochemical, genetic and physiological studies are needed about denitrification process in members of *Haloferax* genus, specially focused in the last steps of this process (reduction of nitric oxide and nitrous oxide), the acquired knowledge since today permits to apply some species such *Hfx. mediterranei* in different biotechnological processes like wastewater treatments or the development of biosensors. In that sense, not only denitrification as pathway but also isolated enzymes involved in denitrification could be used.

6.1 Wastewater Treatments by *Haloferax* Members

Traditional wastewater treatments are based on two essential procedures for nitrogen removal: first, autotrophic nitrification (oxidation of ammonium to nitrate by nitrifying bacteria under aerobic conditions); second, anoxic denitrification (which converts nitrate and nitrite into N_2 gas by denitrifying bacteria under anaerobic conditions) (Zhu et al., 2012).

The most studied microorganisms capable of heterotrophic nitrification and aerobic denitrification are *Paracoccus* (Blaszczyk, 1993), *Thioalkalivibrio* spp. (Sorokin & Kuenen, 2005), *Bacillus licheniformis* (Takenaka et al., 2007), *Halomonas* spp. (Boltianskaia et al., 2007) and *Pseudomonas stutzeri* (Miyahara et al., 2010). These strains can oxidise ammonium to nitrite and, simultaneously, can reduce nitrates and nitrites to N_2 by aerobic denitrification (Guo et al., 2013).

However, this approach is very costly and time consuming because nitrifying bacteria do not grow quickly and require different conditions (Chen et al., 2014). Moreover, in the last years there is an additional problem in urban and industrial wastewater: the increase of salinity.

The fish and seafood processing industry, leather and petroleum industry and the manufacturing of chemicals such as pesticides, herbicides and explosives generates effluents containing complex mixtures of salts and nitrate or nitrite (Nájera-Fernández et al., 2012; Zhang, Zhang, & Quan, 2012). Moreover, the use of seawater as a substitute for freshwater in the municipal sanitation for some human activities such as the maintenance of public urinals, produces saline waste entering in the wastewater circuit and increasing its salinity (Duan, Fang, Su, Chen, & Lin, 2015).

Thus in this context and due to the fact that the bioactivity of denitrification significantly decreased when salt was above 2%, w/v (Guo et al., 2013), traditional denitrifying bacteria used in wastewater treatments may suffer a loss of enzymatic activities and the unbalance of osmotic stress across the cell wall resulting in plasmolysis (Nájera-Fernández et al., 2012; Zhang et al., 2012).

One solution could be some strains from *Halomonas* (Shapovalova, Khijniak, Tourova, Muyzer, & Sorokin, 2008), which are actively denitrifiers under highly halophilic conditions (salt concentrations of 4 M Na^+), but the problem is that they grow and denitrify only aerobically; due to the low dissolved oxygen concentrations in industrial wastewater, these aerobic halophilic species are not suitable for use in the treatment of industrial salt wastewater (Shapovalova et al., 2008).

Therefore, it is very important to use halophilic microorganisms tolerant to a wide range of salinities, resistant to salt shock (Duan et al., 2015) and anoxic conditions. In this sense, organisms belonging to the *Haloferax* genera are perfect candidates for being used in wastewater treatments.

Hfx. mediterranei is an extreme halophilic archaeon able to grow on an unusually large range of concentrations of NaCl (1.0–5.2 M) (Torreblanca et al., 1986). It has been used as model in saline wastewater treatments or brines bioremediation because of its resistance to very high nitrate and nitrite concentrations (Bonete et al., 2008; Martínez-Espinosa, Lledó, Marhuenda-Egea, Díaz, & Bonete, 2009) and tolerance to a broad range of salinities (D'Souza et al., 1997).

In a recent study, Nájera-Fernández et al. grew *Hfx. mediterranei* at 42°C using brines either prepared in the laboratory or collected from wastewater plant treatments with 30, 40 and 50 mM KNO_2. Given that the maximum tolerance of the majority of microorganisms studied to date ranges from 2 to 5 mM NO_2^-, the concentrations in this experiment can be considered very high.

The results showed that *Hfx. mediterranei* consumed 100% of nitrate and nitrite present in media with 30 and 40 mM KNO_2^- and 60% of the nitrate and 75% of the nitrite in medium with 50 mM KNO_2^-. Using nitrogen species by this haloarchaea to grow occurred both in oxic and anoxic conditions, thanks to assimilatory nitrate and nitrite reductases in the first case and to respiratory nitrate and nitrite reductases when oxygen is depleted.

In addition to the advantages of using *Haloferax* species in removing nitrogenous compounds from wastewater, recently it has been discovered that *Hfx. mediterranei* is able to reduce perchlorate and chlorate in anaerobic conditions through respiratory nitrate reductase (NarGH) lacking genes coding for (per) chlorate reductases, as previously mentioned (Martínez-Espinosa et al., 2015). These anions, which have very harmful effects on health, are present in sewage because of many human activities such as the manufacture of propellants, explosives or bleaching paper.

The ability of NarGH to use (per)chlorate as final electron acceptor in the absence of oxygen is due to the location of its active site facing the membrane potential positive face (pNars) (Martinez-Espinosa et al., 2007, 2015). It permits to reduce these anions to chlorite with any intracellular damage. Moreover, this enzyme is capable to reduce bromate too.

Hfx. mediterranei could be a good model to remove perchlorate, chlorate and bromate in brines and wastewater even in the presence of low salt concentrations (Martínez-Espinosa et al., 2015). Due to that many wastewater contain both nitrates and chlorates (Wolterink et al., 2003), the processes based on bioremediation by *Hfx. mediterranei* could replace or even improve protocols where perchlorate and nitrate are removed from brines by ion exchange techniques (Lehman, Badruzzaman, Adham, Roberts, & Clifford, 2008).

Despite the advantages of using denitrifiers to remove toxic anions from sewage, there are some negative effects related to denitrification during biological wastewater treatments such as the emission of nitrous oxide, which is involved in global warming. Although the release of this gas from wastewater treatments is relatively small (3% of the estimated total anthropogenic

N_2O emission), but is a significant factor (26%) in the greenhouse gas footprint of the total water chain (Kampschreur, Temmink, Kleerebezem, Jetten, & van Loosdrecht, 2009).

The emission of N_2O can be due to several factors: first, the use of partial denitrifiers (reducing nitrates and nitrites to nitrous oxide because of the lack of nitrous oxide reductase); second, due to some abiotic conditions such as pH (Čuhel et al., 2010; Huang, Long, Chapman, & Yao, 2014; Liu, Mørkved, Frostegård, & Bakken, 2010; Sánchez-Andrea, Rojas-Ojeda, Amils, & Sanz, 2012), salinity or temperature which can inactivate this enzyme; third, because of some operational parameters in wastewater treatments such as low COD/N ratio in the denitrification stage or the increase of nitrite concentrations (Kampschreur et al., 2009).

A little is known about the relation between the use of *Haloferax* species in wastewater treatments and nitrous oxide emission. In case of *Hfx. mediterranei*, although nitrous oxide reductase has not been purified and characterised yet, it is known that it is a complete denitrifier haloarchaea because of the reduction of N_2O to N_2 observed in anaerobic conditions after the addition of KNO_3 as terminal electron acceptor (Bonete et al., 2008). Concerning other *Hfx.* species such as *Hfx. volcanii*, it is not yet known if it is complete or incomplete denitrifier organism. Despite of this knowledge, there are no studies focused on the nitrous oxide emission during wastewater treatments using members of the *Haloferax* genus.

6.2 Biosensors Based on Denitrification Enzymes

In the last years, a lot of legislative actions for environmental pollution control have been grown due to the social concern in this area (Rodriguez-Mozaz, Lopez De Alda, & Barceló, 2006). Special attention has been given to water quality control in terms of nitrates and nitrites concentrations because of their negative effects on human health. The most important governmental agencies have promulgated rules and directives to restrict the level of these ions in drinking water and food products (Gabriela Almeida, Serra, Silveira, & Moura, 2010).

In order to comply with the new controls, it needs efficient measures of nitrates and nitrites. Their determination using traditional methods such as chromatography, spectrophotometry or polarography is expensive and they have susceptibility to matrix interferences, require pretreatments and long time of analysis (Cosnier, Da Silva, Shan, & Gorgy, 2008; Gabriela Almeida et al., 2010; Mohd Zuki, Suhaity Azmi, & Ling Ling, 2014).

Therefore, it becomes necessary to develop systems which can detect a lot of compounds in environmental samples as quickly and as cheaply as possible (Rodriguez-Mozaz et al., 2006). In these terms, biosensors have become one of the most important devices. They are capable of providing quantitative or semi-quantitative analytical information with the help of two elements: a biological recognition component which detects the analyte connected to a transducing element which transforms the measure in a (semi)quantitative signal (Mohd Zuki et al., 2014; Rodriguez-Mozaz et al., 2006).

Biosensors can be classified into various groups (Fig. 5), according to their two principal components (Rodriguez-Mozaz et al., 2006; Slonczewski, Coker, & DasSarma, 2010): on the basis of the transducing element, they can be named as electrochemical, optical, piezoelectric or thermal sensors; according to the biorecognition principle, biosensors are classified into antibodies and antigens, enzymatic, nonenzymatic, whole-cell and nucleic acids biosensors (Marazuela & Moreno-Bondi, 2002; Rodriguez-Mozaz et al., 2006).

In the context of nitrates and nitrites biosensors, there have been two main lines of development: on the one hand, using whole cells and detecting some products of their nitrogen metabolism; on the other hand, systems based on the immobilisation of denitrification enzymes in a matrix.

In case of using whole cells, bacteria are placed in a reaction chamber where the reduction NO_3^- to N_2O occurs and it is measured by a specific nitrous oxide microelectrode (Larsen, Kjær, & Revsbech, 1997). A new

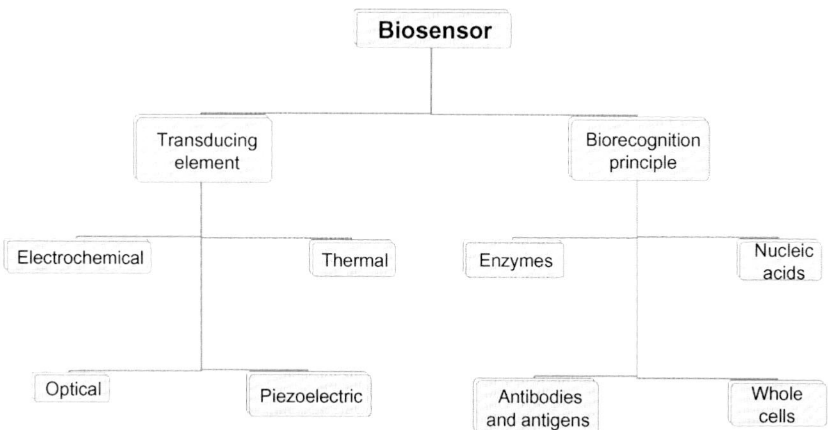

Fig. 5 Classification of biosensors based on the transducing element and the biorecognition principle.

improvement utilises a whole-cell fluorescence system which is based on recombinant *E. coli* bacteria without interference of phosphate, chloride and nitrite (Rodriguez-Mozaz et al., 2006).

But the most important nitrates and nitrites biosensors are based on the immobilisation of nitrate and nitrite reductases from bacteria (*Paracoccus denitrificans*, *Paracoccus pantotrophus*, *Alcaligenes faecalis*, *Desulfovibrio desulfuricans*, etc.) in different materials improving their stability and half-life (Mohd Zuki et al., 2014). The most common methods used to immobilise enzymes are covalent binding and cross-linking (Mohd Zuki et al., 2014). Proteins should be placed on surfaces that do not alter their structure and functions: a good example is the use of inorganic clay nanoparticles as hydrophilic additives in combination with polypyrrole films (Cosnier et al., 2008).

The majority of proposals use redox mediators (viologen derivatives) to shuttle electrons from the protein redox centres to the transducing elements (ammeters or voltmeters). Very recently, new approaches are based on the direct electron transfer between the enzymes and the electrode material (Gabriela Almeida et al., 2010) to simplify the structure of biosensor.

In the context of the genus *Haloferax*, biosensors could be developed based on nitrate and nitrite reductases from *Hfx. mediterranei* because these enzymes are purified and characterised. These biosensors would have several advantages over bacterial biosensors described since today: first, nitrate and nitrite reductases from this haloarchaea tolerate high salt concentrations, allowing their use in salt water and brines; second, their K_m values are quite low so the affinity for their substrates is very high (K_m of nitrite reductase of *Hfx. mediterranei* is 4.04 ± 0.33 mM which is almost 100-fold greater than for other enzymes NirK characterised) (Esclapez et al., 2013). Despite all advantages, it has not yet developed any prototype of biosensor for *Haloferax*.

7. CONCLUSIONS AND FUTURE PERSPECTIVES

Although knowledge about anaerobic metabolism in haloarchaea is improving, more effort should be done to understand how those microorganisms, and in particular, *Haloferax* genus, switch on the biological traits required to be alive in these open and hostile habitats characterised by high salt concentrations and nutrients and oxygen limitation. Currently, huge amount of open questions remain unaddressed about haloarchaea anaerobic metabolism:

– The key signals inducing anaerobic metabolism must be identified (oxygen depletion alone or oxygen depletion concomitant with the presence of specific compounds such as nitrate, chlorate, sulphur, redox balance, etc.).

- The molecular mechanisms involved in anaerobic pathways have to be characterised; no information about regulators involved in anaerobic metabolism induction or inhibition has been reported yet.
- The importance of NO_x emissions as a consequence of haloarchaeal denitrification should be quantified in situ and in vitro.
- Molecules acting as oxygen sensors in haloarchaea must be identified and characterised. Same for nitrate, nitrite, nitric oxide, chlorate, sulphur and nitrous oxide.
- The presence of oxygen alternative terminal electron acceptors is minority in environments with high salt concentrations. Thus, their ecological role is still uncertain in the context of the habitat where *Haloferax* species live.

Another aspect that remains partially known is the capacity of the denitrification from those haloarchaea initially characterised as denitrifiers. This aspect, although the last one in this section, is probably the most important one. When a new taxon is isolated, several microbial and biochemical analysis are performed to characterise the phenotype. Usually, a new isolated is considered denitrifier when nitrate reduction is observed under anoxic conditions. However, this capability does not directly show denitrification activity (denitrification implies complete nitrate reduction to dinitrogen). As a result of this partial characterisation, more work is currently required to properly characterise whether or not one single *Haloferax* species, initially characterised as a denitrifier is in fact a complete or a partial denitrifier.

New insights on gene expression and their regulation as well as protein characterisation will contribute to quantify how important those species are in terms of behaviour as 'microbial weeds', NO_x gases emissions, nitrogen compounds recycling, potential biotechnology applications, etc.

ACKNOWLEDGEMENT

This work was funded by research grant from the MINECO Spain (CTM2013-43147-R).

REFERENCES

Allen, M. A., Goh, F., Leuko, S., Echigo, A., Mizuki, T., Usami, R., et al. (2008). *Haloferax elongans* sp. nov. and *Haloferax mucosum* sp. nov., isolated from microbial mats from Hamelin Pool, Shark Bay, Australia. *International Journal of Systematic and Evolutionary Microbiology*, *58*, 798–802.

Allers, T., Ngo, H. P., Mevarech, M., & Lloyd, R. G. (2004). Development of additional selectable markers for the halophilic archaeon *Haloferax volcanii* based on the *leuB* and *trpA* genes. *Applied and Environmental Microbiology*, *70*, 943–953.

Alvarez-Ossorio, M. C., Muriana, F. J. G., de la Rosa, F. F., & Relimpio, A. V. (1992). Purification and characterization of nitrate reductase from the halophile archaebacterium *Haloferax mediterranei*. *Zeitschrift für Naturforschung*, *47C*, 670–676.

Antón, J., Meseguer, I., & Rodríguez-Valera, F. (1988). Production of an extracellular polysaccharide by *Haloferax mediterranei*. *Applied and Environmental Microbiology, 54*, 2381–2386.

Asker, D., & Ohta, Y. (2002). *Haloferax alexandrinus* sp. nov., an extremely halophilic canthaxanthin-producing archaeon from a solar saltern in Alexandria (Egypt). *International Journal of Systematic and Evolutionary Microbiology, 52*, 729–738.

Bardiya, N., & Bae, J. H. (2005). Bioremediation potential of a perchlorate-enriched sewage sludge consortium. *Chemosphere, 58*, 83–90.

Baymann, F., Schoepp-Cothenet, B., Lebrun, E., van Lis, R., & Nitschke, W. (2012). Phylogeny of Rieske/cytb complexes with a special focus on the Haloarchaeal enzymes. *Genome Biology and Evolution, 4*, 720–729.

Besse, A., Peduzzi, J., Rebuffat, S., & Carré-Mlouka, A. (2015). Antimicrobial peptides and proteins in the face of extremes: Lessons from archaeocins. *Biochimie, 118*, 344–355. pii: S0300-9084(15)00175-3.

Bickel-Sandkotter, S., & Ufer, M. (1995). Properties of a dissimilatory nitrate reductase from the halophilic archaeon *Haloferax volcanii*. *Zeitschrift fur Naturforschung—Section C Journal of Biosciences, 50*, 365–372.

Blasco, F., Santos, J. P. D., Magalon, A., Frixon, C., Guigliarelli, B., Santini, C. L., et al. (1998). NarJ is a specific chaperone required for molybdenum cofactor assembly in nitrate reductase A of *Escherichia coli*. *Molecular Microbiology, 28*, 435–447.

Blaszczyk, M. (1993). Effect of medium composition on the denitrification of nitrate by *Paracoccus denitrificans*. *Applied and Environmental Microbiology, 59*, 3951–3953.

Boltianskaia, I. V., Kevbrin, V. V., Lysenko, A. M., Kolganova, T. V., Turova, T. P., Osipov, G. A., et al. (2007). *Halomonas mongoliensis* sp. nov. and *Halomonas kenyensis* sp. nov., new haloalkaliphilic denitrifiers capable of reducing N_2O, isolated from soda lakes. *Mikrobiologiya, 76*, 834–843.

Bonete, M. J., Martínez-Espinosa, R. M., Pire, C., Zafrilla, B., & Richardson, D. J. (2008). Nitrogen metabolism in haloarchaea. *Saline Systems, 4*, 9.

Brischwein, M., Scharf, B., Engelhard, M., & Mäntele, W. (1993). Analysis of the redox reaction of an archaebacterial copper protein, halocyanin, by electrochemistry and FTIR difference spectroscopy. *Biochemistry, 32*, 13710–13717.

Cabello, P., Roldán, M. D., & Moreno-Vivián, C. (2004). Nitrate reduction and the nitrogen cycle in archaea. *Microbiology, 150*, 3527–3546.

Chen, M., Wang, W., Feng, Y., Zhu, X., Zhou, H., Tan, Z., et al. (2014). Impact resistance of different factors on ammonia removal by heterotrophic nitrification-aerobic denitrification bacterium *Aeromonas* sp. HN-02. *Bioresource Technology, 167*, 456–461.

Cheung, J., Danna, K. J., O'Connor, E. M., Price, L. B., & Shand, R. F. (1997). Isolation, sequence, and expression of the gene encoding halocin H4, a bacteriocin from the halophilic archaeon *Haloferax mediterranei* R4. *Journal of Bacteriology, 179*, 548–551.

Cosnier, S., Da Silva, S., Shan, D., & Gorgy, K. (2008). Electrochemical nitrate biosensor based on poly(pyrrole-viologen) film-nitrate reductase-clay composite. *Bioelectrochemistry, 74*, 47–51.

Čuhel, J., Šimek, M., Laughlin, R. J., Bru, D., Chèneby, D., Watson, C. J., et al. (2010). Insights into the effect of soil pH on N_2O and N_2 emissions and denitrifier community size and activity. *Applied and Environmental Microbiology, 76*, 1870–1878.

DasSarma, S., & Arora, P. (2002). *Halophiles, encyclopedia of life sciences*. London: Nature Publishing Group.

Dassarma, S., Kennedy, S. P., Berquist, B., Ng, V. W., Baliga, N. S., Spudich, J. L., et al. (2001). Genomic perspective on the photobiology of *Halobacterium* species NRC-1, a phototrophic, phototactic, and UV-tolerant haloarchaeon. *Photosynthesis Research, 70*, 3–17.

Dornmayr-Pfaffenhuemer, M., Legat, A., Schwimbersky, K., Fendrihan, S., & Stan-Lotter, H. (2011). Responses of haloarchaea to simulated microgravity. *Astrobiology, 11*(3), 199–205.

D'Souza, S. E., Altekar, W., & D'Souza, S. F. (1997). Adaptive response of *Haloferax mediterranei* to low concentrations of NaCl (<20%) in the growth medium. *Archives of Microbiology, 168*, 68–71.

Duan, J., Fang, H., Su, B., Chen, J., & Lin, J. (2015). Characterization of a halophilic heterotrophic nitrification–aerobic denitrification bacterium and its application on treatment of saline wastewater. *Bioresource Technology, 179*, 421–428.

Eisenberg, H. (1995). Life in unusual environments: Progress in understanding the structure and function of enzymes from extreme halophilic bacteria. *Archives of Biochemistry and Biophysics, 318*(1), 1–5.

Ellington, M. J. K. (2003). *Rhodobacter capsulatus* gains a competitive advantage from respiratory nitrate reduction during light–dark transitions. *Microbiology, 149*, 941–948.

Elshahed, M. S., Najar, F. Z., Roe, B. A., Oren, A., Dewers, T. A., & Krumholz, L. R. (2004). Survey of archaeal diversity reveals an abundance of halophilic archaea in a low-salt, sulfide- and sulfur-rich spring. *Applied and Environmental Microbiology, 70*, 2230–2239.

Enache, M., Itoh, T., Kamekura, M., Teodosiu, G., & Dumitru, L. (2007). *Haloferax prahovense* sp. nov., an extremely halophilic archaeon isolated from a Romanian salt lake. *International Journal of Systematic and Evolutionary Microbiology, 57*, 393–397.

Englert, C., Horne, M., & Pfeifer, F. (1990). Expression of the major gas vesicle protein gene in the halophilic archaebacterium *Haloferax mediterranei* is modulated by salt. *Molecular and General Genetics, 222*, 225–232.

Esclapez, J., Zafrilla, B., Martínez-Espinosa, R. M., & Bonete, M. J. (2013). Cu-NirK from *Haloferax mediterranei* as an example of metalloprotein maturation and exportation via Tat system. *Biochimica et Biophysica Acta—Proteins and Proteomics, 1834*, 1003–1009.

Falb, M., Müller, K., Königsmaier, L., Oberwinkler, T., Horn, P., von Gronau, S., et al. (2008). Metabolism of halophilic archaea. *Extremophiles, 12*, 177–196.

Fendrihan, S., Legat, A., Gruber, C., Pfaffenhuemer, M., Weidler, G., Gerbl, F., et al. (2006). Extremely halophilic archaea and the issue of long term microbial survival. *Reviews in Environmental Sciences and Biotechnology, 5*, 1569–1605.

Gabriela Almeida, M., Serra, A., Silveira, C. M., & Moura, J. J. G. (2010). Nitrite biosensing via selective enzymes—A long but promising route. *Sensors, 10*, 11530–11555.

Gavira, M., Roldan, M. D., Castillo, F., & Moreno-Vivian, C. (2002). Regulation of nap gene expression and periplasmic nitrate reductase activity in the phototrophic bacterium *Rhodobacter sphaeroides* DSM158. *Journal of Bacteriology, 184*, 1693–1702.

Graham, D. E., Overbeek, R., Olsen, G. J., & Woese, C. R. (2000). An archaeal genomic signature. *Proceedings of the National Academy of Sciences, 97*, 3304–3308.

Grant, W. D. (2004). Life at low water activity. *Philosophical transactions of the Royal Society of London Series B, Biological Sciences, 359*(1448), 1249–1266.

Grant, W. D., Kamekura, M., McGenity, T. J., & Ventosa, A. (2001). Class III. Halobacteria class. nov. In D. R. Boone, R. W. Castenholz, & G. M. Garrity (Eds.), *Bergey's manual of systematic bacteriology: Vol. 1.* (2nd ed., pp. 294–301). New York: Springer Verlag.

Guan, Z., Naparstek, S., Calo, D., & Eichler, J. (2012). Protein glycosylation as an adaptive response in archaea: Growth at different salt concentrations leads to alterations in *Haloferax volcanii* S-layer glycoprotein N-glycosylation. *Environmental Microbiology, 3*, 743–753.

Guo, Y., Zhou, X., Li, Y., Li, K., Wang, C., Liu, J., et al. (2013). Heterotrophic nitrification and aerobic denitrification by a novel *Halomonas campisalis*. *Biotechnology Letters, 35*, 2045–2049.

Gupta, R. S., Naushad, S., & Baker, S. (2015). Phylogenomic analyses and molecular signatures for the class *Halobacteria* and its two major clades: A proposal for division of the class *Halobacteria* into an emended order *Halobacteriales* and two new orders, *Haloferacales* ord.

nov. and *Natrialbales* ord. nov., containing the novel families *Haloferacaceae* fam. nov. and *Natrialbaceae* fam. nov. *International Journal of Systematic and Evolutionary Microbiology, 65*, 1050–1069.

Gutierrez, M. C., Kamekura, M., Holmes, M. L., Dyall-Smith, M. L., & Ventosa, A. (2002). Taxonomic characterization of *Haloferax* sp. ("*H. alicantei*") strain Aa 2.2: Description of *Haloferax lucentensis* sp. nov. *Extremophiles, 6*, 479–483.

Han, J., Zhang, F., Hou, J., Liu, X., Li, M., Liu, H., et al. (2012). Complete genome sequence of the metabolically versatile halophilic archaeon *Haloferax mediterranei*, a poly(3-hydroxybutyrate-co-3-hydroxyvalerate) producer. *Journal of Bacteriology, 194*(16), 4463–4464.

Hechler, T., Frech, M., & Pfeifer, F. (2008). Glucose inhibits the formation of gas vesicles in *Haloferax volcanii* transformants. *Environmental Microbiology, 10*, 20–30.

Hechler, T., & Pfeifer, F. (2009). Anaerobiosis inhibits gas vesicle formation in halophilic archaea. *Molecular Microbiology, 71*(1), 132–145.

Hering, O., Brenneis, M., Beer, J., Suess, B., & Soppa, J. (2009). A novel mechanism for translation initiation operates in haloarchaea. *Molecular Microbiology, 71*, 1451–1463.

Heylen, K., Gevers, D., Vanparys, B., Wittebolle, L., Geets, J., Boon, N., et al. (2006). The incidence of *nirS* and *nirK* and their genetic heterogeneity in cultivated denitrifiers. *Environmental Microbiology, 8*, 2012–2021.

Hochstein, L. I., & Lang, F. (1991). Purification and properties of a dissimilatory nitrate reductase from *Haloferax denitrificans*. *Archives of Biochemistry and Biophysics, 288*, 380–385.

Hochstein, L. I., & Tomlinson, G. A. (1988). The enzymes associated with denitrification. *Annual Review of Microbiology, 42*, 231–261.

Huang, Y., Long, X. E., Chapman, S. J., & Yao, H. (2014). Acidophilic denitrifiers dominate the N_2O production in a 100-year-old tea orchard soil. *Environmental Science and Pollution Research, 22*, 4173–4182.

Inatomi, K., & Hochstein, L. I. (1996). The purification and properties of a copper nitrite reductase from *Haloferax denitrificans*. *Current Microbiology, 32*, 72–76.

Juez, G., Rodriguez-Valera, F., Herrero, N., & Mojica, F. J. (1990). Evidence for salt-associated restriction pattern modifications in the archaeobacterium *Haloferax mediterranei*. *Journal of Bacteriology, 172*, 7278–7281.

Juez, G., Rodriguez-Valera, F., Ventosa, A., & Kushner, D. J. (1986). *Haloarcula hispanica* spec. nov. and *Haloferax gibbonsii* spec. nov., two new species of extremely halophilic archaebacteria. *Systematic and Applied Microbiology, 8*, 75–79.

Kampschreur, M. J., Temmink, H., Kleerebezem, R., Jetten, M. S. M., & van Loosdrecht, M. C. M. (2009). Nitrous oxide emission during wastewater treatment. *Water Research, 43*, 4093–4103.

Kletzin, A. (2007). Chapter 2: General characteristics and important model organisms. In *Archaea: Cellular and molecular biology* (pp. 14–92). Washington, DC: ASM Press, ISBN: 978-1-55581-391-8.

Larsen, H. (1962). Halophilism. In I. C. Gunsalus & R. Y. Stanier (Eds.), *The bacteria: Vol. 4.* (pp. 297–342). New York: Academic Press.

Larsen, L. H., Kjær, T., & Revsbech, N. P. (1997). A microscale NO_3^- biosensor for environmental applications. *Analytical Chemistry, 69*(17), 3527–3531.

Lefebvre, O., & Moletta, R. (2006). Treatment of organic pollution in industrial saline wastewater: A literature review. *Water Research, 40*, 3671–3682.

Lehman, S. G., Badruzzaman, M., Adham, S., Roberts, D. J., & Clifford, D. A. (2008). Perchlorate and nitrate treatment by ion exchange integrated with biological brine treatment. *Water Research, 42*, 969–976.

Leigh, J., Albers, S. V., Atomi, H., & Allers, T. (2011). Model organisms for genetics in the domain archaea: Methanogens, halophiles, thermococcales and sulfolobales. *FEMS Microbiology Review, 35*(4), 577–608.

Lestini, R., Duan, Z., & Allers, T. (2010). The archaeal Xpf/Mus81/FANCM homolog Hef and the Holliday junction resolvase Hjc define alternative pathways that are essential for cell viability in *Haloferax volcanii*. *DNA Repair, 9*(9), 994–1002.

Li, R., Zi, X., Wang, X., Zhang, X., Gao, H., & Hu, N. (2013). Marinobacter hydrocarbonoclasticus NY-4, a novel denitrifying, moderately halophilic marine bacterium. *SpringerPlus, 2*, 346.

Liebensteiner, M. G., Oosterkamp, M. J., & Stams, A. J. (2015). Microbial respiration with chlorine oxyanions: Diversity and physiological and biochemical properties of chlorate- and perchlorate-reducing microorganisms. *Annals of the New York Academy of Sciences*. http://dx.doi.org/10.1111/nyas.12806. June 23 [Epub ahead of print].

Lillo, J. G., & Rodriguez-Valera, F. (1990). Effects of culture conditions on poly(beta-hydroxybutyric acid) production by *Haloferax mediterranei*. *Applied and Environmental Microbiology, 56*, 2517–2521.

Liu, X., & DeMoss, J. A. (1997). Characterization of NarJ; A system-specific chaperone required for nitrate reductase biogenesis in *Escherichia coli*. *Journal of Biological Chemistry, 272*, 24266–24271.

Liu, B., Mørkved, P. T., Frostegård, Å., & Bakken, L. R. (2010). Denitrification gene pools, transcription and kinetics of NO, N_2O and N_2 production as affected by soil pH. *FEMS Microbiology Ecology, 72*, 407–417.

Lledó, B., Martínez-Espinosa, R. M., Marhuenda-Egea, F. C., & Bonete, M. J. (2004). Respiratory nitrate reductase from haloarchaeon *Haloferax mediterranei*: Biochemical and genetic analysis. *Biochimica et Biophysica Acta—General Subjects, 1674*, 50–59.

López-García, P., St Jean, A., Amils, R., & Charlebois, R. L. (1995). Genomic stability in the archaeae *Haloferax volcanii* and *Haloferax mediterranei*. *Journal of Bacteriology, 177*(5), 1405–1408.

Lynch, E. A., Langille, M. G., Darling, A., Wilbanks, E. G., Haltiner, C., Shao, K. S., et al. (2012). Sequencing of seven haloarchaeal genomes reveals patterns of genomic flux. *PLoS One*, 7(7), e41389. http://dx.doi.org/10.1371/journal.pone.0041389.

Madern, D., Ebel, C., & Zaccai, G. (2000). Halophilic adaptation of enzymes. *Extremophiles, 4*, 91–98.

Madigan, M. T., & Oren, A. (1999). Thermophilic and halophilic extremophiles. *Current Opinion in Microbiology, 2*(3), 265–269.

Mancinelli, R. L., & Hochstein, L. I. (1986). The occurrence of denitrification in extremely halophilic bacteria. *FEMS Microbiology Letters, 35*, 55–58.

Marazuela, M. D., & Moreno-Bondi, M. C. (2002). Fiber-optic biosensors—An overview. *Analytical and Bioanalytical Chemistry, 372*, 664–682.

Martinez-Espinosa, R. M., Dridge, E. J., Bonete, M. J., Butt, J. N., Butler, C. S., Sargent, F., et al. (2007). Look on the positive side! The orientation, identification and bioenergetics of "archaeal" membrane-bound nitrate reductases. *FEMS Microbiology Letters, 276*, 129–139.

Martínez-Espinosa, R. M., Lledó, B., Marhuenda-Egea, F. C., Díaz, S., & Bonete, M. J. (2009). NO_3^-/NO_2^- assimilation in halophilic archaea: Physiological analysis, *nasA* and *nasD* expressions. *Extremophiles: Life under Extreme Conditions, 13*, 785–792.

Martínez-Espinosa, R. M., Richardson, D. J., & Bonete, M. J. (2015). Characterisation of chlorate reduction in the haloarchaeon *Haloferax mediterranei*. *Biochimica et Biophysica Acta (BBA)—General Subjects, 1850*, 587–594.

Martínez-Espinosa, R. M., Richardson, D. J., Butt, J. N., & Bonete, M. J. (2006). Respiratory nitrate and nitrite pathway in the denitrifier haloarchaeon *Haloferax mediterranei*. *Biochemical Society Transactions, 34*, 115–117.

Mattar, S., Scharf, B., Kent, S. B., Rodewald, K., Oesterhelt, D., & Engelhard, M. (1994). The primary structure of halocyanin, an archaeal blue copper protein, predicts a lipid anchor for membrane fixation. *Journal of Biological Chemistry, 269*(21), 14939–14945.

Mengele, R., & Sumper, M. (1992). Drastic differences in glycosylation of related S-layer glycoproteins from moderate and extreme halophiles. *Journal of Biological Chemistry*, *267*, 8182–8185.

Miyahara, M., Kim, S. W., Fushinobu, S., Takaki, K., Yamada, T., Watanabe, A., et al. (2010). Potential of aerobic denitrification by *Pseudomonas stutzeri* TR2 to reduce nitrous oxide emissions from wastewater treatment plants. *Applied and Environmental Microbiology*, *76*, 4619–4625.

Mohd Zuki, S. N. S., Suhaity Azmi, N., & Ling Ling, T. (2014). Mini review: Nitrite reductase and biosensors development. *Bioremediation Science and Technology Research*, *2*, 5–9.

Mullakhanbhai, M. F., & Larsen, H. (1975). Halobacterium volcanii spec. nov., a Dead Sea halobacterium with a moderate salt requirement. *Archives of Microbiology*, *104*(3), 207–214.

Müller, J. A., & DasSarma, S. (2005). Genomic analysis of anaerobic respiration of the archaeon *Halobacterium* sp. strain NRC-1: Dimethyl sulfoxide and trimethylamine N-oxide as terminal electron acceptors. *Journal of Bacteriology*, *187*, 1659–1667.

Nájera-Fernández, C., Zafrilla, B., Bonete, M. J., & Martínez-Espinosa, R. M. (2012). Role of the denitrifying Haloarchaea in the treatment of nitrite-brines. *International Microbiology*, *15*, 111–119.

Nakara, K., Tanimoto, T., Hatano, K., Usuda, K., & Shoun, H. (1993). Cytochrome P-450 55A1 (P450dNIR) acts as nitric oxide reductase employing NADH as direct electron donor. *Journal of Biological Chemistry*, *268*, 8350–8355.

Naor, A., Yair, Y., & Gophna, U. (2013). A halocin-H4 mutant *Haloferax mediterranei* strain retains the ability to inhibit growth of other halophilic archaea. *Extremophiles*, *17*, 973–979.

Nelson-Sathi, S., Dagan, T., Landan, G., Janssen, A., Steel, M., McInerney, J. O., et al. (2012). Acquisition of 1,000 eubacterial genes physiologically transformed a methanogen at the origin of haloarchaea. *Proceedings of the National Academy of Sciences of the United States of America*, *109*(50), 20537–20542.

Nilsson, T., Rova, M., & Smedja Bäcklund, A. (2013). Microbial metabolism of oxochlorates: A bioenergetic perspective. *Biochimica et Biophysica Acta*, *182*, 189–197.

Offre, P., Spang, A., & Schleper, C. (2013). Archaea in biogeochemical cycles. *Annual Review of Microbiology*, *67*, 437–457.

Oren, A. (1991). Anaerobic growth of halophilic archaeobacteria by reduction of fumarate. *Journal of General Microbiology*, *137*, 1387–1390.

Oren, A. (1999). Bioenergetic aspects of halophilism. *Microbiology and Molecular Biology Reviews*, *63*, 334–348.

Oren, A. (2002). *Halophilic microorganisms and their environments*. Dordrecht, The Netherlands: Kluwer Academic Publishers.

Oren, A. (2005). A hundred years of *Dunaliella* research: 1905–2005. *Saline Systems*, *1*, 2.

Oren, A. (2008). Microbial life at high salt concentrations: Phylogenetic and metabolic diversity. *Saline Systems*, *4*, 2.

Oren, A. (2012). The function of gas vesicles in halophilic archaea and bacteria: Theories and experimental evidence. *Life (Basel, Switzerland)*, *3*(1), 1–20.

Oren, A., Elevi Bardavid, R., & Mana, L. (2014). Perchlorate and halophilic prokaryotes: Implications for possible halophilic life on Mars. *Extremophiles*, *18*(1), 75–80.

Oren, A., & Hallsworth, J. E. (2014). Microbial weeds in hypersaline habitats: The enigma of the weed-like *Haloferax mediterranei*. *FEMS Microbiology Letters*, *359*(2), 134–142.

Oren, A., & Trüper, H. G. (1990). Anaerobic growth of halophilic archaeobacteria by reduction of dimethylsulfoxide and trimethylamine N-oxide. *FEMS Microbiology Letters*, *70*, 33–36.

Palmer, T., Santini, C. L., Iobbi-Nivol, C., Eaves, D. J., Boxer, D. H., & Giordano, G. (1996). Involvement of the *narJ* and *mob* gene products in distinct steps in the biosynthesis of the molybdoenzyme nitrate reductase in *Escherichia coli*. *Molecular Microbiology*, *20*, 875–884.

Papke, R. T., Douady, C. J., Doolittle, W. F., & Rodríguez-Valera, F. (2003). Diversity of bacteriorhodopsins in different hypersaline waters from a single Spanish saltern. *Environmental Microbiology*, *5*(11), 1039–1045.

Pfeifer, F. (2015). Haloarchaea and the formation of gas vesicles. *Life (Basel, Switzerland)*, *5*(1), 385–402.

Philips, S., Laanbroek, H. J., & Verstraete, W. (2002). Origin, causes and effects of increased nitrite concentrations in aquatic environments. *Reviews in Environmental Science and Biotechnology*, *1*, 115–141.

Pinto, L. H., D'Alincourt Carvalho-Assef, A. P., Vieira, R. P., Clementino, M. M., & Albano, R. M. (2015). Complete genome sequence of *Haloferax gibbonsii* strain ARA6, a potential producer of polyhydroxyalkanoates and halocins isolated from Araruama, Rio de Janeiro, Brasil. *Journal of Biotechnology*, *212*, 69–70.

Rascovan, N., Maldonado, J., Vazquez, M. P., & Eugenia Farías, M. (2015). Metagenomic study of red biofilms from Diamante Lake reveals ancient arsenic bioenergetics in haloarchaea. *The ISME Journal*, *10*, 299–309. http://dx.doi.org/10.1038/ismej.2015.109.

Ravishankara, A. R., Daniel, J. S., & Portmann, R. W. (2009). Nitrous oxide (N_2O): The dominant ozone-depleting substance emitted in the 21st century. *Science*, *326*, 123–125.

Remane, A., & Schleper, C. (1971). *Biology of brackish water*. In *Die binnengewasser*, (Vol. 25) Stuttgart: E. Schweizerbart Verlag. 372 pp.

Richardson, D. J., Berks, B. C., Russell, D. A., Spiro, S., & Taylor, C. J. (2001). Functional, biochemical and genetic diversity of prokaryotic nitrate reductases. *Cellular and Molecular Life Sciences*, *58*, 165–178.

Rodrigo-Baños, M., Garbayo, I., Vílchez, C., Bonete, M. J., & Martínez-Espinosa, R. M. (2015). Carotenoids from haloarchaea and their potential in biotechnology. *Marine Drugs*, *13*(9), 5508–5532.

Rodriguez-Mozaz, S., Lopez De Alda, M. J., & Barceló, D. (2006). Biosensors as useful tools for environmental analysis and monitoring. *Analytical and Bioanalytical Chemistry*, *386*, 1025–1041.

Rodriguez-Valera, F., Juez, G., & Kushner, D. J. (1983). *Halobacterium mediterranei* spec, nov., a new carbohydrate-utilizing extreme halophile. *Systematic and Applied Microbiology*, *4*(3), 369–381.

Ruepp, A., & Soppa, J. (1996). Fermentative arginine degradation in *Halobacterium salinarium* (formerly *Halobacterium halobium*): Genes, gene products, and transcripts of the *arcRACB* gene cluster. *Journal of Bacteriology*, *178*, 4942–4947.

Sánchez-Andrea, I., Rojas-Ojeda, P., Amils, R., & Sanz, J. L. (2012). Screening of anaerobic activities in sediments of an acidic environment: Tinto River. *Extremophiles*, *16*, 829–839.

Saralov, A. I., Baslerov, R. V., & Kuznetsov, B. B. (2013). *Haloferax chudinovii* sp. nov., a halophilic archaeon from Permian potassium salt deposits. *Extremophiles*, *17*, 499–504.

Scharf, B., & Engelhard, M. (1993). Halocyanin, an archaebacterial blue copper protein (type I) from *Natronobacterium pharaonis*. *Biochemistry*, *32*(47), 12894–12900.

Shapleigh, W. J., & Payne, W. J. (1985). Differentiation of cd1 cytochrome and copper nitrite reductase production in denitrifiers. *FEMS Microbiology Letters*, *26*, 275–279.

Shapovalova, A. A., Khijniak, T. V., Tourova, T. P., Muyzer, G., & Sorokin, D. Y. (2008). Heterotrophic denitrification at extremely high salt and pH by haloalkaliphilic gammaproteobacteria from hypersaline soda lakes. *Extremophiles*, *12*, 619–625.

Sharma, A. K., Walsh, D. A., Bapteste, E., Rodriguez-Valera, F., Ford Doolittle, W., & Papke, R. T. (2007). Evolution of rhodopsin ion pumps in haloarchaea. *BMC Evolutionary Biology*, *7*, 79.

Slonczewski, J. L., Coker, J. A., & DasSarma, S. (2010). Microbial growth with multiple stressors. *Microbe*, *5*, 110–116.

Soppa, J., Baumann, A., Brenneis, M., Dambeck, M., Hering, O., & Lange, C. (2008). Genomics and functional genomics with haloarchaea. *Archives of Microbiology*, *190*(3), 197–215.

Sorokin, D. Y., & Kuenen, J. G. (2005). Chemolithotrophic haloalkaliphiles from soda lakes. *FEMS Microbiology Ecology*, *52*, 287–295.

Stan-Lotter, H. (2004). Biotic survival over geological times. *Astrobiology*, *4*(3), 325–326.

Sumper, M., Berg, E., Mengele, R., & Strobel, I. (1990). Primary structure and glycosylation of the S-layer protein of *Haloferax volcanii*. *Journal of Bacteriology*, *172*, 7111–7118.

Takenaka, S., Zhou, Q., Kuntiya, A., Seesuriyachan, P., Murakami, S., & Aoki, K. (2007). Isolation and characterization of thermotolerant bacterium utilizing ammonium and nitrate ions under aerobic conditions. *Biotechnology Letters*, *29*, 385–390.

Thomson, A. J., Giannopoulos, G., Pretty, J., Baggs, E. M., & Richardson, D. J. (2012). Biological sources and sinks of nitrous oxide and strategies to mitigate emissions. *Philosophical transactions of the Royal Society of London. Series B, Biological Sciences*, *367*(1593), 1157–1168.

Tindall, B. J., Tomlinson, G. A., & Hochstein, L. I. (1989). Transfer of *Halobacterium denitrificans* (Tomlinson, Jahnke, and Hochstein) to the genus *Haloferax* as *Haloferax denitrificans* comb. nov. *International Journal of Systematic Bacteriology*, *39*(3), 359–360.

Tomlinson, G. A., Jahnke, L. L., & Hochstein, L. I. (1986). *Halobacterium denitrificans* sp. nov., an extremely halophilic denitrifying bacterium. *International Journal of Systematic Bacteriology*, *36*(1), 66–70.

Torreblanca, M., Rodriguez-Valera, F., Juez, G., Ventosa, A., Kamekura, M., & Kates, M. (1986). Classification of non-alkaliphilic *Halobacteria* based on numerical taxonomy and polar lipid composition, and description of *Haloarcula* gen. nov. and *Haloferax* gen. nov. *Systematic and Applied Microbiology*, *8*, 89–99.

Valentine, D. L. (2007). Adaptations to energy stress dictate the ecology and evolution of the archaea. *Nature Reviews Microbiology*, *5*(4), 316–323.

van Ginkel, C. G., van Haperen, A. M., & van der Togt, B. (2005). Reduction of bromate to bromide coupled to acetate oxidation by anaerobic mixed microbial cultures. *Water Research*, *39*, 59–64.

van Leeuwen, F. X. R. (2000). Safe drinking water: The toxicologist's approach. *Food and Chemical Toxicology*, *38*(Suppl. 1), 51–58.

Woese, C. R. (2004). The archaeal concept and the world it lives in: A retrospective. *Photosynthesis Research*, *80*(1–3), 361–372.

Woese, C. R., & Fox, G. E. (1977). Phylogenetic structure of the prokaryotic domain: The primary kingdoms. *Proceedings of the National Academy of Sciences of the United States of America*, *74*(11), 5088–5090.

Woese, C. R., Kandler, O., & Wheelis, M. L. (1990). Towards a natural system of organisms: Proposal for the domains Archaea, Bacteria, and Eucarya. *Proceedings of the National Academy of Sciences of the United States of America*, *87*(12), 4576–4579.

Wolterink, A. F. W. M., Schiltz, E., Hagedoorn, P. L., Hagen, W. R., Kengen, S. W. M., & Stams, A. J. M. (2003). Characterization of the chlorate reductase from *Pseudomonas chloritidismutans*. *Journal of Bacteriology*, *185*, 3210–3213.

Xu, X. W., Wu, Y. H., Wang, C. S., Oren, A., Zhou, P. J., & Wu, M. (2007). *Haloferax larsenii* sp. nov., an extremely halophilic archaeon from a solar saltern. *International Journal of Systematic and Evolutionary Microbiology*, *57*, 717–720.

Yoshimatsu, K., Iwasaki, T., & Fujiwara, T. (2002). Sequence and electron paramagnetic resonance analyses of nitrate reductase NarGH from a denitrifying halophilic euryarchaeote *Haloarcula marismortui*. *FEBS Letters*, *516*, 145–150.

Yoshimatsu, K., Sakurai, T., & Fujiwara, T. (2000). Purification and characterization of dissimilatory nitrate reductase from a denitrifying halophilic archaeon, *Haloarcula marismortui*. *FEBS Letters*, *470*, 216–220.

Zhang, J., Zhang, Y., & Quan, X. (2012). Electricity assisted anaerobic treatment of salinity wastewater and its effects on microbial communities. *Water Research*, *46*, 3535–3543.

Zhu, L., Ding, W., Feng, L. J., Kong, Y., Xu, J., & Xu, X. Y. (2012). Isolation of aerobic denitrifiers and characterization for their potential application in the bioremediation of oligotrophic ecosystem. *Bioresource Technology*, *108*, 1–7.

Zimmermann, P., & Pfeifer, F. (2003). Regulation of the expression of gas vesicle genes in *Haloferax mediterranei*: Interaction of the two regulatory proteins GvpD and GvpE. *Molecular Microbiology*, *49*(3), 783–794.

Zumft, W. G. (1997). Cell biology and molecular basis of denitrification. *Microbiology and Molecular Biology Reviews*, *61*, 533–616.

Zumft, W. G., & Kroneck, P. M. H. (2006). Respiratory transformation of nitrous oxide (N_2O) to dinitrogen by bacteria and archaea. *Advances in Microbial Physiology*, *52*, 107–227.

CHAPTER THREE

Mechanisms of Bacterial Extracellular Electron Exchange

G.F. White, M.J. Edwards, L. Gomez-Perez, D.J. Richardson, J.N. Butt, T.A. Clarke[1]

School of Biological Sciences and School of Chemistry, University of East Anglia, Norwich, United Kingdom
[1]Corresponding author: e-mail address: Tom.Clarke@uea.ac.uk

Contents

Abstract

The biochemical mechanisms by which microbes interact with extracellular soluble metal ions and insoluble redox-active minerals have been the focus of intense research over the last three decades. The process presents two challenges to the microorganism. Firstly, electrons have to be transported at the cell surface, which in Gram-negative bacteria presents an additional problem of electron transfer across the ~6 nm of the outer membrane. Secondly, the electrons must be transferred to or from the terminal electron acceptors or donors. This review covers the known mechanisms that bacteria use to

Advances in Microbial Physiology, Volume 68
ISSN 0065-2911
http://dx.doi.org/10.1016/bs.ampbs.2016.02.002

transport electrons across the cell envelope to external electron donors/acceptors. In Gram-negative bacteria, electron transfer across the outer membrane involves the use of an outer membrane β-barrel and cytochrome. These can be in the form of a porin–cytochrome protein, such as Cyc2 of *Acidithiobacillus ferrooxidans*, or a multiprotein porin–cytochrome complex like MtrCAB of *Shewanella oneidensis* MR-1. For mineral-respiring organisms, there is the additional challenge of transferring the electrons from the cell to mineral surface. For the strict anaerobe *Geobacter sulfurreducens* this requires electron transfer through conductive pili to associated cytochrome OmcS that directly reduces Fe(III)oxides, while the facultative anaerobe *S. oneidensis* MR-1 accomplishes mineral reduction through direct membrane contact, contact through filamentous extensions and soluble flavin shuttles, all of which require the outer membrane cytochromes MtrC and OmcA in addition to secreted flavin.

1. INTRODUCTION

For many years it was considered that the redox cycling of metals within the environment occurred through abiotic processes. It is only within the last three decades that the ability of microorganisms to transform subsurface metals and minerals through oxidation or reduction has been accepted as a globally important phenomenon, while metagenomic analysis has found that different microorganisms influence a substantial proportion of mineral cycling within the environment. As iron is the fourth most abundant element in the earth's crust (after oxygen, silicon and aluminium) and the iron cycle is essential to life on earth, bacterial interactions with iron oxide and hydroxide minerals are the most studied processes. However other environmentally important minerals, such as those containing manganese, arsenic, radioactive contaminants, eg, uranium and technetium or even rare metals such as gold and platinum have been proposed as substrates for these organisms (Lovley, 1993; Nealson & Saffarini, 1994).

Both oxidation and reduction of metal substrates are possible. Many microorganisms undergoing anaerobic respiration generate an excess of electrons through the oxidation of organic substrates. These electrons are transported to the surface of the cell where they are discharged into a diverse range of terminal electron acceptors, these can include soluble electron acceptors such as uranium, technetium, soluble metal chelates, or insoluble acceptors such as iron and manganese oxides or hydroxides (Fredrickson & Zachara, 2008).

In contrast, some microorganisms in environments with limited organic substrate use the oxidation of reduced iron species to liberate electrons that are then transported to oxygen or nitrate species. Microorganisms that utilise metals as electron donors or acceptor represent ancient forms of metabolism that are likely to have evolved on early earth before other anaerobic or aerobic respiratory processes (Ilbert & Bonnefoy, 2013).

These microorganisms have substantial environmental and economic importance. Iron-oxidising bacteria are the cause of acid mine drainage but have also been exploited to profitably extract copper from low-grade ore (Johnson, 2014). Microbes in the subsurface are capable of reducing radionuclides, causing them to precipitate and removing them from the groundwater, while in other areas similar processes cause the solubilisation of arsenic, leading to increased levels of arsenic in the drinking water (Osborne, McArthur, Sikdar, & Santini, 2015; Wilkins, Livens, Vaughan, Beadle, & Lloyd, 2007). There has been substantial interest in the past decade in the ability of these bacteria to interact with electrodes and generate power through biobatteries and microbial fuel cells, with the most recent reported current outputs reaching 6.9 mW m^{-2} (Logan et al., 2015). In addition, microbes also can be grown on cathodes allowing for metabolic control and generation of electrosynthetic metabolites (Desloover, Arends, Hennebel, & Rabaey, 2012).

The increase in the ease of genomic and metagenomic sequencing has lead to the putative respiratory chains of many of these metal-metabolising organisms being identified. However, despite this wealth of information the mechanistic detail by which microorganisms facilitate the reduction or oxidation of different insoluble substrates is poorly understood. In this chapter, we review the current literature underpinning our understanding of how microorganisms achieve this, briefly summarising the processes of iron oxidation and iron reduction, before reviewing the literature on the most well-understood processes, that of the mineral-reducing *Geobacter sulfurreducens* and *Shewanella oneidensis*.

2. DIVERSITY OF MICROBE–MINERAL METABOLISM

Over the past two decades many microorganisms have been identified as having the ability to use metal and metal-containing minerals as metabolic substrates, either as terminal electron donors or acceptors. During mineral respiration microorganisms oxidise carbon substrates such as lactate or acetate within the cytoplasm, resulting in reduced organic electron carriers such as NADH or $FADH_2$. These carriers are oxidised at the cytoplasmic membrane, resulting in the co-transport of electrons and protons across the membrane. The change in proton gradient generates a proton motive force (PMF) that is used to synthesise ATP, while electrons accumulate within the periplasm or intermembrane space. In Gram-negative bacteria, the electrons are transported across the outer membrane to the cell surface before being transferred into terminal electron acceptors, while in Gram-positive bacteria the electrons must be transported through the thick cell wall before being transferred into terminal acceptors. Release of these electrons is not thought to contribute to the PMF, but prevents the build-up of charge inside the cell, this process is known as dissimilatory metal respiration (Fig. 1).

Most metal-reducing bacteria are not true lithotrophs as they typically use organic carbon as an electron donor. Metal-oxidising bacteria use minerals as electron donors by extracting electrons from reduced metal species such as soluble or insoluble Fe(II). These microbes couple the oxidation of iron to the reduction of a terminal electron acceptor by oxidising Fe(II) to Fe(III) on the cell surface, and transporting the electrons through the periplasm to the cytoplasmic membrane where they are used to generate both NADH and a PMF that is used to generate ATP. In order to generate PMF

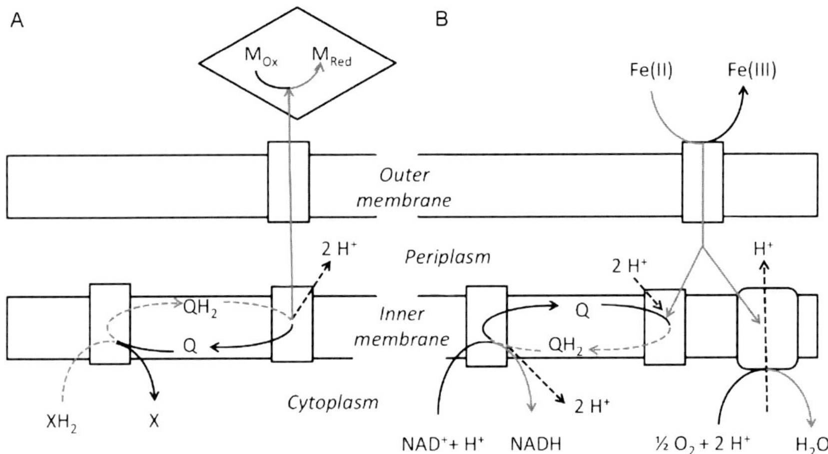

Fig. 1 Electron transfer pathways of energy production for mineral-respiring bacteria. (A) Mineral-reducing bacteria transfer electrons into the quinol pool via a quinone reductase in the cytoplasmic membrane. Quinol dehydrogenases on the periplasmic side of the membrane oxidise quinol and electrons are passed through the periplasm, across the outer membrane to the extracellular electron acceptors (M_{ox}). (B) Pathway of iron-oxidising bacteria, electrons obtained on the cell surface from oxidation of iron are transferred through two separate pathways; either into oxidases for production of a proton motive force or into the quinol pool to power the formation of NADH. *Red lines* (grey in the print version) represent electron transport while *dashed lines* represent proton transport.

transmembrane reductases at the cytoplasmic membrane couple the reduction of a terminal electron acceptor to the net transport of proton across the cytoplasmic membrane. The most common terminal acceptor is oxygen, and the majority of iron oxidisers fall into this category; however, there are examples of anaerobic iron-oxidising bacteria, such as *Dechloromonas* spp., which couples the oxidation of iron to the reduction of nitrate.

2.1 Biology of Iron-Metabolising Bacteria

2.1.1 Iron-Oxidising Bacteria

There are a broad range of different bacteria that are capable of oxidising Fe(II) in order to obtain reducing equivalents for the generation of a PMF and reduction of NAD^+. These include phototrophic organisms (*Rhodobacter* spp.), acidophilic aerobes (*Acidithiobacillus* spp., *Leptospirillum* spp.), neutrophilic aerobes (*Sideroxydans* spp., *Gallionella* spp.) and nitrate respiring anaerobes (*Dechloromonas* spp. Strain UWNR4) (Coby, Picardal, Shelobolina, Xu, & Roden, 2011; Hedrich, Schlomann, & Johnson, 2011). Several of the aerobic iron-oxidising bacteria are autotrophic, requiring Fe(II) as the sole source of energy and electrons for carbon fixation, while the phototrophic and anaerobic iron oxidisers are heterotrophic and use electrons from Fe(II) as a supplement.

One of the most important and best studied of the iron-oxidising bacteria is the microbe *Acidithiobacillus ferrooxidans* (Roger et al., 2012). This microbe is well known for its environmental impact and generates energy by coupling the oxidation of Fe(II) to the reduction of O_2. The best known form of this reaction is the oxidation of the Fe(II)S mineral known as pyrite, which plays a central role in acid mine drainage, the cause of acidic lakes and rivers that contain excessive levels of iron. In the presence of water and O_2 the abiotic and biotic reactions below occur. Fe(II) is oxidised by *A. ferrooxidans* to form Fe(II) and the liberated electrons are used by the bacteria to reduce O_2 to H_2O. The resulting Fe(III) then reacts spontaneously with FeS_2 (pyrite) to produce Fe(III), thiosulphate and an excess of protons. This, coupled with the further oxidation of thiosulphate to sulphate by sulphur-respiring bacteria, has the accelerative effect of lowering pH, which in turn accelerates the dissolution of pyrite and increasing the availability of soluble reduced iron (Roger et al., 2012).

$$4Fe(II) + O_2 + 4H^+ \rightarrow 4Fe(III) + 2H_2O \ (\textit{A. ferrooxidans}\ \text{biotic reaction})$$

$$6Fe(III) + FeS_2 + 3H_2O \rightarrow 7Fe(III) + S_2O_3^{2-} + 6\,H^+ \ (\text{abiotic reaction})$$

Under neutral pH conditions ferrous iron can also be used as an electron donor. However, at this pH soluble ferrous iron reacts more quickly with oxygen, so under oxygen levels typically found in the environment the spontaneous chemical oxidation of iron outcompetes the biological oxidation of most bacteria (Emerson, Fleming, & McBeth, 2010). This limits the environmental niche of neutrophilic iron-oxidising bacteria to the microoxic zone, where the concentration of oxygen is so low that the rate of chemical iron oxidation is outcompeted by the rate of biological iron oxidation. The second challenge facing neutrophilic iron-oxidising bacteria is that Fe(III) is rapidly precipitated by oxygen into insoluble Fe(III)oxides at neutral pH, meaning that these bacteria must also deal with insoluble iron oxides that are generated by the cell during respiration.

Neutrophilic iron-oxidising bacteria that have been studied in isolation include bacteria of the genera *Sideroxydans* and *Gallionellacea*. These organisms grow very slowly (typical doubling times of ~8 h) and to low cell culture densities, limiting studies to genome analysis, growth conditions and electron microscopy (Emerson et al., 2013; Emerson & Moyer, 1997; Neubauer, Emerson, & Megonigal, 2002). Like *A. ferrooxidans*, these organisms couple the extracellular oxidation of Fe(II) to the reduction of O_2 on the cytoplasmic membrane but the oxidised Fe(III) product rapidly precipitates into insoluble Fe(III)oxides that form on the surface of the cell. The *Sideroxydans* genus only contains two known species currently: *Sideroxydans lithotrophicus* ES-1 (Emerson & Moyer, 1997) and *Sideroxydans paludicola* (Weiss et al., 2007). During aerobic respiration by *S. lithotrophicus* ES-1 the Fe(III) has been shown to precipitate as nanoparticles on the surface of the cell, these nanoparticles are proposed to separate from the cell during the bacterial life cycle of the cell (Emerson et al., 2013).

The *Gallionella* genus is composed of several known species and subspecies: *Gallionella ferruginea* is one of the earliest recognised iron-oxidising bacteria, having been identified in the 19th century together with associated iron precipitates that under a microscope appeared as twisted stalks (Ehrenberg, 1836). Like *S. lithotrophicus*, *G. ferruginea* oxidises iron at the surface of the cell, but the precipitation of the iron is tightly controlled. Rather than generating iron oxide particles these bacteria process the iron into twisted 'stalks' that are made from a mixture of microbial expolymer and Fe(III)hydroxide that extend from one side of the bacteria (Chan, Fakra, Emerson, Fleming, & Edwards, 2011; Ghiorse, 1984). These stalks have been proposed to have a number of roles, including tethering the microorganism to sediment in flowing water, preventing the bacteria from becoming encrusted in iron oxide and protection against reactive oxygen species produced through the oxidation iron (Chan, Fakra, Edwards, Emerson, & Banfield, 2009; Hallbeck & Pedersen, 1995).

Iron oxidation is often used as a supplemental source of electrons by metabolically diverse bacteria. For example, phototrophic purple bacteria can use iron oxidation to obtain electrons for carbon fixation (Ehrenreich & Widdel, 1994). The oxidised iron is thought to precipitate away form the cell surface as Fe(III)(hydr)oxides, which then slowly convert to lepidocrocite and goethite, although the mechanism is unclear (Kappler & Newman, 2004). There have been reports of anaerobic iron-oxidising bacteria that use ferrous iron as a supplement rather than being lithoautotrophic anaerobes (Roden, 2012). Most require supplements of acetate or other organics in order to stimulate iron oxidation (Chakraborty & Picardal, 2013), and it has been proposed that the oxidation of nitrate (and in some instances chlorate) is part of a detoxification strategy as well as having a metabolic role (Carlson, Clark, Melnyk, & Coates, 2012).

2.1.2 Mineral-Reducing Bacteria

Mineral-reducing bacteria are a globally disperse group of bacteria that are typically found in the anoxic subsurface and can range in depth from a few centimetres to several kilometres (Nealson, 1997). These microbes can utilise a broad range of substrates as electron donors, including organics such as lactate, acetate or glycerol, or inorganic such as hydrogen, with most organisms able to utilise a broad range of substrates. The number of electron acceptors is also broad, ranging from soluble radionuclides, through insoluble metal oxides to synthetic anodes (Guo, Prevoteau, Patil, & Rabaey, 2015; Lovley, 1993; Nealson, Belz, & McKee, 2002; Wei, Liang, & Huang, 2011). Many bacteria have the capability of extracellular reduction, although many of these require the use of exogenous mediators, such as neutral red, to facilitate electron transfer (Taskan, Ozkaya, & Hasar, 2015). However, many microorganisms already present in the environment, both Gram-negative and Gram-positive, are capable of reducing metal oxides or synthetic electrodes without the addition of mediators (Chabert, Amin Ali, & Achouak, 2015). The most studied mineral-reducing organisms are the genera of *Shewanella* and *Geobacter*. Bacteria of the *Shewanella* genus are typically facultative anaerobes that can often be found in the shallow sediments of lakes, rivers and oceans (Fredrickson et al., 2008). The strict anaerobes of *Geobacter* have also been isolated from a range of different geographical locations and are often identified in the deeper subsurface (Nevin & Lovley, 2002).

The environmental impact of *Shewanella* spp. was first identified in the sediments of lake Oneida, where the concentrations of manganese in the lake waters were observed to increase in the summer and decrease in the winter as manganese nodules were deposited within the sediment layer. Enrichment of the sediments identified an organism initially known as *Alteromonas putrefaciens* MR-1 (later known as *S. oneidensis* MR-1) as the biotic component in the sediment responsible for manganese reduction (Myers & Nealson, 1988). This organism explained the observed manganese cycle within the lake. During the summer months microbial activity increased, leading to an increase in the concentration of soluble, reduced manganese Mn(II) in the lakewater through the process of mineral respiration. During the winter months microbial activity decreases, and the soluble Mn(II) is chemically oxidised back into insoluble manganese Mn(IV) nodules on the surface of the lake sediment, as the lake warms during summer the cycle begins again (Aguilar & Nealson, 1998).

Within the same decade, the environmental capability of *Geobacter* to reduce Fe(III)oxide to the semi-reduced mineral magnetite had also been observed. The ability of bacteria to generate and utilise magnetite as a sensor had already been studied but this was the first evidence that bacteria were transforming Fe(III) minerals to Fe(III/II) minerals during respiration (Lovley et al., 1993). Both *Geobacter* and *Shewanella* are genetically tractable, can be grown as single colonies on plates and can utilise a range of different organic substrates. They have been the source of careful study for almost 30 years, and as a consequence the mechanisms by which each bacteria performs mineral respiration are better understood than for many other mineral-respiring organism.

2.2 Model Iron Oxides Used for Measurement of Microbial Biochemistry

One of the principle challenges in the study of microbe–mineral respiration is defining the metabolic substrates used by the bacteria. The subsurface environment is comprised of an almost infinite variety of different substances, including minerals, metal chelates, complex organics such as humic acids, as well as mineral clay substrates. In order to survive in the environment microorganisms must be able to interact with these. The morphology of minerals and complex organics such as humic acids tend to vary according to environmental conditions and the process whereby they were made. In light of this, attempts to understand the mechanisms of microbial mineral reduction have relied on a relatively small range of soluble metal chelates and insoluble metal oxides and hydroxides.

Unlike conventional substrates used in biochemical and enzymatic analysis, insoluble metal oxides and hydroxides are heterogeneous and cannot be defined by conventional enzyme–substrate kinetic models. There is also competition between microbial catalysis and conventional chemical redox processes. For example, the neutrophilic iron-oxidising *S. lithotrophicus* survives at the oxic/anoxic interface by outcompeting oxygen in ferrous iron oxidation. At atmospheric oxygen concentrations iron oxidation is predominantly a spontaneous chemical process, however at micromolar oxygen concentrations chemical iron oxidation becomes slow enough for *S. lithotrophicus* to oxidise ferrous iron at the cell surface through a biologically driven reaction. Likewise, the *Shewanella* outer membrane cytochrome (OMC) MtrC was shown to become saturated by increasing Fe(III) citrate on the surface of the cell.

There are two approaches used to study biological iron reduction. The first, and simplest, is the reduction of soluble chelates of iron; the second is reduction of iron minerals, which is potentially more physiologically relevant but requires more careful definition. Soluble iron chelates commonly used include Fe(III)citrate, Fe(III)NTA and Fe(III)EDTA. Fe(III)EDTA is the simplest, comprising a single ferric iron atom chelated by an single ethylenediaminetetraacetate (EDTA), however this is an artificial substrate that is unlikely to exist in the environment. Both Fe(III)NTA and Fe(III)citrate will form chelates with different structures depending on the Fe(III):chelate ratio. At stoichiometric ratios complexes containing $Fe(III)_2NTA_2$ and $Fe(III)_2citrate_2$ were observed to bind to defined sites on the surface of the OMC UndA. Studies using these soluble cytochromes have indicated that the redox potential of the chelate largely dictates the observed rate of reduction.

For insoluble minerals, both physical and chemical properties must be considered when determining the availability of Fe(III) and how readily it will reduce. The chemical formula and arrangement of iron and oxygen in the crystal lattice define the mineral and its associated chemical and physical properties. There are 15 different types of iron oxides and hydroxides found in soils and sediments (Schwertmann & Cornell, 2000). The most commonly studied in microbial reduction are the iron hydroxides ferrihydrite, goethite and lepidocrocite and the iron oxide hematite. These range in chemical formula, crystallinity, shape, size and particle surface area (Table 1). The coordination of Fe(III) in the mineral crystals is octahedral. In goethite and hematite, the O and OH ions form layers that are hexagonally close packed (α-phase) and in lepidocrocite they are cubic close packed (γ-phase). This affects the crystallinity, surface texture and general reactivity of the mineral. Goethite is found in most soils, hematite occurs in more tropical regions where there are higher temperatures and lower water activities and lepidocrocite occurs in water-logged environments that are deficient in oxygen. Ferrihydrite is a commonly found, less ordered substance that varies in composition. It has a higher redox potential and consists of smaller particles making it generally more reactive than the other minerals. Ferrihydrite is unstable and over time transforms into the mineral oxides goethite and lepidocrocite. Lepidocrocite, while stable, has a more positive redox potential than goethite or hematite, making it the next most reactive in the series.

For particulate substrates, only surface-exposed metal ions will be accessible to electron transfer. Transmission electron microscopy can be used to measure the average dimensions of the mineral nanoparticles and, from this, the surface area per gram of mineral estimated. However, this does not take account of the surface texture. To account for the surface the Brunauer–Emmet–Teller (BET) method determines the 'specific surface area' by measuring the number of gas molecules, near their condensation temperature, that form a monolayer on a specific mass of particles (Schwertmann & Cornell, 2000). Rates of iron reduction can then be defined in terms of moles Fe reduced $m^{-2}\,s^{-1}$ that normalises the rates in terms of the physical properties of the minerals.

It has been shown that for abiotic mineral dissolution, the rate of Fe(III) reduction depends not only on the physical properties of the mineral particles but also on the thermodynamic properties of the ferric oxide phase (Roden, 2003). For reduction by a soluble electron donor, such as ascorbate, electron transfer is rapid and detachment of Fe(II) ions from the mineral surface is the rate-limiting step as Fe(II) produced by chemical reduction can reassociate with the mineral surface. However, for biological iron oxide reduction the process requires recognition of the mineral surface by catalytic sites on the surface of the bacterium before microbe to mineral electron transfer can take place. In this case, electron transfer is the rate-limiting step rather than detachment of Fe(II). From the examples shown in Table 1, it can be observed that, for equivalent surface areas, ascorbate dissolution rates are typically 10-fold faster for lepidocrocite than goethite; whereas for

Table 1 Properties of Insoluble Iron Oxides Commonly Used as Model Mineral Substrates

Mineral	Goethite	Hematite	Lepidocrocite	Ferrihydrite	References
Formula	$\alpha FeOOH$	αFe_2O_3	$\gamma FeOOH$	Varied	1
Crystal system	Orthorhombic	Trigonal	Orthorhombic	Amorphous	1
Typical size and shape	20–100 nm × 200–500 nm (needles)	30–50 nm (hexagonal)	10–50 nm × 100–300 nm (lath-like)	<50 nm (spheres)	2
Typical BET surface area ($m^2\ g^{-1}$)	38	34	130	230	2
Redox potential typical[a] (*V*)	−0.157	−0.121	−0.103	+0.61	2
Dissolution rate by ascorbate[b] (10^{-9} mol $m^{-2}\ s^{-1}$)	0.4	–	5–7	4–40	3
Dissolution rate by ascorbate[b] (10^{-9} mol $m^{-2}\ s^{-1}$)	0.1	0.1	4	0.2	4
Dissolution rate by *Shewanella*[c] (10^{-9} mol $m^{-2}\ s^{-1}$)	0.06	0.08	0.1	0.01	4

[a]Redox potential (*V* vs SHE at pH 7) 1 μM total dissolved Fe.
[b]Dissolution rate: Reduction of iron oxides by 10 mM ascorbic acid at pH 3, 25°C, anoxic conditions, determined by concentration of Fe^{2+} released into solution.
[c]Dissolution rate: Reduction of iron oxides by *Shewanella putrefaciens* strain CN32, pH 7, 25°C, anoxic conditions, determined by concentration of Fe^{2+} released into solution.
References: 1, Schwertmann and Cornell (2000); 2, White et al. (2013); 3, Larsen and Postma (2001) and 4, Roden (2003).

microbial reduction with *Shewanella* there is less dependence on mineral type and a more direct correlation with specific surface area. Reviewing literature reports of bacterial reduction rates, Roden (2003) commented that similar rates of mineral reduction and the same trend with respect to mineral type has been observed for both *Shewanella* and *Geobacter* (*S. putrefaciens CN32, S. alga BrY, G. sulfurreducens, G. metallireducens*). It must be noted that the results for ferrihydrite do not fit the expected rates of abiotic and biotic reduction. Although the chemical and physical properties predict that it should be the most reactive mineral studied, literature values reported for abiotic reduction of ferrihydrite vary greatly and the production of Fe(II) from ferrihydrite by *Shewanella* is significantly slower than that found with the other minerals. This can be explained by the varied composition of ferrihydrite, its instability and the fact that the small ferrihydrite particles have a tendency to aggregate.

3. BIOLOGICAL ELECTRON TRANSPORT ACROSS THE CELL ENVELOPE

The majority of organisms that perform the metabolic oxidation or reduction of metal species transport electrons to the cell surface where the catalytic redox reaction takes place. For bacteria that utilise minerals as terminal electron acceptors, this requires the transport of electrons generated in the cytoplasm to be transported to the surface of the cell. For bacteria that use minerals as an electron source, the electrons must be brought into the cell either to the cytoplasmic membrane in order to reduce terminal electron accepters or to cytoplasmic NADH. The outer membrane of Gram-negative bacteria is an insulating barrier that requires a conduit for electron passage, while in Gram-positive bacteria the cell outer wall acts as a barrier between cell and mineral surface. Consequently in order to transfer electrons from inside the cell to the cell surface bacteria use a diverse range of cofactor-rich proteins. Fig. 2 shows examples of currently proposed electron transport systems for both mineral-oxidising and mineral-reducing bacteria. All organisms use the quinol pool, either in the form of menaquinol or ubiquinol, as mediators of electrons across the cytoplasmic membrane, and the current understanding of the structural mechanisms that these bacteria use to transport electrons across the cell envelope and exchange electrons exchange with iron (hydr)oxides will be discussed in the remainder of this review.

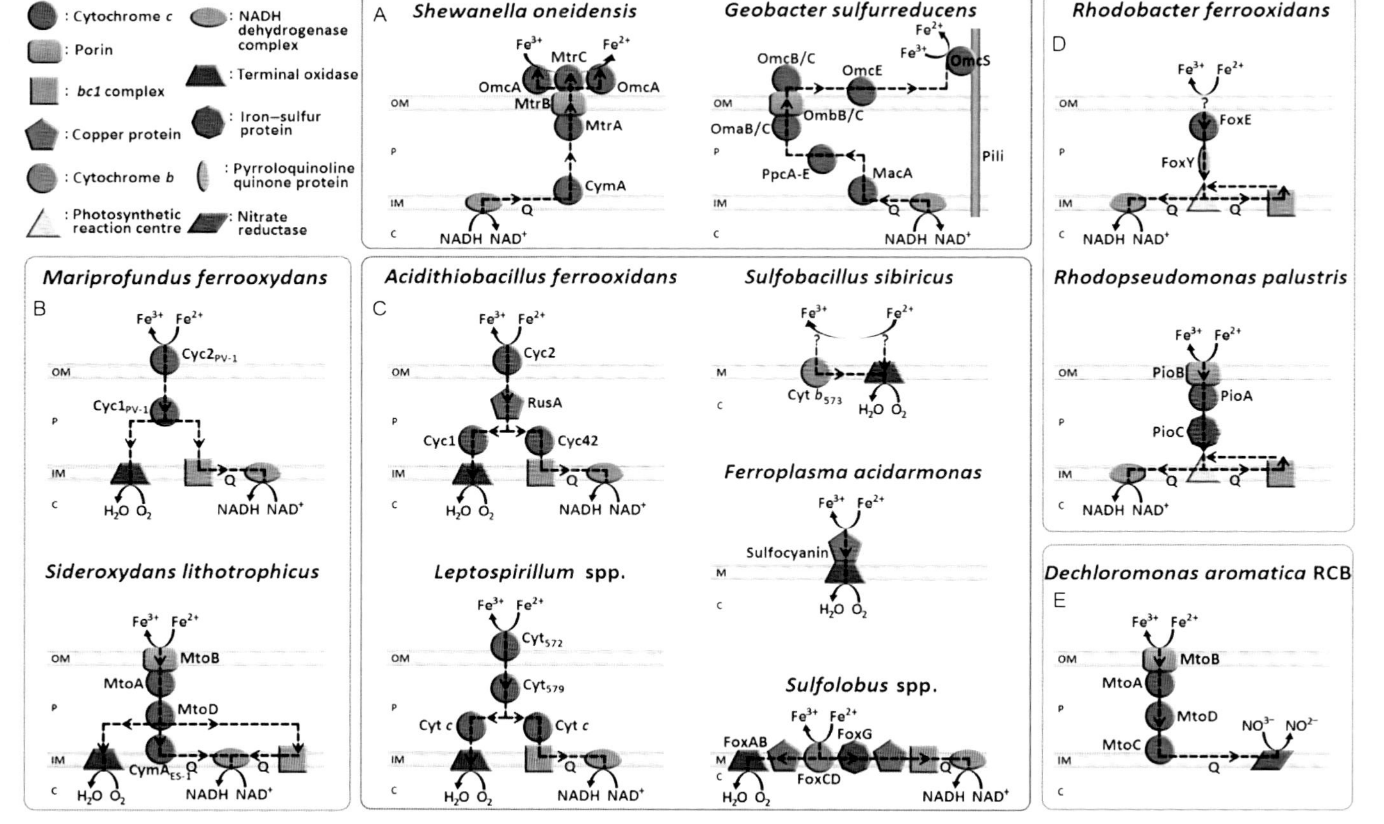

Fig. 2 Iron reduction and oxidation pathways in microorganisms. (A) Anaerobic iron reducers *G. sulfurreducens* and *S. oneidensis* (Liu et al., 2014; Lovley, 2006; Shi, Rosso, Zachara, & Fredrickson, 2012). (B) Neutrophilic aerobic iron oxidisers *M. ferrooxydans* and *S. lithotrophicus* (Barco et al., 2015; Beckwith et al., 2015). (C) Acidophilic aerobic iron oxidisers *A. ferrooxidans*, *Leptospirillum* spp., *S. sibiricus*, *F. acidarmonas*, *Sulfolobus* spp. (Ilbert & Bonnefoy, 2013). (D) Neutrophilic photosynthetic iron oxidisers *R. ferrooxidans* and *R. palustris* (Jiao & Newman, 2007; Saraiva, Newman, & Louro, 2012). (E) Neutrophilic anaerobic iron oxidisers dependent on nitrate, perchlorate or chlorate reduction (Shi, Rosso, et al., 2012). The electron flow is indicated as a *dashed line*. (See the colour plate.)

3.1 Extracellular Electron Transfer in Gram-Positive Bacteria and Archae

A number of Gram-positive bacteria have been shown to be capable of reducing insoluble metal oxides, either as isolated strains or as part of a bacterial consortium. As Gram-positive bacteria do not have an outer membrane, there is no requirement for electron transfer through porin–cytochrome systems, however the thicker cell walls limit the ability of the cell to transfer electrons directly from the cytoplasmic membrane and into insoluble extracellular mineral. The Gram-positive *Sulfobacillus sibiricus* is a moderately thermophilic acidophile that oxidises soluble Fe(II) to soluble Fe(III) (Dinarieva, Zhuravleva, Pavlenko, Tsaplina, & Netrusov, 2010; Ilbert & Bonnefoy, 2013). Fe(II) is oxidised at the cytoplasmic membrane and energy is obtained through the reduction of oxygen on the same membrane. It is not clear which enzyme is responsible for iron oxidation, whether Fe(II) is oxidised directly by the terminal oxidase or if a cytochrome *b* electron shuttle mediates electron transfer between Fe(II) and the terminal oxidase (Fig. 2), but for the Gram-positive acidophilic iron oxidisers, it is not necessary to have an electron conduit between the cytoplasmic membrane and cell surface as both substrate and product can diffuse through the cell wall.

The acidophilic archaea *Ferroplasma acidarmonas* and *Sulfolobus* spp. have also been show to be capable of iron oxidation through direct reduction of iron at the cytoplasmic membrane. *F. acidarmonas* oxidises iron using a copper protein, sulphocyanin, that directly transfers electrons to a terminal aa_3 oxidase in order to supplement the PMF (Fig. 2) (Castelle et al., 2015; Dopson, Baker-Austin, & Bond, 2007, 2005). In *Sulfolobus* spp. a more complex pathway appears to have evolved. A cytochrome *b* complex (FoxCD) oxidises ferrous iron to ferric iron and sends the electrons to an iron–sulphur protein with (Fe-S) domain (FoxG) and then the electrons are transported to a multicopper oxidase (Mco) before they arrive to the terminal oxidase and the NADH dehydrogenase (Fig. 2) (Bathe & Norris, 2007).

In contrast, the Gram-positive thermophilic iron-reducing *Carboxydothermus ferrireducens* (originally called *Thermoterrabacterium ferrireducens*) has been shown to reduce ferrihydrite through direct contact between microbe and mineral (Gavrilov, Lloyd, Kostrikina, & Slobodkin, 2012; Gavrilov, Slobodkin, Robb, & de Vries, 2007). This pathway requires an electron transfer pathway to transport electrons directly from the cytoplasmic membrane to the extracellular minerals on the cell surface. The pathway has yet

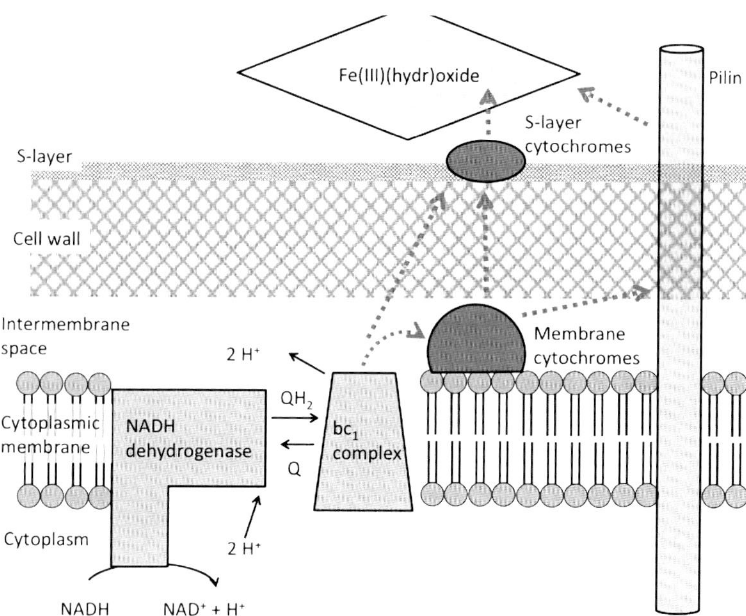

Fig. 3 Mechanism of extracellular electron transport in the Gram-positive *Carboxydothermus ferrireducens*. Electrons generated through the cytochrome bc_1 complex are transferred to extracellular iron oxides through direct electron transfer pathways that are proposed to involve cytochromes associated with the cytoplasmic membrane, the microbe–mineral interface involves cytochromes associated with the S-layer of the cell wall. The electron transfer mechanisms indicated by *dashed arrows* have not yet been determined.

to be fully defined but it is reported to involve the reduction of Fe(III)oxides using cytochromes embedded within the S-layer of the cell wall, and pili that are expressed during respiration on insoluble minerals. At least one cytochrome Fe-(EDTA) reductase has been identified that attached to the cytoplasmic membrane, although it is not clear whether this is the only protein that is reduced by the cytochrome bc_1 complex, or how electrons from this complex are transferred either to pili or through the cell wall into the cytochromes embedded in the S-layer (Gavrilov et al., 2012) (Fig. 3).

3.2 The Porin–Cytochrome Complex as a Transmembrane Electron Conduit

The most commonly studied, and best understood, of the complexes that allow electron transfer across the outer membrane of Gram-negative bacteria is the MtrCAB porin–cytochrome complex of *S. oneidensis* MR-1, which is composed of three proteins encoded by the *mtrCAB* operon

(Hartshorne et al., 2009; Ross et al., 2007). *S. oneidensis* also produces a second outer membrane cytochrome OmcA and has an *mtrDEF* gene cluster paralogous to *mtrCAB*. OmcA is co-expressed with MtrCAB, while the expression of MtrDEF is under a separate promotor and appears to be preferentially expressed in biofilms or aggregated cells (McLean et al., 2008).

Both *mtrA* and *mtrC* encode for two decahaem cytochromes, with molecular weights of 37 and 75 kDa for MtrA and MtrC, respectively. *mtrB* encodes MtrB, a transmembrane β-barrel with a molecular weight of 70 kDa. MtrA is periplasmic, while MtrC is transported by the type-II secretion system to the surface of the cell where it associates with the outer membrane through an N-terminal lipid anchor (Ross et al., 2007). The three proteins come together as a transmembrane complex that allows electrons from MtrA to be transferred to MtrC through MtrB (Fig. 2). It is proposed that MtrB functions as a porin into which both MtrA and MtrC insert far enough to allow direct haem-to-haem electron exchange (Hartshorne et al., 2009; Richardson et al., 2012). In support of the proposed 'porin–cytochrome mechanism' there is no evidence for co-factors in MtrB that would mediate electron transfer across the ~50 nm of the hydrophobic lipid bilayer. Recombinant MtrB cannot be expressed in the folded state without co-expression of MtrA, however, if the genes for expression of outer membrane proteases are deleted, isolated MtrB can be expressed but the protein is produced in an unfolded state (Hartshorne et al., 2009; Schicklberger, Bucking, Schuetz, Heide, & Gescher, 2011).

It is proposed MtrA extends into the periplasm where it contacts other periplasmic electron transfer proteins. Once electrons enter MtrA, they are transported through MtrA and into the adjoining MtrC with MtrB acting as an insulating sheath. Because electrons move through the haem chain by tunnelling between adjacent haems the porin–cytochrome complex is more an 'electron hopping conduit' than a true wire, with electrons being able to move in both directions across the membrane depending on the potential difference across the membrane (Hartshorne et al., 2009; White et al., 2012, 2013). Several groups have exploited this by growing *S. oneidensis* on an electrode and lowering the potential to drive electrons back into the bacterium, thereby altering the metabolic properties of the organism and causing the formation of various intermediates, most notably succinate (Grobbler et al., 2015; Ross, Flynn, Baron, Gralnick, & Bond, 2011).

Structurally much of the MtrCAB porin–cytochrome complex is still unresolved. The best structurally resolved component is MtrC, where a structure of the soluble MtrC protein from *S. oneidensis* MR-1 has been solved to a resolution of 1.8 Å. The dimensions of the protein are approximately 80 × 60 × 30 Å, giving the protein a disc-like appearance (Edwards et al., 2015). Much less is known about MtrB beyond theoretical modelling; MtrB contains up to 28 β-strands as predicted from topology modelling software, and a short N-terminal domain that contains a CXXC amino acid motif. The role of the two cysteine at the N-terminus are unclear, but they are known to be on the surface of the cell, are redox active and at least one is required for successful assembly of the MtrCAB complex. In contrast to the orientation of most known β-barrel porins the MtrB porin is inverted, with the long soluble loops facing the periplasmic side of the outer membrane, and short loops facing the surface of the protein (Huysmans, Baldwin, Brockwell, & Radford, 2010; White et al., 2013). This inverted state is suggestive of an alternative assembly mechanism and is consistent with the need for MtrA to stabilise MtrB in the barrel—it is possible that MtrB assembles around MtrA in or on the outer membrane before formation of the stable complex.

The MtrA protein has been spectroscopically studied by several groups and the haems within the protein are known to cluster around two groups of standard electrode potentials, a high-potential group around −200 and a low potential group around −100 mV; the split of haems between each group is approximately equal, with five haems in each group (Bewley et al., 2012; Pitts et al., 2003). There is no high resolution structural information available, so nothing is known about the way that the haems arrange within the MtrA structure, but a low resolution SAXS envelope of MtrA has been proposed by Firer-Sherwood and colleagues, which has an approximate length of 100 Å, a diameter of 40 Å and tapers to a narrow 30 Å at one end, suggesting that it is capable of inserting into MtrB (Firer-Sherwood, Ando, Drennan, & Elliott, 2011). MtrA appears modular, as insertion of a stop codon into the *mtrA* gene after the sequence encoding the first five CXXCH haem motifs results in the expression of a truncated MtrA cytochrome with five bis-his ligated haems (Clarke et al., 2008). The structure of a lipopolysaccharide transporter has been resolved and shown to be a 26-strand β-barrel that contains a second protein as a 'plug' (Dong et al., 2014). This porin–plug complex is the closest structurally resolved protein to MtrB and has approximate dimensions of 60 × 40 nm, suggesting that MtrA would be small enough to insert completely to MtrB while the larger MtrC may be restricted from entering fully (Fig. 4). These predictions are supported by proteinase K experiments revealing the complete digestion

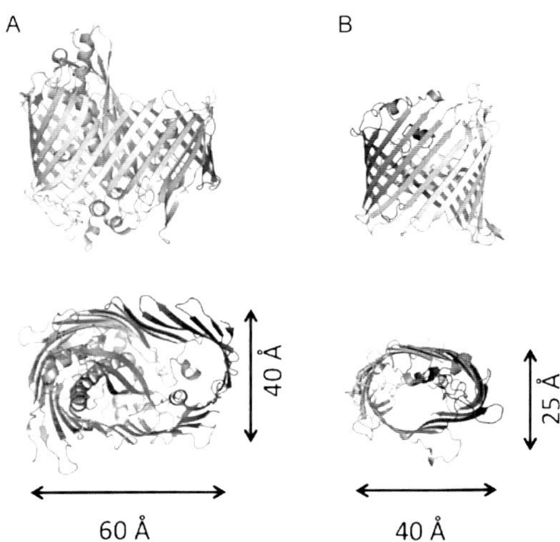

Fig. 4 Crystal structures of porins similar in size to Cyc2 and MtrB. (A) Lipopolysaccharide translocon consisting of a 26 β-strand porin LptD and α-helical LptE plug (*light grey*) (PDB ID: 4N4R). (B) 18 β-strand monomer of the trimeric sucrose transporter ScrY (PDB ID: 1A0T). (See the colour plate.)

of MtrC from MtrCAB embedded in liposomes or the cell membrane of *S. oneidensis* (Edwards et al., 2015; White et al., 2013).

Taken together, these data suggest that the MtrCAB complex functions by allowing electron transfer from one side of the lipid bilayer to the other through a chain of 20 haems that is formed between two cytochromes stabilised by a large porin. The much smaller diameter of MtrA would mean that this protein inserts further into MtrB than MtrC, such that the majority of MtrC remains exposed on the surface of the cell.

Genes homologous to *mtrCAB* are found in several other bacterial families, including the related genus *Rhodoferax ferrireducens*, *Aeromonas hydrophila*, *Halorhodospira halophila*, and *Vibrio* spp. These bacteria have the full complement of *mtrCAB* genes, but the proposed role of the genes vary (Shi, Rosso, et al., 2012). *Rhodoferax* is a known reducer of iron oxides and has been shown to grow using iron oxides as a terminal electron acceptor, it therefore seems likely the role of *mtrCAB* in this organism is the same as for *S. oneidensis*: to facilitate the transfer of electrons to extracellular iron oxides (Finneran, Johnsen, & Lovley, 2003). Many have not been shown to respire on iron oxides and it has been suggested that these bacteria actually use the MtrCAB pathway to sequester iron from the environment (Bücking, Schicklberger, & Gescher, 2012). Many other bacteria have also been shown to contain homologues of *mtrAB* (Shi, Rosso, et al., 2012) but relatively few of these have been characterised and the roles of the genes are not known.

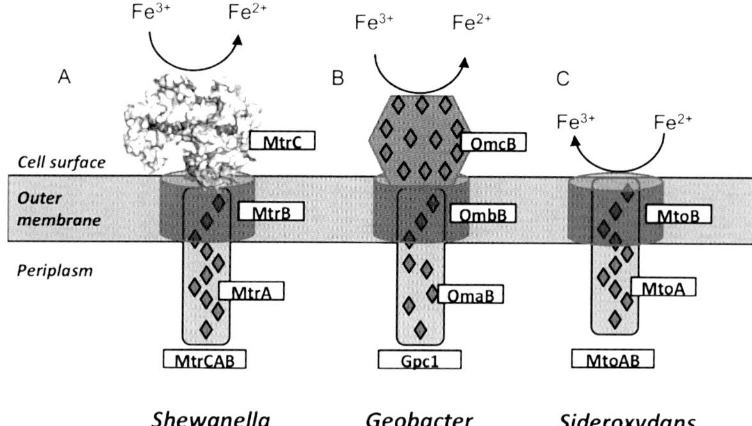

Fig. 5 Theoretical structures of known outer membrane porin–cytochromes complexes. (A) MtrCAB complex of *Shewanella oneidensis* consisting of the decahaem MtrA, the 28 β-strand porin MtrB and the decahaem cytochrome MtrC (PDB ID: 4LMB). (B) Gpc1/Gpc2 complex of *Geobacter sulfurreducens* consisting of the octahaem OmaB, 20 β-strand porin OmbB and dodecahaem OmcB. (C) MtoAB complex from *Sideroxydans lithotrophicus ES-1* consisting of the decahaem MtoA and 28 β-strand porin MtoB. *Red diamonds* represent c-type haems. (See the colour plate.)

Outer membrane porin–cytochrome complexes with a similar modular configuration to MtrCAB have been identified in *Geobacter* (Fig. 5). These complexes (Gpc1 and Gpc2) have been demonstrated to be capable of transferring electrons across a membrane and their deletion has been shown to be detrimental to the reduction of extracellular electron acceptors by *Geobacter sulfurreducens* (Liu et al., 2014). The Gpc1 gene cluster encodes for OmaB, OmbB and OmcB and the paralogous Gpc2 gene cluster encodes for OmaC, OmbC and OmcC. Both OmaB:OmbB:OmcB and OmaC:OmbC:OmcC have been purified as single complexes from the outer membrane of *G. sulfurreducens* and, using proteoliposome models, have been shown to be capable of transferring electrons through a lipid bilayer (Liu et al., 2014). OmaB/OmaC are octahaem periplasmic proteins, OmbB/OmbC are transmembrane porins and OmcB/OmcC are dodecameric extracellular cytochromes. The proposed functions and cellular localisations of OmaB/OmaC, OmbB/OmbC and OmcB/OmcC are analogous to those of MtrA, MtrB and MtrC from *S. oneidensis*. However, although both contain haem-binding motifs typical of c-type cytochromes, there is little sequence homology between Gpc1/Gpc2 proteins and the proteins of the *Shewanella* MtrCAB complex and the distribution of the c-type haems is not conserved.

The porin–cytochrome complexes of mineral-respiring *Geobacter* and *Shewanella* are three-component complexes that have been clearly identified as facilitating extracellular electron transfer across a lipid bilayer. In contrast, the porin–cytochrome complexes found in iron-oxidising bacteria contain only two proteins, which are homologues of MtrA and MtrB (Shi, Rosso, et al., 2012). Few of these two-component complexes have been characterised so far. The genome of *S. lithotrophicus* ES-1 contains the *mtoA* gene, a decahaem cytochrome with 42–44% homology to the *mtrA* genes of *Shewanella* spp. The *mtoA* gene is next to *mtoB* in the *S. lithotrophicus* genome, which is predicted to encode a porin. To date, only the MtoA protein has been characterised, and shown to be a soluble decahaem cytochrome with a redox potential range slightly higher than MtrA. It has been proposed that together with MtoB it forms a decahaem MtoAB porin–cytochrome complex in the outer membrane that facilitates oxidation of soluble Fe(II) to insoluble Fe(III) (Liu et al., 2012) (Fig. 5). There is no MtrC homologue in the gene cluster, but there is a periplasmic monohaem c-type cytochrome, MtoD, that has been purified and structurally characterised (Beckwith et al., 2015). MtoD is proposed to transfer electrons across the periplasm into the cytoplasmic membrane either to bc_1 oxidase or to a CymA homologue, presumably to insert electrons into the quinol pool where they will ultimately be coupled to NAD^+ reduction (Fig. 2). Homologous MtoAB systems have been identified in the genomes of *D. aromatica* and *G. capsiferriformans*, where extracellular electrons are taken from Fe(II), across the outer membrane through the decahaem porin–cytochrome complex and passed to intracellular cytochromes for reduction pathways on the cytoplasmic membrane (Emerson et al., 2013; Shi, Rosso, et al., 2012). The PioAB iron-oxidising system of *Rhodopseudomonas palustris TIE-1* also has high homology to MtoAB and MtrAB, composed of the decahaem PioA and porin PioB. This phototrophic organism can use soluble Fe(II) as a electron donor for carbon fixation, with the periplasmic PioC, an iron–sulphur-containing protein as the periplasmic shuttle between the outer and inner membrane (Jiao & Newman, 2007) (Fig. 2).

This family of porin–cytochrome complexes seems to provide the major route of electron transfer across the outer membranes of a broad variety of Gram-negative bacteria, with known examples found in the α-, β-, γ- and ζ-proteobacteria. The number of haems found within the porin–cytochrome complex of iron-oxidising organisms is 10, while the porin–cytochrome complexes of iron-reducing organisms contain a total of 20 haems. The 10 haems therefore appear to represent the optimum number of haems required to traverse the outer membrane through the porin and allow electron transfer on each side of the membrane. The porins within the complexes are predicted to be composed from between 20 and 28 β-strands which would give pore diameters ranging from 30 to 50 Å, sufficient to allow the insertion of a cytochrome through the centre (Fig. 4).

3.3 The Cyc2 Outer Membrane Fused Porin–Cytochrome

While the porin–cytochrome complex system is the most commonly identified system in iron-respiring bacteria, there is at least one other system identified that regulates electron transfer across the outer membrane. The OMC Cyc2 from *A. ferrooxidans* was the first outer membrane electron transfer protein to be identified (Appia-Ayme et al., 1998), and was shown to be a monohaem c-type cytochrome localised to the outer membrane. Its proposed role is to obtain electrons from Fe(II) on the cell surface and transfer them to rusticyanin (RusA), a high-potential periplasmic copper protein (Fig. 2) (Yarzabal et al., 2002). The electron transfer network then diverges as RusA transfers these electrons to either Cyc1 or Cyc42, which are two dihaem periplasmic cytochromes *c*. Electrons on the haems of Cyc1 are transferred to terminal oxidases, while Cyc42 will transfer electrons through the cytochrome bc_1 complex and the quinol pool to NADH dehydrogenases on the inner membrane (Roger et al., 2012) (Fig. 2). Supercomplexes of Cyc2, RusA, Cyc1 and cytochrome oxidase have been isolated, suggesting that there are stable networks of electron transfer proteins from the cell surface to the inner membrane in these bacteria, although it is unclear how Cyc42 would accept electrons in these systems (Castelle et al., 2008). The structure of Cyc2 is poorly understood; it contains a single c-type haem with a measured redox potential of +560 mV vs SHE at pH 4.8, making it possible to accept electrons from Fe(II) (Ilbert & Bonnefoy, 2013). Secondary structure predictions suggest that the N-terminal of Cyc2 contains a single CXXCH c-haem motif in an N-terminal region of approximately 30 amino acids, followed by a porin domain of 18 β-strands (Yarzabal et al., 2002). Similar sized porins have been structurally resolved using X-ray crystallography, such as the trimeric ScrY. The monomers of this sucrose transporter have 18 β-strands and form a channel with approximate dimensions of 25 × 40 Å (Fig. 3). The most obvious way that the 30 amino acid haem domain could fold is with the 18 β-strands of the porin domain wrapping around an N-terminal haem domain (as is the typical folding pattern for most outer membrane β-barrel proteins). Cyc2 could therefore be representative of a family of fused porin–cytochromes where a monohaem cytochrome domain forms the 'plug' within the β-barrel. It is currently unclear whether the N-terminal haem domain faces the periplasm or the surface, a periplasmic facing haem domain would require soluble Fe(II) to diffuse and be oxidised within the barrel. If the cytochrome is exposed on the surface then it might be expected that rustacyanin would have to insert into the porin from the periplasmic side in order to form an electron conduit (Fig. 6). However, rustacyanin only contains a single copper atom on the surface of the protein and it is unclear how this could orient in a stable complex to exchange electrons between the haems Cyc2 and Cyc1A/Cyc42.

The *Leptospirillum* spp. are a second group of acidophilic aerobes that utilise a system with homology to Cyc2. The outer membrane contains a large amount of an outer membrane protein called cytochrome 572 (Cyt_{572}) that is also predicted to contain a small monohaem N-terminal domain followed by a 15 β-strand C-terminal domain. A periplasmic monohaem cytochrome called cytochrome 579 is predicted to mediate electron transfer from the outer membrane to the inner membrane (Fig. 2). Despite the low sequence homology between Cyt_{572} the secondary structure analysis is consistent formation of a fused porin–cytochrome similar to the *A. ferrooxidans*

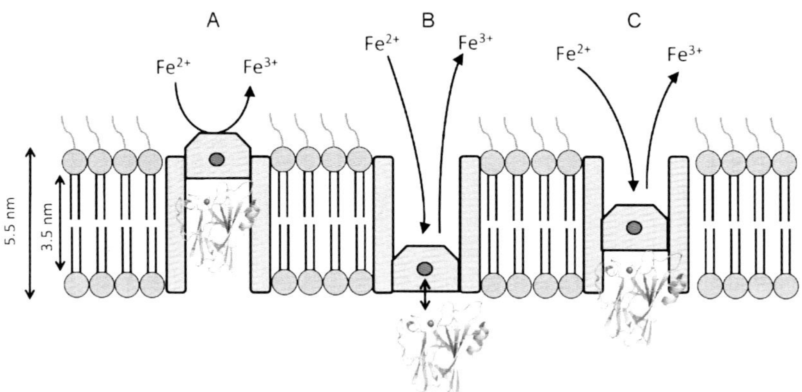

Fig. 6 Possible configurations of a Cyc2 fused porin–cytochrome. (A) The haem domain is located on the cell surface, allowing access to Fe^{2+} and insertion of rustacyanin into the barrel. (B) The haem domain located on the periplasmic face of the membrane, allowing reversible association with rustacyanin and Fe^{2+} diffusion through the extracellular-facing barrel entrance. (C) The haem domain is in the centre of the barrel with Fe^{2+} diffusing into the extracellular side and electrons being transferred to a bound rustacyanin at the periplasmic face of the membrane. (See the colour plate.)

porin–cytochrome, allowing *Leptospirillum* spp. to obtain electrons from Fe(II) in the acidic extracellular environment. Both Cyc2 and cytochrome 572 have been shown to oxidise reduced iron under acidic conditions (Castelle et al., 2008; Jeans et al., 2008) and have periplasmic electron transfer partners that mediate electron transfer across the membrane. Unlike the porin–cytochrome complexes of *Shewanella* and *Geobacter*, Cyc2 has been isolated from the cellular membranes of both *A. ferrooxidans* and *Leptospirillum* and shown to be a single subunit protein (Yarzabal et al., 2002). This suggests that neither rustacyanin nor cytochrome 579 bind tightly to Cyc2, but suggests that they interact transiently with the base of the porin and that the haem domain lies at the periplasmic side of the barrel as shown in Fig. 6.

The first marine iron oxidiser to be characterised was *Mariprofundus ferrooxydans*. Found in iron-rich mats in hydrothermal fields it has recently been studied and its genome and proteomes sequenced. Two of the most common cytochromes expressed by *M. ferrooxydans* are homologues of Cyc2 and Cyc1 of *A. ferrooxidans*, suggesting that this neutrophilic microorganism utilises the same pathway as the acidophilic iron-oxidising organisms (Barco et al., 2015). Perhaps surprisingly the genomes of both *S. lithotrophicus* and *Gallionella capsiferriformans* contain genes homologous to *cyc2* of *M. ferrooxydans*, suggesting that these organisms contain both porin–cytochrome complex and the porin–cytochrome fusion systems. It is currently unclear which of these two systems would be responsible for neutrophilic iron oxidation for these organisms.

3.4 Electron Transfer Through the Outer Membrane of Gram-Negative Bacteria

Current research has identified two separate systems by which electrons might pass through the outer membrane of Gram-negative bacteria. The first, the MtrAB porin–cytochrome complex, represents the best characterised of the electron transfer conduits, with representative complexes identified from the iron-reducing *Shewanella*, *Geobacter* and lithotrophic iron-oxidising *Sideroxydans* and *Gallionella*. It appears a key feature of respiration using the transfer of electrons to and from iron under conditions where iron oxides develop and is important when dealing with insoluble metal oxides. The second system is the Cyc2 fused porin–cytochrome system that is common amongst microorganisms that utilise extracellular soluble Fe(II), such as *A. ferrooxidans*, this much simpler system is used under acidic conditions when insoluble iron oxides are unlikely to develop. There is also overlap within these systems, *S. lithotrophicus* contains homologues of both the MtrAB porin–cytochrome complex and the Cyc2 fused porin–cytochrome while certain strains of *Geobacter* contain homologues of MtrAB, Cyc2 and also generate conductive pili. It is not known whether these systems have overlapping functionality or perhaps work together to create an efficient iron metabolic pathway.

Despite the observation of these conserved pathways for outer membrane electron transfer there are also several organisms that have, as yet, unknown mechanisms for electron transfer through their outer membranes. For example, *Rhodobacter ferrooxidans* does not contain the genes for porin–cytochromes in their outer membranes, and yet can oxidise metal on the cell surface under neutrophilic conditions (Hegler, Posth, Jiang, & Kappler, 2008; Saraiva et al., 2012) (Fig. 2).

4. STRUCTURES AT THE INTERFACE OF MICROBE–MINERAL INTERACTION

In order to utilise extracellular electron acceptors and donors as substrates during respiration bacteria have generated molecular conduits, detailed in the previous section, to transport electrons to the cell surface. From here it is necessary to then transfer the accumulating electrons to terminal electron donors and acceptors. This step is perhaps the most complicated and variable step of all, as it results in the transformation of a wide range of different substrates into different products and intermediates. For acidophilic iron oxidisers this is not a substantial challenge as both Fe(II) and Fe(III) are soluble and readily diffuse around the cell. However, for neutrophilic iron-oxidising organisms the challenge is to prevent the accumulation of insoluble iron oxides on the surface of the cell, effectively entombing the bacteria, while for iron-reducing organisms the challenge is to move the electrons from the surface to an insoluble, non-diffusive iron oxide particle.

In addition to the challenges posed by the transformation of iron between different mineral states, there is the challenge of access for bacteria attempting to utilise insoluble substrates as electron acceptors. Bacteria typically live either in planktonic suspension or as an aggregate biofilm. For bacteria living planktonically, there is a challenge of obtaining regular access to the surface of insoluble substrates when such interactions might only occur transiently. For mineral-respiring bacteria in a biofilm the challenge of nutrient access is on two fronts. Aggregated bacteria directly on the surface of the mineral have the greatest access to terminal electron acceptors but are furthest away for nutrients accessible from the media while bacteria on the surface of the biofilm have access to nutrients, but not to an electron donor (Bond, Strycharz-Glaven, Tender, & Torres, 2012). It is essential that all bacteria have access to both nutrients and the mineral oxide buried at the base of the biofilm to survive.

In order for bacteria to transfer electrons to minerals surface there have been three possible pathways postulated for iron (hydr)oxide-reducing bacteria. The possible mechanisms by which electrons can transfer from cell to terminal acceptor have been loosely defined into three separate categories: direct, indirect and mediated. Direct mechanisms involve contact between the cell and the acceptor such that cytochromes and other electron transfer proteins on the outer membrane surface could make contact and transfer electrons directly. Indirect electron transfer involves the production of extracellular wire-like appendages that allow electron flux from the cell surface and into an electron acceptor several cell lengths distant. Mediated electron transfer is a broad category where soluble iron chelators and siderophores, secreted organics and environmental mediators might be utilised as soluble electron shuttles from electron transfer proteins on the cell surface and into the terminal acceptors. There is evidence for all in the literature and it is likely that all three mechanisms can contribute under different environmental stresses (Brutinel & Gralnick, 2012; Lovley, 2008; Nealson et al., 2002; Richardson et al., 2012).

4.1 Extracellular Electron Transfer Through Conductive Filaments

In a biofilm actively respiring on a mineral or electrode surface only the first layer of cells will directly contact the mineral or electrode, meaning that the remaining cells require an alternative method of transferring electrons into the acceptor surface (Bond et al., 2012; Lovley, 2008). Many bacteria generate conductive filaments that can be seen, by electron microscopy, to extend from the surface of bacteria and are long enough to make contact either with a terminal electron acceptor such as a mineral or electrode, or another bacteria (Fig. 7). Studies on these filaments have been greatly helped by the application of conductive atomic force microscopy, which allows the conductivity of individual filaments to be measured (Gorby et al., 2006; Reguera et al., 2005).

For *S. oneidensis MR-1* two types of filament have been described, a type IV pilin that has been structurally determined (Gorgel et al., 2015) and a

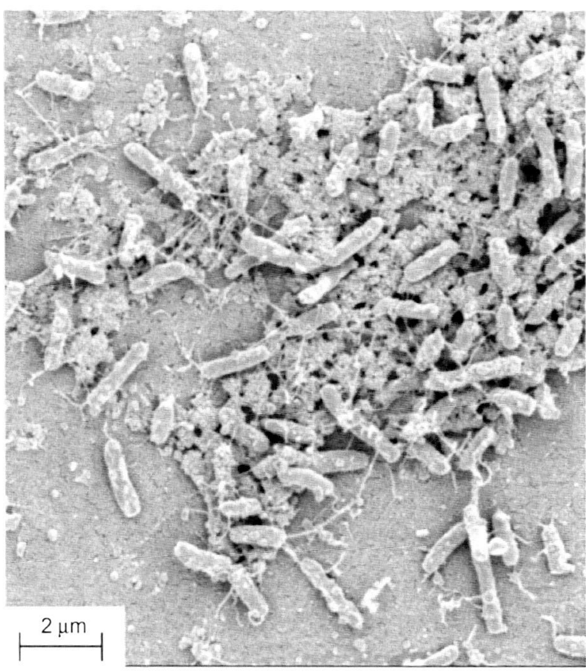

Fig. 7 Scanning electron microscopy image of *S. oneidensis* grown on a vitreous carbon anode. Filaments similar to the nanowires observed by Pirbadian et al. (2014) are extending from the microorganisms. *Image courtesy of Saad Ibrahim (UEA) and Kim Findlay (John Innes Centre).*

second type of filament that is an extension of the outer membrane. These membrane extensions are known to be conductive with reported values of 1 S cm^{-1} and lengths measured of up to 9 μm (Pirbadian et al., 2014). The conductive properties of the *S. oneidensis* MR-1 nanowires have also been shown to be non-metallic, meaning that electrons are likely to travel through redox centres along the filament, a process known as super-exchange conductivity (Bond et al., 2012; Pirbadian & El-Naggar, 2012).

Type IV pili in Gram-negative bacteria are associated with a range of different properties, including adhesion, motility, DNA transfer and electrical conductivity (Giltner, Nguyen, & Burrows, 2012). The filaments that extend from the surface of *G. sulfurreducens* were originally shown to be a member of the type IVa family of pili, have a conductivity of 5 mS cm^{-1} and be essential for Fe(III)oxide reduction (Malvankar et al., 2011; Reguera et al., 2005). The structure and mechanism by which the pili transfer electrons from *Geobacter* to the surface of the cell remains the subject of intense research by several groups. Wild type pili that are expressed by

G. sulfurreducens are up to 20 μm long and are coated in an extracellular cytochrome called OmcS. OmcS is a hexahaem OMC that is essential for reduction of Fe(III)oxides or electrodes (Holmes et al., 2006; Mehta, Coppi, Childers, & Lovley, 2005). However, the distance between adjacent OmcS cytochromes on the pili surface is not sufficiently close to allow for electron transfer between cytochromes, suggesting that electron transfer occurs through the pili, and not through the associated cytochromes (Aklujkar et al., 2013).

The intrinsic conductivity of the pilin is proposed to be metallic like, with electrons travelling through of aromatic residues with closely packed pi orbitals, a process known as metal-like conductivity (Boesen & Nielsen, 2013). A substantial challenge has been to prove the packing of aromatic amino acids at the centre of the pilin. The PilA subunit of the pilin from *G. sulfurreducens* contains five conserved aromatic amino acids at the carboxy terminus that, when substituted for alanines substantially decreases the conductivity of the pilin without noticeably interfering with the structure (Vargas et al., 2013). This is promising work implicating aromatic amino acid residues in the conductivity of pili, but it is still difficult to rationalise how the PilA subunits, which contain 5 aromatic residues in a total of 66 amino acids would be able to come together to form a continuous chain of pi-stacked orbitals over a distance of several cell lengths. In helping to understand how these five aromatic residues support electron transfer the NMR structure of PilA subunit has been determined and shown to have an α-helical content of approximately 85% with the five aromatic residues clustered at the carboxy terminus of PilA (Reardon & Mueller, 2013) (Fig. 8). Arrangement of this NMR structure into a pilin superstructure based on the pilin arrangement of *Neisseria gonorrhoeae* did not generate a conformation that would allow a continuous chain of stacked aromatic amino acids, but clustered the aromatic amino acids within a 15 Å sphere that was separated from adjacent spheres by a zone devoid of aromatic amino acids. The alternative model, based on the pilus assembly of *Pseudomonas aeruginosa* suggested a continuous chain of aromatic residues within the centre of the pilin with the aromatic amino acids arranged in a helical pattern on the within 3.6 Å of each other that could allow for metallic-like electron transfer (Malvankar et al., 2015) (Fig. 8C and D). This helical arrangement required three aromatic amino acids to form a continuous chain, and in principle disruption of any of these three should break the chain. It is surprising that five aromatic residues required substitution in order to create the observed decrease in conductivity when deletion of just one of the three aromatic residues should have an effect.

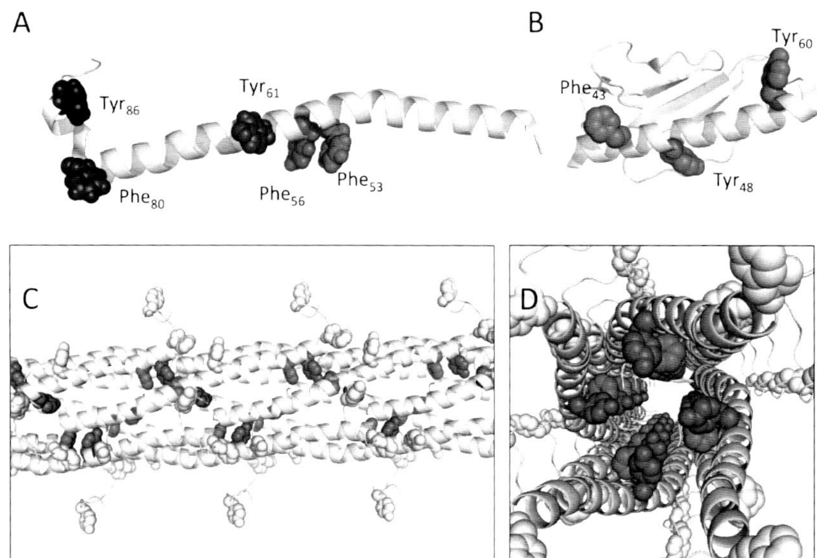

Fig. 8 The structures and configuration of pilin associated with mineral-reducing bacteria. (A) The *G. sulfurreducens* PilA structure (Reardon & Mueller, 2013, PDB ID: 2M7G) with aromatic residues implicated in electron transfer shown as *spheres*. (B) PilA structure from *S. oneidensis* (Gorgel et al., 2015, PDB ID: 4D40) with aromatic residues associated with the pilin centre shown as *spheres*. (C) Side view of proposed assembly of conductive pilin based on homology modelling using *P. aeruginosa* pilin assembly as a template. Aromatic residues shown as *spheres* with proposed stacking residues highlighted in *red*. (D) End-on view of *G. pilin* assembly. *Structural coordinates for pilin assembly obtained from Malvankar et al. (2015).* (See the colour plate.)

While the pili of *G. sulfurreducens* are the most studied of all the conductive nanowires, there are a host of other microorganisms that have been shown to produce conductive filaments, including *Synechocystis* sp. *PCCC6803*, *Pelotomaculum thermopropionicum* (Gorby et al., 2006) and *C. ferrireducens* (Gavrilov et al., 2012). For most of these bacteria, the composition of the filaments is not yet known and so it is unclear whether superexchange or metal-like conductive mechanism is favoured by the majority of other bacterial species.

Although the pili of *Geobacter* and several other microorganisms have been shown to be conductive, it has been less clear as to whether the *S. oneidensis* pili are also conductive. Genetic studies showed that deletion of the genes that encode the biosynthetic expression system *mshH-Q* decreased the current generated by a *S. oneidensis* fuel cell, however deletion of the genes *mshA-D* that actually form the pili only caused a 20% decrease in current, suggesting that the assembly system, rather than the pili were required for extracellular electron transfer (Fitzgerald et al., 2012). The structure of the *S. oneidensis* type IV pilin has also been solved and, like *G. sulfurreducens* contains a number of aromatic amino acids that could be modelled to generate two clusters of parallel aromatic residues approximately 4–7 Å apart, however there was a maximal distance of 11 Å between the two clusters, which would make electron transfer across the cluster gap unlikely without the presence of a mediator (Gorgel et al., 2015). As a consequence it is still unclear as to whether the type IV pili of *S. oneidensis* are conductive.

In mixed microbial communities the nanowires can allow the transfer of electron equivalents between different species, for example co-cultures of *G. sulfurreducens* and *G. metallireducens* were able to grown on a mixture of ethanol and fumarate, while isolated cultures were not. The growth was attributed to direct electron transfer between the two species through the exchange of electrons via the *Geobacter* pili (Summers et al., 2010). Further studies have shown that *Geobacter* is also capable of growing syntrophically with a range of different bacterial species, including *Methanosarcina barkeri* or *Hydrogenophaga* spp. (Kimura & Okabe, 2013; Rotaru et al., 2014), surprisingly these bacteria are not known to produce pili that could conduct electrons between species so the mechanism that allows microbes to obtain electrons from *Geobacter* pili is not known.

4.2 Direct Contact: The Structures of the Outer Membrane Cytochromes of *Shewanella* spp.

The majority of extracellular electron transfer proteins that have been structurally characterised thus far are the OMCs of the *Shewanella* genus (Clarke et al., 2011; Edwards, Hall, et al., 2012; Edwards et al., 2014, 2015; Fredrickson et al., 2008). Phylogenetic analysis revealed that these OMCs could be grouped into four separate clades, called the OmcA, MtrC, MtrF and UndA clades (Edwards, Fredrickson, Zachara, Richardson, & Clarke, 2012; Edwards et al., 2015). The OmcA and UndA clades are more closely related than to either the MtrC or MtrF clades, consistent with their positions within the gene clusters. OmcA and UndA are interchangeable between different *Shewanella* species while MtrC and MtrF are associated with the porin and periplasmic cytochromes. There is often little sequence homology between the different clade members with pairwise alignments of representative members of each clade giving values for sequence identity of 23–30%. However, the arrangement of CXXCH c-type cytochrome-binding motifs shows significant conservation (Edwards et al., 2014;

Edwards, Fredrickson, et al., 2012). The first approximately 200 N-terminal amino acids typically that contain the LXXC lipid-binding motif and, in OmcA, UndA, MtrF and half of the MtrC family a $CX_{2-5}C$ motif that forms a disulphide bond. There then follows a CXXCH rich region containing five CXXCH cytochrome-binding motifs within 200 amino acids before a gap of around 200 amino acids that contains a completely conserved $CX_{8-15}C$ motif that forms a disulphide bond. The C-terminal ~200 amino acids then contains the final 5 (or 6 in the case of UndA) CXXCH-binding motifs. This pattern of interspersed cytochrome-binding regions is specific to this family of OMCs and resolution of the structures of representatives of each of the four clades revealed the significance of this conserved organisation (Clarke et al., 2011; Edwards et al., 2014, 2015; Edwards, Hall, et al., 2012) (Fig. 9).

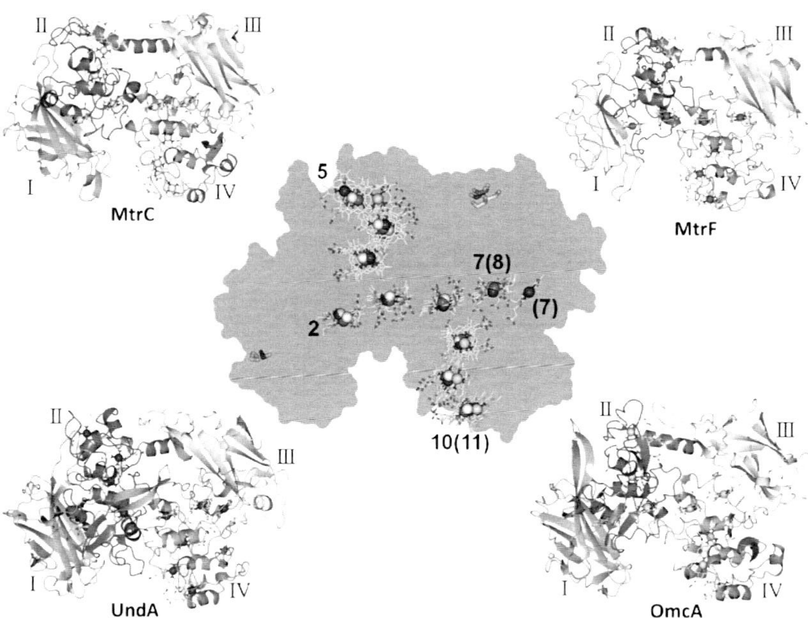

Fig. 9 Cartoon representations of the crystal structures of outer membrane cytochromes. Structures of MtrC (PDB ID: 4LM8), MtrF (PDB ID: 3PMQ), OmcA (PDB ID: 4LMH) and UndA (PDB ID:3UCP) isolated from *Shewanella* spp. Domains are numbered according to their position in the amino acid sequence (*roman numerals*). (Centre) Superposition of the haems of each cytochrome shows conservation haem configuration within the cytochrome structure. Haems are numbered according to the position of the CXXCH-binding motif in the amino sequence. *Numbers* refer to MtrC, MtrF and OmcA with *numbers in parenthesis* referring to UndA. The position of resolved disulphide bonds in each structure are shown as *sticks*. (See the colour plate.)

Despite the low sequence homology the structures of these cytochromes are markedly conserved. Each cytochrome is comprised of four domains, two alternating β-barrel domains and two multihaem domains. The domains are arranged in a loop so that the distance between the two porphyrin rings of adjacent haems in the two multihaem domains is within electron transfer distance of each other, in this way rapid electron exchange between the two domains and across the entire protein is allowed. The arrangement of the four domains results in the formation of a cross-like arrangement of the haems. This 'staggered-cross' is so far unique to structures of the OMCs of *Shewanella* with the minimum distance between the porphyrin rings of adjacent haems being less than 7 Å, allowing for rapid electron transfer between the four terminal haem groups (Breuer, Zarzycki, Blumberger, & Rosso, 2012). All the haems in the OMC crystal structures are *bis*-histidine coordinated, making it difficult to predict which sites are used for entry/egress. Haems 5 and 10 (11 in UndA) are exposed at opposite ends of the structure and are obvious sites for electron entrance/exit, but haems 2 and 7 are oriented towards the β-barrels that scaffold the two pentahaem domains and may direct electrons into potential binding sites for soluble metal ions or shuttles.

Overlay of the haems of the four OMCs reveals that the highest positional change is observed in haem 5, where the MtrC and MtrF haems are close in spatial arrangement, but the OmcA and UndA haems are widely different. The orientation of haem 5 of UndA is flipped relative to MtrC/MtrF/OmcA and haem 5 of OmcA is displaced. It is tempting to suggest that the high conservation in MtrC and MtrF might be due to their interaction with the MtrDE and MtrAB porin–cytochrome complexes, leaving haems 10 as the site of direct electron exchange. As OmcA and UndA do not form isolatable complexes with either porin–cytochrome complex, the requirement for a conserved terminal haem may not be necessary. In contrast, deletion of MtrC still allows electron exchange through OmcA, so a contrary argument could be made for haem 10, which is conserved. The other obvious area of difference between the homology of the other proteins is the position of the extra haem in UndA, which is located at the interface of domain 2 and the β-barrel. This extra haem could be responsible for the accelerated rates of Fe chelate reduction observed for Fe(III)NTA and Fe(III)EDTA as crystal soaks revealed that both Fe(III)citrate and FE(III) NTA associated within electron transfer of this haem (Edwards, Hall, et al., 2012; Shi et al., 2011).

The redox potential windows of the haems of the members of the OMCs from *S. oneidensis* have been measured (Clarke et al., 2011; Firer-Sherwood, Pulcu, & Elliott, 2008; Hartshorne et al., 2007) and range in value from −500 to +100 mV vs standard hydrogen electrode. The cytochromes have also been tested for various soluble and insoluble substrates, by measuring the rate at which a fully reduced cytochrome is oxidised in the presence of substrate. The OMCs of *Shewanella* spp. characterised so far are rapidly oxidised by a range of soluble substrates including Fe(III)citrate, Fe(III) NTA, Fe(III)EDTA, flavin mononucleotide (FMN) and riboflavin, as well as insoluble Fe(III)(hydr)oxides such as goethite, hematite, lepidocrocite or ferrihydrite.

Initial in vitro experiments on the OMCs of *S. oneidensis* revealed that soluble chelated iron species such as Fe(III)citrate and Fe(III)NTA as well as flavins such as FMN and riboflavin, could rapidly oxidise reduced samples of MtrC and OmcA. The oxidation of these OMCs was dependent on the redox potential of the electron acceptor, with Fe(III) chelates fully oxidising OMCs, and FMN only partially reducing OMCs. Surprisingly, the rates of reduction by reduced cytochromes purified from *S. oneidensis* are too low to support physiological respiration. However, when soluble reduced OMCs were mixed with iron oxides such as Fe(III)hydroxide, the rate of oxidation was less than $0.005\ s^{-1}$, suggesting that the OMCs were not capable of reducing iron oxides fast enough to support respiration (Clarke et al., 2011; Ross, Brantley, & Tien, 2009). By incorporating the MtrCAB complex into proteoliposomes containing the membrane impermeable methyl viologen, a system was established whereby each MtrC could catalyse the reduction of iron oxides from the reduced methyl viologen inside the liposome. This method had two advantages over the use of soluble proteins. (1) Each MtrCAB could transfer >1000 electrons from the liposome interior, thereby functioning as an electron conduit. (2) The MtrCAB complex was correctly oriented, with MtrC on the membrane surface and so would be optimised for correct interaction with the iron oxides, rather than perhaps interacting in an orientation that did not favour electron transfer. These proteoliposome experiments proved for the first time that direct electron transfer from the cell surface to an iron oxide or electrode was possible and that the rate of electron transfer appeared to be proportional to the driving force between electron donor and acceptor (White et al., 2013).

The remarkable structural conservation, both in domain organisation and haem arrangement, suggests that both the β-barrel domains and the staggered-haem cross are essential features of these mineral-reducing cytochromes, however, it still not clear what area on the surface of these elliptical proteins would interact with insoluble substrates.

Previously, Lower and coworkers used phage-display technology to enrich for peptides that bind to hematite. This work identified a hematite-binding motif, with a conserved sequence of Ser/Thr-hydrophobic/aromatic-Ser/Thr-Pro-Ser/Thr (Lower et al., 2008, 2007). Molecular dynamic simulations with the peptide Ser-Pro-Ser indicated that hydrogen bonding occurs between two serine amino acids and the hydroxylated hematite surface and that the proline induces a structure-binding motif by limiting the peptide flexibility. The location of the residues comprising the proposed hematite-binding motif was subsequently identified to be adjacent to haem 10 in the crystal structures of MtrC and OmcA (Edwards et al., 2014, 2015).

The current evidence is indicative of an orientation of the MtrC cytochrome with haem 5 interacting with MtrAB and haem 10 capable of interacting with the extracellular environment, however this is still far from certain. The orientation of the cytochromes on the surface of the membrane, along with the interactions between different cytochromes on the membrane surface are still significant questions that need to be addressed before a molecular understanding of the interaction between cytochrome and extracellular environment can be understood.

4.2.1 The Role of Shuttles in Shewanella *Extracellular Electron Transfer*

The use of electron shuttles as mediators during mineral respiration by *Shewanella* spp. has been widely researched. The first experiments separated *S. putrefaciens* spp. 200 from the insoluble mineral goethite by a dialysis membrane and found that the goethite could not be reduced (Arnold, DiChristina, & Hoffmann, 1986). A similar observation was made using *S. putrefaciens* MR-4 and manganite (MnOOH), where dialysis tubing prevented reduction of the insoluble manganese hydroxide (Larsen et al., 1998). Subsequent experiments encased iron oxide particles in microporous alginate beads. The beads had a diameter of 5 mm and pores of 12 kDa with Fe(III)oxide evenly distributed throughout. While *G. metallireducens* was only able to reduce the Fe(III)oxide exposed on the surface of the beads, *S. alga* BR1Y was capable of reducing much of the internalised Fe(III) as well (Nevin & Lovley, 2002). These later results suggested that certain species of *Shewanella* might be able to reduce insoluble iron oxides indirectly through the use of redox mediators. The first studies on indirect reduction by *S. oneidensis* used Fe(III)oxides encased in porous glass beads, these beads were approximately 50 μm in diameter, with the glass shell around the Fe(III)oxide being 0.3 μm thick. In these experiments, *S. oneidensis*, *S. putrefaciens* CN-32 and *Shewanella* spp. strain ANA-3 were all shown to reduce Fe(III) through the 0.3 μm porous barrier of the glass bead.

S. oneidensis was further shown to form a biofilm over the surface of the beads, causing them to cluster together (Lies et al., 2005). Together these results using encapsulated Fe(III)oxides suggested that, under certain conditions, *Shewanella* respiration of insoluble metal oxide was at least partly due to the involvement of soluble mediators that could diffuse through the permeable glass or alginate and reduce the encapsulated Fe(III).

Initially *S. oneidensis* was reported to secrete quinol-based compounds that were able to restore viability to menaquinone biosynthesis mutants (Newman & Kolter, 2000). These compounds were later identified as being released by lysed cells, rather than intentionally secreted shuttles (Myers & Myers, 2004). Eventually the search for secreted compounds in the extracellular environment in batch culture experiments by *S. oneidensis* revealed the presence of the flavin compounds FMN and riboflavin (Marsili et al., 2008; von Canstein, Ogawa, Shimizu, & Lloyd, 2008). These two compounds were determined not to be the products of cell lysis and a dedicated transport pathway was identified. FADH is secreted into the periplasm and hydrolysed into FMN by a 5′-nucleotidase called UshA. The FMN is then secreted into the extracellular matrix by a flavin transporter encoded by the *bfe* gene (Covington, Gelbmann, Kotloski, & Gralnick, 2010; Kotloski & Gralnick, 2013). Deletion of the *bfe* gene decreased the ability of *Shewanella* to reduce iron oxides or transfer electrons to carbon anodes but the phenotype could be restored through the addition of riboflavin to the culture mixture. *S. oneidensis* mutants lacking either *mtrC* or *omcA* were limited in their ability to generate current or reduce ferrihydrite, and addition of flavin did not improve this, showing the electron transfer pathway from the *S. oneidensis* cell surface to terminal electron acceptor required both MtrC/OmcA and FMN/riboflavin (Coursolle, Baron, Bond, & Gralnick, 2010) Taken together, these results clearly demonstrated that extracellular flavin in the forms of FMN or riboflavin were essential for *Shewanella* to respire on either insoluble Fe(III) or electrode surfaces, and that MtrC and OmcA were responsible for reduction (Brutinel & Gralnick, 2012).

Surprisingly, experiments mixing reduced flavins and insoluble iron oxides revealed that the ability of flavins to reduce insoluble Fe(III) was variable depending on the composition of the Fe(III) species. Mixing ferrihydrite or lepidocrocite with fully reduced FMN or riboflavin resulted in flavin oxidation coupled to the release of soluble ferrous iron (Shi, Zachara, et al., 2012). In contrast experiments involving the reduction of the iron oxides goethite and hematite by FMN reveal almost no electron exchange, indicating that *S. oneidensis* respiration on these minerals was unlikely to occur using soluble flavin shuttles (Wang et al., 2015).

S. oneidensis has been shown to be fully capable of respiring and reducing both goethite and hematite (Learman, Bose, Wigginton, Brown, & Hochella, 2007; Lower, Hochella, & Beveridge, 2001; Neal, Rosso, Geesey, Gorby, & Little, 2003; Ruebush, Brantley, & Tien, 2006), in these instances an alternative electron transfer system must be applicable if FMN or riboflavin is not capable of reducing them.

5. SUMMARY OF ELECTRON TRANSPORT MODELS ACROSS THE OUTER MEMBRANE

5.1 The Extracellular Electron Transfer Systems of Neutrophilic Iron-Oxidising Organisms

S. lithotrophicus, *G. capsiferriformans* and *R. palustris* all appear to utilise a two-component porin–cytochrome complex to transfer electrons from Fe(II) at the cell surface to cytochrome electron shuttles in the periplasm. The mechanism by which electrons are abstracted from soluble Fe(II) at the cell surface is not known but the resulting formation of insoluble Fe(III)(hydr)oxides is well documented. In *S. lithotrophicus* the insoluble Fe(III) species precipitate on the surface of the cell as nanoparticles that appear to separate from the cell rather than accumulate (Emerson & Moyer, 1997), while in *Gallionella* and related species the iron is oxidised and incorporated into twisted stalks. These stalks have been shown to be predominantly iron and organic carbon and are typically have diameters of 0.4 μm and can be up to 400 μm in length. The stalks are only produced under growth on iron and are composed of fibres that have nanometre diameters. The number of both stalks and fibres within the stalks varies on species and growth conditions.

The mechanism by which electrons are abstracted from Fe(II) in this process is still not understood. It is unclear if the Fe(II) is oxidised to Fe(III) and then precipitated to Fe(III)OOH, or whether the steps of electron abstraction and oxide formation occur simultaneously. By coupling the oxidation of Fe(II) with the formation of an Fe(III)oxide it would be possible to harvest the electron from a Fe-complex at a much lower redox potential. The reduction potential of the haems from MtoA from *S. lithotrophicus* have been measured and shown to vary between +100 and −400 mV (Liu et al., 2012), and the ability of MtoA to oxidise a range of different iron chelates, including Fe(II)citrate and Fe(II)EDTA, was shown to occur at rates between 1×10^{-3} and 6.3×10^{-3} $\mu M^{-1}\, s^{-1}$. It is likely that Fe(II) exists in a chelated form in the environment, attached to humic or organic acids, and this is the physiologically relevant form of iron utilised by iron-oxidising bacteria such as *S. lithotrophicus*. *S. lithotrophicus* was shown to grow on a range of different

Fe(II) species, including FeS, $FeCO_3$ (siderite), $FeCl_2$ and $FeSO_4$ suggesting that the species and consequently the redox potential of the Fe(II) source is not important (Emerson et al., 2013).

5.2 The Mechanism of Mineral Reduction in *Geobacter sulfurreducens*

While conductive pili are responsible for electron transfer from the surface of *G. sulfurreducens*, the mechanisms by which electrons are transferred through the outer membrane and into the mineral surfaces are likely to involve the expression of cytochromes. The genome of *G. sulfurreducens* contains over 100 c-type cytochromes, some containing up to 64 CXXCH motifs, the canonical c-type cytochrome-binding motifs (Methe et al., 2003). These genes appear to have formed through multiple rounds of duplication, suggesting that the expressed cytochromes will form chains of multihaem domains. The structurally resolved cytochromes from *Geobacter* so far include the periplasmic monohaem PccH, dihaem proteins MacA (DCH2), dodecahaem GSU1996 and a trihaem cytochrome *c*7 family of PpcA–F (Dantas, Campelo, Duke, Salgueiro, & Pokkuluri, 2015; Heitmann & Einsle, 2005; Pokkuluri et al., 2011, 2010). The only cytochrome associated with the outer membrane that has been structurally characterised is the outer membrane monohaem cytochrome OmcF. Many other key cytochromes of *G. sulfurreducens*, including OmaB, OmcB and OmcS have yet to be structurally resolved.

There are excellent reviews describing the role of the periplasmic cytochromes in transferring electrons from the cytoplasmic membrane to the outer membrane (Richter, Schicklberger, & Gescher, 2012; Santos, Silva, Morgado, Dantas, & Salgueiro, 2015). The interactions between electron transfer proteins within the periplasm appear to be transient, allowing pathways to alter in response to changes in the periplasmic redox environment during homeostasis. The electrons are then proposed to leave the cell through the OmaB–OmbB–OmcB porin–cytochrome complex with electrons emerging on the *G. sulfurreducens* surface through OmcB (Liu et al., 2014), rather than electrons being transported through the cell surface via the pili. There is little evidence for electron transfer through pi-orbital stacking within bacterial cells, with the majority of electrons being carefully transported through protein cofactors within 14 Å of each other (Moser, Anderson, & Dutton, 2010). It is interesting to note that the conductivity of *G. sulfurreducens* pili is sensitive to environmental changes such as pH, and as consequence their properties could be expected to change in the hydrophobic environment of the lipid bilayer (Malvankar et al., 2015). If the aromatic stacking is disrupted within the core of the lipid bilayer then pilin electron transfer through the outer membrane will not occur, ensuring the pilin only conduct electrons occurs on the cell exterior and preventing electrons re-entering the bacterium.

The dodecahaem OmcB is an extracellular c-type cytochrome associated to the outer membrane by a N-terminal lipid anchor. OmcB is expressed during Fe(III) respiration and its repression limits the ability of *G. sulfurreducens* to respire on Fe(III)oxides (Leang, Coppi, & Lovley, 2003). There is currently little experimental biochemical or structural information on OmcB. The amino acid sequence reveals that there is little homology to any other characterised cytochrome apart from OmcC, a 71% homologous protein that is part of a tandem chromosomal repeat on the *G. sulfurreducens* genome (Leang & Lovley, 2005; Liu et al., 2014). The arrangement of CXXCH motifs in the 722 amino acids of the OmcB sequence shows that seven CXXCH motifs are clustered in the first 285 amino acids, while the remaining five are found in the next 284 residues, leaving the final 153 amino acids free. This possibly indicates a tighter clustering of haems around the N-terminal domain of OmcB, and could indicate an electron transfer pathway from the OmbA–OmbB outer membrane complex that involves partial embedding in the OmbB porin. In support of this hypothesis is previous research that showed that OmcB was partially, but not wholly, exposed on the surface of *G. sulfurreducens* (Qian, Reguera, Mester, & Lovley, 2007).

The other important OMCs of *G. sulfurreducens* are the hexahaem OmcS and tetrahaem OmcE. These cytochromes are loosely associated to the surface and deletion of either gene causes *G. sulfurreducens* to lose the ability to respire on insoluble Fe(III)oxides, despite being able to reduce extracellular soluble metals (Mehta et al., 2005). Of these OmcS has provoked the most interest as it is localised to the conductive pili, is known to contain six *bis*-histidine coordinated c-type haems that have midpoint redox potentials of −40 mV to −360 mV vs SHE. OmcS is also the only cytochrome of *G. sulfurreducens* that has been shown to reduce insoluble iron oxides, although no rates have been measured (Qian et al., 2011). Deletion of *omcS* also prevents respiration on Fe(III)oxides although addition of magnetite nanoparticles to *G. sulfurreducens* cells can restore Fe(III)oxide respiration (Liu et al., 2015), it has been reasonable to deduce from these experiments that OmcS therefore is a catalytically important cytochrome that functions as the final electron donor, from the conductive pili, to insoluble iron oxides.

The number of OMCs that have been shown genetically to be important for Fe(III)oxide respiration suggests that there is a specific electron transfer pathway that runs from the periplasm to the surface of an iron oxide. A recent proposed mechanism for electron transfer across the surface of *G. sulfurreducens* suggested that electrons would pass through the outer membrane via the OmaB–OmbB–OmcB porin–cytochrome, then electrons would pass from OmcB to OmcE across the surface of the cell, electrons would then enter at the base of the pili and pass through the conductive pili to the OmcS cytochromes that coat the pili (Santos et al., 2015) (Fig. 10).

This mechanism provides a plausible explanation for the number of different cytochromes required for electron transfer between the outer membrane and Fe(III)oxide surface, but there is still a number of other proteins that are somehow involved in anaerobic respiration, for example, a multicopper protein OmpB has been shown to be important somehow in mineral

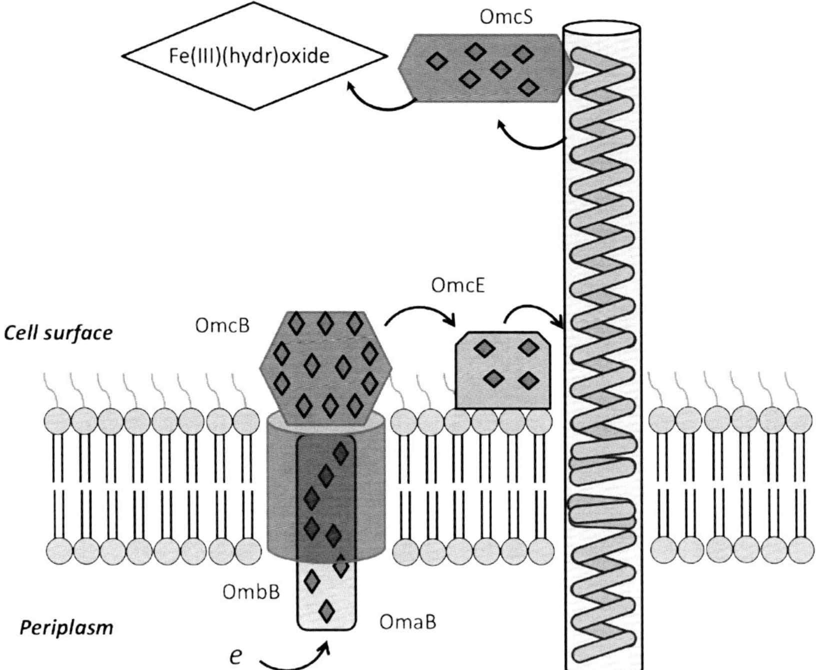

Fig. 10 Possible mechanism of insoluble mineral reduction by *Geobacter sulfurreducens*. Electrons from the periplasm are proposed to pass through a porin–cytochrome complex to OmcB on the cell surface, OmcB passes electrons to OmcE with mediates electron transfer to the extended pilin. Electrons transfer from the pilin-to-pilin-associated OmcS cytochromes that directly reduce the insoluble Fe(III)oxide/hydroxide.

reduction, although it is diffusive over the surface of the cell and not localised to the pili (Qian et al., 2007). It is possible that these proteins are not directly involved in Fe(III)oxide respiration but help to maintain the cell under conditions of Fe(III) reduction.

5.3 The Mechanism of Electron Transfer from the Cell Surface of *Shewanella* spp.

One of the primary advantages in research into mineral respiration of the *Shewanella* spp. has been that the genes *mtrCAB* and *omcA* responsible for microbial electron transfer were identified early on as being part of a single gene cluster (Beliaev & Saffarini, 1998). Given that deletion of *mtrA* or *mtrB* fully prevented *S. oneidensis* MR-1 from reducing insoluble metal oxides and deletion of *mtrC* or *omcA* caused only partial loss of Fe(III)oxide reduction, it is likely there is overlap between the roles of MtrC and OmcA (Coursolle & Gralnick, 2010; Myers & Myers, 2001). The role of the MtrCAB porin–cytochrome complex as a conduit for electron transport across the outer membrane will result in the accumulation of electrons within the MtrC cytochromes on the cell surface, but the pathway by which electron are then transferred to insoluble Fe(III) and Mn(VI)oxides is less clear.

The mechanism by which *S. oneidensis* transfers electrons from periplasmic electron donors to extracellular substrates has been experimentally shown to be a multistep process that requires the participation of both MtrA, MtrB, MtrC and OmcA as well as extracellular flavins. *S. oneidensis* produces filaments that have been shown to be conductive, but when examined these conductive filaments were revealed to be extensions of the cell membrane and periplasm that could extend up to 9 μm (Gorby et al., 2006; Pirbadian et al., 2014). These extensions have been shown to be conductive, but have different properties to the metallic-like conductivity of the *G. sulfurreducens* pili, instead they show super-exchange conductivity, where the electrons are passing through a number of co-factors. These outer membrane extensions have been shown to contain large numbers of both MtrC and OmcA spread across the surface, allowing for the possibility of electrons being transferred through the filament and into terminal electron acceptors. It is unclear how electrons could travel through these filaments, either through the internal periplasmic space and then out through the porin–cytochrome complexes, or directly through MtrC and OmcA on the outside. It is also currently unclear how these filaments are assembled, or how their expression is regulated.

The pathway through the outer membrane is well defined, through the MtrCAB porin–cytochrome complex that permits electron transfer to the cell surface. From here electrons can pass from MtrC to OmcA, although it appears possible for electrons to directly move from MtrAB to OmcA as deletion of MtrC still allows for partial Fe(III)reduction and transfer to electrodes (Bretschger et al., 2007; Coursolle & Gralnick, 2010). Both MtrC and OmcA have been shown to interact on the surface of living *S. oneidensis* cells at a ratio of 2 OmcA:1 MtrC, so each porin–cytochrome outer membrane conduit complex, consisting of MtrCAB–OmcA$_2$, could feasibly hold up to 40 electrons within a haem chain that could transfer electrons to an available electron acceptor (Shi et al., 2006).

It is possible for MtrC to directly reduce mineral oxides at rates that could support physiological respiration, given a sufficient driving force in the form of a lower redox potential on the inner side of the membrane (White et al., 2013). However, the observed increase in Fe(II) when either FMN or riboflavin are added to *S. oneidensis* cells respiring on insoluble Fe(III) indicates that the flavins have an important role in supporting anaerobic respiration on iron oxides.

The isolated MtrC and OmcA were capable of reducing flavins at physiologically favourable rates showing that flavins and cytochromes could transiently interact to exchange electrons, however no stable complex or measurable association could be determined, indicating the electron transfer interaction was very weak and short lived (Coursolle et al., 2010). Paquete et al. measured dissociation constants of 29 and 225 μM between FMN and MtrC or OmcA, suggestive of a very weak interaction that, at the submicromolar concentrations of flavin secreted by *S. oneidensis* under batch culture conditions, would favour a transient interaction between cytochrome and soluble shuttle (Paquete et al., 2014). However, an in vivo investigation using whole cell EPR spectroscopy coupled with cyclic voltammetry revealed that evidence for tight FMN–MtrC and riboflavin–OmcA complexes on the surface of the cell (Okamoto, Hashimoto, Nealson, & Nakamura, 2013; Okamoto et al., 2014). A separate study by Wang et al. (2015) using the proteoliposome method of White et al. (2012) revealed that stoichiometric amounts of flavin binding to MtrCAB were sufficient to accelerate the initial rate of electron transfer to hematite, lepidocrocite and goethite (Wang et al., 2015). A study that tracked the movement of *S. oneidensis* in the presence of Mn(IV)oxide particles showed that many cells remained planktonic, making intermittent contact with the mineral surface in a manner that suggested respiration was occurring through direct contact, rather than through electron shuttles diffused through the media (Harris, El-Naggar, & Nealson, 2012).

Taken together the in vivo and in vitro data suggest that these outer membrane flavocytochrome complexes are the dominant form responsible for mineral reduction and electron transfer to electrodes, rather than cytochromes transferring electrons to soluble shuttles. Further studies on MtrC and OmcA revealed that the binding of flavins to either cytochrome was enhanced when the protein was reduced. Rather than the haems, flavin binding was revealed to be regulated by a conserved disulphide on the surface of the second β-barrel domain observed in the four available OMC structures (Fig. 9). Reduction of this disulphide caused either riboflavin or FMN to associate so tightly to the OMC that it was possible to isolate the flavin–cytochrome complex through gel filtration. Oxidation through exposure to air caused the disulphide to reform and flavin to dissociate (Edwards et al., 2015). It has been suggested that this control of flavin binding by the redox-active disulphide may be a mechanism to protect *S. oneidensis* against the formation of reactive oxygen species, as *S. oneidensis* has a very low tolerance for oxygen species such as peroxide (Ghosal et al., 2005).

MtrC, OmcA, MtrF and UndA were all reported to bind a single flavin, with no observable preference for FMN or riboflavin, in contrast to Okamoto et al. (2013) who observed preferential interactions for MtrC with FMN, and OmcA for riboflavin. These data suggest the observed differences between cytochromes and flavins may possibly be due to accessibility on the cell surface, rather than the affinity between cytochrome and flavin.

While it is still not possible to rule out the potential role of flavin as a soluble shuttle between *S. oneidensis* and the more reactive minerals such as ferrihydrite and lepidocrocite, it is clear that the mechanism of direct contact between *S. oneidensis* OMC and mineral oxide will dominate for the more stable minerals such as goethite and hematite. The *S. oneidensis* nanowires first reported by Gorby et al. in 2006 utilise the same MtrC/OmcA catalytic mechanism for mineral reduction as the rest of the cell surface (Gorby et al., 2006; Pirbadian et al., 2014), so it is possible to suggest a flexible mechanism by which *S. oneidensis* can interact with insoluble metals under different levels of oxygen (Fig. 11). Under anoxic conditions the disulphides on the surface of the OMCs becomes reduced, possibly through proximity with reduced haem groups of the cytochrome or reduced environmental species. The cytochrome then binds flavin to give a flavocytochrome with enhanced electron transfer activity. This

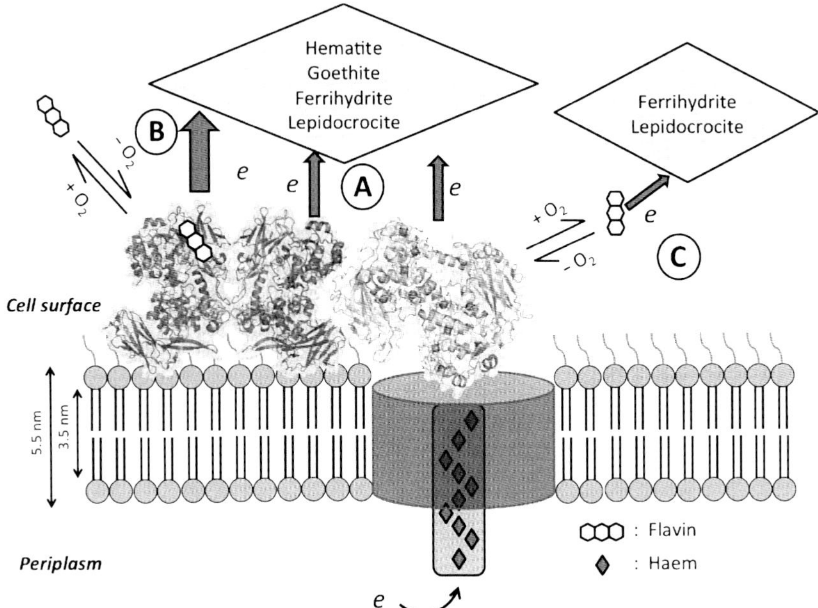

Fig. 11 Proposed mechanisms of extracellular electron transfer to insoluble Fe(III)oxide/hydroxide minerals by *Shewanella oneidensis* MR-1. (A) Periplasmic electrons are passed through the MtrCAB porin–cytochrome complex to the surface-exposed MtrC cytochromes. From here they can either be directly transferred to the surface of the mineral or to cell membrane-associated OmcA. (B) In the absence of oxygen, secreted flavins bind to OmcA and MtrC, which enhances the mineral reductase activity of these cytochromes. Exposure to oxygen causes the flavin to dissociate. (C) Reduction of ferrihydrite and lepidocrocite by soluble flavin shuttles can occur but the relative contributions of direct or mediated reduction are not known. (See the colour plate.)

flavocytochrome is capable of reducing minerals, metals and electrodes at much faster rates than the cytochrome form, but when the environment changes from anoxic to microoxic the disulphide reforms and the mechanisms switches to a mediated process where flavins may become soluble shuttles to other electron acceptors, such as ferrihydrite.

6. FUTURE PERSPECTIVES

The past decades have brought huge insight into the biochemical relationship between microorganisms and the inorganic environment. In particular, our understanding of how the model mineral-respiring organisms *S. oneidensis* and *G. sulfurreducens* transfer electrons from the cytoplasmic membrane to extracellular Fe(III)oxides has greatly increased. However, there are still substantial areas in these systems where our understanding in these systems is limited. For *Shewanella* the interactions between mineral, cytochrome and flavin are still unclear, and it appears likely that the microbe–mineral interface is dynamic and adaptive, although how this adaptation is driven is unclear. For *Geobacter*, the OMCs are poorly characterised, and a mechanistic understanding of their roles, particularly for OmcS, is an important step in understanding how these complexes work. There is evidence for metallic-like conductivity in *Geobacter* pili, but better understanding of the roles of the aromatic residues, as well as improved models for pilin assembly, are still required.

Due to their low growth rates and poor biomass yields, substantially less is known about the iron-oxidising bacteria, and this is the area where more experimental biochemical evidence is needed. Better growth conditions for these bacteria, or expression of iron-oxidising systems in suitable bacteria is required before the electron transfer pathways and metabolic pathways of these bacteria can be understood to the same level as the mineral-respiring strains.

ACKNOWLEDGEMENTS

This research was supported by the Biotechnology and Biological Sciences Research Council (BB/H007288/1, BB/K00929X/1 and BB/K009885/1) and a UEA studentship to L.G.-P. We are grateful to Profs. Jim Fredrickson and John Zachara for useful discussion.

REFERENCES

Aguilar, C., & Nealson, K. H. (1998). Biogeochemical cycling of manganese in Oneida lake, New York: Whole lake studies of manganese. *Journal of Great Lakes Research*, *24*(1), 93–104.

Aklujkar, M., Coppi, M. V., Leang, C., Kim, B. C., Chavan, M. A., Perpetua, L. A., et al. (2013). Proteins involved in electron transfer to Fe(III) and Mn(IV) oxides by Geobacter sulfurreducens and Geobacter uraniireducens. *Microbiology*, *159*(Pt. 3), 515–535.

Appia-Ayme, C., Bengrine, A., Cavazza, C., Giudici-Orticoni, M. T., Bruschi, M., Chippaux, M., et al. (1998). Characterization and expression of the co-transcribed cyc1 and cyc2 genes encoding the cytochrome c4 (c552) and a high-molecular-mass cytochrome c from Thiobacillus ferrooxidans atcc 33020. *FEMS Microbiology Letters*, *167*(2), 171–177.

Arnold, R. G., DiChristina, T. J., & Hoffmann, M. R. (1986). Inhibitor studies of dissimilative Fe(III) reduction by Pseudomonas sp. Strain 200 ("Pseudomonas ferrireductans"). *Applied and Environmental Microbiology*, *52*(2), 281–289.

Barco, R. A., Emerson, D., Sylvan, J. B., Orcutt, B. N., Jacobson Meyers, M. E., Ramirez, G. A., et al. (2015). New insight into microbial iron oxidation as revealed by the proteomic profile of an obligate iron-oxidizing chemolithoautotroph. *Applied and Environmental Microbiology*, *81*(17), 5927–5937.

Bathe, S., & Norris, P. R. (2007). Ferrous iron- and sulfur-induced genes in sulfolobus metallicus. *Applied and Environmental Microbiology*, *73*(8), 2491–2497.

Beckwith, C. R., Edwards, M. J., Lawes, M., Shi, L., Butt, J. N., Richardson, D. J., et al. (2015). Characterization of mtod from Sideroxydans lithotrophicus: A cytochrome c electron shuttle used in lithoautotrophic growth. *Frontiers in Microbiology*, *6*, 332.

Beliaev, A. S., & Saffarini, D. A. (1998). Shewanella putrefaciens mtrb encodes an outer membrane protein required for Fe(III) and Mn(IV) reduction. *Journal of Bacteriology*, *180*(23), 6292–6297.

Bewley, K. D., Firer-Sherwood, M. A., Mock, J. Y., Ando, N., Drennan, C. L., & Elliott, S. J. (2012). Mind the gap: Diversity and reactivity relationships among multi-haem cytochromes of the mtra/dmse family. *Biochemical Society Transactions*, *40*(6), 1268–1273.

Boesen, T., & Nielsen, L. P. (2013). Molecular dissection of bacterial nanowires. *mBio*, *4*(3), e00270.

Bond, D. R., Strycharz-Glaven, S. M., Tender, L. M., & Torres, C. I. (2012). On electron transport through geobacter biofilms. *ChemSusChem*, *5*(6), 1099–1105.

Bretschger, O., Obraztsova, A., Sturm, C. A., Chang, I. S., Gorby, Y. A., Reed, S. B., et al. (2007). Current production and metal oxide reduction by Shewanella oneidensis mr-1 wild type and mutants. *Applied and Environmental Microbiology*, *73*(21), 7003–7012.

Breuer, M., Zarzycki, P., Blumberger, J., & Rosso, K. M. (2012). Thermodynamics of electron flow in the bacterial deca-heme cytochrome mtrf. *Journal of the American Chemical Society*, *134*(24), 9868–9871.

Brutinel, E. D., & Gralnick, J. A. (2012). Shuttling happens: Soluble flavin mediators of extracellular electron transfer in Shewanella. *Applied Microbiology and Biotechnology*, *93*(1), 41–48.

Bücking, C., Schicklberger, M., & Gescher, J. (2012). The biochemistry of dissimilatory ferric iron and manganese reduction in Shewanella oneidensis. In J. Gescher & A. Kappler (Eds.), *Microbial metal respiration from geochemistry to potential applications: Vol. 8* (pp. 49–82). Berlin, Heidelberg: Springer.

Carlson, H. K., Clark, I. C., Melnyk, R. A., & Coates, J. D. (2012). Toward a mechanistic understanding of anaerobic nitrate-dependent iron oxidation: Balancing electron uptake and detoxification. *Frontiers in Microbiology*, *3*, 57.

Castelle, C., Guiral, M., Malarte, G., Ledgham, F., Leroy, G., Brugna, M., et al. (2008). A new iron-oxidizing/o-2-reducing supercomplex spanning both inner and outer membranes, isolated from the extreme acidophile Acidithiobacillus ferrooxidans. *Journal of Biological Chemistry*, *283*(38), 25803–25811.

Castelle, C. J., Roger, M., Bauzan, M., Brugna, M., Lignon, S., Nimtz, M., et al. (2015). The aerobic respiratory chain of the acidophilic archaeon ferroplasma acidiphilum: A membrane-bound complex oxidizing ferrous iron. *Biochimica et Biophysica Acta: Bioenergetics*, *1847*(8), 717–728.

Chabert, N., Amin Ali, O., & Achouak, W. (2015). All ecosystems potentially host electrogenic bacteria. *Bioelectrochemistry*, *106*(Pt. A), 88–96.

Chakraborty, A., & Picardal, F. (2013). Neutrophilic, nitrate-dependent, fe(ii) oxidation by a Dechloromonas species. *World Journal of Microbiology and Biotechnology*, *29*(4), 617–623.

Chan, C. S., Fakra, S. C., Edwards, D. C., Emerson, D., & Banfield, J. F. (2009). Iron oxyhydroxide mineralization on microbial extracellular polysaccharides. *Geochimica et Cosmochimica Acta*, *73*(13), 3807–3818.

Chan, C. S., Fakra, S. C., Emerson, D., Fleming, E. J., & Edwards, K. J. (2011). Lithotrophic iron-oxidizing bacteria produce organic stalks to control mineral growth: Implications for biosignature formation. *ISME Journal*, *5*(4), 717–727.

Clarke, T. A., Edwards, M. J., Gates, A. J., Hall, A., White, G. F., Bradley, J., et al. (2011). Structure of a bacterial cell surface decaheme electron conduit. *Proceedings of the National Academy of Sciences of the United States of America*, *108*(23), 9384–9389.

Clarke, T. A., Holley, T., Hartshorne, R. S., Fredrickson, J. K., Zachara, J. M., Shi, L., et al. (2008). The role of multihaem cytochromes in the respiration of nitrite in Escherichia coli and Fe(III) in Shewanella oneidensis. *Biochemical Society Transactions*, *36*, 1005–1010.

Coby, A. J., Picardal, F., Shelobolina, E., Xu, H., & Roden, E. E. (2011). Repeated anaerobic microbial redox cycling of iron. *Applied and Environmental Microbiology*, 77(17), 6036–6042.

Coursolle, D., Baron, D. B., Bond, D. R., & Gralnick, J. A. (2010). The Mtr respiratory pathway is essential for reducing flavins and electrodes in Shewanella oneidensis. *Journal of Bacteriology*, *192*(2), 467–474.

Coursolle, D., & Gralnick, J. A. (2010). Modularity of the mtr respiratory pathway of Shewanella oneidensis strain mr-1. *Molecular Microbiology*, 77(4), 995–1008.

Covington, E. D., Gelbmann, C. B., Kotloski, N. J., & Gralnick, J. A. (2010). An essential role for usha in processing of extracellular flavin electron shuttles by Shewanella oneidensis. *Molecular Microbiology*, *78*(2), 519–532.

Dantas, J. M., Campelo, L. M., Duke, N. E., Salgueiro, C. A., & Pokkuluri, P. R. (2015). The structure of pcch from Geobacter sulfurreducens—A novel low reduction potential monoheme cytochrome essential for accepting electrons from an electrode. *FEBS Journal*, *282*(11), 2215–2231.

Desloover, J., Arends, J. B. A., Hennebel, T., & Rabaey, K. (2012). Operational and technical considerations for microbial electrosynthesis. *Biochemical Society Transactions*, *40*, 1233–1238.

Dinarieva, T. Y., Zhuravleva, A. E., Pavlenko, O. A., Tsaplina, I. A., & Netrusov, A. I. (2010). Ferrous iron oxidation in moderately thermophilic acidophile Sulfobacillus sibiricus n1(t). *Canadian Journal of Microbiology*, *56*(10), 803–808.

Dong, H., Xiang, Q., Gu, Y., Wang, Z., Paterson, N. G., Stansfeld, P. J., et al. (2014). Structural basis for outer membrane lipopolysaccharide insertion. *Nature*, *511*(7507), 52–56.

Dopson, M., Baker-Austin, C., & Bond, P. (2007). Towards determining details of anaerobic growth coupled to ferric iron reduction by the acidophilic archaeon 'Ferroplasma acidarmanus' fer1. *Extremophiles*, *11*(1), 159–168.

Dopson, M., Baker-Austin, C., & Bond, P. L. (2005). Analysis of differential protein expression during growth states of ferroplasma strains and insights into electron transport for iron oxidation. *Microbiology*, *151*, 4127–4137.

Edwards, M. J., Baiden, N. A., Johs, A., Tomanicek, S. J., Liang, L., Shi, L., et al. (2014). The X-ray crystal structure of Shewanella oneidensis omca reveals new insight at the microbe–mineral interface. *FEBS Letters*, *588*(10), 1886–1890.

Edwards, M. J., Fredrickson, J. K., Zachara, J. M., Richardson, D. J., & Clarke, T. A. (2012). Analysis of structural MtrC models based on homology with the crystal structure of MtrF. *Biochemical Society Transactions*, *40*(6), 1181–1185.

Edwards, M. J., Hall, A., Shi, L., Fredrickson, J. K., Zachara, J. M., Butt, J. N., et al. (2012). The crystal structure of the extracellular 11-heme cytochrome unda reveals a conserved 10-heme motif and defined binding site for soluble iron chelates. *Structure*, *20*(7), 1275–1284.

Edwards, M. J., White, G. F., Norman, M., Tome-Fernandez, A., Ainsworth, E., Shi, L., et al. (2015). Redox linked flavin sites in extracellular decaheme proteins involved in microbe–mineral electron transfer. *Scientific Reports*, *5*, 11677.

Ehrenberg, C. G. (1836). Vorläufige mitteilungen über das wirkliche vorkommen fossiler infusorien und ihre große verbreitung. *Poggendorff's Ann Phys Chem*, *38*, 213–227.

Ehrenreich, A., & Widdel, F. (1994). Anaerobic oxidation of ferrous iron by purple bacteria, a new type of phototrophic metabolism. *Applied and Environmental Microbiology*, *60*(12), 4517–4526.

Emerson, D., Field, E. K., Chertkov, O., Davenport, K. W., Goodwin, L., Munk, C., et al. (2013). Comparative genomics of freshwater fe-oxidizing bacteria: Implications for physiology, ecology, and systematics. *Frontiers in Microbiology*, *4*, 254.

Emerson, D., Fleming, E. J., & McBeth, J. M. (2010). Iron-oxidizing bacteria: An environmental and genomic perspective. *Annual Review of Microbiology*, *64*, 561–583.

Emerson, D., & Moyer, C. (1997). Isolation and characterization of novel iron-oxidizing bacteria that grow at circumneutral pH. *Applied and Environmental Microbiology*, *63*(12), 4784–4792.

Finneran, K. T., Johnsen, C. V., & Lovley, D. R. (2003). Rhodoferax ferrireducens sp. nov., a psychrotolerant, facultatively anaerobic bacterium that oxidizes acetate with the reduction of Fe(III). *International Journal of Systematic and Evolutionary Microbiology*, *53*(Pt. 3), 669–673.

Firer-Sherwood, M., Pulcu, G. S., & Elliott, S. J. (2008). Electrochemical interrogations of the mtr cytochromes from Shewanella: Opening a potential window. *Journal of Biological Inorganic Chemistry*, *13*(6), 849–854.

Firer-Sherwood, M. A., Ando, N., Drennan, C. L., & Elliott, S. J. (2011). Solution-based structural analysis of the decaheme cytochrome, mtra, by small-angle X-ray scattering and analytical ultracentrifugation. *Journal of Physical Chemistry B*, *115*(38), 11208–11214.

Fitzgerald, L. A., Petersen, E. R., Ray, R. I., Little, B. J., Cooper, C. J., Howard, E. C., et al. (2012). Shewanella oneidensis mr-1 msh pilin proteins are involved in extracellular electron transfer in microbial fuel cells. *Process Biochemistry*, *47*(1), 170–174.

Fredrickson, J. K., Romine, M. F., Beliaev, A. S., Auchtung, J. M., Driscoll, M. E., Gardner, T. S., et al. (2008). Towards environmental systems biology of Shewanella. *Nature Reviews. Microbiology*, *6*(8), 592–603.

Fredrickson, J. K., & Zachara, J. M. (2008). Electron transfer at the microbe-mineral interface: A grand challenge in biogeochemistry. *Geobiology*, *6*(3), 245–253.

Gavrilov, S. N., Lloyd, J. R., Kostrikina, N. A., & Slobodkin, A. I. (2012). Fe(III) oxide reduction by a Gram-positive thermophile: Physiological mechanisms for dissimilatory reduction of poorly crystalline Fe(III) oxide by a thermophilic Gram-positive bacterium Carboxydothermus ferrireducens. *Geomicrobiology Journal*, *29*(9), 804–819.

Gavrilov, S. N., Slobodkin, A. I., Robb, F. T., & de Vries, S. (2007). Characterization of membrane-bound Fe(III)-EDTA reductase activities of the thermophilic Gram-positive dissimilatory iron-reducing bacterium Thermoterrabacterium ferrireducens. *Microbiology*, *76*(2), 139–146.

Ghiorse, W. C. (1984). Biology of iron- and manganese-depositing bacteria. *Annual Review of Microbiology*, *38*, 515–550.

Ghosal, D., Omelchenko, M. V., Gaidamakova, E. K., Matrosova, V. Y., Vasilenko, A., Venkateswaran, A., et al. (2005). How radiation kills cells: Survival of Deinococcus radiodurans and Shewanella oneidensis under oxidative stress. *FEMS Microbiology Reviews*, *29*(2), 361–375.

Giltner, C. L., Nguyen, Y., & Burrows, L. L. (2012). Type iv pilin proteins: Versatile molecular modules. *Microbiology and Molecular Biology Reviews*, *76*(4), 740–772.

Gorby, Y. A., Yanina, S., McLean, J. S., Rosso, K. M., Moyles, D., Dohnalkova, A., et al. (2006). Electrically conductive bacterial nanowires produced by Shewanella oneidensis strain MR-1 and other microorganisms. *Proceedings of the National Academy of Sciences of the United States of America*, *103*(30), 11358–11363.

Gorgel, M., Ulstrup, J. J., Boggild, A., Jones, N. C., Hoffmann, S. V., Nissen, P., et al. (2015). High-resolution structure of a type iv pilin from the metal-reducing bacterium Shewanella oneidensis. *BMC Structural Biology*, *15*, 4.

Grobbler, C., Virdis, B., Nouwens, A., Harnisch, F., Rabaey, K., & Bond, P. L. (2015). Use of SWATH mass spectrometry for quantitative proteomic investigation of Shewanella oneidensis MR-1 biofilms grown on graphite cloth electrodes. *Systematic and Applied Microbiology, 38*(2), 135–139.

Guo, K., Prevoteau, A., Patil, S. A., & Rabaey, K. (2015). Engineering electrodes for microbial electrocatalysis. *Current Opinion in Biotechnology, 33*, 149–156.

Hallbeck, L., & Pedersen, K. (1995). Benefits associated with the stalk of Gallionella ferruginea, evaluated by comparison of a stalk-forming and a non-stalk-forming strain and biofilm studies in situ. *Microbial Ecology, 30*(3), 257–268.

Harris, H. W., El-Naggar, M. Y., & Nealson, K. H. (2012). Shewanella oneidensis MR-1 chemotaxis proteins and electron-transport chain components essential for congregation near insoluble electron acceptors. *Biochemical Society Transactions, 40*(6), 1167–1177.

Hartshorne, R. S., Jepson, B. N., Clarke, T. A., Field, S. J., Fredrickson, J., Zachara, J., et al. (2007). Characterization of Shewanella oneidensis mtrc: A cell-surface decaheme cytochrome involved in respiratory electron transport to extracellular electron acceptors. *Journal of Biological Inorganic Chemistry, 12*(7), 1083–1094.

Hartshorne, R. S., Reardon, C. L., Ross, D., Nuester, J., Clarke, T. A., Gates, A. J., et al. (2009). Characterization of an electron conduit between bacteria and the extracellular environment. *Proceedings of the National Academy of Sciences of the United States of America, 106*(52), 22169–22174.

Hedrich, S., Schlomann, M., & Johnson, D. B. (2011). The iron-oxidizing proteobacteria. *Microbiology, 157*(Pt. 6), 1551–1564.

Hegler, F., Posth, N. R., Jiang, J., & Kappler, A. (2008). Physiology of phototrophic iron(ii)-oxidizing bacteria: Implications for modern and ancient environments. *FEMS Microbiology Ecology, 66*(2), 250–260.

Heitmann, D., & Einsle, O. (2005). Structural and biochemical characterization of dhc2, a novel diheme cytochrome c from Geobacter sulfurreducens. *Biochemistry, 44*(37), 12411–12419.

Holmes, D. E., Chaudhuri, S. K., Nevin, K. P., Mehta, T., Methe, B. A., Liu, A., et al. (2006). Microarray and genetic analysis of electron transfer to electrodes in Geobacter sulfurreducens. *Environmental Microbiology, 8*(10), 1805–1815.

Huysmans, G. H., Baldwin, S. A., Brockwell, D. J., & Radford, S. E. (2010). The transition state for folding of an outer membrane protein. *Proceedings of the National Academy of Sciences of the United States of America, 107*(9), 4099–4104.

Ilbert, M., & Bonnefoy, V. (2013). Insight into the evolution of the iron oxidation pathways. *Biochimica et Biophysica Acta: Bioenergetics, 1827*(2), 161–175.

Jeans, C., Singer, S. W., Chan, C. S., Verberkmoes, N. C., Shah, M., Hettich, R. L., et al. (2008). Cytochrome 572 is a conspicuous membrane protein with iron oxidation activity purified directly from a natural acidophilic microbial community. *ISME Journal, 2*(5), 542–550.

Jiao, Y., & Newman, D. K. (2007). The pio operon is essential for phototrophic Fe(ii) oxidation in Rhodopseudomonas palustris TIE-1. *Journal of Bacteriology, 189*(5), 1765–1773.

Johnson, D. B. (2014). Biomining—Biotechnologies for extracting and recovering metals from ores and waste materials. *Current Opinion in Biotechnology, 30*, 24–31.

Kappler, A., & Newman, D. K. (2004). Formation of Fe(III)-minerals by Fe(II)-oxidizing photoautotrophic bacteria. *Geochimica et Cosmochimica Acta, 68*(6), 1217–1226.

Kimura, Z., & Okabe, S. (2013). Acetate oxidation by syntrophic association between Geobacter sulfurreducens and a hydrogen-utilizing exoelectrogen. *ISME Journal*, 7(8), 1472–1482.

Kotloski, N. J., & Gralnick, J. A. (2013). Flavin electron shuttles dominate extracellular electron transfer by Shewanella oneidensis. *mBio, 4*(1), e00553.

Larsen, I., Little, B., Nealson, K. H., Ray, R., Stone, A., & Tian, J. H. (1998). Manganite reduction by Shewanella putrefaciens MR-4. *American Mineralogist, 83*(11–12), 1564–1572.

Larsen, O., & Postma, D. (2001). Kinetics of reductive bulk dissolution of lepidocrocite, ferrihydrite, and goethite. *Geochimica et Cosmochimica Acta, 65*(9), 1367–1379.

Leang, C., Coppi, M. V., & Lovley, D. R. (2003). Omcb, a c-type polyheme cytochrome, involved in Fe(III) reduction in Geobacter sulfurreducens. *Journal of Bacteriology, 185*(7), 2096–2103.

Leang, C., & Lovley, D. R. (2005). Regulation of two highly similar genes, omcB and omcC, in a 10 kb chromosomal duplication in Geobacter sulfurreducens. *Microbiology, 151*(Pt. 6), 1761–1767.

Learman, D. R., Bose, S., Wigginton, N. S., Brown, S. D., & Hochella, M. F. (2007). Reduction of hematite nanoparticles by Shewanella oneidensis MR-1. *Geochimica et Cosmochimica Acta, 71*(15), A551.

Lies, D. P., Hernandez, M. E., Kappler, A., Mielke, R. E., Gralnick, J. A., & Newman, D. K. (2005). Shewanella oneidensis mr-1 uses overlapping pathways for iron reduction at a distance and by direct contact under conditions relevant for biofilms. *Applied and Environmental Microbiology, 71*(8), 4414–4426.

Liu, F., Rotaru, A. E., Shrestha, P. M., Malvankar, N. S., Nevin, K. P., & Lovley, D. R. (2015). Magnetite compensates for the lack of a pilin-associated c-type cytochrome in extracellular electron exchange. *Environmental Microbiology, 17*(3), 648–655.

Liu, J., Wang, Z., Belchik, S. M., Edwards, M. J., Liu, C., Kennedy, D. W., et al. (2012). Identification and characterization of mtoa: A decaheme c-type cytochrome of the neutrophilic Fe(ii)-oxidizing bacterium Sideroxydans lithotrophicus ES-1. *Frontiers in Microbiology, 3*, 37.

Liu, Y., Wang, Z., Liu, J., Levar, C., Edwards, M. J., Babauta, J. T., et al. (2014). A trans-outer membrane porin–cytochrome protein complex for extracellular electron transfer by Geobacter sulfurreducens PCA. *Environmental Microbiology Reports, 6*(6), 776–785.

Logan, B. E., Wallack, M. J., Kim, K. Y., He, W. H., Feng, Y. J., & Saikaly, P. E. (2015). Assessment of microbial fuel cell configurations and power densities. *Environmental Science and Technology Letters, 2*(8), 206–214.

Lovley, D. R. (1993). Dissimilatory metal reduction. *Annual Review of Microbiology, 47*, 263–290.

Lovley, D. R. (2006). Bug juice: Harvesting electricity with microorganisms. *Nature Reviews Microbiology, 4*(7), 497–508.

Lovley, D. R. (2008). Extracellular electron transfer: Wires, capacitors, iron lungs, and more. *Geobiology, 6*(3), 225–231.

Lovley, D. R., Giovannoni, S. J., White, D. C., Champine, J. E., Phillips, E. J., Gorby, Y. A., et al. (1993). Geobacter metallireducens gen. nov. sp. nov., a microorganism capable of coupling the complete oxidation of organic compounds to the reduction of iron and other metals. *Archives of Microbiology, 159*(4), 336–344.

Lower, B. H., Lins, R. D., Oestreicher, Z., Straatsma, T. P., Hochella, M. F., Jr., Shi, L., et al. (2008). In vitro evolution of a peptide with a hematite binding motif that may constitute a natural metal-oxide binding archetype. *Environmental Science & Technology, 42*(10), 3821–3827.

Lower, B. H., Shi, L., Yongsunthon, R., Droubay, T. C., McCready, D. E., & Lower, S. K. (2007). Specific bonds between an iron oxide surface and outer membrane cytochromes MtrC and OmcA from Shewanella oneidensis MR-1. *Journal of Bacteriology, 189*(13), 4944–4952.

Lower, S. K., Hochella, M. F., & Beveridge, T. J. (2001). Bacterial recognition of mineral surfaces: Nanoscale interactions between Shewanella and alpha-feooh. *Science, 292*(5520), 1360–1363.

Malvankar, N. S., Vargas, M., Nevin, K., Tremblay, P. L., Evans-Lutterodt, K., Nykypanchuk, D., et al. (2015). Structural basis for metallic-like conductivity in microbial nanowires. *mBio*, *6*(2), e00084.

Malvankar, N. S., Vargas, M., Nevin, K. P., Franks, A. E., Leang, C., Kim, B. C., et al. (2011). Tunable metallic-like conductivity in microbial nanowire networks. *Nature Nanotechnology*, *6*(9), 573–579.

Marsili, E., Baron, D. B., Shikhare, I. D., Coursolle, D., Gralnick, J. A., & Bond, D. R. (2008). Shewanella secretes flavins that mediate extracellular electron transfer. *Proceedings of the National Academy of Sciences of the United States of America*, *105*(10), 3968–3973.

McLean, J. S., Pinchuk, G. E., Geydebrekht, O. V., Bilskis, C. L., Zakrajsek, B. A., Hill, E. A., et al. (2008). Oxygen-dependent autoaggregation in Shewanella oneidensis MR-1. *Environmental Microbiology*, *10*(7), 1861–1876.

Mehta, T., Coppi, M. V., Childers, S. E., & Lovley, D. R. (2005). Outer membrane c-type cytochromes required for Fe(III) and Mn(IV) oxide reduction in Geobacter sulfurreducens. *Applied and Environmental Microbiology*, *71*(12), 8634–8641.

Methe, B. A., Nelson, K. E., Eisen, J. A., Paulsen, I. T., Nelson, W., Heidelberg, J. F., et al. (2003). Genome of Geobacter sulfurreducens: Metal reduction in subsurface environments. *Science*, *302*(5652), 1967–1969.

Moser, C. C., Anderson, J. L., & Dutton, P. L. (2010). Guidelines for tunneling in enzymes. *Biochimica et Biophysica Acta*, *1797*(9), 1573–1586.

Myers, C. R., & Myers, J. M. (2004). Shewanella oneidensis MR-1 restores menaquinone synthesis to a menaquinone-negative mutant. *Applied and Environmental Microbiology*, *70*(9), 5415–5425.

Myers, C. R., & Nealson, K. H. (1988). Bacterial manganese reduction and growth with manganese oxide as the sole electron acceptor. *Science*, *240*(4857), 1319–1321.

Myers, J. M., & Myers, C. R. (2001). Role for outer membrane cytochromes OmcA and OmcB of Shewanella putrefaciens MR-1 in reduction of manganese dioxide. *Applied and Environmental Microbiology*, *67*(1), 260–269.

Neal, A. L., Rosso, K. M., Geesey, G. G., Gorby, Y. A., & Little, B. J. (2003). Surface structure effects on direct reduction of iron oxides by Shewanella oneidensis. *Geochimica et Cosmochimica Acta*, *67*(23), 4489–4503.

Nealson, K. H. (1997). Sediment bacteria: Who's there, what are they doing, and what's new? *Annual Review of Earth and Planetary Sciences*, *25*, 403–434.

Nealson, K. H., Belz, A., & McKee, B. (2002). Breathing metals as a way of life: Geobiology in action. *Antonie Van Leeuwenhoek*, *81*(1–4), 215–222.

Nealson, K. H., & Saffarini, D. (1994). Iron and manganese in anaerobic respiration: Environmental significance, physiology, and regulation. *Annual Review of Microbiology*, *48*, 311–343.

Neubauer, S. C., Emerson, D., & Megonigal, J. P. (2002). Life at the energetic edge: Kinetics of circumneutral iron oxidation by lithotrophic iron-oxidizing bacteria isolated from the wetland-plant rhizosphere. *Applied and Environmental Microbiology*, *68*(8), 3988–3995.

Nevin, K. P., & Lovley, D. R. (2002). Mechanisms for Fe(III) oxide reduction in sedimentary environments. *Geomicrobiology Journal*, *19*(2), 141–159.

Newman, D. K., & Kolter, R. (2000). A role for excreted quinones in extracellular electron transfer. *Nature*, *405*(6782), 94–97.

Okamoto, A., Hashimoto, K., Nealson, K. H., & Nakamura, R. (2013). Rate enhancement of bacterial extracellular electron transport involves bound flavin semiquinones. *Proceedings of the National Academy of Sciences of the United States of America*, *110*(19), 7856–7861.

Okamoto, A., Kalathil, S., Deng, X., Hashimoto, K., Nakamura, R., & Nealson, K. H. (2014). Cell-secreted flavins bound to membrane cytochromes dictate electron transfer reactions to surfaces with diverse charge and pH. *Scientific Reports*, *4*, 5628.

Osborne, T. H., McArthur, J. M., Sikdar, P. K., & Santini, J. M. (2015). Isolation of an arsenate-respiring bacterium from a redox front in an arsenic-polluted aquifer in West Bengal, Bengal Basin. *Environmental Science & Technology*, *49*(7), 4193–4199.

Paquete, C. M., Fonseca, B. M., Cruz, D. R., Pereira, T. M., Pacheco, I., Soares, C. M., et al. (2014). Exploring the molecular mechanisms of electron shuttling across the microbe/metal space. *Frontiers in Microbiology*, *5*, 318.

Pirbadian, S., Barchinger, S. E., Leung, K. M., Byun, H. S., Jangir, Y., Bouhenni, R. A., et al. (2014). Shewanella oneidensis MR-1 nanowires are outer membrane and periplasmic extensions of the extracellular electron transport components. *Proceedings of the National Academy of Sciences of the United States of America*, *111*(35), 12883–12888.

Pirbadian, S., & El-Naggar, M. Y. (2012). Multistep hopping and extracellular charge transfer in microbial redox chains. *Physical Chemistry Chemical Physics*, *14*(40), 13802–13808.

Pitts, K. E., Dobbin, P. S., Reyes-Ramirez, F., Thomson, A. J., Richardson, D. J., & Seward, H. E. (2003). Characterization of the Shewanella oneidensis MR-1 decaheme cytochrome mtra: Expression in Escherichia coli confers the ability to reduce soluble Fe(III) chelates. *Journal of Biological Chemistry*, *278*(30), 27758–27765.

Pokkuluri, P. R., Londer, Y. Y., Duke, N. E., Pessanha, M., Yang, X., Orshonsky, V., et al. (2011). Structure of a novel dodecaheme cytochrome c from Geobacter sulfurreducens reveals an extended 12 nm protein with interacting hemes. *Journal of Structural Biology*, *174*(1), 223–233.

Pokkuluri, P. R., Londer, Y. Y., Yang, X., Duke, N. E., Erickson, J., Orshonsky, V., et al. (2010). Structural characterization of a family of cytochromes c(7) involved in Fe(III) respiration by Geobacter sulfurreducens. *Biochimica et Biophysica Acta*, *1797*(2), 222–232.

Qian, X., Mester, T., Morgado, L., Arakawa, T., Sharma, M. L., Inoue, K., et al. (2011). Biochemical characterization of purified OmcS, a c-type cytochrome required for insoluble Fe(III) reduction in Geobacter sulfurreducens. *Biochimica et Biophysica Acta*, *1807*(4), 404–412.

Qian, X., Reguera, G., Mester, T., & Lovley, D. R. (2007). Evidence that OmcB and OmpB of Geobacter sulfurreducens are outer membrane surface proteins. *FEMS Microbiology Letters*, *277*(1), 21–27.

Reardon, P. N., & Mueller, K. T. (2013). Structure of the type IVa major pilin from the electrically conductive bacterial nanowires of Geobacter sulfurreducens. *Journal of Biological Chemistry*, *288*(41), 29260–29266.

Reguera, G., McCarthy, K. D., Mehta, T., Nicoll, J. S., Tuominen, M. T., & Lovley, D. R. (2005). Extracellular electron transfer via microbial nanowires. *Nature*, *435*(7045), 1098–1101.

Richardson, D. J., Butt, J. N., Fredrickson, J. K., Zachara, J. M., Shi, L., Edwards, M. J., et al. (2012). The 'porin–cytochrome' model for microbe-to-mineral electron transfer. *Molecular Microbiology*, *85*(2), 201–212.

Richter, K., Schicklberger, M., & Gescher, J. (2012). Dissimilatory reduction of extracellular electron acceptors in anaerobic respiration. *Applied and Environmental Microbiology*, *78*(4), 913–921.

Roden, E. E. (2003). Fe(III) oxide reactivity toward biological versus chemical reduction. *Environmental Science & Technology*, *37*(7), 1319–1324.

Roden, E. E. (2012). Microbial iron-redox cycling in subsurface environments. *Biochemical Society Transactions*, *40*(6), 1249–1256.

Roger, M., Castelle, C., Guiral, M., Infossi, P., Lojou, E., Giudici-Orticoni, M. T., et al. (2012). Mineral respiration under extreme acidic conditions: From a supramolecular organization to a molecular adaptation in Acidithiobacillus ferrooxidans. *Biochemical Society Transactions*, *40*(6), 1324–1329.

Ross, D. E., Brantley, S. L., & Tien, M. (2009). Kinetic characterization of OmcA and MtrC, terminal reductases involved in respiratory electron transfer for dissimilatory iron reduction in Shewanella oneidensis MR-1. *Applied and Environmental Microbiology, 75*(16), 5218–5226.

Ross, D. E., Flynn, J. M., Baron, D. B., Gralnick, J. A., & Bond, D. R. (2011). Towards electrosynthesis in Shewanella: Energetics of reversing the mtr pathway for reductive metabolism. *PLoS ONE, 6*(2), e16649.

Ross, D. E., Ruebush, S. S., Brantley, S. L., Hartshorne, R. S., Clarke, T. A., Richardson, D. J., et al. (2007). Characterization of protein–protein interactions involved in iron reduction by Shewanella oneidensis MR-1. *Applied and Environmental Microbiology, 73*(18), 5797–5808.

Rotaru, A. E., Shrestha, P. M., Liu, F., Markovaite, B., Chen, S., Nevin, K. P., et al. (2014). Direct interspecies electron transfer between Geobacter metallireducens and Methanosarcina barkeri. *Applied and Environmental Microbiology, 80*(15), 4599–4605.

Ruebush, S. S., Brantley, S. L., & Tien, M. (2006). Reduction of soluble and insoluble iron forms by membrane fractions of Shewanella oneidensis grown under aerobic and anaerobic conditions. *Applied and Environmental Microbiology, 72*(4), 2925–2935.

Santos, T. C., Silva, M. A., Morgado, L., Dantas, J. M., & Salgueiro, C. A. (2015). Diving into the redox properties of Geobacter sulfurreducens cytochromes: A model for extracellular electron transfer. *Dalton Transactions, 44*(20), 9335–9344.

Saraiva, I. H., Newman, D. K., & Louro, R. O. (2012). Functional characterization of the FoxE iron oxidoreductase from the photoferrotroph Rhodobacter ferrooxidans SW2. *Journal of Biological Chemistry, 287*(30), 25541–25548.

Schicklberger, M., Bucking, C., Schuetz, B., Heide, H., & Gescher, J. (2011). Involvement of the Shewanella oneidensis decaheme cytochrome MtrA in the periplasmic stability of the beta-barrel protein MtrB. *Applied and Environmental Microbiology, 77*(4), 1520–1523.

Schwertmann, U., & Cornell, R. M. (2000). *Iron oxides in the laboratory: Preparation and characterisation*. Weinheim: Wiley-VCH.

Shi, L., Belchik, S. M., Wang, Z., Kennedy, D. W., Dohnalkova, A. C., Marshall, M. J., et al. (2011). Identification and characterization of UndAHRCR-6, an outer membrane endecaheme c-type cytochrome of Shewanella sp. strain HRCR-6. *Applied and Environmental Microbiology, 77*(15), 5521–5523.

Shi, L., Chen, B., Wang, Z., Elias, D. A., Mayer, M. U., Gorby, Y. A., et al. (2006). Isolation of a high-affinity functional protein complex between OmcA and MtrC: Two outer membrane decaheme c-type cytochromes of Shewanella oneidensis MR-1. *Journal of Bacteriology, 188*(13), 4705–4714.

Shi, L., Rosso, K. M., Zachara, J. M., & Fredrickson, J. K. (2012). Mtr extracellular electron-transfer pathways in Fe(III)-reducing or Fe(II)-oxidizing bacteria: A genomic perspective. *Biochemical Society Transactions, 40*(6), 1261–1267.

Shi, Z., Zachara, J. M., Shi, L., Wang, Z., Moore, D. A., Kennedy, D. W., et al. (2012). Redox reactions of reduced flavin mononucleotide (FMN), riboflavin (RBF), and anthraquinone-2,6-disulfonate (AQDS) with ferrihydrite and lepidocrocite. *Environmental Science & Technology, 46*(21), 11644–11652.

Summers, Z. M., Fogarty, H. E., Leang, C., Franks, A. E., Malvankar, N. S., & Lovley, D. R. (2010). Direct exchange of electrons within aggregates of an evolved syntrophic coculture of anaerobic bacteria. *Science, 330*(6009), 1413–1415.

Taskan, E., Ozkaya, B., & Hasar, H. (2015). Combination of a novel electrode material and artificial mediators to enhance power generation in an mfc. *Water Science and Technology, 71*(3), 320–328.

Vargas, M., Malvankar, N. S., Tremblay, P. L., Leang, C., Smith, J. A., Patel, P., et al. (2013). Aromatic amino acids required for pili conductivity and long-range extracellular electron transport in Geobacter sulfurreducens. *mBio, 4*(2), e00105.

von Canstein, H., Ogawa, J., Shimizu, S., & Lloyd, J. R. (2008). Secretion of flavins by Shewanella species and their role in extracellular electron transfer. *Applied and Environmental Microbiology, 74*(3), 615–623.

Wang, Z. M., Shi, Z., Shi, L., White, G. F., Richardson, D. J., Clarke, T. A., et al. (2015). Effects of soluble flavin on heterogeneous electron transfer between surface-exposed bacterial cytochromes and iron oxides. *Geochimica et Cosmochimica Acta*, *163*, 299–310.

Wei, J., Liang, P., & Huang, X. (2011). Recent progress in electrodes for microbial fuel cells. *Bioresource Technology*, *102*(20), 9335–9344.

Weiss, J. V., Rentz, J. A., Plaia, T., Neubauer, S. C., Merrill-Floyd, M., Lilburn, T., et al. (2007). Characterization of neutrophilic Fe(II)-oxidizing bacteria isolated from the rhizosphere of wetland plants and description of Ferritrophicum radicicola gen. nov. sp. nov., and Sideroxydans paludicola sp. nov. *Geomicrobiology Journal*, *24*(7–8), 559–570.

White, G. F., Shi, Z., Shi, L., Dohnalkova, A. C., Fredrickson, J. K., Zachara, J. M., et al. (2012). Development of a proteoliposome model to probe transmembrane electron-transfer reactions. *Biochemical Society Transactions*, *40*(6), 1257–1260.

White, G. F., Shi, Z., Shi, L., Wang, Z., Dohnalkova, A. C., Marshall, M. J., et al. (2013). Rapid electron exchange between surface-exposed bacterial cytochromes and Fe(III) minerals. *Proceedings of the National Academy of Sciences of the United States of America*, *110*(16), 6346–6351.

Wilkins, M. J., Livens, F. R., Vaughan, D. J., Beadle, I., & Lloyd, J. R. (2007). The influence of microbial redox cycling on radionuclide mobility in the subsurface at a low-level radioactive waste storage site. *Geobiology*, *5*(3), 293–301.

Yarzabal, A., Brasseur, G., Ratouchniak, J., Lund, K., Lemesle-Meunier, D., DeMoss, J. A., et al. (2002). The high-molecular-weight cytochrome c cyc2 of Acidithiobacillus ferrooxidans is an outer membrane protein. *Journal of Bacteriology*, *184*(1), 313–317.

CHAPTER FOUR

Cooperation of Secondary Transporters and Sensor Kinases in Transmembrane Signalling: The DctA/DcuS and DcuB/DcuS Sensor Complexes of *Escherichia coli*

G. Unden[1], S. Wörner, C. Monzel

Institute for Microbiology and Wine Research, University of Mainz, Mainz, Germany

[1]Corresponding author: e-mail address: unden@uni-mainz.de

Contents

Abstract

Many membrane-bound sensor kinases require accessory proteins for function. The review describes functional control of membrane-bound sensors by transporters. The C_4-dicarboxylate sensor kinase DcuS requires the aerobic or anaerobic

Advances in Microbial Physiology, Volume 68
ISSN 0065-2911
http://dx.doi.org/10.1016/bs.ampbs.2016.02.003

C_4-dicarboxylate transporters DctA or DcuB, respectively, for function and forms DctA/DcuS or DcuB/DcuS sensor complexes. Free DcuS is in the permanent (ligand independent) ON state. The DctA/DcuS and DcuB/DcuS complexes, on the other hand, control expression in response to C_4-dicarboxylates. In DctA/DcuS, helix 8b of DctA and the PAS_C domain of DcuS are involved in interaction. The stimulus is perceived by the extracytoplasmic sensor domain (PAS_P) of DcuS. The signal is transmitted across the membrane by a piston-type movement of TM2 of DcuS which appears to be pulled (by analogy to the homologous citrate sensor CitA) by compaction of PAS_P after C_4-dicarboxylate binding. In the cytoplasm, the signal is perceived by the PAS_C domain of DcuS. PAS_C inhibits together with DctA the kinase domain of DcuS which is released after C_4-dicarboxylate binding.

DcuS exhibits two modes for regulating expression of target genes. At higher C_4-dicarboxylate levels, DcuS is part of the DctA/DcuS complex and in the C_4-dicarboxylate-responsive form which stimulates expression of target genes in response to the concentration of the C_4-dicarboxylates (catabolic use of C_4-dicarboxylates, mode I regulation). At limiting C_4-dicarboxylate concentrations ($\leq$0.05 mM), expression of DctA drops and free DcuS appears. Free DcuS is in the permanent ON state (mode II regulation) and stimulates low level (C_4-dicarboxylate independent) DctA synthesis for DctA/DcuS complex formation and anabolic C_4-dicarboxylate uptake.

Bacterial two-component systems that basically consist of a sensor kinase and a response regulator (Krell et al., 2010; Mascher, Helmann, & Unden, 2006) often require accessory or auxiliary proteins for function (Buelow & Raivio, 2010; Tetsch & Jung, 2009). Accessory proteins have been described for membrane bound and for cytoplasmic sensor kinases. Well-characterized examples are represented by the PII protein that regulates the activity of the NtrB kinase of the NtrB–NtrC two-component system. In many bacteria, NtrB–NtrC controls the expression of genes for nitrogen assimilation in response to the N- (and the C-) status of the bacteria (reviewed by Leigh & Dodsworth, 2007; Ninfa & Jiang, 2005). PII represents the actual sensor of the system, whereas the 'sensor' kinase NtrB is not able to respond to the stimulus on its own. In the phosphorelay of *Bacillus subtilis* that controls onset of endospore formation, proteins like KipI and SdaA function as inhibitors of the His kinase KinA and of the phosphor-transfer to Spo0F (Cunningham & Burkholder, 2009; Wang, Grau, Perego, & Hoch, 1997). KipI and Sda respond to various environmental stimuli and control in this way the sporulation process.

In membrane-bound two-component systems, the auxiliary proteins often control sensing, transmembrane signalling or the kinase domain of the sensor kinase (Buelow & Raivio, 2010). The control of membrane-bound sensor kinases is sometimes realized by transporters (reviewed by Tetsch & Jung, 2009). ABC transporters, extracytoplasmic binding proteins that can be part of ABC and tripartite transporters, and secondary transporters have been shown to control the activity of sensor kinases (Fig. 1). Before discussing the DcuS–DcuR two-component system in detail that uses secondary transporters DctA or DcuB as coregulators, sensor kinases using transporters or extracytoplasmic binding proteins as accessory proteins will be described briefly for reference.

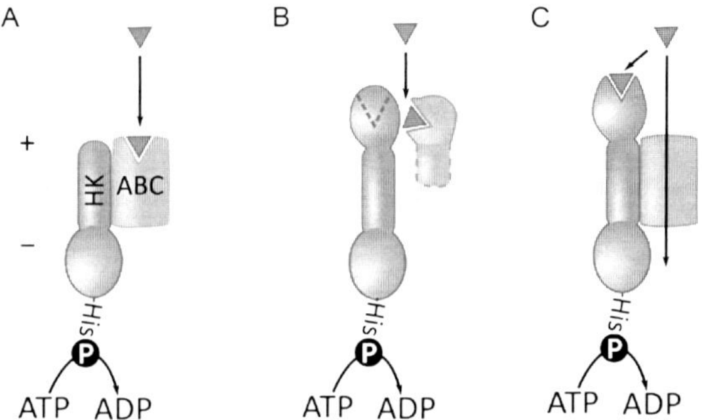

Fig. 1 Coregulation of bacterial sensor kinases (histidine kinases HK) by ABC transporters (A), binding proteins (B) or secondary transporters (C). The scheme shows basic principles for coregulation of sensor kinases (*blue*) by transporters or components of transporters (*orange*). (A) The sensor kinase forms a complex with the transporter. The sensor kinase contains no sensing domain. The ligand (*red triangle*) binds to the transporter which transmits a signal to HK stimulating autokinase activity. Transporters serving this type of sensing are often ABC transporters (examples: transporter $PstSCAB_2$ and PhoR–PhoB phosphate sensor system of *E. coli*; BceAB transporter and BceS–BceR two-component system for bacitracin detoxification by *B. subtilis*). (B) Activation of a sensor kinase by an extracytoplasmic binding protein. HK may contain a binding site for the ligand as well (*broken line* in HK). Examples: Activation of the citrate sensor kinase BctE of the BctE–BctB two-component system by binding of citrate to the binding protein BctC binding protein of the tripartite citrate transporter BctCAB of *Bordetella pertussis*, or the TorS–TorR TMAO-sensing system of *E. coli* interacting with the TMAO-binding protein TorT which is involved only in sensing and does not cooperate with transporters. The binding proteins can be attached to the membrane by protein or lipid anchors. (C) Coregulation of two-component systems by secondary carriers. The DcuS–DcuR two-component system for C_4-dicarboxylate-dependent regulation forms with the C_4-dicarboxylate transporters DctA or DcuB permanent complexes. Binding of the stimulus occurs by the sensor kinase (DcuS). References and more detailed description in the text. + and – indicate the periplasmic and cytoplasmic sides of the membrane. (See the colour plate.)

1. TRANSPORTERS AS COREGULATORS OF SENSOR KINASES

1.1 ABC Transporters

Sensor kinases that are devoid of a sensing domain have been shown to cooperate with ABC transporters (Fig. 1A) as represented by the two-component system PhoR–PhoB of *Escherichia coli*. PhoR–PhoB responds to external phosphate and regulates expression of the genes of the phosphate regulon. A recent model for the function of the system is presented by Hsieh and Wanner (2010). The sensor kinase PhoR comprises two transmembrane (TM) helices with a small periplasmic loop, but lacks a periplasmic sensing domain (Scholten & Tommassen, 1993). Sensing of external phosphate depends on the high-affinity phosphate transporter $PstSCAB_2$ (Hsieh & Wanner, 2010; Makino et al., 1989). The transporter binds phosphate by the PstS binding protein and communicates the information to PhoR by an additional protein, PhoU (Baek & Lee, 2006; Gardner, Johns, Tanner, & McCleary, 2014; Rice, Pollard, Lewis, & McCleary, 2009). Interaction studies show that PhoU interacts with both the sensor kinase and the $PstSCAB_2$ transporter in the cytoplasm, suggesting a seven-component complex of the transporter, the sensor kinase and PhoU which is according to sequence prediction a chaperone protein. The cytoplasmic PAS domain of PhoR is important for PhoR/PhoU interaction (Gardner et al., 2014). In the absence of external phosphate, PhoR is in the active (or default) state that catalyses autophosphorylation and PhoB phosphorylation. Binding of phosphate to the PstS binding protein results in inhibition of PhoR–PhoB phosphorylation and deactivation of the system by PhoR-P dephosphorylation. In this sensing system, therefore presence of phosphate is only perceived by the transporter but not by the sensor kinase (Fig. 1A). The transporter has a dual role in transport and sensing, but sensing appears to be independent of transport and the transport process (Cox, Webb, & Rosenberg, 1989).

Firmicutes like *Bacillus*, *Staphylococcus* and *Enterococcus* contain two-component systems that form together with ABC transporters, a detoxification system against antimicrobial peptides (Gebhard & Mascher, 2011; Rietkötter, Hoyer, & Mascher, 2008). Transporters and two-component systems form modules that have coevolved (Dintner et al., 2011). The BceS–BceR two-component system cooperates with the BceAB ABC transporter in bacitracin detoxification. The BceS sensor kinase is membrane integral by two TM helices. BceS has no extracytoplasmic domain, and the TM helices are directly linked without a recognizable ligand binding domain. The senor kinase forms a sensory complex with the transporter (Dintner, Heermann, Fang, Jung, & Gebhard, 2014). In contrast to the Pho system of *E. coli*, the activity of BceS–BceR responds to changes in the activity of the transporters and the amount of antibiotic transported, rather than to changes in the ambient antibiotic concentration (Fritz et al., 2015). Therefore both PhoR–PhoB and BceS–BceR two-component systems have no site or domain for sensing on their own and their sensor kinases rely on interacting ABC transporters for stimulus perception. However, the mode of sensing appears to be different. The former transporter monitors the ambient concentration of the stimulus (phosphate) whereas the latter regulates the kinase by transmitting the information on the transport activity of the individual transporter to the attached sensor kinase (flux sensing).

1.2 Extracytoplasmic Binding Proteins

Some sensor kinases require extracytoplasmic binding proteins as coregulators for function (Fig. 1B). The systems are at variance to the PhoR–PhoB phosphate system where the binding protein PstS is part of the cosensing ABC transporter $PstSCAB_2$. In the sensors discussed here, the binding proteins interact with sensor kinase and are responsible for stimulus perception by the sensor kinase. Thus the monosaccharide binding protein ChvE is part of the ABC-type sugar transporter GguAB of *Agrobacterium tumefaciens*. ChvE also interacts with the sensor kinase of the VirA–VirG two-component system controlling expression of the *vir* regulon (Kemner, Liang, & Nester, 1997; Shimoda et al., 1990). Similarly, the binding protein BctC that is part of the tripartite citrate transporter BctCAB, serves additionally in transmitting the stimulus citrate to the BctE sensor kinase of the BctE–BctD two-component system. BctC is supposed to interact directly with BctE and to represent a part of the signalling cascade. BctE–BctD induces in the presence of citrate the *bctCBA* transporter genes (Antoine et al., 2005). Other sensor kinases use small extracellular or periplasmic substrate binding proteins that are not part of transport systems. The binding proteins bind the ligand and interact with the sensor kinase. The TorT protein of *E. coli* represents a new family of periplasmic binding proteins (Baraquet et al., 2006). TorT binds TMAO and interacts with the periplasmic domain of TorS His kinase from the TorS–TorR two-component system that controls gene expression in response to TMAO. It is assumed that TMAO binds only via TorT inducing in this way conformational changes and activation of TorS. The senor kinase DctS of the DctS–DctR two-component system of *B. subtilis* requires the extracytoplasmic binding protein DctB for inducing gene expression in the presence of C_4-dicarboxylates (Asai, Baik, Kasahara, Moriya, & Ogasawara, 2000; Graf, Schmieden, Tschauner, Hunke, & Unden, 2014). DctB is homologous in sequence and structure prediction to the DctP-type binding proteins of the TRAP transporter family (Graf et al., 2014). DctB has no role in C_4-dicarboxylate transport in *B. subtilis*. It binds to the sensor kinase DctS, and mutant strains deficient of DctB are not able to stimulate genes of the DctS–DctR regulon in the presence of C_4-dicarboxylates. The sensor kinase DctS contains a C_4-dicarboxylate binding motif (Graf et al., 2016, 2014), and it can be speculated that the C_4-dicarboxylate is transmitted from DctB to DctS.

1.3 Secondary Transporters: UhpA–UhpB and C_4-Dicarboxylate Two-Component Systems

In yeast, plant and mammals secondary transporters with regulatory function have been identified (Antoine et al., 1997; Didion, Regenberg, Jorgensen, Kielland-Brandt, & Andersen, 1998; Lalonde et al., 1999; Meyer et al., 2000; Özcan, Dover, & Johnson, 1996). In bacteria, some ToxR-like transcriptional activators (Rauschmeier, Schüppel, Tetsch, & Jung, 2014; Tetsch & Jung, 2009; Tetsch, Koller, Haneburger, & Jung, 2008) and some two-component systems require secondary transporters for function (Fig. 1C). Here the function of secondary transporters as coregulators or cosensors of two-component systems will be discussed.

1.3.1 The Interplay Between Transporter UhpC and the UhpB Hexose-6-P Sensor Kinase

UhpABC represents the classical system of bacteria for secondary transporters in the regulation of sensor kinases. UhpABC induces in the presence of external hexose-6-phosphate the adjacent *uhpT* gene. *uhpT* encodes the transporter UhpT for the uptake of external hexose-6-phosphate in antiport against phosphate (Island & Kadner, 1993; Kadner, Island, Dahl, & Webber, 1994). The sensor kinase UhpB of the UhpB–UhpA two-component system is membrane integral by eight predicted TM helices which include a significant periplasmic loop region of approx. 50 AA (Island & Kadner, 1993). Bioinformatic analysis clusters the TM helices of the sensor kinase

UhpB within the MASE1 (membrane-associated sensor1) family of sensor domains (http://pfam.xfam.org/; Finn et al., 2014). MASE1 sensory domains are found in many bacterial sensing or signalling proteins but the nature of their signal and functional details of this class of sensor kinases are unknown (Anantharaman & Aravind, 2003; Nikolskaya, Mulkidjanian, Beech, & Galperin, 2003). The accessory membrane protein UhpC is required for the activation and phosphorylation of UhpB and UhpA when hexose-6-phosphates are present. UhpB–UhpA interacts in an unknown manner with the accessory protein UhpC. UhpC has sequence similarity with UhpT and displays significant, though low activity for hexose-phosphate/phosphate antiport (Schwöppe, Winkler, & Neuhaus, 2002), and it has been suggested that UhpC has evolved from a transporter like UhpT. Transport by UhpC is, however, not required for sensing (Schwöppe, Winkler, & Neuhaus, 2003). Although the data suggest that UhpB and UhpC act jointly, it is not known whether a (permanent or transient) UhpB/UhpC complex is formed, and which protein (UhpC or UhpB) represents the actual sensor for the ligand.

1.3.2 C_4-Dicarboxylate Sensor Kinases Are Negatively Regulated by Transporters DctA or DcuB

In *E. coli* and other bacteria, the enzyme systems for degradation of external C_4-dicarboxylates (fumarate, L-malate, succinate) require in aerobic growth the expression of a C_4-dicarboxylate transporter which is typically DctA (*dctA* gene) at neutral pH. Additionally, PEP carboxykinase (*pck* gene) or malic enzyme (*maeB* or *sfcA* genes) for the production of pyruvate from C_4-dicarboxylates is required. Pyruvate is then fed via acetyl-CoA to the citric acid cycle (Bologna, Andreo, & Drincovich, 2007; Goldie & Sanwal, 1980). Under anaerobic conditions, fumarate, L-malate or aspartate is metabolized by fumarate respiration, which produces succinate as the product that has to be excreted (Kröger, Geisler, Lemma, Theis, & Lenger, 1992; Unden & Kleefeld, 2004). The *dctA*, *dcuB*, *frdABCD* and *fumB* genes encoding the aerobic and anaerobic transporters, fumarate reductase FrdABCD and fumarase FumB, respectively (Guest, 1992; Unden & Kleefeld, 2004), are induced by external C_4-dicarboxylates under aerobic and anaerobic conditions, respectively. In *E. coli* and *B. subtilis*, the DcuS–DcuR (and $DctS_{Bs}$–$DctR_{Bs}$, respectively) two-component systems induce expression of these genes in the presence of the C_4-dicarboxylates (Asai et al., 2000; Golby, Davies, Kelly, Guest, & Andrews, 1999; Janausch, Zientz, Tran, Kröger, & Unden, 2002; Zientz, Bongaerts, & Unden, 1998). *Rhizobium* (or *Sinorhizobium*) and *Pseudomonas* degrade C_4-dicarboxylates aerobically, requiring then transporters DctA or DctPQM for C_4-dicarboxylate uptake. In (*Sino*) *Rhizobium* and *Pseudomonas* the DctB–DctD C_4-dicarboxylate two-component system performs induction under corresponding conditions (Reid & Poole, 1998; Valentini, Storelli, & Lapouge, 2011). In rhizobia, and later in *E. coli*, *B. subtilis* and *Pseudomonas aeruginosa*, PAO1 it was shown that DctA controls its own synthesis and the synthesis of other genes that are subject to induction by C_4-dicarboxylates (Asai et al., 2000; Davies et al., 1999; Jording, Uhde, Schmidt, & Pühler, 1994; Reid & Poole, 1998; Valentini et al., 2011; Watson, 1990; Yarosh, Charles, & Finan, 1989; Yurgel & Kahn, 2004). Deletion of *dctA* causes constitutive expression of the corresponding genes, suggesting that in all systems the sensor kinases DctB, DcuS or DctS, respectively, are in the permanent active state when no DctA is present and that DctA negatively affects the sensor kinases. In *E. coli* under anaerobic growth conditions when no DctA is produced, the transporter DcuB shows a similar effect on DcuS function (Kleefeld, Ackermann, Bauer, Krämer, & Unden, 2009). Deletion of *dcuB* (but not of the related *dcuA* or *dcuC*) causes constitutive expression of the anaerobically expressed target genes. The transporter DauA that replaces DctA at acidic growth conditions has an effect on DctA expression as well and is suggested to affect *dctA* transcription in an unknown manner (Karinou et al., 2013).

The bacterial C_4-dicarboxylate sensor kinases can be grouped in the CitA/DcuS, the DctB and the $DctS_{Rc}$ families (Graf et al., 2016). The need for transporters as accessory proteins or coregulators for the DcuS- and DctB-type sensor kinases represents a common feature for this type of sensor kinases. In contrast, the CitA citrate sensor kinases that are closely related to the DcuS sensors apparently do not require coregulation by transporters (Scheu et al., 2012).

2. POLAR LOCALIZATION OF DcuS AND OF THE DcuS/DctA SENSOR COMPLEX

Two-component systems controlling metabolic reactions are expected to show no specific subcellular localization in the cell membrane. Thus the citrate-sensing CitS from *B. subtilis*, the quorum sensing kinase LuxQ from *Vibrio harveyi*, and others display either a heterogeneous localization at unspecified positions or a homogeneous distribution over the cell membrane (Meile, Wu, Ehrlich, Errington, & Noirot, 2006; Neiditch et al.,

2006; other references in Scheu, Steinmetz, Dempwolff, Graumann, & Unden, 2014).

Sensor kinase DcuS, in contrast, reveals polar localization (Scheu et al., 2008, 2014). In fluorescence microscopy, DcuS fused to GFP derivatives has a dynamic and preferential polar localization, even at very low expression levels (Fig. 2). Single assemblies of DcuS show high mobility in fast time-lapse acquisitions and fast recovery in FRAP experiments, demonstrating that the polar localization is physiological and no artefact. DctA and DcuR fused to derivatives of the YFP protein are dispersed in the membrane or in the cytosol, respectively, when expressed without DcuS. When DcuS was coexpressed at appropriate levels, both proteins colocalize with DcuS. Thus DcuS is responsible for location of DctA and DcuR at the poles and for the formation of tripartite DctA/DcuS/DcuR complexes. DctA functions in the free state as a transporter for the uptake of C_4-dicarboxylates and is present at high levels that surpass those of the senor DcuS. Therefore, only the portion of DctA participating in the DctA/DcuS sensor complex is expected to colocalize with DcuS. On the other hand, DctA, DcuR and the alternative succinate transporter DauA were not essential for polar localization of DcuS, suggesting that polar trapping occurs by DcuS. Cellular factors like

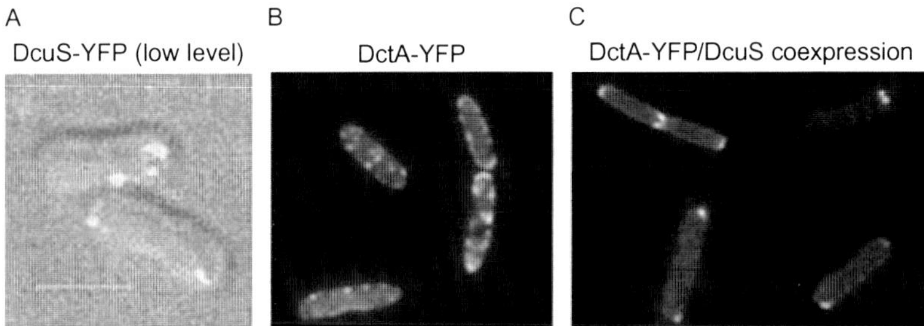

Fig. 2 Polar localization of DcuS (or DcuS-YFP) (A), colocalization of DctA in the absence (B) and the presence of equivalent levels of DcuS (C). For (A) stationary-phase cells of *E. coli* W3110pMW407-expressing DcuS-YFP at low level were analysed by confocal microscopy. For colocalization of DctA-YFP and DcuS, DctA-YFP (pMW526) was either produced alone (B) or coexpressed (C) with DcuS in strain IMW262. The fluorescence of DctA-YFP showed a spotty appearance that is dispersed over the cell surface (B), or concentrated at the cell poles when DcuS is overproduced at the same time (C). The scale bar in (A) corresponds to 2 μm. *Panel (A) modified from Scheu, P. D., Steinmetz, P. A., Dempwolff, F., Graumann,P. L.,& Unden, G. (2014). Polar localization of a tripartite complex of the two-component system DcuS/DcuR and the transporter DctA in* Escherichia coli *depends on the sensor kinase DcuS.* PLoS One, 9, *e115534. http://dx.doi.org/10.1371/journal.pone.0115534.* (See the colour plate.)

cardiolipin, high curvature at the cell poles, and the cytoskeletal protein MreB were not required for polar localization. Polar localization of DcuS required the presence of the cytoplasmic PAS_C and the kinase domains of DcuS.

Formation of clusters was also described for other sensor kinases. Recently, TorS and EvgS of *E. coli* were reported to form clusters predominantly at the cell poles (Sommer, Koler, Frank, Sourjik, & Vaknin, 2013). The role of supramolecular organization of sensors, cosensors and components of signal transducing pathways in bacterial sensor complexes has been suggested to increase the sensitivity, specificity and versatility of sensors systems (Sourjik & Berg, 2000; Tetsch et al., 2008; Witan, Monzel, Scheu, & Unden, 2012) by allowing signal exchange and integration.

3. DcuS AS MEMBRANE-BOUND SENSOR KINASE FOR TRANSMEMBRANE SIGNALLING

DcuS is a typical extracytoplasmic sensing histidine kinase that is composed of an extracytoplasmic PAS (PER-ARNT-SIM) domain (PAS_P), two TM helices, a cytoplasmic PAS domain (PAS_C) and the kinase domain (Scheu, Kim, Griesinger, & Unden, 2010) (Fig. 3A). Whereas PAS_P contains the ligand binding site (Cheung & Hendrickson, 2008; Kneuper et al., 2005; Pappalardo et al., 2003), the role of PAS_C is in signal transduction from TM2 to the kinase (Etzkorn et al., 2008; Monzel, Degreif-Dünnwald, Gröpper, Griesinger, & Unden, 2013). DcuS is a dimer or a higher oligomer in the membrane (Scheu, Liao, et al., 2010).

3.1 Stimulus Perception and Compaction of the Extracytoplasmic PAS_P Domain

DcuS is a member of the CitA/DcuS family comprising the citrate (CitA) and the C_4-dicarboxylate (DcuS) sensor kinases (Bott, Meyer, & Dimroth, 1995; Graf et al., 2016; Scheu, Kim, et al., 2010; Zientz et al., 1998). The extracytoplasmic sensing domain of DcuS is framed by two TM helices (Fig. 3A). The sensory domain has a modified PAS-fold and was termed PAS_P (periplasmic PAS) or PDC (Pho/DcuS/DctB/CitA) domain (Cheung & Hendrickson, 2008; Pappalardo et al., 2003; Reinelt, Hofmann, Gerharz, Bott, & Madden, 2003). In DcuS and other sensor kinases of the CitA/DcuS family, TM2 is followed by a cytoplasmic PAS domain (PAS_C) and the C-terminal kinase domain of the HisKA/HATPase type (Grebe & Stock, 1999; Scheu, Kim, et al., 2010).

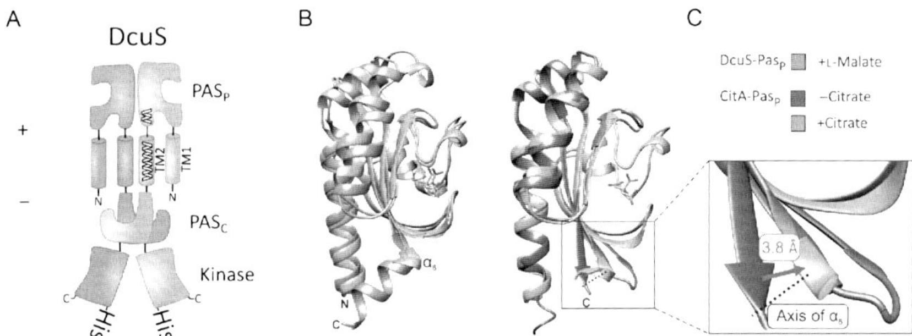

Fig. 3 Domain structure and topology of the DcuS sensor kinase (A) and compaction (B) of the periplasmic citrate/C_4-dicarboxylate binding domains of CitA upon citrate or of DcuS upon L-malate binding. (C) Uplift of the C-terminal region of PAS$_P$ after compaction of the domain. (A) DcuS is membrane embedded by transmembrane helices 1 and 2 (TM1, TM2) that are flanked extracytoplasmically by the PER-ARNT-SIM domain PAS$_P$ and by the PAS domain PAS$_C$, and a C-terminal HisKA/HATPase-type kinase. DcuS is a homodimer, and the monomers are presented in *grey* and *light blue*. In the *blue* monomer, the α-helical structure of TM2 and of the C-terminal helix α_6 of PAS$_P$ is indicated. + and − indicate the periplasmic and cytoplasmic sides of the membrane. (B) Structure comparison of the periplasmic PAS$_P$ domains of DcuS with L-malate (*grey*; #3BY8 (10)) and CitA$_{Kp}$ with citrate (*orange*; #2J80) and without citrate (*grey*, #2V9A) (Sevvana et al., 2008). Structures were superimposed using the software Chimera (Pettersen et al., 2004). The structure was derived from isolated domains of PAS$_P$ with α_6 representing the C-terminal end (Cheung & Hendrickson, 2008; Sevvana et al., 2008). The orientation of α_6 is presumably not the same as in full-length DcuS where α_6 appears to be an extension and linear continuation of TM2 (Monzel & Unden, 2015). (C) The C-terminus of CitA$_{Kp}$ and its uplift in parallel to the helix axis is zoomed in (presented for DcuS-PAS$_P$ helix α_6). The uplift is shown by a *red arrow* and a *dotted line*, respectively. For details, see text. (See the colour plate.)

DcuS of *E. coli* controls expression of target genes in response to all types of C_4-dicarboxylates (like fumarate, L- and D- malate, aspartate, D-, L-, meso-tartrate, mesaconate) and of citrate (Kneuper et al., 2005). However, L-malate and fumarate are the preferred substrates (S. Wörner & G. Unden, unpublished). The approx. K_m for the induction of the DcuS-dependent genes is in the mM range whereas CitA has high affinity and specificity for citrate and few tricarboxylates with a K_D value in the low μM range (Kaspar & Bott, 2002; Kaspar et al., 1999; Kneuper et al., 2005). Only extracellular C_4-dicarboxylates cause DcuS-dependent induction which is in agreement with the role of DcuS–DcuR in regulating genes that are required for degradation of external C_4-dicarboxylates, in particular the genes for the C_4-dicarboxylate transporters. Fumarate reductase that serves in glucose fermentation and in fumarate respiration with external fumarate, on the other hand, is only slightly induced by C_4-dicarboxylates and DcuS–DcuR.

When citrate is bound by PAS_P of CitA, the central β-sheet of the domain encloses the ligand, resulting in compaction of the domain and an uplift of the C-terminal part of the β-sheet, away from the membrane surface (Sevvana et al., 2008) (Fig. 3B). The uplift amounts to approx. 3.6 Å at the C-terminal α_6-helix of PAS_P (Sevvana et al., 2008). Helix α_6 directly extends into TM2. For DcuS, a high-resolution structure is available only for the malate-bound form that is almost identical to PAS_P from $CitA_{Kp}$ with an overall r.m.s.d. of 1.6 Å (Cheung & Hendrickson, 2008; Fig. 3B). Based on the similar structures of the ligand-bound forms of CitA- and DcuS-PAS_P, a similar compaction and structural re-organization is predicted for DcuS-PAS_P upon ligand binding (Scheu, Kim, et al., 2010).

3.2 Transmembrane Signalling by DcuS: A Piston-Type Displacement of TM2

For direct information on transmembrane signalling and related movement of the TM helices, the water accessibility of the amino acid residues of DcuS at the water–membrane interface was probed during activation of DcuS by fumarate in vivo, using cysteine accessibility scanning (Scan-SCAM; Monzel & Unden, 2015). According to the data, TM1 was inserted in the membrane with amino acid residues 21–41 in both the fumarate-activated (ON) and inactive (OFF) states. In contrast, membrane insertion of TM2 showed significant changes in response to the presence of fumarate. TM2 was located in the membrane with residues 181–201 in the OFF state and residues 185–205 in the ON state (Fig. 4).

Trp residues often flank and position TM helices by their amphiphilic and aromatic head group at the membrane interface (Draheim, Bormans, Lai, & Manson, 2005). Remarkably, TM2 carries Trp residues at positions 181 and 185 which corresponds to the position of the water/membrane interface of TM2 in the OFF and the ON state, respectively (Monzel & Unden, 2015). Positively charged Arg residues in a TM helix tend to move to the hydrophilic region (Draheim et al., 2005; Li, Vorobyov, & Allen, 2013). Artificial displacement of TM2 by introducing Arg residues at positions 181 and 185 is able to activate DcuS. Overall, the data show a displacement of TM2 to the periplasm upon activation by physiological (fumarate) or nonphysiological (Trp/Arg replacement in TM2) conditions that are compatible in their changes of the accessibility pattern with a piston-type shift.

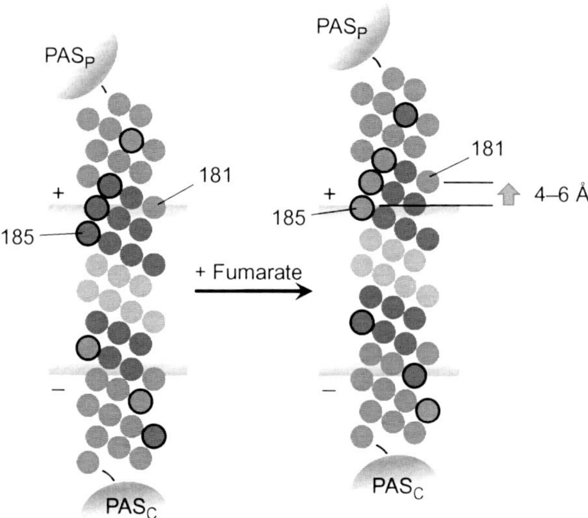

Fig. 4 Accessibility changes of Cys residues around the second transmembrane helix (TM2) of DcuS without (*left*) and with (*right*) activation by fumarate. The figure shows a schematic representation of the accessibility of Cys residues engineered at different positions into TM2 by PEG-mal using the Scan-SCAM method (Monzel & Unden, 2015). The inaccessibility of residues was deduced from a modified mobility after labelling in SDS-PAGE and Western blotting (Monzel & Unden, 2015). Accessible (unPEGylated) cysteine residues are shown in *green* and inaccessible (PEGylated) ones in *red*. The residues showing a change in accessibility are framed in *bold lining*. + and − indicate the periplasmic and cytoplasmic sides of the membrane. The positions of Trp residues 181 and 185 that are at the membrane water interface in the inactive and the fumarate-activated state, respectively, are indicated. (See the colour plate.)

The data from accessibility changes of TM2 upon activation suggest a piston-type shift of TM2 by four residues to the periplasm upon activation (or fumarate binding). A piston-type mode of function is supported by energetic calculations of preferred positions for TM2 insertion in the membrane. The calculations that are based on the aminoacid composition of the corresponding helices (Hessa et al., 2007, Monzel & Unden, 2015) suggest a preferred membrane insertion of TM2 at two positions (residues 181–201, and 185–205, respectively). There is a continuum of low apparent free energy of insertion (ΔG^{app}) for helices in both positions and the region in between, and an increase of ΔG^{app} beyond these positions (Monzel & Unden, 2015). This indicates that membrane insertion of TM2 at position 181–201 (corresponding to the OFF state) and 185–205 (ON state) represents energetically preferred positions, and there is only a low energy barrier for the transition between both positions. For TM1, the same calculations give only one preferred position (residues 21–41) that corresponds closely to the experimentally determined position of TM1.

3.2.1 Additional Modes for TM Signalling in DcuS?

Alternative modes for transmembrane signalling include rotation of TM helices along their axis, or a scissors movement (Ottemann, Xiao, Shin, & Koshland, 1999). In the scissors movement, tilting of TM helices (or part of the helices) can be induced by a compression of unstably arranged TM helices after ligand binding (Falke, 2014); vice versa, loss of tilting can happen after decompression of kinked helices by pulling at the TM after ligand binding.

Transmembrane helices TM2/TM2′ of the DcuS homodimer can be linked via Cys residues in a 4-3-4-3 distance pattern in the same way in the absence and the presence of fumarate, suggesting cross-linking and the same mirror-image arrangement of the helices under both conditions (Monzel & Unden, 2015). The cross-linked region covers four connected helical turns within α_6 of PAS_P and the periplasmic part of TM2. The data argue against a rotational movement in transmembrane signalling of DcuS. However, contribution of scissors-type movements to transmembrane signalling cannot be excluded.

3.3 Signal Transfer from the Membrane (TM2) by PAS_C to the Kinase Domain

At the cytoplasmic side, the transmembrane signal from TM2 is perceived by the cytoplasmic PAS_C domain and transmitted to the kinase (see Fig. 3A). The N-terminal region of PAS_C has a high plasticity that is important for signal perception and seems to be used for converting the linear movement from TM2 to a change in PAS_C homodimerization (Etzkorn et al., 2008). Deletion or mutational inactivation of PAS_C causes activation of the kinase (Monzel et al., 2013). Therefore, the activity of the kinase is inhibited by the PAS_C dimer, and the default state of the kinase is the ON state. PAS_C represents one of the dimerization sites within DcuS, and as the major interaction site with the transporter DctA that serves as a coregulator of DcuS.

PAS_C controls the activity of the kinase domains, and various factors contribute to the regulation. PAS_C basically functions as an inhibitor of the kinase (Monzel et al., 2013). Thus, destabilization of the PAS_C/PAS_C dimerization by mutation (type I-ON mutation) of the contact region results in permanent activation of the kinase (permanent ON state). In a second type of ON mutations (type IIA-ON), the interaction with the coregulator DctA is affected. This finding is in line with experiments showing that DctA is an inhibitor of DcuS function (Steinmetz, Wörner, & Unden, 2014), and that PAS_C is important for interaction with DctA (Witan, Bauer, et al., 2012, Witan, Monzel, et al., 2012). A third class of ON mutations (type IIB-ON) neither affects PAS_C dimer stability nor interaction with DctA, a situation that was similar to physiological activation by fumarate.

The experiments suggest (Monzel et al., 2013) that dimerization of PAS_C is required to inhibit the kinase, which is in agreement with structural studies (Etzkorn et al., 2008). Disturbing the PAS_C/PAS'_C homodimer activates the kinase in a fumarate-independent manner. Finally, physiological activation by fumarate causes a subtle structural modification in PAS_C that alleviates kinase inhibition under the respective conditions. It is assumed that the subtle structural and functional changes of type IIB-ON represent the response of PAS_C to the piston-type movement of TM2 that is communicated to the N-terminus of PAS_C. As will be shown in the following section, the transporters DctA and DcuB play an important role in controlling DcuS function which might be via PAS_C.

4. THE DctA/DcuS (OR DcuB/DcuS) SENSOR COMPLEXES AND THEIR ROLE FOR DcuS FUNCTION

4.1 DctA (and DcuB) Form Permanent Sensor Complexes with DcuS

The transporters DctA or DcuB serve as coregulators of DcuS under aerobic and anaerobic conditions, respectively, and their loss causes constitutive activity of DcuS (Davies et al., 1999; Janausch et al., 2002; Kleefeld et al., 2009). So far no DctA/DcuS complexes were isolated. In vivo, however, interaction between DctA and DcuS was demonstrated by FRET, a bacterial two-hybrid system and cochromatography (Witan, Bauer, et al., 2012). In addition, DcuS traps DctA that forms a complex with it at the cell poles (Scheu et al., 2014). In the same way, DcuB interacts with DcuS under anaerobic conditions. The DctA/DcuS and DcuB/DcuS sensor complexes appear to being formed permanently with and without fumarate (S. Wörner & G. Unden, unpublished).

DctA is composed of eight TM helices in an arrangement similar to that of Glt_{Ph}, the Na^+/glutamate symporter of *Pyrococcus horikoshii*. The structure of Glt_{Ph} has been solved (Yernool, Boudker, Jin, & Gouaux, 2004). DctA is homologous to Glt_{Ph} and has a similar predicted structure (Witan, Bauer,

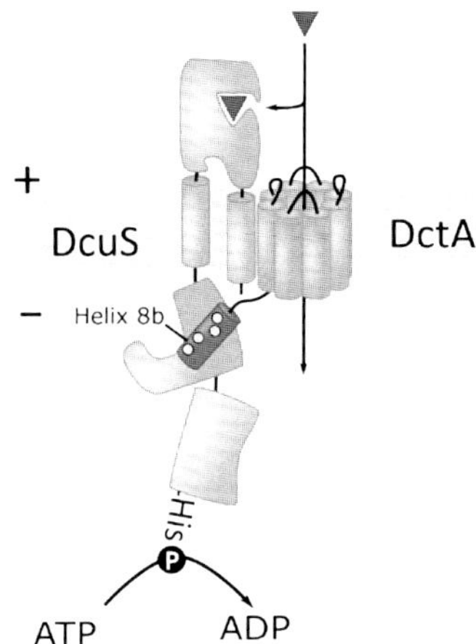

Fig. 5 Molecular details of DctA/DcuS interaction. For DctA and DcuS, only monomers of the proteins are shown. DcuS is preferentially dimeric (Scheu, Liao, et al., 2010), whereas DctA is presumably a trimer (Slotboom, Konings, & Lokema, 1999; J. Witan & G. Unden, unpublished). The C-terminal cytoplasmic helix 8b of DctA plays a central role in the interaction with DcuS (Witan, Bauer, et al., 2012). The amphipathic helix comprises three helical turns that contain a LDX_3LX_3L signature sequence (*red* (*dark grey* in the print version) *tube*, with the three L and the D residues presented as *yellow* (*white* in the print version) *dots*). The three Leu residues that are separated by four and three residues and the Asp residue were important for function and interaction of DctA (Witan, Bauer, et al., 2012). Helix 8b interacts with the PAS_C domain of DcuS and controls by the interaction the kinase activity of DcuS (see text for details).

et al., 2012). DctA contains, however, an additional C-terminal amphipathic helix 8b on the cytoplasmic side of the membrane that is required for interaction with membrane-embedded DcuS (Fig. 5). Mutations in H8b diminish coregulation and interaction with DcuS, and a genetic construct of isolated H8b showed strong interaction with DcuS (Witan, Bauer, et al., 2012) as well as with separate PAS_C (Monzel et al., 2013). The helix contains a LX_4LX_3L signature with three Leu residues that is required for interaction with DcuS. On the DcuS side, deletion and mutation of the cytoplasmic PAS_C domain affected the interaction between DctA and DcuS (Monzel et al., 2013; Witan, Bauer, et al., 2012). The data therefore suggest that DctA and DcuS form a sensor complex, and H8b and PAS_C represent specific sites for the interaction.

4.2 DcuS as the Site for Sensing in the DctA/DcuS or DcuB/DcuS Sensor Complexes

Presence of a transporter (DctA or DcuB) in the sensor complex was taken as an indication that the transporters may present the site, or an additional site, for sensing due to their capability to interact with the same substance. However, the K_m values for succinate or fumarate transport by DctA at pH 7 are 30 μM (Janausch, Kim, & Unden, 2001; Kay & Kornberg, 1971) and approx. 100 μM for DcuB (Engel, Krämer, & Unden, 1994), whereas the apparent K_m for induction of *dcuB-lacZ* by succinate, L-tartrate or L-aspartate was in the range of 2.0–3.0 mM (Kneuper et al., 2005). The difference in affinity for both processes by nearly two orders of magnitude suggests that the transporters do not represent the site for sensing. This was proven directly by showing that (a) binding of the effectors at the transporters is not required for the function of DctA/DcuS or DcuB/DcuS complexes in sensing, (b) the specificity for detecting the stimulus relies on DcuS and (c) no transport of the substrates is required for coregulation (Fig. 6).

(a) Both, DcuS and the transporters DcuB and DctA have a broad substrate range for C_4-dicarboxylates and related compounds. Most

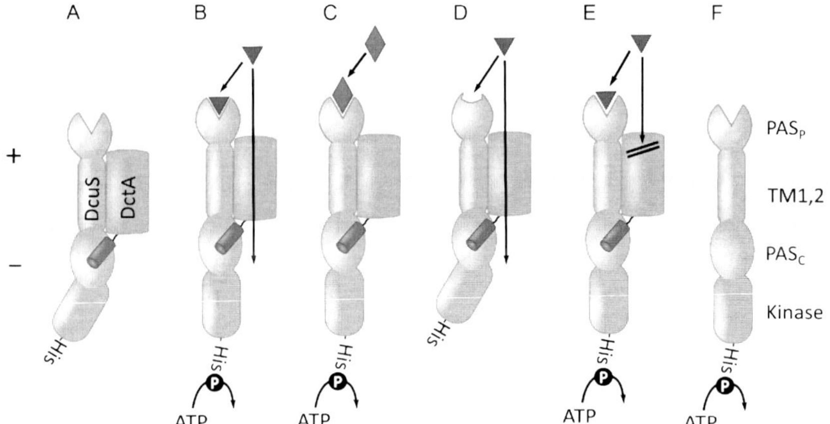

Fig. 6 Control of DcuS OFF/ON conversion in the DctA/DcuS complex: role of ligands, DctA, PAS$_P$ and PAS$_C$ for the conversion. The functional state of DcuS is indicated by the capability for autophosphorylation. (A) Functional DctA/DcuS complex, no activation by ligands and (B) functional DctA/DcuS complex in the presence of effectors. The effectors (typically C_4-dicarboxylates (*blue* (*dark grey* in the print version) *triangle*) or citrate) activate DcuS. The C_4-dicarboxylates can be transported by DctA, but the transport activity of DctA in the DctA/DcuS complex has not been tested. (C) Effectors like citrate (*green* (*dark grey* in the print version) *rhomb*) activate DcuS, but are no substrate for DctA. (D) By mutation of the ligand binding site in PAS$_P$ of DcuS, the ligand binding or specificity of sensing can be altered, whereas substrate specificity and transport by DctA is retained. (E) In strains with DctA(S380D) (or DcuB(E79A)), transport activity of the carriers is lost, but sensing by DctA/DcuS (or DcuB/DcuS) is retained. (F) DcuS without DctA is in the permanent ON state due to the absence of the transporter that functions as an inhibitor of DcuS. The same phenotype is obtained by DcuS-ON mutations in PAS$_C$ (PAS$_C$(N248D) or PAS$_C$(N204D)). Details and references in the text.

C_4-dicarboxylates are substrates for the transporters and for DcuS (Engel et al., 1994; Kay & Kornberg, 1971; Kneuper et al., 2005; Six, Andrews, Unden, & Guest, 1994). However, in aerobic and anaerobic growth, citrate induces DcuS-dependent regulation, but is no substrate or competitive inhibitor for DctA or DcuB (Steinmetz et al., 2014; S. Wörner & G. Unden, unpublished). In the same way under anaerobic conditions, substrates like maleate or L-tartrate are substrates for DcuS-dependent regulation, but were no substrates or competitive inhibitors of DcuB. Therefore there are stimuli that activate DcuS but are no substrate for the transporters (Fig. 6C), which demonstrates that ligand binding at the transporter is not required for sensing.

(b) Mutation of the ligand site of DcuS was sufficient to change the ligand specificity of sensing (S. Wörner & G. Unden, unpublished) (Fig. 6D). In this way, DcuS variants were produced that were specific for L-malate, or for mesaconate and L-malate, or that lost substrate response completely (Kneuper et al., 2005; S. Wörner & G. Unden, unpublished). Therefore DcuS alone is responsible for recognition of the stimuli, and DcuB (or DctA) do not contribute to this process.

(c) Variants of DcuB and DctA were isolated (eg, DcuB(E79A), DcuB(R83A) or DctA(S380D)) that lost the capacity for transport of C_4-dicarboxylates (Fig. 6E) but retained the capacity for coregulation (Table 1) (Kleefeld et al., 2009; Steinmetz et al., 2014). Therefore there is no need that the transporters are active in transport for functioning as coregulators of DcuS. In contrast to the situation with transport inactive carriers (Fig. 6E), deletion of DcuB or DctA (Fig. 6F) also results in the loss of transport. Under these conditions DcuS is in the permanent ON state (Davies et al., 1999; Kleefeld et al., 2009). Comparison of both situations demonstrates the deregulation of DcuS function by the loss of the transporters is caused by the loss of the corresponding proteins but not of the transport function.

Altogether the situations of mutants summarized in Fig. 6 demonstrate that DctA and DcuB exert their function in coregulation without having a sensory function neither in the sense of concentration sensing (or binding) of C_4-dicarboxylates nor in coupling sensing to transport or transport measurement (flux sensing). Rather it appears that the transporters function as coregulators and have to be present to form a functional sensor or sensor complex.

The transfer of DcuS to the permanent ON state by deletion of DctA or DcuB resembles very much the situation when PAS_C of DcuS is deleted

Table 1 Effect of Various Mutations in the Transporters DctA or DcuB on C_4-Dicarboxylate Transport Activity and Regulatory Activity in the Expression of *dcuB-lacZ*

Transporter	Property or Phenotype	Induction by Fumarate (-Fold) 'Control Factor'	Transport (% Wild Type)
DctA	Wild type	5.2	100
DctA(S380D)	Transport deficient	4.1	11
DcuB	Wild type	17.4	100
DcuB(T394I)	Regulation deficient	0.85	115
DcuB(E79A)	Transport deficient	18.0	13

Transport was determined as the uptake of [^{14}C]fumarate into washed cells, expression of *dcuB-lacZ* was determined as the β-galactosidase activity. The data of the mutants were related to strains with wild-type situation.

Original data from Kleefeld, A., Ackermann, B., Bauer, J. Krämer, J., & Unden, G. (2009). The fumarate/succinate antiporter DcuB of *Escherichia coli* is a bifunctional protein with sites for regulation of DcuS-dependent gene expression. *Journal of Biological Chemistry, 284*, 265–275 and Steinmetz, P. A., Wörner, S., & Unden, G. (2014). Differentiation of DctA and DcuS function in the DctA/DcuS sensor complex of *Escherichia coli*: Function of DctA as an activity switch and of DcuS as the C_4-dicarboxylate sensor. *Molecular Microbiology, 94*, 218–229.

(Monzel et al., 2013). Both the transporters and PAS_C are obviously inhibitors of the kinase domain of DcuS, and their deletion causes constitutive expression of DcuS-dependent target genes. The conclusion was confirmed by titrating the levels of DctA or DcuB: Increase of DctA or DcuB from very low-to-intermediate levels was required to convert DcuS from the noninhibited (or permanent active) to the C_4-dicarboxylate-responsive state (Steinmetz et al., 2014; S. Wörner & G. Unden, unpublished).

4.3 C_4-Dicarboxylate-Dependent Induction by the DcuS Sensor Complexes: A Model for DctA/DcuS (or DcuB/DcuS) Function

4.3.1 Function of DcuS in Signal Perception, Transmembrane Signalling and Control of the Kinase

Combining data for DcuS function allows to propose a model for stimulus perception and signalling by DcuS (Fig. 7). The DctA/DcuS complex functions as a sensor that controls expression of the target genes in response to the concentration of extracellular C_4-dicarboxylates, which are detected by binding to the PAS_P domain of DcuS. The model describes (Fig. 7B and C) (i) stimulus perception and conversion by PAS_P, (ii) transmembrane signalling by TM2 and (iii) signal transfer to the kinase via PAS_C by overcoming the inactivation of the kinase. For the first step (i) the data rely mostly on

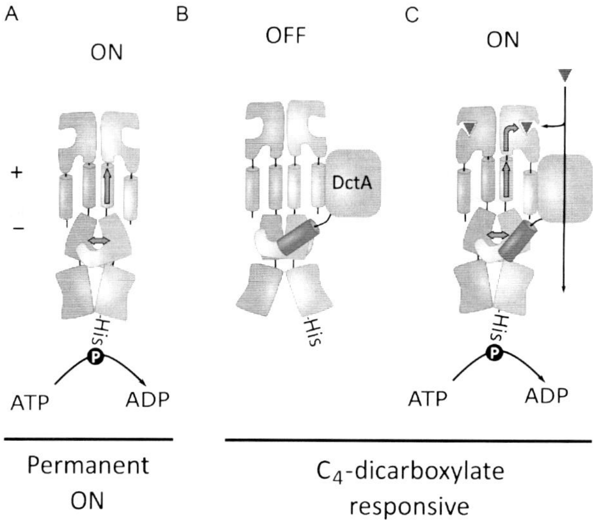

Fig. 7 Transmembrane signalling by DcuS: Control of the kinase activity by C_4-dicarboxylates and the transporter proteins DctA (or DcuB). The functional state of DcuS is shown for the DctA/DcuS sensor complex where DcuS is able to respond to C_4-dicarboxylates (B and C), and for DcuS without DctA (A) where DcuS is in the permanent ON state. In the C_4-dicarboxylate-responsive state, binding of C_4-dicarboxylates at PAS_P causes compaction of PAS_P with an uplift of α_6 in PAS_P and of TM2 (*red* (*dark grey* in the print version) *arrows*) by one helical turn in TM2. Movement of TM2 is transmitted at the cytoplasmic side to PAS_C and results in relieved PAS_C dimerization and of kinase inhibition. DctA (via helix 8b) and PAS_C are both required for kinase inhibition: PAS_C is only able to inhibit the kinase when properly positioned by DctA (compare the model and the activity state in the absence of ligand in the DctA-less form of DcuS (A) and in the DctA/DcuS complex (B)). DctA functions as a coinhibitor of PAS_C, and inhibition of the kinase is abolished artificially both by deletion of PAS_C and DctA. In this model, the activity of the kinase is 'silenced' by PAS_C in analogy to the 'silencing' of the kinase of ArcB by the oxidation of Cys residues (Malpica, Franco, Rodriguez, Kwon, & Georgellis, 2004), and DctA would be a cosilencer. Protein silencing (Malpica et al., 2004; Watt, Heinrich, & Thomas, 2006) functions in analogy to gene or RNA silencing by inhibiting or inactivation protein function. See text for detail and references.

the closely related citrate sensor $CitA_{Kp}$ of *Klebsiella pneumoniae*. Due to the low affinity of DcuS, the induction by the C_4-dicarboxylates starts at concentrations >0.1 mM. Binding of the ligand causes a compaction of PAS_P resulting in an uplift of the C-terminal region around helix α_6 by approx. 3.8 Å (see Fig. 3 for details). For DcuS a similar uplift is suggested upon fumarate or L-malate binding due to the structural similarity (Scheu, Kim, et al., 2010). (ii) The uplift of α_6 induces the displacement of TM2 towards the periplasm which has the characteristics of a piston-type shift of the helix.

The shift comprises four amino acid residues of TM2 corresponding to 1.1 turns or 4–6 Å at the Cα-atoms of an α-helix. This dimension fits to the uplift of the C-terminal region by 3.8 Å. Transmembrane signalling may include additional modes of transmembrane signalling in addition to the piston-type movement. (iii) The displacement of TM2 is further transmitted towards PAS_C by a short linker of nine amino acid residues (residues 203–211) that connects both regions within DcuS. PAS_C serves as the transmitter and converts the TM2 shift to a conformational movement controlling kinase activity. Signal transmission by PAS_C depends on a modification of PAS_C dimerization. Default kinase is in the functional ON state which is inactivated by the dimeric PAS_C. For activation of the kinase, the inhibition by PAS_C has to be relieved. The relief in inhibition is achieved by a partial relieve of PAS_C dimerization after signal input from TM2 (fumarate activation).

4.3.2 Role of the Transporters DctA or DcuB as Coregulators of DcuS

DcuS forms a sensor complex with the transporters DctA (DctA/DcuS, aerobic growth) or DcuB (DcuB/DcuS, anaerobic growth). Formation of the complex is permanent under the respective conditions and independent from the presence of C_4-dicarboxylates. The transporters have no role in sensing of the C_4-dicarboxylates, neither by sensing the concentration of the ligand nor by metabolic or flux sensing. The transporters convert, however, DcuS structurally from the permanent ON state (Fig. 7A) to the C_4-dicarboxylate-responsive state (Fig. 7B and C). DctA interacts by the cytoplasmic helix 8b with PAS_C of DcuS. PAS_C together with DctA inhibits the activity of the kinase domain, and both partners are required for the inhibition: deletion of either PAS_C or DctA causes permanent activity of the kinase without control by the C_4-dicarboxylate availability. The molecular basis for the inhibition of PAS_C by DctA might be structural changes in PAS_C after DctA binding that allow proper response of PAS_C to shifting of TM2, or appropriate positioning of TM2. It is assumed that DcuB takes over the role of DctA in coregulation of DcuS in a similar way.

4.4 Negative Autoregulation of DcuS Function by DctA Under C_4-Dicarboxylate Limitation

In addition to 'normal' concentration-dependent regulation in the preceding paragraph and the functional model for the DctA/DcuS complex (mode I regulation), DcuS is also capable of negative autoregulation (mode II regulation). Combination of both modes results in a biphasic response of DcuS to inducer concentration (Fig. 8).

As described earlier, in mode I regulation DcuS prevails in complex with a transporter (DctA or DcuB). The transporters are coregulators and part of the sensor complex and transform DcuS to the C_4-dicarboxylate-responsive form. The DctA/DcuS (or DcuB/DcuS) sensor complex induces expression of target genes with increasing concentrations of (external) C_4-dicarboxylates (concentration sensing when the concentration of the substrates increase and higher metabolic rates are required). This type of regulation represents a typical mode of transcriptional regulation for catabolic systems.

At low concentration of C_4-dicarboxylates ($\leq$0.1 mM or below, which is a factor of 20 or more below the approx. K_m of DcuS for the substrates (Kneuper et al., 2005)), transcriptional regulation of *dctA* or *dcuB* occurs

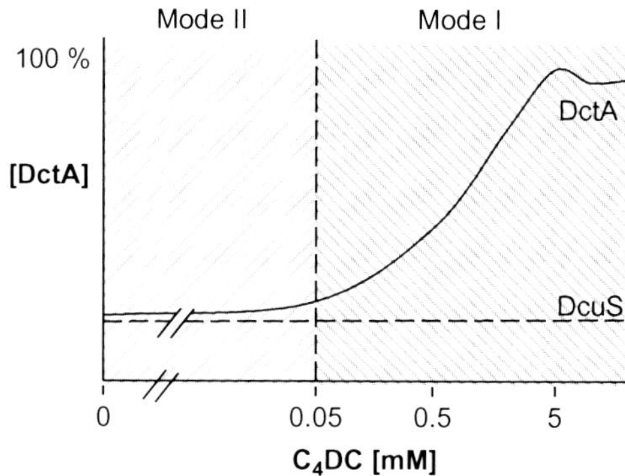

Fig. 8 Function of DcuS in two different modes (modes I and II) at high and low substrate concentrations. Under *Mode I condition*, DcuS prevails as the DctA/DcuS complex which induces expression of target genes (*dctA* and others) in response to the C_4-dicarboxylate concentration, resulting in high levels of DctA. DctA contents are sufficient for DctA/DcuS complex formation, and formation of DctA for catabolic C_4-dicarboxylate uptake. At low concentrations of C_4-dicarboxylates, *Mode II type of regulation* sets in. The levels of DctA drop due to weak or lack of induction. When not sufficient DctA is produced for DctA/DcuS complex formation, free DcuS is present which is active without C_4-dicarboxylate, resulting in resumed DctA production until sufficient DctA for DctA/DcuS complex is produced. After DctA/DcuS complex formation further expression of *dctA* stops since DctA/DcuS is inactive in the absence of sufficient C_4-dicarboxylates. By the autoregulation under mode II condition, the expression of DctA is maintained at significant and constant level even in the absence of C_4-dicarboxylate. The absolute levels of DctA and DcuS are not known but expression data show that in the absence of C_4-dicarboxylates significant expression of *dctA* and *dcuB* is retained (Steinmetz et al., 2014; S. Wörner & G. Unden, unpublished).

according to a different mechanism (Fig. 8, mode II). Under these conditions, substrate concentration is no longer sufficient to induce expression of *dctA* or *dcuB*. However, as shown in Fig. 8, the expression of *dctA* or levels of DctA are low but significant and stay constant with further decreasing C_4-dicarboxylate concentrations (Steinmetz et al., 2014). When the concentration of DctA decreases to a level that is no longer sufficient for DctA/DcuS complex formation, the uncomplexed DcuS is autophosphorylated independently of C_4-dicarboxylates and stimulates *dctA* expression. DcuS on the other hand is not autoregulated by C_4-dicarboxylates, and the cellular contents stay constant with respect to C_4-dicarboxylate levels (Goh et al., 2005; Oyamada, Yokoyama, Morinaga, Suzuki, & Makino, 2007). When the contents of DctA drop below levels required for complexing DcuS according to this scenario, the noncomplexed DcuS autophosphorylates independent of C_4-dicarboxylates and stimulates *dctA* expression. The resulting DctA production and DctA/DcuS complex formation is supposed to halt *dctA* expression in the absence of C_4-dicarboxylates. This mode of regulation ensures presence of minimal contents of DctA at very low or lacking concentrations of C_4-dicarboxylates in order to maintain basic levels of transport and formation of a functional sensor complex.

Regulation by mode II and involvement of transporters as coregulators is apparently of broad significance, and most bacteria use the DcuS–DcuR two-component systems in combination with transporters as coregulators (G. Unden et al., unpublished observation). C_4-dicarboxylates like aspartate, L-malate and fumarate serve two different metabolic roles in most bacteria. On the one hand, they are used in anabolism as precursors for the synthesis of various amino acids, or as electron acceptors in the synthesis of heme or the pyrimidines. When present at higher concentrations, however, the major role of the C_4-dicarboxylates is catabolic, for example, in fumarate respiration. *E. coli* growing in the gut or rumen encounters low but significant levels of C_4-dicarboxylates (mostly ≤ 0.1 mM), in particular L-malate and L-aspartate. Under these conditions, essentially mode II regulation is predicted to apply, and the C_4-dicarboxylates are used for anabolic but also catabolic purpose. When the levels of the C_4-dicarboxylates increase significantly, either after a specific diet of the host or after transfer of the bacteria to a different environment, mode I regulation comes into play with increased induction of DctA and DcuB and metabolic systems for growth on C_4-dicarboxylates. Under such conditions, the C_4-dicarboxylates are substrates for catabolism and their role in anabolism is decreased. The biphasic regulation therefore allows use of the C_4-dicarboxylates under different environmental conditions with low and high C_4-dicarboxylate supply, and for anabolism as well as catabolism.

ACKNOWLEDGEMENTS

We are grateful to P. Steinmetz and J. Witan for many important contributions and Deutsche Forschungsgemeinschaft for financial support.

REFERENCES

Anantharaman, V., & Aravind, L. (2003). Application of comparative genomics in the identification and analysis of novel families of membrane-associated receptors in bacteria. *BMC Genomics*, *4*, 34.

Antoine, R., Huvent, I., Chemial, K., Deray, I., Raze, D., Locht, C., et al. (2005). The periplasmic binding protein of a tripartite tricarboxylate transporter is involved in signal transduction. *Journal of Molecular Biology*, *351*, 799–809.

Antoine, B., Lefrancois-Martinez, A. M., Le Guillou, G., Leturgue, A., Vandervalle, A., & Kahn, A. (1997). Role of the GLUT 2 glucose transporter in the response of the 1-type pyruvate kinase gene to glucose in liver-derived cells. *Journal of Biological Chemistry*, *272*, 17937–17943.

Asai, K., Baik, S. H., Kasahara, Y., Moriya, S., & Ogasawara, N. (2000). Regulation of the transport system for C_4-dicarboxylic acids in *Bacillus subtilis*. *Microbiology*, *146*, 263–271.

Baek, J. H., & Lee, S. Y. (2006). Novel gene members in the Pho regulon of *Escherichia coli*. *FEMS Microbiology Letters*, *264*, 104–109.

Baraquet, C., Théraulaz, L., Guiral, M., Lafitte, D., Méjean, V., & Jourlin-Castelli, C. (2006). TorT, a member of a new periplasmic binding protein family, triggers induction of the Tor respiratory system upon trimethylamine N-oxide electron-acceptor binding in *Escherichia coli*. *Journal of Biological Chemistry*, *281*, 38189–38199.

Bologna, F. P., Andreo, C. S., & Drincovich, M. F. (2007). *Escherichia coli* malic enzymes: Two isoforms with substantial differences in kinetic properties, metabolic regulation, and structure. *Journal of Bacteriology*, *189*, 5937–5946.

Bott, M., Meyer, M., & Dimroth, P. (1995). Regulation of anaerobic citrate metabolism in *Klebsiella pneumoniae*. *Molecular Microbiology*, *18*, 533–546.

Buelow, D. R., & Raivio, T. L. (2010). Three (and more) component regulatory systems—Auxiliary regulators of bacterial histidine kinases. *Molecular Microbiology*, *75*, 547–566.

Cheung, J., & Hendrickson, W. A. (2008). Crystal structures of C_4-dicarboxylate ligand complexes with sensor domains of histidine kinases DcuS and DctB. *Journal of Biological Chemistry*, *283*, 30256–30265.

Cox, G. B., Webb, D., & Rosenberg, H. (1989). Specific amino acid residues in both the Pst B and Pst C proteins are required for phosphate transport by the *Escherichia coli* Pst system. *Journal of Bacteriology*, *171*, 1531–1534.

Cunningham, K. A., & Burkholder, W. F. (2009). The histidine kinase inhibitor Sda binds near the site of autophosphorylation and may sterically hinder autophosphorylation and phosphotransfer to Spo0F. *Molecular Microbiology*, *71*, 659–677.

Davies, S., Golby, P., Omrani, D., Broad, S. A., Harrington, V. L., Guest, J. R., et al. (1999). Inactivation and regulation of the aerobic C_4-dicarboxylate transport (*dctA*) gene of *Escherichia coli*. *Journal of Bacteriology*, *181*, 5624–5635.

Didion, T., Regenberg, B., Jorgensen, M. U., Kielland-Brandt, M. C., & Andersen, H. A. (1998). The permease homologue SSy1p controls the expression of amino acid and peptide transporter genes in *Saccharomyces cerevisiae*. *Molecular Microbiology*, *27*, 643–650.

Dintner, S., Heermann, R., Fang, C., Jung, K., & Gebhard, S. (2014). A sensory complex consisting of an ATP-binding cassette transporter and a two-component regulatory system controls bacitracin resistance in *Bacillus subtilis*. *Journal of Biological Chemistry*, *289*, 27899–27910.

Dintner, S., Staron, A., Berchtold, E., Petri, T., Mascher, T., & Gebhard, S. (2011). Coevolution of ABC transporters and two-component regulatory systems as resistance modules against antimicrobial peptides in *Firmicutes* bacteria. *Journal of Bacteriology*, *193*, 3851–3862.

Draheim, R. R., Bormans, A. F., Lai, R. Z., & Manson, M. D. (2005). Tryptophan residues flanking the second transmembrane helix (TM2) set the signaling state of the Tar chemoreceptor. *Biochemistry*, *44*, 1268–1277.

Engel, P., Krämer, R., & Unden, G. (1994). Transport of C4-dicarboxylates by anaerobically grown *Escherichia coli*: Energetics and mechanism of exchange, uptake and efflux. *European Journal of Biochemistry*, *222*, 605–614.

Etzkorn, M., Kneuper, H., Dünnwald, P., Vijayan, V., Krämer, J., Griesinger, C., et al. (2008). Plasticity of the PAS domain and a potential role for signal transduction in the histidine kinase DcuS. *Nature Structural and Molecular Biology*, *15*, 1031–1039.

Falke, J. J. (2014). Piston versus scissors: Chemotaxis receptors versus sensor His-kinase receptors in two-component signaling pathways. *Structure*, *22*, 1219–1220.

Finn, R. D., Bateman, A., Clements, J., Coggill, P., Eberhardt, R. Y., Eddy, S. R., et al. (2014). Pfam: The protein families database. *Nucleic Acids Research*, *42*(Database Issue), D222–D230.

Fritz, G., Dintner, S., Treichel, N. S., Radeck, J., Gerland, U., Mascher, T., et al. (2015). A new way of sensing: Need-based activation of antibiotic resistance by a flux-sensing mechanism. *mBio*, *6*, e00915–e00975.

Gardner, S. G., Johns, K. D., Tanner, R., & McCleary, W. R. (2014). The PhoU protein from *Escherichia coli* interacts with PhoR, PstB, and metals to form a phosphate-signaling complex at the membrane. *Journal of Bacteriology*, *196*, 1741–1752.

Gebhard, S., & Mascher, T. (2011). Antimicrobial peptide sensing and detoxification modules: Unravelling the regulatory circuitry of *Staphylococcus aureus*. *Molecular Microbiology*, *81*, 581–587.

Goh, E. B., Bledsoe, P. J., Chen, L. L., Gyaneshwar, P., Stewart, V., & Igo, M. M. (2005). Hierarchical control of anaerobic gene expression in *Escherichia coli* K-12: The nitrate-responsive NarX-NarL regulatory system represses synthesis of the fumarate-responsive DcuS-DcuR regulatory system. *Journal of Bacteriology*, *187*, 4890–4899.

Golby, P., Davies, S., Kelly, D. J., Guest, J. R., & Andrews, S. C. (1999). Identification and characterization of a two-component sensor-kinase and response regulator system (DcuS-DcuR) controlling gene expression in response to C_4-dicarboxylates in *Escherichia coli*. *Journal of Bacteriology*, *181*, 1238–1248.

Goldie, A. H., & Sanwal, B. D. (1980). Genetic and physiological characterization of *Escherichia coli* mutants deficient in phosphoenolpyruvate carboxykinase activity. *Journal of Bacteriology*, *141*, 1115–1121.

Graf, S., Broll, C., Wissig, J., Strecker, A., Parowatkin, M., & Unden, G. (2016). CitA/DcuS-type sensor kinases from the thermophilic *Geobacillus kaustophilus* and *G. thermodenitrificans* for sensing C4-dicarboxylates and citrate. *Microbiology*, *162*, 127–137. http://dx.doi.org/10.1099/mic.0.000171.

Graf, S., Schmieden, D., Tschauner, K., Hunke, S., & Unden, G. (2014). The sensor kinase DctS forms a tripartite sensor unit with DctB DctA for sensing C_4-dicarboxylates in *Bacillus subtilis*. *Journal of Bacteriology*, *196*, 1084–1093.

Grebe, T. W., & Stock, J. B. (1999). The histidine protein kinase superfamily. *Advances in Microbial Physiology*, *41*, 139–227.

Guest, J. R. (1992). Oxygen-regulated gene expression in *Escherichia coli*. *Journal of General Microbiology*, *138*, 2253–2263.

Hessa, T., Meindl-Beinker, N. M., Bernsel, A., Kim, H., Sato, Y., Lerch-Bader, M., et al. (2007). Molecular code for transmembrane-helix recognition by the Sec61 translocon. *Nature*, *450*, 1026–1030.

Hsieh, Y. J., & Wanner, B. L. (2010). Global regulation by the seven-component Pi signaling system. *Current Opinion in Microbiology*, *13*, 198–203.

Island, M. D., & Kadner, R. J. (1993). Interplay between the membrane associated UhpB and UhpC regulatory proteins. *Journal of Bacteriology*, *175*, 5028–5034.

Janausch, I. G., Kim, O. B., & Unden, G. (2001). DctA- and Dcu-independent transport of succinate in *Escherichia coli*: Contribution of diffusion and of alternative carriers. *Archives of Microbiology*, *176*, 224–230.

Janausch, I. G., Zientz, E., Tran, Q. H., Kröger, A., & Unden, G. (2002). C_4-dicarboxylate carriers and sensors in bacteria. *Biochimica et Biophysica Acta*, *1553*, 39–56.

Jording, D., Uhde, C., Schmidt, R., & Pühler, A. (1994). The C_4-dicarboxylate transport system of *Rhizobium meliloti* and its role in nitrogen-fixation during symbiosis. *Experientia*, *50*, 874–883.

Kadner, R. J., Island, M. D., Dahl, J. L., & Webber, C. A. (1994). A transmembrane signaling complex controls transcription of the Uhp sugar phosphate transport system. *Research in Microbiology*, *145*, 381–387.

Karinou, E., Compton, E. L. R., Morel, M., & Javelle, A. (2013). The *Escherichia coli* SLC26 homologue YchM (DauA) is a C4-dicarboxylic acid transporter. *Molecular Microbiology*, *87*, 623–640.

Kaspar, S., & Bott, M. (2002). The sensor kinase CitA (DpiB) of *Escherichia coli* functions as a high-affinity citrate receptor. *Archives of Microbiology*, *177*, 313–321.

Kaspar, S., Perozzo, R., Reinelt, S., Meyer, M., Pfister, K., Scapozza, L., et al. (1999). The periplasmic domain of the histidine autokinase CitA functions as a highly specific citrate receptor. *Molecular Microbiology*, *33*, 858–872.

Kay, W. W., & Kornberg, H. L. (1971). The uptake of C4-dicarboxylic acids by *Escherichia coli*. *European Journal of Biochemistry*, *18*, 274–281.

Kemner, J. M., Liang, X., & Nester, E. W. (1997). The *Agrobacterium tumefaciens* virulence gene *chvE* is part of a putative ABC-type sugar transport operon. *Journal of Bacteriology*, *179*, 2452–2458.

Kleefeld, A., Ackermann, B., Bauer, J., Krämer, J., & Unden, G. (2009). The fumarate/succinate antiporter DcuB of *Escherichia coli* is a bifunctional protein with sites for regulation of DcuS-dependent gene expression. *Journal of Biological Chemistry*, *284*, 265–275.

Kneuper, H., Janausch, I. G., Vijayan, V., Zweckstetter, M., Bock, V., Griesinger, C., et al. (2005). The nature of the stimulus and the fumarate binding site of the fumarate sensor DcuS of *Escherichia coli*. *Journal of Biological Chemistry*, *280*, 20596–20603.

Krell, T., Lacal, J., Busch, A., Silva-Jiménez, H., Guazzaroni, M. E., & Ramos, J. L. (2010). Bacterial sensor kinases: Diversity in the recognition of environmental signals. *Annual Reviews of Microbiology*, *64*, 539–559.

Kröger, A., Geisler, V., Lemma, E., Theis, F., & Lenger, R. (1992). Bacterial fumarate respiration. *Archives of Microbiology*, *158*, 311–314.

Lalonde, S., Boles, E., Hellmann, H., Barker, L., Patrick, J. W., Frommer, W. B., et al. (1999). The dual function of sugar carriers: Transport and sugar sensing. *The Plant Cell*, *11*, 707–726.

Leigh, J. A., & Dodsworth, J. A. (2007). Nitrogen regulation in bacteria and archaea. *Annual Reviews of Microbiology*, *61*, 349–377.

Li, L., Vorobyov, I., & Allen, T. W. (2013). The different interactions of lysine and arginine side chains with lipid membranes. *Journal of Physical Chemistry B*, *117*, 11906–11920.

Makino, K., Shinagawa, H., Amemura, M., Kawamoto, T., Yamada, M., & Nakate, A. (1989). Signal transduction in the phosphate regulon of *Escherichia coli* involves phosphotransfer between PhoR and PhoB proteins. *Journal of Molecular Biology, 210*, 551–559.

Malpica, R., Franco, B., Rodriguez, C., Kwon, O., & Georgellis, D. (2004). Identification of a quinone-sensitive redox switch in the ArcB sensor kinase. *Proceedings of the National Academy of Sciences of the United States of America, 101*, 13318–13323.

Mascher, T., Helmann, J. D., & Unden, G. (2006). Stimulus perception in bacterial signal transducing histidine kinases. *Microbiology and Molecular Biology Reviews, 70*, 910–938.

Meile, J. C., Wu, L. J., Ehrlich, S. D., Errington, J., & Noirot, P. (2006). Systematic localization of proteins fused to the green fluorescent protein in *Bacillus subtilis*: Identification of new proteins at the DNA replication factory. *Proteomics, 6*, 2135–2146.

Meyer, S., Melzer, M., Truernit, E., Hummer, C., Besenbeck, R., Stadler, R., et al. (2000). AtSUC3, a gene encoding a new Arabidopsis sucrose transporter, is expressed in cells adjacent to the vascular tissue and in a carpel cell layer. *The Plant Journal, 24*, 869–882.

Monzel, C., Degreif-Dünnwald, P., Gröpper, C., Griesinger, C., & Unden, G. (2013). The cytoplasmic PAS_C domain of the sensor kinase DcuS of *Escherichia*: Role in signal transduction, dimer formation and DctA interaction. *MicrobiologyOpen, 2*, 912–927.

Monzel, C., & Unden, G. (2015). Transmembrane signaling in the sensor kinase DcuS of *Escherichia coli*: A long-range piston-type displacement of transmembrane helix 2. *Proceedings of the National Academy of Sciences of the United States of America, 112*, 11042–11047.

Neiditch, M. B., Federle, M. J., Pompeani, A. J., Kelly, R. C., Swem, D. L., Jeffrey, P. D., et al. (2006). Ligand-induced asymmetry in histidine sensor kinase complex regulates quorum sensing. *Cell, 126*, 1095–1108.

Nikolskaya, A. N., Mulkidjanian, A. Y., Beech, I. B., & Galperin, M. Y. (2003). MASE1 and MASE2: Two novel integral membrane sensory domains. *Journal of Molecular Microbiology and Biotechnology, 5*, 11–16.

Ninfa, A. J., & Jiang, P. (2005). PII signal transduction proteins: Sensors of α-ketoglutarate that regulate nitrogen metabolism. *Current Opinion in Microbiology, 8*, 168–173.

Ottemann, K. M., Xiao, W., Shin, Y. K., & Koshland, D. E. (1999). A piston model for transmembrane signaling of the aspartate receptor. *Science, 285*, 1751–1754.

Oyamada, T., Yokoyama, K., Morinaga, M., Suzuki, M., & Makino, K. (2007). Expression of *Escherichia coli* DcuS-R two-component regulatory system is regulated by the secondary internal promoter which is activated by CRP-cAMP. *Journal of Microbiology, 45*, 234–240.

Özcan, S., Dover, J., & Johnson, M. (1996). Glucose sensing and signaling by two glucose receptors in the yeast *Saccharomyces cerevisiae*. *EMBO Journal, 17*, 2566–2573.

Pappalardo, L., Janausch, I. G., Vijayan, V., Zientz, E., Junker, J., Peti, W., et al. (2003). The NMR structure of the sensory domain of the membranous two-component fumarate sensor (histidine protein kinase) DcuS of *Escherichia coli*. *Journal of Biological Chemistry, 278*, 39185–39188.

Pettersen, E. F., Goddard, T. D., Huang, C. C., Couch, G. S., Greenblatt, D. M., Meng, E. C., et al. (2004). UCSF Chimera—A visualization system for exploratory research and analysis. *Journal of Computational Chemistry, 25*, 1605–1612.

Rauschmeier, M., Schüppel, V., Tetsch, L., & Jung, K. (2014). New insights into the interplay between the lysine transporter LysP and pH sensor CadC in *Escherichia coli*. *Journal of Molecular Biology, 426*, 215–229.

Reid, C. J., & Poole, P. S. (1998). Roles of DctA and DctB in Signal detection by the dicarboxylic acid transport system of *Rhizobium leguminosarum*. *Journal of Bacteriology, 180*, 2660–2669.

Reinelt, S., Hofmann, E., Gerharz, T., Bott, M., & Madden, D. R. (2003). The structure of the periplasmic ligand-binding domain of the sensor kinase CitA reveals the first extracellular PAS domain. *Journal of Biological Chemistry*, *278*, 39189–39196.

Rice, C. D., Pollard, J. E., Lewis, Z. T., & McCleary, W. R. (2009). Employment of a promoter-swapping technique shows that PhoU modulates the activity of the $PstSCAB_2$ ABC transporter in *Escherichia coli*. *Applied and Environmental Microbiology*, *75*, 573–582.

Rietkötter, E., Hoyer, D., & Mascher, T. (2008). Bacitracin sensing in *Bacillus subtilis*. *Molecular Microbiology*, *68*, 768–785.

Scheu, P. D., Kim, O. B., Griesinger, C., & Unden, G. (2010). Sensing by the membrane-bound sensor kinase DcuS: Exogenous versus endogenous sensing of C_4-dicarboxylates in bacteria. *Future Microbiology*, *5*, 1383–1402.

Scheu, P. D., Liao, Y. F., Bauer, J., Kneuper, H., Basché, T., Unden, G., et al. (2010). Oligomeric sensor kinase DcuS in the membrane of *Escherichia coli* and in proteoliposomes: Chemical cross-linking and FRET spectroscopy. *Journal of Bacteriology*, *192*, 3474–3483.

Scheu, P., Sdorra, S., Liao, Y. F., Wegner, M., Basché, T., Unden, G., et al. (2008). Polar accumulation of the metabolic sensory histidine kinases DcuS and CitA in *Escherichia coli*. *Microbiology*, *154*, 2463–2472.

Scheu, P. D., Steinmetz, P. A., Dempwolff, F., Graumann, P. L., & Unden, G. (2014). Polar localization of a tripartite complex of the two-component system DcuS/DcuR and the transporter DctA in *Escherichia coli* depends on the sensor kinase DcuS. *PLoS One*. *9*. http://dx.doi.org/10.1371/journal.pone.0115534. e115534.

Scheu, P. D., Witan, J., Rauschmeier, M., Graf, S., Liao, Y. F., Ebert-Jung, A., et al. (2012). The CitA/CitB two-component system regulating citrate fermentation in *Escherichia coli* and its relation to the DcuS/DcuR system in vivo. *Journal of Bacteriology*, *194*, 636–645.

Scholten, M., & Tommassen, J. (1993). Topology of the PhoR protein of *Escherichia coli* and functional analysis of internal deletion mutants. *Molecular Microbiology*, *8*, 269–275.

Schwöppe, C., Winkler, H. H., & Neuhaus, H. E. (2002). Properties of the glucose-6-phosphate transporter from *Chlamydia pneumonia* (HPTcp) and the glucose-6-phosphate sensor from *Escherichia coli* (UhpC). *Journal of Bacteriology*, *184*, 2108–2115.

Schwöppe, C., Winkler, H. H., & Neuhaus, H. E. (2003). Connection of transport and sensing by UhpC, the sensor for external glucose-6-phosphate in *Escherichia coli*. *European Journal of Biochemistry*, *270*, 1450–1457.

Sevvana, M., Vijayan, V., Zweckstetter, M., Reinelt, S., Madden, D., Herbst Irmer, R., et al. (2008). A ligand-induced switch in the periplasmic domain of sensor histidine kinase CitA. *Journal of Molecular Biology*, *377*, 512–523.

Shimoda, N., Toyoda-Yamamoto, A., Nagamine, J., Usami, S., Katayama, M., Sakagami, Y., et al. (1990). Control of expression of *Agrobacterium vir* genes by synergistic actions of phenolic signal molecules and monosaccharides. *Proceedings of the National Academy of Sciences of the United States of America*, *87*, 6684–6688.

Six, S., Andrews, S. C., Unden, G., & Guest, J. R. (1994). *Escherichia coli* possesses two homologous anaerobic C_4-dicarboxylate membrane transporters (DcuA and DcuB) distinct from the aerobic dicarboxylate transport system (Dct). *Journal of Bacteriology*, *176*, 6470–6478.

Slotboom, D. J., Konings, W. N., & Lokema, J. S. (1999). Structural features of the glutamate transporter family. *Microbiology and Molecular Biology Reviews*, *63*, 293–307.

Sommer, E., Koler, M., Frank, V., Sourjik, V., & Vaknin, A. (2013). The sensory histidine kinases TorS and EvgS tend to form clusters in *Escherichia coli* cells. *PLoS One*, *8*. e77708.

Sourjik, V., & Berg, H. C. (2000). Localization of components of the chemotaxis machinery of *Escherichia coli* using fluorescent protein fusions. *Molecular Microbiology*, *37*, 740–751.

Steinmetz, P. A., Wörner, S., & Unden, G. (2014). Differentiation of DctA and DcuS function in the DctA/DcuS sensor complex of *Escherichia coli*: Function of DctA as an activity switch and of DcuS as the C_4-dicarboxylate sensor. *Molecular Microbiology, 94*, 218–229.

Tetsch, L., & Jung, K. (2009). The regulatory interplay between membrane-integrated sensors and transport proteins in bacteria. *Molecular Microbiology, 73*, 982–991.

Tetsch, L., Koller, C., Haneburger, I., & Jung, K. (2008). The membrane-integrated transcriptional activator CadC of *Escherichia coli* senses lysine indirectly via the interaction with the lysine permease LysP. *Molecular Microbiology, 67*, 570–583.

Unden, G., & Kleefeld, A. (2004). C_4-Dicarboxylate degradation in aerobic and anaerobic growth. Module 3.4.5. in: R. Curtiss, III (Editor in Chief), *EcoSal—Escherichia coli and Salmonella: Cellular and molecular biology*. Washington, D.C.: ASM Press (Online) http://www.ecosal.org.

Valentini, M., Storelli, N., & Lapouge, K. (2011). Identification of C_4-dicarboxylate transport systems in *Pseudomonas aeruginosa* PAO1. *Journal of Bacteriology, 193*, 4307–4316.

Wang, L., Grau, R., Perego, M., & Hoch, J. A. (1997). A novel histidine kinase inhibitor regulating development in *Bacillus subtilis*. *Genes and Development, 11*, 2569–2579.

Watson, R. J. (1990). Analysis of C_4-dicarboxylate transport genes of *Rhizobium meliloti*: Nucleotide sequence and deduced products of *dctA*, *dctB* and *dctD*. *Molecular Plant-Microbe Interactions, 3*, 174–181.

Watt, P. M., Heinrich, T. K., & Thomas, W. R. (2006). Protein silencing with phylomers: A new tool for target validation and generating lead biologicals targeting protein interactions. *Expert Opinion on Drug Discovery, 1*, 491–502.

Witan, J., Bauer, J., Wittig, I., Steinmetz, P. A., Erker, W., & Unden, G. (2012). Interaction of the *Escherichia coli* transporter DctA with the sensor kinase DcuS: Presence of functional DctA/DcuS sensor units. *Molecular Microbiology, 85*, 846–861.

Witan, J., Monzel, C., Scheu, P., & Unden, G. (2012). The sensor kinase DcuS of *Escherichia coli*: Two stimulus input sites and a merged signal pathway in the DctA/DcuS sensor unit. *Biological Chemistry, 393*, 1291–1297.

Yarosh, O. K., Charles, T. C., & Finan, T. M. (1989). Analysis of C_4-dicarboxylate transport genes in *Rhizobium meliloti*. *Molecular Microbiology, 3*, 813–823.

Yernool, D., Boudker, O., Jin, Y., & Gouaux, E. (2004). Structure of a glutamate transporter homologue from *Pyrococcus horikoshii*. *Nature, 431*, 811–818.

Yurgel, S. N., & Kahn, M. L. (2004). Dicarboxylate transport by rhizobia. *FEMS Microbiology Reviews, 28*, 489–501.

Zientz, E., Bongaerts, J., & Unden, G. (1998). Fumarate regulation of gene expression in *Escherichia coli* by the DcuSR (*dcuSR* genes) two-component regulatory system. *Journal of Bacteriology, 180*, 5421–5425.

CHAPTER FIVE

Pivotal Role of Iron in the Regulation of Cyanobacterial Electron Transport

A. González, E. Sevilla, M.T. Bes, M.L. Peleato, M.F. Fillat[1]
University of Zaragoza, Zaragoza, Spain
[1]Corresponding author: e-mail address: fillat@unizar.es

Contents

Abstract

Iron-containing metalloproteins are the main cornerstones for efficient electron transport in biological systems. The abundance and diversity of iron-dependent proteins in

Advances in Microbial Physiology, Volume 68
ISSN 0065-2911
http://dx.doi.org/10.1016/bs.ampbs.2016.02.005

cyanobacteria makes those organisms highly dependent of this micronutrient. To cope with iron imbalance, cyanobacteria have developed a survey of adaptation strategies that are strongly related to the regulation of photosynthesis, nitrogen metabolism and other central electron transfer pathways. Furthermore, either in its ferrous form or as a component of the haem group, iron plays a crucial role as regulatory signalling molecule that directly or indirectly modulates the composition and efficiency of cyanobacterial redox reactions. We present here the major mechanism used by cyanobacteria to couple iron homeostasis to the regulation of electron transport, making special emphasis in processes specific in those organisms.

1. INTRODUCTION

Iron is an essential micronutrient for nearly all living beings with the exceptions of some species of lactobacilli, *Borrelia* and *Treponema pallidum*, the later being an obligate intracellular parasite that is entirely dependent on the iron-dependent metabolic processes of their hosts (Andrews, Robinson, & Rodriguez-Quinones, 2003). Iron is present in haem proteins, iron–sulphur cluster proteins, and diiron and mononuclear enzymes, among others. The reduction of ferric iron (Fe^{3+}) to the ferrous state (Fe^{2+}) requires a high redox potential $\left(E_0' = +770\,\text{mV}\right)$. However, in the variety of iron-containing proteins so far characterized, iron displays a wide range of redox potentials ranging from the most oxidizing of oxyhaemoglobin $\left(E_0' = +330\,\text{mV}\right)$ to the reducing one of ferric enterobactin $\left(E_0' = -700\,\text{mV}\right)$. This feature confers iron the ability to participate in a wide range of cellular redox processes, being the most ubiquitous metal present in the different cofactors of the proteins involved in the electron transport chains. On the other hand, iron contributes to the generation of deleterious reactive oxygen species through the Fenton reaction. Therefore, in order to keep cell performance, iron homeostasis must be strictly controlled in all types of cells. Moreover, iron and some iron-containing cofactors, such as haem, have emerged as important signalling molecules involved in the modulation of transcription factors or the activation of regulatory RNAs among other processes (Faller, Matsunaga, Yin, Loo, & Guo, 2007; Girvan & Munro, 2013; Hernández-Prieto et al., 2012; Oglesby-Sherrouse & Murphy, 2013; Pellicer et al., 2012). Thus, despite the importance of iron metabolism in the environment, our knowledge of its physiology and biochemistry is still very limited. In particular, oxygenic photosynthesis exerts unique stresses on photosynthetic organisms. The photosynthetic apparatus is composed of a number of membrane-embedded protein supercomplexes that contain many cofactors. Among them are Fe cofactors such as Fe–S clusters, cytochromes and nonhaem Fe. As a result, the demand for this metal far exceeds that of other, non-photosynthetic organisms. The Fe quota of *Synechocystis* sp. PCC 6803 cells is one order of magnitude higher than that of the nonphotosynthetic bacterium *Escherichia coli* (Keren, Aurora, & Pakrasi, 2004; Shcolnick & Keren, 2006). In the case of diazothropic cyanobacteria, nitrogen fixation is even more iron demanding, probably requiring >200 μmol Fe/mol C (Morel & Price, 1991). Furthermore, iron is a major cofactor in other cyanobacterial electron transport carriers such as those involved in oxidative phosphorylation and nitrate assimilation. Notably, most metalloproteins involved in all those routes contain iron–sulphur clusters that are sensitive to reactive oxygen species generated in the Mehler reaction as by-products of the photosynthetic metabolism, namely, ferredoxins and peroxiredoxins (Latifi, Ruiz, & Zhang, 2009). In addition, the nitrogenase complex has proved to be extremely unstable in the presence of oxygen evolved by photoactive PSII (Fay, 1992). Consequently, cyanobacterial iron homeostasis is coordinately regulated with the oxidative stress response and nitrogen metabolism through an intricate regulatory network in which the ferric uptake regulator (FurA) protein plays a major role (González, Angarica, Sancho, & Fillat, 2014; González, Bes, Peleato, & Fillat, 2011; González, Valladares, Peleato, & Fillat, 2013; López-Gomollón, Hernández, Pellicer, et al., 2007). This network involves among many others a source of key genes in the photosynthetic process, allowing an optimal functionality of the cell. Cyanobacterial responses to iron unbalance include cell retrenchment, as well as the adaptation of photosynthetic antennae and the substitution of ferredoxin I for flavodoxin in the electron transport chain involved in $NADP^+$. In the case of filamentous nitrogen-fixing cyanobacteria, the array of responses to iron deprivation includes transcriptional activation of *nifH*, coding for the nitrogenase Fe protein or dinitrogenase reductase (Razquin, Schmitz, Fillat, Peleato, & Bohme, 1994). However, a fully functional nitrogenase complex is only accomplished in mature heterocysts, the specialized cells that differentiate and develop as response to nitrogen deprivation (Flores & Herrero, 2010).

This chapter deals with the pivotal role of iron in cyanobacterial metabolism as component of most metalloproteins that conform the major electron transport processes but also as regulatory signalling molecule that directly or indirectly modulates the composition and efficiency of cyanobacterial redox reactions.

2. A SURVEY OF DIFFERENT ROLES OF IRON-CONTAINING PROTEINS IN THE PHOTOSYNTHETIC PROCESS

Iron is the most common transition metal in living organisms. The biological use of this essential micronutrient is as an enzyme cofactor, predominantly in electron transfer and catalysis. The main forms of iron cofactor are, in order of decreasing abundance, iron–sulphur clusters, haem and diiron or mononuclear iron, with a wide functional range (Dokmanic, Sikic, & Tomic, 2008). Before we can discuss the different roles of iron-containing proteins in photosynthesis, it is necessary to briefly go over the chemical properties of Fe, either on its own or in complex with sulphur (S) or organic structures.

Iron belongs to the group of transition metals, which can give rise to cations with an incompleted subshell of electrons. The more exposed electrons can form π bonds with organic ligands. The accessible oxidation states are Fe^{2+} with six *d* electrons and Fe^{3+} with five *d* electrons. Furthermore, because of the wide range of organic structures, ligands and protein folds containing iron, the redox potentials of iron cofactors can span an enormous range. Generally, Fe–S enzymes tend to have more negative redox potentials, and haem and other Fe proteins tend to have more positive redox potentials (Beinert, 2000).

In haem, Fe is coordinated in a tetrapyrrole ring with four nitrogens as ligands. Haem is a flat, rigid structure that is highly stable. The function of Fe in haem depends on the axial ligands perpendicular to the tetrapyrrole ring. If one axial ligand is unoccupied, Fe can bind oxygen and serve as an oxygen carrier, as in haemoglobin and myoglobin. In cytochrome P450 enzymes, the free site binds water to catalyse hydroxylations and other reactions. In cytochromes with an electron transfer function, the axial ligands are commonly two histidines or one histidine and one methionine. In the sirohaem proteins sulfite reductase and nitrite reductase, a cysteine ligand bridges to an Fe_4S_4 cluster. Haem can occur as free haem, but it is highly insoluble and its concentration inside the cell is extremely low. Haem is transported across membranes, but the current knowledge of haem transporters is fragmented and in some cases ambiguous (Hamza & Dailey, 2012).

In Fe–S clusters (Beinert, Holm, & Munck, 1997), $Fe^{2+/3+}$ is combined with S^{2-}, which is inserted in the sulfane (S^0) form and subsequently reduced. Once incorporated in a cluster, the S^{2-} does not undergo redox transitions, but the *d* electrons of Fe become delocalized. Therefore, the valency of the cluster as a whole is indicated using square brackets: $[2Fe–2S]^{2+} + e \leftrightarrow [2Fe–2S]^{+}$. The Fe atoms are usually bound by cysteinyl ligands of the protein. Other ligands also occur, such as histidine in the Rieske-type cluster, increasing the redox potential to +300 mV. Most common are rhombic Fe_2S_2 and cubane Fe_4S_4 clusters. Most Fe–S clusters have an electron transfer function (Beinert, 2000). A string of clusters with individual distances of less than 14 Å can transfer electrons across a protein complex, such as in photosystem I (PSI) and respiratory complex I. Fe–S clusters can also have catalytic functions. For example, the Fe_4S_4 cluster in aconitase acts as a Lewis acid: one of the nonliganded Fe atoms of the cluster binds the substrate citrate and removes a proton and a hydroxyl group, which is followed by rehydration of the substrate to form isocitrate. In radical *S*-adenosyl methionine enzymes such as biotin synthase and lipoate synthase, an Fe_4S_4 cluster is employed to stabilize an *S*-adenosyl radical, and a second cluster serves as an S donor. In helicases and DNA-binding proteins, Fe_4S_4 clusters appear to provide stability to a protein fold (Wu & Brosh, 2012).

In photosynthetic organisms, the primary light-driven reactions of photosynthesis occur in the thylakoid membranes and are mediated by two photochemical reaction centres, termed photosystem II (PSII) and PSI. The coupling of the two reaction centres in a linear electron transfer chain from water to the reductant NADPH forms the basis of the Z-scheme, which was proposed by Hill and Bendall in 1960 (Hill & Bendall, 1960) and it still prevails today (Nelson & Ben-Shem, 2004). In plants and green algae, both photosystems are composed of a core complex of proteins and pigments and are energetically linked to peripheral antenna pigment–protein complexes, the light-harvesting complexes (LHCs). The core complexes are embedded and specifically oriented within thylakoid membranes, which are folded upon themselves to create an inner-membrane space known as lumen and are surrounded on the outside by the stroma. In cyanobacteria and red algae, thylakoid membranes are located adjacent to the plasma membrane and the peripheral antennae appears as external structures on the stromal side forming the phycobilisome (PBS). The PBS is an extra-membrane supramolecular complex composed of many chromophore (phycobilin)-binding proteins (phycobiliproteins) and linker proteins (Sidler, 1994). PBS collects light energy of a wide range of wavelengths, funnels it to the central core and then transfers it to photosystems. Although phycobiliproteins are evolutionarily related to each other, the binding of different pycobilin pigments ensures the ability to collect light energy over a wide range of wavelengths, and then transfer this energy to chlorophyll. Phycobilins are especially efficient at absorbing red, orange, yellow and green light, wavelengths that are not efficiently absorbed by chlorophyll *a* (Brown, Houghton, & Vernon, 1990; Watanabe & Ikeuchi, 2013).

The initial step in photosynthesis is the absorption of light by chlorophylls and other pigments attached to proteins in the thylakoid membranes. Like cytochromes, chlorophylls consist of a porphyrin ring attached to a long hydrocarbon side chain. They differ from cytochromes (and haem) in containing a central Mg^{2+} ion (rather than Fe atom) and having an additional five-membered ring. The energy of the absorbed light is used to remove electrons from an unwilling donor (water, in green plants and cyanobacteria), forming dioxygen. In PSII, the initial photochemical event results in an electron transfer from the pair of core chlorophyll molecules (P680) to a primary quinone acceptor (Q_A). The oxidized P680 is reduced through electron donation from the PSII manganese cluster on the lumenal side of the membrane. Four such photochemical events drive the oxidation of two H_2O molecules, releasing one O_2 and four H^+ ions to the lumen. From Q_A, electrons are transferred to the secondary and mobile quinone (Q_B). Two electrons are transferred to Q_B, which is protonated from the stroma and released from PSII as plastoquinol (PQH_2) to migrate within the membrane and dock with the transmembrane cytochrome (Cyt) $b_6 f$ complex (Nelson & Ben-Shem, 2004). At Cyt $b_6 f$, the two protons are released into the lumen and one electron is transferred to a mobile Cyt b_{553}, which migrates in the lumen to the donor side of PSI. The second electron is transferred to another plastoquinone (PQ) molecule on Cyt $b_6 f$ and used to increase proton translocation through a pathway termed the Q cycle. The final steps of the electron transfer chain involve light absorption by PSI, photochemical conversion at or near its special pair (P700), and electron transfer through a series of bound intermediates to a mobile ferredoxin (Fd 1) that is replaced in most cyanobacteria by flavodoxin under iron-deficient conditions (Sandmann, Peleato, Fillat, Lázaro, & Gómez-Moreno, 1990). Ferredoxin or flavodoxin reduce $NADP^+$ as ultimate electron acceptor to form NADPH via ferredoxin-$NADP^+$ oxidoreductase (FNR). P700 is rereduced through electron donation from Cyt b_{553}. The electron transfer along the photosynthetic electron transport chain is accompanied by alkalization of stroma and acidification of lumen, thus generating the transthylakoid difference in electrochemical potentials of hydrogen ions which serves as the driving force for operation of the ATP synthase (Junge & Nelson, 2015). The net outcome of the photosynthetic electron transfer chain is that four photons utilized by PSII and four photons utilized by PSI yield one O_2, two NADPHs, and the accumulation of $12H^+$ ions in the lumen, which are subsequently used by the ATP synthase complex to form approximately three ATPs. The products of the light-induced stages of photosynthesis (ATP and NADPH) are used in reductive biosynthetic reactions of the Bassham–Benson–Calvin cycle (Bassham, Benson, & Calvin, 1950).

The consequences of iron deficiency in cyanobacterial photosynthesis have been studied for long time (Fillat, Peleato, Razquin, & Gómez-Moreno, 1994; Sandmann, 1985). Below a critical concentration of this nutrient, photosynthetic electron transport is depressed mainly due to a decrease in the synthesis of soluble cytochromes and the Fe–S centres of PSI and the cytochrome complex. Table 1 shows that nearly every

Table 1 Iron-Containing Proteins of the Photosynthetic Electron Transport Chain in Cyanobacteria

Complex	Fe-containing Cofactor	Number of Fe Atoms	Fe-binding Protein	Gene
PSI		12		
	1 F_x ([4Fe–4S])	4	Iron–sulphur protein	*psaA/B*
	1 F_A ([4Fe–4S])	4	Iron–sulphur protein	*psaC*
	1 F_B ([4Fe–4S])	4	Iron–sulphur protein	*psaC*
Cyt b_6f		6		
	3 haem	3	Cyt b_6	*petB*
	1 haem	1	Cyt f	*petA*
	1 [2Fe–2S]	2	Rieske iron–sulphur protein	*petC*
PSII		4		
	1 nonhaem iron	1	D1/D2	*psbA/D*
	2 haem	2	Cyt b_{559}	*psbE/F*
	1 haem	1	Cyt c_{550}	*psbV*
Cyt c_{553}	1 haem	1	Cyt c_{553}	*petJ*
Fd	1 [2Fe–2S]	2	Fd	*petF*

component of the photosynthetic electron transport chain is iron dependent. PSI represents the largest photosynthetic requirement for iron; each PSI trimer contains 36 Fe atoms in 9 [4Fe–4S] clusters (Jordan et al., 2001). Thus, more than 25% of the Fe quota in *Synechocystis* cells is in PSI alone (Keren et al., 2004).

Isolated cyanobacterial PSI exists as a trimer (relative molecular mass Mr $3 \times 356{,}000$); this form is also present in vivo. One monomer consists of at least 11 different protein subunits coordinating more than 100 cofactors comprising 96 chlorophylls, 22 carotenoids, 2 phylloquinones, 3 [4Fe–4S] clusters, 4 lipids and a putative Ca^{2+} ion. PSI captures light energy by a large internal antenna system and guides it to the core of the reaction centre with high efficiency. After primary charge separation initiated by excitation of the chlorophyll dimer P700, the electron passes along the electron transfer chain consisting of the spectroscopically identified cofactors A_0 (Chl*a*), A1 (phylloquinone) and the [4Fe–4S] clusters F_X, F_A and F_B (Brettel, 1997). At the stromal (cytoplasmic) side, the electron is donated by F_B to ferredoxin and thence transferred to $NADP^+$ reductase. PSI contains nine protein subunits featuring transmembrane α-helices (PsaA, PsaB, PsaF, PsaI, PsaJ, PsaK, PsaL, PsaM and PsaX) and three stromal subunits (PsaC, PsaD and PsaE). PsaC harbours the two [4Fe–4S] clusters F_A and F_B. It exhibits pseudo-twofold symmetry similar to bacterial 2[4Fe–4S] ferredoxins, but contains an insertion of 10 amino acids in the loop connecting the iron–sulphur cluster-binding motifs and extensions of the N and C termini by 2 and 14 amino acids, respectively. As the insertion extrudes as a large loop, it may be engaged in docking of ferredoxin. The long C terminus of PsaC interacts with PsaA/B/D and appears to be important for the proper assembly of PsaC into the PSI complex. F_X is coordinated to both PsaA and PsaB in strictly conserved loop segments A/B-hi containing two cysteines each, as proposed from the amino acid sequences. The arrangement of the clusters with F_A being closer to F_X than F_B suggests a sequence $F_X \rightarrow F_A \rightarrow F_B$ in electron transfer, in agreement with spectroscopic data (Brettel, 1997).

The Cyt b_6f is a heterooligomeric integral membrane–protein complex which provides the electronic connection between the PSII and PSI reaction centres (Baniulis, Yamashita, Zhang, Hasan, & Cramer, 2008; Kurisu, Zhang, Smith, & Cramer, 2003). Three-dimensional structures of the b_6f complex obtained from X-ray diffraction show that the b_6f complex contains eight polypeptide subunits with 13 transmembrane helices in each monomer of a functional dimer. Four of the eight subunits, PetA, B,

C and D are "large" (16–31 kDa), and contain or confine the redox prosthetic groups Cyt *f*, Cyt b_6, the Rieske iron–sulphur protein, and subunit IV, respectively. The remaining four small (3.3–4.1 kDa) hydrophobic subunits, PetG, L, M and N form a "fence" at the outside periphery of each monomer, with each small subunit containing one transmembrane helix. In total, the Cyt $b_6 f$ complex contains five Fe atoms: two haem prosthetic groups associated with Cyt b_6, one haem associated with Cyt *f*, and a [2Fe–2S] Rieske protein involved in electron transport to Cyt c_{553}.

PSII is a huge, multisubunit membrane–protein complex responsible for the photooxidation of water into dioxygen and the reduction of plastoquinone (Shen, 2015; Suga et al., 2015). It operates as a water:plastoquinone oxidoreductase where the transformation of light into electrochemical Gibbs energy occurs. PSII is composed of at least 20 protein subunits and about 100 cofactor molecules, including the antenna and reaction centre chlorophylls, pheophytins, cytochromes, carotenoids, quinones, lipids and the Mn_4CaO_5 cluster. The Mn_4CaO_5 cluster is the catalytic centre for water oxidation forming the oxygen evolving complex (OEC). The PSII reaction centre core consists of two homologous proteins, D1 (PsbA) and D2 (PsbD), and two further closely related chlorophyll (Chl)-containing proteins CP43 (PsbC) and CP47 (PsbB). The three extrinsic proteins PsbO, PsbU and PsbV form a cap over the catalytic site where oxygen evolution occurs, preventing access by reductants other than water, while some of the low-molecular-weight subunits are located on the peripheral of the CP43/D1/D2/CP47 cluster, where they probably help to stabilize the binding of Chl and β-carotene molecules contained within the complex. The exceptions to this are the PsbE and PsbF proteins, which provide histidine ligands for the haem of Cyt b_{559}, and the PsbL, PsbM and PsbT proteins located at the monomer–monomer interface, where they possibly play a role in stabilizing the dimeric nature of the PSII complex. All of the small subunits have a single transmembrane helix except for PsbZ, which has two (Barber, 2008). PSII employs a nonhaem Fe^{2+} ion for electron transport between Q_A and Q_B (Muh & Zouni, 2013). Besides, PSII contains two haems associated with the reaction centre core Cyt b_{559}, and has one additional haem in cyanobacteria associated with an extrinsic reaction centre Cyt c_{550} (four Fe atoms total).

Beyond these three membrane–protein complexes (PSI, PSII and Cyt $b_6 f$), some mobile electron carriers also depend of iron cofactor to its proper function (Table 1). Thus, Cyt c_{553} has one haem, and ferredoxin contains a [2Fe–2S] centre in their structures. Ferredoxins are small, mostly acidic soluble proteins, which possess a highly negative redox potential and use iron–sulphur clusters to act as electron distributors in various metabolic pathways (Cassier-Chauvat & Chauvat, 2014). Like other photosynthetic organisms, cyanobacteria contain several ferredoxin proteins that can be classified according to the nature of their iron–sulphur centre ([2Fe–2S], [3Fe–4S] or [4Fe–4S]) and the organisms in which they were isolated for the first time. The most abundant ferredoxin, designated as Fd1, is recognized primarily as the protein that mediates electron transfer from iron–sulphur centres of PSI to FNR, though Fd1 is also involved in other redox processes such as cyclic photophosphorylation, nitrogen assimilation, biosynthesis of glutamate and chlorophyll, etc. As an example, *Synechocystis* sp. PCC 6803 expresses nine Fd species: Fd1–6 which contain a [2Fe–2S] cluster, Fd7 with a [4Fe–4S] cluster, Fd8 with [3Fe–4S] and [4Fe–4S] clusters and Fd9 containing two [4Fe–4S] clusters (Cassier-Chauvat & Chauvat, 2014).

Furthermore, contribution of iron to photosynthesis not only implicates the components of the electron transport chain themselves, but also other iron-containing proteins like flavodiiron proteins (FDP) (Allahverdiyeva et al., 2013; Aro et al., 2005; Bersanini et al., 2014; Zhang, Allahverdiyeva, Eisenhut, & Aro, 2009). FDP, also known as flavoproteins, are modular enzymes widely present in prokaryotes. All FDP share two conserved structural domains: the N-terminal metallo-β-lactamase-like domain harbouring a nonhaem diiron centre where O_2 and/or NO reduction take place, and the C-terminal flavodoxin-like domain containing a flavin mononucleotide (FMN) moiety (Vicente, Carrondo, Teixeira, & Frazao, 2008). In addition to the common sequence core, some FDP also have C-terminal extensions. Based on these extensions, FDP can be grouped into four classes: A, B, C and D. Particularly, the class C appears to be specific to cyanobacteria (Aro et al., 2005; Vicente et al., 2008). In this class, an additional NADPH: flavin reductase-like domain makes it possible for $NADP^+$ to be directly used as an electron donor. Analysis of sequenced cyanobacterial genomes reveals the presence of several genes encoding distinct FDP in the same organism. Thus, the genomes of the unicellular *Synechocystis* sp. PCC 6803 and the filamentous *Anabaena* sp. PCC 7120 contain four FDP genes, coding for proteins denoted as Flv1, Flv2, Flv3 and Flv4. Several analyses have shown that Flv1 and Flv3 enable cyanobacterial growth and photosynthesis under fluctuating light. Flv1 and Flv3 function as NADPH: oxygen oxidoreductases donating electrons directly to O_2 without production of reactive oxygen species (ROS). Thus, Flv1/Flv3 heterodimers provide protection for PSI under fluctuating growth light maintaining the redox balance on the electron transport chain (Allahverdiyeva et al., 2013; Aro et al., 2005). Flv2/Flv4 heterodimers participate in an alternative electron transfer pathway from PSII, which dissipates excitation pressure under high light conditions by channelling up to 30% of PSII-originated electrons, contributing to photoprotection of PSII (Aro et al., 2005; Bersanini et al., 2014; Zhang et al., 2009).

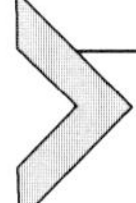

3. CONTROL OF IRON HOMEOSTASIS IN CYANOBACTERIA

Because of the importance of iron to a proper functioning of photosynthesis and other essential physiological processes, but also its potential deleterious effect due to the Fenton reaction (Latifi et al., 2009), iron homeostasis must be tightly regulated by all photosynthetic organisms. Since iron is scarcely soluble in aqueous environments at neutral pH, cyanobacteria have evolved different strategies to efficiently scavenge (Goldman, Lammers, Berman, & Sanders-Loehr, 1983; Jeanjean et al., 2008), incorporate (Mirus, Strauss, Nicolaisen, von Haeseler, & Schleiff, 2009; Nicolaisen et al., 2008), and store this essential micronutrient in the cell (Keren et al., 2004; Shcolnick, Summerfield, Reytman, Sherman, & Keren, 2009). As most Gram-negative and several Gram-positive bacteria, the effective balance between iron acquisition and protection against oxidative stress is controlled in cyanobacteria mainly by a global transcriptional regulator known as Fur, which stands for *f*erric *u*ptake *r*egulator (Andrews et al., 2003). Fur typically acts as a transcriptional repressor, which senses intracellular free iron and modulates the transcription of target genes in response to iron availability. This is accomplished by binding Fur–Fe^{2+} complexes to *cis*-acting regulatory elements known as Fur boxes, located in the promoter regions of iron-responsive genes (Escolar, Pérez-Martín, & de Lorenzo, 1999). Under iron-restricted conditions, the metal corepressor is released and the repressor becomes inactive, allowing the transcription of target genes. More recently, Fur-mediated direct and indirect activation of transcription involving a variety of mechanisms has been established (Masse & Gottesman, 2002; Teixido, Carrasco, Alonso, Barbe, & Campoy, 2011; Yu & Genco, 2012).

As a global regulator, Fur not only controls the expression of iron acquisition and storage systems, but also a plethora of genes and operons belonging to a broad range of functional categories (Deng et al., 2012; Gilbreath et al., 2012; Lin et al., 2011; Teixido et al., 2011). All these genes constitute the Fur regulon. In the filamentous nitrogen-fixing cyanobacterium *Anabaena* sp. PCC 7120, FurA is the master regulator of iron homeostasis and modulates both, directly and indirectly, the expression of several genes involved in photosynthesis as response to iron availability (González et al., 2014; González, Bes, Barja, Peleato, & Fillat, 2010; González et al., 2011; González, Bes, Valladares, Peleato, & Fillat, 2012). Thus, FurA directly regulates the transcription of genes *isiA* (Leonhardt & Straus, 1994) and *isiB* (Bes, Hernández, Peleato, & Fillat, 2001), encoding the iron-stress-induced protein IsiA (Kouril et al., 2005) and flavodoxin (Ferreira & Strauss, 1994; Setif, 2001), respectively. Likewise, FurA modulates the transcription of the photosystems subunits PsbA (González et al., 2010), PsaL and PsbZ (López-Gomollón, Hernández, Pellicer, et al., 2007), while predicted FurA boxes were observed in the promoter region of gene *psaK* (González et al., 2014). Other photosynthesis-related genes such as *all0865* encoding the CO_2 concentrating mechanism protein CcmM (González et al., 2011) and *rbcL* (López-Gomollón, Hernández, Pellicer, et al., 2007), coding for the large subunit of enzyme ribulose 1,5-bisphosphate carboxylase/oxygenase (RuBisCO) (Badger & Price, 2003), are also FurA direct targets.

In addition, FurA appears to modulate indirectly the amount of active iron-containing proteins involved in the photosynthetic processes since this metalloregulator controls the synthesis of iron cofactors. Thus, FurA modulates the expression of at least five genes coding for enzymes involved in the tetrapyrrole biosynthesis pathway, mediating at early stages by controlling porphobilinogen synthase (*hemB*) and porphobilinogen deaminase (*hemC*) levels, but also at the final steps of haem biosynthesis branch, controlling the expression of protoporphyrinogen oxidase (*hemK*) and ferrochelatase (*hemH*). In addition, FurA appears to regulate the haem-degrading enzyme haem oxygenase 1 (*ho1*), the first step of the phycobilin synthesis branch (González et al., 2012). Thereby FurA modulates not only the amount of haem as response to iron availability, but also influences the synthesis of cyanobacterial photosynthetic pigments like chlorophyll and phycobilins (Beale, 1994). In addition, recent studies have shown that FurA regulates transcription of *sufS* (our unpublished results), coding for the enzyme cysteine desulphurase implicated in the biosynthesis of Fe–S clusters (Wang et al., 2005).

Another transcriptional regulator involved in iron homeostasis, PfsR (Cheng & He, 2014), has been recently described in *Synechocystis* PCC 6803. The results of several analyses suggest a role of PfsR in transcriptional regulation of iron-responsive genes involved in photosynthesis. Compared to the wild type, the *pfsR* deletion mutant displayed stronger tolerance to iron limitation and accumulated significantly more chlorophyll *a*, carotenoid and phycocyanin under iron-limiting conditions. The mutant also maintained higher levels of major photosystem proteins including PsaC, PsaD, PsbA and PsbB than the wild type after iron deprivation. In addition, the activities of PSI and PSII were much higher in a *pfsR* deletion mutant than in wild-type cells under iron-limiting conditions (Cheng & He, 2014).

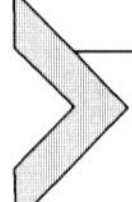

4. IRON-RESPONSIVE PROTEINS AND PHOTOSYSTEM PERFORMANCE

Oxygen-producing photosynthesis must have evolved before the pervasive oxidation of the atmosphere around 2.4 billion years ago (Great Oxidation Event), and there are evidences of generated oxygen at least 3 billion years ago (Kaufman, 2014). The consequences of the changes affected the survival of the producers itself in many aspects. Among other aspects, oxygen in the atmosphere jeopardizes the availability of iron for the structure or synthesis of many photosynthetic elements. Moreover, the presence of oxygen put in danger the cells due to Fenton reactions, and also photooxidation processes endanger the oxygenic photosynthesis machinery. In fact, Ionescu, Voss, Oren, Hess, and Muro-Pastor (2010) proposed that oxygenic photosynthesis provides a protection mechanism for cyanobacteria against iron-encrustation in environments with high Fe^{2+} concentrations. For those reasons, cyanobacteria developed many strategies to cope with the new environment. Among those mechanisms, plasticity to adapt to an increasing iron limitation, iron uptake strategies to scavenge iron from iron-scarce environments, synthesis of new proteins capable of replace iron-proteins, and many other responses geared to maintain the photosynthetic efficiency (Ferreira & Strauss, 1994) and avoid possible photooxidation events. In fact, the iron-dependent synthesis of chlorophyll is a mechanism that harmonizes the configuration of the photosystems with iron availability.

4.1 Plasticity of the Photosynthetic Electron Transport Chains to Adapt to Iron Availability

It is well known that the quality and intensity of light determine how photosynthetic membranes are organized and their operation. On the other hand, light and iron are specifically linked as the ability to convert photons to chemical energy, due to the requirements of iron-rich photosystems and other iron-dependent components as chlorophyll. In this way, there is a close relationship between the photosynthetic capacity and the iron uptake, described to be light dependent in several cyanobacteria, as *Microcystis aeruginosa* (Alexova et al., 2011; Utkilen & Gjolme, 1995). An interesting case is the cyanobacterium *Fremyella diplosiphon* that undergoes complementary chromatic adaptation. During this process, phycobiliprotein composition of light-harvesting antennae from *Fremyella* is altered in response to green and red light to optimize the use of light energy for photosynthesis (Pattanaik, Busch, Hu, Chen, & Montgomery, 2014). Moreover, iron limitation triggers light-regulated responses that differ depending of the quality of light. Significant reductions in growth and pigment levels, alterations in iron-associated proteins and accumulation of reactive oxygen species were observed under red light, whereas green light-grown cells exhibited partial resistance to iron limitation.

Hence, as a general response in cyanobacteria, in case of shortage of the metal, the configuration of the membranes and therefore their function is limited by iron availability. Thus, the need for finely tuned regulation to adapt to environment, leads to mechanisms that regulated the interplay between iron homeostasis and light responses. Dynamic changes in the photosynthetic membranes expected to modify the energy distribution balance have been described: under iron stress, *Synechococcus* sp. PCC 7942 exhibits a dynamic ability to uncouple PSII and PSI electron transport, showing enhanced cyclic electron transport (Ivanov et al., 2000). This reorganization mimics responses induced to maintain the functional balance between PSII and PSI upon changes in light quality (Allen, Mullineaux, Sanders, & Melis, 1989; Mekala, Suorsa, Rantala, Aro, & Tikkanen, 2015) and quantity (Mekala et al., 2015), and to avoid photooxidation events. The lack of iron affects the photosynthetic machinery, and there is a light-induced oxidative stress associated with iron limitation, inducing cell changes similar to the chromatic adaptation observed in the cyanobacterium *F. diplosiphon* (Pattanaik et al., 2014).

Among the first notable responses of cyanobacteria to iron stress are the substitution of Cyt c_{553} with the copper-dependent plastocyanin and the partial replacement of Fd1 by flavodoxin (Ferreira & Strauss, 1994). Although flavodoxin expression varies between species, it has been traditionally used as a molecular indicative of iron stress in field studies (Geiss et al., 2001). Several analyses suggest that Fd1 is regulated by iron at the level of differential mRNA stability (Bovy, de Kruif, de Vrieze, Borrias, & Weisbeek, 1993; Bovy, de Vrieze, et al., 1993). Other Fd proteins in cyanobacteria like Fd4, Fd5 and Fd8 are also downregulated under iron-stress conditions (Cassier-Chauvat & Chauvat, 2014). Likewise, iron starvation led to significant changes in gene expression of major membrane components of the photosynthetic electron chain (Nodop et al., 2008; Shi, Sun, & Falkowski, 2007; Singh, McIntyre, & Sherman, 2003). Thus, *Synechocystis* sp. PCC 6803 substantially downregulated the expression of *psaA* and *psaC* from PSI; *psbB*, *psbT* and *psbH* from PSII, as well as *petG* from Cyt $b_6 f$ complex after 24 h under iron deprivation (Singh et al., 2003). If iron limitation persists, overall photosynthetic capacity decreases and major stoichiometric changes occur that are reflective of unsupplied iron requirements. The more iron is required, the more the system is suppressed by iron stress. For example, moderate iron stress decreases the electron transport capacity by 83% for PSI, 69% for Cyt $b_6 f$ and 46% for PSII in the cyanobacterium *Aphanocapsa* (*Synechocystis* sp. strain PCC 6714) (Sandmann, 1985). This preferential downregulation of PSI relative to PSII and Cyt $b_6 f$ appears as a common iron-stress response across both prokaryotic and eukaryotic photoautotrophs (Fraser et al., 2013; Strzepek & Harrison, 2004).

4.2 Fitness of Photosynthetic Machinery Under Iron-Limited Environments: Advantage of isiA Gene

Iron deficiency causes oxidative stress, a process not observed in heterotrophic bacteria (Latifi, Jeanjean, Lemeille, Havaux, & Zhang, 2005). Light incident on deficient photosystems results on photooxidative damage, with potential disassemble of the photosynthetic machinery and in any case, affecting the photosynthetic efficiency and lowering the cell performance. One of the most relevant adaptive processes showing the plasticity of photosynthetic membranes to overcome iron deficiency is the expression of the *isi* (iron stress induced) genes that in some cyanobacteria are arranged as a dicistronic operon (Nakao et al., 2010). The product of *isiA* was described in iron-starved *Anacystis nidulans*, later renamed as *Synecchococcus* PCC 7942 (Laudenbach & Straus, 1988), as an induced chlorophyll-binding protein (Sherman & Sherman, 1983). The IsiA protein was previously named CP43′ due its similarity to CP43, located at the PSII (Leonhardt & Straus, 1992). Initially it was proposed to play a role as an additional LHC (Riethman & Sherman, 1988), and over the years, it has been suggested that it plays several functions, summarized by (Junge & Nelson, 2015): (i) IsiA is a chlorophyll storage protein for fast recovery of cyanobacteria after stress (Tetenkin, Golitsin, & Gulyaev, 1998), (ii) IsiA acts as an excitation energy dissipater, protecting photosystems from photoinhibition (Park, Sandstrom, Gustafsson, & Oquist, 1999), (iii) IsiA servers as a LHC potentially for both photosystems (Pakrasi, Goldenberg, & Sherman, 1985; Riethman & Sherman, 1988) and (iv) IsiA replaces CP43 in PSII and permits a cyclic electron transfer pathway involving PSII and the cytochrome $b_6 f$ complex (De Las Rivas & Barber, 2004; Nogi & Miki, 2001).

IsiA binds an average of 16 chlorophyll *a* molecules and 4 carotenoids (β-carotene, zeaxanthin and echinenone in a stoichiometry 2:1:1) (Berera, van Stokkum, Kennis, van Grondelle, & Dekker, 2010) and has been described associated either with PSI and PSII (Sun & Golbeck, 2015). IsiA forms supercomplexes in the thylakoids leading to a reorganization of photosynthetic components with 18 IsiA subunits surrounding PSI trimers (Bibby, Nield, & Barber, 2001; Burnap, Troyan, & Sherman, 1993; Kouril et al., 2005; Latifi et al., 2005; Pakrasi et al., 1985). IsiA complexes increase the LHC size by about 70% (Bibby et al., 2001). Two novel IsiA supercomplexes in addition to the initially described were observed by sucrose gradient ultracentrifugation in *Synechocystis* PCC 6803 under extensive iron starvation (Wang, Hall, Al-Adami, & He, 2010). One of them was identified as an IsiA–PSI–PSII supercomplex, while the other was assigned as an IsiA–PSI supercomplex (Wang et al., 2010). The initial observations were confirmed by different authors and it seems that the oligomeric complexes of IsiA around PSI increase the absorptional cross-section of PSI, serving as a LHC for the PSI complexes during iron starvation (Melkozernov, Barber, & Blankenship, 2006; Yeremenko et al., 2004), with enhanced PSI electron throughput (Sun & Golbeck, 2015). In addition, the complexes seem to act quenching chlorophyll singlet excited states to cope with the deleterious effects of excess of light for the iron-deficient membranes, protecting the photosynthetic machinery (Berera et al., 2010). It is noticeable that the *isiA* gene is also transcribed under oxidative stress (Yousef, Pistorius, & Michel, 2003) and other environmental stresses, including salt stress, heat stress, and high or limited light conditions (Havaux et al., 2005; Jeanjean et al., 2003; Vinnemeier & Hagemann, 1999; Vinnemeier, Kunert, & Hagemann, 1998). A Fur binding site in the promoter region of *isiAB* in *Synechocystis* PCC 6803 has been identified (Vinnemeier et al., 1998), though repression by Fur seems not to be the only mechanism for the control of the iron-inducible *isiAB* operon in the cyanobacterium *Synechocystis* PCC sp. PCC 6803 (Kunert, Vinnemeier, Erdmann, & Hagemann, 2003). Additionally, IsrR (iron stress-repressed RNA), a *cis*-encoded antisense RNA transcribed from the *isiA* noncoding strand has been described to be involved in such regulation (Duhring, Axmann, Hess, & Wilde, 2006).

It is interesting to notice that *isiA* is not present in all cyanobacteria, and no homologs of *isiA* were found in plants. In fact, the occurrence of *isiA* in cyanobacteria found in the iron-limited, high-nutrient low-chlorophyll regions of the equatorial Pacific led to the suggestion that the presence of this gene can be a natural biomarker for iron limitation in oceanic environments (Bibby, Zhang, & Chen, 2009).

4.3 *isiB* Encoding Flavodoxin Permits Efficient Electron Transport in Iron-Deficient Environments

Usually downstream of *isiA*, the *isiB* gene encodes for the small FMN-flavoprotein flavodoxin. In this case, flavodoxin is not exclusive of cyanobacteria, and may be also present in heterotrophic bacteria as well as in a few cases of algae (Peleato, Ayora, Inda, & Gomez-Moreno, 1994). The versatile electron transfer shuttle ferredoxin, an iron–sulphur protein, is particularly affected when iron is scarce, and its downregulation under adverse conditions severely compromises survival of phototrophs (Lodeyro, Ceccoli, Pierella Karlusich, & Carrillo, 2012). Replacement of ferredoxin by flavodoxin is a common strategy employed by cyanobacteria to overcome environmental adversities and under iron deficiency, flavodoxin replaces ferredoxin in many electron transfer chains, including $NADP^+$ photoreduction (Lodeyro et al., 2012; Sandmann et al., 1990; Vigara, Inda, Vega, Gomez-Moreno, & Peleato, 1998). An exhaustive list of ferredoxin and flavodoxin-dependent reactions and redox partners in plastids and cyanobacteria is provided in (Lodeyro et al., 2012). Surprisingly, flavodoxin is not able to functionally replace heterocyst ferredoxin, even though electron transfer chain to nitrogenase is also an iron-dependent process (Razquin et al., 1995). Ferredoxin and flavodoxin are isofunctional, but they do not share any significant similarity in primary, secondary or tertiary structures, and yet they can interact productively with the same redox partners and exhibit kinetic constants in the same range (Lodeyro et al., 2012; Vigara et al., 1998). Notably, flavodoxin resulted more efficient than ferredoxin I in $NADP^+$ photoreduction (Fig. 1) (Razquin et al., 1995), in contrast with its lower activity as electron donor in inorganic nitrogen assimilation (Vigara et al., 1998).

Both ferredoxin and flavodoxin proteins evolved in the anaerobic environment preceding the appearance of oxygenic photosynthesis. The new oxygen-rich atmosphere proved to compromise ferredoxin synthesis and many

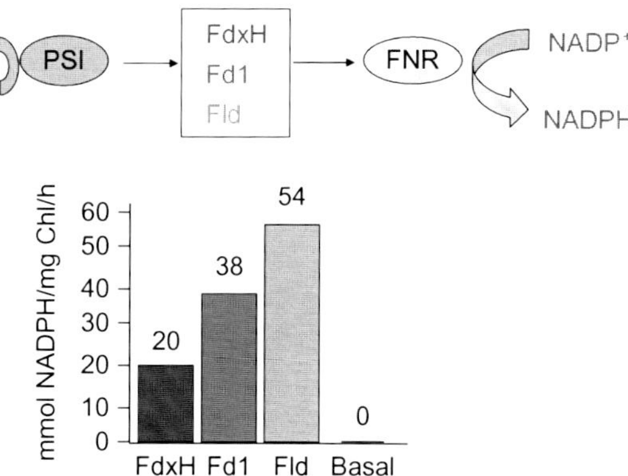

Fig. 1 Ability of equimolar amounts of different electron carriers from *Anabaena* sp. PCC 7119 to support $NADP^+$ photoreduction using cyanobacterial thylakoids. *Fd1*, vegetative cell ferredoxin; *FdxH*, heterocyst ferredoxin; *Fld*, flavodoxin (Razquin et al., 1995). (See the colour plate.)

microorganisms used the stress responsive flavodoxin expression to replace ferredoxin. Phylogenetic analyses reveal that the evolutionary history of flavodoxin implies several horizontal gene transfer events between distant organisms, including *Eukarya*, *Bacteria* and *Archaea* (Karlusich, Lodeyro, & Carrillo, 2014; Lodeyro et al., 2012). However, the flavodoxin gene was lost in many cases in cyanobacteria, and in most of the algae and all the plants (Karlusich et al., 2014) due to environmental selection pressure concerning iron availability (Pierella Karlusich, Ceccoli, Grana, Romero, & Carrillo, 2015).

Flavodoxin expression is also induced under an ample range of several environmental stresses that result in ferredoxin downregulation, especially by oxidative stress (Fulda & Hagemann, 1995; Laudenbach & Straus, 1988; Lodeyro et al., 2012). In relation to photosynthesis, flavodoxin behaved as an alternative intermediate for the photosynthetic electron transfer chain in vivo, acting, as ferredoxin does, as the main distributor of the reducing power (Lodeyro et al., 2012). Under iron limitation, reduced flavodoxin also signals for the whole cell the presence of an active photosynthetic electron transfer chain. Flavodoxin fuels a key process, the thioredoxin electron transfer pathway. Reduced thioredoxin, via thioredoxin reductase, regenerates through reduction of their cysteine residues the active forms of many target enzymes like peroxiredoxins, Calvin cycle enzymes and $NADP^+$-malate dehydrogenase, among others. Light signal prevents that the Calvin cycle work in dark conditions, and also implements mechanisms aimed to harmonize the pool of redox power and its utilization. Flavodoxin allows this key process still to proceed under iron-deficient conditions. Iron deficiency and other environmental stresses in photosynthetic organisms share a common feature: oxidative stress, resulting in damage to the cell structures and modification of signalling responses (Latifi et al., 2009). Flavodoxin could have initially evolved as an oxidative stress defence, and later been recruited to play the ferredoxin role in the electron transfer chains when the synthesis of ferredoxin was diminished. It is a fact that the presence of flavodoxin increases the tolerance of cells, even not photosynthetic ones, exposed to several oxidants and stresses, though the mechanism is still under discussion (Li, Singh, McIntyre, & Sherman, 2004; Yousef et al., 2003; Zheng, Doan, Schneider, & Storz, 1999). An abundant amount of data has been derived from work with plants transformed with flavodoxin (Coba de la Pena, Redondo, Manrique, Lucas, & Pueyo, 2010; Tognetti, Monti, Valle, Carrillo, & Smania, 2007; Tognetti et al., 2006; Tognetti, Zurbriggen, et al., 2007; Zurbriggen et al., 2008). Expression of a chloroplast-targeted cyanobacterial flavodoxin in tobacco plants led to a considerable level of stress tolerance and resistance to strong oxidants. This protection, that was described to be dose-dependent, was not mediated by a general induction of the antioxidant defence (Tognetti, Monti, et al., 2007; Tognetti et al., 2006).

Moreover, the recycling of NADP(H) through the Fld/sFNR couple relieves the accumulation of the reduced forms in the electron transfer chain, and prevents excessive reduction of the NADP(H) pool under adverse situations (Lodeyro et al., 2012), allowing photosynthetic electron transport to occur under nonoptimal conditions. Furthermore, maintenance of high levels of reduced thioredoxins could fuel several dissipative and scavenging pathways, mainly the activation of peroxiredoxins, eliminating H_2O_2 and peroxides. If the mechanism of the observed oxidative stress protection of the flavodoxin relays in the ferredoxin substitution, one could hypothesize that overexpression of ferredoxin would also confer analogous protection. However, transgenic tobacco plants overexpressing cyanobacterial ferredoxin does not exhibit enhanced tolerance to oxidants (Ceccoli, Blanco, Medina, & Carrillo, 2011). According to the authors, the failure of such lines to display enhanced stress tolerance can be explained by the stress-dependent decline of the foreign cyanobacterial ferredoxin in the cells, likely due to the high sensitivity of Fe–S clusters to oxidants. Nevertheless, it is not discarded that flavodoxin can act differently than ferredoxin, and not only as a replace in alleviate oxidative stress of photosynthetic organisms.

4.4 Roles idiABC Protecting Photosynthetic Machinery from Photooxidation

In addition to the induction of IsiA and flavodoxin, under iron and manganese limitation some cyanobacterial strains synthesize the iron deficiency-induced protein IdiA (Michel, Thole, & Pistorius, 1996); No counterpart of IdiA seems to exist in green algae and higher plants (Michel & Pistorius, 2004). The transcriptional regulator IdiB regulates the expression of *idiA*, with *idiB* expression controlled by iron availability (Michel, Kruger, Puhler, & Pistorius, 1999). IdiA plays an important role in protecting the acceptor side of PSII against oxidative damage, especially under iron-limiting growth conditions (Exss-Sonne, Tolle, Bader, Pistorius, & Michel, 2000).

IdiA shows considerable sequence similarity to a family of bacterial periplasmic ABC transporter complexes involved in iron import known as FutA, SfuA, FbpA or HitA (http://genome.microbedb.jp/cyanobase/). Even though some IdiA-similar proteins have been found in the periplasm (Tolle et al., 2002), in cyanobacteria IdiA is predominantly associated to thylakoids (Michel et al., 1998), suggesting different functions for the distinct IdiA-similar proteins (Tolle et al., 2002). IdiA propiciates prominent structural changes upon iron deficiency and forms a tight and specific complex with dimeric PSII by interaction with CP43 and D1 (Lax et al., 2007), suggesting that IdiA protects the acceptor side of PSII, which is more exposed under iron limitation due to ongoing PBS degradation (Lax et al., 2007).

In the *idi* operon, IdiB positively regulates transcription of *idiA* under iron starvation. IdiB belongs to the family of Crp/Fnr transcriptional regulators (Michel & Pistorius, 2004). It is transcribed under iron limitation and oxidative stress, and controlled itself by iron-responsive Fur family members (Yousef et al., 2003). A third iron-regulated gene is *idiC*, belonging to the thioredoxin-like (2Fe–2S) ferredoxin family. Even though the synthesis of IdiC is constitutive, iron limitation induces a strongly enhanced expression of *idiC*. IdiC is loosely attached to the thylakoid and to other membranes and its expression is enhanced during conditions of iron starvation and during the late growth phase (Pietsch et al., 2011). Even though its role is still unclear, based on the similarity of IdiC to NuoE of the respiratory *E. coli* NDH-1 complex, the authors suggested that IdiC is a component of the NDH-1 complex in *S. elongatus* and, thus, has a function in the electron donation from NAD(P)H to plastoquinone. Under stress conditions, when PS II is damaged, IdiC would prevent or reduce the oxidative stress deviating electron transport via alternative dehydrogenases, increasing PSI cyclic flow interconnected with respiratory routes (Pietsch et al., 2011).

4.5 Microcystin Is Synthesized as Consequence of Iron Starvation. Is Its Production Another Adaptative Response to Survive in Iron-Scarce Environments?

Metabolic plasticity of cyanobacteria includes the synthesis of a broad variety of secondary metabolites, some of them potentially toxic for eukaryotic organisms, the so called cyanotoxins (Carmichael et al., 2001). Some strains of cyanobacteria are able to synthesize microcystins, which are the most abundant and ubiquitous of those cyanotoxins. Microcystins are a family of cyclic heptapeptides, synthesized in a mixed polyketide synthase/nonribosomal peptide synthetase system called microcystin synthetase or *mcy* operon (Tillett et al., 2000). The role of microcystins in cyanobacteria is still unclear, but there are evidences that confer to the strain-forming advantages for survivance under iron-limited conditions (our unpublished results). Martin-Luna et al. (2006) showed that the *mcy* operon is likely regulated by Fur, being *mcy* transcription and microcystin synthesis enhanced under iron-limited conditions (Sevilla et al., 2008). On the other hand, an active photosynthetic electron chain is necessary for *mcy* transcription and microcystin synthesis (Sevilla, Martin-Luna, Bes, Fillat, & Peleato, 2011) and mutants defective in microcystin production exhibit a clearly increased sensitivity under high light conditions and presence of oxidants (Zilliges et al., 2011). Therefore, microcystin production might be another mechanism evolved by cyanobacteria related to iron homeostasis, on track to survive in iron-limited conditions.

5. IRON-DEPENDENT PROTEINS INVOLVED IN NITROGEN METABOLISM: ROLES AND REGULATION

Nitrogen can constitute as much as about 11% of the dry weight of a cyanobacterial cell. The nitrogen sources most commonly used by cyanobacteria are nitrate, ammonium and dinitrogen, but urea and other organic sources such as amino acids can also be assimilated (Flores & Herrero, 1994). Considering all the metabolic pathways related to nitrogen assimilation, cyanobacteria might contain, at least, five iron-containing enzymes involved in nitrogen metabolism (Table 2).

The uptake of nitrate, nitrite, urea and ammonium is mainly mediated by permeases located in the cytoplasmic membrane; most of them use ATP to

Table 2 Iron-Containing Enzymes Involved in Nitrogen Assimilation in Cyanobacteria

Enzyme	Prosthetic Groups	Physiological Substrate
Nitrogenase		
Dinitrogenase reductase	[4Fe–4S] cluster	Fd, ATP
Dinitrogenase	4× [4Fe–4S] clusters 2× Mo-Fe cofactors	N_2
Nitrate reductase	[4Fe–4S] cluster Mo cofactor	Nitrate, Fd/flavodoxin
Nitrite reductase	[4Fe–4S] cluster Sirohaem	Nitrite, Fd/flavodoxin
Glutamate synthase	[3Fe–4S] cluster FMN	Glutamine, 2-OG, Fd

drive an active, concentrative transport of their substrates (Olmedo-Verd, Flores, Herrero, & Muro-Pastor, 2005). Intracellular nitrate is sequentially reduced to nitrite and ammonium by the iron-containing enzymes nitrate reductase and nitrite reductase, which are the products of the *narB* and *nir* genes, respectively. Cyanobacterial nitrate reductase is homologous with Mo-containing bacterial oxidoreductases but it is unique in that it uses ferredoxin as an electron donor (Rubio, Herrero, & Flores, 1996). The enzyme contains a [4Fe–4S] cluster and a Mo-*bis*-molybdopterin guanine dinucleotide type cofactor (Jepson et al., 2004; Rubio, Flores, & Herrero, 2002). In this system, electrons flow from reduced ferredoxin to the iron–sulphur cluster and then to the Mo cofactor, where nitrate is reduced to nitrite. In a second step, nitrite is reduced to ammonium by nitrite reductase, with reduced ferredoxin serving as the electron donor. Cyanobacterial nitrite reductase contains a [4Fe–4S] cluster and sirohaem as prosthetic groups (Luque, Flores, & Herrero, 1993). Electrons from reduced ferredoxin are transferred to the iron–sulphur cluster and then to sirohaem, where nitrite is reduced to ammonium. The *narB* and *nir* genes are clustered together with the nitrate/nitrite permease-encoding genes in numerous cyanobacteria forming an operon with the structure of *nir*-permease genes-*narB* (Olmedo-Verd et al., 2005).

Intracellular urea in cyanobacteria is degraded to ammonium and CO_2 by a standard bacterial Ni^{2+}-dependent urease, whereas amino acids are catabolized by different pathways rendering ammonium as final product (Flores & Herrero, 1994). Whatever the nitrogen source used for growth, intracellular ammonium is incorporated into carbon skeletons by the sequential action of the enzymes glutamine synthetase (GS) and glutamate synthase (GOGAT), in a cycle commonly known as the GS-GOGAT pathway (Muro-Pastor, Reyes, & Florencio, 2005). GS catalyses the ATP-dependent amidation of glutamate to yield glutamine, while GOGAT catalyses the reductive transfer of the amide group from glutamine to 2-oxoglutarate (2-OG) to yield two molecules of glutamate. This pathway is also iron dependent in cyanobacteria, since in those organisms GOGAT is a monomeric iron-containing enzyme of about 180 kDa molecular mass, carrying a flavin FMN and a [3Fe–4S] cluster that uses ferredoxin as electron donor (Navarro, Martin-Figueroa, Candau, & Florencio, 2000). The operation of the GS-GOGAT cycle requires two direct photosynthetic products, ATP and reducing power, but a supply of carbon skeletons in the form of 2-OG is also needed. Thus, the GS-GOGAT cycle represents an important pathway in the utilization of 2-OG and the connecting step between carbon and nitrogen metabolisms (Muro-Pastor, Reyes, & Florencio, 2001).

In cyanobacteria, ammonium is the preferred nitrogen source, causing the repression of other nitrogen assimilation systems. Thus, the operon *nir*-permease genes-*narB* is expressed in high levels only when ammonium is not present in the growth medium. Likewise, genes encoding permeases for urea, as well as some ureases are repressed by growth in the presence of ammonium (Herrero, Muro-Pastor, & Flores, 2001). 2-OG acts as an indicator of the C to N ratio of the cells. When the cyanobacterial cells are incubated in the presence of a limiting concentration of ammonium, but with an adequate supply of carbon, they sense a high C to N ratio that determines expression of genes encoding permeases and enzymes required for an efficient assimilation of ammonium or for the assimilation of alternative nitrogen sources. This activation of gene expression is mediated by the global regulator of nitrogen metabolism NtcA (Herrero et al., 2001).

When combined nitrogen sources become depleted, certain cyanobacteria perform biological nitrogen fixation reducing the N_2 to ammonium through a multimeric enzyme complex, nitrogenase. Nitrogenase is one of the most iron-rich enzymes in nature, containing up to 50 Fe atoms per enzyme complex (Kupper et al., 2008). Nitrogenase is irreversibly inhibited by molecular oxygen and ROS. Because cyanobacteria are the only diazotrophs that actually produce oxygen as a by-product of the photosynthetic process, they must negotiate the inevitable presence of molecular oxygen with an essentially anaerobic enzyme. Some filamentous diazotrophic cyanobacteria spatially separate photosynthesis and nitrogen fixation by differentiating highly specialized cells called heterocysts (Flores & Herrero, 2010; Kumar, Mella-Herrera, & Golden, 2010). Heterocysts deposit glycolipid and polysaccharide layers outside of their cell walls to limit the entry of atmospheric oxygen. They also lack PSII activity and increase their respiration rate to consume O_2 that enters the cell, providing thereby a microoxic compartment for the expression of the oxygen-sensitive nitrogenase. Heterocyst development and its pattern formation are consequences of multiple external and internal signals, the action of several positive and negative regulators, the communication between cells in a filament, and the spatial–temporal regulation of gene expression and cellular processes (Zhang, Laurent, Sakr, Peng, & Bedu, 2006). Heterocyst differentiation is triggered by a metabolic signal, the accumulation of 2-OG, which is sensed by the global regulator of nitrogen metabolism NtcA. NtcA activates the expression of the master regulator of heterocyst differentiation HetR and a subsequent complex regulatory cascade is initiated (Flores & Herrero, 2010; Kumar et al., 2010).

The nitrogenase complex consists of two different iron-containing protein components: dinitrogenase (the Mo-Fe protein) and dinitrogenase reductase (the Fe protein). Dinitrogenase is an $\alpha_2\beta_2$ tetramer, and its α and β subunits are encoded by the *nifD* and *nifK* genes, respectively (Flores & Herrero, 1994). This tetramer binds four [4Fe–4S] clusters, organized into two "P cluster" and two Mo-Fe cofactors that bind N_2 used by the enzyme to catalyse its reduction to ammonia (Flores & Herrero, 1994). Dinitrogenase reductase is a dimer of identical subunits, that are encoded by the *nifH* gene and that together bind one intersubunit [4Fe–4S] centre. Dinitrogenase reductase mediates the ATP-dependent transference of electrons from external electron donors, such as Fd or flavodoxin to the P clusters of dinitrogenase. In addition to the set of ferredoxins present in vegetative cells, heterocystous cyanobacteria possess a heterocyst-specific ferredoxin (FdxH) that is expressed in these differentiated cells (Schrautemeier & Böhme, 1985). The vegetative (PetF or Fd1) and heterocyst ferredoxins isolated from *Anabaena* sp. PCC 7120 differ in midpoint oxidation–reduction potentials, amino acid compositions, immunological crossreactivities and electron spin resonance spectra (Bohme & Haselkorn, 1989). Although the amino acid sequences of Fd1 and FdxH exhibit a 51% identity, heterocysts ferredoxin contains a group of positively charged residues clustered in its surface that are important for its interaction with dinitrogenase reductase (Schmitz, Schrautemeier, & Böhme, 1993). Actually, in vitro tests comparing the efficiency of flavodoxin and both, vegetative cell (Fd1) and heterocyst ferredoxin in the different potential electron transport chains to nitrogenase, showed that heterocyst ferredoxin was the most effective electron carrier to nitrogenase reductase regardless of the transport route assayed (Fig. 2) (Razquin et al., 1995). It is remarkable that, unlike vegetative cell ferredoxin, that is substituted by flavodoxin in vegetative cells and heterocyst under iron deficiency, the synthesis of heterocyst ferredoxin was not affected (Razquin et al., 1994).

Thus, the architecture of the electron transport chains to nitrogenase and these of nitrogenase itself, imposes additional iron demands for growth beyond those attributed to photosynthesis and respiration. In fact, iron requirement for diazotrophy is ~5 times higher than that for ammonia assimilation (Sohm, Webb, & Capone, 2011). Models predict that the distribution of nitrogen fixation in the modern ocean may be constrained by the availability of iron, derived mainly from the deposition of dust from adjacent desert areas. Thus, North Atlantic Ocean, which receives a very high deposition of dust from Sahara and Sahel deserts shows larger nitrogen fixation rates than other oceanic areas (Sohm et al., 2011). Iron limitation caused a fast downregulation of nitrogenase activity and protein levels (Kupper et al., 2008). To maintain viability under iron plus nitrogen limitation, the marine diazotrophic cyanobacterium *Trichodesmiun erythraeum* IMS101 selectively sacrifices nitrogen fixation to conserve iron for photosynthetic

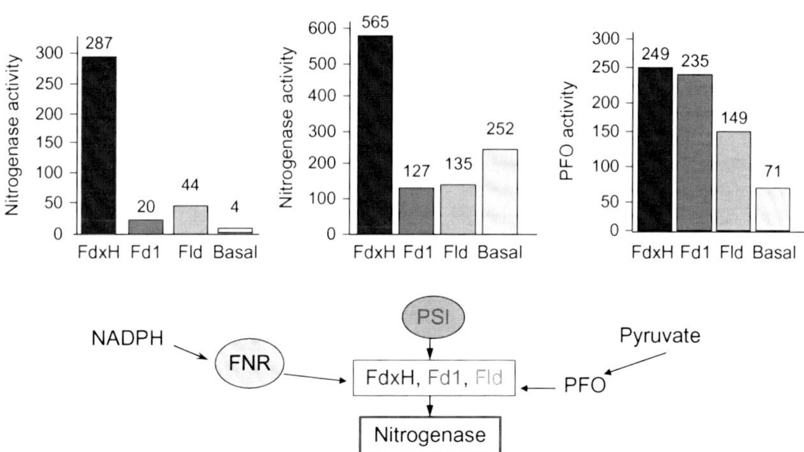

Fig. 2 Efficiency of equimolar amounts of different electron carriers from *Anabaena* sp. PCC 7119 as electron carriers to nitrogenase reductase. *Fd1*, vegetative cell ferredoxin; *FdxH*, heterocyst ferredoxin; *Fld*, flavodoxin; *FNR*, ferredoxin-NADP$^+$ reductase; *PFO*, pyruvate flavodoxin oxidoreductase (Razquin et al., 1995). (See the colour plate.)

and respiratory electron transport (Kupper et al., 2008; Shi et al., 2007). Likewise, nitrogen fixation and heterocyst differentiation are severely impaired in *Anabaena* sp. PCC 7120 grown under iron deficiency (Narayan, Kumari, & Rai, 2011; Saxena, Raghuvanshi, Singh, & Bisen, 2006; Wen-Liang, Yong-Ding, & Cheng-Cai, 2003).

Different studies have gave evidences about a tight connection between iron availability and regulation of nitrogen metabolism in cyanobacteria (González et al., 2013, 2014; López-Gomollón, Hernández, Pellicer, et al., 2007; López-Gomollón, Hernández, Wolk, Peleato, & Fillat, 2007). The expression of FurA, the master regulator of iron homeostasis (González et al., 2012), appeared activated by NtcA and strongly induced in proheterocysts during the first 15 h after nitrogen step-down, remaining stably expressed in mature heterocysts (López-Gomollón, Hernández, Wolk, et al., 2007). In contrast, in vitro and in vivo analyses have shown that FurA acts as a transcriptional repressor of the *ntcA* expression (González et al., 2013). Taken together, the data appeared to suggest that FurA might function as an NtcA shutoff switch, which in conjunction with other signals regulates the timing of NtcA induction during the heterocyst development.

The identification of common elements overlapping the NtcA and FurA regulons, including glutamine synthetase (*glnA*), glutamate synthase (*gltS*) and dinitrogenase reductase (*nifH*), additionally suggested a cross-talk between iron and nitrogen regulatory networks (López-Gomollón, Hernández, Pellicer, et al., 2007). Recent studies indicated a direct activating role of FurA on the expression of other players involved in heterocyst differentiation, such as those encoding by genes *hetC*, *patA* and *alr1728*, while also suggest the modulation of other predicted targets like *asr1734* or *patS* (González et al., 2014). These data further support the connection between iron homeostasis and nitrogen metabolism via FurA in cyanobacteria.

6. RELATIONSHIP BETWEEN IRON AND RESPIRATORY ELECTRON TRANSPORT IN CYANOBACTERIA

Respiration in cyanobacteria is much more complex than photosynthesis and some aspects remain to be fully understood. Respiratory electron transport mainly occurs in the thylakoid membranes, where the respiratory chain shares a quinone/quinol pool (Schmetterer, 1994), as well as several components, such as plastoquinone, cytochrome b_6f and plastocyanin/cytochrome *c*, with the photosynthetic electron transport chain (Lea-Smith et al., 2013; Peschek, Obinger, & Paumann, 2004; Schmetterer, 1994). A second complete respiratory chain is present in the cyanobacterial cell membrane that also uses the same mobile quinone pool mediating electrons in the photosynthetic and thylakoidal respiratory processes. The intrincate landscape of cyanobacterial respiration has explored with more detail in the unicellular model cyanobacterium *Synechocystis* sp. PCC 6803 (Cooley & Vermaas, 2001; Lea-Smith et al., 2013; Pils & Schmetterer, 2001). This bacterium contains three iron-rich terminal oxidases with different functional features. The importance of cyanide-sensitive haem-copper cytochrome *c* oxidase COX (aa_3-Type) and cytochrome bd quinol oxidase (Cyd) in cellular respiration is well established (Alge & Peschek, 1993; Howitt & Vermaas, 1998; Schmetterer, 1994). However, the function of the alternative respiratory terminal oxidase ARTO is still controversial. Cytochrome *c* oxidase COX occurs in both, thylakoidal and cell membranes seems to be the main type of oxidase in cyanobacteria (Peschek, 1996). Encoded by the *cox* operon (*coxBAC*), subunit II ligates Cu_A and is the main docking site for soluble donors. Subunit I ligates haem *a* and the haem *a*3-Cu_B binuclear centre, while subunit III lacks metals. It is noticeable that some cyanobacteria, such as *Anabaena* sp. PCC 7120, have two copies of the *cox* operon (*coxBAC1* and *coxBAC2*) (Jones & Haselkorn, 2002). Expression of *cox* genes varies with growth conditions, such as nitrogen status in nitrogen-fixing organisms (Valladares, Herrero, Pils, Schmetterer, & Flores, 2003) and salt concentrations (Hart, Schlarb-Ridley, Bendall, & Howe, 2005). Furthermore, a holistic analysis of the FurA regulon in *Anabaena* sp. PCC 7120 unveils *cox*B as a target of this master regulator of iron homeostasis, linking cellular iron status with the respiratory process (González et al., 2014). Additionally, the study of iron reduction in an ΔARTO mutant of *Synechocystis* sp. PCC 6803 suggest that this alternative oxidase in involved in reductive iron uptake (Kranzler et al., 2014). Those results are consistent with the localization of ARTO in the cyanobacterial outer membrane, nearby the periplasmic space, where the uptake-associated iron reduction is likely to take place (Kranzler et al., 2014).

Although it has been proposed that ARTO is regulated by an oxygen responsive transcriptional regulator (Howitt & Vermaas, 1998), it is reasonable to think that somehow, iron availability will also influence its assembly and functionality.

Finally, transcription of *alr0869* (*ndhF*) and the subunit 5 of NADH dehydrogenase encoded by *all1127* is regulated by FurA as response to iron availability (González et al., 2014). All those observations are consistent with the depletion in oxygen consumption observed when *Anabaena* sp. PCC 7120 exhibits a reduced internal pool of iron either due to removal of this nutrient from the medium or to disruption of the *alr1690-α-furA* operon (Hernández et al., 2010).

7. IRON-REGULATED RNAs RELATED TO THE CONTROL OF CYANOBACTERIAL ELECTRON TRANSPORT

The availability of fast and cheaper technical approaches for transcriptomic studies has allowed to identify a plethora of potential regulatory small RNAs (sRNAs) in cyanobacteria (Kopf & Hess, 2015). As an example, differential RNA sequencing in the model cyanobacteria *Synechocystis* sp. PCC 6803 indicates that its transcriptome contains more than 4000 transcriptional units, and a half of them are sRNAs. Surprisingly, this study also reveals a massive antisense transcription resulting in a high level of antisense sRNAs (Mitschke et al., 2011). In cyanobacteria the expression of sRNAs is highly regulated by environmental or nutritional conditions (Gierga, Voss, & Hess, 2012). One of the key conditions in such regulation is iron availability what is reflected in several iron-regulated sRNAs. For instance, in *Synechocystis* sp. PCC 6803, 10 sRNAs have been found as novel compounds of the iron regulatory network (Hernández-Prieto et al., 2012). Indeed, the presence of iron-regulated sRNAs involved in the regulation of iron homeostasis is frequent also in other organisms such as bacteria and animals (Billenkamp, Peng, Berghoff, & Klug, 2015; Brantl, 2012; Theil, 1987). Some of the best studied iron-related sRNAs in bacteria are RhyB of *E. coli* (Masse & Gottesman, 2002; Salvail & Masse, 2012; Zhang et al., 2003) and its homologs in several nonenterobacteria such as PrrF1 and PrrF2 in *Pseudomonas aeruginosa* (Wilderman et al., 2004).

In cyanobacteria, the best characterized iron-regulated sRNAs are IsaR1 in the unicellular cyanobacterium *Synechocystis* sp. PCC 6803, *α-fur* in the filamentous, nitrogen-fixing cyanobacterium *Anabaena* sp. PCC 7120 and its homolog *α-fur* in the toxic unicellular cyanobacterium *Microcystis* sp. PCC 7806. All of these sRNAs are directly or indirectly connected with the control of cyanobacterial electron transport chain.

As we have already mentioned, one of the responses of cyanobacteria to iron stress consists of the expression of IsiA that forms a giant ring structure around PSI. Transcription of *isiA* is regulated by Fur but it is also post-transcriptionally regulated by the antisense sRNA IsrR. IsrR is expressed in iron-replete conditions and represses the synthesis of IsiA under these conditions. Duplexes formed between the *isiA* mRNA and its antisense sRNA IsrR are probably targeted for degradation (Duhring et al., 2006).

In *Anabaena* sp. PCC 7120 *α-fur* RNA covers the complete transcript of *furA* but also holds the transcript of the membrane protein Alr1690 (Hernández et al., 2006). When *α-furA-alr1690* was disrupted, an enhanced expression of FurA expression was observed triggering an iron-deficient phenotype (Hernández et al., 2010). As a consequence, all the genes involved in the regulation of electron transport chain regulated by FurA are indirectly regulated by *α-furA-alr1690*. Phenotypic analyses of *α-furA-alr1690* mutant showed that this sRNA is necessary for a proper thylakoid arrangement and optimal yield of the photosynthetic machinery.

The presence of *α-fur* is not limited to the filamentous cyanobacteria *Anabaena* sp. PCC 7120 but it is present among different cyanobacterial genera although the genetic context of *α-fur* is not well conserved (Martin-Luna et al., 2011). In *Synechocystis* sp. PCC 6803, the *furA* orthologue is located between two hypothetical proteins. In this cyanobacterium the *α-fur* is smaller than that of *Anabaena* sp. PCC 7120 and covers part of the *fur* gene. However, in *Microcystis* sp. PCC 7806 *furA* is flanked between *dnaJ* and *sufE* genes. This locus shows a complex organization since comprises several overlapping transcripts. *α-fur* RNA contains the complete *fur* gene and part of the flanking gene sequences *dnaJ* and *sufE*. Moreover, *fur* seems to be part of a dicistronic operon harbouring *fur* and an *α-sufE* RNA. In *E. coli*, the *sufABCDSE* operon is responsible for the synthesis of Fe–S clusters under iron starvation and oxidative stress conditions. In fact the expression of the *suf* operon is regulated by Fur (Outten, Djaman, & Storz, 2004). Among the ORF belonging to the *suf* operon, SufE is a novel sulphur transfer protein that triggers the cysteine desulphurase SulfS activity together with the SufBCD complex (Outten, Wood, Munoz, & Storz, 2003). Thus, *α-fur* and *α-sufE* sRNAs in *Microcystis* sp. PCC 7806 could be indirectly involved in the regulation of Fe–S cluster synthesis.

Expression of both *fur-α-sufE* and *α-fur* RNAs of *Microcystis* sp. PCC 7806 was analysed in response to light and oxidative stress (Martin-Luna et al., 2011). The results indicated that *α-fur* plays an important role regulating *fur* expression. In addition, *α-fur* was upregulated in response to oxidative stress whereas it was not detected in presence of DCMU that blocks the electron transport chain at the Q_b site in the PSII. These results again suggest a link between the *α-fur/fur-α-sufE* system and the photosynthetic apparatus in *Microcystis* sp. PCC 7806.

Other sRNAs such as PsbA2R, PsbA3R and As1-Flv4 are involved in regulating important photosynthetic proteins. PsbA2R and PsbA3R antisense sRNAs are positive post-transcriptional regulators of *psbA2* and *psbA3* genes. These genes encode the D1 reaction centre protein of PSII. The expression of PsbA2R and PsbA3R antisense sRNAs as well as psbA2 and psbA3 mRNAs is regulated by light. When cells are transfer to high light thresholds, PsbA2R and PsbA3R antisense sRNAs overlap with psbA2 and psbA3 mRNAs and protect them to the attack of RNaseE avoiding their degradation (Sakurai, Anzai, & Furukawa, 2014). On the contrary, the as1-Flv4 RNA has a repressive effect on the expression of Flv4 protein. Flv4 and Flv2 are FDP that perform a pivotal role in photoprotection of PSII under low carbon conditions (Zhang et al., 2009). Finally the sRNA PsrR1 controls photosynthesis by targeting multiples mRNAs (Georg et al., 2014). Recently a comparative transcriptomic study of the closely related *Synechocystis* sp. PCC 6714 and *Synechocystis* sp. PCC 6803 strains showed significant differences in their repertoires of sRNAs. Surprisingly a lower degree of conservation was found in antisense sRNAs present in both strains what is reflected in those sRNAs involved in the regulation of photosynthetic machinery. As it could be expected IsiR is well conserved in both *Synechocystis* spp. strains but as1-FLv4, PsbA2R and PsbA3R are absent in *Synechocystis* sp. PCC 6714 (Kopf, Klähn, Scholz, Hess, & Voß, 2015). In view of the results, the authors concluded that fluctuations on the sRNAs composition in bacterial transcriptomes can contribute to inter-strain divergence and bacterial evolution (Kopf et al., 2015). Anyway, it seems clear that the regulation mediated by sRNAs and photosynthesis are deeply intertwined in cyanobacteria as well as in other organisms. However, future work must be done in order to clarify the role of all of these sRNAs in regulating the photosynthesis machinery.

8. INSIGHTS INTO THE MECHANISMS OF GENETIC REGULATION OF CYANOBACTERIAL ELECTRON TRANSPORT OPERATED BY FurA

Lack of iron or haem regulation has serious effects on electron transport chains in cyanobacteria since both play a central role in electron transfer reactions due to their ability to donate and accept electrons in haem and iron–sulphur proteins and to act as a cofactor. Moreover, the cyclic tetrapyrrole haem contains a central iron atom and originates phycobilins as products of haem metabolism in which the tetrapyrrole ring has been reopened and the centrally chelated iron has been lost. Upon excitation, haem and linear tetrapyrrole molecules may interact with a variety of substrates to yield radicals in a hydrogen atom and/or electron transfer reaction that react with oxygen and generate highly reactive oxidized intermediates that oxidize many biomolecules with deleterious consequences (Vavilin & Vermaas, 2002). On the other hand, the level of the major photosynthetic complexes PSII, cyt b_6f and PSI is reduced under iron limitation. As a consequence a decrease in electron transport and carbon fixation rate occurs (Rochaix, 2011). In addition, under conditions in which cyanobacteria exhibit iron disequilibrium, redox imbalance can occur. A direct evidence of this relation is obtained in studies of adaptation of cyanobacteria to iron starvation (Latifi et al., 2005). It is remarkable that many genes whose transcription is triggered by low iron are also induced under oxidative challenge (Nodop et al., 2008; Singh & Sherman, 2007). Therefore, a set of switches in cyanobacteria directly influence the pathways of electron transport by controlling the stoichiometry, interactions or proximity of the different electron transport components in and around the thylakoid membrane. They can act to control the balance of linear and cyclic electron transport, which is critical for the redox balance of the cell (Mullineaux, 2014). One of the routes for cyclic electron transport around PSI involves electron transfer from ferredoxin to plastoquinone via PGR5 protein (Yeremenko et al., 2005). Since flavodoxin expression in substitution of ferredoxin is controlled at the genetic level by the Fur, it would be reasonably to attribute a role of Fur in the control pathways of cyclic vs linear electron flow apart of its involvement in coordinating the responses to iron imbalance and oxidative stress (Latifi et al., 2009). Although there is much information about Fur target genes and the mechanisms for iron uptake and storage in cyanobacteria, little is known about the molecular mechanisms for controlling Fur activity. A recent study, that directly links thiol redox metabolism with iron sensing and regulation in *Anabaena* sp. PCC 7120, sheds light on this open question (Botello-Morte et al., 2014).

8.1 An Iron Sensing Mechanism Based on a Thiol Redox Switch in FurA from *Anabaena* sp. PCC 7120

In many organisms, reactive oxygen species target the thiol group of specific cysteine residues in sensory proteins that by means of thiol-based switches coordinate the expression of appropriate adaptative responses. For instance, OxyR is a DNA-binding protein that positively regulates a peroxide inducible regulon in response to peroxide stress. Activation of *E. coli* OxyR is mediated by disulfide-bond formation. Also the OhrR family of regulators sense organic hydroperoxides (OHP) and other ROS by oxidation of a critical highly conserved Cys residue (Hillion & Antelmann, 2015). Many of these redox regulators contain reactive cysteines that remain in their thiolate form to increase redox sensitivity towards harmful oxidant species. In particular, reactive cysteines arranged into a redox CXXC motif are considered to play an important role in sensing those redox signals by thiol-based regulators. The sequence of the XX dipeptide located between the cysteines of the motif is very important in controlling the redox properties of the protein in which it is found, so it has been termed a redox rheostat (Quan, Schneider, Pan, Von Hacht, & Bardwell, 2007). On the other hand, iron is used as a sensor of cellular iron status (Kobayashi & Nishizawa, 2015) and iron-based sensors incorporate Fe–S clusters, haem, and mononuclear iron sites to act as switches to control protein activity in response to changes in cellular redox balance. Some of them utilize thiolate ligands to coordinate the metal (Outten & Theil, 2009).

FurA from *Anabaena* sp. PCC 7120 is a protein that contains five cysteine residues. Four of these cysteines form part of two highly conserved $C_{101}XXC_{104}$ and $C_{141}XXC_{144}$ motifs in Fur homologues, both of them located in the C-terminal domain of the protein. The redox sensitivity of FurA activity and the presence in its primary sequence of two highly conserved CXXC motifs displaying disulfide reductase activity, according to in vitro assays (Botello-Morte et al., 2014), suggest that FurA activity may be associated to a disulfide formation. Furthermore, in vivo experiments indicate that FurA is mainly a monomer with a single free cysteine in the cytoplasm of *Anabaena* sp. PCC 7120 culture at the stationary phase, suggesting that it is capable of forming two disulfide bonds. Moreover, single cysteine mutations introduced in FurA have demonstrated that individual mutation of C^{101} led to the complete inactivation of FurA. In this sense, isothermal titration calorimetry (ITC) and electrophoretic mobility shift assay (EMSA) experiments indicate that C^{101} and its particular redox state play an essential role in the coordination of the metal corepressor, which ultimately controls FurA-DNA-binding activity in vitro. Since the redox state of C^{101} varies with the presence or absence of C^{104} or C^{133} it is suggested that the environments of these cysteines are mutually interdependent and a mechanism of FurA functioning based on a thiol/disulfide redox switch involving these cysteines and controlling the redox state of C^{101}, which directly coordinates the corepressor metal, is proposed (Botello-Morte et al., 2016). According to this mechanism, C^{133} would be responsible for maintaining

C^{104} in the oxidized state, avoiding the formation of a C^{101}–C^{104} disulfide bridge that would lead to inactivation of FurA. The reverse process would be necessary to activate FurA again (Fig. 3). The proposed mechanism is probably specific for cyanobacterial Fur since it relies on the presence of C^{133}, a residue only conserved in Fur homologues from cyanobacteria and not in Fur homologues from heterotrophic bacteria (Botello-Morte et al., 2016). This mechanism explains why purified recombinant FurA from *Anabaena* sp. PCC 7120 lacks structural zinc unlike other Fur homologues where the cysteine residues of the CXXC motifs coordinate the so called structural zinc atom (Butcher, Sarvan, Brunzelle, Couture, & Stintzi, 2012; Fillat, 2014; Hernández, Bes, Fillat, Neira, & Peleato, 2002; Vitale et al., 2009). Since C^{101} and C^{104} are part of the redox switch they cannot be found simultaneously reduced to coordinate zinc.

In photosynthetic organisms reversible protein thiol oxidation is an essential regulatory mechanism of photosynthesis, metabolism and gene expression. For instance, a recent proteome-wide quantitative and site-specific profiling of in vivo thiol oxidation modulated by light/dark in the cyanobacterium *Synechocystis* sp. PCC 6803 has revealed broad changes in thiol oxidation in many key biological processes, including photosynthetic electron transport (Guo et al., 2014). Also, a thiol-based redox modulation has been described for a eukaryotic-type serine/threonine kinase SpkB in *Synechocystis* sp. PCC 6803 that is inhibited by oxidation and reactivated by thioredoxin-catalysed reduction (Mata-Cabana, Garcia-Dominguez, Florencio, & Lindahl, 2012).

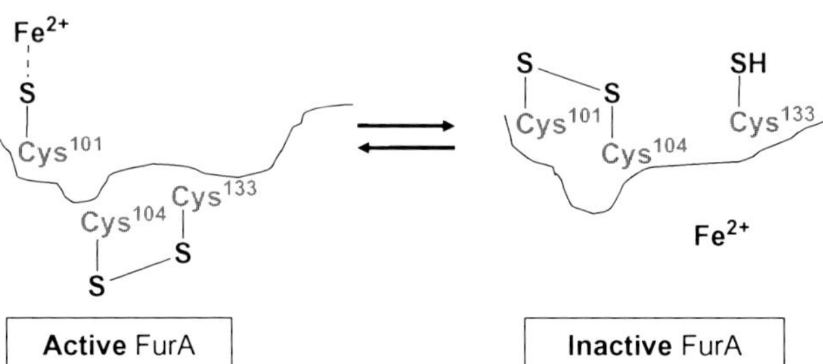

Fig. 3 Proposed model of FurA operating mechanism. When the protein is transcriptionally active residues C^{133} and C^{104} form an intramolecular disulfide bond and C^{101} can coordinate the corepressor metal. Upon an intracellular redox change, a thiol/disulfide interchange would generate a new intramolecular disulfide bond between residues C^{101} and C^{104}. This oxidation event would simultaneously occur with metal release and FurA inactivation. The reverse process would be necessary to activate FurA again. (See the colour plate.)

8.2 The Haem-FurA Interaction: A Potential Link Between Iron and Haem Metabolism in a Changing Redox Ambiance in Cyanobacterial Cytoplasm

The pathways of photosynthetic and respiratory electron transport in cyanobacteria are directly controlled by stoichiometry, localization and interaction of photosynthetic and respiratory electron transfer complexes. They can also be indirectly regulated by the control of the transfer of excitation energy from the LHCs (Mullineaux, 2014). For most cyanobacteria the principal LHCs are PBSs that contain phycobiliproteins with phycobilins acting as chromophores. Phycobilins consist of an open-chain tetrapyrrole that is covalently attached to phycobiliproteins and synthesized from haem. As mentioned previously, in the cyanobacterium *Anabaena* sp. PCC 7120 iron deprivation transcriptionally induces enzymes involved in haem metabolism (González et al., 2012). In particular, iron limitation results in up-regulation of transcription of the enzymes haem oxygenase I (*ho1*) and haem oxygenase II (*ho2*), both of them involved in haem degradation and subsequent release of iron. Additionally, iron limitation also triggers transcriptional induction of all haem biosynthesis enzymes although the increase of transcript abundance of haem metabolism enzymes observed is not homogeneous in all genes. According to EMSA experiments, FurA specifically binds to the promoter regions of four haem biosynthesis enzymes encoding genes (*hemB*, *hemC*, *hemK* and *hemH*) and to the promoter of haem-degrading enzyme haem oxygenase I (*ho1*). Notably, the binding of FurA to DNA in vitro is affected by the presence of haem. Purified recombinant FurA from *Anabaena* sp. PCC 7120 interacts strongly with haem in the micromolar range of concentration and this interaction affects the in vitro ability of FurA to bind DNA, inhibiting that process in a concentration-dependent manner (Hernández, Peleato, Fillat, & Bes, 2004). Thiolate form of residue C^{141} coordinates the Fe^{3+} haem. This residue is part of a $C^{141}P^{142}$ sequence analogous to the CP motif that is present in the haem binding/sensing sites of many haem sensor proteins where this motif is considered to be important for its function (Shimizu, 2012). Upon reduction the axial ligand for the Fe^{2+} haem complex is other than C^{141} thiolate. In consequence, cysteine coordination to the Fe^{3+} haem complex as the axial ligand is altered to another amino acid in the Fe(II) haem complex (Fig. 4). So that, the cysteine residue appears to play a critical role functioning as redox switch to coordinate the haem molecule (Pellicer et al., 2012).

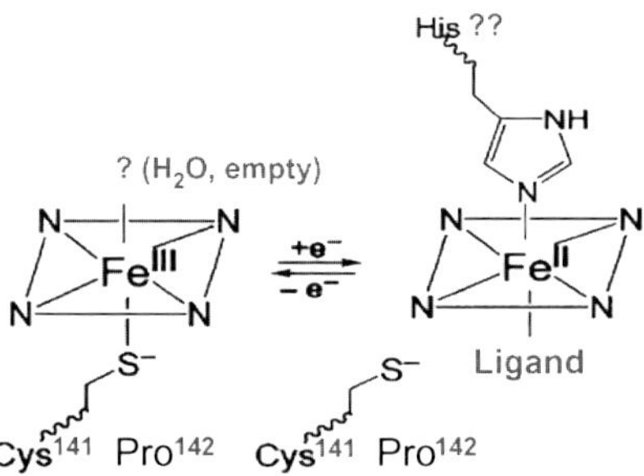

Fig. 4 Haem redox-dependent axial ligand switching in FurA. When the Fe(III) haem is reduced to Fe(II) haem, cysteine thiolate dissociates and is replaced by another residue such as histidine imidazol as the axial ligand. ? and ??: Ligands to be determined. (See the colour plate.)

Moreover, association/dissociation of the haem iron complex to/from FurA is relatively weak. The dissociation rate constant of the iron haem complex from FurA determined by difference absorption spectrometry is 0.35 ± 0.18 μM (Hernández et al., 2004). This value is higher than that of prototype haem proteins as myoglobin since apomyoglobin has an affinity for haem in the picomolar range of concentration (Hargrove, Barrick, & Olson, 1996). The above physico-chemical features of the binding of haem to FurA, namely: (i) fast haem dissociation rate constant, (ii) haem redox-dependent ligand switching and (iii) presence of a CP (Cys-Pro) motif in the amino acid sequence, match those described for haem sensor proteins (Shimizu, 2012). Recent studies have suggested the implication of thiol disulfide redox dependence of haem binding and haem ligand switching in haem sensor proteins. Thereby, the haem oxygenase HO2, constitutively expressed in mammals, contains three Cys-Pro signatures, known as haem regulatory motifs (HRMs). The HRM consists of a conserved Cys-Pro core sequence that is usually flanked at the N terminus by basic amino acids and at the C terminus by a hydrophobic residue (Kuhl et al., 2013). The C-terminal HRMs in HO2 constitute a thiol/disulfide redox switch that regulates affinity of the enzyme for haem (Yi & Ragsdale, 2007). The redox potential of this thiol/disulfide switch was measured to be −200 mV, which is near the ambient intracellular redox potential. For bacterial and human cells expressing HO2 under normal growth conditions, the HRMs are 60–70% reduced, whereas oxidative stress conditions convert most (86–89%) of the HRMs to the disulfide state. Treatment with reductants converts the HRMs largely (81–87%) to the reduced state (Yi et al., 2009). Consequently, the thiol/disulfide switch in HO2 responds to cellular oxidative stress and reductive conditions, although the HRMs in HO2 do not bind haem per se but instead form a reversible thiol/disulfide redox switch that indirectly regulates the affinity of HO2 for haem (Yi & Ragsdale, 2007). Notably, C^{141} from *Anabaena* sp. PCC 7120 FurA is involved in the formation of a S–S bond that shows disulfide reductase activity (Pellicer et al., 2012). The use of DTT to modulate the redox state of a FurA triple cysteine mutant containing the $C^{141}XXC^{144}$ redox pair followed by a combination of thiol alkylation and sodium dodecyl sulphate polyacrylamide gel electrophoresis (SDS-PAGE) to monitor the thiol redox state indicate that the redox potential of the redox pair is −238 mV, similar to that described for the bacterial cytoplasm, which is assumed to be between −220 and −240 mV (Wouters, Fan, & Haworth, 2010). Therefore, this redox pair could be responsive to slight changes affecting cytosolic redox homeostasis that could modulate its ability to bind haem in the sense than other haem sensing proteins do (Gupta & Ragsdale, 2011; Hu, Wang, Xia, & Varshavsky, 2008; Yi & Ragsdale, 2007). In these cases, under oxidative or normoxic conditions cysteine thiolate shows tendency to form intra- or intermolecular S–S bridges hindering Fe(III) haem coordination by thiolate cysteine. On the contrary, when conditions are reductive or hypoxic cysteine in its free thiolate form is generated and it coordinates Fe(III) haem more easily. Since an active photosynthetic electron transport chain is required for the transcription of the FurA orthologue in cyanobacteria (Martin-Luna et al., 2011), a light-driven modulation of the activity of FurA could be envisaged that would link redox homeostasis in cyanobacterial cytoplasm and haem synthesis and degradation to avoid alteration in tetrapyrrole homeostasis leading to reactive oxygen species generation (Busch & Montgomery, 2015; López-Gomollón, Sevilla, Bes, Peleato, & Fillat, 2009). Unlike the case of other haem sensors, the way to reduce the disulfide bridge once oxidized constitutes an unresolved issue for *Anabaena* sp. PCC 7120 FurA thiol/disulfide redox switches. Thiol/disulfide couples in FurA respond to changes in intracellular redox conditions in vivo. In fact when cells in early exponential phase are subjected to oxidative stress generated by addition of 0.5 mM H_2O_2 the proportion of reduced cysteines decreases (Botello-Morte et al., 2016). However, attempts to identify the reducing partner of FurA have failed. In the chloroplasts of higher plants, algae and cyanobacteria, thioredoxins convert disulphides to dithiols in their respective target enzymes by dithiol/disulfide exchange thereby modulating their activities. Cross-linking experiments aimed to analyse the ability of FurA to interact with thioredoxin indicate that this regulator is not a target of thioredoxin since the complex necessary for electron transfer was not detected (Botello-Morte et al., 2014). This is in accord with the results of a previous study using an immobilized cysteine to serine site-directed mutant of the *Synechocystis* sp. PCC 6803 thioredoxin TrxA as a bait aimed to find proteins interacting with TrxA (Lindahl & Florencio, 2003). The putative *Synechocystis* sp. PCC 6803 TrxA target proteins identified in this study were mainly enzymes participating in anabolic processes.

While much remains to be learned regarding the biological conditions governing the redox status of Fur, evidences point to a possible role of this transcriptional regulator in the redox control of electron transfer processes in cyanobacteria through a thiol/disulfide switch mechanism governing not only its ability to coordinate iron but also haem.

ACKNOWLEDGEMENTS

The authors acknowledge the Spanish Ministry of Economy and Competitivity (MINECO) for grant BFU2012-31458 cofunded with the help of FEDER funds and Gobierno de Aragón for project B18. We also would like to thank all the people that have been working at the "Genetic regulation and physiology of cyanobacteria" lab for their contribution to the knowledge on this field.

REFERENCES

Alexova, R., Fujii, M., Birch, D., Cheng, J., Waite, T. D., Ferrari, B. C., et al. (2011). Iron uptake and toxin synthesis in the bloom-forming *Microcystis aeruginosa* under iron limitation. *Environmental Microbiology, 13*(4), 1064–1077.

Alge, D., & Peschek, G. A. (1993). Identification and characterization of the *ctaC* (*coxB*) gene as part of an operon encoding subunits I, II, and III of the cytochrome c oxidase (cytochrome aa3) in the cyanobacterium *Synechocystis* PCC 6803. *Biochemical and Biophysical Research Communications, 191*(1), 9–17.

Allahverdiyeva, Y., Mustila, H., Ermakova, M., Bersanini, L., Richaud, P., Ajlani, G., et al. (2013). Flavodiiron proteins Flv1 and Flv3 enable cyanobacterial growth and photosynthesis under fluctuating light. *Proceedings of the National Academy of Sciences of the United States of America, 110*(10), 4111–4116.

Allen, J. F., Mullineaux, C. W., Sanders, C. E., & Melis, A. (1989). State transitions, photosystem stoichiometry adjustment and non-photochemical quenching in cyanobacterial cells acclimated to light absorbed by photosystem I or photosystem II. *Photosynthesis Research, 22*(2), 157–166.

Andrews, S. C., Robinson, A. K., & Rodriguez-Quinones, F. (2003). Bacterial iron homeostasis. *FEMS Microbiology Reviews, 27*(2–3), 215–237.

Aro, E. M., Suorsa, M., Rokka, A., Allahverdiyeva, Y., Paakkarinen, V., Saleem, A., et al. (2005). Dynamics of photosystem II: A proteomic approach to thylakoid protein complexes. *Journal of Experimental Botany, 56*(411), 347–356.

Badger, M. R., & Price, G. D. (2003). CO2 concentrating mechanisms in cyanobacteria: Molecular components, their diversity and evolution. *Journal of Experimental Botany, 54*(383), 609–622.

Baniulis, D., Yamashita, E., Zhang, H., Hasan, S. S., & Cramer, W. A. (2008). Structure-function of the cytochrome b6f complex. *Photochemistry and Photobiology, 84*(6), 1349–1358.

Barber, J. (2008). Crystal structure of the oxygen-evolving complex of photosystem II. *Inorganic Chemistry, 47*(6), 1700–1710.

Bassham, J. A., Benson, A. A., & Calvin, M. (1950). The path of carbon in photosynthesis. *The Journal of Biological Chemistry, 185*(2), 781–787.

Beale, S. I. (1994). Biosynthesis of cyanobacterial tetrapyrrole pigments: Hemes, chlorophylls, and phycobilins. In D. A. Bryant (Ed.), *The Molecular Biology of Cyanobacteria* (pp. 519–558). Dordrecht: Kluwer Academic Publishers.

Beinert, H. (2000). Iron-sulfur proteins: Ancient structures, still full of surprises. *Journal of Biological Inorganic Chemistry, 5*(1), 2–15.

Beinert, H., Holm, R. H., & Munck, E. (1997). Iron-sulfur clusters: Nature's modular, multipurpose structures. *Science, 277*(5326), 653–659.

Berera, R., van Stokkum, I. H. M., Kennis, J. T. M., van Grondelle, R., & Dekker, J. P. (2010). The light-harvesting function of carotenoids in the cyanobacterial stress-inducible IsiA complex. *Chemical Physics, 373*(1–2), 65–70.

Bersanini, L., Battchikova, N., Jokel, M., Rehman, A., Vass, I., Allahverdiyeva, Y., et al. (2014). Flavodiiron protein Flv2/Flv4-related photoprotective mechanism dissipates excitation pressure of PSII in cooperation with phycobilisomes in Cyanobacteria. *Plant Physiology, 164*(2), 805–818.

Bes, M. T., Hernández, J. A., Peleato, M. L., & Fillat, M. F. (2001). Cloning, overexpression and interaction of recombinant Fur from the cyanobacterium *Anabaena* PCC 7119 with *isiB* and its own promoter. *FEMS Microbiology Letters, 194*(2), 187–192.

Bibby, T. S., Nield, J., & Barber, J. (2001). Iron deficiency induces the formation of an antenna ring around trimeric photosystem I in cyanobacteria. *Nature, 412*(6848), 743–745.

Bibby, T. S., Zhang, Y. A., & Chen, M. (2009). Biogeography of photosynthetic light-harvesting genes in marine phytoplankton. *PloS One, 4*(2), e4601.

Billenkamp, F., Peng, T., Berghoff, B. A., & Klug, G. (2015). A cluster of four homologous small RNAs modulates C1 metabolism and the pyruvate dehydrogenase complex in *Rhodobacter sphaeroides* under various stress conditions. *Journal of Bacteriology, 197*(10), 1839–1852.

Bohme, H., & Haselkorn, R. (1989). Expression of *Anabaena* ferredoxin genes in *Escherichia coli*. *Plant Molecular Biology, 12*(6), 667–672.

Botello-Morte, L., Bes, M. T., Heras, B., Fernandez-Otal, A., Peleato, M. L., & Fillat, M. F. (2014). Unraveling the redox properties of the global regulator FurA from *Anabaena* sp. PCC 7120: Disulfide reductase activity based on its CXXC motifs. *Antioxidants & Redox Signaling, 20*(9), 1396–1406.

Botello-Morte, L., Pellicer, S., Sein-Echaluce, V. C., Contreras, L. M., Neira, J. L., Abian, O., et al. (2016). Cysteine mutational studies provide insight into a thiol-based redox switch mechanism of metal and DNA binding in FurA from *Anabaena* sp. *Antioxidants & Redox Signaling, 24*(4), 173–185.

Bovy, A., de Kruif, J., de Vrieze, G., Borrias, M., & Weisbeek, P. (1993). Iron-dependent protection of the *Synechococcus* ferredoxin I transcript against nucleolytic degradation requires cis-regulatory sequences in the 5′ part of the messenger RNA. *Plant Molecular Biology, 22*(6), 1047–1065.

Bovy, A., de Vrieze, G., Lugones, L., van Horssen, P., van den Berg, C., Borrias, M., et al. (1993). Iron-dependent stability of the ferredoxin I transcripts from the cyanobacterial strains *Synechococcus* species PCC 7942 and *Anabaena* species PCC 7937. *Molecular Microbiology, 7*(3), 429–439.

Brantl, S. (2012). Acting antisense: Plasmid- and chromosome-encoded sRNAs from Gram-positive bacteria. *Future Microbiology*, 7(7), 853–871.

Brettel, K. (1997). Electron transfer and arrangement of the redox cofactors in photosystem I. *Biochimica et Biophysica Acta (BBA)—Bioenergetics*, *1318*(3), 322–373.

Brown, S. B., Houghton, J. D., & Vernon, D. I. (1990). Biosynthesis of phycobilins. Formation of the chromophore of phytochrome, phycocyanin and phycoerythrin. *Journal of Photochemistry and Photobiology, B*, *5*(1), 3–23.

Burnap, R. L., Troyan, T., & Sherman, L. A. (1993). The highly abundant chlorophyll-protein complex of iron-deficient *Synechococcus* sp. PCC7942 (CP43′) is encoded by the *isiA* gene. *Plant Physiology*, *103*(3), 893–902.

Busch, A. W., & Montgomery, B. L. (2015). Interdependence of tetrapyrrole metabolism, the generation of oxidative stress and the mitigative oxidative stress response. *Redox Biology*, *4*, 260–271.

Butcher, J., Sarvan, S., Brunzelle, J. S., Couture, J. F., & Stintzi, A. (2012). Structure and regulon of *Campylobacter jejuni* ferric uptake regulator Fur define apo-Fur regulation. *Proceedings of the National Academy of Sciences of the United States of America*, *109*(25), 10047–10052.

Carmichael, W. W., Azevedo, S. M., An, J. S., Molica, R. J., Jochimsen, E. M., Lau, S., et al. (2001). Human fatalities from cyanobacteria: Chemical and biological evidence for cyanotoxins. *Environmental Health Perspectives*, *109*(7), 663–668.

Cassier-Chauvat, C., & Chauvat, F. (2014). Function and regulation of ferredoxins in the cyanobacterium, *Synechocystis* PCC 6803: Recent advances. *Life (Basel)*, *4*(4), 666–680.

Ceccoli, R. D., Blanco, N. E., Medina, M., & Carrillo, N. (2011). Stress response of transgenic tobacco plants expressing a cyanobacterial ferredoxin in chloroplasts. *Plant Molecular Biology*, *76*(6), 535–544.

Cheng, D., & He, Q. (2014). PfsR is a key regulator of iron homeostasis in *Synechocystis* PCC 6803. *PloS One*, *9*(7)e101743.

Coba de la Pena, T., Redondo, F. J., Manrique, E., Lucas, M. M., & Pueyo, J. J. (2010). Nitrogen fixation persists under conditions of salt stress in transgenic *Medicago truncatula* plants expressing a cyanobacterial flavodoxin. *Plant Biotechnology Journal*, *8*(9), 954–965.

Cooley, J. W., & Vermaas, W. F. (2001). Succinate dehydrogenase and other respiratory pathways in thylakoid membranes of *Synechocystis* sp. strain PCC 6803: Capacity comparisons and physiological function. *Journal of Bacteriology*, *183*(14), 4251–4258.

De Las Rivas, J., & Barber, J. (2004). Analysis of the structure of the PsbO protein and its implications. *Photosynthesis Research*, *81*(3), 329–343.

Deng, X., Sun, F., Ji, Q., Liang, H., Missiakas, D., Lan, L., et al. (2012). Expression of multidrug resistance efflux pump gene *norA* is iron responsive in *Staphylococcus aureus*. *Journal of Bacteriology*, *194*(7), 1753–1762.

Dokmanic, I., Sikic, M., & Tomic, S. (2008). Metals in proteins: Correlation between the metal-ion type, coordination number and the amino-acid residues involved in the coordination. *Acta Crystallographica. Section D, Biological Crystallography*, *64*(Pt. 3), 257–263.

Duhring, U., Axmann, I. M., Hess, W. R., & Wilde, A. (2006). An internal antisense RNA regulates expression of the photosynthesis gene *isiA*. *Proceedings of the National Academy of Sciences of the United States of America*, *103*(18), 7054–7058.

Escolar, L., Pérez-Martín, J., & de Lorenzo, V. (1999). Opening the iron box: Transcriptional metalloregulation by the Fur protein. *Journal of Bacteriology*, *181*(20), 6223–6229.

Exss-Sonne, P., Tolle, J., Bader, K. P., Pistorius, E. K., & Michel, K. P. (2000). The IdiA protein of *Synechococcus* sp PCC 7942 functions in protecting the acceptor side of photosystem II under oxidative stress. *Photosynthesis Research*, *63*(2), 145–157.

Faller, M., Matsunaga, M., Yin, S., Loo, J. A., & Guo, F. (2007). Heme is involved in microRNA processing. *Nature Structural & Molecular Biology*, *14*(1), 23–29.

Fay, P. (1992). Oxygen relations of nitrogen fixation in cyanobacteria. *Microbiological Reviews*, *56*(2), 340–373.

Ferreira, F., & Strauss, N. A. (1994). Iron deprivation in cyanobacteria. *Journal of Applied Phycology*, *6*, 199–210.

Fillat, M. F. (2014). The FUR (ferric uptake regulator) superfamily: Diversity and versatility of key transcriptional regulators. *Archives of Biochemistry and Biophysics*, *546*, 41–52.

Fillat, M. F., Peleato, M. L., Razquin, P., & Gómez-Moreno, C. (1994). Effects of iron deficiency in photosynthetic electron transport and nitrogen fixation in the cyanobacterium *Anabaena*: flavodoxin as adaptative response. In J. Abadía (Ed.), *Iron nutrition in soils and plants. Developments in plant and soil sciences: vol. 59.* (pp. 315–322). Dordrecht: Kluwer Academic Publishers.

Flores, E., & Herrero, A. (1994). Assimilatory nitrogen metabolism and its regulation. In D. A. Bryant (Ed.), *The molecular biology of cyanobacteria* (pp. 487–517). Dordrecht: Kluwer Academic Publishers.

Flores, E., & Herrero, A. (2010). Compartmentalized function through cell differentiation in filamentous cyanobacteria. *Nature Reviews. Microbiology*, *8*(1), 39–50.

Fraser, J. M., Tulk, S. E., Jeans, J. A., Campbell, D. A., Bibby, T. S., & Cockshutt, A. M. (2013). Photophysiological and photosynthetic complex changes during iron starvation in *Synechocystis* sp. PCC 6803 and *Synechococcus elongatus* PCC 7942. *PLoS One*, *8*(3), e59861.

Fulda, S., & Hagemann, M. (1995). Salt treatment induces accumulation of flavodoxin in the cyanobacterium *Synechocystis* sp. PCC6803. *Journal of Plant Physiology*, *146*(4), 520–526.

Geiss, U., Vinnemeier, J., Kunert, A., Lindner, I., Gemmer, B., Lorenz, M., et al. (2001). Detection of the *isiA* gene across cyanobacterial strains: Potential for probing iron deficiency. *Applied and Environmental Microbiology*, *67*(11), 5247–5253.

Georg, J., Dienst, D., Schurgers, N., Wallner, T., Kopp, D., Stazic, D., et al. (2014). The small regulatory RNA SyR1/PsrR1 controls photosynthetic functions in cyanobacteria. *Plant Cell*, *26*(9), 3661–3679.

Gierga, G., Voss, B., & Hess, W. R. (2012). Non-coding RNAs in marine *Synechococcus* and their regulation under environmentally relevant stress conditions. *The ISME Journal*, *6*(8), 1544–1557.

Gilbreath, J. J., West, A. L., Pich, O. Q., Carpenter, B. M., Michel, S., & Merrell, D. S. (2012). Fur activates expression of the 2-oxoglutarate oxidoreductase genes (*oorDABC*) in *Helicobacter pylori*. *Journal of Bacteriology*, *194*(23), 6490–6497.

Girvan, H. M., & Munro, A. W. (2013). Heme sensor proteins. *The Journal of Biological Chemistry*, *288*(19), 13194–13203.

Goldman, S. J., Lammers, P. J., Berman, M. S., & Sanders-Loehr, J. (1983). Siderophore-mediated iron uptake in different strains of *Anabaena* sp. *Journal of Bacteriology*, *156*(3), 1144–1150.

González, A., Angarica, V. E., Sancho, J., & Fillat, M. F. (2014). The FurA regulon in *Anabaena* sp. PCC 7120: In silico prediction and experimental validation of novel target genes. *Nucleic Acids Research*, *42*(8), 4833–4846.

González, A., Bes, M. T., Barja, F., Peleato, M. L., & Fillat, M. F. (2010). Overexpression of FurA in *Anabaena* sp. PCC 7120 reveals new targets for this regulator involved in photosynthesis, iron uptake and cellular morphology. *Plant & Cell Physiology*, *51*(11), 1900–1914.

González, A., Bes, M. T., Peleato, M. L., & Fillat, M. F. (2011). Unravelling the regulatory function of FurA in *Anabaena* sp. PCC 7120 through 2-D DIGE proteomic analysis. *Journal of Proteomics*, *74*(5), 660–671.

González, A., Bes, M. T., Valladares, A., Peleato, M. L., & Fillat, M. F. (2012). FurA is the master regulator of iron homeostasis and modulates the expression of tetrapyrrole biosynthesis genes in *Anabaena* sp. PCC 7120. *Environmental Microbiology*, *14*(12), 3175–3187.

González, A., Valladares, A., Peleato, M. L., & Fillat, M. F. (2013). FurA influences heterocyst differentiation in *Anabaena* sp. PCC 7120. *FEBS Letters*, *587*(16), 2682–2690.

Guo, J., Nguyen, A. Y., Dai, Z., Su, D., Gaffrey, M. J., Moore, R. J., et al. (2014). Proteome-wide light/dark modulation of thiol oxidation in cyanobacteria revealed by quantitative site-specific redox proteomics. *Molecular & Cellular Proteomics*, *13*(12), 3270–3285.

Gupta, N., & Ragsdale, S. W. (2011). Thiol-disulfide redox dependence of heme binding and heme ligand switching in nuclear hormone receptor rev-erb{beta}. *The Journal of Biological Chemistry*, *286*(6), 4392–4403.

Hamza, I., & Dailey, H. A. (2012). One ring to rule them all: Trafficking of heme and heme synthesis intermediates in the metazoans. *Biochimica et Biophysica Acta*, *1823*(9), 1617–1632.

Hargrove, M. S., Barrick, D., & Olson, J. S. (1996). The association rate constant for heme binding to globin is independent of protein structure. *Biochemistry*, *35*(35), 11293–11299.

Hart, S. E., Schlarb-Ridley, B. G., Bendall, D. S., & Howe, C. J. (2005). Terminal oxidases of cyanobacteria. *Biochemical Society Transactions*, *33*(Pt. 4), 832–835.

Havaux, M., Guedeney, G., Hagemann, M., Yeremenko, N., Matthijs, H. C., & Jeanjean, R. (2005). The chlorophyll-binding protein IsiA is inducible by high light and protects the cyanobacterium *Synechocystis* PCC6803 from photooxidative stress. *FEBS Letters*, *579*(11), 2289–2293.

Hernández, J. A., Alonso, I., Pellicer, S., Luisa Peleato, M., Cases, R., Strasser, R. J., et al. (2010). Mutants of *Anabaena* sp. PCC 7120 lacking *alr1690* and alpha-*furA* antisense RNA show a pleiotropic phenotype and altered photosynthetic machinery. *Journal of Plant Physiology*, *167*(6), 430–437.

Hernández, J. A., Bes, M. T., Fillat, M. F., Neira, J. L., & Peleato, M. L. (2002). Biochemical analysis of the recombinant Fur (ferric uptake regulator) protein from *Anabaena* PCC 7119: Factors affecting its oligomerization state. *The Biochemical Journal*, *366*(Pt. 1), 315–322.

Hernández, J. A., Muro-Pastor, A. M., Flores, E., Bes, M. T., Peleato, M. L., & Fillat, M. F. (2006). Identification of a *furA* cis antisense RNA in the cyanobacterium *Anabaena* sp. PCC 7120. *Journal of Molecular Biology*, *355*(3), 325–334.

Hernández, J. A., Peleato, M. L., Fillat, M. F., & Bes, M. T. (2004). Heme binds to and inhibits the DNA-binding activity of the global regulator FurA from *Anabaena* sp. PCC 7120. *FEBS Letters*, *577*(1–2), 35–41.

Hernández-Prieto, M. A., Schon, V., Georg, J., Barreira, L., Varela, J., Hess, W. R., et al. (2012). Iron deprivation in *Synechocystis:* Inference of pathways, non-coding RNAs, and regulatory elements from comprehensive expression profiling. *G3 (Bethesda)*, *2*(12), 1475–1495.

Herrero, A., Muro-Pastor, A. M., & Flores, E. (2001). Nitrogen control in cyanobacteria. *Journal of Bacteriology*, *183*(2), 411–425.

Hill, R., & Bendall, F. (1960). Function of the 2 cytochrome components in chloroplast—Working hypothesis. *Nature*, *186*(4719), 136–137.

Hillion, M., & Antelmann, H. (2015). Thiol-based redox switches in prokaryotes. *Biological Chemistry*, *396*(5), 415–444.

Howitt, C. A., & Vermaas, W. F. (1998). Quinol and cytochrome oxidases in the cyanobacterium *Synechocystis* sp. PCC 6803. *Biochemistry*, *37*(51), 17944–17951.

Hu, R. G., Wang, H., Xia, Z., & Varshavsky, A. (2008). The N-end rule pathway is a sensor of heme. *Proceedings of the National Academy of Sciences of the United States of America*, *105*(1), 76–81.

Ionescu, D., Voss, B., Oren, A., Hess, W. R., & Muro-Pastor, A. M. (2010). Heterocyst-specific transcription of NsiR1, a non-coding RNA encoded in a tandem array of direct repeats in cyanobacteria. *Journal of Molecular Biology*, *398*(2), 177–188.

Ivanov, A. G., Park, Y. I., Miskiewicz, E., Raven, J. A., Huner, N. P., & Oquist, G. (2000). Iron stress restricts photosynthetic intersystem electron transport in *Synechococcus* sp. PCC 7942. *FEBS Letters*, *485*(2–3), 173–177.

Jeanjean, R., Talla, E., Latifi, A., Havaux, M., Janicki, A., & Zhang, C. C. (2008). A large gene cluster encoding peptide synthetases and polyketide synthases is involved in production of siderophores and oxidative stress response in the cyanobacterium *Anabaena* sp. strain PCC 7120. *Environmental Microbiology*, *10*(10), 2574–2585.

Jeanjean, R., Zuther, E., Yeremenko, N., Havaux, M., Matthijs, H. C., & Hagemann, M. (2003). A photosystem 1 *psaFJ*-null mutant of the cyanobacterium *Synechocystis* PCC 6803 expresses the *isiAB* operon under iron replete conditions. *FEBS Letters*, *549*(1–3), 52–56.

Jepson, B. J., Anderson, L. J., Rubio, L. M., Taylor, C. J., Butler, C. S., Flores, E., et al. (2004). Tuning a nitrate reductase for function. The first spectropotentiometric characterization of a bacterial assimilatory nitrate reductase reveals novel redox properties. *The Journal of Biological Chemistry*, *279*(31), 32212–32218.

Jones, K. M., & Haselkorn, R. (2002). Newly identified cytochrome c oxidase operon in the nitrogen-fixing cyanobacterium *Anabaena* sp. strain PCC 7120 specifically induced in heterocysts. *Journal of Bacteriology*, *184*(9), 2491–2499.

Jordan, P., Fromme, P., Witt, H. T., Klukas, O., Saenger, W., & Krauss, N. (2001). Three-dimensional structure of cyanobacterial photosystem I at 2.5 A resolution. *Nature*, *411*(6840), 909–917.

Junge, W., & Nelson, N. (2015). ATP synthase. *Annual Review of Biochemistry*, *84*, 631–657.

Karlusich, J. J. P., Lodeyro, A. F., & Carrillo, N. (2014). The long goodbye: The rise and fall of flavodoxin during plant evolution. *Journal of Experimental Botany*, *65*(18), 5161–5178.

Kaufman, A. J. (2014). Early earth: Cyanobacteria at work. *Nature Geoscience*, 7(4), 253–254.

Keren, N., Aurora, R., & Pakrasi, H. B. (2004). Critical roles of bacterioferritins in iron storage and proliferation of cyanobacteria. *Plant Physiology*, *135*(3), 1666–1673.

Kobayashi, T., & Nishizawa, N. K. (2015). Intracellular iron sensing by the direct binding of iron to regulators. *Frontiers in Plant Science*, *6*, 155.

Kopf, M., & Hess, W. R. (2015). Regulatory RNAs in photosynthetic cyanobacteria. *FEMS Microbiology Reviews*, *39*(3), 301–315.

Kopf, M., Klähn, S., Scholz, I., Hess, W. R., & Voß, B. (2015). Variations in the non-coding transcriptome as a driver of inter-strain divergence and physiological adaptation in bacteria. *Scientific Reports*. *5*(9560). http://dx.doi.org/10.1038/srep09560.

Kouril, R., Arteni, A. A., Lax, J., Yeremenko, N., D'Haene, S., Rogner, M., et al. (2005). Structure and functional role of supercomplexes of IsiA and photosystem I in cyanobacterial photosynthesis. *FEBS Letters*, *579*(15), 3253–3257.

Kranzler, C., Lis, H., Finkel, O. M., Schmetterer, G., Shaked, Y., & Keren, N. (2014). Coordinated transporter activity shapes high-affinity iron acquisition in cyanobacteria. *The ISME Journal*, *8*(2), 409–417.

Kuhl, T., Wissbrock, A., Goradia, N., Sahoo, N., Galler, K., Neugebauer, U., et al. (2013). Analysis of Fe(III) heme binding to cysteine-containing heme-regulatory motifs in proteins. *ACS Chemical Biology*, *8*(8), 1785–1793.

Kumar, K., Mella-Herrera, R. A., & Golden, J. W. (2010). Cyanobacterial heterocysts. *Cold Spring Harbor Perspectives in Biology*, *2*(4), a000315.

Kunert, A., Vinnemeier, J., Erdmann, N., & Hagemann, M. (2003). Repression by Fur is not the main mechanism controlling the iron-inducible *isiAB* operon in the cyanobacterium *Synechocystis* sp. PCC 6803. *FEMS Microbiology Letters*, *227*(2), 255–262.

Kupper, H., Setlik, I., Seibert, S., Prasil, O., Setlikova, E., Strittmatter, M., et al. (2008). Iron limitation in the marine cyanobacterium *Trichodesmium* reveals new insights into regulation of photosynthesis and nitrogen fixation. *The New Phytologist*, *179*(3), 784–798.

Kurisu, G., Zhang, H., Smith, J. L., & Cramer, W. A. (2003). Structure of the cytochrome b6f complex of oxygenic photosynthesis: Tuning the cavity. *Science*, *302*(5647), 1009–1014.

Latifi, A., Jeanjean, R., Lemeille, S., Havaux, M., & Zhang, C. C. (2005). Iron starvation leads to oxidative stress in *Anabaena* sp. Strain PCC 7120. *Journal of Bacteriology, 187*(18), 6596–6598.

Latifi, A., Ruiz, M., & Zhang, C. C. (2009). Oxidative stress in cyanobacteria. *FEMS Microbiology Reviews, 33*(2), 258–278.

Laudenbach, D. E., & Straus, N. A. (1988). Characterization of a cyanobacterial iron stress-induced gene similar to *psbC*. *Journal of Bacteriology, 170*(11), 5018–5026.

Lax, J. E., Arteni, A. A., Boekema, E. J., Pistorius, E. K., Michel, K. P., & Rogner, M. (2007). Structural response of photosystem 2 to iron deficiency: Characterization of a new photosystem 2-IdiA complex from the cyanobacterium *Thermosynechococcus elongatus* BP-1. *Biochimica et Biophysica Acta, 1767*(6), 528–534.

Lea-Smith, D. J., Ross, N., Zori, M., Bendall, D. S., Dennis, J. S., Scott, S. A., et al. (2013). Thylakoid terminal oxidases are essential for the cyanobacterium *Synechocystis* sp. PCC 6803 to survive rapidly changing light intensities. *Plant Physiology, 162*(1), 484–495.

Leonhardt, K., & Straus, N. A. (1992). An iron stress operon involved in photosynthetic electron transport in the marine cyanobacterium *Synechococcus* sp. PCC 7002. *Journal of General Microbiology, 138*(Pt. 8), 1613–1621.

Leonhardt, K., & Straus, N. A. (1994). Photosystem II genes *isiA*, *psbDI* and *psbC* in *Anabaena* sp. PCC 7120: Cloning, sequencing and the transcriptional regulation in iron-stressed and iron-repleted cells. *Plant Molecular Biology, 24*(1), 63–73.

Li, H., Singh, A. K., McIntyre, L. M., & Sherman, L. A. (2004). Differential gene expression in response to hydrogen peroxide and the putative PerR regulon of *Synechocystis* sp. Strain PCC 6803. *Journal of Bacteriology, 186*(11), 3331–3345.

Lin, C. T., Wu, C. C., Chen, Y. S., Lai, Y. C., Chi, C., Lin, J. C., et al. (2011). Fur regulation of the capsular polysaccharide biosynthesis and iron-acquisition systems in *Klebsiella pneumoniae* CG43. *Microbiology, 157*(Pt. 2), 419–429.

Lindahl, M., & Florencio, F. J. (2003). Thioredoxin-linked processes in cyanobacteria are as numerous as in chloroplasts, but targets are different. *Proceedings of the National Academy of Sciences of the United States of America, 100*(26), 16107–16112.

Lodeyro, A. F., Ceccoli, R. D., Pierella Karlusich, J. J., & Carrillo, N. (2012). The importance of flavodoxin for environmental stress tolerance in photosynthetic microorganisms and transgenic plants. Mechanism, evolution and biotechnological potential. *FEBS Letters, 586*(18), 2917–2924.

López-Gomollón, S., Hernández, J. A., Pellicer, S., Angarica, V. E., Peleato, M. L., & Fillat, M. F. (2007a). Cross-talk between iron and nitrogen regulatory networks in *Anabaena* (*Nostoc*) sp. PCC 7120: Identification of overlapping genes in FurA and NtcA regulons. *Journal of Molecular Biology, 374*(1), 267–281.

López-Gomollón, S., Hernández, J. A., Wolk, C. P., Peleato, M. L., & Fillat, M. F. (2007b). Expression of *furA* is modulated by NtcA and strongly enhanced in heterocysts of *Anabaena* sp. PCC 7120. *Microbiology, 153*(Pt. 1), 42–50.

López-Gomollón, S., Sevilla, E., Bes, M. T., Peleato, M. L., & Fillat, M. F. (2009). New insights into the role of Fur proteins: FurB (All2473) from *Anabaena* protects DNA and increases cell survival under oxidative stress. *The Biochemical Journal, 418*(1), 201–207.

Luque, I., Flores, E., & Herrero, A. (1993). Nitrite reductase gene from *Synechococcus* sp. PCC 7942: Homology between cyanobacterial and higher-plant nitrite reductases. *Plant Molecular Biology, 21*(6), 1201–1205.

Martin-Luna, B., Sevilla, E., Gonzalez, A., Bes, M. T., Fillat, M. F., & Peleato, M. L. (2011). Expression of fur and its antisense alpha-*fur* from *Microcystis aeruginosa* PCC7806 as response to light and oxidative stress. *Journal of Plant Physiology, 168*(18), 2244–2250.

Martin-Luna, B., Sevilla, E., Hernandez, J. A., Bes, M. T., Fillat, M. F., & Peleato, M. L. (2006). Fur from *Microcystis aeruginosa* binds in vitro promoter regions of the microcystin biosynthesis gene cluster. *Phytochemistry*, *67*(9), 876–881.

Masse, E., & Gottesman, S. (2002). A small RNA regulates the expression of genes involved in iron metabolism in *Escherichia coli*. *Proceedings of the National Academy of Sciences of the United States of America*, *99*(7), 4620–4625.

Mata-Cabana, A., Garcia-Dominguez, M., Florencio, F. J., & Lindahl, M. (2012). Thiol-based redox modulation of a cyanobacterial eukaryotic-type serine/threonine kinase required for oxidative stress tolerance. *Antioxidants & Redox Signaling*, *17*(4), 521–533.

Mekala, N. R., Suorsa, M., Rantala, M., Aro, E. M., & Tikkanen, M. (2015). Plants actively avoid state transitions upon changes in light intensity: Role of light-harvesting complex II protein dephosphorylation in high light. *Plant Physiology*, *168*(2), 721–734.

Melkozernov, A. N., Barber, J., & Blankenship, R. E. (2006). Light harvesting in photosystem I supercomplexes. *Biochemistry*, *45*(2), 331–345.

Michel, K. P., Exss-Sonne, P., Scholten-Beck, G., Kahmann, U., Ruppel, H. G., & Pistorius, E. K. (1998). Immunocytochemical localization of IdiA, a protein expressed under iron or manganese limitation in the mesophilic cyanobacterium *Synechococcus* PCC 6301 and the thermophilic cyanobacterium *Synechococcus elongatus*. *Planta*, *205*(1), 73–81.

Michel, K. P., Kruger, F., Puhler, A., & Pistorius, E. K. (1999). Molecular characterization of *idiA* and adjacent genes in the cyanobacteria *Synechococcus* sp. strains PCC 6301 and PCC 7942. *Microbiology*, *145*(Pt. 6), 1473–1484.

Michel, K. P., & Pistorius, E. K. (2004). Adaptation of the photosynthetic electron transport chain in cyanobacteria to iron deficiency: The function of IdiA and IsiA. *Physiologia Plantarum*, *120*(1), 36–50.

Michel, K. P., Thole, H. H., & Pistorius, E. K. (1996). IdiA, a 34 kDa protein in the cyanobacteria *Synechococcus* sp. Strains PCC 6301 and PCC 7942, is required for growth under iron and manganese limitations. *Microbiology*, *142*(Pt. 9), 2635–2645.

Mirus, O., Strauss, S., Nicolaisen, K., von Haeseler, A., & Schleiff, E. (2009). TonB-dependent transporters and their occurrence in cyanobacteria. *BMC Biology*, *7*, 68.

Mitschke, J., Georg, J., Scholz, I., Sharma, C. M., Dienst, D., Bantscheff, J., et al. (2011). An experimentally anchored map of transcriptional start sites in the model cyanobacterium *Synechocystis* sp. PCC6803. *Proceedings of the National Academy of Sciences of the United States of America*, *108*(5), 2124–2129.

Morel, F. M., & Price, N. M. (1991). Iron nutrition of phytoplankton and its possible importance in the ecology of ocean regions with high nutrient and low biomass. *Oceanography*, *4*(2), 56–61.

Muh, F., & Zouni, A. (2013). The nonheme iron in photosystem II. *Photosynthesis Research*, *116*(2–3), 295–314.

Mullineaux, C. W. (2014). Electron transport and light-harvesting switches in cyanobacteria. *Frontiers in Plant Science*, *5*, 7.

Muro-Pastor, M. I., Reyes, J. C., & Florencio, F. J. (2001). Cyanobacteria perceive nitrogen status by sensing intracellular 2-oxoglutarate levels. *Journal of Biological Chemistry*, *276*(41), 38320–38328.

Muro-Pastor, M. I., Reyes, J. C., & Florencio, F. J. (2005). Ammonium assimilation in cyanobacteria. *Photosynthesis Research*, *83*(2), 135–150.

Nakao, M., Okamoto, S., Kohara, M., Fujishiro, T., Fujisawa, T., Sato, S., et al. (2010). CyanoBase: The cyanobacteria genome database update 2010. *Nucleic Acids Research*, *38*(Database issue), D379–D381.

Narayan, O. P., Kumari, N., & Rai, L. C. (2011). Iron starvation-induced proteomic changes in *Anabaena* (*Nostoc*) sp. PCC 7120: Exploring survival strategy. *Journal of Microbiology and Biotechnology*, *21*(2), 136–146.

Navarro, F., Martin-Figueroa, E., Candau, P., & Florencio, F. J. (2000). Ferredoxin-dependent iron-sulfur flavoprotein glutamate synthase (GlsF) from the cyanobacterium *Synechocystis* sp. PCC 6803: Expression and assembly in *Escherichia coli*. *Archives of Biochemistry and Biophysics, 379*(2), 267–276.

Nelson, N., & Ben-Shem, A. (2004). The complex architecture of oxygenic photosynthesis. *Nature Reviews. Molecular Cell Biology, 5*(12), 971–982.

Nicolaisen, K., Moslavac, S., Samborski, A., Valdebenito, M., Hantke, K., Maldener, I., et al. (2008). Alr0397 is an outer membrane transporter for the siderophore schizokinen in *Anabaena* sp. strain PCC 7120. *Journal of Bacteriology, 190*(22), 7500–7507.

Nodop, A., Pietsch, D., Hocker, R., Becker, A., Pistorius, E. K., Forchhammer, K., et al. (2008). Transcript profiling reveals new insights into the acclimation of the mesophilic fresh-water cyanobacterium *Synechococcus elongatus* PCC 7942 to iron starvation. *Plant Physiology, 147*(2), 747–763.

Nogi, T., & Miki, K. (2001). Structural basis of bacterial photosynthetic reaction centers. *Journal of Biochemistry, 130*(3), 319–329.

Oglesby-Sherrouse, A. G., & Murphy, E. R. (2013). Iron-responsive bacterial small RNAs: Variations on a theme. *Metallomics, 5*(4), 276–286.

Olmedo-Verd, E., Flores, E., Herrero, A., & Muro-Pastor, A. M. (2005). HetR-dependent and -independent expression of heterocyst-related genes in an *Anabaena* strain overproducing the NtcA transcription factor. *Journal of Bacteriology, 187*(6), 1985–1991.

Outten, F. W., Djaman, O., & Storz, G. (2004). A suf operon requirement for Fe-S cluster assembly during iron starvation in *Escherichia coli*. *Molecular Microbiology, 52*(3), 861–872.

Outten, F. W., & Theil, E. C. (2009). Iron-based redox switches in biology. *Antioxidants & Redox Signaling, 11*(5), 1029–1046.

Outten, F. W., Wood, M. J., Munoz, F. M., & Storz, G. (2003). The SufE protein and the SufBCD complex enhance SufS cysteine desulfurase activity as part of a sulfur transfer pathway for Fe-S cluster assembly in *Escherichia coli*. *The Journal of Biological Chemistry, 278*(46), 45713–45719.

Pakrasi, H. B., Goldenberg, A., & Sherman, L. A. (1985). Membrane development in the cyanobacterium *Anacystis nidulans* during recovery from iron starvation. *Plant Physiology, 79*(1), 290–295.

Park, Y. I., Sandstrom, S., Gustafsson, P., & Oquist, G. (1999). Expression of the *isiA* gene is essential for the survival of the cyanobacterium *Synechococcus* sp. PCC 7942 by protecting photosystem II from excess light under iron limitation. *Molecular Microbiology, 32*(1), 123–129.

Pattanaik, B., Busch, A. W., Hu, P., Chen, J., & Montgomery, B. L. (2014). Responses to iron limitation are impacted by light quality and regulated by RcaE in the chromatically acclimating cyanobacterium *Fremyella diplosiphon*. *Microbiology, 160*(Pt. 5), 992–1005.

Peleato, M. L., Ayora, S., Inda, L. A., & Gomez-Moreno, C. (1994). Isolation and characterization of two different flavodoxins from the eukaryote *Chlorella fusca*. *The Biochemical Journal, 302*(Pt. 3), 807–811.

Pellicer, S., González, A., Peleato, M. L., Martinez, J. I., Fillat, M. F., & Bes, M. T. (2012). Site-directed mutagenesis and spectral studies suggest a putative role of FurA from *Anabaena* sp. PCC 7120 as a heme sensor protein. *The FEBS Journal, 279*(12), 2231–2246.

Peschek, G. A. (1996). Structure-function relationships in the dual-function photosynthetic-respiratory electron-transport assembly of cyanobacteria (blue-green algae). *Biochemical Society Transactions, 24*(3), 729–733.

Peschek, G. A., Obinger, C., & Paumann, M. (2004). The respiratory chain of blue-green algae (cyanobacteria). *Physiologia Plantarum, 120*(3), 358–369.

Pierella Karlusich, J. J., Ceccoli, R. D., Grana, M., Romero, H., & Carrillo, N. (2015). Environmental selection pressures related to iron utilization are involved in the loss of the flavodoxin gene from the plant genome. *Genome Biology and Evolution, 7*(3), 750–767.

Pietsch, D., Bernat, G., Kahmann, U., Staiger, D., Pistorius, E. K., & Michel, K. P. (2011). New insights into the function of the iron deficiency-induced protein C from *Synechococcus elongatus* PCC 7942. *Photosynthesis Research, 108*(2–3), 121–132.

Pils, D., & Schmetterer, G. (2001). Characterization of three bioenergetically active respiratory terminal oxidases in the cyanobacterium *Synechocystis* sp. strain PCC 6803. *FEMS Microbiology Letters, 203*(2), 217–222.

Quan, S., Schneider, I., Pan, J., Von Hacht, A., & Bardwell, J. C. (2007). The CXXC motif is more than a redox rheostat. *The Journal of Biological Chemistry, 282*(39), 28823–28833.

Razquin, P., Schmitz, S., Fillat, M. F., Peleato, M. L., & Bohme, H. (1994). Transcriptional and translational analysis of ferredoxin and flavodoxin under iron and nitrogen stress in *Anabaena* sp. strain PCC 7120. *Journal of Bacteriology, 176*(23), 7409–7411.

Razquin, P., Schmitz, S., Peleato, M. L., Fillat, M. F., Gomezmoreno, C., & Bohme, H. (1995). Differential activities of heterocyst ferredoxin, vegetative cell ferredoxin, and flavodoxin as electron carriers in nitrogen-fixation and photosynthesis in *Anabaena* sp. *Photosynthesis Research, 43*(1), 35–40.

Riethman, H. C., & Sherman, L. A. (1988). Purification and characterization of an iron stress-induced chlorophyll-protein from the cyanobacterium *Anacystis nidulans* R2. *Biochimica et Biophysica Acta, 935*(2), 141–151.

Rochaix, J. D. (2011). Reprint of: Regulation of photosynthetic electron transport. *Biochimica et Biophysica Acta, 1807*(8), 878–886.

Rubio, L. M., Flores, E., & Herrero, A. (2002). Purification, cofactor analysis, and site-directed mutagenesis of *Synechococcus* ferredoxin-nitrate reductase. *Photosynthesis Research, 72*(1), 13–26.

Rubio, L. M., Herrero, A., & Flores, E. (1996). A cyanobacterial *narB* gene encodes a ferredoxin-dependent nitrate reductase. *Plant Molecular Biology, 30*(4), 845–850.

Sakurai, Y., Anzai, I., & Furukawa, Y. (2014). A primary role for disulfide formation in the productive folding of prokaryotic Cu, Zn-superoxide dismutase. *The Journal of Biological Chemistry, 289*(29), 20139–20149.

Salvail, H., & Masse, E. (2012). Regulating iron storage and metabolism with RNA: An overview of posttranscriptional controls of intracellular iron homeostasis. *Wiley Interdisciplinary Reviews RNA, 3*(1), 26–36.

Sandmann, G. (1985). Consequences of iron deficiency on photosynthetic and respiratory electron transport in blue-green algae. *Photosynthesis Research, 6*(3), 261–271.

Sandmann, G., Peleato, M. L., Fillat, M. F., Lázaro, M. C., & Gómez-Moreno, C. (1990). Consequences of the iron-dependent formation of ferredoxin and flavodoxin on photosynthesis and nitrogen fixation on Anabaena strains. *Photosynthesis Research, 26*(2), 119–125.

Saxena, R. K., Raghuvanshi, R., Singh, S., & Bisen, P. S. (2006). Iron induced metabolic changes in the diazotrophic cyanobacterium *Anabaena* PCC 7120. *Indian Journal of Experimental Biology, 44*(10), 849–851.

Schmetterer, G. (1994). Cyanobacterial respiration. In D. A. Bryant (Ed.), *The molecular biology of cyanobacteria* (pp. 409–435). Dordrecht: Kluwer Academic.

Schmitz, S., Schrautemeier, B., & Böhme, H. (1993). Evidence from directed mutagenesis that positively charged amino acids are necessary for interaction of nitrogenase with the [2Fe-2S] heterocyst ferredoxin (FdxH) from the cyanobacterium *Anabaena* sp. PCC 7120. *Molecular and General Genetics, 240*(3), 455–460.

Schrautemeier, B., & Böhme, H. (1985). A distinct ferredoxin for nitrogen fixation isolated from heterocysts of the cyanobacterium *Anabaena variabilis*. *FEBS Letters, 184*(2), 304–308.

Setif, P. (2001). Ferredoxin and flavodoxin reduction by photosystem I. *Biochimica et Biophysica Acta, 1507*(1–3), 161–179.

Sevilla, E., Martin-Luna, B., Bes, M. T., Fillat, M. F., & Peleato, M. L. (2011). An active photosynthetic electron transfer chain required for mcyD transcription and microcystin synthesis in *Microcystis aeruginosa* PCC7806. *Ecotoxicology, 21*(3), 811–819.

Sevilla, E., Martin-Luna, B., Vela, L., Bes, M. T., Fillat, M. F., & Peleato, M. L. (2008). Iron availability affects *mcyD* expression and microcystin-LR synthesis in *Microcystis aeruginosa* PCC7806. *Environmental Microbiology, 10*(10), 2476–2483.

Shcolnick, S., & Keren, N. (2006). Metal homeostasis in cyanobacteria and chloroplasts. Balancing benefits and risks to the photosynthetic apparatus. *Plant Physiology, 141*(3), 805–810.

Shcolnick, S., Summerfield, T. C., Reytman, L., Sherman, L. A., & Keren, N. (2009). The mechanism of iron homeostasis in the unicellular cyanobacterium *Synechocystis* sp. PCC 6803 and its relationship to oxidative stress. *Plant Physiology, 150*(4), 2045–2056.

Shen, J. R. (2015). The structure of photosystem II and the mechanism of water oxidation in photosynthesis. *Annual Review of Plant Biology, 66*, 23–48.

Sherman, D. M., & Sherman, L. A. (1983). Effect of iron deficiency and iron restoration on ultrastructure of *Anacystis nidulans*. *Journal of Bacteriology, 156*(1), 393–401.

Shi, T., Sun, Y., & Falkowski, P. G. (2007). Effects of iron limitation on the expression of metabolic genes in the marine cyanobacterium *Trichodesmium erythraeum* IMS101. *Environmental Microbiology, 9*(12), 2945–2956.

Shimizu, T. (2012). Binding of cysteine thiolate to the Fe(III) heme complex is critical for the function of heme sensor proteins. *Journal of Inorganic Biochemistry, 108*, 171–177.

Sidler, W. A. (1994). Phycobilisome and phycobiliprotein structures. In D. A. Bryant (Ed.), *The molecular biology of cyanobacteria* (pp. 139–216). Dordrecht: Kluwer Academic Publishers.

Singh, A. K., McIntyre, L. M., & Sherman, L. A. (2003). Microarray analysis of the genome-wide response to iron deficiency and iron reconstitution in the cyanobacterium *Synechocystis* sp. PCC 6803. *Plant Physiology, 132*(4), 1825–1839.

Singh, A. K., & Sherman, L. A. (2007). Reflections on the function of IsiA, a cyanobacterial stress-inducible, Chl-binding protein. *Photosynthesis Research, 93*(1–3), 17–25.

Sohm, J. A., Webb, E. A., & Capone, D. G. (2011). Emerging patterns of marine nitrogen fixation. *Nature Reviews Microbiology, 9*(7), 499–508.

Strzepek, R. F., & Harrison, P. J. (2004). Photosynthetic architecture differs in coastal and oceanic diatoms. *Nature, 431*(7009), 689–692.

Suga, M., Akita, F., Hirata, K., Ueno, G., Murakami, H., Nakajima, Y., et al. (2015). Native structure of photosystem II at 1.95 A resolution viewed by femtosecond X-ray pulses. *Nature, 517*(7532), 99–103.

Sun, J., & Golbeck, J. H. (2015). The presence of the IsiA-PSI supercomplex leads to enhanced photosystem I electron throughput in iron-starved cells of *Synechococcus* sp. PCC 7002. *J Phys Chem B, 119*(43), 13549–13559.

Teixido, L., Carrasco, B., Alonso, J. C., Barbe, J., & Campoy, S. (2011). Fur activates the expression of *Salmonella enterica* pathogenicity island 1 by directly interacting with the *hilD* operator in vivo and in vitro. *PloS One, 6*(5), e19711.

Tetenkin, V. L., Golitsin, V. M., & Gulyaev, B. A. (1998). Stress protein of cyanobacteria CP36: Interaction with photoactive complexes and formation of supramolecular structures. *Biochemistry (Mosc), 63*(5), 584–591.

Theil, E. C. (1987). Ferritin: Structure, gene regulation, and cellular function in animals, plants, and microorganisms. *Annual Review of Biochemistry, 56*, 289–315.

Tillett, D., Dittmann, E., Erhard, M., von Dohren, H., Borner, T., & Neilan, B. A. (2000). Structural organization of microcystin biosynthesis in *Microcystis aeruginosa* PCC7806: An integrated peptide-polyketide synthetase system. *Chemistry & Biology*, 7(10), 753–764.

Tognetti, V. B., Monti, M. R., Valle, E. M., Carrillo, N., & Smania, A. M. (2007). Detoxification of 2,4-dinitrotoluene by transgenic tobacco plants expressing a bacterial flavodoxin. *Environmental Science & Technology, 41*(11), 4071–4076.

Tognetti, V. B., Palatnik, J. F., Fillat, M. F., Melzer, M., Hajirezaei, M. R., Valle, E. M., et al. (2006). Functional replacement of ferredoxin by a cyanobacterial flavodoxin in tobacco confers broad-range stress tolerance. *Plant Cell, 18*(8), 2035–2050.

Tognetti, V. B., Zurbriggen, M. D., Morandi, E. N., Fillat, M. F., Valle, E. M., Hajirezaei, M. R., et al. (2007). Enhanced plant tolerance to iron starvation by functional substitution of chloroplast ferredoxin with a bacterial flavodoxin. *Proceedings of the National Academy of Sciences of the United States of America, 104*(27), 11495–11500.

Tolle, J., Michel, K. P., Kruip, J., Kahmann, U., Preisfeld, A., & Pistorius, E. K. (2002). Localization and function of the IdiA homologue Slr1295 in the cyanobacterium *Synechocystis* sp. strain PCC 6803. *Microbiology, 148*(Pt. 10), 3293–3305.

Utkilen, H., & Gjolme, N. (1995). Iron-stimulated toxin production in *Microcystis aeruginosa*. *Applied and Environmental Microbiology, 61*(2), 797–800.

Valladares, A., Herrero, A., Pils, D., Schmetterer, G., & Flores, E. (2003). Cytochrome c oxidase genes required for nitrogenase activity and diazotrophic growth in *Anabaena* sp. PCC 7120. *Molecular Microbiology, 47*(5), 1239–1249.

Vavilin, D. V., & Vermaas, W. F. (2002). Regulation of the tetrapyrrole biosynthetic pathway leading to heme and chlorophyll in plants and cyanobacteria. *Physiologia Plantarum, 115*(1), 9–24.

Vicente, J. B., Carrondo, M. A., Teixeira, M., & Frazao, C. (2008). Structural studies on flavodiiron proteins. *Methods in Enzymology, 437*, 3–19.

Vigara, A. J., Inda, L. A., Vega, J. M., Gomez-Moreno, C., & Peleato, M. L. (1998). Flavodoxin as an electronic donor in photosynthetic inorganic nitrogen assimilation by iron-deficient *Chlorella fusca* cells. *Photochemistry and Photobiology, 67*(4), 446–449.

Vinnemeier, J., & Hagemann, M. (1999). Identification of salt-regulated genes in the genome of the cyanobacterium *Synechocystis* sp. strain PCC 6803 by subtractive RNA hybridization. *Archives of Microbiology, 172*(6), 377–386.

Vinnemeier, J., Kunert, A., & Hagemann, M. (1998). Transcriptional analysis of the *isiAB* operon in salt-stressed cells of the cyanobacterium *Synechocystis* sp. PCC 6803. *FEMS Microbiology Letters, 169*(2), 323–330.

Vitale, S., Fauquant, C., Lascoux, D., Schauer, K., Saint-Pierre, C., & Michaud-Soret, I. (2009). A ZnS(4) structural zinc site in the *Helicobacter pylori* ferric uptake regulator. *Biochemistry, 48*(24), 5582–5591.

Wang, Q., Hall, C. L., Al-Adami, M. Z., & He, Q. (2010). IsiA is required for the formation of photosystem I supercomplexes and for efficient state transition in *Synechocystis* PCC 6803. *PloS One, 5*(5), e10432.

Wang, Y., Wu, S. L., Hancock, W. S., Trala, R., Kessler, M., Taylor, A. H., et al. (2005). Proteomic profiling of *Escherichia coli* proteins under high cell density fed-batch cultivation with overexpression of phosphogluconolactonase. *Biotechnology Progress, 21*(5), 1401–1411.

Watanabe, M., & Ikeuchi, M. (2013). Phycobilisome: Architecture of a light-harvesting supercomplex. *Photosynthesis Research, 116*(2–3), 265–276.

Wen-Liang, X., Yong-Ding, L., & Cheng-Cai, Z. (2003). Effect of iron deficiency on heterocyst differentiation and physiology of the filamentous cyanobacterium *Anabaena* sp. PCC 7120. *Wuhan University Journal of Natural Sciences, 8*(3A), 880–884.

Wilderman, P. J., Sowa, N. A., FitzGerald, D. J., FitzGerald, P. C., Gottesman, S., Ochsner, U. A., et al. (2004). Identification of tandem duplicate regulatory small RNAs in *Pseudomonas aeruginosa* involved in iron homeostasis. *Proceedings of the National Academy of Sciences of the United States of America, 101*(26), 9792–9797.

Wouters, M. A., Fan, S. W., & Haworth, N. L. (2010). Disulfides as redox switches: From molecular mechanisms to functional significance. *Antioxidants & Redox Signaling*, *12*(1), 53–91.

Wu, Y., & Brosh, R. M., Jr. (2012). DNA helicase and helicase-nuclease enzymes with a conserved iron-sulfur cluster. *Nucleic Acids Research*, *40*(10), 4247–4260.

Yeremenko, N., Jeanjean, R., Prommeenate, P., Krasikov, V., Nixon, P. J., Vermaas, W. F., et al. (2005). Open reading frame *ssr2016* is required for antimycin A-sensitive photosystem I-driven cyclic electron flow in the cyanobacterium *Synechocystis* sp. PCC 6803. *Plant & Cell Physiology*, *46*(8), 1433–1436.

Yeremenko, N., Kouril, R., Ihalainen, J. A., D'Haene, S., van Oosterwijk, N., Andrizhiyevskaya, E. G., et al. (2004). Supramolecular organization and dual function of the IsiA chlorophyll-binding protein in cyanobacteria. *Biochemistry*, *43*(32), 10308–10313.

Yi, L., Jenkins, P. M., Leichert, L. I., Jakob, U., Martens, J. R., & Ragsdale, S. W. (2009). Heme regulatory motifs in heme oxygenase-2 form a thiol/disulfide redox switch that responds to the cellular redox state. *The Journal of Biological Chemistry*, *284*(31), 20556–20561.

Yi, L., & Ragsdale, S. W. (2007). Evidence that the heme regulatory motifs in heme oxygenase-2 serve as a thiol/disulfide redox switch regulating heme binding. *The Journal of Biological Chemistry*, *282*(29), 21056–21067.

Yousef, N., Pistorius, E. K., & Michel, K. P. (2003). Comparative analysis of *idiA* and *isiA* transcription under iron starvation and oxidative stress in *Synechococcus elongatus* PCC 7942 wild-type and selected mutants. *Archives of Microbiology*, *180*(6), 471–483.

Yu, C., & Genco, C. A. (2012). Fur-mediated global regulatory circuits in pathogenic *Neisseria species*. *Journal of Bacteriology*, *194*(23), 6372–6381.

Zhang, P., Allahverdiyeva, Y., Eisenhut, M., & Aro, E. M. (2009). Flavodiiron proteins in oxygenic photosynthetic organisms: Photoprotection of photosystem II by Flv2 and Flv4 in *Synechocystis* sp. PCC 6803. *PLoS One*, *4*(4), e5331.

Zhang, C. C., Laurent, S., Sakr, S., Peng, L., & Bedu, S. (2006). Heterocyst differentiation and pattern formation in cyanobacteria: A chorus of signals. *Molecular Microbiology*, *59*(2), 367–375.

Zhang, A., Wassarman, K. M., Rosenow, C., Tjaden, B. C., Storz, G., & Gottesman, S. (2003). Global analysis of small RNA and mRNA targets of Hfq. *Molecular Microbiology*, *50*(4), 1111–1124.

Zheng, M., Doan, B., Schneider, T. D., & Storz, G. (1999). OxyR and SoxRS regulation of *fur*. *Journal of Bacteriology*, *181*(15), 4639–4643.

Zilliges, Y., Kehr, J. C., Meissner, S., Ishida, K., Mikkat, S., Hagemann, M., et al. (2011). The cyanobacterial hepatotoxin microcystin binds to proteins and increases the fitness of microcystis under oxidative stress conditions. *PloS One*, *6*(3), e17615.

Zurbriggen, M. D., Tognetti, V. B., Fillat, M. F., Hajirezaei, M. R., Valle, E. M., & Carrillo, N. (2008). Combating stress with flavodoxin: A promising route for crop improvement. *Trends in Biotechnology*, *26*(10), 531–537.

CHAPTER SIX

Bacterial Electron Transfer Chains Primed by Proteomics

H.J.C.T. Wessels*, N.M. de Almeida[†], B. Kartal[†,‡], J.T. Keltjens[†,1]

*Nijmegen Center for Mitochondrial Disorders, Radboud Proteomics Centre, Translational Metabolic Laboratory, Radboud University Medical Center, Nijmegen, The Netherlands

[†]Institute of Water and Wetland Research, Radboud University Nijmegen, Nijmegen, The Netherlands

[‡]Laboratory of Microbiology, Ghent University, Ghent, Belgium

[1]Corresponding author: e-mail address: j.keltjens@science.ru.nl

Contents

Abstract

Electron transport phosphorylation is the central mechanism for most prokaryotic species to harvest energy released in the respiration of their substrates as ATP. Microorganisms have evolved incredible variations on this principle, most of these we perhaps do not know, considering that only a fraction of the microbial richness is known. Besides these variations, microbial species may show substantial versatility in using respiratory systems. In connection herewith, regulatory mechanisms control the expression of these respiratory enzyme systems and their assembly at the translational and posttranslational levels, to optimally accommodate changes in the supply of their energy substrates. Here, we present an overview of methods and techniques

Advances in Microbial Physiology, Volume 68
ISSN 0065-2911
http://dx.doi.org/10.1016/bs.ampbs.2016.02.006

from the field of proteomics to explore bacterial electron transfer chains and their regulation at levels ranging from the whole organism down to the Ångstrom scales of protein structures. From the survey of the literature on this subject, it is concluded that proteomics, indeed, has substantially contributed to our comprehending of bacterial respiratory mechanisms, often in elegant combinations with genetic and biochemical approaches. However, we also note that advanced proteomics offers a wealth of opportunities, which have not been exploited at all, or at best underexploited in hypothesis-driving and hypothesis-driven research on bacterial bioenergetics. Examples obtained from the related area of mitochondrial oxidative phosphorylation research, where the application of advanced proteomics is more common, may illustrate these opportunities.

ABBREVIATIONS

16-BAC 16-benzyldimethyl-*n*-hexadecylammonium chloride
2DE two-dimensional gel electrophoresis
AP-MS, q-AP-MS (quantitative) affinity purification
AQUA absolute quantitation
BNE blue native gel electrophoresis
CBB Coomassie Brilliant blue
CE capillary electrophoresis
CID collision-induced dissociation
CNE clear (or colourless) native gel electrophoresis
CTAB cetyltrimethyl-ammonium bromide
CX-MS chemical cross-linking coupled with MS
Da Dalton
DIA data-independent analysis
DIGE difference gel electrophoresis
EI electron ionisation
emPAI exponentially modified protein abundance index
ESI electrospray ionisation
ETC electron transfer chain
ETD electron transfer dissociation
ETP electron transport phosphorylation
ETS electron transfer system
FAB fast atom bombardment
FT-ICR Fourier transform ion-cyclotron-resonance
FWHM full width at half maximum
GRAVY grand average of hydropathicity
HDX-MS hydrogen/deuterium exchange coupled with MS
ICAT isotope-coded affinity tags
IEF isoelectric focusing
IM-MS ion-mobility MS
IT ion trap
iTRAQ isobaric tags for relative and absolute quantitation
LC liquid chromatography

LC-MS/MS liquid chromatography online tandem mass spectrometry
LIT linear ion trap
LMW low-molecular weight
m relative mass
m/z mass-to-charge ratio
MALDI-TOF matrix-assisted laser desorption ionisation time-of-flight
MRM multiple reaction monitoring
MS mass spectrometry
MS^n higher order mass spectrometry experiment
MudPIT multidimensional protein identification technology
NDH NADH:quinone oxidoreductase (complex I)
ORF open-reading frame
OXPHOS oxidative phosphorylation, more specifically in mitochondria
PAGE polyacrylamide gel electrophoresis
p*I* isoelectric point
pmf proton-motive force
PTMs posttranslational modifications
Q analytical quadrupole
q nonanalytical quadrupole
QqQ triple quadrupole
Q-TOF quadrupole time of flight
RCS, RES, RNS, ROS reactive chlorine, electrophilic, nitrogen, oxygen species
RMP (bacterial) respiratory membrane-bound protein complex
RP reversed-phase
SC supercomplex
SCX strong cation-exchange chromatography
SDH succinate dehydrogenase (complex II)
SDS sodium dodecyl sulphate
SILAC stable isotope labelling with amino acids in cell culture
SRM selected reaction monitoring
TAP tandem affinity purification
Th thomson
TMH transmembrane helix
TMT tandem mass tags
TOF time-of-flight
u atomic mass unit
z charge number

1. INTRODUCTION

Respiration is a hallmark of life. By this principle, also known as electron transport phosphorylation (ETP) or oxidative phosphorylation (OXPHOS), electrons derived from the oxidation of organic or inorganic substrates are transferred to a terminal electron acceptor (Al-Attar & de Vries, 2013). Substrate oxidation and reduction, and intermediary electron transfer are catalysed by respiratory membrane-bound protein complexes (RMPs). Low-molecular-weight compounds dissolved either in the membrane, such as quinones, or in the cytoplasmic (NADH, ferredoxins) and periplasmic (*c*-type cytochromes, blue copper proteins) solutes assist as electron shuttles between the RMPs. These systems are organised such that the energy released in these redox processes is conserved as a proton-motive force (*pmf*) or a sodium-motive force (*smf*) across the cell membrane. This force drives the synthesis of ATP by the membrane-bound, H^+ (or Na^+)-translocating ATP synthase. The textbook example of an electron transfer system (ETS) is the mitochondrion of eukaryotes. In mitochondria, electrons from NADH or succinate oxidation enter the chain via NADH:quinone oxidoreductase (NDH, complex I) or succinate dehydrogenase (SDH, complex II), respectively, pass via a quinol:cytochrome *c* oxidoreductase (complex III) to end in the terminal oxidase (complex IV) that reduces O_2 to H_2O. Energy is conserved by the ATP synthase (complex V).

Eukaryotes generally respire only oxygen by using electrons generated from the oxidation of organic substrates (sugars, fatty acids, amino acids). Aerobic microorganisms use oxygen as well, but they have a diversity of oxidases at their disposal that can cope with an enormous range of environmental oxygen concentrations (Borisov, Gennis, Hemp, & Verkhovsky, 2011; Han et al., 2011; Hemp & Gennis, 2008). Prokaryotes also can use many other (in)organic electron acceptors in anaerobic respiration in species-dependent ways (Croal, Gralnick, Malasarn, & Newman, 2004; Grein, Ramos, Venceslau, & Pereira, 2013; Heimann et al., 2007; Kraft, Strous, & Tegetmeyer, 2011; Matias, Pereira, Soares, & Carrondo, 2005; Richter, Schicklberger, & Gescher, 2012; Shan, Lai, & Yan, 2012; Simon & Kern, 2008; Suh, Kuntumalla, Yu, & Pieper, 2014; Tosha & Shiro, 2013; Zumft & Kroneck, 2007), whereas an almost unlimited number of (in)organic substrates serve as electron donors, provided energy is gained from the redox processes. The conversion of each specific set of substrates requires its dedicated RMPs that are organised as linear or branched electron transfer chains (ETCs). Besides this versatility, prokaryotes may display a tremendous respiratory flexibility to cope with changes in the supply of energy substrates. Many different ETSs from model organisms have been studied by now in varying levels of detail (see above references). Still, we most likely have only a limited notion of these systems: only a minor fraction of the microbial species richness is known. However, our insight into this richness increases at an incredible pace, supported by the power of (meta)genomics and (meta)transcriptomics (Brown et al., 2015; Caro-Quintero & Konstantinidis, 2012; Faust, Lahti, Gonze, de Vos, & Raes, 2015; Gerber, 2014; Godzik, 2011; Land et al., 2015; Méndez-García et al., 2015; Oren & Garrity, 2014; Ponomarova & Patil, 2015; Suenaga, 2015; Temperton & Giovannoni, 2012; van Schaik, 2015). From metagenome assemblies, it can be inferred that microorganisms have evolved overwhelming variations of already known RMPs, permitting a species to occupy a specific niche. Many of the genome assemblies deal with species of completely new phyla, lacking cultured representatives with known physiology and metabolism. These new species may metabolise their substrates by new pathways whose nature is hidden in DNA sequences as genes coding for 'hypothetical' or 'conserved' proteins, or genes that have been annotated erroneously. Moreover, even if a certain gene product is highly homologous to a protein of known function, this does not imply that the gene product is functional or has the predicted function. Functional expression may be highly complicated requiring the association of other partners that are not immediately identified by DNA analyses, no matter how advanced. This is particularly the case for RMPs.

RMPs are usually composed of different subunits in fixed ratios. In prokaryotes, for instance, NDH consists of 13 subunits, while the same complex (I) in eukaryotic mitochondria comprises about 45 subunits (Mimaki, Wang, McKenzie, Thorburn, & Ryan, 2012). After translation, each of the subunits undergoes modifications, resulting in mature functional complexes (Section 5.2.1). Next, posttranslational modifications (PTMs) may involve the addition of specific groups (phosphorylation, acetylation, methylation, etc.) to specific amino acids for structural integrity or to control enzyme activity (Section 4.2). All in all, the assembly and PTMs must be well timed and carefully controlled events, which are regulated in the way that the living cell can immediately and properly respond to environmental changes.

By now, it has become clear that the above-depicted view of ETCs as being composed of individual RMPs linked by mobile electron carriers is more complicated. RMPs appear to combine into supercomplexes (SCs) (Section 5.1). The modes of SC formation and factors directing their formation have become one of the most exciting recent advances in the research on ETSs.

All of the phenomena outlined earlier boil down to the fact that proteins are composed of specific amino acid sequences that are subjected to cleavage, insertion of (nonpeptide) cofactors, the addition of functional groups, small molecules or other chemical modifications: In the end, it is all about chemistry. This chemical a priori makes their study amenable to one of the most powerful, sensitive and accurate methods in the field, mass spectrometry (MS). In the last two decades, we have seen incredible developments in MS instrumentation and methodologies, as well as in the generation of massive amounts of high-quality data and computer software-assisted handling of these data. Hereby, MS-supported protein research has become a field on its own: proteomics. In this chapter, we will present an overview of the way proteomics has added to the fundamental understanding of electron transport systems and their regulation, and particularly in the way applications of the proteomics toolbox may contribute to the future understanding of bacterial respiration. We will do so from the global and enzyme-targeted perspectives. The global approach aims at mapping the complete protein complement and all its modifications (proteoforms; Smith, Kelleher, & Consortium for Top Down Proteomics, 2013) of an organism, and how this protein complement is modified in response to environmental changes. This approach is mainly discovery based and can be used to generate new hypotheses. In contrast, the enzyme-targeted, hypothesis-driven research is directed towards the insight in the structure and functioning of an enzyme or enzyme system as isolated, taking advantage of the methodology offered by proteomics. Proteomics supports both approaches by bottom-up and top-down methods. In the bottom-up direction, which we will also refer to as peptide-centric, proteins in a sample are first proteolytically or chemically cleaved. Hereafter, peptide fragments are analysed by MS as to identify (and quantify) the proteins present in the sample or to get structural information in case of an enzyme-targeted research question. The top-down, protein-centred approach takes the whole protein (complex) as the starting point. Both in bottom-up and in top-down directions, proteomics offers an arsenal of MS analytical methods. To make a nonexpert reader familiar with these methods, we will start our review with an introduction to proteomics.

2. METHODOLOGY OF PROTEOMICS

Unambiguous protein identification, characterisation and quantitation by MS are invaluable to gain molecular insight into the complicated protein complexes of the electron transport system. The possibility to fully dissect protein complexes at the molecular level using only small sample amounts with high throughput is unique to MS and often crucial in protein research. MS is used to accurately determine a (poly)peptide's mass, amino acid sequence (or identity), quantity, PTMs and interaction partners. Even more elaborate experiments can be performed to unravel fine structural details such as disulphide bridge arrangements, cross section and protein–protein interaction sites. Continuous advancements in hardware, software and method development rapidly expand the capabilities of MS in protein research but also make it difficult for nonexperts to grasp the possibilities and limitations of this technique. In this section, we will provide some basics of MS and proteomics methodology that are relevant for the analysis of ETS protein complexes (see for a more detailed discussion: de Hoffmann & Stroobant, 2007).

2.1 Introduction to Mass Spectrometry

To understand how proteomics methodology works, it is important to have a minimal understanding of MS hardware and how experiments are conducted. We will first explain what information can be obtained from a mass spectrum and introduce the units that are used in this review.

2.1.1 The Mass Spectrum

In contrast to its name, a mass spectrometer does not measure the mass of a (poly)peptide directly per se. Rather, it measures the mass-to-charge ratio (m/z) in thomson (Th) of ions from which the relative mass (m) can be derived if the charge number of the ion (z) is known. The relative mass (m) is a dimensionless number which is defined as the mass of an atom, a molecule or an ion divided by 1/12th of the mass of the ^{12}C carbon atom. The charge number (z) is the total charge on an ion divided by the elementary charge (e), 1.602177×10^{-19} C. The thomson is defined as 1 Th = 1 $u/e = 1.036426 \times 10^{-8}$ kg C^{-1}. The mass of a (poly)peptide in proteomics is described by its monoisotopic mass, average mass or nominal mass in atomic mass units (u or amu), which is called the Dalton (Da). The monoisotopic mass of an ion is calculated using the 'exact' mass of the predominant isotope of each element. The average mass (or chemical mass) is a weighted average of the natural isotopes for the atomic mass of each element. The nominal mass is the mass of the predominant isotope of each element rounded to the nearest integer value. At first sight, these mass definitions seem to differ only marginally. However, for large biopolymers such as (poly)peptides the differences become significant, especially when taking the mass accuracy of modern mass spectrometers into account. For example, if we consider a carbamidomethylated (CAM) tryptic peptide from bovine serum albumin (BSA), $YIC_{(cam)}DNQDTISSK$ (chemical formula: $C_{59}H_{94}N_{16}O_{24}S_1$), significant mass differences are observed for the different mass calculations: monoisotopic mass, 1442.6348 u; average mass, 1443.5376 u; nominal mass, 1442.0000 u. The monoisotopic mass is typically used in proteomics experiments to describe the mass of a (poly)peptide if its isotopic peaks are resolved in the mass spectrum. A mass spectrum plots the intensity of an ion vs its mass-to-charge ratio. Fig. 1 shows the isotopic peaks in an electrospray ionisation (ESI) mass spectrum for ions of the tryptic BSA peptide $YIC_{(cam)}DNQDTISSK$ with $z=2$ (Fig. 1A) and $z=3$ (Fig. 1B). From both charge states of the ion, we derive the same monoisotopic mass: Mr 1442.6358 Da from m/z 722.3257 $[M+2H]^{2+}$ and Mr 1442.6381 Da from m/z 481.8872 $[M+3H]^{3+}$. The charge state of an ion can be determined from the mass spectrum on the basis of the observed distance in Th between the different isotopic peaks. Since the isotopic peaks differ by 1 Da, the observed distance will be $1/z$. Therefore, the charge state of an ion can be calculated as $z=1/[\text{observed distance in Th}]$. For the ion in Fig. 1A, we observe a mass difference of 0.5 Th between the isotopic peaks, which determines the charge number of the ion: $z=1/0.5=2$. Similarly, for the ion in Fig. 1B, the observed mass difference of 0.33 Th establishes its charge number: $z=1/0.33=3$. Computer algorithms can also generate a 'deconvoluted' spectrum in which the spectral information from each charge state peak of a single compound (identical m but different z) is combined into a single spectral peak with mass [M] or $[M+H]^{1+}$. This is particularly useful for large biomolecules such as intact proteins, which are usually detected with a wide range of different charges in a single ESI mass spectrum. Fig. 2A shows the ESI mass spectrum (as recorded) of bovine cytochrome *c*, which was detected with charge states in the range of $z=7$ up to $z=21$. The deconvoluted spectrum shown in Fig. 2B is much easier to interpret as it shows only a single spectral peak for cytochrome *c* with a neutral mass of Mr 12,351.3059 Da (calculated neutral mass for $C_{560}H_{873}N_{148}O_{156}S_4Fe_1$ is Mr 12,351.3161 Da; measured mass error: 0.0102 Da/0.8 ppm). The isotopic distribution observed in mass spectra contains even more information than the mere charge number of an ion. The natural occurrence of isotopes differs between elements. Therefore,

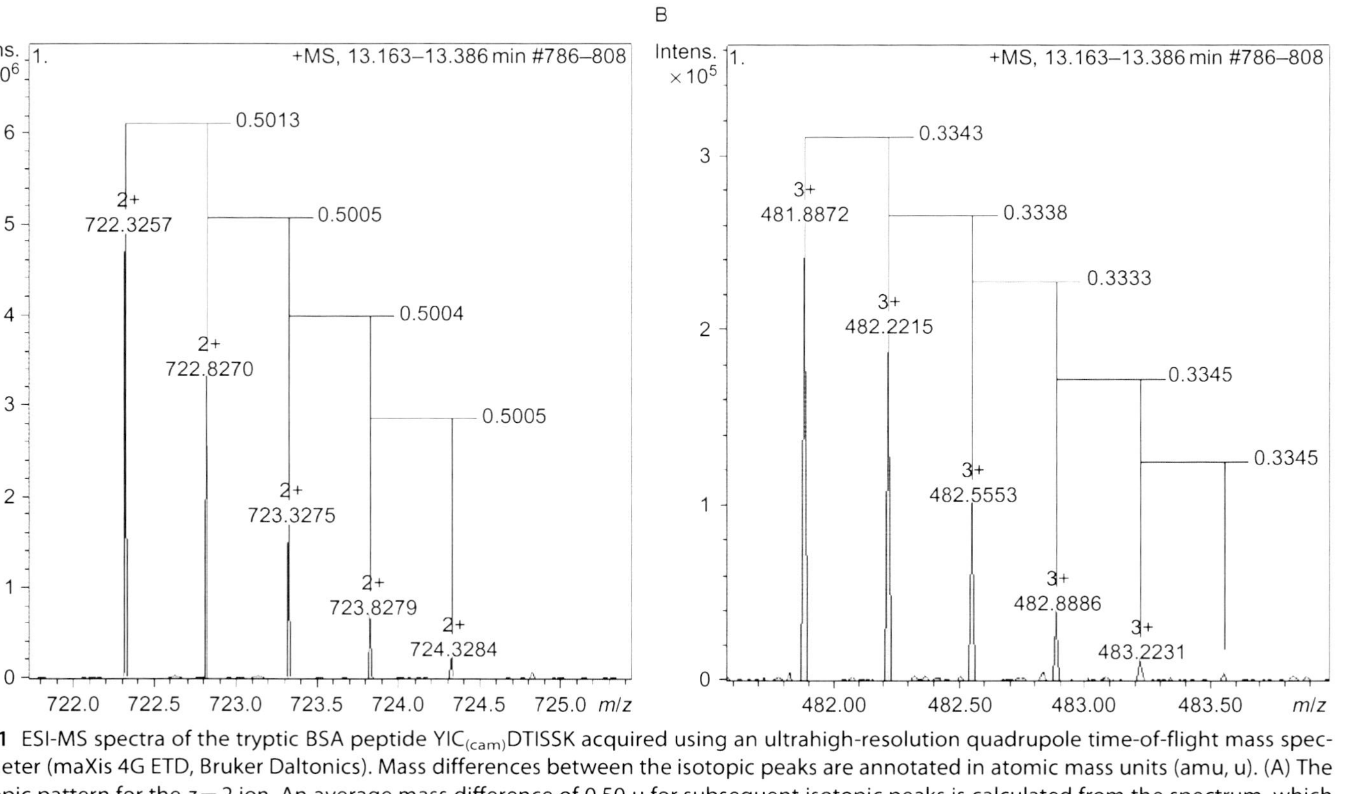

Fig. 1 ESI-MS spectra of the tryptic BSA peptide $YIC_{(cam)}DTISSK$ acquired using an ultrahigh-resolution quadrupole time-of-flight mass spectrometer (maXis 4G ETD, Bruker Daltonics). Mass differences between the isotopic peaks are annotated in atomic mass units (amu, u). (A) The isotopic pattern for the $z=2$ ion. An average mass difference of 0.50 u for subsequent isotopic peaks is calculated from the spectrum, which can be used to determine the charge state of the ion: $z=1/0.5=2$. (B) The isotopic pattern for the $z=3$ ion. The average mass difference of 0.33 u for subsequent isotopic peaks establishes $z=1/0.33=3$.

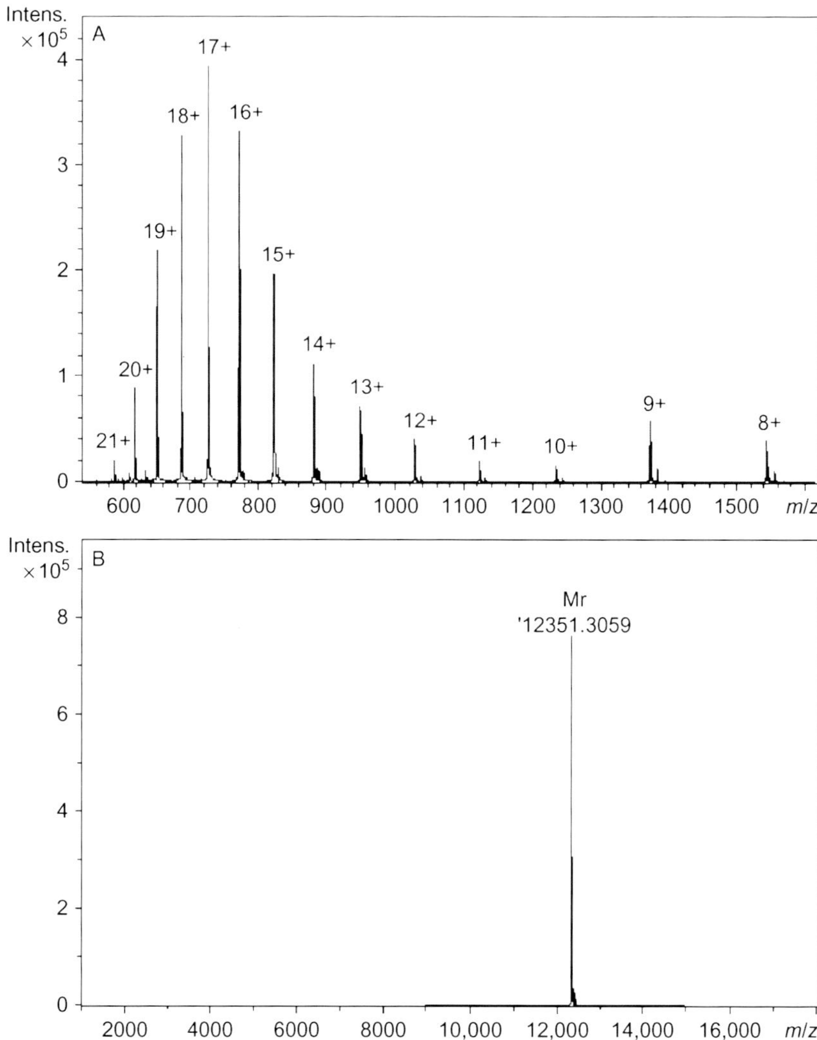

Fig. 2 ESI-MS and deconvoluted MS spectra of horse cytochrome *c*. (A) The ESI-MS spectrum in which a range of different charge state peaks are observed for horse cytochrome *c*. (B) The deconvoluted neutral mass spectrum for cytochrome *c*. The mass measurement error for cytochrome *c* was only 0.0102 Da or 0.826 ppm. Spectra were acquired using an ultrahigh-resolution quadrupole time-of-flight mass spectrometer (maxis 4G ETD, Bruker Daltonics).

the relative intensities from the different isotopic peaks of a molecule reflect directly its elemental composition. Thus far, besides accurate mass, the isotopic pattern is mostly used to verify the identity of small molecules by MS. To summarise, a mass spectrum contains the mass-to-charge ratio (m/z) of an ion from which *m* and *z* can be derived, information about the elemental composition of the ion from its isotopic pattern, and the detected intensity of the ion in arbitrary units.

2.1.2 The Ionisation Process

Any MS experiment starts with the ionisation process, which takes place in the ion source. Ionisation of (poly)peptides is essential in MS since electrical and/or magnetic fields are applied by the mass analyser to subsequently transport, manipulate and detect molecules with a net positive or negative charge (ions). During ionisation, a (poly)peptide acquires a net positive or negative charge by gaining or losing protons. Besides charging, the ionisation process must ensure that ions are available in the gas phase for the mass analyser. This may seem trivial, but efficient ionisation of nonvolatile biomolecules such as (poly)peptides has been a challenge that was solved by the introduction of soft ionisation techniques of 'ESI' and 'matrix-assisted laser desorption ionisation' (MALDI) in 1984 (Yamashita & Fenn, 1984) and 1985 (Karas, Bachmann, & Hillenkamp, 1985), respectively. Till then, harsh ionisation techniques such as 'electron ionisation' (EI) (de Hoffmann & Stroobant, 2007) and the softer 'fast atom bombardment' (FAB) (Barber, Bordoli, Sedgwick, & Tyler, 1981) methods were used to ionise (poly)peptides, but these techniques were often inefficient or induced undesired fragmentation of the analyte prior to detection. Both ESI and MALDI are nondestructive ionisation methods for (poly)peptides that achieve high ionisation efficiency to enable routine mass spectrometric analysis of (poly) peptides.

In MALDI, (poly)peptides are cocrystallised with a small organic compound on the surface of a sample plate. Following crystallisation, the sample plate is inserted via a vacuum lock into the mass spectrometer where it is placed onto a *XYZ* stage and transported into the MALDI source chamber. On the *XYZ* stage, the target is aligned with a laser beam and ion optics of the mass analyser. Once positioned, the laser is pulsed at the sample spot at frequencies of up to 10,000 Hz to generate ions. The sample plate is meticulously shifted after a fixed number of laser shots to prevent signal decrease due to sample depletion. Each shift thus presents 'fresh' sample to the mass analyser. The pulsed light of the laser is absorbed by the cocrystallised matrix molecules. Following ablation, a phase explosion takes place in which clusters of analyte and matrix molecules fly away from the target towards the entrance of the mass analyser. At the same time, protons are transferred from the matrix to the (poly)peptide or vice versa to yield positively or negatively charged ions, respectively. Classical MALDI matrices such as α-cyano-4-hydroxycinnamic acid (CHCA), sinapinic acid and 2,5-dihydroxybenzoic acid (Juhasz, Costello, & Biemann, 1993) typically yield singly charged ions. Other 'cooler' matrices such as 1,5-diaminonaphthalene (1,5-DAN) (Fukuyama, Iwamoto, & Tanaka, 2006), 2,6-dihydroxyacetophenone (DHAP) (Gorman, Ferguson, & Nguyen, 1996) and 5-aminosalicylic acid (5-ASA) (Sakakura & Takayama, 2010) are available for MALDI, which allow the generation of multiply charged ions and can be used for in-source decay tandem MS experiments.

In ESI, (poly)peptides in solution are passed through a capillary that ends in a narrow opening to produce small droplets (electrospray emitter). Ionisation is induced at atmospheric pressure by applying a distal voltage directly to the solvent or to the capillary outlet in which the mass spectrometer acts as ground, or by applying an electric potential difference between the mass spectrometer inlet capillary and a spray cap. During ESI, protons are transferred from the solvent to the (poly)peptide or vice versa to yield positively or negatively charged ions, respectively. Desolvation occurs prior to entrance into the vacuum of the mass spectrometer to prevent loss of vacuum and coulomb explosions of droplets that contaminate the inlet capillary or ion optics of the mass spectrometer. Contrary to MALDI, ESI typically generates multiply charged ions for peptides and proteins.

The choice for MALDI or ESI ionisation depends on the sample, mass analyser and desired MS experiment that needs to be performed. For instance, one should consider the mass and expected charge of the (poly) peptide in relation to the upper mass range limit of the mass analyser. If we would analyse a peptide of 5999 Da by MALDI using CHCA matrix on an ion trap with an upper mass range limit of 3000 m/z, the ion would not be detected since its mass-to-charge ratio likely is m/z 6000 $[M+1H]^{1+}$. On the other hand, a mass analyser with a higher upper mass range limit, such as a time-of-flight (TOF) analyser, would be able to detect the $z=1$ ion generated by MALDI. Conversely, if the peptide acquires three protons in ESI to form an ion of m/z 2000.7 $[M+3H]^{3+}$, it is detectable by the ion trap.

2.1.3 Mass Analysers

Once ionisation and desolvation have taken place, ions are introduced into the high vacuum region of the mass spectrometer where ion optics focus and transport the ion beam to the mass analyser. A variety of mass analysers are currently used in proteomics, each of which has their own specifications, advantages and limitations. One such specification is the resolution of an instrument. In MS, resolution is defined by the full width at half maximum (FWHM; Δm at 50% height) of a spectral peak. Resolution for mass spectrometers is expressed as the mass resolving power (R) which is calculated as $m/\Delta m_{50\%}$.

Quadrupole mass analysers scan through a selected m/z range using oscillating electric fields to create a stable trajectory for ions of a specific m/z ratio to reach the detector (Kero, Pedder, & Yost, 2004). Ions with a stable trajectory produce a signal that scales proportionally to the number of ions that reach the detector. A single spectrum is thus the combined product of a few up to tens of thousands of trajectories at scan speeds of up to 1000 Th s^{-1} and more. Here, the number of trajectories that are being used per spectrum depends on the resolving power of the instrument and selected mass range. Quadrupole analysers are the lowest-resolution analysers in proteomics and are mainly employed in tandem to selectively quantify ions of interest. Isotopic resolution can generally be achieved only for up to doubly charged ions. Still, quadrupoles are frequently employed in combination with other mass analysers, such as TOF or orbitrap mass analysers, to select ions of interest for multiple stage mass spectrometric analyses (MS^n).

Ion trap mass analysers can be classified into two types that essentially follow the same principle: the 3D ion trap (Paul trap) (Brancia, 2006; March, 1997; McLuckey, Van Berkel, Goeringer, & Glish, 1994) or the 2D ion trap (linear ion trap—LIT) (Douglas, Frank, & Mao, 2005). Conceptually, an ion trap is based on the inverse principle of a quadrupole analyser. First, ions generated at the ion source are trapped together within a small radius inside the trap in two or three dimensions using a radiofrequency quadrupolar field. Inert helium gas inside the ion trap removes excess energy from trapped ions to ensure a stable trajectory. Once trapped, potentials are adjusted to scan through the selected mass range by ejecting only ions of a specific m/z ratio out of the trap towards a detector. Similar to the quadrupole analyser, a single mass spectrum is the product of a few up to tens of thousands of scans at scan speeds as high as 52,000 Th s^{-1}. Modern ion traps are considered high-resolution mass spectrometers and may achieve sufficient resolution to resolve the individual isotopes of ions up to charge state $z=6$. Similar to the quadrupole analyser, ion traps can be combined with other mass analysers such as ion-cyclotron-resonance or orbitrap mass analysers for MS^n experiments. Some notable characteristics of ion traps are that slower scan speeds result in higher resolution mass spectra and that FWHM is constant throughout the mass range. In addition, ion traps are able to perform multiple consecutive stages of MS, up to 11 (MS^{11}) on commercial instruments.

TOF mass analysers (Mamyrin, 2001) first accelerate ions after which they drift according to their velocities through a free-field region, until they reach a detector located at the end of the flight tube. The TOF that an ion requires to reach the detector after the initial acceleration depends on its mass-to-charge ratio. Ions with a low mass-to-charge ratio travel faster than ions with a high mass-to-charge ratio. It is important that ions acquire the same kinetic energy during acceleration to achieve a high-resolution mass spectrum. By now, different methods have been established to minimise the kinetic energy distribution during acceleration and its negative effect on resolution. Both MALDI-TOF and ESI-TOF instruments may be equipped with an ion mirror or 'reflectron' to focus ions (Cotter, Griffith, & Jelinek, 2007). Ions drift into the reflectron where they are deflected by an electric field and repelled towards the detector. The reflectron is situated at the end of the flight tube and the detector is placed off-axis near the beginning of the flight tube. Ions with the same m/z ratio but with higher kinetic energy enter the reflectron deeper than ions with lower kinetic energy. As a result, ions with higher kinetic energy travel a longer flight path to arrive at the detector at approximately the same time as ions with a lower kinetic energy. Another advantage is that a reflectron nearly doubles the flight path of ions without physically extending the flight tube, which also enhances resolution. TOF instruments with continuous ion sources such as ESI use 'orthogonal extraction' of ions into the flight tube to reduce the kinetic energy dispersion during acceleration (Chernushevich, Loboda, & Thomson, 2001). 'Delayed extraction' is an alternative for TOF instruments with pulsed ion sources such as MALDI to minimise the kinetic energy dispersion (Brown & Lennon, 1995). Some unique features of TOF analysers include the potentially unlimited upper mass range limit in linear mode, spectra acquisition rate-independent resolution, constant ultrahigh mass resolving power across the entire mass range (up to $R \geq 80{,}000$) and highest possible dynamic range (10^5).

Fourier transform ion-cyclotron-resonance (FT-ICR) mass analysers (Marshall, Hendrickson, & Jackson, 1998) determine the mass-to-charge ratio of ions from their circular motion (detected frequency) when trapped in a high magnetic field. An ICR cell is located at the centre (bore) of a high magnetic field superconducting magnet, which is in the range of 3–21 Tesla (T) (Hendrickson et al., 2015). First, ions are injected into the ICR cell and trapped by applying a trapping voltage to the front and back plates. Once trapped, ions have a small radius circular motion perpendicular to the magnet field along the z-axis. Next, a radiofrequency wave emitted by the excitation plates raises the kinetic energy of trapped ions in order to increase the radius of their circular motion. This results in ions orbiting close to the walls of the ICR cell, which are equipped with excitation and detector plates. At each passage, an ion package induces a current on the detector plates. Ions with low mass-to-charge ratio will have a fast circular motion and are detected at a high frequency, whereas ions with a high mass-to-charge ratio will have a slow circular motion detected as a low-frequency signal. The excitation frequency wave has the same frequency as the ion in the cyclotron to allow resonance absorption of this wave. Importantly, all ions of the same m/z need to be in phase or grouped together in an orbit to produce an induction current. If ions are located randomly along the orbit, the induced current will be null. In Fourier transform MS all ions present in the cyclotron are excited at the same time by applying a rapid scan of a large frequency range. This enables simultaneous detection of all ions present in the ICR cell in a single measurement. As a result, a highly complex wave will be recorded which is then converted by Fourier transformation into a frequency-dependent intensity function from which the mass spectrum is obtained. FT-ICR instruments offer the highest possible spectral resolution that can be achieved by MS, over $R \geq 10^7$ (at m/z 400 using a 7 T magnet; Bruker Daltonics, www.bruker.com). Resolution increases with longer spectra acquisition time and higher magnetic field strength. It is, however, important to note that resolution is not constant across the mass range, but is inversely proportional to m/z, which means that resolution decreases significantly at increasing m/z. To date, FT-ICR instruments provide the widest range of fragmentation methods available in commercial instruments for structural characterisation of ions.

Orbitrap or electrostatic ion trap analysers (Makarov, 2000; Perry, Cooks, & Noll, 2008) measure the frequency of oscillating ions along the central electrode of the trap from which the individual frequencies and intensities are derived via Fourier transformation of the broadband signal to produce the mass spectrum. An electrostatic ion trap is capable of achieving ultrahigh-resolution spectra (up to $R = 240{,}000$ at m/z 200 with isotopic fidelity (Thermo Scientific, Waltham, MA; www.thermoscientific.com)), next to what is possible by FT-ICR. However, loss of resolution is inversely proportional to $(m/z)^{1/2}$, which results in a less dramatic loss of resolution at increasing m/z compared to FT-ICR MS.

As mentioned before, each mass analyser has its own advantages and limitations. In addition, there is a significant difference in acquisition and maintenance costs between mass analysers. It is therefore important to choose the best suitable mass analyser for a specific experiment to ensure optimal results at lowest possible costs. Table 1 summarises relevant system specifications for commercially available instruments (2015).

2.1.4 Peptide Sequencing: Tandem Mass Spectrometry

Although a mass spectrum contains highly informative data about (poly)peptides, one should realise that proteomics applications also require structural information about ions of interest. Hence, accurate mass and isotope distribution cannot differentiate between peptides with identical amino acid composition but different amino acid sequence. For example, peptides SHAPE and HEAPS are composed of the same amino acids and consequently have an identical mass and isotope distribution. Hence, structural

Table 1 System Specifications of State-of-the-Art Instruments for Each Type of Mass Analyser[a]

Mass Analyser	Mass Accuracy	Resolving Power (k)[b]	Dynamic Range (k)[b]	Mass Range (m/z)
FT-ICR	<1 ppm	260	10	100–10,000
Orbitrap	<1 ppm	107	5	50–6000
Time-of-flight	<1 ppm	80	100	50–20,000[c] 50–400,000[d]
Ion trap	<0.35 Da	10	10	50–6000
Quadrupole	<0.2 Da	2	10	50–2000

[a]Specifications were obtained from: Bruker Daltonics, www.bruker.com; Thermo Scientific, http://www.thermoscientific.com/; Waters, www.waters.com; AB Sciex, www.sciex.com; Agilent Technologies, www.agilent.com; Shimadzu, www.shimadzu.com; JEOL USA, www.jeolusa.com.
[b]Resolving power for each instrument is specified at 1000 m/z at ~1 Hz spectra acquisition rate. Resolving power is calculated as $m/\Delta m_{50\%}$.
[c]ESI orthogonal time of flight in reflectron mode.
[d]MALDI time-of-flight mass analysers in linear mode.

information from which the amino acid sequence can be deduced is required to differentiate between them. To this end, most mass spectrometers can produce mass spectra of fragment ions that are generated in a fragmentation experiment on an isolated ion. Such MS experiments are referred to as 'tandem mass spectrometry', since the mass of the (poly)peptide is measured in a first round of MS (MS or MS^1), followed by a second round of MS (MS/MS or MS^2) in which the 'precursor' ion is isolated, fragmented, and its fragment ions are analysed. Instruments that have the capability to perform higher order (MS^n) experiments are also denoted 'tandem mass spectrometers', even though they are not limited to only one or two successive MS experiments per se; commercial ion traps can perform up to MS^{11} experiments. Tandem mass spectrometers are often 'hybrid instruments' that combine different mass analysers for various stages of MS^n experiments. Examples are ESI Q-TOF (quadrupole time-of-flight) (Chernushevich et al., 2001) and LIT FT-ICR MS (Peterman, Dufresne, & Horning, 2005; Syka et al., 2004) instruments that use an analytical quadrupole (Q) or LIT for the isolation or/and fragmentation of precursor ions, respectively.

Most tandem MS experiments begin with the isolation of a precursor ion from a mixture of available ions inside the instrument. This isolation step ensures that exclusive fragment ions from a single ion species are recorded in the MS^2 spectrum. Mixed MS^2 spectra that contain fragment ions from multiple precursors complicate the interpretation of the spectrum severely, since any of the fragment ions might derive from different precursor ions. Following isolation, the precursor ions need to be fragmented in some way to produce sequence-specific fragment ions. For this sake, most fragmentation methods aim at producing a complete set of unique fragment ions for which the mass difference between the consecutive fragments corresponds with the known (adjusted) mass of a single amino acid residue. It is important to realise that each unique fragment ion is produced by a single peptide backbone cleavage of a single precursor ion. This means that instead of a single precursor ion, multiple precursor ions, ie, with identical mass-to-charge ratio, are used to generate a single fragmentation spectrum. Thus, the term 'precursor ion' refers to the total complement of identical ions that undergo fragmentation.

Cleavage of the peptide backbone may occur at three different sites between amino acid residues to form peptide fragments: alkyl carbonyl bond (CHR–CO), peptide amide bond (CO–NH) and amino alkyl bond (NH–CHR) cleavage. A widely accepted nomenclature for peptide fragment ions is shown in its concise form in Fig. 3. This nomenclature was first introduced

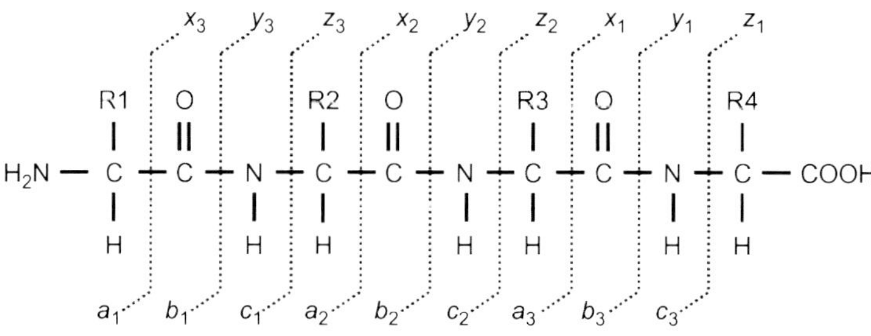

Fig. 3 Roepstorff–Fohlmann–Biemann nomenclature for fragment ions. The generic chemical structure of a peptide backbone is shown with the designation of fragment ions. The *a*-, *b*- and *c*-fragment ions extend from the N-terminus of the peptide and are numbered accordingly. Similarly, the *x*-, *y*- and *z*-fragment ions extend from the C-terminus of the peptide. The *a*- and *x*-fragment ions result from alkyl carbonyl bond cleavage, *b*- and *y*-ions result from peptide amide bond cleavage, and *c*- and *z*-ions result from amino alkyl bond cleavage. *Adapted from Steen, H., & Mann, M. (2004). The ABC's (and XYZ's) of peptide sequencing.* Nature Reviews. Molecular Cell Biology, *5, 699–711.*

by Roepstorff and Fohlman (1984) and later modified by Biemann (1990) to accommodate unambiguous annotation of fragment ion type and fragment order. Fragment ions that extend from the N-terminus of the (poly)peptide are referred to as *a*-, *b*- or *c*-ions, whereas their complementary C-terminal fragment ions are called *z*-, *y*- or *x*-ions. Thus, the numbering of the N-terminal fragment ions starts at the N-terminus and the numbering of C-terminal fragment ions starts at the C-terminus. Different fragmentation methods generally induce only one of the three possible peptide backbone cleavages. For example, collision-induced dissociation (CID) typically produces *b*- and *y*-ion series, whereas electron transfer dissociation (ETD) produces *c*- and *z*-ions. A range of methods are available on modern instruments for the controlled fragmentation of (poly)peptide ions in the gas phase. It is beyond the scope of this review to discuss all available fragmentation methods and we will limit to the two most common fragmentation methods for (poly)peptides: CID and ETD.

To date, CID is the preferred fragmentation method for peptides because of its high efficiency, sensitivity and short duty cycle time. In CID, precursor ions collide with an uncharged gas atom—usually nitrogen or argon—by which the kinetic energy is partially converted to vibrational energy until a critical threshold is reached and the precursor ion breaks into two fragments (Seidler, Zinn, Boehm, & Lehmann, 2010). CID is commonly used to analyse proteolytical peptides fragments of proteins (eg, tryptic peptides) or small proteins. A significant limitation of CID is its relatively 'slow' fragmentation reaction, which allows energised ions to fragment through their weakest links. This may severely limit the amount of amino acid sequence information in the fragmentation spectrum because only few fragment ions are produced. In addition, labile PTMs may be difficult to analyse by CID (Zhou, Dong, & Vachet, 2011). Phosphorylated peptides are a notorious example. In this case, CID tends to induce a dominant neutral loss of the phosphoryl group. As a consequence, peptide fragmentation occurs with very low efficiency since most of the energy is dissipated by the cleavage of the phosphate ester bond. In addition, the neutral loss reaction impairs the assignment of the phosphorylation site since the phosphoryl group is not attached anymore to any of the fragment ions.

ETD is a 'fast' fragmentation method, which induces near-instant fragmentation of precursor ions. In ETD, electrons are transferred from a reagent anion to a multiple-charged precursor cation via ion/ion reactions. Upon capturing an electron, a multiply protonated peptide or protein is transformed into a hypervalent species. The resulting odd-electron species undergo fragmentation pathways driven by radical chemistry, where the N–Cα backbone bonds are cleaved to generate *c*- and *z*-type fragment ions (Huang & McLuckey, 2010). ETD generally provides more extensive sequence information for larger molecules and preserves labile modifications such as phosphoryl groups to allow assignment of the modification site (Zhou, Dong, et al., 2011; Zhou, Morgner, et al., 2011). Both CID and ETD can provide complementary information (Guthals & Bandeira, 2012) that can be employed to fully elucidate both the structure of the PTM and the amino acid sequence and modification site (Hogan, Pitteri, Chrisman, & McLuckey, 2005). An example is the analysis of glycopeptides by CID and ETD (Fig. 4). Here, CID (Fig. 4A) provides structural information on the glycan moiety, while the amino acid sequence and modification site are elucidated by ETD (Fig. 4B).

2.1.5 Separation of Complex Samples: LC-MS/MS

The high complexity of most samples does not allow for direct tandem MS analysis by ESI-MS/MS or MALDI-MS/MS. In general, direct analysis of complex samples by MS will only produce signals for the most abundant ions due to dynamic range limitations and ion suppression. The highest possible dynamic range of Q-TOF instruments (10^5) is still orders of magnitude lower than the dynamic range of protein abundances in biological samples, which can vary over 10 orders of magnitude in eukaryotic cells. In addition,

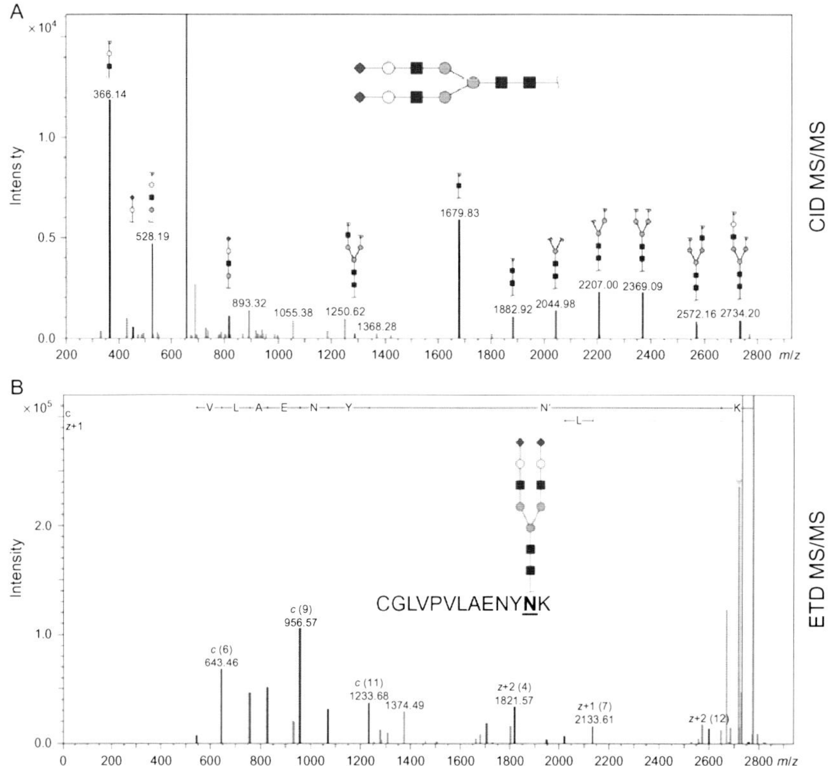

Fig. 4 Complementary structural information obtained by CID and ETD fragmentation of glycopeptides. (A) Deconvoluted CID MS/MS spectrum of the glycosylated tryptic peptide CGLVPVLAENYNK from human transferrin by ultrahigh-resolution quadrupole time-of-flight tandem mass spectrometry (maXis 4G ETD, Bruker Daltonics). Extensive structural information for the glycan moiety of the glycopeptide obtained by CID fragmentation was used to identify the corresponding glycan structure by the GlycoQuest software programme (Bruker Daltonics). GlycoQuest correctly identified the glycan moiety as $Hex_5HexNAc_4NeuAc_2$ in a glycan database search via its annotated fragment ions in the spectrum. (B) Deconvoluted ETD MS/MS spectrum of the same glycopeptide from human transferrin acquired by high-resolution ion trap tandem mass spectrometry (amaZon speed ETD, Bruker Daltonics). Following elucidation of the glycan moiety, an ion trap ETD MS/MS experiment was performed on the same glycopeptide and the acquired ETD spectrum was used in a MASCOT database search that specified the $Hex_5HexNAc_4NeuAc_2$ as variable modification. MASCOT search results were imported into ProteinScape 3.1 (Bruker Daltonics) for visualisation. MASCOT correctly identified the peptide moiety as CGLVPVLAENYNK via its annotated *c*- and *z*-fragment ions (MASCOT score: 125). In addition, the glycosylation site of the peptide was readily identified from the deconvoluted spectrum as the glycan was preserved during ETD fragmentation of the peptide backbone. Symbols for the sugars are the following: hexose, *circle*; *N*-acetylneuramic acid, *diamond*; *N*-acetylhexosamine, *square*.

significant ion suppression would take place during ionisation due to competition between (poly)peptides for ionisation efficiency. Finally, diverse peptides with identical mass but distinct amino acid sequence would be detected as a single spectral peak and would produce a mixture of fragment ions in tandem MS experiments. Therefore, mass spectrometers are often connected to other separation methods that partition the (poly)peptides from the sample at which resolved peaks are fed online into the ion source. For (poly)peptides, liquid chromatography (LC) is by far the most convenient technique for hyphenation with MS via an ESI source. Capillary electrophoresis (CE) has unique advantages over LC for the separation of peptides and proteins (Haselberg, de Jong, & Somsen, 2013; Robledo & Smyth, 2014) but is less common. We will therefore focus on LC as separation technique. The combination of liquid chromatography with online tandem MS analysis is referred to as LC-MS/MS. Out of available chromatographic separation methods, reversed-phase (RP) chromatography is mostly applied because of its reproducibility, robustness, high separation efficiency and its compatibility with ESI. To maximise sensitivity, columns preferably have small inner diameters of 75 or 100 μm that require flow rates of only 150–300 nL/min (Gama, Collins, & Bottoli, 2013; Sestak, Moravcova, & Kahle, 2015). These low flow rates minimise the dilution of trace-level (poly)peptides during separation, but they also provide better signal to noise for detected analytes: the ionisation efficiency in ESI increases proportionally with lower flow rates (Gama et al., 2013). Selection of chromatographic particles depends on various factors such as the size of the (poly) peptide and operational back pressure of the columns that can be handled by the liquid chromatograph. In general, smaller diameter particles provide higher resolution at the expense of substantial increase of the backpressure in the system. Presently, particles of less than 2 μm are available for the routine separation of (poly)peptides. Such small particles require newer generation liquid chromatographs that can withstand operational pressures of 10,000 PSI (690 bar) up to 15,000 PSI (1034 bar) (Sestak et al., 2015). Pore sizes for peptide separations are in the range of 100–300 Å and for protein separations pore sizes may range between 300 and 8000 Å. Correspondingly, RP columns for peptide or protein separations are packed with C18 or C4 materials, respectively. Because of the low flow rate compared to the relatively large injection volume, which is in the range of 0.1–20 μL, most nanoflow LC setups employ trap columns to minimise the amount of time that would be needed to load the sample onto the column. For example, injection of a 10-μL sample using a 10-μL sample loop and two sample loop volumes of solvent would require at least 67 min of loading time at a flow rate of 0.3 μL min^{-1}, without taking the dead volume of the system into account. By comparison, applying the same injection and sample loop volumes, the loading time by a trap setup at a flow rate of 10 μl min^{-1} is approximately 2 min.

2.2 Protein Identification

In MS, protein identification can be performed directly on the intact protein (top-down proteomics) (Catherman, Skinner, & Kelleher, 2014) or on peptides of the protein following enzymatic or chemical digestion (bottom-up proteomics) (Zhang, Fonslow, Shan, Baek, & Yates, 2013). Bottom-up proteomics is by far the most common methodology thanks to its robust and sensitive performance, and its relatively low demands on instrumentation and experience with MS technology.

2.2.1 Peptide-Centric (Bottom-Up) Proteomics

From a generic perspective, three different approaches can be distinguished in bottom-up proteomics for protein identification: peptide mass fingerprinting, MS/MS database searches and de novo sequencing. In the first two methods, MS data are matched to in silico calculated data for all available protein sequences in a database to find the proteins or peptides that best explain the experimental data. In peptide mass fingerprinting (Henzel, Watanabe, & Stults, 2003), the collection of measured peptide masses is compared with the in silico calculated peptide masses for each protein sequence and subsequently scored. The calculated protein identification scores determine which protein sequence(s) in the database provide the best, ie, statistically significant, match with the experimental data. This method relies on mass accuracy of the measurement and sequence-specific proteolysis to assign protein identifications. Peptide mass fingerprinting is commonly performed on tryptic digests of (nearly) pure proteins in combination with MALDI-TOF MS. These (nearly) pure proteins are obtained by standard purification methods or after one- or two-dimensional gel electrophoresis (1DE, 2DE) of a sample (see Section 3.1.1). The method is time demanding, but it is well suited for low-complexity samples of only a few proteins.

Contrary to peptide mass fingerprinting, MS/MS analysis allows the identification of a theoretically unlimited number of proteins in a sample through identification of the individual proteolytical peptides combined with protein database searches. As outlined earlier, in tandem MS the mass of the proteolytical peptide can be determined from the acquired MS^1 spectrum, whereas the sequence-specific fragment ion masses are derived from the MS^2 spectrum. Database search algorithms first compare the precursor ion mass with those of all peptides from the in silico digested protein sequences in the database to find a match with a specified mass tolerance. For each extracted peptide, the algorithm will score the matching quality between the collections of measured and calculated fragment ions. The highest scoring (significant) spectrum-peptide match is considered as peptide identification. Finally, the collection of identified proteolytical peptides infers protein identifications. Fig. 5 shows a schematic representation of the typical workflow in an LC-MS/MS-based bottom-up (shotgun) proteomics experiment. For database searches of MS^2 spectra, various software programmes are available, such as MASCOT (Matrix Science) (Perkins, Pappin, Creasy, & Cottrell, 1999), MaxQuant (Cox & Mann, 2008), Trans-Proteomic Pipeline (Pedrioli, 2010), X!Tandem (Craig & Beavis, 2004), SEQUEST (Eng, McCormack, & Yates, 1994), OMSSA (Geer et al., 2004) and PEAKS (Zhang et al., 2012). Each software package has its advantages and limitations, and a number of these have been critically evaluated by Dagda, Sultana, and Lyons-Weiler (2010). Regardless of the software or database search algorithm used, care must be taken to control the false discovery rate (FDR) for both peptide and protein identifications when analysing large tandem MS data sets (Aggarwal & Yadav, 2016; Brosch, Yu, Hubbard, & Choudhary, 2009; Cox & Mann, 2008; Weatherly et al., 2005; Wright et al., 2012; Zhang et al., 2012). FDR is determined by performing a search against the protein sequence database and a decoy sequence database that contains nonsense protein sequences. Any spectrum-peptide match in the decoy database is considered a false positive result. The false discovery rate is then controlled by adjusting thresholds to include spectrum-peptide matches until the maximum-allowed FDR is reached (typically 1%).

Database searches to identify peptide and protein sequences are not always possible, or may not retrieve a significant spectrum-peptide match, irrespective of the data quality. The finding of a match between a peptide sequence and a fragmentation spectrum presumes that the measured peptide sequence is actually available in the database. This is obviously a problem in proteomics of species for which there is no comprehensive genomic sequence available. The application of de novo sequencing provides an alternative to database search strategies since the amino acid sequence is

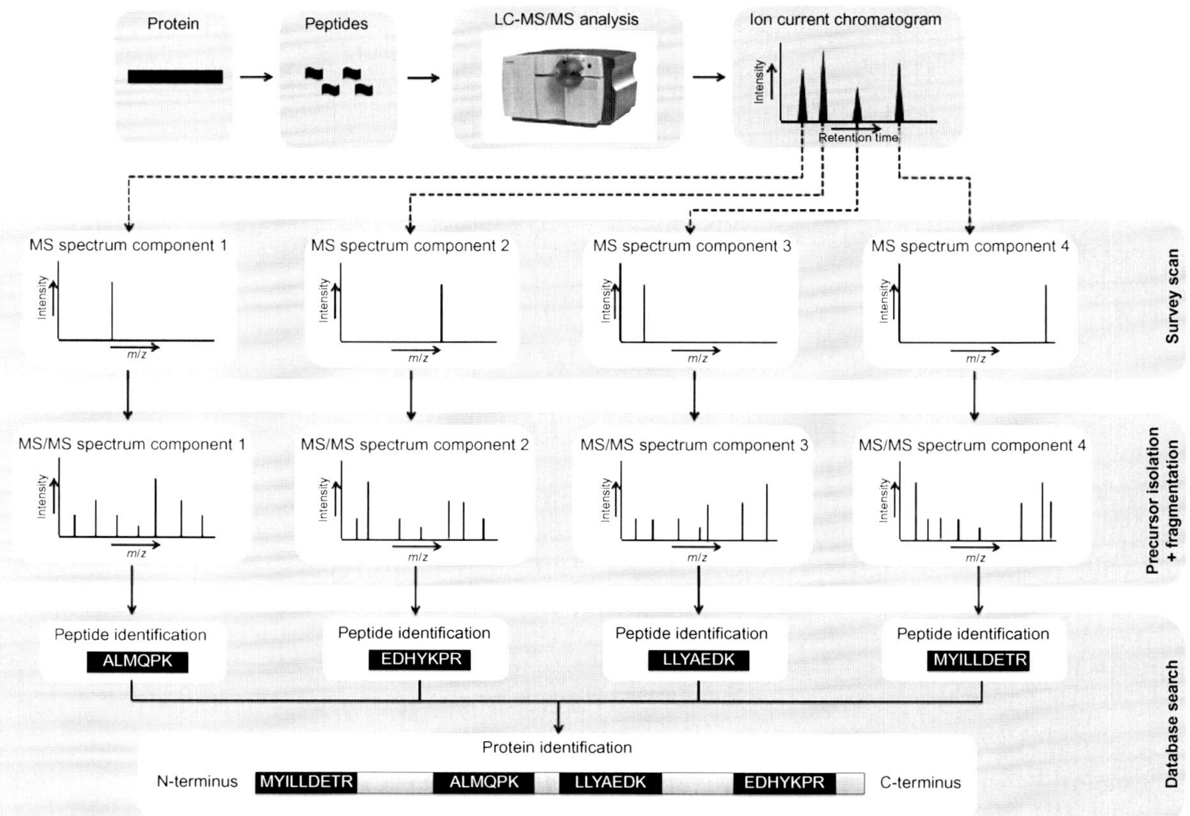

Fig. 5 See legend on opposite page.

determined directly from the MS^2 spectrum (Seidler et al., 2010). Here, the mass difference between consecutive fragment ions establishes the amino acid sequence as shown in Fig. 6 for peptide 'YIC$_{(CAM)}$DNQDTISSK' from BSA. The annotated deconvoluted spectrum reveals a dominant *y*-ion series from which part of the peptide sequence can be read directly. This spectrum is easily interpreted, but manual interpretation of fragmentation spectra is challenging and time-consuming in general since each spectral peak in the MS^n spectrum can be any of the expected fragment ion types (eg, *b*- or *y*-ion). The processing of large tandem MS data sets is facilitated by de novo sequencing software programmes (Allmer, 2011), including PEAKS (Zhang et al., 2012), Lutefisk (Taylor & Johnson, 1997), PepNovo (Frank & Pevzner, 2005), NovoHMM (Pevtsov, Fedulova, Mirzaei, Buck, & Zhang, 2006), AUDENS (Grossmann et al., 2005), Novor (Ma, 2015), Sherenga (Dancik, Addona, Clauser, Vath, & Pevzner, 1999), UniNovo (Jeong, Kim, & Pevzner, 2013) and pNovo$^+$ (Chi, Chen, et al., 2013). In these programmes, high-throughput automated interpretation of MS^2 spectra is supported by different algorithms. Applicability of these programmes specifically depends on the type of mass analyser used to generate tandem MS data as concluded by Pevtsov et al. (2006), who compared the performance of established de novo sequencing algorithms for Q-TOF and ion trap tandem MS data.

Fig. 5 Schematic representation of the typical steps in shotgun proteomics. Proteins are cut into proteolytic peptides by peptidases or chemicals (eg, trypsin). The resulting peptide mixture is desalted and concentrated by solid-phase extraction prior to liquid chromatography with online tandem mass spectrometry analysis. Peptides are separated by reversed-phase chromatography and eluting peptides are subjected to mass spectrometric analysis. In time, the mass spectrometer will first acquire a survey scan (MS^1 level) to measure the precursor ion mass-to-charge ratio which is used to calculate the peptide mass. Next, the mass spectrometer isolates a precursor ion, which is fragmented and its fragment ions are analysed in a second round of mass spectrometry (MS/MS or MS^2 level). Once a set number of MS^2 spectra are recorded, the cycle of MS^1 and related MS^2 spectra acquisition is repeated till the end of the analysis to collect tandem mass spectrometry data for as many peptides as possible. After the LC-MS/MS analysis, vendor-specific software is used to extract data for the precursor- and fragment ions, which is used in subsequent protein sequence database searches. For each acquired fragmentation spectrum, the database search algorithm will attempt to match the experimental data with in silico data for all available peptides in the sequence database. Ideally, each MS/MS spectrum provides a unique peptide identification, which is used to infer protein identifications in the final stage of MS/MS database searches. *Adapted from Wessels, H. J. (2015).* Mitochondrial proteomics: Method development and application. *PhD thesis. Radboud University Nijmegen.*

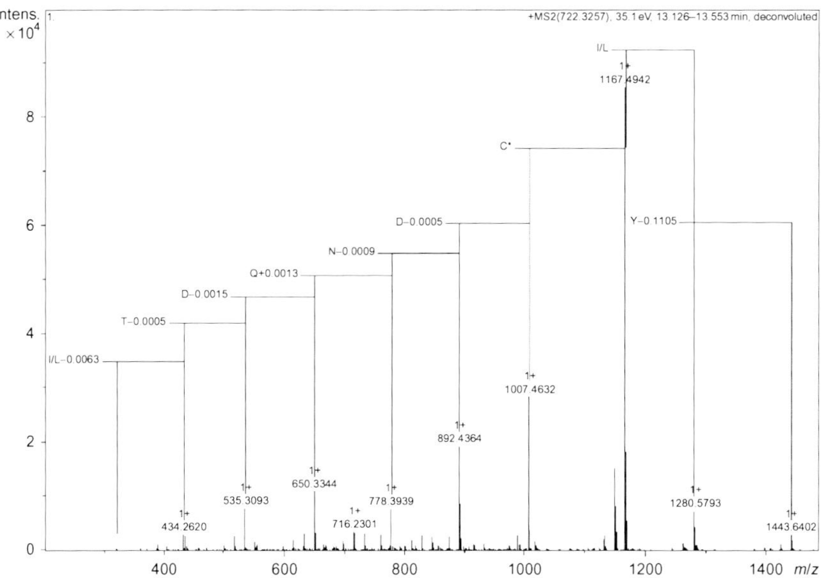

Fig. 6 Manual de novo sequencing result for a CID MS/MS spectrum of the tryptic BSA peptide $YIC_{(CAM)}DNQDTISSK$ acquired by ultrahigh-resolution quadrupole time-of-flight tandem mass spectrometry (maXis 4G ETD, Bruker Daltonics). The mass difference between successive fragment ion peaks was used to determine part of the sequence for this peptide in Data Analysis 4.2 software (Bruker Daltonics). Each amino acid that could be determined using subsequent fragment ions is annotated in the figure with its according mass measurement error in u. Only dominant *y*-fragment ions were annotated for simplicity.

2.2.2 Protein Centric (Top-Down) Proteomics

The direct analysis of intact proteins by MS, referred to as 'top-down' proteomics, is preferred over peptide-centric methods when information about the exact molecular composition of a protein (proteoform) is required. The term proteoform was proposed by Smith et al. (2013) to designate all of the different molecular forms in which the protein product of a single gene can be found, including changes due to genetic variations, alternatively spliced RNA transcripts and PTMs. In bottom-up proteomics, information on proteoforms in a sample is irreversibly lost after digestion of a protein into smaller peptide fragments (Rappsilber & Mann, 2002). Analysis of intact proteins by top-down proteomics preserves the true biological complexity of proteins since every unique proteoform is analysed directly 'as is' (Catherman et al., 2014; Tipton et al., 2011). One may wonder why bottom-up approaches are still generally used in proteomics, if top-down proteomics is capable of providing exact information at the level of proteoforms. The answer to this question relates to the fact that top-down proteomics is very challenging to perform due to the high complexity of the mass spectral data and technical limitations (Smith et al., 2013). In general, top-down proteomics is still unable to achieve the high sensitivity and throughput that bottom-up proteomics provides. The application of bottom-up approaches is thus preferred in research that does not require proteoform-specific information. However, recent hardware and software developments have been making a start to close the gap between top-down and bottom-up proteomics with respect to sensitivity, throughput and robustness. The lower sensitivity for whole proteins as compared to peptides in LC-ESI MS^n experiments supported by modern instruments is partly explained by 'charge state dilution' and 'isotope dilution' effects. Unlike peptides, a single protein species typically acquires a variable number of charges during ionisation by ESI. This implies that the total signal for a single protein is divided over all the different charge state peaks in the mass spectrum (charge state dilution), as observed in Fig. 2A. In addition, proteins have a broad isotopic distribution that again lowers the signal-to-noise ratio (isotope dilution). An example of such isotopic distribution pattern for proteins is presented in Fig. 7 showing a deconvoluted mass spectrum of horse cytochrome *c* (12.3 kDa). One may note that the monoisotopic peak for

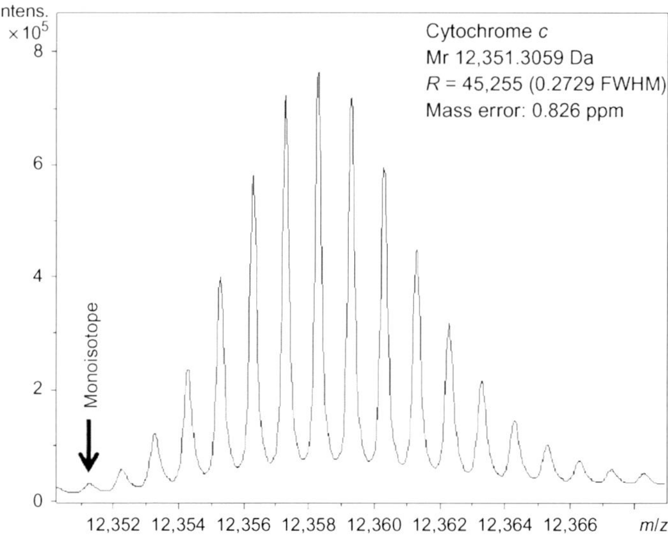

Fig. 7 Isotopic distribution of horse cytochrome *c* in a deconvoluted ESI-MS spectrum acquired by ultrahigh-resolution quadrupole time-of-flight tandem mass spectrometry (maXis 4G ETD, Bruker Daltonics). The *monoisotopic peak* is indicated in the spectrum.

cytochrome *c* is barely detectable and that the monoisotopic peak may not be observed in deconvoluted spectra of larger proteins. In fact, the monoisotopic mass is calculated by a peak-detection algorithm (SNAP 2.0, Bruker Daltonics), which fits an in silico isotopic distribution model, based on the average amino acid 'averagine', to the experimental data. The high complexity of data in top-down proteomics generally requires ultrahigh-resolution MS in combination with sophisticated peak picking and/or deconvolution algorithms to produce information-rich data. Thus far, top-down proteomics primarily relied on FT-ICR and electrostatic ion trap mass analysers. However, the latest generation of ultrahigh-resolution Q-TOF instruments provides an interesting alternative to FT-based mass analysers. These new Q-TOF spectrometers have unique characteristics, such as a significantly higher dynamic range, wider mass range and constant mass resolving power over the full mass range (Table 1).

Protein identification in top-down proteomics by tandem MS is in essence similar to bottom-up proteomics. Database searches supported by software algorithms permit the identification of proteoforms, but de novo sequencing can be applied as well. However, database search strategies are adapted for top-down proteomics, since any possible combination of PTMs as well as sequence variants needs to be considered to successfully match an acquired fragmentation spectrum to a specific proteoform. Software programmes capable of handling top-down proteomics data include ProSightPTM2 (LeDuc et al., 2004; Zamdborg et al., 2007), MASCOT top-down (Karabacak et al., 2009) and MS-Align+ (Liu et al., 2012). CID or ETD fragmentation methods are both applicable in MS^n experiments on whole proteins. Generally, ETD fragmentation will provide higher sequence coverage than CID, but the former method requires more precursor ions to produce reasonable signal-to-noise MS^n spectra.

Top-down proteomics commonly find application in MS measurements on intact proteins in their denatured state. Developments in 'native mass spectrometry' will enable the study of intact proteins as well, and even of protein complexes in which the higher order complexity is preserved. Eventually, native MS may allow the most detailed analysis of protein complexes composed of unique combinations of proteoforms that form distinct complexes together. The opportunities of native MS are illustrated by an experiment with Trastuzumab (Fig. 8). Trastuzumab is a therapeutic monoclonal antibody and it is a protein complex composed of two heavy and two light chains (Bernard-Marty, Lebrun, Awada, & Piccart, 2006). Each heavy chain may be substituted by different N-glycans (Listinsky, Siegal, & Listinsky, 2013); the light chains are present as one proteoform. The different

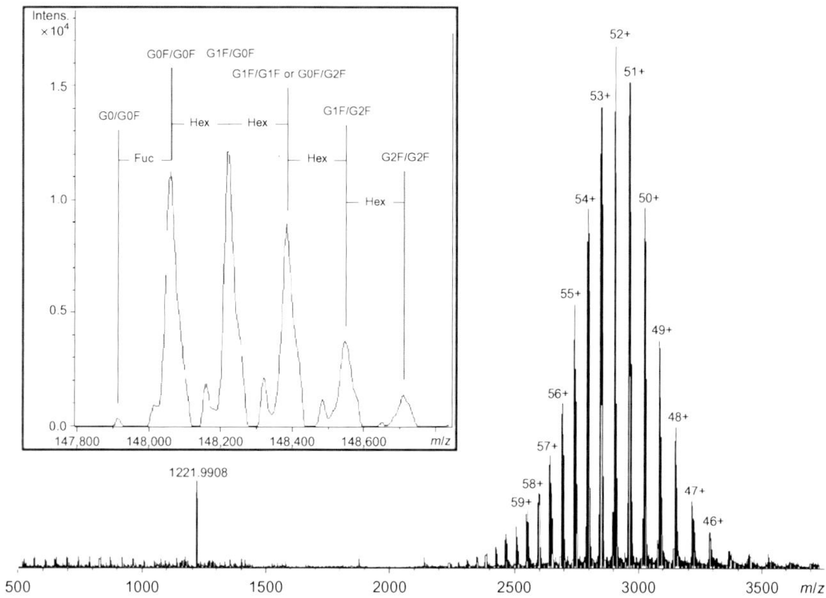

Fig. 8 Native ESI-MS and deconvoluted MS (inset) spectra of the intact monoclonal antibody Trastuzumab by ultrahigh-resolution quadrupole time-of-flight mass spectrometry (maXis 4G ETD, Bruker Daltonics). The ESI-MS spectrum shows a wide charge state distribution for Trastuzumab ($z = 41$ up to $z = 66$), which is typical for large proteins or complexes. The inset presents the deconvoluted spectrum of Trastuzuman from which the different configurations of heavy chains in the monoclonal antibody complex can be detected. The mass differences between subsequent peaks correspond with the mass of either a fucose (Fuc) or a hexose (Hex) as annotated in the figure. Glycan nomenclature to label spectral peaks is as follows. *G0*, $GlcNAc_2Man_3GlcNAc_2$; *G0F*, $GlcNAc_2Man_3GlcNAc_2Fuc_1$; *G1F*, $GalGlcNAc_2Man_3GlcNAc_2Fuc_1$; *G2F*, $Gal_2GlcNAc_2Man_3GlcNAc_2Fuc_1$.

combinations of heavy chain proteoforms in a single antibody molecule yield unique protein complexes of distinct masses, except for combinations of N-glycans with identical masses (eg, G0F/G2F and G1F/G1F combinations in Fig. 8). The multiple spectral peaks in the deconvoluted mass spectrum of Trastuzumab result from the unique combinations of heavy chain proteoforms. Within the context of this review, it should be noted that native MS is still notoriously challenging for membrane complexes.

2.3 Protein Quantitation by Mass Spectrometry

2.3.1 Protein Quantitation: Principles

As mentioned previously, the intensity in acquired mass spectra correlates with the abundance of a (poly)peptide in the sample. However, intensities for equimolar amounts of different peptides may vary over several orders of magnitude in mass spectra due to differences in ionisation efficiency. Alternatively, the integrated peak area of a peptide eluting from the LC column serves as a measure for its abundance. Spectral intensity and peak area are thus meaningful to determine the relative abundance of a chemically identical (poly)peptide between samples. However, these data are of less value in making quantitative comparisons between different (poly)peptides. An exception to this rule holds for stable isotope-labelled variants of the same peptide.

In profiling experiments, the chromatographic peak area of extracted ion chromatograms from MS^1 spectra is used to compare protein abundances between samples. Quantitation software programmes for shotgun proteomics such as MaxQuant (Cox & Mann, 2008), MASCOT Distiller (Matrix Science; www.matrixscience.com), MSQuant (Gouw & Krijgsveld, 2012), TPP (Pedrioli, 2010) or IDEAL-Q (Tsou et al., 2010) extract MS^1 intensity-based quantitative information for identified peptides into LC-MS/MS data sets. Most of these programmes accurately match mass and retention time to robustly retrieve quantitative information for peptides that were not identified (but present) in some of the parallel measurements. This is particularly important in shotgun proteomics where precursors are on-the-fly and data-dependently selected by the mass spectrometer for subsequent MS^2 experiments. Since part of the precursors selected for MS^2 analysis differs between measurements, data gaps are introduced for proteins that were present in a sample but were excluded from MS^2 analysis. To minimise those data gaps, software programmes construct a database that contains the accurate mass and retention time for every identified peptide within a proteomics experiment to subsequently find a peptide ion in the MS^1 spectra of every individual measurement without the need for a positive peptide identification.

2.3.2 Label-Free and Stable Isotope-Labelling Quantitation Methods

For intensity-based quantitation in shotgun proteomics both label-free and stable isotope-labelling quantitation methods have been developed. In label-free quantitation, each sample is analysed in a separate LC-MS/MS run at which the intensity of a peptide from each run is used to calculate the relative abundance of proteins between two or more samples. Data are normalised to correct for experimental artefacts that may result from differences in, for instance, protein amount or sample injection volume.

The so-called spectral count methods provide an alternative to intensity-based quantitation (Lundgren, Hwang, Wu, & Han, 2010). In spectral counting, the number of MS^2 spectra that are matched to a protein sequence is taken as a measure of its abundance. Spectral count methods have much lower accuracy and linear dynamic range than intensity-based quantitation methods, but quantitative data are easily calculated from experimental database search results. A common spectral count procedure to assess protein abundance differences between samples uses the exponentially modified protein abundance index (emPAI) (Ishihama et al., 2005). The emPAI value is calculated by the following formula: $emPAI = 10\,\hat{}\,[N_{observed\ peptides}]/[N_{observable\ peptides}]$, where $N_{observed\ peptides}$ is the number of unique peptide matches or total number of matched MS^2 spectra for a given protein sequence and $N_{observable\ peptides}$ is the number of unique observable peptides from an in silico digest of the protein sequence. Since emPAI calculations compensate for the number of detectable tryptic peptides between protein sequences, emPAI values allow comparisons between different proteins.

Stable isotope labelling of peptides can be achieved by either in vitro or in vivo labelling approaches. Common in vitro labelling methods include stable isotope dimethyl labelling (Boersema, Aye, van Veen, Heck, & Mohammed, 2008; Boersema, Raijmakers, Lemeer, Mohammed, & Heck, 2009), isotope-coded affinity tags (ICAT) (Gygi et al., 1999), tandem mass tags (TMT) (Thompson et al., 2003) or isobaric tags for relative and absolute quantitation (iTRAQ) (Ross et al., 2004). By the introduction of labels that differ in isotope composition (^1H vs ^2H, ^{12}C vs ^{13}C, ^{14}N vs ^{15}N, ^{16}O vs ^{18}O), a mass offset is realised for peptides between two or more samples. This allows samples to be mixed together and analysed in a single LC-MS/MS experiment using the mass offset to differentiate between the samples ("multiplexing"). Importantly, the different isotope-labelled versions of a peptide exactly coelute in reversed-phase LC, enabling unambiguous relative quantitation. Stable isotope dimethyl labelling and ICAT allow multiplexing of two to three different samples at which the signal intensity in MS^1 spectra serves quantitation. Dimethyl labelling is the most popular in vitro method due to its simple, efficient and robust derivatisation of peptides at very low cost. Also TMT and iTRAQ have gained rapidly increasing interest. TMT and iTRAQ reagents bind N-terminal amine groups (and the ε amine of lysine residues) of all peptides. In contrast to dimethyl and ICAT labelling, TMT and iTRAQ reagents are isobaric, being composed of a reporter molecule (of variable mass) combined with a balancer (also of variable mass) such that the molecular masses of reporter and balancer remain the same in the different reagents. During the tandem MS step, the bound iTRAQ label is fragmented, at which the low-mass reporter is used for peptide quantitation, whereas the remaining peaks in the MS/MS spectrum serve peptide identification. By the constant mass of the labels, the complexity of the sample is not increased, which allows for multiplexing of up to 10 different samples in a single analysis (McAlister et al., 2012; Werner et al., 2014). Here, the ratio between the various reporter ions in the MS^2 or MS^3 spectrum is used for relative quantitation.

For in vivo metabolic labelling of (poly)peptides with stable isotopes, cells are grown in the presence of stable isotope-coded amino acids (SILAC) (Ong et al., 2002). This method requires organisms that are auxotrophic for the pertinent amino acid or organisms in which its biosynthesis is fully arrested, when supplied with the labelled amino acid. If not available, ^{15}N-ammonium may serve as the substrate for the in vivo labelling (Krijgsveld et al., 2003; Wang, Ma, Quinn, & Fu, 2002). Care must be taken to achieve near complete metabolic incorporation of the stable isotopes to avoid labelling artefacts (unless pulse labelling is applied to study protein synthesis or protein degradation kinetics).

Label-free quantitation and stable isotope-labelling approaches have their own advantages and limitations. First of all, the multiplexed analysis of 2–10 samples in a single LC-MS/MS analysis effectively removes run-to-run variation in signal intensities. In case of in vivo isotope labelling, cells can even be mixed prior to protein extraction to eliminate any experimentally introduced variation between samples. Label-free quantitation on the other hand does not increase the complexity of the sample and permits the (retrospective) use of LC-MS/MS data from different experiments.

There is growing interest in the field of proteomics to acquire multiplexed tandem mass spectra using data-independent analysis (DIA). In DIA, all precursor ions within the full m/z range, or in multiple overlapping narrow m/z windows of f.i. 25–50 Th, are fragmented at the same time and fragment ions measured by high-resolution mass analysers. Different approaches are subsequently used to deconvolute the multiplexed MS^n spectra as to enable identification and quantitation of cofragmented peptides. DIA is used to bypass the technical limitations of tandem MS instruments to ultimately identify and quantify the entire proteome. DIA is still in development and considering its main application (proteome-wide analysis) is not key to analyse ETC complexes; we have confined ourselves to data-dependent analysis (DDA) methods that rely on precursor isolation for MS^n experiments. Interested readers are encouraged to read the excellent review on DIA methodology by Chapman, Goodlett, and Masselon (2014).

2.3.3 Untargeted and Targeted Approaches

The above quantitation methods are employed in profiling experiments in which up to thousands of proteins are analysed in an untargeted fashion. Unfortunately, proteins of interest can go beyond detection and quantitation due to technical reasons such as the quasi-random nature of precursor selection for MS^2 analysis by the mass spectrometer or overlapping isotope patterns of coeluting peptides. This is a major limitation for untargeted experiments. Another disadvantage of untargeted measurements is that statistical analysis and biological interpretation of acquired data can be highly challenging for large data sets. In addition, why should one measure thousands of proteins if only data from a handful of a priori known proteins are needed to test a hypothesis?

Targeted proteomics offers an attractive alternative to profiling in which only a subset of peptides from proteins of interest are measured selectively (Gallien, Duriez, & Domon, 2011; Gillette & Carr, 2013). Advantages of targeted proteomics over profiling methods are its superior sensitivity, robustness, higher sample throughput (by using short LC gradients) and simplicity of the data. Targeted proteomics is usually performed on triple quadrupole (QqQ) mass spectrometers using 'multiple reaction monitoring' (MRM). In selected reaction monitoring (SRM) on QqQ instruments, a precursor ion is selected by the first quadrupole, dissociated in the second quadrupole, and one of the generated fragment ions is exclusively selected by the third quadrupole for detection. The precursor-ion and fragment-ion pair is referred to as SRM 'transition'. Ideally, a transition is selected to be unique for a single peptide to allow straightforward quantitation via a single chromatographic peak in its constructed SRM chromatogram. Unfortunately, it is often impossible to find a single transition that is unique for a given peptide in highly complex peptide mixtures. The problem of nonspecific transitions is solved by MRM. In MRM, multiple fragment ions of one precursor ion are measured by separate SRM measurements. Although each separate transition might detect several peptides, leading to multiple peaks in its chromatogram, the combination of transitions is chosen to be specific for only a single peptide. Therefore, the correct chromatographic peak in each SRM chromatogram that corresponds with the peptide of interest is readily identified by searching for a chromatographic peak with the exact same retention time and peak shape in all SRM chromatograms of the target peptide. Different software programmes are available for the selection of protein-specific peptides with optimal quantitative characteristics and subsequent selection of MRM transitions (Colangelo, Chung, Bruce, & Cheung, 2013). The software programme Skyline (MacLean, Tomazela, Shulman, et al., 2010) is at present the most commonly used software tool to develop MRM assays in proteomics. It also supports empirical refinement of SRM methods to achieve optimal quantitative results (Bereman, MacLean, Tomazela, Liebler, & MacCoss, 2012; MacLean, Tomazela, Abbatiello, et al., 2010). Absolute quantitation (AQUA) is also possible by mixing a known amount of synthesised stable isotope-labelled peptide standard (AQUA peptide; Stemmann, Zou, Gerber, Gygi, & Kirschner, 2001) into the sample prior to injection. The mass spectrometer is set to acquire MRM data for both the native (light) and stable isotope-labelled (heavy) versions of the peptide such that the absolute amount of the native peptide can be calculated on the basis of an a priori-prepared standard curve of the heavy standard peptide.

3. BACTERIAL ELECTRON TRANSFER CHAINS AND THEIR REGULATION: GLOBAL APPROACHES

In principle, taking advantage of the power of proteomics, the full protein complement of a bacterial cell can be analysed and quantified in a single experiment. However, practice is more refractory for several reasons: the enormous complexity of even relatively 'simple' microorganisms; physicochemical properties of proteins that pose specific demands as to their detection and identification; differences in the abundance of the different protein species that can vary over six orders of magnitude; and efforts and costs associated with MS analyses and data handling. In Section 3.1, we will cover these aspects in a general way, starting from two approaches that are aimed at analysing proteomes from the global, discovery-based perspective, (1) proteomics supported by two-dimensional gel electrophoresis (2DE) (Section 3.1.1) and (2) shotgun proteomics (Section 3.1.2). Hereafter, we will continue with membrane proteins, a subset of proteins of which proteomic analyses are challenging, but that are key components in bacterial ETS (Section 3.2). We will close this part of our review with an outline of proteomic methods to quantify protein abundance and differences in this, which is essentially the study of differential expression, at both the translational and posttranslational levels (Section 3.3).

3.1 2DE-Based Methods and Shotgun Proteomics

3.1.1 Methods Based on 2D IEF SDS-PAGE, the Classical Approach

The success of this method was founded on two major experimental breakthroughs: 2DE separation of large numbers of proteins (Neidhardt, 2011;

O'Farrell, 1975) and their identification by MS (Henzel et al., 1993; Hillenkamp & Karas, 1990; Hillenkamp, Karas, Beavis, & Chait, 1991; James, Quadroni, Carafoli, & Gonnet, 1993; Pappin, Hojrup, & Bleasby, 1993; Yates, Speicher, Griffin, & Hunkapiller, 1993). Separation relies on two unrelated properties of a protein, its isoelectric point (p*I*) and molecular mass. Upon electrophoresis through a pH gradient gel, a protein in a mixture migrates until it reaches that part in the gel that equals its p*I* value. A prerequisite for the proper separation is the presence of isoelectric focusing (IEF) gels or gel strips with stable, immobilised pH gradients. Gel strips that either cover a wide range of pH values (pH 3–10) or are more narrow ranged (gradients across 2, 1, 0.5 or even less than 0.5 pH units) are commercially available. The latter enables the better resolution of complex protein mixtures in the pH range applied. Following IEF, the gel strip is positioned unto a large (eg, 20×20 cm^2) denaturing gel and proteins are separated in the second dimension by sodium dodecyl sulphate-polyacrylamide gel electrophoresis (SDS-PAGE), based on their molecular masses. This technique enables the separation of several thousands of proteins from a sample on a single gel. Next, proteins are visualised by staining, and proteins of interest can be individually picked and digested by a proteolytic enzyme (usually trypsin) or a cocktail of proteolytic enzymes. Cleaved proteins are then analysed by MALDI-TOF MS, FAB or ESI-MS. Proteolytic cleavage results in a mixture of peptide fragments that is unique for a certain protein (fingerprint). The comparison of this fingerprint against an in silico-generated peptide database derived from the genome of the organism under investigation allows the identification of the analysed protein with a certain amount of confidence (Sections 2.1.2 and 2.2.1). Thus, this technique of peptide mass fingerprinting starts from the complete and resolved polypeptide, unlike the shotgun approach described later.

The 2DE-MS methodology has been applied to many prokaryotic species. Laboratories have developed all sorts of variations on basic techniques in order to improve proteome analysis of their model organisms, often consisting of species-specific preparation of the protein samples. This methodology has been used for whole-cell proteomics as well as for the proteomic analyses of subcellular fractions, such as the cytoplasmic proteins, the protein complement associated with the cytoplasmic membrane, with the cell surface outer membrane in Gram-negative bacteria or the cell envelope of Gram-positive species. All these aspects have been described in numerous publications and have been covered in reviews in which the merits and future perspectives of 2DE-MS are critically evaluated (eg, Brewis & Brennan, 2010; Chevalier, 2010; Cordwell, 2004; Curreem, Watt,

Lau, & Woo, 2012; Görg, Weiss, & Dunn, 2004; Issaq & Veenstra, 2008; Mathy & Sluse, 2008; Rabilloud, Chevallet, Luche, & Lelong, 2008a, 2008b). Here, we would like to illustrate the opportunities and intrinsic shortcomings of 2DE-MS with our work on *Methanothermobacter thermautotrophicus* strain ΔH (Farhoud, 2011).

M. thermautotrophicus is a methane-forming (methanogenic) Archaeon that gains its energy for autotrophic growth from the oxidation of hydrogen with CO_2 as the terminal electron acceptor. It was one of the first organisms whose genome was sequenced (Smith et al., 1997). This genome codes for 1922 open-reading frames (ORFs) and comprises 1874 protein-coding sequences, which had been divided over 49 functional categories next to 3 categories with unknown function (unclassified, conserved and unknown, hypothetical proteins). At the time of publication (1995), 844 of these protein-coding ORFs could be given an annotated function (46%), a situation that still holds in the genomic database (NCBI accession number, PRJNA289). In the years that followed the publication of the genome, our knowledge on protein functions has advanced greatly, and keeping up with the literature, we now can assign a function to 1594 protein ORFs (83%) (Farhoud, 2011). To get a better understanding of which genes are expressed at the protein level and under which conditions, we performed a comprehensive 2DE analysis, employing a variety of IEF gradient gels. An example of this is shown in Fig. 9.

Fig. 9 immediately reveals the strength of 2DE analysis, visibility in what one is doing, but this strength is a weakness at the same time. After gel electrophoretic separations and gel staining, over 3000 protein spots were analysed by MALDI-TOF MS, which were unambiguously identified with a >95% success rate: 'you can get whatever you can see'. However, the drawback is: 'you won't get what you don't see'. Small-sized proteins (<10 kDa) generally went unassigned, mainly due to the lack of sufficient and appropriate proteolytic cleavage sites for reliable identification. To quite an extent, successful MS-supported identification depends on high-quality genome assembly of the organism under investigation and its high-quality annotation. No matter how trivial, these factors may compromise future proteomic studies based on metagenomes that have been assembled and annotated automatically without proper, laborious and time-demanding curating by experts. For instance, even though it was well assembled, we noticed that in the *M. thermautotrophicus* genome 34 pairs of adjacent ORFs that were translated into N-terminal and C-terminal parts of known or conserved proteins. By single nucleotide substitution, deletion or insertion all pairs could be merged into full-length proteins. Six of these pairs comigrated

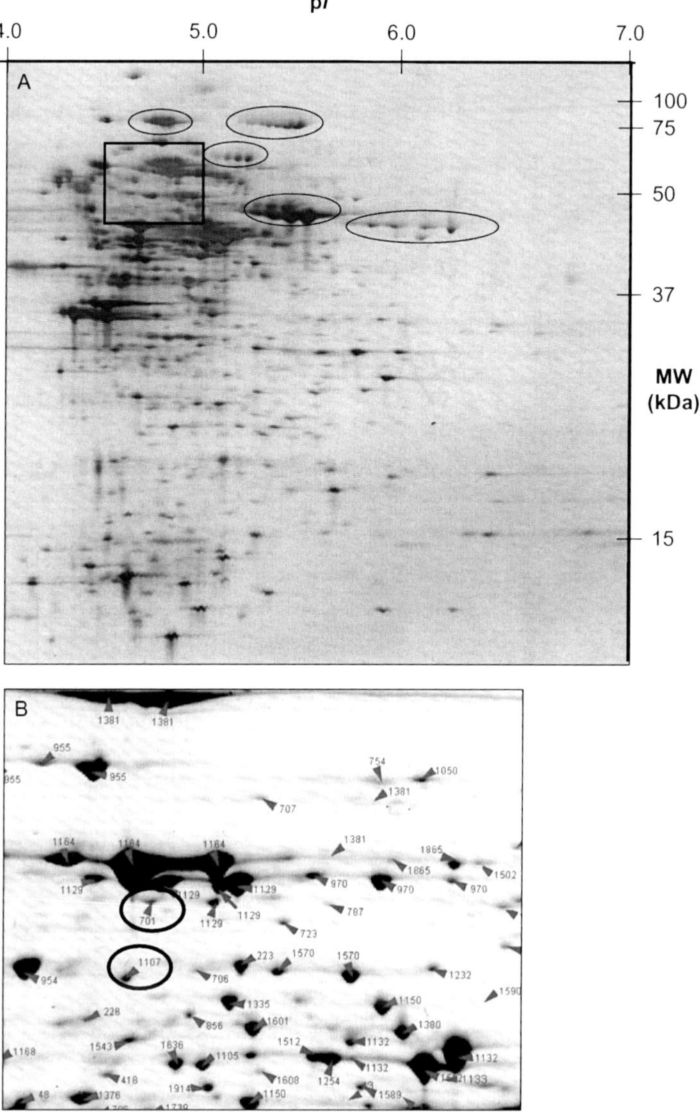

Fig. 9 Two-dimensional isoelectric focusing SDS-PAGE reference map of the soluble cell fraction of *Methanothermobacter thermautotrophicus*. (A) Soluble proteins were separated in the first dimension (*horizontal*) by IEF (pH 4–7), followed by SDS-PAGE in the second dimension (*vertical*). Proteins were visualised by *silver staining*. A number of 'trains' representing proteoforms of a same protein are encircled. (B) Detail of the reference map obtained from a 2D IEF (pH 4.5–5.5) SDS-PAGE gel. This section is framed in Fig. 9A. *Encircled* are MTH701, which is a combination of MTH701 and MTH702, annotated as the N- and C-terminal parts, respectively, of acetyl-CoA synthase 2, and MTH1107, annotated as the biotin-carrying subunit of pyruvate carboxylase (PycB). However, MALDI-TOF MS reveals that biotin is absent in this subunit. Please note the various proteoforms of MTH1164 and MTH1129, representing the alpha subunits of methylcoenzyme M reductase type I and type II, respectively. Protein extract was prepared from cells grown under hydrogen-limited conditions. (See the colour plate.)

into the same spot upon 2DE (see f.i. Fig. 9B), indicating that the gene splitting was rather the result of sequencing errors than of genetic events.

Protein staining also provides the means for protein quantification. In 2D gels this can be done by gel scanning combined with software programmes that convert the size of a spot and its colour intensity into a protein amount. However, like in each protein staining method the practical applicability depends on two factors, detection limit and the dynamic range in protein amounts, expressed as log values, giving a linear response. Conventional staining dyes such as Coomassie Brilliant blue (CBB), colloidal Coomassie and silver nitrate have detection limits of 50, 8–16 and 0.5–4 ng of protein, respectively (Chevalier, 2010; Curreem et al., 2012). (Note that only pico- or femtomoles of proteins are needed for MS analysis.) The dynamic range of the more sensitive silver staining is very low (1.5 log) and this method is not easily compatible with MS. As alternatives, differently fluorescent dyes with detection limits in the nanogram range, like Sypro Ruby, Flamingo and Deep Purple, have been developed (Poland, Rabilloud, & Sihna, 2005). Albeit not particularly sensitive, the advantage of these fluorescent dyes is their dynamic range (4 log). Furthermore, proteins can be selectively stained on the basis of the presence of haem groups or iron (Ferrer, Golyshina, Beloqui, Golyshin, & Timmis, 2007), which could be of relevance in studies on respiratory complexes. Nevertheless, lowly expressed proteins will go unseen (and undetected) in 2DE. In addition, such proteins can be obscured in the gels by highly expressed proteins. The use of narrow-ranged IEF strips may obviate the latter problem to some degree (Fig. 9B).

In 2DE gels, a protein species may be represented by multiple spots that mainly differ in their p*I* values (proteoforms, 'trains'). Proteoforms could be the result of physiological relevant PTMs (phosphorylation, substitution of charged amino acids with other groups), but this is not necessarily the case. Proteoforms also might represent folding isomers or other electrophoresis artefacts (Berven, Karlsen, Murrell, & Jensen, 2003; Lutter et al., 2001; Nag et al., 1994; Paton, Gerrard, & Bryson, 2008). In Fig. 9, several examples of protein trains in the proteome of *M. thermautotrophicus* are highlighted. Two such examples are the alpha subunits of methane-forming methylcoenzyme M reductase type I (McrABC) and type II (MrtABC) that were observed as three to five proteoforms (Fig. 9B). The high-resolution crystal structure of Mcr I, indeed, reveals five PTMs in this alpha subunit, being methylated amino acid residues in close proximity to the F_{430} catalytic site (Ermler, Grabarse, Shima, Goubeaud, & Thauer, 1997;

Grabarse et al., 2001). 2DE suggests various proteoforms, but a directed MS method would be needed to substantiate if these proteoforms are related to different degrees of methylation. In general, PTMs easily evade identification because modified peptides are hardly ionised by MALDI, peptide coverage of in-gel trypsin digestion is relatively low, substitution of amino acids may go unnoticed in database searches and phosphopeptides, for example, tend to decompose prior to detection the MALDI-TOF reflector mode. The identification of PTMs requires its dedicated MS analysis methods (Section 4.2).

Altogether, 2DE of *M. thermautotrophicus* cell extracts allowed the identification of 633 unique proteins (33.8% of the proteome), including 50 out of 109 proteins (45.9%) involved in the process of methanogenesis. However, recovery was biased. High-molecular weight proteins (MW > 50 kDa) and basic proteins (p*I* > 10) were not well represented in the data set, which is due to the poor resolution properties of IEF for such proteins. Most notably, only 26 of the predicted 406 membrane-bound proteins (6.4%) could be found. Apart from other reasons discussed in Section 3.2.1, the low yield of membrane-bound proteins is caused by their insolubility: these proteins precipitate in the IEF gel before reaching their p*I* (Braun, Kinkl, Beer, & Ueffing, 2007).

The above considerations do not apply to *M. thermautotrophicus* specifically, but hold in general ways for 2DE proteomic analyses (see the reviews cited earlier). Apart from that the method is quite laborious and time demanding. Still, 2DE-MS could be very well the method of choice to address research questions with a more limited and focused scope than the analysis of the total proteome of a prokaryote.

3.1.2 Shotgun Proteomics

To overcome the limitations of 2DE-based proteomics, shotgun proteomics has been implemented as an alternative method for the high-throughput analysis of complex protein samples (Aebersold & Mann, 2003; Washburn, Wolters, & Yates, 2001; Wolters, Washburn, & Yates, 2001). In this method, proteins in a sample are proteolytically cleaved before tandem MS. Next, all peptide fragments are computationally matched to the protein sequences in the database (see Section 2.2.2) (Washburn, 2004; Wei et al., 2005; Yates, Ruse, & Nakorchevsky, 2009). Before MS analysis, the peptide mixture is resolved by RP (C18) LC on a nanocapillary column that is connected online with the tandem mass spectrometer (LC-MS/MS). This whole procedure is fully automatised, and to achieve a better resolution, LC can be performed in two or even more dimensions ("MudPIT") by use of strong cation-exchange, immobilised-metal affinity or TiO columns placed before the hydrophobic column (Block et al., 2009; Cheung, Wong, & Ng, 2012; Fränzel & Wolters, 2011; Kislinger, Gramolini, MacLennan, & Emili, 2005; Sun, Chiu, & He, 2008; Whitelegge, 2002). For further reduction of the sample complexity, proteins can be preseparated by conventional SDS-PAGE or LC. After SDS-PAGE, the gel is cut into multiple pieces, proteins in the pieces are digested in-gel and digests are subjected to LC-MS/MS separately. This preseparation may also involve offline IEF, common protein separation methods such as anion exchange or affinity column chromatography, CE as well as methods that have been specifically designed for tandem MS proteomic analyses (Cologna, Russell, Lim, Vigh, & Russell, 2010; Freeman & Ivanov, 2011; Gilmore & Washburn, 2010; Lam, Antonioli, Righetti, Citterio, & Girault, 2007). Obviously, this method can be applied to lysates of whole cells as well as of subcellular fractions, which have been enriched or fully purified by designated procedures. A variety of in vitro and in vitro labelling methods are then available to quantify relative or absolute protein amounts, as described in Section 2.3.2.

Shotgun proteomics has been broadly used to make an inventory of the proteome of bacterial model species such as the Gram-negative *Escherichia coli* (Han & Lee, 2006; Krug et al., 2013), Gram-positive *Bacillus subtilis* (Becher, Büttner, Moche, Hessling, & Hecker, 2011; Völker & Hecker, 2005; Wolff et al., 2007), the industrially relevant *Corynebacterium glutamicum* (Fränzel et al., 2010; Fränzel & Wolters, 2011), pathogens such as *Streptococcus pneumoniae* (Sun et al., 2011), *Mycobacterium tuberculosis* (Calder, Soares, de Kock, & Blackburn, 2015; de Souza & Wiker, 2011; Mawuenyega et al., 2005) and *Mycoplasma pneumomiae* (Catrein & Herrmann, 2011), environmentally relevant *Shewanella oneidensis* (Brown, Romine, Schepmoes, Smith, & Lipton, 2010; Elias et al., 2005), the extreme thermophilic Archaeon *Pyrococcus furiosus* (Lee, Sevinsky, Bundy, Grunden, & Stephenson, 2009) and various others. As the result of extensive analyses of whole-cell lysates or subcellular fractions of these organisms, 50–60% of all proteins encoded in their genomes could be recovered. In case of the 'relatively simple' *M. pneumomiae*, no less than 620 gene products out of the 700 protein-coding genes (89%) could be identified (Catrein & Herrmann, 2011). Here, tandem affinity chromatography was extensively utilised to selectively enrich proteins from the cytoplasmic protein complement. Quite recently, a similar recovery (88%, viz., 2300 out of the approximately 2600 protein-encoding ORFs) has been established for *E. coli* by the application of high-resolution MS combined with Super-SILAC protein quantification (Soares, Spät, Krug, & Macek, 2013; Soufi, Krug, Harst, & Macek, 2015). By applying LC-MS/MS combined with SDS-PAGE prefractionation, we could identify 894 unique proteins (48%) from the soluble and membrane fractions of *M. thermautotrophicus* grown under high and low concentrations of its energy source hydrogen (Farhoud, 2011). Of these 894 proteins, 372 were not found in the 2DE proteome, whereas 111 proteins retrieved by the latter method were not detected by the shotgun approach. Apparently, both approaches have their flaws. Together, both methods allowed the identification of 1005 different polypeptide gene products (54% of the expected proteome), including 83 proteins (76%) involved in primary methanogenic metabolism. Importantly, LC-MS/MS gave a far better yield of membrane proteins (124; 30%); still this type of proteins remained underrepresented in the data set. Indeed, most of the proteins of the methanogenesis pathway that remained beyond detection were membrane proteins. In line herewith, proteins that were obtained with <50% efficiency belonged to functional categories dominated by membrane proteins, viz., glycerolipid metabolism (10 out of 22 expected proteins; 45%), cell envelope and membrane metabolism (51/129; 39.5%) and transport of inorganic (47/95; 49.5%) and organic (12/38; 31.5%) compounds. Obviously, these percentages may be underestimations, since only part of the genome may be expressed under the conditions at which *M. thermautotrophicus* was grown. Next to this, membrane proteins might have been observed, but with too low peptide scores to permit reliable identification.

It goes without saying that shotgun proteomics has definite benefits above classical 2DE-MS, especially regarding its opportunities for high-throughput analysis. Nevertheless, the former method has its limitations. No matter how welcomed, the comprehensive in-depth analysis of the flood of data is timely. Regulatory systems and protein–protein associations and protein–protein interactions remain unresolved. In addition, the method still has a bias against membrane-bound proteins. Lastly, information on PTMs is lost, unless directed approaches are applied to identify and quantify these protein modifications (see Section 4.2). Despite these shortcomings, shotgun proteomics provides us with a most powerful tool in hypothesis-generating research or to answer questions by well-designed and well-focused experiments.

3.2 Membrane-Bound Systems

3.2.1 Analysis of Membrane-Bound Proteins

Earlier, we noticed that membrane-bound proteins are recovered only poorly by 2DE-MS, whereas they remained underrepresented in shotgun proteins. In their review on early status of proteomics, Santoni, Molloy, and Rabilloud (2000) even wondered if proteomics and membrane proteins would not be 'un amour impossible'. In the years that followed, lots of efforts and ingenuity have been invested to enhance the detection and identification of membrane proteins (see for reviews: Bernsel & Daley, 2009; Bunai & Yamane, 2005; Braun et al., 2007; Gilmore & Washburn, 2010; Helbig, Heck, & Slijper, 2010; Lu, McClatchy, Kim, & Yates, 2008; Poetsch & Wolters, 2008; Rabilloud et al., 2008a, 2008b; Savas, Stein, Wu, & Yates, 2011; Soufi & Macek, 2015; Vuckovic, Dagley, Purcell, & Emili, 2013; Weiner & Li, 2008). Reviewing the outcome of these pursuits 9 years later, Rabilloud (2009) was less pessimistic: 'love is possible, but so difficult'.

The reasons why proteomic analysis of membrane proteins is often—but not always—difficult may become clear if we look at the nature of these proteins, thereby defining what is implied by a membrane protein, or more specifically a cytoplasmic membrane protein (Fig. 10). These proteins come in a number of classes with different topologies. Simple ones cross the membrane by one transmembrane helix (TMH), their N- and C-terminal parts being localised on opposite sides of the membrane. Terminal parts are comprised of hydrophilic peptide chains of which lengths may vary substantially. More

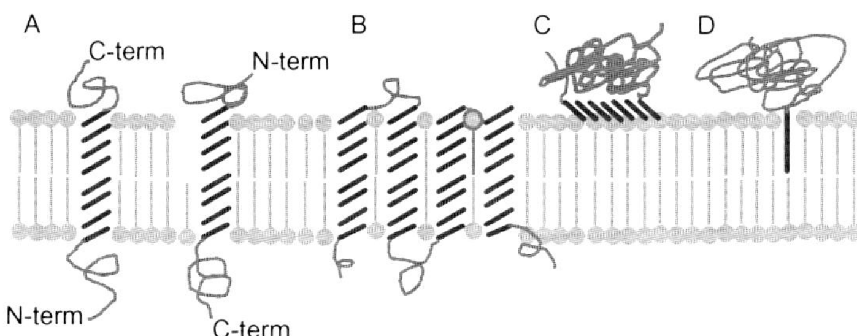

Fig. 10 Schematic representation of the various types of membrane proteins. (A) Proteins that span the lipid bilayer with one TMH having opposite orientations of their N- and C-termini. (B) Multipass membrane protein (here, four TMHs). (C) Protein attached to the membrane by hydrophobic and/or ionic interactions. (D) Protein anchored in the membrane by a lipid chain. *Adapted from Helbig, A. O., Heck, A. J., & Slijper, M. (2010). Exploring the membrane proteome—Challenges and analytical strategies.* Journal of Proteomics, *73, 868–878.*

complex membrane proteins have multiple TMHs. One such TMH is composed of a stretch of 15–25 mostly hydrophobic amino acids that fold into an alpha helix. The hydrophobic helices make the protein to reside inside the membrane bilayer. In this bilayer, proteins are packed by annular lipids that act nonspecifically as 'lubricant' and 'solvent' to allow rotational or lateral motions of the protein, or by nonannular lipids, like cardiolipin (see Section 4.1.4) that may be determinants in the structure and activity of the membrane protein (Lee, 2003; Paradies, Paradies, De Benedictis, Ruggiero, & Petrosillo, 2014). By the presence of both hydrophilic and hydrophobic regions, membrane proteins are amphipathic to variable degrees. This amphipathic character can be expressed by the grand average of hydropathicity (GRAVY) index (Kyte & Doolittle, 1982). In general, the more THMs are present, the more hydrophobic the protein becomes and the more difficult the protein is detected by proteomic analyses. Proteins may also be just anchored to the membrane by hydrophobic and/or ionic interactions (Fig. 10). Alternatively, contact is facilitated by lipids or by glycosylphosphatidylinositol moieties that are covalently bound to the protein. These proteins lack any TMH, which may hamper their immediate identification as membrane proteins by genomic analyses. In case of multi-subunit respiratory enzymes, complexes can be composed from different types of membrane proteins depicted in Fig. 10. Here, association may also result from subunit–subunit rather than from protein–lipid interactions.

The main problem in proteomic analysis of membrane proteins is how to cope with hydrophobicity. In the absence of lipids, membrane proteins readily precipitate or aggregate in an aqueous environment having a pH close to the p*I* of the proteins (Eichacker et al., 2004; Wu, MacCoss, Howell, & Yates, 2003). This is less of a problem for proteins with low GRAVY indices (<0.1–0.2), ie, proteins that harbour only one or two THMs and/or contain hydrophilic regions of sufficient length. Generally, such proteins can be identified by conventional 2DE or shotgun approaches. Membrane lipids may be replaced by ionic detergents that pack the protein in a micelle, making the protein soluble in water. SDS has superior qualities in this respect and, in fact, SDS-PAGE is most suitable for the separation and analysis of membrane proteins. However, the presence of SDS is incompatible with IEF. A second problem regarding the proteomic analysis of membrane proteins is that TMHs lack sufficient (hydrophilic) lysine and arginine residues for proteolytic cleavage by trypsin, yielding peptide fragments that are out of range for conventional peptide-centric MS analysis (Helbig et al., 2010). Still, appropriate trypsin cleavage sites may be present in the hydrophilic domains, permitting the identification of the protein. Third, membrane proteins may be present in too low amounts or are masked by highly abundant proteins. Obviously, low abundance may not apply to respiratory membrane proteins related with the primary energy metabolism. In this case, the questions would be, which proteins are detected and which ones are not, and why the latter ones were not detected. The listed problems have been tackled at two levels: the development (1) of procedures to enrich membrane proteins or proteins of interest and (2) of methods that enhance their detection and identification by MS.

Proteins residing at the membrane can be enriched by differential (ultra) centrifugation steps, sucrose gradient ultracentrifugation or gel filtration column chromatography. Taking advantage of membrane hydrophobicity, a biphasic partitioning system might be advantageous in separating the hydrophobic (membrane) from the hydrophilic (cytoplasmic) protein complement (Everberg, Gustavasson, & Tjerned, 2008; Everberg, Leiding, Schiöth, Tjerneld, & Gustavsson, 2006; Masuda, Saito, Tomita, & Ishihama, 2009; Poetsch & Wolters, 2008). In this procedure, the hydrophobic fraction is extracted into the more hydrophobic polyethylene glycol (PEG) phase, while hydrophilic proteins partition into the hydrophilic dextran phase. The applicability of this method in prokaryotic research does not seem to be tested as yet. Besides, the presence of PEG is extremely incompatible with MS and the compound has to be carefully removed before LC-MS/MS or MALDI-MS. Another tool for the enrichment of the membrane subcellular fraction, which has not been exploited in this research, is free flow electrophoresis (Eichacker, Weber, Sukop-Köppel, & Wildgruber, 2015; Zischka et al., 2006). However, highly abundant cytoplasmic proteins can still contaminate membrane-bound protein fractions. These contaminants may be removed by incubation in alkaline solvent systems, for instance, a high concentration of carbonate (Fujiki, Hubbard, Fowler, & Lazarow, 1982; Molloy, 2008) or of NaBr (Schluesener, Fischer, Kruip, Rögner, & Poetsch, 2005). Under alkaline conditions, membranes spread as sheets (Fujiki et al., 1982; Wu et al., 2003) and nonspecifically associated proteins are dissociated, perhaps including ones of which membrane association is functional. Further enrichment may involve hydrophilic interaction chromatography (Hendrickx, Adams, & Cabooter, 2015). After tagging peripheral lysines with biotin, proteins present in membrane fractions can be further enriched by the using of avidine/streptavidin beads or columns (Rybak, Scheurer, Neri, & Elia, 2004; Scheurer et al., 2005; Schiapparelli et al., 2014). As far as relevant in bacterial research, lectin affinity column chromatography could provide a tool to selectively enrich glycosylated membrane proteins (McDonald, Yang, Marathe, Yen, & Macher, 2009; Zhou, Aebersold, & Zhang, 2007). If investigations address one or a small subset of membrane proteins and their binding partners, antibody affinity chromatography can also be a method of choice (Section 4.1.1) (Banks, Kong, & Washburn, 2012).

Procedures to enhance the MS-based detection of membrane proteins developed along the two lines described before: 2DE-MS (Braun et al., 2007; Bunai & Yamane, 2005; Kashino, Harayama, Pakrasi, & Satoh, 2007) and shotgun proteomics (Lu et al., 2008; Rabilloud, 2009; Vuckovic et al., 2013). Regarding 2DE, the issue is to dissolve the proteins from the membrane and transfer them without precipitation to a detergent mimicking the membrane environment, so that gel electrophoresis becomes feasible. Dissolution with 1% SDS, an anionic detergent, provides these opportunities, but membrane protein complexes will dissociate and the possibility of a second dimension will be excluded: Upon SDS-PAGE in the second dimension, proteins are just distributed across the gel along a diagonal, which is not very informative. To an extent, off-diagonal resolution can be improved if separations in the first and second dimensions are carried out under different conditions with respect to PAGE (gradient) concentrations and the composition of the gel electrophoresis buffer systems (Braun et al., 2007). An alternative for SDS has been found in the cationic detergents 16-benzyldimethyl-*n*-hexadecylammonium chloride (16-BAC) (Bisle et al., 2006; Nothwang & Schindler, 2009; Zahedi, Meisinger, & Sickmann, 2005; Zahedi, Moebius, & Sickmann, 2007) and cetyltrimethyl-ammonium bromide (CTAB) (Helling et al., 2006). In these methods, membrane proteins are dissolved by treatment with a mild detergent and packed by a layer of 16-BAC or CTAB molecules, which gives charge for PAGE separation in a discontinuous acidic gradient. As for SDS-PAGE, separation proceeds on the basis of the size of the proteins. Hence, after 16-BAC and SDS electrophoreses in the first and second dimensions, respectively, proteins are again resolved along a diagonal. Future improvements have to be directed towards getting more evenly distributed protein patterns across the whole 2D gel. Two such, very robust methods are blue native (BNE) and clear (or colourless) native (CNE) gel electrophoresis, which will be described hereafter (Section 3.2.2). While detergents are essential to resolve proteins from the membrane and their presence is needed during gel electrophoresis, these compounds may hamper MS analysis severely, an aspect that also holds for shotgun proteomics. Consequently, detergents have to be carefully removed before those analyses. Otherwise, this issue may be overcome by the application of laser-induced liquid bead ion-desorption (LILBID)-MS (Sokolova et al., 2010).

Shotgun proteomics-supported methods aimed at the improved identification of membrane proteins have been focused on protein cleavage procedures (Helbig et al., 2010). Trypsic digestion of intact membrane proteins usually yields only the subset of peptides that are present in the hydrophilic domains (membrane shaving). Although this information is important for determining the topology of a membrane protein (see next), it may be insufficient for an unambiguous identification, creating the need to access the complete protein sequence embedded in the membrane. Again, SDS would be a perfect agent to dissolve membrane proteins, but this compound compromises protease activity. The presence of 50% methanol dissolves the membranes, but keeps the proteins intact, enabling the targeting of intramembrane cleavage sides. As the hydrophobic parts of membrane proteins contain only few cleavage sites for trypsin, various methods have been tested in order to obtain a more comprehensive peptide coverage, including combined chemical (CNBr) and proteolytic cleavages by trypsin or other proteases (Fischer & Poetsch, 2006; Fischer, Wolters, Rögner, & Poetsch, 2005; Poetsch & Wolters, 2008). One of these proteases is Proteinase K (ProtK) (Wu et al., 2003). This protease cleaves a protein ultimately in dipeptides, which are irrelevant for MS analyses. However, if carried out under alkaline conditions, ProtK activity is strongly inhibited, yielding peptide fragments that are sufficiently sized for those analyses.

We just indicated that membrane shaving might give important topological, structural information, indicating on which side of the membrane the identified peptides occur. For instance, when performed on protoplasts only those peptides that face the periplasm will be recovered. In addition, the absence of peptides in the N- or C-terminal part of the protein may hint at proteolytic cleavage related with protein export and protein assemblage.

As mentioned in the beginning of this section, many efforts have been invested in elucidating membrane proteomes, including those of a variety of prokaryotic species. Table 2 presents a selected overview of investigated species and of the methods employed. Results were satisfactory or even very promising, especially if investigations had been performed by complementary approaches. In retrospect (2015), one could even state that membrane proteins and proteomics are becoming to like each other at second sight, but it is somewhat early to speak of love.

Table 2 Cytoplasmic Membrane Proteomes of Selected Prokaryotic Species

Species	Method	References
Bacteria, general	2DE	Wilmes and Bond (2004)
Bacillus subtilis	2DE	Eymann et al. (2004)
Campylobacter jejuni	2DE + shotgun	Cordwell et al. (2008)
Chlorobium tepidum	2DE + shotgun	Kouyianou, Aivaliotis, Gevaert, Karas, and Tsiotis (2010)
Corynebacterium glutamicum	Shotgun	Schluesener et al. (2005)
Enterococcus faecalis	BNE	Maddalo et al. (2011)
Escherichia coli	2DE	Lai, Nair, Phadke, and Maddock (2004)
	BNE	Lasserre et al. (2006)
	BNE	Stenberg et al. (2005)
Francisella tularensis	BNE	Dresler, Klimentova, and Stulik (2011)
Helicobacter pylori	BNE	Pyndiah et al. (2007)
Kuenenia stuttgartiensis	Shotgun	Neumann et al. (2014)
Methanothermobacter thermautotrophicus	BNE	Farhoud et al. (2005)
Mycobacterium bovis	2DE + shotgun	Målen et al. (2008)
Mycobacterium tuberculosis	Shotgun	Sinha et al. (2005)
Paracoccus denitrificans	BNE	Stroh et al. (2004)
Porphyromonas gingivalis	BNE	Glew et al. (2014)
Staphylococcus aureus	2DE + shotgun	Scherl et al. (2005)
Streptococcus pneumoniae	Shotgun	Choi et al. (2010)

3.2.2 Membrane-Bound Protein Complexes

Respiratory RMPs are usually composed of several subunits. In the global approaches based on 2D IEF SDS-PAGE and shotgun MS, the information regarding subunit composition is lost. In the first method, RMPs are resolved as separate polypeptides, whereas in the second method peptide fragments from enzyme complexes are immersed in an ocean of other peptides, which are only allocated to single subunits. Most elegant and powerful methods that overcome these restrictions are BNE (Schägger, Aquila, & Von Jagow, 1988; Schägger & von Jagow, 1991, Wittig, Braun, & Schägger, 2006) and the related CNE (Wittig, Karas, & Schägger, 2007; Wittig & Schägger, 2008a, 2009a). In BNE, membrane proteins are resolved from the membranes by mild neutral detergents, like Triton X-100, dodecyl-β-D-maltoside or digitonin, and in the presence of 6-aminohexanoic acid, a zwitterionic compound that improves membrane solubilisation. Hereafter, the anionic dye CBB G-250 is added. The critical step is to find the conditions at which subunit interactions and higher order protein interactions (SCs) are preserved. Extracted protein complexes get embedded in a layer of CBB molecules. The dye binds hydrophobic proteins and gives them a negative charge allowing their solution in aqueous systems and their separation by gel electrophoresis. Because the dye also binds cationic amino acids of soluble proteins, BNE is also suitable for this type of proteins. After extraction, the dissolved protein complexes are separated by PAGE using a designated buffer system containing CBB. Separation occurs on the basis of protein sizes. By varying the concentrations of polyacrylamide or by applying different gradients, protein complexes with molecular masses as low 10 kDa and as high as 10 MDa are resolved as blue bands (Strecker, Wumaier, Wittig, & Schägger, 2010; Wittig, Beckhaus, Wumaier, Karas, & Schägger, 2010; Wittig & Schägger, 2008a, 2009a). Upon separation in the second dimension by SDS-PAGE, subunits of protein complexes or the constituents of SCs are diverted into their individual polypeptides that in turn can be identified by MS. Importantly, polypeptides derived from a single protein complex migrate below to each other at the same width of the SDS gel, which facilitates their immediate allocation to a certain protein (super)complex. CNE works in principal the same, except that CBB is omitted from the electrophoresis buffer (Wittig et al., 2007). In this method, protein complexes (first dimension) or individual subunits (second dimension) are colourless and can be stained by fluorescent dyes. Such staining is not possible in case of BNE since CBB quenches fluorescence. Due to the absence of CBB in the buffer system of CNE, this method is restricted to acidic proteins and their intrinsic migration through the electric field. As a result, protein resolution in these gels is far less than for BNE, and proteins are often observed as smears rather than as defined bands (Wittig & Schägger, 2005). This problem has been addressed by including colourless mixed micelles of neutral and anionic detergents to the gel electrophoresis buffer (high-resolution CNE; Wittig et al., 2007).

A rapidly increasing number of studies on bacterial ETS have benefitted from the method of BNE (Table 2); our work on *M. thermautotrophicus* was one of the first (Farhoud et al., 2005). Here, we investigated the membrane-bound and soluble protein complexes of this organism grown under high and low hydrogen concentrations. Fig. 11 presents an example of one of the two-dimensional BN gels obtained. Protein spots were analysed by MALDI-TOF MS and LC-MS/MS. Although only part (1550) of these protein spots was investigated, 361 unique proteins could be identified, including all key enzymes involved in methanogenesis, the soluble and membrane-bound ones alike. Also H^+-translocating F_1F_0-ATP synthase was recovered. This enzyme was found as monomeric and dimeric complexes, and as higher order aggregates (Fig. 11B and C), which never has been shown for prokaryotes, but has firmly been established for mitochondrial ATP synthase (see Section 5.1.1) (Wittig & Schägger, 2008b, 2009b). Similarly, the multisubunit membrane-bound Na^+-dependent N^5-methyl-tetrahydro-methanopterin:coenzymeM methyltransferase (MTR) occurred as monomeric and dimeric complexes, and part of it was even associated with ATP synthase (Fig. 11C). Importantly, association was not an experimental artefact, but it was related to the growth conditions applied. Besides this, the change in growth conditions revealed that soluble benzyl viologen-reducing hydrogenase and soluble heterodisulphide reductase differentially formed complexes (Fig. 11D). This observation could be of relevance in the further understanding of how the exergonic terminal step of methane formation, methylcoenzyme M reduction, is coupled to the endergonic first step of CO_2 reduction under hydrogen limitation by an electron bifurcation mechanism (Kaster, Moll, Parey, & Thauer, 2011; Thauer, Kaster, Seedorf, Buckel, & Hedderich, 2008).

Recently, we used the new BNE-based method of complexome profiling to make an inventory of the enzymes and respiratory complexes that play a role in the energy metabolism and cell carbon synthesis of anammox bacteria (our unpublished results). The particular method was pioneered by Wessels et al. (2009) to identify new assembly factors of mitochondrial complex I and was later on optimised (Heide et al., 2012; Huynen, Mühlmeister, Gotthardt, Guerrero-Castillo, & Brandt, 2015; Nijtmans, Huynen, & Thorburn, 2013; Remmerie et al., 2011; Weber et al., 2013). In this method, proteins and their complexes isolated from the membrane fraction are separated on a BN gel as described earlier. Instead of the separation in the second dimension by SDS-PAGE, the gel is cut in multiple pieces, each of which is proteolytically cleaved and subjected to label-free quantitative LC-MS/MS. Such separation is shown in Fig. 12 for a membrane

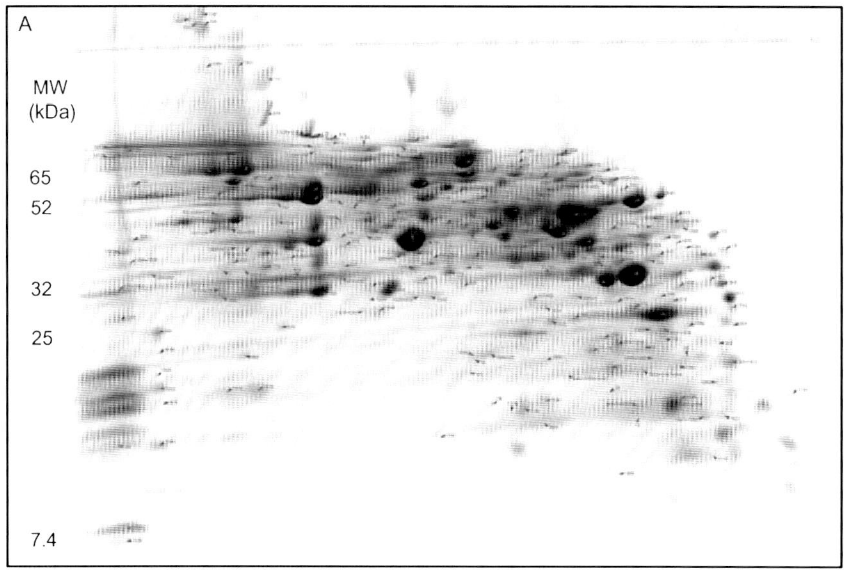

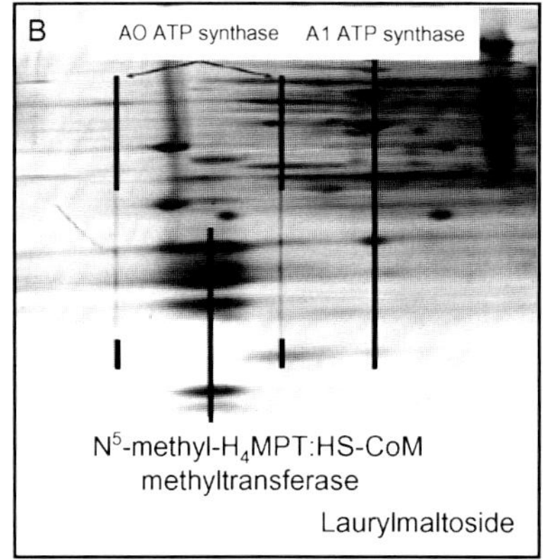

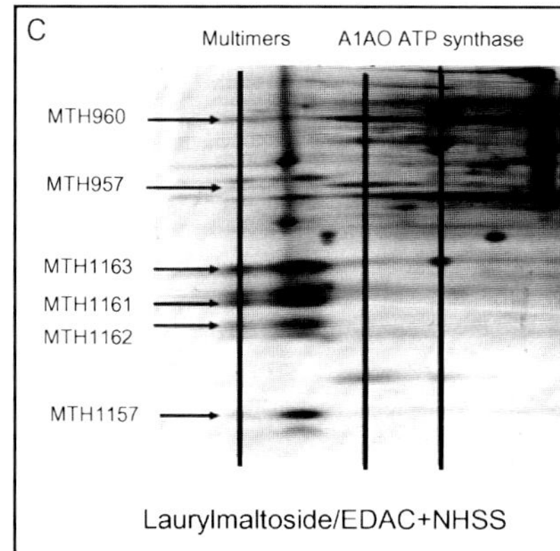

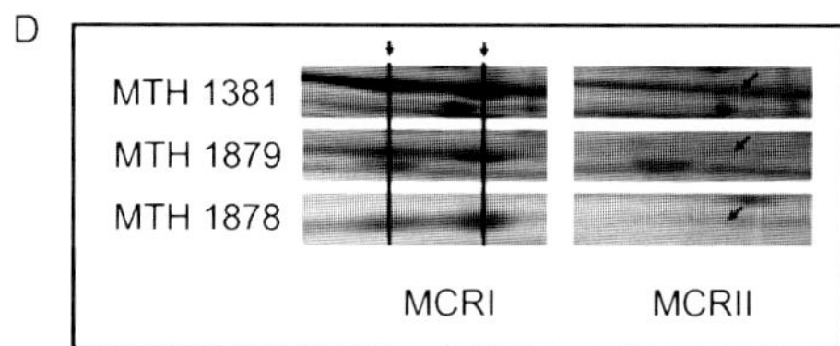

Fig. 11 Blue Native gel electrophoresis of soluble and membrane fractions of *Methanothermobacter thermautotrophicus*. (A) Two-dimensional reference (2D-BNE) map of the soluble fraction. This fraction was separated in the first dimension (*horizontal*) by 5–10% gradient BNE, followed by 10% SDS-PAGE in the second dimension (*vertical*) and subsequent staining with colloidal Coomassie. The same chromatographic conditions and staining method were applied in (B)–(D). (B) Detail of a 2D-BNE gel of the membrane fraction solubilised with β-laurylmaltoside, showing monomers, dimers and higher-order

preparation of the anammox bacterium *Kuenenia stuttgartiensis*. LC-MS/MS not only identifies the proteins that are found in each gel slice but also quantifies their relative amounts on the basis of the intensity of the mass spectral peaks (Section 2.3.2). The distribution if a certain protein across the whole gel can then be deduced based on the protein migration patterns. Finally, these protein profiles are hierarchically clustered, eliciting which proteins comigrate as protein complexes through the BN gel. The application of various polyacrylamide gradients thereby assures reproducibility of the migration profiles. After analysis of the *K. stuttgartiensis* membrane preparations, nearly 1400 different proteins could be identified, representing 34.5% of the transcribed genome. 44.4% of the proteins with one predicted TMH were recovered. This value might be an underestimation, since a single TMH could be a signal sequence that is cleaved during protein translocation, making the protein a soluble one in situ. Proteins with two or more predicted TMHs were recovered with >50% efficiency, irrespective of the number of TMHs (2–15). Most importantly, most if not all of the subunits of NADH:quinone dehydrogenase (complex I), membrane-bound formate dehydrogenase, three different types of bc_1 complexes and four different types of ATP synthase/ATPase, and various other respiratory RMPs known from other species were retrieved. Subunits were consistently assembled into protein complexes as predicted from genome analyses (Kartal et al., 2013). In addition, the presence of a number of predicted, new anammox-specific RMPs could be verified as well.

supercomplexes of the archaeal A_0A_1-type ATP synthase, and of the sodium-pumping N^5-methyltetrahydro-methanopterin:coenzyme M methyltransferase (MTR) complex. (C) Detail of a 2D-BNE gel of the membrane fraction solubilised with β-laurylmaltoside followed by cross-linking with 'zero-length' 1-ethyl-3-(3-dimethylaminopropyl) carbodiimide (EDAC) in the presence of N-hydroxysulphosuccinimide (NHSS). The figure shows monomeric and dimeric forms of the A_0A_1-type ATP synthase (MTH957, MTH960), the MTR complex (MTH1157–1163) as well as a putative supercomplex of the ATP synthase and MTR. (D) Culture condition-dependent association between the heterodisulphide reductase (HDR) subunits A (MTH1878), B (MTH1879), C (MTH1381) and of the viologen-reducing hydrogenase (MVH) (not specified). HDR is found as two species associated with MVH, sizing approximately 400 and 300 kDa (*arrows*) when cells were grown under hydrogen limitation (MCR I condition). These species were hardly seen when growth had occurred under hydrogen excess (MCR II condition). *Adapted from Farhoud, M. H., Wessels, H. J., Steenbakkers, P. J., Mattijssen, S., Wevers, R. A., van Engelen, B. G., et al. (2005). Protein complexes in the archaeon* Methanothermobacter thermautotrophicus *analyzed by blue native/SDS-PAGE and mass spectrometry.* Molecular & Cellular Proteomics, *4, 1653–1663.* (See the colour plate.)

Fig. 12 See legend on opposite page.

Results obtained with BNE in the research on eukaryotic and prokaryotic ETS (Table 2), and as illustrated by the above two examples, underscore the broad and novel potentials of this technique. This method appears to be particular useful in studies on the composition and assembly of RMPs (Section 5). Presently, BNE has been given one more dimension by the introduction of metal isotope native radio autography in gel electrophoresis (MIRAGE) (Sevcenco, Hagen, & Hagedoorn, 2012). In this application, cells are grown in the presence of radioactive isotopes of metals (Fe, Zn, Cu, Mo, W) that play a key role in respiratory complexes. The subsequent analysis of these cells by BNE in combination with radio autography establishes which metal is incorporated into which protein. As an alternative method to map the 'metalloproteome' (Tainer & Adams, 2010) of an organism without the use of radioactive metals, prefractionation of whole-cell lysates by conventional column chromatography has been suggested (Cvetkovic et al., 2010; Lancaster et al., 2011; Maret, 2010; Shi & Chance, 2011). Column fractions are subjected to inductively coupled plasma mass spectrometry (ICP-MS) to detect the presence of metals and relevant fractions are further analysed by BNE combined with tandem LC-MS/MS. Instead of conventional column chromatography, immobilised-metal affinity chromatography may serve the enrichment of specific metalloproteins from complex samples (Block et al., 2009; Cheung et al., 2012; Shi & Chance, 2011; Sun et al., 2008).

3.3 Differential Expression

The bacterial cell is a highly dynamic system that has to be optimally tuned for changes in its environment, such as the supply in its energy sources. This

Fig. 12 Complexome profiling of membrane preparations of the anammox bacterium *Kuenenia stuttgartiensis*. Membrane preparations were subjected to 4–16% (*top, left*) and 3–12% gradient (*top, right*) BNE. After electrophoresis and gel staining with colloidal Coomassie, gels were each cut into about 65 equally sized pieces and each of these pieces was tryptically digested and analysed by LC-MS/MS, yielding in total ~1400 unique proteins. Each of these proteins was label-free quantified for each gel piece. The *large central part* of the figure shows the gel migration profile (*horizontal* direction) of each of these ~1400 proteins (*vertical* direction) displayed as heat maps. Next, migration profiles were hierarchically clustered (*right-hand side* of the *central part* of the figure) as to reveal proteins that comigrated in the gels. In the *bottom part* a detail of the central figure is displayed, establishing the comigration of all subunits of NADH:quinone oxidoreductase (complex I; kuste2660–2672), except for the very hydrophobic subunits NuoK (kuste2669; 3 TMHs, MW 11.2 kDa) and NuoN (kuste1172; 12 TMHs, MW 52.9 kDa) that remained undetected. All observed subunits were found in a protein complex of about 620 kDa, which favourably compares the theoretical MW 575 kDa of complex I. (See the colour plate.)

tuning relies on the adaptation of the enzymatic machinery, both by the differential expression at the transcriptional level and, more directly and elegantly, by PTMs of proteins to control their activities. Proteomics might provide the toolbox to investigate these adaptive processes on a global scale and at both levels simultaneously while taking into account the limitations of global approaches (see Section 3.1). Hereafter, we will outline the proteomics-based methods and applications to map the regulation of bacterial ETSs, following the two routes along which proteomics has evolved, 2DE and shotgun proteomics. Regulation at the posttranslational level will be described separately (Section 4.2).

The strength of 2DE is its capacity to resolve and visualise a vast amount of proteins on a 2D gel. Already in the preproteomics era, these merits were appreciated. By changing growth conditions or implementing a stress situation together with the addition of ^{35}S- or ^{14}C-labelled substrates, the proteins that were synthesised in response to the particular change could be revealed by autoradiography (see f.i. VanBogelen, Olson, Wanner, & Neidhardt, 1996). At that time, the challenge was to identify these proteins. The introduction of MS has solved the issue. The close inspection of two or more 2D gels of lysates from cells grown under varied conditions immediately demonstrates changes in expression patterns. Fig. 9B shows a simple, but puzzling, example of this. *M. thermautotrophicus* contains one protein with a biotin prosthetic group, the beta subunit of pyruvate carboxylase (PycB, MTH1107). This enzyme introduces fixed CO_2 into the TCA cycle for further distribution across anabolic pathways. *M. thermautotrophicus* perfectly grows without biotin administration, but the organism lacks the pathway for its biosynthesis (J. Keltjens, unpublished results). In agreement herewith, MTH1107 found on 2D gels (Fig. 9B) is not biotinylated. Strikingly, after growth in the presence of this cofactor the MT1107 spot is moved to another position and here the protein harbours covalently bound biotin. The puzzling aspects are, why pyruvate carboxylase is expressed in an inactive form and how fixed CO_2 is fed into the TCA cycle.

The close comparison of highly complicated protein distribution patterns among different gels is facilitated by gel imaging and special software programmes. By measuring the sizes and intensities of the stained spots (spot volumes), differences in protein expression levels can be evaluated quantitatively. The problem, however, is that proteins never migrate in exactly the same way due to minute differences in sample and gel preparation and electrophoresis conditions. In principle, this problem can be computationally addressed by the application of programmes that resize and reshape gel images, so that a set of marker proteins overlay each other (warping), but

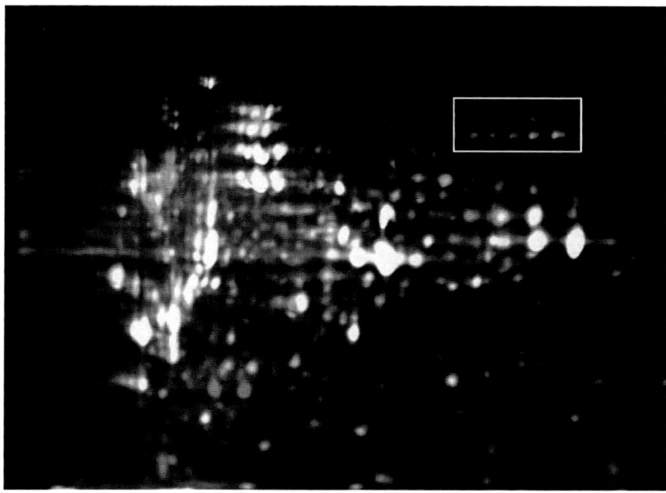

Fig. 13 Two-dimensional difference gel electrophoresis (DIGE) of *Methanothermobacter thermautotrophicus* cells grown under hydrogen-depleted and hydrogen-excess conditions. The cytoplasmic proteins from the hydrogen-excess and hydrogen-limited cultured conditions were tagged with *red* and *green* fluorescent dyes, respectively. Equal amounts of protein from both cultures were mixed and tagged with a *yellow* fluorescent dye serving as an internal control. Prior to 2D IEF SDS-PAGE, same protein amounts of these three preparations were combined. Multiple spots (proteoforms) derived from translation initiation factor 2α (aIF2α, MTH1872) are accentuated by the *white frame*. (See the colour plate.)

this warping is not without faults, resulting in a substantial rate of false positives or false negatives. Difference gel electrophoresis (DIGE) addresses this problem (Larbi & Jefferies, 2009; Minden, 2012; Sapra, 2009; Viswanathan, Unlü, & Minden, 2006). In this method, protein samples prepared from two differently grown cell cultures are labelled with two different fluorescent markers. Equal amounts of proteins from both samples are mixed and labelled with a third marker to serve as an internal control. Before 2DE, these three samples are combined and separation is performed in one gel. Subsequent gel imaging that exploits the high dynamic range (4 log) of the fluorescent markers reveals which proteins are differentially expressed and to what degree. Fig. 13 shows an example from our own work on *M. thermautotrophicus* (Farhoud, 2011). While the translated genome predicted the presence of 877 different proteins in the pH 3.5–5.5 range covered by the IEF strip, more than 2500 fluorescent protein spots could be detected in mixed preparations from cells grown with excess and limiting hydrogen concentrations. Eight of these were differentially expressed more than 5-fold, 18 of these 3–5-fold, 69 by a factor 2–3 and 214 by a significant

1.15–2-fold. Importantly, differential expression comprises a number of protein proteoforms (trains). This concerned, among others, MTH1872 (Fig. 13), a homolog of the eukaryotic translation initiation factor 2α (aIF2α). In Eukarya and Archaea (Andaya, Villa, Jia, Fraser, & Leary, 2014; Schmitt, Naveau, & Mechulam, 2010; Tahara, Ohsawa, Saito, & Kimura, 2004), aIF2α activity is known to be controlled by phosphorylation. This could be the case for *M. thermautotrophicus*, but a dedicated MS method would be needed to substantiate this prediction.

Shotgun proteomics is a convenient tool for the global inventory of expressed proteins (Section 3.1.2), but the approach is even well suited to determine and quantify differences in protein expression levels. The prerequisite for its application is the availability of highly reproducible protocols for the preparation of samples obtained from whole-cell lysates or subcellular fractions of interest. For differential expression quantification, protein samples of cells cultured under the varied conditions are labelled in vitro or in vivo as explained in Section 2.3.2. Like in the DIGE methodology, labelled and unlabelled preparations are mixed before 1DE, LC or MudPIT separations and subsequent multiplexed tandem MS, in order to avoid analytical artefacts and to facilitate mass spectral comparisons. As before (Section 3.1.2), a drawback in shotgun MS protein identification and quantification is that PTMs may go unnoticed. Interference of PTMs with peptide ionisation and the complete loss of information regarding protein proteoforms may even compromise the correct interpretation of data.

Hitherto, a large number of studies took advantage of 2DE- and/or shotgun-based proteomics to map differential expression patterns of bacteria in response to a change in environmental conditions. Table 3 presents a selected overview of studies that focused on the effect of changes in the supply of primary electron donors or terminal electron acceptors. In general, expression profiles are complex and involve numerous proteins. Besides this complexity, it should be taken into account that views may be biased because of flaws in protein detection, such as membrane proteins. Nevertheless, in concert with genomics, transcriptomics and metabolomics, proteomic analyses may contribute significantly to systems-biology endeavours. This is especially true if protein expression levels are profiled on the basis of series of experiments performed under carefully controlled conditions (see f.i. Kohlmann et al., 2011, 2014; Soufi et al., 2015; Wegener et al., 2010). Moreover, findings can be completely unexpected, thereby raising hypotheses for further, more directed investigations aimed at the elucidation of underlying regulatory systems. The role of proteomics in this, however, may be restricted, since regulatory proteins that exert their function in

Table 3 Differential Expression Analyses of Selected Prokaryotic Species in Response to Change in Energy Metabolism

Species	Condition	Method; Quantitation	References
Escherichia coli	Ethanol stress	Shotgun; SILAC	Soufi et al. (2015)
	Amino acid supply	DIGE	Lopez-Campistrous et al. (2005)
Geobacter metallireducens	Aromatic e-donors	Shotgun; label free	Heintz et al. (2009)
	Electron donor excess	Shotgun; label free	Marozava et al. (2014)
Geobacter sulfurreducens	U(VI) e-acceptor	Shotgun; label free	Orellana et al. (2014)
Metallosphaera cuprina	S-Oxidation + C-metabolism	Shotgun; label free	Jiang et al. (2014)
Neisseria gonorrhoeae	Biofilm/anaerobiosis	Shotgun; label free	Phillips et al. (2012)
Pseudoalteromonas tunicata	Planktonic/biofilm	2DE/BNE	Hoke et al. (2011)
Psychroflexus torquis	Light	Shotgun; label free	Feng, Powell, Wilson, and Bowman (2015)
Purple photobacteria	Light	CNE	Niederman (2013)
Ralstonia eutropha	Chemolithoautotrophy	2DE/1DE	Kohlmann et al. (2011)
	Denitrification	2DE/1DE	Kohlmann et al. (2014)
Shewanella decolorationis	Azo/Fe^{3+} e-acceptor	2DE	Wang, Xu, and Sun (2010) and Wang, Zhang, et al. (2010)
Shewanella oneidensis	Aerobic/microaerobic	Shotgun; label free	Fang et al. (2006)
Synechocystis sp.	C/N assimilation	Shotgun; label free	Wegener et al. (2010)
Syntrophomonas wolfei	Butyrate oxidation	2DE	Schmidt, Müller, Schink, and Schleheck (2013) and Schmidt, Zhou, et al. (2013)
	Synthrophy	Shotgun; label free	Sieber et al. (2015)

modified forms are usually present in low copy numbers, precluding their detection by MS-based methods. If approached in a targeted way (Section 2.3.3), proteomics could provide a most valuable complementary tool. The comparison of expression patterns between mutant and wild-type strains immediately shows the effect of a mutation (see f.i. Section 5.2.2). Herewith, known observations are substantiated by another technique, but the comparison also may disclose new proteins whose expression is affected by particular mutation, which opens directions for further research.

4. RESPIRATORY PROTEIN COMPLEXES IN ASSEMBLY

RMPs represent a notoriously difficult class of proteins to work with within the field of structural biology. Usually, these proteins are fairly resistant towards crystallisation, and a load of trial and errors (and tricks) is required to get crystals of sufficient refractory quality that yield high-resolution X-ray images (Loll, 2014; Moraes, Evans, Sanchez-Weatherby, Newstead, & Stewart, 2014). The other method to explore protein structures, NMR spectroscopy, does not depend on protein crystals, but this approach is still limited in its scope regarding the size of the proteins. Even though X-ray crystallography and NMR give the most detailed structural information, this information may not necessarily reflect a physiological state. Crystallisation trials and NMR studies need a vast amount of purified protein by often elaborate column chromatographic methods. During purification, essential components could be lost. In addition, purified proteins are investigated in a nonnatural environment, which is the cell lipid membrane. Again, lipids that play a key role in structural integrity may get lost and may be replaced by artificial detergents. To speed up large-scale purification, enzymes are often (heterologously) overexpressed in tagged forms, but overexpression represents a nonphysiological condition. Furthermore, heterologously overexpressed proteins may be devoid of PTMs that could be structurally or functionally important. Lastly, crystal structures and NMR spectra present only a static view. Enzyme action is highly dynamic and it involves a multiplicity of conformational states, substrate–protein and protein–protein interactions.

In recent years, we witnessed a boom of mass spectroscopic methods and techniques aimed at the resolution of the structural properties of RMPs and dynamic processes related with their catalytic action, including PTMs. The common property of these methods and techniques is that they take the whole protein as the starting point, ushering the fields of top-down and structural proteomics. For quite some time, the development and application of these expensive and often home-built apparatus have been the playground of specialists. Regarding the study of electron transfer processes, these specialists took as their 'toys' mitochondrial OXPHOS complexes from human tissues or, to simplify matters, from animal or yeast model organisms. The focus on these OXPHOS complexes is understandable, considering the manifold of metabolic diseases that are caused by the malfunctioning of mitochondrial respiration. All in all, application of these new methods to advance our knowledge on bacterial ETSs and their regulation has lagged behind. Hereafter, we will present an overview of top-down proteomics that allowed, or may allow, the better comprehension of the structure of bacterial respiratory RMPs and their regulation at the level of PTMs. We will exemplify progress in this knowledge mainly with achievements in research on eukaryotic systems, which just reflects the current status of the field, and on prokaryotic systems, as much as available. The examples of the work on OXPHOS complexes permit an illustration of the promises that new proteomics-based methods offer, but also of their limitations, as to invite their critical implementation in the investigations on bacterial systems. For the interested reader who likes to learn more about new developments and their applications, we would like to refer to a number of recent reviews (Boersema, Kahraman, & Picotti, 2015; Catherman et al., 2014; Cui, Rohrs, & Gross, 2011; Hyung & Ruotolo, 2012; Marcoux & Cianférani, 2015; Owens, 2011; Petrotchenko & Borchers, 2014; Zhou & Robinson, 2010).

4.1 Structural Aspects of Respiratory Complexes

4.1.1 Protein Purification

An enormous advantage in the study of RMPs by MS techniques is that analyses need only minor amounts of protein (less than a microgram). Nevertheless, preparations need to be as pure as possible and protein complexes should be catalytically active, implying that they should be composed of all subunits in the correct stoichiometry at which the overall structure is maintained as found in the original membranous milieu. The latter may comprise the presence in the right quantities of native structural lipids. Obviously, purification can be achieved by conventional means. Because only minute protein amounts are required, efforts have been and are being directed at scaling down the repertoire of column chromatographic methods and of new separation methodologies to the nanolitre volumes, such as nanoparticle-based monoliths (Catherman et al., 2014; Tang et al., 2014). Otherwise, purification by BNE and elution from the gels might even provide a sufficient amount of pure protein for MS analyses.

An approach that has been especially designed for proteomics is affinity purification combined with MS (AP-MS), or variants of it like FLAG-tag (Terpe, 2003) and tandem affinity purification (TAP-MS) (Puig et al., 2001; Rigaut et al., 1999; Saada et al., 2009; Vogel et al., 2005). In the original TAP-MS procedure, the target protein is fused in frame with two tags separated by a tobacco etch virus cleavage site and expressed. The proximal tag is a calmodulin-binding peptide and the second one is composed of the IgG-binding part of the *Staphylococcus aureus* protein A. This setup allows a two-step purification of the target protein and its binding partners, but the method has its faults. The size of protein A (~21 kDa) may affect the folding, activity of the protein (complex) or interaction with the physiological binding partners. Purification is rather harsh and these partners may get lost. Moreover, despite the stringent conditions, the protein (complex) of interest may not be pure. Sometimes, up to 95% of all proteins that copurify are contaminants and their presence requires a tedious investment in MS and data analyses time to identify the right candidate(s) (Meyer & Selbach, 2015). In addition, data may be ambiguous in establishing which protein belongs to what complex. A reason for the high detection rate of contaminants is the extreme sensitivity of modern MS apparatus. To address these issues, several variants of the original method have been proposed, including the use of smaller tags (Gingras, Gstaiger, Raught, & Aebersold, 2007; Trinkle-Mulcahy, 2012; Völkel, Le Faou, & Angrand, 2010). Another solution is quantitative affinity purification (q-AP-MS) (Meyer & Selbach, 2015; Vermeulen, Hubner, & Mann, 2008). Herein, differentially SILAC-labelled cells are transfected with the tagged protein to be investigated, whereas a control vector contains the tag only. After immunoprecipitation, elution, protein digestion and shotgun proteins, lowly abundant contaminant proteins are always recovered in a 1:1 ratio and can be ignored. On the basis of the relative changes in peptide intensities, the subunit composition of the protein complex to be characterised is derived.

4.1.2 Primary Structure Information

Once purified, the identity of the subunits and subunit stoichiometry of a protein complex can be analysed. Subunits can be identified in the traditional way using SDS gel electrophoresis or 2D-BNE and subsequent peptide mass fingerprinting or shotgun MS at which the absence of specific peptides is indicative of N- or C-terminal cleavage events. Top-down proteomics explores the whole protein or protein complex inside the mass spectrometer. Here, the sample is volatised under atmospheric pressure and by soft ionisation (Fig. 14). Conditions are chosen such that subunit interactions

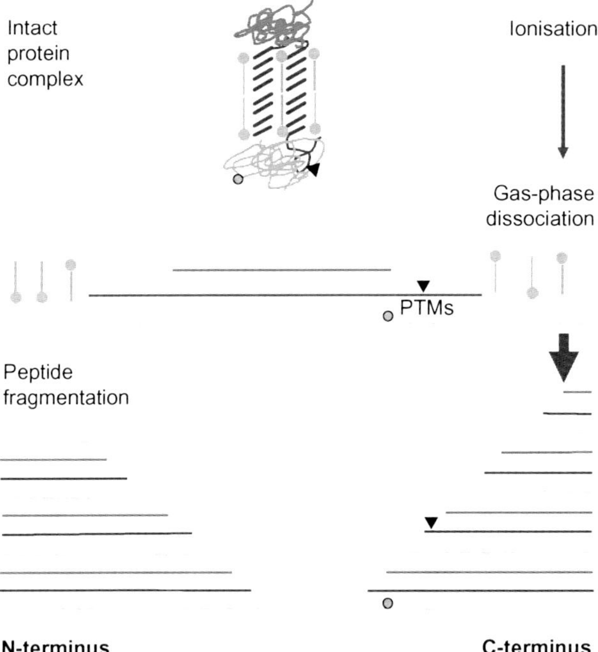

Fig. 14 Schematic representation of top-down MS analysis of a hypothetical membrane-bound protein complex composed of three subunits localised in the periplasmic (*grey*), membrane (*dark grey*) and cytoplasmic (*light grey*) aspects of the bacterial cell, respectively. The membrane-inserted subunit (2 TMHs) is associated with lipid molecules (*filled circles* and *sticks*) and contains a PTM (*black triangle*). The cytoplasmic subunit harbours another PTM (*grey circle*). Upon volatilisation of the as-isolated native complex inside the mass spectrometer, the gradual increase of ion activation/ionisation results in gas-phase dissociation of the complex, followed by fragmentation of the subunits. Hereafter, peptide fragments are sequenced and the PTMs are localised and identified.

and other noncovalent bindings are preserved. Herewith, PTMs remain intact and are subjected to further identification (see Section 4.2). After increase of ion activation (collision energy), the complex disintegrates into its constituent subunits of which the molecular masses can be determined, simultaneously enabling the establishment of the subunit stoichiometry (Hopper & Robinson, 2014). Further increase of the ion activation results in the fragmentation of the polypeptide chains, which can be fully sequenced. Interestingly, first fragments observed stem from membrane-embedded regions. Hence, by the deliberate tuning of ionisation energies, the relation between higher order structure and protein sequence can be probed in a single experiment (Konijnenberg et al., 2015).

Precise native MW measurements pose high demands on the resolution of the mass spectrometer, since it would have to distinguish between S–H groups and disulphide bridges ($\Delta m = 2$ Da), deamidation ($\Delta m = 1$ Da), acetylation vs trimethylation ($\Delta m = 0.04$ Da, sic) and phosphorylation vs sulphation ($\Delta m = 10$ Da). An inherent problem of molecular mass measurements is that the parent ion [M] is distributed across multiple peaks as the result of natural ^{1,2}H and ^{12,13}C isotope ratios (Section 2.1.1; Figs 2, 7 and 8).

Quite a few studies on respiratory systems have been supported by MS-based methods to obtain information on the subunit composition and subunit primary sequences. As already alluded, most of these were related to mitochondrial OXPHOS complexes. Studies were directed either at the full respiratory repertory of mitochondria from humans (Catherman et al., 2013; Wessels et al., 2013, 2009), plants (Klodmann, Lewejohann, & Braun, 2011) and yeasts (Helbig et al., 2009; Lemaire & Dujardin, 2008; Nübel, Wittig, Kerscher, Brandt, & Schägger, 2009), or specifically at eukaryotic complex I (Batista, Franco, Mendes, Coelho, & Pereira, 2010; Bridges, Fearnley, & Hirst, 2010; Carroll, Fearnley, Shannon, Hirst, & Walker, 2003; Dröse et al., 2011; Fearnley, Carroll, & Walker, 2007; Klodmann & Braun, 2011; Klodmann, Sunderhaus, Nimtz, Jänsch, & Braun, 2010; Pocsfalvi et al., 2006), complex III (Marín-Buera et al., 2015) or ATP synthase (Hoffmann et al., 2010; Wittig & Schägger, 2008b). Hitherto, only few investigations have explored bacterial respiratory systems from the proteomics perspective at some depth (Battchikova & Aro, 2007; Castelle et al., 2015; Menon et al., 2009), albeit with most promising results regarding SC formation. The latter work will be discussed in Section 5.1.

4.1.3 On Track to Secondary, Tertiary and Quaternary Structures of Respiratory Complexes

Proteomics does not only provide invaluable data on protein sequences, but nowadays approaches also allow an insight in the three-dimensional organisation of a protein complex, lending the field a firm place within structural biology (Benesch, Aquilina, Ruotolo, Sobott, & Robinson, 2006; Benesch & Robinson, 2006; Hyung & Ruotolo, 2012; Petrotchenko & Borchers, 2014; Zhou & Robinson, 2010). To this end, a number of tools and methods have been established that in combination permit the proposal of structural models, even though of as-yet low resolution (Konermann, Tong, & Pan, 2008; Owens, 2011). Most methods are still peptide-centric and are directed at the immediately accessible part of a protein, its surface. These methods are schematically represented in Fig. 15.

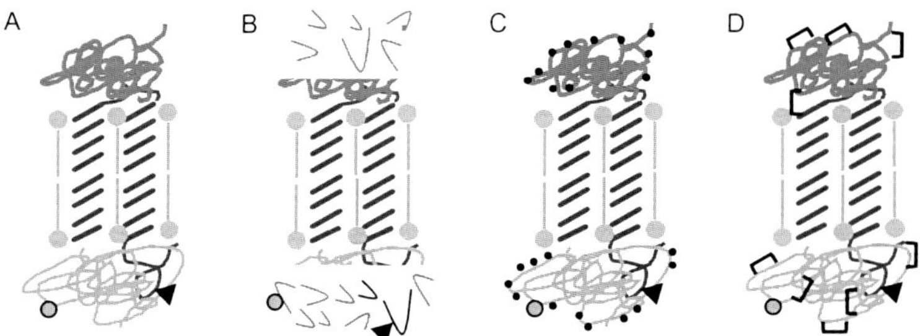

Fig. 15 Schematic representation of different MS-based methods to get structural information of a protein complex. The figure uses the same hypothetical protein as in Fig. 14. (A) As-isolated protein. (B) After limited proteolysis. Peptide fragments can be sequenced and PTMs identified and localised. (C) After probing of specific amino acids with chemical reagents, with low-molecular-weight dyes (protein *painting*), with in situ-generated hydroxyl radicals, or after deuterium exchange of amino acids amide hydrogen atoms. (D) After chemical cross-linking of neighbouring amino acid residues within or between the different subunits.

A method that actually has been used for decades to probe conformational features of proteins is limited chemical or enzymatic proteolysis (Feng et al., 2014; Fontana, de Laureto, Spolaore, & Frare, 2012; Fontana et al., 2004). We already referred to it in the context of 'membrane shaving' (Section 3.2.1). The first peptides that are released by this treatment are the ones exposed to the (aqueous) environment and their identification provides information on the topology of an RMP. If applied under different physiological conditions, limited proteolysis also may inform on conformational changes, and the dynamics of an enzyme. Amino acids and peptides are susceptible to chemical modifications. This property is utilised by the treatment of whole proteins with reagents that bring about known modifications (see for a review: Bennett, Matthiesen, & Roepstorff, 2000). Again, proteolysis and subsequent tandem MS disclose those parts of the enzyme complex that have been modified (or have resisted modification). 'Protein painting' is a recent further development of the approach in which small chemical dyes are used to probe the protein (Luchini, Espina, & Liotta, 2014). Alternatively, surface-exposed amino acids are oxidatively modified by exposure to hydroxyl radicals, once more permitting the elucidation of targeted peptides (Konermann & Pan, 2012). Whereas chemical modifications address specific amino acids, hydrogen/deuterium exchange coupled with MS (HDX-MS) exploits the property that all amino acids (except for proline) can exchange amide hydrogen atoms for deuterium, which permits more homogeneous protein labelling (Mandell, Baerga-Ortiz, Croy, Falick, & Komives, 2005; Mandell, Baerga-Ortiz,

Falick, & Komives, 2005). When incubated in a D_2O-containing buffer, the H/D exchange rates are not the same for all amides. Rates will generally be highest for the D_2O-exposed parts of a protein. This difference in rates is advantageous in identifying these parts or to investigate kinetics of a protein system (Tsutsui & Wintrode, 2007). HDX-MS has been widely used in bottom-up proteomics (see for reviews: Boersema et al., 2015; Iacob & Engen, 2012; Marcoux & Cianférani, 2015). After incubation in the D_2O buffer and quenching the exchange reaction at low pH 2–4 and 0–4°C, followed by pepsin cleavage, tryptic fragments are analysed by LC-MS/MS as usual. This method, however, also has its flaws. Peptide fragments may go unnoticed or unidentified, spectra are extremely complicated and LC separation may cause H/D scrambling and back reactions. Nevertheless, HDX-MS is fully compatible with a top-down approach where these drawbacks can be overcome (Bache, Rand, Roepstorff, & Jørgensen, 2008; Burns, Sarpe, Wagenbach, Wordeman, & Schriemer, 2015; Konermann, Pan, & Liu, 2011; Rand, Zehl, & Jørgensen, 2014).

Another chemical modification procedure that provides structural information is chemical cross-linking (CX-MS) introduced by Rappsilber, Siniossoglou, Hurt, and Mann (2000) and Young et al. (2000), and recently reviewed in Petrotchenko and Borchers (2010), Rappsilber (2011), Leitner et al. (2014, 2010), Ramisetty and Washburn (2011), Sinz, Arlt, Chorev, and Sharon (2015) and Tran, Goodlett, and Goo (2015). In this method, two amino acids are connected to each other through a reagent with two (or more) reactive side groups. Nowadays, a range of cross-linkers are available like N-hydroxysuccinimide derivatives that covalently bind to the N-terminus and the ε amine group of lysine residues, sulphydryl reactive cross-linkers, cross-linkers directed at the C-terminus and the carboxylic group of aspartates and glutamates, and photoreactive cross-linkers. Their main difference is the spacing between the reactive groups (5–15 Å). Since these lengths are known, the reagents act as rulers measuring the maximal distance between the targeted amino acids. Application of CX-MX, however, has its challenges. The number of suitable cross-link sites is limited or too many (in large proteins). MS spectra are quite convoluted and additional purification steps are required to recover large polypeptides that have evaded proteolytic cleavage for further investigation. These issues have been solved to an extent by cleavable linkers. In addition, software programmes have been developed that can handle cross-linked peptides (see Tran et al., 2015).

The simplest cross-linker is formaldehyde that is water soluble, permeates the cell membrane and forms reversible short-linked bonds (2.3–2.7 Å) (Ramisetty & Washburn, 2011). Despite recent promising results, the use of formaldehyde still needs thorough testing: this compound not only acts as a cross-linking agent, but it is capable of various other complicating chemical modifications as well (Srinivasa, Ding, & Kast, 2015).

Most studies that employed earlier methods were focused on soluble proteins. Research on respiratory RMPs is still in its infancy (Zhou & Robinson, 2014). The reason may be clear: RMPs are cumbersome to handle and difficult to analyse by MS (see Section 3.2.1). In addition, membrane-embedded, hydrophobic regions may be inaccessible for reagents used for peptide modification. For instance, amino acids required for cross-linking are rather rare in hydrophobic domains. Yet, cross-linking may stabilise the interaction between soluble and membrane-embedded subunits, supporting their combined isolation (Puts, Lenoir, Krijgsveld, Williamson, & Holthuis, 2010). Anyway, an examination of at least the hydrophilic stretches of membrane subunits and of the associated solvent-exposed subunits should be feasible (Fig. 15). Indeed, cross-linking was fruitfully applied to investigate the effect of serine/threonine/tyrosine (de)phosphorylation on the structure of chloroplast F_1F_0-ATP synthase (Schmidt, Zhou, et al., 2013). By successive dephosphorylations combined with cross-linking, the authors could accurately probe the changes in subunit interactions and distances resulting from (de)phosphorylation, providing strong evidence for their role in the structural *integrity* and catalytic action of ATP synthase.

4.1.4 Protein–Lipid Interactions

The lipid bilayer is vital for life. It not only separates the inside of the cell from the outside world, but it is also the site where respiratory processes are localised. It has become increasingly clear that close interaction between respiratory RMPs and lipid molecules is essential for the proper structuring and activity of these RMPs (Martfeld, Rajagopalan, Greathouse, & Koeppe, 2015; Stangl & Schneider, 2015; Yeagle, 2014). Cardiolipin and other anionic lipids play a key role in this respect, both in mitochondria (Claypool, 2009; Hasan, Yamashita, Ryan, Whitelegge, & Cramer, 2011; Paradies et al., 2014) and in bacterial respiratory systems (Arias-Cartin, Grimaldi, Arnoux, Guigliarelli, & Magalon, 2012; Arias-Cartin et al., 2011). It has been known for a long time that the loss of cardiolipin results in the loss of activity of bacterial membrane-bound enzymes such as NDH, nitrate reductase NarGHI, formate dehydrogenase, SDH, alpha-ketoglutarate dehydrogenase, quinol:cytochrome *c* oxidoreductase (complex III) and cytochrome bo_3:ubiquinol oxidase (Arias-Cartin et al., 2012; van Gestel et al., 2010). Supplementation with cardiolipin restores these activities. Furthermore, tight, noncovalent, yet highly specific spatial associations between membrane-embedded subunits and cardiolipin are found in crystal structures of RMPs (Arias-Cartin et al., 2012, 2011; Planas-Iglesias et al., 2015).

Cardiolipin is not evenly distributed across the cell membrane, but it is dynamically accumulated at the poles and the division centre of rod-shaped bacterial cells (Oliver et al., 2014). Intriguingly, nitrate reductase is mainly found at the poles as well, consistent with the partnership between protein and lipid molecules (Alberge et al., 2015). Obviously, this uneven placement has its consequences for the spatial organisation of respiratory systems and concurrent bioenergetic processes (Alberge et al., 2015).

The immediate implication of the above considerations for structural proteomics on respiratory RMPs is that the field has to take protein–lipid interactions into account. That is to say, investigations should not only be restricted to peptides but should also include the identification and quantification of lipid molecules. This requires a close collaboration between proteomics and lipidomics, which means lipid analysis by MS (Kliman, May, & McLean, 2011). In the ideal case, methods should become available to precisely probe the interaction sides between protein and lipid. One such way is the ground-breaking methodology described in the next section.

4.1.5 Membrane Protein Complexes Studied in Thin Air

One of the most amazing and counterintuitive recent discoveries in proteomics of RMPs was that these proteins may be studied not only in their native lipid environment (or once purified, in artificial solvent systems) but also in gas phase and under vacuum (Laganowsky, Reading, Hopper, & Robinson, 2013). Just like soluble proteins, an RMP present in appropriate micelles (Barrera, Di Bartolo, Booth, & Robinson, 2008; Calabrese, Watkinson, Henderson, Radford, & Ashcroft, 2015) can get volatised and ionised, despite the low charge state associated with hydrophobicity. In this volatilised state, the native complex remains intact with all interactions between subunits, substrates, lipids and its other partners (Chorev, Ben-Nissan, & Sharon, 2015; Hopper & Robinson, 2014; Landreh & Robinson, 2015; Morgner, Montenegro, Barrera, & Robinson, 2012; Zhou & Robinson, 2014). And just like for soluble proteins, upon increase of ionisation energy all binding partners consecutively disintegrate into subcomplexes, individual subunits still carrying PTMs, lipid molecules and cleaved protein products. Subsequently, all of these components can be identified, quantified and sequenced (Fig. 14). By combined peptide and lipid analyses, the foundation has been paved for the in-depth elucidation of the RMP–lipid interplay (Barrera, Zhou, & Robinson, 2013; Borysik, Hewitt, & Robinson, 2013; Hopper et al., 2013).

In addition to the rapid development of other top-down native MS techniques, another recently introduced powerful tool is ion-mobility MS (IM-MS) (Allison et al., 2015; Lanucara, Holman, Gray, & Eyers, 2014; May & McLean, 2015). In this method, ionised proteins are introduced into a cell filled with inert gas. Across this cell an electric field is maintained. Herein, proteins (and other molecules present in the sample) collide. The collision frequency depends on surface size: the larger the surface, the higher the number of collisions. Driven by the electric field, charged molecules exit the cell, somewhat resembling liquid CE. This diffusion rate (drift time) is directly related to the size and charge of the molecule: molecules with the highest mass migrate slowest. The result is a series of elution profiles of the individual molecule species. After exit, molecules are analysed in the MS part of the instrument and quantified. When calibrated with proteins of known structures, collected data allow the calculation of the average collisional cross section of a protein (in Å^2) providing important 3D structural information. In addition, these data enable the calculation of dissociation constants of binding partners as well as the revelation of subtle conformational changes and ligand binding-induced unfolding patterns (Laganowsky et al., 2014; Niu & Ruotolo, 2015; Rabuck et al., 2013; Stojko et al., 2015). Thus, by IM-MS information is gathered on the 3D structure and on the dynamics of a protein. A particular nice example of the opportunities offered by native MS and IM-MS in combination with lipidomics is the work by Zhou, Dong, et al. (2011) and Zhou, Morgner, et al. (2011). These authors investigated the V-type ATPases of *Thermus thermophilus* and *Enterococcus hirae* by these methods, which enabled the establishment of the subunit stoichiometries and the identification of the lipids that were associated with the rotor ring. Additionally, from analyses of subcomplexes formed in solution and in the gas phase, the effect of nucleotide binding on both ATP hydrolysis and proton translocation was assessed. In the later context of RMP assembly and maturation (Section 5.2.1) the application of IM-MS in unravelling the conformational dynamics between the NarJ chaperone and the NarG catalytic subunit of nitrate reductase should be noted (Lorenzi et al., 2012).

4.1.6 Putting Structural Information in a Timed Perspective

The application of bottom-up and top-down MS approaches and their variants, especially when performed by well-designed combinations, may give information at all four structural levels of a protein complex to the extent that a 3D model of it can be proposed (Boersema et al., 2015; Marcoux & Cianférani, 2015; Walzthoeni, Leitner, Stengel, & Aebersold, 2013). The MS-based model will not be as refined as one deduced from X-ray crystallography or NMR. However, these three approaches can be complementary (see f.i. Nyon et al., 2015). If refractory to X-ray or NMR analyses, MS structural data of a protein complex can be modelled onto cryo-electron microscopy (EM) maps, as has been done to determine the molecular architecture of the proteasome of *Schizosaccharomyces pombe* (Lasker et al., 2012). Unlike X-ray crystallography, NMR and EM, structural proteomics brings one more dimension into play: time. Cooperative movements and structural changes are revealed over a timescale from milliseconds to minutes (Zhou & Robinson, 2014). Moreover, proteomics offers two more benefits. Firstly, purification or BNE under mild conditions may detect more interaction partners like SCs, assembly factors, regulatory proteins and other compounds that direct enzyme activity (see Section 5). Secondly, the MS-supported approach is strong in identifying PTMs (see Section 4.2). The challenge is to apply these proteomics tools on bacterial respiratory systems of which many known ones remain to be investigated in more detail and even more new ones may be discovered in the near future. The above examples of the work on the F_1F_0- and V-type ATPases, and nitrate reductase indicate that such an approach could be feasible.

4.2 Posttranslational Modifications

Proteins are composed of a standard collection of 20 different amino acids, but currently more than 450 ways are known in which the primary structure of these amino acids is modified in biological systems. Modifications are made after translation to serve proper folding, assembly and function (Ryšlavá, Doubnerová, Kavan, & Vaněk, 2013). Generally, a modification results from the action of dedicated enzymes, whereas other, even so dedicated enzymes may remove PTMs, recovering the original amino acid. Reversible PTM provides a means to change enzyme activity instantaneously. PTMs originate also chemically from exogenous or endogenous sources like reactive oxygen species (ROS) and reactive nitrogen species (RNS) that destroy proteins, requiring the controlled action of repair or protein degradation systems. Next to this, PTM lies at the basis of signalling pathways by which an organism responds to changes in its environment. Living systems have evolved a plethora of relatively simple to extremely complex signalling pathways that may communicate with each other (crosstalk) to enable the appropriate response of the whole cell to an environmental change. Crosstalk may also occur at the level of the enzyme itself. A substantial percentage of enzymes has more than one and even many sites for a same PTM or for different PTMs (Soufi, Soares, Ravikumar, & Macek, 2012; Venne, Kollipara, & Zahedi, 2014; Young, Plazas-Mayorca, & Garcia, 2010). By differential modification, enzyme activity can be fine-tuned in response to different stimuli.

The process of (reversible) protein modification has been known for a long time, especially in eukaryotes that have exploited its possibilities in every conceivable variant. The true magnitude of this PTM universe has been revealed by proteomics (Olsen & Mann, 2013). Currently, tens of thousands of PTMs have been identified and localised within protein sequences. Proteomics made it even so clear that prokaryotes utilise a same arsenal of PTMs as Eukarya to control their cell processes (Grangeasse, Stülke, & Mijakovic, 2015).

On several occasions, we pointed out difficulties in analysing and identifying PTMs by (standard) proteomics, especially since identification of each type of PTM requires its own strategy (see for reviews on methodological aspects: Černý, Skalák, Cerna, & Brzobohatý, 2013; Kettenbach, Rush, & Gerber, 2011; Young et al., 2010; Zhao & Jensen, 2009). Consequently, identification of PTMs could only be addressed within the field of proteomics after the development of methods that enabled (1) the routine enrichment of modified peptides from complex biological samples, (2) robust MS analysis and (3) analyses on a high-throughput basis. In the following, we will give a general overview of prokaryotic research on protein phosphorylation, acetylation, methylation and redox-related PTMs, and we will refer to recent reviews that will give entry to the extensive primary literature. Specific examples are related to the topic of this volume, the regulation of bacterial ETCs and of mitochondrial OXPHOS complexes where the impact of the different types of PTM is better comprehended.

4.2.1 Protein Phosphorylation

For a long time, it was believed that the role of phosphorylation in the regulation of bacterial cell processes was limited to the phosphorelay of two-component systems, the phosphotransfer system in the uptake of sugars (Stülke & Hillen, 1998) and serine phosphorylation of the catabolite control protein A (ccpA) (Fujita, 2009), and in the regulation of the isocitrate dehydrogenase by the bifunctional AceK kinase/phosphatase controlling the entry into the glyoxylate shunt (Zheng & Jia, 2010; Zheng, Yates, & Jia, 2012). However, the introduction and application of MS-based methods (Gerber, Rush, Stemman, Kirschner, & Gygi, 2003; Macek,

Mann, & Olsen, 2009; Olsen & Macek, 2009; Villén & Gygi, 2008; Wang, Pan, & Tao, 2014) for the global identification of eubacterial and archaeal phosphoproteomes demonstrated that this view was too narrow (Macek & Mijakovic, 2011; Mijakovic & Macek, 2012).

Even though histidine and aspartate phosphorylation acting in the two-component regulatory and phosphotransfer systems remained undetected because of the labile nature of the phosphoryl bonds, proteomics showed extensive phosphorylation of serine (S), threonine (T) and tyrosine (Y) residues in all investigated prokaryotes (Table 4). This table shows that the extent of phosphorylation is highly variable among different species with respect to the number of proteins and the number of verified phosphorylation sites. Although microbial species most certainly differ from each other in the degree and ways protein phosphorylation is employed in the regulation of cellular processes, these differences are also due to methodological flaws. Initially, phosphoproteomes were analysed by 2DE by which protein phosphorylation was visualised by fluorescent labelling by Pro-Q Diamond (Agrawal & Thelen, 2005). Unfortunately, the original method was rather imprecise, both regarding visual detection and MS identification. (Please note the recent improvement by Wang et al., 2014.) An up to 10-fold higher recovery of phosphorylated proteins and identified phosphopeptides (events) was achieved by the use of shotgun proteomics (Macek et al., 2007) (Table 4). Herein, phosphopeptides from tryptic lysates are subsequently separated by strong cation-exchange chromatography (SCX), enriched by binding to TiO_2 beads or by immobilised-metal affinity chromatography (IMAC) and analysed by high-accuracy MS to identify phosphopeptides in the low-femtomole range (Macek & Mijakovic, 2011). Alternatively, phosphopeptides are enriched by immunoprecipitation by phosphoserine, phosphothreonine or phosphotyrosine antibodies (Černý et al., 2013). Despite continual methodological improvements, due to differences in sample (pre)preparation phosphopeptide recoveries vary substantially even for the same organism (see f.i. *E. coli* in Table 4) (Lin et al., 2015). While microbial species differ in the extent of protein phosphorylation, they also differ in phosphorylation targets and quite interesting differences may be seen in the ratios between serine:threonine:tyrosine phosphorylation (Table 4), likely reflecting species-related variations in regulatory strategies.

The phosphorylation degree of a protein is determined by the antagonistic activities between ATP-dependent protein kinases targeting S, T or Y residues and phosphatases. For serine and threonine phosphorylation, prokaryotes employ Hanks-type kinases as also known for eukaryotes (Cousin et al., 2013; Macek & Mijakovic, 2011). Tyrosine phosphokinases belong to

Table 4 Bacterial Phosphoproteomes[a]

Species (Proteins)[d]	Method	P-Proteins[b] (P-Events)	S:T:Y (%)[c]	References
Bacillus subtilis (4175)	2DE	28 (9)	NR	Eymann et al. (2007)
	Shotgun	78 (103)	69:20.5:10	Macek et al. (2007)
	Shotgun	175 (339)	74.8:17.7:7.1	Lin, Sugiyama, and Ishihama (2015)
Campylobacter jejuni (1672)	2DE	36 (58)	NR	Voisin et al. (2007)
Escherichia coli (4140)	2DE	24 (80)	NR	Soung, Miller, Koc, and Koc (2009)
	Shotgun	79 (105)	68:23.5:8.5	Macek et al. (2008)
	Shotgun	150 (108)	75.9:16.7:7.4	Soares et al. (2013)
	Shotgun	393 (1088)	69.5:21.8:7.7	Lin et al. (2015)
Halobacterium salinarium (2592)	Shotgun	69 (81)	86:12:1	Aivaliotis et al. (2009)
Helicobacter pylori (1447)	Shotgun	76 (126)	42.8:38.7:18.5	Ge et al. (2011)
Klebsiella pneumoniae (5501)	Shotgun	81 (117)	NR	Lin et al. (2009)
	Shotgun	559 (663)	72.9:13.7:12.9	Lin et al. (2015)
Lactobacillus lactis (1638)	Shotgun	63 (73)	46.5:50.5:3	Soufi et al. (2008)
Mycobacterium tuberculosis (3906)	Shotgun	301 (516)	60:40:NR	Prisic et al. (2010)
Mycoplasma pneumoniae (691)	2DE	63 (16)	NR	Schmidl et al. (2010)
	Shotgun	61 (93)	58:37:5	van Noort et al. (2012)

Continued

Table 4 Bacterial Phosphoproteomes—cont'd

Species (Proteins)	Method	P-Proteins (P-Events)	S:T:Y (%)	References
Pseudomonas aeruginosa (5572)	Shotgun	23 (55)	52.7:32.7:14.5	Ravichandran, Sugiyama, Tomita, Swarup, and Ishihama (2009)
Pseudomonas putida (5350)	Shotgun	40 (53)	52.8:39.6:7.5	Ravichandran et al. (2009)
Sinorhizobium meliloti (6208)	Shotgun	77 (96)	63:28:5	Liu, Tian, and Chen (2015)
Staphylococcus aureus (2767)	2DE	103 (68)	NR	Bäsell et al. (2014)
Streptococcus pneumoniae (1814)	Shotgun	102 (163)	47:44:9	Sun et al. (2010)
Streptomyces coelicolor (8152)	Shotgun	40 (46)	34:52:14	Parker et al. (2010)
Synechocystis sp. PCC 6803 (3233)	2DE	32 (8)	NR	Mikkat, Fulda, and Hagemann (2014)
	Shotgun	183 (301)		Spät, Maček, and Forchhammer (2015)

NR, not reported.
[a]Adapted from Macek and Mijakovic (2011).
[b]P-proteins, number of unique phosphorylated proteins; P-events, number of uniquely identified phosphorylation sites.
[c]S:P:Y, ratio (%) between serine, threonine and tyrosine phosphorylation sites, respectively, in the phosphoproteome.
[d]Number of protein-encoding genes as recorded in the NCBI genome database.

the BY-kinase family and are unrelated to those of Eukarya (Grangeasse, Cozzone, Deutscher, & Mijakovic, 2007), even though their catalytic mechanisms are similar (Alber, 2009). Prokaryotic protein phosphatases are generally homologous to eukaryotic ones (Pereira, Goss, & Dworkin, 2011). However, it is largely unknown how those kinases and phosphatases are activated. The elucidation of these regulatory mechanisms and the identification of protein substrates of the kinases and phosphatases are topics of extensive research (Macek & Mijakovic, 2011; Mijakovic & Deutscher, 2015; Ravikumar et al., 2014). One of the outcomes of these research efforts is that different phosphorylation systems are able of crosstalk (Jers, Kobir, Søndergaard, Jensen, & Mijakovic, 2011; Shi et al., 2014; van Noort et al., 2012). No matter the complete underlying picture, phosphoproteome inventories made it fully clear that protein (de)phosphorylation directs the regulation of many key cellular processes, including central metabolism and respiration (for further details see references listed in Table 4).

Indeed, bacterial respiratory complexes are phosphorylated. In *E. coli*, four phosphorylation sites are found on the alpha subunit of anaerobic respiratory fumarate reductase (FrdA), one each on the NuoB and NuoI subunits of NDH, one on the NarG catalytic subunit of nitrate reductase next to three on NarL, one on the cydA subunit of the high-oxygen affinity alternative terminal oxidase CydAB (cytochrome *d*:ubiquinol oxidase, cytochrome *bd*-I oxidase; Borisov et al., 2011) and seven on the ATP synthase (AtpA, 1; AtpD, 3; AtpF, 3) (Macek et al., 2008). These same respiratory complexes are also targeted in *B. subtilis* (NDH, 1 site; NarG, 1 site; CydAB, 1 site; ATP synthase, 4 sites, including 2 sites each on AtpA and AtpB). Also respiratory complexes of *Klebsiella pneunomiae* undergo phosphorylation, albeit in different fashions. Here, 1 site is seen on FrdA, 2 sites on the NuoI subunit of NDH, 4 sites on AtpB, 4 sites on CydA and no less than 10 sites on CydC, the putative subunit III of the CydAB complex. The structural and functional significance of these bacterial phosphorylations remains to be established, but it is interesting to note that, except for ATP synthase, these events address RMPs that operate under anaerobic or microaerophilic conditions. In mitochondria, protein phosphorylation is known to be of fundamental importance in the regulation of the activities, assembly and maintenance of the structural integrity of the OXPHOS complexes (Kane & Van Eyk, 2009; Pagliarini & Dixon, 2006; Reiland et al., 2009; Reinders et al., 2007; Valsecchi, Ramos-Espiritu, Buck, Levin, & Manfredi, 2013). Notably, multiple phosphorylations have been revealed and partially identified with regard to their roles in the complexes I (NDH), IV (cytochrome *c* oxidase) and V (ATP synthase).

Phosphorylation of complex I concerns the nuclear-encoded accessory subunits NDUFA7, NDUFA10, NDUFC2, NDUFS4, ESSS and GRIM-19, which have been implemented with the assembly of the 45-subunit complex (De Rasmo et al., 2010; Palmisano, Sardanelli, Signorile, Papa, & Larsen, 2007; Vartak, Semwal, & Bai, 2014). Cytochrome *c* oxidase is phosphorylated at no less than 18 positions facing both the matrix and intermembrane space of the mitochondrion (Helling, Hüttemann, Kadenbach, et al., 2012; Helling, Hüttemann, Ramzan, et al., 2012). At least a number of these reversible phosphorylations seem to deal with 'allosteric ATP inhibition' involved in the control of an appropriate *pmf* and in preventing the formation of superoxide radicals (Helling et al., 2008; Helling, Hüttemann, Kadenbach, et al., 2012; Helling, Hüttemann, Ramzan, et al., 2012; Hüttemann et al., 2012). The ATP synthesis machinery itself is phosphoryl-targeted at several subunits, enabling the regulation of its activity (Kane & Van Eyk, 2009; Kane, Youngman, Jensen, & Van Eyk, 2010). In the aforementioned papers by Schmidt, Müller, et al. (2013) and Schmidt, Zhou, et al. (2013), it was shown that the phosphorylation degree had a direct effect on subunit interactions that might play a role in complex assembly and/or enzyme activity. Phosphorylation of the mitochondrial OXPHOS complexes is mediated by protein kinases. At least some of the kinases are allosterically regulated by cyclic-AMP, which involves, among other ones, the action of a novel, mitochondrial-specific soluble adenylate cyclase (De Rasmo et al., 2015; Nowak & Bakajsova, 2015; Valsecchi, Konrad, & Manfredi, 2014; Valsecchi et al., 2013). As mentioned, the role of (reversible) phosphorylation of bacterial respiratory complexes is not understood, no more than possible signalling pathways either.

4.2.2 Protein Acetylation

Proteomics established that besides protein phosphorylation, protein acetylation is fundamental in the regulation of central cellular processes, also in prokaryotes (see for reviews: Bernal et al., 2014; Hentchel & Escalante-Semerena, 2015; Jones & O'Connor, 2011; Kim & Yang, 2011; Soppa, 2010). In order to make a global inventory of acetylation sites and events (the 'acetylome') shotgun proteomics is currently supported by the selective immunoprecipitation of acetylated peptides by specific antibodies (Guan, Yu, Lin, Xiong, & Zhao, 2010; Hentchel & Escalante-Semerena, 2015; Zhao et al., 2010). Using this methodology, the acetylomes of a number of bacterial species have been determined (Table 5). These acetylomes comprised between 60 and over 600 different proteins in species-dependent

Table 5 Bacterial Acetylomes

Species (Proteins)[a]	Ac-Proteins (Ac-Sites)[b]	References
Bacillus subtilis (4175)	185 (332)	Kim, Kim, et al. (2013) and Kim, Yu, et al. (2013)
	629 (1355)	Kosono et al. (2015)
Erwinia amylovora (3298)	96 (141)	Wu, Vellaichamy, et al. (2013) and Wu, Lee, Liao, and Wei (2013)
Escherichia coli (4140)	85 (125)	Yu, Kim, Moon, Ryu, and Pan (2008)
	91 (138)	Zhang, Zheng, et al. (2013)
Geobacillus kaustophilus (3352)	114 (253)	Lee et al. (2013)
Mycobacterium tuberculosis (3908)	137 (226)	Liu et al. (2014)
	658 (1128)	Xie et al. (2015)
Mycoplasma pneumoniae (691)	221 (719)	van Noort et al. (2012)
Rhodopseudomonas palustris (4652)	62 (—)	Crosby, Pelletier, Hurst, and Escalante-Semerena (2012)
Salmonella enterica (4430)	191 (235)	Wang, Xu, et al. (2010) and Wang, Zhang, et al. (2010)
Staphylococcus aureus (2810)	412 (—)	Zhang et al. (2014)
Streptomyces roseosporus (6406)	667 (1143)	Liao, Xie, Li, Cheng, and Xie (2014)
Thermus thermophilus (2173)	128 (197)	Okanishi, Kim, Masui, and Kuramitsu (2013)
Vibrio parahemolyticus (4449)	656 (1413)	Pan et al. (2014)

[a]Number of protein-encoding genes as recorded in the NCBI genome database.
[b]Ac-proteins, number of unique acetylated proteins; Ac-sites, number of unambiguously identified acetylation sites.

ways, and these numbers may very well be underestimates. For example, a recent study on the acetylome of *B. subtilis* (Kosono et al., 2015) reported a 3.5-fold higher coverage than found earlier (Kim, Yu, et al., 2013). Nevertheless, the width and scope of protein acetylation make the phenomenon of at least equal importance as phosphorylation. Unlike Eukarya, irreversible N-terminal acetylation is relatively rare in bacteria

(Hentchel & Escalante-Semerena, 2015). Just as it is the case for eukaryotes (Guan & Xiong, 2011; Kim & Yang, 2011; Shi & Tu, 2015; Tripodi, Nicastro, Reghellin, & Coccetti, 2015; Zhao et al., 2010), acetylation addresses transcription, translation, protein folding, amino acid- and nucleotide biosynthesis and metabolism in prokaryotes (Bernal et al., 2014; Hentchel & Escalante-Semerena, 2015; Jones & O'Connor, 2011). Protein acetylation even appears to be the crucial mechanism to control the central carbon metabolism and the TCA cycle at the posttranslational level (Wang, Zhang, et al., 2010). In agreement herewith, fructose-1,6-bisphosphate aldolase in *E. coli* (Zhang, Fonslow, et al., 2013; Zhang, Zheng, Yang, Chen, & Cheng, 2013) has no less than 11 acetylation sites, besides 6 succinylation sites, which represents another recently discovered PTM (Kosono et al., 2015; Weinert et al., 2013). *E. coli* glyceraldehyde-3-phosphate dehydrogenase harbours even 14 acetylation sites, next to 9 succinylation sites, and in isocitrate lyase, which lacks any of the latter, 11 sites are subjected to acetylation. In contrast, isocitrate dehydrogenase of this organism, the model enzyme of bacterial protein phosphorylation, is acetylated on five positions, whereas phosphorylation occurs at four positions. In fact, a large proportion of bacterial enzymes has both phosphorylation and acetylation sites. The availability of PTMs in different modes and numbers provides the opportunity of crosstalking (Soufi et al., 2012; van Noort et al., 2012). It remains to be experimentally investigated how and to what extent this occurs.

Protein acetylation usually concerns the ε amino group of lysine to which acetyl-coenzyme A is the acetyl donor. The reaction is catalysed by lysine acetyltransferases and in bacterial genomes three different classes of this enzyme are found (Bernal et al., 2014; Hentchel & Escalante-Semerena, 2015; Jones & O'Connor, 2011). It should be noted that lysine acetylation is also performed nonenzymatically by acetyl phosphate, but its physiological relevance is not fully understood (Kuhn et al., 2014; Wagner & Hirschey, 2014). Removal of the acetyl group is achieved by lysine deacetylases that are represented by four different classes of enzymes. By the presence of lysine acetyltransferases and -deacetylases, the process is reversible and is prone to regulation. Bacterial genomes usually harbour a number of candidates of either enzyme (Hentchel & Escalante-Semerena, 2015). Yet, except for some cases (Nambi et al., 2013; Xu, Hegde, & Blanchard, 2011; reviewed in Hentchel & Escalante-Semerena, 2015), it is again far from clear which acetyltransferase or -deacetylase targets which set of substrates.

Concerning respiratory complexes, in *E. coli* NDH is acetylated at the NuoF (one site) and NuoE (two sites) subunits, at terminal oxidase CydAB on one site each of both subunits, at the AtpA (two sites) and AtpB (three sites) subunits of ATP synthase, at one site of nitrate reductase NarG and at four different positions of FrdA (Zhang, Fonslow, et al., 2013; Zhang, Zheng, et al., 2013). One may note that the very same target proteins were found for phosphorylation (see Section 4.2.1), which again might suggest regulatory crosstalking. Once more, the structural, regulatory or functional implications of those acetylations are elusive. For the sake of comparison, more than one-third of all mitochondrial proteins are acetylated, most of these being involved in some aspect of energy metabolism (Anderson & Hirschey, 2012; Papanicolaou, O'Rourke, & Foster, 2014). Regarding mitochondrial complex I, 13 of its subunits, all being nuclear encoded, are acetylated at their N-terminal position (Carroll, Ding, Fearnley, & Walker, 2013). ATP synthase is acetylated at evolutionary conserved positions on the F_1 alpha, beta and gamma subunits as well as on the oligomycin sensitivity-conferring protein (Vassilopoulos et al., 2014). Deacetylation is catalysed by the NAD^+-dependent SIRT3 sirtuin. In cells lacking functional sirtuin, ATP synthase remains acetylated and is impaired in its activity (Vassilopoulos et al., 2014; Wu, Lee, et al., 2013). The role of SIRT3 is not restricted to ATP synthase. If absent or inactive, many mitochondrial proteins become hyperacetylated. Herewith, SIRT3 might turn out to be a key modulator of energy homeostasis (Anderson & Hirschey, 2012; Osborne, Cooney, & Turner, 2014). Sirtuins are also found in genomes of prokaryotes (Hentchel & Escalante-Semerena, 2015), suggesting a universal function of these proteins in PTM by reversible acetylation (Choudhary, Weinert, Nishida, Verdin, & Mann, 2014). In mycobacteria, a homolog already has also been identified as a regulator of a variety of enzymes (Gu et al., 2015; Xu et al., 2011). It is unknown whether sirtuins, or other lysine acetylases and deacetylases, act in the regulation of bacterial respiratory enzyme complexes.

4.2.3 Protein Methylation

Besides acetylation, lysine can be posttranslationally modified by mono-, di- and trimethylation. The other amino acids that can be added methyl groups are arginine, glutamine, aspartate and histidine. Two broad classes of enzymes are able to methylate amino acids and both use S-adenosylmethionine as a methyl donor (Schubert, Blumenthal, & Cheng, 2003). The first class (formally Class V methyltransferases) is characterised by a so-called SET domain and these enzymes predominantly methylate histones in eukaryotes. Members of the second class (formally Class I methyltransferases), the seven β-strand methyltransferases, methylate DNA, RNA and the aforementioned amino acids. Lysine methylation is reversible, but bacterial demethylases are less well defined (Lanouette, Mongeon, Figeys, & Couture, 2014). Protein methylation plays a diverse role in biosynthesis, signal transduction, protein repair, chromatin regulation and gene silencing (Lanouette et al., 2014; Schubert et al., 2003), even though the chemistry of a group is only little effected by methylation. However, the presence of a methyl group may protect a protein from being modified by another PTM, like ubiquitination in eukaryotes (Pang, Gasteiger, & Wilkins, 2010). Next, lysine methylation creates a binding surface for the recruitment of other proteins or peptides. Amino acid methylation can also have a structural role in a protein itself. In methylcoenzyme M reductase (MCR) from methanogens, for example, histidine, arginine, glutamine and cysteine moieties close to the factor F_{430} catalytic centre in the alpha subunit are all methylated (Ermler et al., 1997). However, 2DE analysis (Fig. 9B) shows that the alpha subunit in both type I and type II MCR occurs in different proteoforms, possibly representing variations in these methylations.

While protein methylation addresses many cellular functions, application of proteomics to explore these functions on a genome-wide basis has lagged behind (Lanouette et al., 2014). There are several reasons for this delay. Firstly, it took a while before antibodies were found to enrich low-abundant methylated peptides specifically and with high affinity (Carlson, Moore, Green, Martín, & Gozani, 2014; Moore et al., 2013). Secondly, methylated peptides separate poorly on SCX-LC columns. Thirdly, the peptides are highly charged under ESI, which limits the number of sequence-informative products produced by CID, and loss of the labile methylation moieties during CID precludes effective fragmentation of the peptide backbone (Snijders, Hung, Wilson, & Dickman, 2010; Wang et al., 2009). The latter problem has been solved by high-resolution ETD and MS^3 methods (Snijders et al., 2010; Wang et al., 2009). By the application of one of these techniques, the identification of mono-, di- or trimethylated amino acids in purified or highly enriched protein preparations that harbour those species structurally should not be a problem, considering mass shifts of 14, 28 and 42 Da. (Note, however, the very small mass difference between trimethyllysine, 42.05 Da, and acetyllysine, 42.01 Da.) We will illustrate these applications in view of mitochondrial OXPHOS complexes.

The analysis of whole complex I revealed a complex PTM pattern of subunit B12 that forms part of the membrane arm (Carroll et al., 2013). Its N-terminus is a mixture of acetylated and free amino acids, whereas one, two or three methyl moieties are attached to histidine residues 4, 6 and 8, respectively, in various combinations. However interesting, the physiological significance of this complexity is not understood. Next, arginine 82 of the 42-kDa subunit (NDUFS2) is symmetrically dimethylated on the ω-N^{G} and ω-$N^{G\prime}$ nitrogen atoms of the guanidine group (Carroll et al., 2013). Arginine 82 is localised near the quinone binding site and its position is conserved in yeast and in *Paracoccus denitrificans*, suggesting that this PTM is functionally important. Dimethylation is catalysed by the assembly factor NDUAF7 (Rhein, Carroll, Ding, Fearnley, & Walker, 2013) and this methylation step occurs early in complex I assembly. Another interesting example is ATP synthase. One of its key elements is the rotor ring composed of 75 amino acid c subunits, each crossing the membrane twice. TMHs are linked by a short loop that contains a lysine (K43). In metazoans, this lysine is trimethylated, which might provide a site for cardiolipin binding (Walpole et al., 2015). Importantly, in metazoans the rotor ring is comprised of eight c subunits. In unicellular Eukarya and prokaryotes, the lysine is either not methylated or substituted for another amino acid. The rotor ring of these organisms harbours 9–15 c subunits. This small difference in lysine substitution has very important bioenergetic consequences for the H^+ translocation/ATP synthesis ratio: each c subunit takes one proton across the membrane per rotor rotation at which three molecules of ATP are synthetised. Hence, the more c subunits are present in the rotor ring, the less ATP is made per rotation. A direct implication for the latter is that ATP is still feasible at lower *pmf*.

4.2.4 Oxidative Stress-Related Posttranslational Modifications

In their natural environment or during infections, bacteria are exposed to what is collectively called oxidative stress. This stress is caused by ROS, RNS, chlorine (RCS) or electrophilic species (RES) (Antelmann & Helmann, 2011; Loi, Rossius, & Antelmann, 2015). Hosts use these species to bombard and kill their infectious invaders. However, oxidative stress is also endogenous during of respiration. ROS include superoxide (O_2^-), hydrogen peroxide (H_2O_2) and the most detrimental of all, the hydroxyl radical ($OH^\cdot$) (Bleier et al., 2015; Imlay, 2013). These species are formed by successive one-electron reductions of dioxygen, which are the result of side reactions of respiratory enzymes or are catalysed by enzymes that are intended to remove ROS, such as superoxide dismutase, catalase, cytochrome *c* peroxidase and NADPH-dependent alkyl hydroperoxide reductase (Fig. 16A). In respiratory enzymes, reactions of oxygen with iron–sulphur clusters, (semi)reduced flavins or quinones are a major source

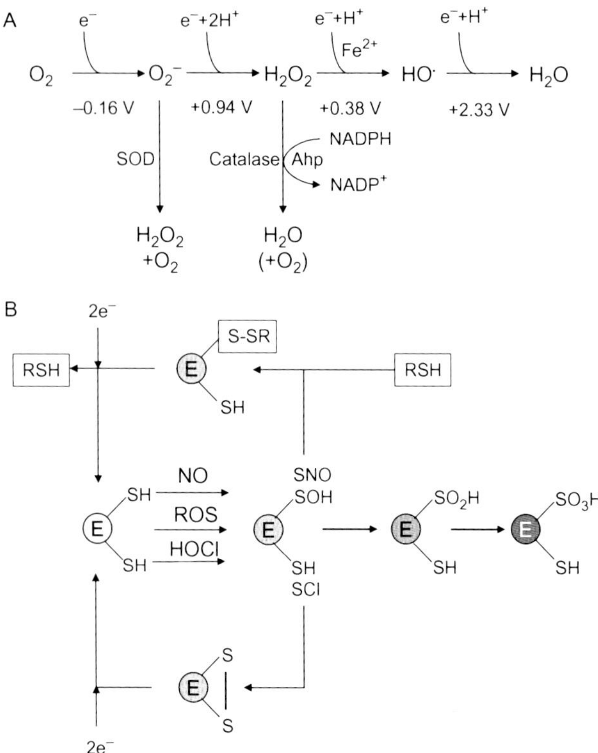

Fig. 16 Oxidative stress-related phenomena and protein repair. (A) Generation of reactive oxygen species (ROS) by subsequent one-electron reductions of O_2. The superoxide radical ($O_2^{\cdot}$) is converted into hydrogen peroxide (H_2O_2) and O_2 by superoxide dismutase (SOD). Catalase converts H_2O_2 into water and O_2, whereas NADPH-dependent alkyl hydroperoxide reductase (Ahp) or cytochrome *c* peroxidase (not shown) reduces H_2O_2 to water. The hydroxyl radical ($OH^{\cdot}$) is produced from H_2O_2 by chemical reaction with Fe^{2+} (Fenton reaction). (B) Reaction of thiol groups in the protein with ROS, reactive nitrogen species (NO) or hypochlorite (HOCl) converting a thiol in sulphenic acid (–SOH), a nitrosyl derivative (–SNO) or a sulphenylchloride (–SCl), respectively. The latter species are irreversibly oxidised to sulphinic ($-SO_2^-$) and sulphonic ($-SO_3^-$) acid. Alternatively, these species react with other thiols in the protein to form disulphides (lower branch) or with low-molecular-weight 'redox buffers' (RSH; glutathione, bacillithiol, mycothiol) to form heterodisulphides (*upper branch*). Enzymatic reduction of these disulphides regenerates the original thiols, establishing protein repair.

of superoxide and H_2O_2. Hydroxyl radicals are nonenzymatically produced by the reaction of H_2O_2 with Fe^{2+}, better known as the Fenton reaction (Imlay, 2013). The major RNS are nitric oxide (NO) and peroxynitrite, which is formed by the reaction between NO and oxygen. RNS are generated endogenously during anaerobic nitrate respiration or they are actively produced as an antimicrobial agent by activated neutrophils. These neutrophils are also producers of the predominant RCS, hypochloric acid (HOCl, bleach), which is a strong oxidant ($E'_0 = 1.28$ V) (Hillion & Antelmann, 2015; Loi et al., 2015). RES are comprised of broad collection of compounds, like quinones, methylglyoxal and diamide, that react with suitable nucleophiles, most notably with the thiol (SH) group of cysteine, via (irreversible) thiol-S-alkylation chemistry.

ROS and other species have a detrimental effect on DNA, lipids and proteins. The compounds cause lesions in proteins and the destruction of mononuclear iron centres and iron–sulphur clusters involved in catalysis or electron transfer (Imlay, 2013). Cysteine is particularly sensitive towards these species. Upon reaction with ROS, its thiol group is first oxidised to the unstable cysteine sulphenic acid (R-SOH), which reacts with other thiols to form either intra- and intermolecular protein disulphides (RS–SR′) or heterodisulphides with low-molecular-weight (LMW) thiols in the cell (Antelmann & Helmann, 2011; Hillion & Antelmann, 2015; Loi et al., 2015) (Fig. 16B). Disulphide formation is reversible and this reversibility is the basis of protein repair. Cysteine sulphenic acid readily oxidises to sulphinic (RSO_2^-) and sulphonic (RSO_3^-) acid and these oxidations are irreversible. Also NO and HOCl rapidly react with cysteine, causing cysteine S-nitrosation ($RS–NO^-$) and cysteine sulphenylchloride (RS–Cl) formation, respectively. The latter compounds are also unstable and reaction with other thiols forms disulphides. In the absence of these thiols, nitrosated cysteines and cysteine sulphenylchlorides are further irreversibly oxidised to sulphinic and sulphonic acid.

Microorganisms have developed ingenious systems to cope with oxidative stress and they do this by a diversity of regulatory systems that have been appreciated only recently (see for reviews: Antelmann & Helmann, 2011; Fu, Yuan, & Gao, 2015; Hillion & Antelmann, 2015; Loi et al., 2015). Best known are the OxyR and SoxR systems of *E. coli*, which protect the organism from peroxide and superoxide stress, respectively (Chiang & Schellhorn, 2012; Imlay, 2013). The common property of all those systems is the presence of transcriptional regulators that both sense specific stressing agents at very low concentrations, and in response activate enzyme machineries that clear those agents and repair the damage. In order to sense, the regulators exploit the same chemistry just described. Reaction with a stressing compound elicits a conformational change of the regulator, which is translated in protein expression. For instance, after reaction with hydrogen peroxide or with NO two specific cysteines in OxyR make a disulphide bond, causing the concomitant structural change of the protein (Jo et al., 2015). In SorR, a Fe_2S_2 cluster is oxidised upon reaction with superoxide, leading to dimerisation and activation of this regulator (Kobayashi, Fujikawa, & Kozawa, 2014). Key components in the protection against oxidative stressors are LMW thiol compounds, familiarly termed redox buffers. The roles of these thiols are dual. Firstly, by chemical reaction with a stressing agent, this agent is eliminated at which the thiol is oxidised to its homodisulphide. After reduction catalysed by a (NADPH-dependent) disulphide reductase, the thiol is regenerated for a next reaction. Secondly, thiols react with cysteine sulphenic acids in oxidatively damaged proteins (Fig. 16B) resulting in the formation of cysteine heterodisulphides (protein-S-thiolation), but protecting these cysteines from further oxidation. Enzymatic reduction of the disulphide bond recovers the original cysteine(s) in the protein. Bacteria harbour a number of these thiols, the simplest one being cysteine itself (Fahey, 2013). However, the disadvantage of cysteine is that it undergoes a heavy metal-catalysed autooxidation at considerable rate and at which ROS are formed. Such autooxidation of the tripeptide glutathione (GSH, γ-L-glutamyl-L-cysteinyl-glycine) is two orders of magnitude slower and GSH is used by Gram-negative bacteria and in eukaryotes as a redox buffer. Firmicutes lack GSH, but have bacillithiol (BSH, L-cysteinyl-D-glucosamine-malate) instead, whereas *Actinobacteria* employ mycothiol (MSH, N-acetyl-L-cysteinyl-D-glucosamine-inositol). It cannot be ruled out completely that also coenzyme A serves this purpose to an extent. Thus, oxidative damage of proteins is observed as protein-S-thiolation by GSH, BSH, MSH or just cysteine, and this thiolation supports for the global mapping of oxidative stress. For this mapping proteomics comes into play.

Several methods have been developed to make a whole-cell analysis of proteins that underwent PTMs due to oxidative stress. These methods follow the 2DE and shotgun methodologies described earlier (see Sections 3.1–3.3) with specific modifications (reviewed in Černý et al., 2013; Leichert & Jakob, 2006; Lindahl, Mata-Cabana, & Kieselbach, 2011; Loi et al., 2015). Briefly, a protein extract of ROS-exposed cells is reduced by an appropriate reagent (f.i. ascorbate for nitrosation, arsenite for sulphenylation, tris(carboxyethyl)phosphine for general disulphides, or specific enzymes that reduce RS–SGH, RS–BSH RS–MSH disulphides). Cysteine moieties in proteins formed after reduction are fluorescently labelled and the enzyme mixture is separated by 2DE for further MS analyses (Leichert & Jakob, 2004). By using two different fluorescent dyes, DIGE immediately reveals the (quantitative) difference in protein patterns between controls and ROS-exposed cells. Alternatively, reduced extracts are labelled with thiol-reactive biotin and biotinylated proteins are visualised by antibiotin immunoblotting. Biotin labelling provides also the means for selective enrichment by streptavidin affinity chromatography. Proteolytic cleavage before or after enrichment enables bottom-up shotgun proteomics of the protein preparations; furthermore, there are also methods to differentially quantify PTMs (Lindemann & Leichert, 2012). These procedures allow the identification of modified proteins, but the information regarding the nature of protein-S-thiolation is obviously lost. Consequently, shotgun analyses have to be run in parallel on preparations that have not been biotinylated, hoping the find candidate proteins with molecular masses that account for a substation by a LMW thiol.

The above proteomics methods have been applied to uncover the global thiol-based redox regulation in a number of bacterial species (see f.i. Chi et al., 2011; Chi, Chen, et al., 2013; Chi, Roberts, et al., 2013; Chi et al., 2014; Hochgräfe et al., 2007; Kim, Kim, et al., 2013; Kim, Yu, et al., 2013; Lindemann et al., 2013; Pohl et al., 2011; Schroeter et al., 2011). Recently, Ansong et al. (2013) used a top-down approach to map a S-thiolation switch on whole proteins from *Salmonella typhimurium* in response to infection-like conditions. Protein preparations obtained from cells grown under different conditions were separated on a long (80 cm) nanocapillary column and an extended (250 min) gradient to prevent undersampling for subsequent orbitrap MS. By this method, 1665 S-thiolation sites on 563 unique proteins were detected. Intriguingly, the S-thiolation mode switched between S-cysteinylation and S-glutathionylation depending on the growth condition. Also respiratory enzymes were targeted by S-thiolation and these comprised the CydAB terminal oxidase (one site each on CydA and CydB), three sites on the DmsB subunit of dimethylsulphoxide oxidoreductase and no less the eight positions on the cyt b_{562} subunit of quinone:cytochrome *c* oxidoreductase (complex III). It may be worth mentioning that in *C. glutamicum* CydB expression stands under control of the OxyR system (Teramoto, Inui, & Yukawa, 2013). ATP synthase was particularly sensitive towards PTM resulting from oxidative stress, being modified at AtpA (6 sites), AtpC (4 sites), AtpD (7 sites), AtpG (3 sites), AtpF (1 site) and AtpH (1 site).

In mitochondria, 0.2–2% of all oxygen respired ends up in ROS as the result of the activities of complexes I–III. Therefore, it will be no surprise that many enzymes in this cell organelle are targeted by ROS and that elaborate regulatory systems are available to protect and repair proteins in response to oxidative stress (see for a review: Dröse, Brandt, & Wittig, 2014). However, proteomics revealed that protection could proceed in quite unexpected manners. Several of subunits of complex I are targets of oxidative thiol-modification with one being of special interest. In vitro and in vivo, vertebrate complex I occurs as two conformers, so-called active (A) and deactive (D) states (Brandt, 2011). The A-state is catalytically competent and the D-state is a dormant one, which is taken when the complex is devoid of its substrates (NADH, ubiquinone). The presence of these substrates induces the transition of the D- into the A-state: So, A/D transition is reversible (Babot, Birch, Labarbuta, & Galkin, 2014). This transition is also known for complex I of unicellular Eukarya, but this has not been established for prokaryotic NDHs thus far. Only in the D-state, a specific cysteine is accessible for artificial modification by N-ethylmaleimide or iodoacetamide, or physiologically by nitrosation by NO (Galkin et al., 2008). This particular cysteine (Cys-39) is localised in the ND3 subunit, close to the quinone binding site. If Cys-39 is S-nitrosated, complex I is kept in the D-state and is consequently unable of turnover, preventing the formation of excess ROS. By the removal of NO (or reduction of an S-glutathionylated form) catalytic activity is restored eventually (Chouchani et al., 2013; Prime et al., 2009); nevertheless, it is still elusive how this is done.

5. ASSEMBLAGES OF RESPIRATORY PROTEIN COMPLEXES

5.1 Supercomplexes

5.1.1 Mitochondrial Supercomplexes

The introduction of BNE into the research on respiratory systems by Schägger and colleagues (Schägger, 2002; Schägger et al., 1988; Schägger & Pfeiffer, 2000; Schägger & von Jagow, 1991) changed our thinking on the way these systems are organised. Until these publications, the accepted idea was that respiratory protein complexes moved laterally and independently from each other along the membranes (liquid-state model),

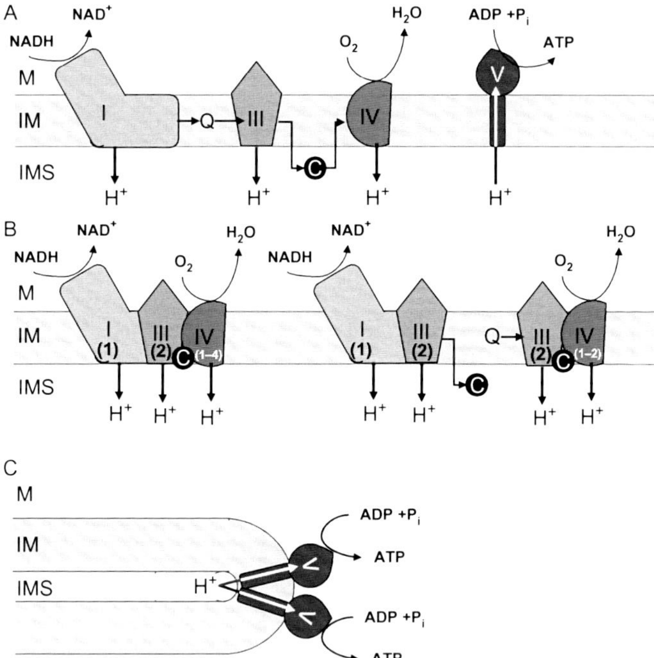

Fig. 17 Schematic representation of the organisation of mitochondrial OXPHOS complexes. The figures schematically show complexes I, III–V. Complex II (succinate dehydrogenase) that does not form part of supercomplexes (SCs) is omitted for better clarity. (A) The 'liquid-state model' in which the different OXPHOS complexes act separately. Electron transfer between NADH-oxidising complexes I and III is mediated by ubiquinone (Q), whereas cytochrome *c* (C) transfers electrons from complex III to oxygen-reducing complex IV. (B) SCs observed in mitochondria from different eukaryotic species. The figure displays from *left* to *right* the SC of complexes I + III + IV catalysing NADH oxidation with oxygen reduction (respirasome), the SC of complexes I + III coupling NADH oxidation with cytochrome *c* reduction and the SC of complexes III + IV, which mediates ubiquinol oxidation and oxygen reduction. The number of individual complexes found in preparations of mitochondria from eukaryotic representatives is shown in *parentheses* below the roman numberings. (C) ATP synthase dimers residing on top of the cristae that are found in long strings perpendicular the plane of the figure. The direction of proton (H^+) movements is indicated by *arrows* without any specification of proton–electron stoichiometry ($H^+/2e^-$). Abbreviations: *M*, matrix; *IM*, inner membrane; *IMS*, intermembrane space.

a view that is still common in many textbooks (Fig. 17A). However, BNE analyses revealed that mitochondrial OXPHOS complexes may aggregate as large functional 'supercomplexes' (SCs), as already had been suggested earlier (solid-state model) (Schägger, 2002). Initially, the SC concept was met with scepticism. It was argued that aggregation was the result of experimental artefacts that arise from membrane resolution and subsequent BNE of respiratory complexes. This objection is valid and should always be kept in mind when performing such experiments. Nevertheless, original results could be repeated by many labs and by application of BNE and CNE as outlined in Section 3.2.2, and SCs have since been detected in mitochondria of animals, plants (Bultema, Braun, Boekema, & Kouril, 2009; Eubel, Heinemeyer, Sunderhaus, & Braun, 2004; Klodmann et al., 2011; Krause et al., 2004) and yeasts (Helbig et al., 2009; Lemaire & Dujardin, 2008; Matus-Ortega et al., 2015; Nübel et al., 2009), as well as in respiratory systems of prokaryotes (see Section 5.1.2). Furthermore, SC formation is also observed for photosystems (Cartron et al., 2014; Chang et al., 2015; Watanabe, Kubota, Wada, Narikawa, & Ikeuchi, 2011; Watanabe et al., 2014). Importantly, ideas regarding their organisation based on proteomics could be confirmed independently by high-resolution electron microscopic studies (Chaban, Boekema, & Dudkina, 2014; Dudkina, Kouril, Peters, Braun, & Boekema, 2010; Schäfer, Dencher, Vonck, & Parcej, 2007; Vonck & Schäfer, 2009). By now, there is little doubt that SCs play somehow a critical role in respiratory processes (Acin-Perez & Enriquez, 2014; Shoubridge, 2012; Vartak, Porras, & Bai, 2013; Wittig, Carrozzo, Santorelli, & Schägger, 2006).

In mitochondria, complexes I–IV form SCs in different combinations (Fig. 17B). Complexs II (SDH) and V (ATP synthase) do not take part these SCs. In fact, ATP synthase molecules assemble as SCs by themselves (Chaban et al., 2014; Dudkina et al., 2010; Schäfer et al., 2007; Vonck & Schäfer, 2009; Wittig & Schägger, 2009b). The basic unit is the ATP synthase dimer (Wittig & Schägger, 2009b; Wittig, Velours, Stuart, & Schägger, 2008) at which the monomeric forms are arranged such that they make angles between 30 and 145 degrees with respect to each other (Chaban et al., 2014) (Fig. 17C). Angles vary between different organisms and they may change when the protein is resolved from the mitochondrial membrane. In yeast, dimerisation is controlled by the phosphorylation state of the ATP synthase (Bornhövd, Vogel, Neupert, & Reichert, 2006; Kane et al., 2010; Reinders et al., 2007). In turn, ATP synthase dimers are arranged in parallel to make long strings that sit on top of the cristae and bend the membrane, thus shaping this inner membrane system (Davies, Anselmi, Wittig, Faraldo-Gómez, & Kühlbrandt, 2012; Dudkina, Heinemeyer, Keegstra, Boekema, & Braun, 2005). This parallel arrangement is facilitated by dimer-specific, noncatalytic subunits e and g (Bornhövd et al., 2006; Saddar, Dienhart, & Stuart, 2008; Wittig & Schägger, 2009b). The planar parts of the cristae thus provide space for the respiratory complexes (Cogliati et al., 2013). BNE and CNE analyses of mitochondria of representatives of all Eukarya conclusively showed that these complexes formed SCs composed of individual complexes in more or less fixed ratios, viz., $I + III_2$, $III_2 + IV_{1-2}$ and $I + III_2 + IV_{1-4}$ (Fig. 17B) (Chaban et al., 2014). The different SCs have been purified in their native states and they all showed the expected activities (Acín-Pérez, Fernández-Silva, Peleato, Pérez-Martos, & Enriquez, 2008). SC $I + III_2$, for instance, is capable of cytochrome *c* reduction by NADH, and $I + III_2 + IV_{1-4}$ (respirasome) catalyses the NADH-dependent reduction of oxygen. The SC $I + III_2 + IV_{1-4}$ seems to be the building unit of long strings or patched strings that stretch and spread along the cristae (Dudkina et al., 2010; Muster et al., 2010; Wittig & Schägger, 2009b). The resulting straightforward questions are: (1) what is the function of the SCs and (2) how are these SCs assembled, assuming that assembly is a directed process.

Based on ample experimental evidence, a number of functions have been proposed for SCs, but none of them provide a definite answer as yet. By the tight control of electron carrier interchange (viz. ubiquinone and cyt *c*) in between the individual complexes, formation of ROS would be minimised (Gómez, Monette, Chavez, Maier, & Hagen, 2009; Maranzana, Barbero, Falasca, Lenaz, & Genova, 2013; Rosca et al., 2008). The close association between the individual complexes would decrease the distances that electron carriers have to travel, thereby increasing the electron flux (substrate tunnelling) (Acin-Perez & Enriquez, 2014; Chaban et al., 2014; Vartak et al., 2013). Indeed, this electron flux is governed by SC assembly (Lapuente-Brun et al., 2013). Moreover, structural models made on basis of electron microscopy and known crystal structures indicate a short distance between the cytochrome *c* binding sites of complexes III and IV (Schäfer et al., 2007), and an SC composed of the latter two was shown to contain a stoichiometric amount of two cytochrome *c* molecules (Moreno-Beltrán et al., 2015). However, such a short distance is less apparent for the quinone binding sites of complexes I and III (Dudkina, Eubel, Keegstra, Boekema, & Braun, 2005), and quinone was shown to move freely through the mitochondrial inner membrane during respiration (Blaza, Serreli, Jones, Mohammed, & Hirst, 2014; Genova & Lenaz, 2013, 2014; Lenaz & Genova, 2009; Trouillard, Meunier, & Rappaport, 2011). Another possibility is that individual complexes cooperatively assist and stabilise each other's assemblies during SC formation and maintenance (Calvaruso et al., 2012; Chojnacka, Gornicka, Oeljeklaus, Warscheid, & Chacinska, 2015; Cui,

Conte, Fox, Zara, & Winge, 2014; Duarte & Videira, 2009; Habersetzer et al., 2013; Marques, Dencher, Videira, & Krause, 2007; Saddar et al., 2008; Stroh et al., 2004; see however: Maas, Krause, Dencher, & Sainsard-Chanet, 2009). Indeed, deletion or mutation of factors known to be involved in the assembly of one complex affected assembly of other complexes as well and of SCs as a whole. Presently, three factors are known to assist in SC formation as such, stomatin-like protein 2 (Mitsopoulos et al., 2015), metalloprotease OPA1 (Bohovych et al., 2015) and dynamin-related protein OMA1 (Cogliati et al., 2013) at which genetic deletion of the latter results in the complete disorganisation of the cristae (Frezza et al., 2006). Besides these proteinaceous factors, one more component is essentially for at least the structural integrity of mitochondrial SCs, cardiolipin(Bazán et al., 2013; Desmurs et al., 2015; Mileykovskaya & Dowhan, 2014; Wenz et al., 2009; Zhang, Mileykovskaya, & Dowhan, 2002, 2005).

Thus, it is obvious that the function of mitochondrial SCs is not fully comprehended and that assemblage is a highly complicated and even so poorly understood interplay of a multitude of assembly factors. In addition, SCs are found as different aggregates (Fig. 17B) and their composition may change depending on the growth conditions applied (Ramírez-Aguilar et al., 2011; Schägger & Pfeiffer, 2000). Perhaps, transitions in between the various SCs are dynamic, as proposed in the plasticity model by Acin-Perez and Enriquez (2014), providing an additional level of control of respiratory processes and at which control could even place at a very local scale.

5.1.2 Bacterial Supercomplexes

BNE/CNE and other biochemical studies established that SC formation is also common to prokaryotic ETCs (Magalon, Arias-Cartin, & Walburger, 2012; Melo & Teixeira, 2016), where the composition of the SCs reflects the respiratory status. In *P. denitrificans*, the presumed progenitor of mitochondria, the main SC is comprised of NDH (complex I), the bc_1 complex (complex III) and the cytochrome aa_3 oxidase (complex IV), just like the respirasomes in mitochondria (Stroh et al., 2004). The thermophilic bacterium *Aquifex aeolicus*, which is able of aerobic growth with H_2S as the electron donor, harbours an SC composed of one or two sulphide:quinone oxidoreductases, one dimeric bc_1 complex and one ba_3-type terminal oxidase, next to traces of quinones and monohaem cytochrome c_{555} (Prunetti et al., 2010). As-isolated this SC is able indeed to couple H_2S oxidation with oxygen reduction. Similarly, a supermolecular structure isolated from the acidiphilic bacterium *Acidithiobacillus ferroxidans*, which spans two membranes and is constituted of iron-oxidising proteins and cytochrome *c* oxidase aa_3, catalyses Fe^{2+} oxidation with oxygen reduction when supplemented with the physiological electron carrier cytochrome c_2 (Castelle et al., 2008). By strep-tag-affinity chromatography an SC has been purified from aerobic Gram-positive *C. glutamicum*, which is assembled from a bc_1 complex, a dihaem cytochrome c_1 electron carrier and cytochrome $c\ aa_3$ oxidase, containing a hitherto elusive fourth subunit (Niebisch & Bott, 2003). The complex had quinol oxidase activity as expected. A similar SC has been obtained from *B. subtilis* (Sousa, Videira, Santos, et al., 2013). Here, the bc: aa_3 complex was observed upon BNE in different ratios, $(bc)_4{:}(caa_3)_2$, $(bc)_2{:}(caa_3)_4$ and $2[(bc)_4{:}(caa_3)_2]$, indicative of string formation. BNE allowed the detection of one more interesting SC, a succinate:quinone oxidoreductase: nitrate reductase. One may note that SDH (complex II) never formed part of mitochondrial SCs. The presence of the succinate:nitrate oxidoreductase and $bc{:}aa_3$ SCs in *B. subtilis* was substantiated in another study (García Montes de Oca et al., 2012), where it was shown that also membrane-bound cytochrome c_{550} was a component and the electron carrier of the $bc{:}aa_3$ SC. Intriguingly, this SC was associated with ATP synthase monomers. Next, a stable hybrid $bc{:}aa_3$-type SC has been artificially assembled from two different *Mycobacterium* species (Kim, Jang, et al., 2015). The presence of $bc{:}aa_3$ SCs in Gram-positive bacteria is highly reminiscent to the complex III:complex IV SCs found in mitochondria (Fig. 17B). Some Gram-negative bacteria, like *E. coli*, lack complex III, but these organisms have their share in SCs as well, supporting the various respiratory pathways that enable these organisms to grow under varied environmental conditions (Sousa et al., 2011, 2012; Sousa, Videira, & Melo, 2013; Sousa, Videira, Santos, et al., 2013). Using BNE and CNE combined with kinetic data of wild-type and mutant strains and transcription analyses, these authors identified a number of SCs, including a trimer of SDH, a direct association between the two types of NDH found in *E. coli*, viz. the canonical multisubunit NDH-1 and more simple flavin NDH-2, and SCs composed of the aerobically stable formate dehydrogenase (fdo) together with two oxidases that are found in *E. coli*, cytochrome bo_3 quinol:oxygen reductase and cytochrome *bd* quinol:oxygen reductase (CydAB).

Like in mitochondrial research, BNE and CNE supported by MS provide a strong method to analyse prokaryotic SCs. Accessory subunits and LMW electron carriers are preserved in RMPs by mild detergent treatments of the membrane. They had gone unnoticed in classic biochemical purifications, as exemplified earlier. The oxygen-stable, membrane-bound [NiFe]-hydrogenase (MBH) from the 'Knallgas' bacterium *Ralstonia eutropha* exemplifies such loss very well (Frielingsdorf, Schubert, Pohlmann, Lenz, & Friedrich, 2011). The enzyme has been well characterised, but the as-isolated protein always lacked one candidate subunit found in the genome, which would bind MBH to the membrane and shuttle electrons into the ETS, viz., a haem *b* subunit. Preparations only contained this haem *b* subunit after mild extraction with digitonin and, unexpectedly, MBH appeared to be a homotrimeric SC. The lipids phosphatidylethanolamine and phosphatidyl glycerol were indispensable to keep interaction between the soluble hydrogenase module and the membrane-bound haem *b* subunit intact. Hitherto, one mitochondrial SC has evaded detection in prokaryotes: dimeric ATP synthase and its string-like multimeric forms. This should not be a surprise, considering that bacteria do not have cristae. However, we did find dimeric and multimeric ATP synthase in the Archaeon *M. thermautotrophicus* under certain growth conditions (Fig. 11B and C) (Farhoud et al., 2005), and it cannot be ruled out a priori that such dimers and multimers could be found in other prokaryotes, especially when equipped with convolute laminar, tubular or highly curved membrane systems. ATP synthase formed part of the $bc{:}aa_3$ SC in *B. subtilis* (García Montes de Oca et al., 2012). For *Bacillus pseudofirmus* OF4, an SC has been described consisting of cytochrome aa_3 and ATP synthase (Liu et al., 2007). Such an assembly could be physiologically essential. *B. pseudofirmus* is an alkaliphilic bacterium and an alkaline environment opposes the building up of a *pmf*. The association between cytochrome aa_3 that pumps protons out the cell and ATP synthase that is driven by the import of protons brings proton exchange in immediate proximity.

The above findings present firm indications that SC formation is relevant in microbial respiration, although their function is not completely clear, neither for bacterial nor mitochondrial SCs. However, the eventual picture may be more diffuse and complex. By labelling respiratory RMPs with green fluorescent protein while maintaining activity after expression, confocal laser microscopy applied to living cells of *B. subtilis* and *E. coli* showed that RMPs are not evenly distributed across the cell membrane (Erhardt et al., 2014; Johnson, van Horck, & Lewis, 2004; Lenn, Leake, & Mullineaux, 2008). RMPs were accumulated in slowly moving small patches sizing roughly 300 nm in diameter. Single-molecule studies with *E. coli* on NDH-1, SDH, cytochrome *bd* oxidase and cytochrome *c* aa_3 oxidase labelled with different fluorophores confirmed that all these RMPs were consistently confined to the patches, albeit in various amounts (NDH-1, 10–20 molecules; SDH, 20–40; CydAB, 70–180; cyt *caa*$_3$, 24–45; ATP synthase, 40–60) (Lenn & Leake, 2016; Llorente-Garcia et al., 2014). In contrast, a fluorescent ubiquinone derivative was evenly distributed across the whole membrane. Within these patches, different RMPs moved at different rates, but they did not comigrate, as far as limits in resolution allow such conclusion. Obviously, lack of comigration strongly argues against SC formation. However, it should be noted that the fdo:cydAB and fdo: cyt *caa*$_3$ SCs observed by BNE (see earlier) were not considered specifically. It also cannot be excluded that fluorescent labelling prevented aggregation of RMPs. Yet, taking SCs for granted, regulation of respiratory activities may include one more level, membrane transertion/segregation resulting in the local and polar distribution of respiratory complexes (Fishov and Norris, 2012; Lenn et al., 2008; Magalon & Alberge, 2015; Matsumoto, Hara, Fishov, Mileykovskaya, & Norris, 2015).

5.2 The Interactome

In order to become active, respiratory RMPs have to be assembled (Luirink, Yu, Wagner, & de Gier, 2012; Price & Driessen, 2010; Sargent, 2007), which requires the precisely timed action of generic and enzyme-specific machineries that insert catalytic and electron transfer cofactors into the subunits or add, for structural purposes, other molecules to specific amino acids. Such an assembly includes the directed placement of these subunits inside or outside the membrane, N- and/or C-terminal cleavage events related with peptide translocation across the membrane, proper folding, subunit interactions or further maturing of enzyme complexes. Next, catalytic activity may be controlled by PTM, which requires an association, no matter how brief, with enzymes that add or remove these molecules. Also other associations with LMW compounds or proteins may affect structures of RMPs, resulting in the inhibition or stimulation of their activities. The sum total of all permanent or temporary interactions of an enzyme with other proteinaceous or nonpeptide partners is called its interactome.

Hereafter we will give, on the basis of recent reviews, a bird's eye view of what is known about the assembly of bacterial RMPs, and how proteomics has contributed to this knowledge or may contribute to a better understanding in the near future. In addition, we will briefly deal with the toolboxes that proteomics provides to pinpoint the interactome of an RMP and to unravel the dynamics of such interaction networks.

5.2.1 Assembly of Bacterial Respiratory Complexes

As alluded above, the assembly of RMPs involves the action of generic machineries that translocate polypeptides through the membrane. The common systems in bacteria are the Sec, YidC and TAT translocons. The Sec (secretory) system threads translated polypeptide chains in an unfolded form to the periplasm, which is associated with the cleavage of an N-terminal leader sequence by a specific peptidase (du Plessis, Nouwen, & Driessen, 2011; Paetzel, 2014; Sala, Bordes, & Genevaux, 2014). The Sec system is assisted by the YidC protein, which correctly inserts TMHs (Dalbey, Kuhn, Zhu, & Kiefer, 2014). In contrast, in the 'twin-arginine translocation' (TAT) pathway proteins are taken across the lipid bilayer in a preassembled and prefolded manner (Cline, 2015; Fröbel, Rose, & Müller, 2012; Goosens, Monteferrante, & van Dijl, 2014a, 2014b; Goosens et al., 2013; Hopkins, Buchanan, & Palmer, 2014; Robinson et al., 2011). The pathway is named after two adjacent arginines that typically are located about 20 positions after the translation start. Preassembly within the cytoplasm may include the insertion of iron–sulphur clusters synthesised by the Suf and ISC systems (Blanc, Gerez, & Ollagnier de Choudens, 2015; Kim, Bothe, Alderson, & Markley, 2015; Kim, Jang, et al., 2015; Wayne Outten, 2015; Yokoyama & Leimkühler, 2015) or of the molybdopterin cofactor, which is the catalytic heart of a wide variety of respiratory enzymes, including formate dehydrogenase and nitrate reductase (Grimaldi, Schoepp-Cothenet, Ceccaldi, Guigliarelli, & Magalon, 2013). Translation at the ribosomes and transport through the membrane can be intimately coupled by action of the signal recognition particle, trigger factor and of the DnaK and GroeEL protein-folding apparatus (Castanié-Cornet, Bruel, & Genevaux, 2014; Saraogi & Shan, 2014). In case of molybdopterin proteins, members of the NarJ family perform a proofreading to check for correct folding before the peptide is directed to the TAT system (Bay, Chan, & Turner, 2015; Chan et al., 2014; Lanciano, Vergnes, Grimaldi, Guigliarelli, & Magalon, 2007). Haems that act in electron transfer, such as *c*- and *b*-type haems, or in catalysis by terminal oxidases and nitric oxide reductases (*b*-, *a*- and *o*-type haems) are localised at the outer aspect of the membranes. So, before incorporation into a subunit(s), these molecules have to be translocated as well, which is realised by the cytochrome *c* maturation machinery (Kranz, Richard-Fogal, Taylor, & Frawley, 2009; Verissimo & Daldal, 2014). Periplasmic subunits can be attached to the membrane by covalently bound lipid molecules. This covalent binding is realised in the lipoprotein modification pathway (Buddelmeijer, 2015; Hutchings, Palmer, Harrington, & Sutcliffe, 2009). Besides the prosthetic groups mentioned, RMPs depend on specific catalytic metals (Cu, Fe). The acquirement of these metals, usually at the periplasm, and insertion into the subunits is mediated by enzyme-specific chaperones (Abicht et al., 2014; Dash, Alles, Bundschuh, Richter, & Ludwig, 2015; Hannappel, Bundschuh, & Ludwig, 2012; Robinson & Winge, 2010). All in all, enzyme maturation is highly complex, wherein each RMP depends on its dedicated set of assembly factors acting in concert. Assembly processes of the major RMPs, including quinol: cytochrome *c* oxidoreductases (Hasan, Proctor, Yamashita, Dokholyan, & Cramer, 2014; Hasan, Yamashita, & Cramer, 2013; Smith, Fox, & Winge, 2012), terminal oxidases (Bühler et al., 2010; Ekici, Pawlik, Lohmeyer, Koch, & Daldal, 2012; Gurumoorthy & Ludwig, 2015), nitrite-/nitric oxide-/nitrous oxide reductases (Adamczack et al., 2014; Barth, Isabella, & Clark, 2009; Hatzixanthis, Richardson, & Sargent, 2005; Nicke et al., 2013; Spiro, 2012), hydrogenases (Peters et al., 2015), formate dehydrogenases (Hartmann, Schwanhold, & Leimkühler, 2015) and other molybdopterin-containing RMPs (Arnoux et al., 2015; Magalon & Mendel, 2015), have been studied in more or less detail. Notoriously, the assembly of complex I with its about 45 subunits in unicellular and higher eukaryotes is puzzling (Mimaki et al., 2012; Vartak et al., 2014; Vogel, Smeitink, & Nijtmans, 2007). Assembly should be much simpler for prokaryotic NDHs that harbour a mere 13 subunits, but virtually nothing is known about it. Most of our knowledge on RMP assemblages has been gained from tedious genetic and biochemical studies, but this knowledge is often still fragmentary and far from complete.

Hitherto, relatively few studies aimed at unravelling assembly pathways of RMPs have taken full advantage of the potentials offered by proteomics. Most of these studies applied proteomics methods to detect the effect of gene deletion, thereby making differential expression analyses by 2DE or shotgun methods outlined in Section 3.3. Moreover, the majority of the latter studies were directed at membrane transport processes of polypeptides, more specifically at the role in these processes of Sec components (Baars et al., 2006, 2008), YidC protein (Wickström, Wagner, Simonsson, et al., 2011), trigger factor (Wickström, Wagner, Baars, et al., 2011), TAT proteins (Hitchcock et al., 2010) and of periplasmic chaperones (Götzke et al., 2015). In a broader context, established and new proteomics methods were used to make inventories of secreted proteins (secretome) by mapping polypeptides that were transported across the cytoplasmic membrane or beyond (see f.i. Johnson, Sikora, Zielke, & Sandkvist, 2013; Renier et al., 2015; Zijnge, Kieselbach, & Oscarsson, 2012). These maps supported the evaluation of signal peptide algorithms (Leversen et al., 2009) and even suggested the presence of yet-unknown peptide transport systems (Wilson, Anderson, & Bernstein, 2015).

To identify binding partners of chaperones that assist in the insertion of copper into the catalytic subunits of cytochrome aa_3 oxidase from *P. denitrificans*, Gurumoorthy and Ludwig (2015) used these chaperones as baits in pull-down and cross-link experiments. Wickström, Wagner, Simonsson, et al. (2011) employed BNE (Section 3.2.2) to assess folding intermediates and partially assembled RMPs that accumulated in the cytoplasm as the result of YidC depletion. In order to start understanding how NDH (NuoA-N) is assembled in prokaryotes, Erhardt et al. (2012) also applied BNE. The authors explored the effect of the disruption of various *nuo* genes by analysing partially assembled NDH complexes.

Indeed, BNE may turn out to become a powerful instrument to elucidate the assembly of RMPs (see for OXPHOS complexes: Calvaruso, Smeitink, & Nijtmans, 2008; Gil Borlado et al., 2010; McKenzie, Lazarou, Thorburn, & Ryan, 2007; Leary, 2012), especially by complexome profiling (Section 3.2.2; Fig. 12). Here, RMPs from cells grown under different conditions and/or isolated from the membranes by different detergents are separated by BNE and comigrating polypeptides are profiled using recently developed software (Giese et al., 2015). This analysis may establish already known, fully or incompletely assembled RMPs, but also may reveal new RMPs. Next, known or yet unknown comigrating partners, ie, other than subunits, may be found, opening a way for further exploration of their function. The potential of this BNE profiling method is exemplified by several studies on mitochondrial OXPHOS complexes, which enabled the identification of new assembly factors (Heide et al., 2012; Wessels et al., 2013, 2009).

5.2.2 The Far from Complete Picture of Interaction Partners and Interaction Dynamics

Proteomics offers the means to investigate the interaction of proteins on all organisational levels and on different timescales, ranging between stable subunit associations, temporary assembly intermediates and contacts with assembly factors, enzymes involved in PTMs or other high- or low-molecular-weight compounds that affect activity. In fact, all available methods have been described in previous sections. Here we will only point out from the global and enzyme-directed perspectives, how these methods can be employed to unravel interactome dynamics.

The method of choice for a global analysis of protein interactions is BNE (see Sections 3.2.2 and 5.2.1) (Heide & Wittig, 2013; Krause, 2006). Even though such global analysis provides a wealth of data, these data may lack detail, but BNE is applicable in the study of individual complexes as well. Actually, most MS-based approaches take isolated enzymes as the starting point to explore their interactomes (Dengjel, Kratchmarova, & Blagoev, 2010; Gingras et al., 2007; Ramisetty & Washburn, 2011; Wepf, Glatter, Schmidt, Aebersold, & Gstaiger, 2009). Proteins are generally purified in their tagged forms and partners that coelute from affinity columns are rigorously screened for physiologically relevance (see Section 4.1.1) (Formosa et al., 2015; Nouws et al., 2010; Saada et al., 2009; Tucker et al., 2013; van den Ecker et al., 2010; Vogel, Dieteren, et al., 2007; Vogel, Janssen, et al., 2007). In principle, this screening can even be done by high-throughput robotics (Hakhverdyan et al., 2015). Time series at which proteins become labelled in vivo (SILAC) or at which isolated protein samples are labelled by in vitro methods (see Section 2.3) may detect temporary interactions in multiplexed tandem MS analyses (Ramisetty & Washburn, 2011). These interactions can be 'frozen' by in vivo or in vitro cross-linking (Tran et al., 2015). The detailed inspection of cross-linked peptides permits the pinpointing of contact sides between the different proteins. In H/D exchange experiments or by probing proteins by hydroxyl radicals or other reagents (see Section 4.1.3), conformational dynamics due to protein–protein interactions or induced by PTMs, ligand or metal ion bindings may be unravelled (Ramisetty & Washburn, 2011; Venable et al., 2015). Whereas the above approaches are shotgun MS oriented, novel MS methods allow interaction studies to be made on whole protein complexes and other associates (see Section 4.1.5). Like in many other occasions, only little use has been made of MS-based methods in research on RMP interactomes. Experimental challenges posed by membrane-bound proteins and inaccessibility to the constantly developing field of proteomics may be reasons for this.

6. PERSPECTIVES

Within the wide field of proteomics, an area of methods and techniques has been established to study protein chemistry in unprecedented detail and accuracy. Proteomics offers the means to make an inventory of all expressed proteins in all their proteoforms in an organism and to quantitatively map changes in this inventory that are made by the organism in response to changes in its environment. From the perspective of the isolated enzyme (system), proteomics enables the collection of information on protein structures and dynamics, lending the field a central place in structural biology.

By now, over 80% of all proteins that are coded by a bacterial genome can be detected on a high-throughput basis, if whole-cell preparations are analysed by appropriate and complementary prefractionation methods (Section 3.1) (Catrein & Herrmann, 2011; Soares et al., 2013; Soufi et al., 2015). This percentage may even be an underestimation, since not all proteins are expressed under the investigated growth conditions. Also membrane-bound proteins that had been a real nuisance for a long time can be identified with a high success rate (Section 3.2.1). Routine methods are available to identify and quantify PTMs (phosphorylation, acetylation, methylation, oxidative stress-induced structural changes) in protein sequences at large scale (Section 4.2). By label-free or in vivo or in vitro labelling techniques (Section 2.3), differential expression can be profiled quantitatively in multiplexed experiments (Section 3.3), providing invaluable data for systems-biology endeavours (see f.i. Hui et al., 2015; Kohlmann et al., 2011, 2014; Soufi et al., 2015; Wegener et al., 2010).

The earlier outcome primarily relies on shotgun proteomics. Due to its success, it seems that classical gel-based proteomics to have taken a place on the back seat. Still, this second-rank position may be unjustified (Oliveira, Coorssen, & Martins-de-Souza, 2014; Oliveira & Picotti, 2014). As pointed out, the strengths of 2DE are the visibility of proteins (Figs 9, 11 and 13) and its relatively easy access by nonexperts (Sections 3.1.1, 3.2.2 and 3.3). On the other hand, the disadvantages of 2DE are its insensitivity and, if not robotised, the workload that is associated with the identification of gel spots: each protein spot has to be extracted from the gel and analysed by MS individually. Despite these obvious shortcomings, gel-based methods still have their undeniable advantages as exemplified by BNE and CNE (Section 3.2.2). By their assistance in the elucidation of respiratory SCs, BNE and CNE even caused a paradigm shift in our understanding of ETSs (Section 5.1). The stronghold of these methods is that enzyme systems, also in very complex protein mixtures, are investigated as their native complexes, even associated with other interaction partners like assembly factors (Section 5.2). This information is lost in classical shotgun approaches. BNE and CNE combined with complexome profiling allow the detection of completely new RMPs and their subunit composition (Section 3.2.2; Fig. 12). In this respect, BNE and CNE may become powerful tools in eliciting new respiratory systems in poorly studied microorganisms that come to us at an incredible pace. Here, gel-based and shotgun proteomics may go hand in hand by clever combinations of top-down and bottom-up approaches to get the maximum structural information out of minimal amounts of proteins (Millea, Krull, Cohen, Gebler, & Berger, 2006; Strader et al., 2004; Wu et al., 2009).

Indeed, a definite advantage of proteomics is that only minimal amounts of proteins are required for structural studies. Miniaturised liquid column chromatography, affinity binding and other methods are available or are being developed for the targeted purification of enzymes in their native form (Section 4.1.1). Purified preparations can be analysed online or offline by MS. In Sections 4.1.3 (Fig. 15) and 4.1.5, we presented an overview of shotgun-based techniques that inform us on secondary, tertiary and quaternary aspects of protein complexes, and of the progress that has been made in studying whole complexes and their dynamics from the top-down perspective. It was stated (Section 4.1.6) that the combined information obtained from proteomics experiments would allow the proposal of 3D structures, although they currently can achieve only a modest resolution (~5 Å). This could be highly relevant in the case of RMPs that are notoriously difficult to investigate by X-ray crystallography and NMR spectroscopy.

Finishing this review, the reader may have wondered that its topic, bacterial ETSs, may have been underexposed. This is definitely true, but it reflects the present status of the application of proteomics to understand bacterial ETS. Nevertheless, we hope that we may have given an impression of how proteomics may contribute to come to a full understanding of these systems in all their variations that emerged during evolution.

ACKNOWLEDGEMENTS

The authors would like to thank Dr Alain van Gool, Dr Dirk Lefeber, Dr Monique van Scherpenzeel and Ms Esther Willems for their contribution in generating the glycopeptide tandem mass spectra shown in Fig. 4. This work was supported by the Netherlands Organization for Scientific Research (NWO) by ALW Grant ALW2PJ/08021 to N.M.A. and VENI Grant 863.11.003 to B.K.

REFERENCES

Abicht, H. K., Schärer, M. A., Quade, N., Ledermann, R., Mohorko, E., Capitani, G., et al. (2014). How periplasmic thioredoxin TlpA reduces bacterial copper chaperone ScoI and cytochrome oxidase subunit II (CoxB) prior to metallation. *The Journal of Biological Chemistry*, *289*, 32431–33244.

Acin-Perez, R., & Enriquez, J. A. (2014). The function of the respiratory supercomplexes: The plasticity model. *Biochimica et Biophysica Acta*, *1837*, 444–450.

Acín-Pérez, R., Fernández-Silva, P., Peleato, M. L., Pérez-Martos, A., & Enriquez, J. A. (2008). Respiratory active mitochondrial supercomplexes. *Molecular Cell, 21*, 529–539.

Adamczack, J., Hoffmann, M., Papke, U., Haufschildt, K., Nicke, T., Bröring, M., et al. (2014). NirN protein from *Pseudomonas aeruginosa* is a novel electron-bifurcating dehydrogenase catalyzing the last step of heme d1 biosynthesis. *The Journal of Biological Chemistry, 289*, 30753–30762.

Aebersold, R., & Mann, M. (2003). Mass spectrometry-based proteomics. *Nature, 422*, 198–207.

Aggarwal, S., & Yadav, A. K. (2016). False discovery rate estimation in proteomics. *Methods in Molecular Biology, 1362*, 119–128.

Agrawal, G. K., & Thelen, J. J. (2005). Development of a simplified, economical polyacrylamide gel staining protocol for phosphoproteins. *Proteomics, 5*, 4684–4688.

Aivaliotis, M., Macek, B., Gnad, F., Reichelt, P., Mann, M., & Oesterhelt, D. (2009). Ser/Thr/Tyr protein phosphorylation in the archaeon *Halobacterium salinarum*—A representative of the third domain of life. *PLoS One, 4*, e4777.

Al-Attar, S., & de Vries, S. (2013). Energy transduction by respiratory metallo-enzymes: From molecular mechanism to cell physiology. *Coordination Chemistry Reviews, 257*, 64–80.

Alber, T. (2009). Signaling mechanisms of the *Mycobacterium tuberculosis* receptor Ser/Thr protein kinases. *Current Opinion in Structural Biology, 19*, 650–657.

Alberge, F., Espinosa, L., Seduk, F., Sylvi, L., Toci, R., Walburger, A., et al. (2015). Dynamic subcellular localization of a respiratory complex controls bacterial respiration. *Elife, 16*(4), e05357.

Allison, T. M., Reading, E., Liko, I., Baldwin, A. J., Laganowsky, A., & Robinson, C. V. (2015). Quantifying the stabilizing effects of protein-ligand interactions in the gas phase. *Nature Communications, 6*, 8551.

Allmer, J. (2011). Algorithms for the *de novo* sequencing of peptides from tandem mass spectra. *Expert Review of Proteomics, 8*, 645–657.

Andaya, A., Villa, N., Jia, W., Fraser, C. S., & Leary, J. A. (2014). Phosphorylation stoichiometries of human eukaryotic initiation factors. *International Journal of Molecular Sciences, 15*, 11523–11538.

Anderson, K. A., & Hirschey, M. D. (2012). Mitochondrial protein acetylation regulates metabolism. *Essays in Biochemistry, 52*, 23–35.

Ansong, C., Wu, S., Meng, D., Liu, X., Brewer, H. M., Deatherage Kaiser, B. L., et al. (2013). Top-down proteomics reveals a unique protein S-thiolation switch in *Salmonella typhimurium* in response to infection-like conditions. *Proceedings of the National Academy of Sciences of the United States of America, 110*, 10153–10158.

Antelmann, H., & Helmann, J. D. (2011). Thiol-based redox switches and gene regulation. *Antioxidants & Redox Signaling, 14*, 1049–1063.

Arias-Cartin, R., Grimaldi, S., Arnoux, P., Guigliarelli, B., & Magalon, A. (2012). Cardiolipin binding in bacterial respiratory complexes: Structural and functional implications. *Biochimica et Biophysica Acta, 1817*, 1937–1949.

Arias-Cartin, R., Grimaldi, S., Pommier, J., Lanciano, P., Schaefer, C., Arnoux, P., et al. (2011). Cardiolipin-based respiratory complex activation in bacteria. *Proceedings of the National Academy of Sciences of the United States of America, 108*, 7781–7786.

Arnoux, P., Ruppelt, C., Oudouhou, F., Lavergne, J., Siponen, M. I., Toci, R., et al. (2015). Sulphur shuttling across a chaperone during molybdenum cofactor maturation. *Nature Communications, 6*, 6148.

Baars, L., Wagner, S., Wickström, D., Klepsch, M., Ytterberg, A. J., van Wijk, K. J., et al. (2008). Effects of SecE depletion on the inner and outer membrane proteomes of *Escherichia coli*. *Journal of Bacteriology, 190*, 3505–3525.

Baars, L., Ytterberg, A. J., Drew, D., Wagner, S., Thilo, C., van Wijk, K. J., et al. (2006). Defining the role of the *Escherichia coli* chaperone SecB using comparative proteomics. *The Journal of Biological Chemistry, 281*, 10024–10034.

Babot, M., Birch, A., Labarbuta, P., & Galkin, A. (2014). Characterisation of the active/de-active transition of mitochondrial complex I. *Biochimica et Biophysica Acta*, *1837*, 1083–1092.

Bache, N., Rand, K. D., Roepstorff, P., & Jørgensen, T. J. (2008). Gas-phase fragmentation of peptides by MALDI in-source decay with limited amide hydrogen (1H/2H) scrambling. *Analytical Chemistry*, *80*, 6431–6435.

Banks, C. A., Kong, S. E., & Washburn, M. P. (2012). Affinity purification of protein complexes for analysis by multidimensional protein identification technology. *Protein Expression and Purification*, *86*, 105–119.

Barber, M., Bordoli, R. S., Sedgwick, R. D., & Tyler, A. N. (1981). Fast atom bombardment of solids (F.A.B.): A new ion source for mass spectrometry. *Journal of the Chemical Society, Chemical Communications*, 325–327.

Barrera, N. P., Di Bartolo, N., Booth, P. J., & Robinson, C. V. (2008). Micelles protect membrane complexes from solution to vacuum. *Science*, *321*, 243–246.

Barrera, N. P., Zhou, M., & Robinson, C. V. (2013). The role of lipids in defining membrane protein interactions: Insights from mass spectrometry. *Trends in Cell Biology*, *23*, 1–8.

Barth, K. R., Isabella, V. M., & Clark, V. L. (2009). Biochemical and genomic analysis of the denitrification pathway within the genus *Neisseria*. *Microbiology*, *155*, 4093–4103.

Bäsell, K., Otto, A., Junker, S., Zühlke, D., Rappen, G. M., Schmidt, S., et al. (2014). The phosphoproteome and its physiological dynamics in *Staphylococcus aureus*. *International Journal of Medical Microbiology*, *304*, 121–132.

Batista, A. P., Franco, C., Mendes, M., Coelho, A. V., & Pereira, M. M. (2010). Subunit composition of *Rhodothermus marinus* respiratory complex I. *Analytical Biochemistry*, *407*, 104–110.

Battchikova, N., & Aro, E. M. (2007). Cyanobacterial NDH-1 complexes: Multiplicity in function and subunit composition. *Physiologia Plantarum*, *131*, 22–32.

Bay, D. C., Chan, C. S., & Turner, R. J. (2015). NarJ subfamily system specific chaperone diversity and evolution is directed by respiratory enzyme associations. *BMC Evolutionary Biology*, *15*, 110.

Bazán, S., Mileykovskaya, E., Mallampalli, V. K., Heacock, P., Sparagna, G. C., & Dowhan, W. (2013). Cardiolipin-dependent reconstitution of respiratory supercomplexes from purified *Saccharomyces cerevisiae* complexes III and IV. *The Journal of Biological Chemistry*, *288*, 401–411.

Becher, D., Büttner, K., Moche, M., Hessling, B., & Hecker, M. (2011). From the genome sequence to the protein inventory of *Bacillus subtilis*. *Proteomics*, *11*, 2971–2980.

Benesch, J. L., Aquilina, J. A., Ruotolo, B. T., Sobott, F., & Robinson, C. V. (2006). Tandem mass spectrometry reveals the quaternary organization of macromolecular assemblies. *Chemistry & Biology*, *13*, 597–605.

Benesch, J. L., & Robinson, C. V. (2006). Mass spectrometry of macromolecular assemblies: Preservation and dissociation. *Current Opinion in Structural Biology*, *16*, 245–251.

Bennett, K. L., Matthiesen, T., & Roepstorff, P. (2000). Probing protein surface topology by chemical surface labeling, crosslinking, and mass spectrometry. *Methods in Molecular Biology*, *146*, 113–131.

Bereman, M. S., MacLean, B., Tomazela, D. M., Liebler, D. C., & MacCoss, M. J. (2012). The development of selected reaction monitoring methods for targeted proteomics via empirical refinement. *Proteomics*, *12*, 1134–1141.

Bernal, V., Castaño-Cerezo, S., Gallego-Jara, J., Écija-Conesa, A., de Diego, T., Iborra, J. L., et al. (2014). Regulation of bacterial physiology by lysine acetylation of proteins. *Nature Biotechnology*, *31*, 586–595.

Bernard-Marty, C., Lebrun, F., Awada, A., & Piccart, M. J. (2006). Monoclonal antibody-based targeted therapy in breast cancer: Current status and future directions. *Drugs*, *66*, 1577–1591.

Bernsel, A., & Daley, D. O. (2009). Exploring the inner membrane proteome of *Escherichia coli*: Which proteins are eluding detection and why? *Trends in Microbiology, 17*, 444–449.

Berven, F. S., Karlsen, O. A., Murrell, J. C., & Jensen, H. B. (2003). Multiple polypeptide forms observed in two-dimensional gels of *Methylococcus capsulatus* (Bath) polypeptides are generated during the separation procedure. *Electrophoresis, 24*, 757–761.

Biemann, K. (1990). Appendix 5. Nomenclature for peptide fragment ions (positive ions). *Methods in Enzymology, 193*, 886–887.

Bisle, B., Schmidt, A., Scheibe, B., Klein, C., Tebbe, A., Kellermann, J., et al. (2006). Quantitative profiling of the membrane proteome in a halophilic archaeon. *Molecular & Cellular Proteomics, 5*, 1543–1558.

Blanc, B., Gerez, C., & Ollagnier de Choudens, S. (2015). Assembly of Fe/S proteins in bacterial systems: Biochemistry of the bacterial ISC system. *Biochimica et Biophysica Acta, 1853*, 1436–1447.

Blaza, J. N., Serreli, R., Jones, A. J., Mohammed, K., & Hirst, J. (2014). Kinetic evidence against partitioning of the ubiquinone pool and the catalytic relevance of respiratory-chain supercomplexes. *Proceedings of the National Academy of Sciences of the United States of America, 111*, 15735–15740.

Bleier, L., Wittig, I., Heide, H., Steger, M., Brandt, U., & Dröse, S. (2015). Generator-specific targets of mitochondrial reactive oxygen species. *Free Radical Biology & Medicine, 78*, 1–10.

Block, H., Maertens, B., Spriestersbach, A., Brinker, N., Kubicek, J., Fabis, R., et al. (2009). Immobilized-metal affinity chromatography (IMAC): A review. *Methods in Enzymology, 463*, 439–473.

Boersema, P. J., Aye, T. T., van Veen, T. A., Heck, A. J., & Mohammed, S. (2008). Triplex protein quantification based on stable isotope labeling by peptide dimethylation applied to cell and tissue lysates. *Proteomics, 8*, 4624–4632.

Boersema, P. J., Kahraman, A., & Picotti, P. (2015). Proteomics beyond large-scale protein expression analysis. *Current Opinion in Biotechnology, 34*, 162–170.

Boersema, P. J., Raijmakers, R., Lemeer, S., Mohammed, S., & Heck, A. J. (2009). Multiplex peptide stable isotope dimethyl labeling for quantitative proteomics. *Nature Protocols, 4*, 484–494.

Bohovych, I., Fernandez, M. R., Rahn, J. J., Stackley, K. D., Bestman, J. E., Anandhan, A., et al. (2015). Metalloprotease OMA1 fine-tunes mitochondrial bioenergetic function and respiratory supercomplex stability. *Scientific Reports, 5*, 13989.

Borisov, V. B., Gennis, R. B., Hemp, J., & Verkhovsky, M. I. (2011). The cytochrome *bd* respiratory oxygen reductases. *Biochimica et Biophysica Acta, 1807*, 1398–1413.

Bornhövd, C., Vogel, F., Neupert, W., & Reichert, A. S. (2006). Mitochondrial membrane potential is dependent on the oligomeric state of F1F0-ATP synthase supracomplexes. *The Journal of Biological Chemistry, 281*, 13990–13998.

Borysik, A. J., Hewitt, D. J., & Robinson, C. V. (2013). Detergent release prolongs the lifetime of native-like membrane protein conformations in the gas-phase. *Journal of the American Chemical Society, 135*, 6078–6083.

Brancia, F. L. (2006). Recent developments in ion-trap mass spectrometry and related technologies. *Expert Review of Proteomics, 3*, 143–151.

Brandt, U. (2011). A two-state stabilization-change mechanism for proton-pumping complex I. *Biochimica et Biophysica Acta, 1807*, 1364–1369.

Braun, R. J., Kinkl, N., Beer, M., & Ueffing, M. (2007). Two-dimensional electrophoresis of membrane proteins. *Analytical and Bioanalytical Chemistry, 389*, 1033–1045.

Brewis, I. A., & Brennan, P. (2010). Proteomics technologies for the global identification and quantification of proteins. *Advances in Protein Chemistry and Structural Biology, 80*, 1–44.

Bridges, H. R., Fearnley, I. M., & Hirst, J. (2010). The subunit composition of mitochondrial NADH: Ubiquinone oxidoreductase (complex I) from *Pichia pastoris*. *Molecular & Cellular Proteomics*, *9*, 2318–2326.

Brosch, M., Yu, L., Hubbard, T., & Choudhary, J. (2009). Accurate and sensitive peptide identification with Mascot Percolator. *Journal of Proteome Research*, *8*, 3176–3181.

Brown, C. T., Hug, L. A., Thomas, B. C., Sharon, I., Castelle, C. J., Singh, A., et al. (2015). Unusual biology across a group comprising more than 15% of domain bacteria. *Nature*, *523*, 208–211.

Brown, R. S., & Lennon, J. J. (1995). Mass resolution improvement by incorporation of pulsed ion extraction in a matrix-assisted laser desorption/ionization linear time-of-flight mass spectrometer. *Analytical Chemistry*, *67*, 1998–2003.

Brown, R. N., Romine, M. F., Schepmoes, A. A., Smith, R. D., & Lipton, M. S. (2010). Mapping the subcellular proteome of *Shewanella oneidensis* MR-1 using sarkosyl-based fractionation and LC-MS/MS protein identification. *Journal of Proteome Research*, *9*, 4454–4463.

Buddelmeijer, N. (2015). The molecular mechanism of bacterial lipoprotein modification—How, when and why? *FEMS Microbiology Reviews*, *39*, 246–261.

Bühler, D., Rossmann, R., Landolt, S., Balsiger, S., Fischer, H. M., & Hennecke, H. (2010). Disparate pathways for the biogenesis of cytochrome oxidases in *Bradyrhizobium japonicum*. *The Journal of Biological Chemistry*, *285*, 15704–15713.

Bultema, J. B., Braun, H. P., Boekema, E. J., & Kouril, R. (2009). Megacomplex organization of the oxidative phosphorylation system by structural analysis of respiratory supercomplexes from potato. *Biochimica et Biophysica Acta*, *1787*, 60–67.

Bunai, K., & Yamane, K. (2005). Effectiveness and limitation of two-dimensional gel electrophoresis in bacterial membrane protein proteomics and perspectives. *Journal of Chromatography, B: Analytical Technologies in the Biomedical and Life Sciences*, *815*, 227–236.

Burns, K. M., Sarpe, V., Wagenbach, M., Wordeman, L., & Schriemer, D. C. (2015). HX-MS^2 for high performance conformational analysis of complex protein states. *Protein Science*, *24*, 1313–1324.

Calabrese, A. N., Watkinson, T. G., Henderson, P. J., Radford, S. E., & Ashcroft, A. E. (2015). Amphipols outperform dodecylmaltoside micelles in stabilizing membrane protein structure in the gas phase. *Analytical Chemistry*, *87*, 1118–1126.

Calder, B., Soares, N. C., de Kock, E., & Blackburn, J. M. (2015). Mycobacterial proteomics: Analysis of expressed proteomes and post-translational modifications to identify candidate virulence factors. *Expert Review of Proteomics*, *12*, 21–35.

Calvaruso, M. A., Smeitink, J., & Nijtmans, L. (2008). Electrophoresis techniques to investigate defects in oxidative phosphorylation. *Methods*, *46*, 281–287.

Calvaruso, M. A., Willems, P., van den Brand, M., Valsecchi, F., Kruse, S., Palmiter, R., et al. (2012). Mitochondrial complex III stabilizes complex I in the absence of NDUFS4 to provide partial activity. *Human Molecular Genetics*, *21*, 115–120.

Carlson, S. M., Moore, K. E., Green, E. M., Martín, G. M., & Gozani, O. (2014). Proteome-wide enrichment of proteins modified by lysine methylation. *Nature Protocols*, *9*, 37–50.

Caro-Quintero, A., & Konstantinidis, K. T. (2012). Bacterial species may exist, metagenomics reveal. *Environmental Microbiology*, *14*, 347–355.

Carroll, J., Ding, S., Fearnley, I. M., & Walker, J. E. (2013). Post-translational modifications near the quinone binding site of mammalian complex I. *The Journal of Biological Chemistry*, *288*, 24799–24808.

Carroll, J., Fearnley, I. M., Shannon, R. J., Hirst, J., & Walker, J. E. (2003). Analysis of the subunit composition of complex I from bovine heart mitochondria. *Molecular & Cellular Proteomics*, *2*, 117–126.

Cartron, M. L., Olsen, J. D., Sener, M., Jackson, P. J., Brindley, A. A., Qian, P., et al. (2014). Integration of energy and electron transfer processes in the photosynthetic membrane of *Rhodobacter sphaeroides*. *Biochimica et Biophysica Acta, 1837*, 1769–1780.

Castanié-Cornet, M. P., Bruel, N., & Genevaux, P. (2014). Chaperone networking facilitates protein targeting to the bacterial cytoplasmic membrane. *Biochimica et Biophysica Acta, 1843*, 1442–1456.

Castelle, C., Guiral, M., Malarte, G., Ledgham, F., Leroy, G., Brugna, M., et al. (2008). A new iron-oxidizing/O_2-reducing supercomplex spanning both inner and outer membranes, isolated from the extreme acidophile *Acidithiobacillus ferrooxidans*. *The Journal of Biological Chemistry, 283*, 25803–25811.

Castelle, C. J., Roger, M., Bauzan, M., Brugna, M., Lignon, S., Nimtz, M., et al. (2015). The aerobic respiratory chain of the acidophilic archaeon *Ferroplasma acidiphilum*: A membrane-bound complex oxidizing ferrous iron. *Biochimica et Biophysica Acta, 1847*, 717–728.

Catherman, A. D., Li, M., Tran, J. C., Durbin, K. R., Compton, P. D., Early, B. P., et al. (2013). Top down proteomics of human membrane proteins from enriched mitochondrial fractions. *Analytical Chemistry, 85*, 1880–1888.

Catherman, A. D., Skinner, O. S., & Kelleher, N. L. (2014). Top down proteomics: Facts and perspectives. *Biochemical and Biophysical Research Communications, 445*, 683–693.

Catrein, I., & Herrmann, R. (2011). The proteome of *Mycoplasma pneumoniae*, a supposedly "simple" cell. *Proteomics, 11*, 3614–3632.

Černý, M., Skalák, J., Cerna, H., & Brzobohatý, B. (2013). Advances in purification and separation of posttranslationally modified proteins. *Journal of Proteomics, 92*, 2–27.

Chaban, Y., Boekema, E. J., & Dudkina, N. V. (2014). Structures of mitochondrial oxidative phosphorylation supercomplexes and mechanisms for their stabilisation. *Biochimica et Biophysica Acta, 1837*, 418–426.

Chan, C. S., Bay, D. C., Leach, T. G., Winstone, T. M., Kuzniatsova, L., Tran, V. A., et al. (2014). 'Come into the fold': A comparative analysis of bacterial redox enzyme maturation protein members of the NarJ subfamily. *Biochimica et Biophysica Acta, 1838*, 2971–2984.

Chang, L., Liu, X., Li, Y., Liu, C. C., Yang, F., Zhao, J., et al. (2015). Structural organization of an intact phycobilisome and its association with photosystem II. *Cell Research, 25*, 726–737.

Chapman, J. D., Goodlett, D. R., & Masselon, C. D. (2014). Multiplexed and data-independent tandem mass spectrometry for global proteome profiling. *Mass Spectrometry Reviews, 33*, 452–470.

Chernushevich, I. V., Loboda, A. V., & Thomson, B. A. (2001). An introduction to quadrupole-time-of-flight mass spectrometry. *Journal of Mass Spectrometry, 36*, 849–865.

Cheung, R. C., Wong, J. H., & Ng, T. B. (2012). Immobilized metal ion affinity chromatography: A review on its applications. *Applied Microbiology and Biotechnology, 96*, 1411–1420.

Chevalier, F. (2010). Highlights on the capacities of "Gel-based" proteomics. *Proteome Science, 8*, 23.

Chi, B. K., Busche, T., Van Laer, K., Bäsell, K., Becher, D., Clermont, L., et al. (2014). Protein S-mycothiolation functions as redox-switch and thiol protection mechanism in *Corynebacterium glutamicum* under hypochlorite stress. *Antioxidants & Redox Signaling, 20*, 589–605.

Chi, H., Chen, H., He, K., Wu, L., Yang, B., Sun, R. X., et al. (2013). pNovo+: De novo peptide sequencing using complementary HCD and ETD tandem mass spectra. *Journal of Proteome Research, 12*, 615–625.

Chi, B. K., Gronau, K., Mäder, U., Hessling, B., Becher, D., & Antelmann, H. (2011). S-Bacillithiolation protects against hypochlorite stress in *Bacillus subtilis* as revealed by transcriptomics and redox proteomics. *Molecular & Cellular Proteomics*, *10*. M111.009506.

Chi, B. K., Roberts, A. A., Huyen, T. T., Bäsell, K., Becher, D., Albrecht, D., et al. (2013). S-Bacillithiolation protects conserved and essential proteins against hypochlorite stress in firmicutes bacteria. *Antioxidants & Redox Signaling*, *18*, 1273–1295.

Chiang, S. M., & Schellhorn, H. E. (2012). Regulators of oxidative stress response genes in *Escherichia coli* and their functional conservation in bacteria. *Archives of Biochemistry and Biophysics*, *525*, 161–169.

Choi, C. W., Yun, S. H., Kwon, S. O., Leem, S. H., Choi, J. S., Yun, C. Y., et al. (2010). Analysis of cytoplasmic membrane proteome of *Streptococcus pneumoniae* by shotgun proteomic approach. *Journal of Microbiology*, *48*, 872–876.

Chojnacka, M., Gornicka, A., Oeljeklaus, S., Warscheid, B., & Chacinska, A. (2015). Cox17 protein is an auxiliary factor involved in the control of the mitochondrial contact site and cristae organizing system. *The Journal of Biological Chemistry*, *290*, 15304–15312.

Chorev, D. S., Ben-Nissan, G., & Sharon, M. (2015). Exposing the subunit diversity and modularity of protein complexes by structural mass spectrometry approaches. *Proteomics*, *15*, 2777–2791.

Chouchani, E. T., Methner, C., Nadtochiy, S. M., Logan, A., Pell, V. R., Ding, S., et al. (2013). Cardioprotection by *S*-nitrosation of a cysteine switch on mitochondrial complex I. *Nature Medicine*, *19*, 753–759.

Choudhary, C., Weinert, B. T., Nishida, Y., Verdin, E., & Mann, M. (2014). The growing landscape of lysine acetylation links metabolism and cell signalling. *Nature Reviews Molecular Cell Biology*, *15*, 536–550.

Claypool, S. M. (2009). Cardiolipin, a critical determinant of mitochondrial carrier protein assembly and function. *Biochimica et Biophysica Acta*, *1788*, 2059–2068.

Cline, K. (2015). Mechanistic aspects of folded protein transport by the twin arginine translocase (Tat). *The Journal of Biological Chemistry*, *290*, 16530–16538.

Cogliati, S., Frezza, C., Soriano, M. E., Varanita, T., Quintana-Cabrera, R., Corrado, M., et al. (2013). Mitochondrial cristae shape determines respiratory chain supercomplexes assembly and respiratory efficiency. *Cell*, *155*, 160–171.

Colangelo, C. M., Chung, L., Bruce, C., & Cheung, K. H. (2013). Review of software tools for design and analysis of large scale MRM proteomic datasets. *Methods*, *61*, 287–298.

Cologna, S. M., Russell, W. K., Lim, P. J., Vigh, G., & Russell, D. H. (2010). Combining isoelectric point-based fractionation, liquid chromatography and mass spectrometry to improve peptide detection and protein identification. *Journal of the American Society for Mass Spectrometry*, *21*, 1612–1619.

Cordwell, S. J. (2004). Exploring and exploiting bacterial proteomes. *Methods in Molecular Biology*, *266*, 115–135.

Cordwell, S. J., Len, A. C., Touma, R. G., Scott, N. E., Falconer, L., Jones, D., et al. (2008). Identification of membrane-associated proteins from *Campylobacter jejuni* strains using complementary proteomics technologies. *Proteomics*, *8*, 122–139.

Cotter, R. J., Griffith, W., & Jelinek, C. (2007). Tandem time-of-flight (TOF/TOF) mass spectrometry and the curved-field reflectron. *Journal of Chromatography, B: Analytical Technologies in the Biomedical and Life Sciences*, *855*, 2–13.

Cousin, C., Derouiche, A., Shi, L., Pagot, Y., Poncet, S., & Mijakovic, I. (2013). Protein-serine/threonine/tyrosine kinases in bacterial signaling and regulation. *FEMS Microbiology Letters*, *346*, 11–19.

Cox, J., & Mann, M. (2008). MaxQuant enables high peptide identification rates, individualized p.p.b.-range mass accuracies and proteome-wide protein quantification. *Nature Biotechnology*, *26*, 1367–1372.

Craig, R., & Beavis, R. C. (2004). TANDEM: Matching proteins with tandem mass spectra. *Bioinformatics*, *20*, 1466–1467.

Croal, L. R., Gralnick, J. A., Malasarn, D., & Newman, D. K. (2004). The genetics of geochemistry. *Annual Review of Genetics*, *38*, 175–202.

Crosby, H. A., Pelletier, D. A., Hurst, G. B., & Escalante-Semerena, J. C. (2012). System-wide studies of N-lysine acetylation in *Rhodopseudomonas palustris* reveal substrate specificity of protein acetyltransferases. *The Journal of Biological Chemistry*, *287*, 15590–15601.

Cui, T. Z., Conte, A., Fox, J. L., Zara, V., & Winge, D. R. (2014). Modulation of the respiratory supercomplexes in yeast: Enhanced formation of cytochrome oxidase increases the stability and abundance of respiratory supercomplexes. *The Journal of Biological Chemistry*, *289*, 6133–6141.

Cui, W., Rohrs, H. W., & Gross, M. L. (2011). Top-down mass spectrometry: Recent developments, applications and perspectives. *Analyst*, *136*, 3854–3864.

Curreem, S. O., Watt, R. M., Lau, S. K., & Woo, P. C. (2012). Two-dimensional gel electrophoresis in bacterial proteomics. *Protein & Cell*, *3*, 346–363.

Cvetkovic, A., Menon, A. L., Thorgersen, M. P., Scott, J. W., Poole, F. L., 2nd., Jenney, F. E., Jr., et al. (2010). Microbial metalloproteomes are largely uncharacterized. *Nature*, *466*, 779–782.

Dagda, R. K., Sultana, T., & Lyons-Weiler, J. (2010). Evaluation of the consensus of four peptide identification algorithms for tandem mass spectrometry based proteomics. *Journal of Proteomics and Bioinformatics*, *3*, 39–47.

Dalbey, R. E., Kuhn, A., Zhu, L., & Kiefer, D. (2014). The membrane insertase YidC. *Biochimica et Biophysica Acta*, *1843*, 1489–1496.

Dancik, V., Addona, T. A., Clauser, K. R., Vath, J. E., & Pevzner, P. A. (1999). *De novo* peptide sequencing via tandem mass spectrometry. *Journal of Computational Biology*, *6*, 327–342.

Dash, B. P., Alles, M., Bundschuh, F. A., Richter, O. M., & Ludwig, B. (2015). Protein chaperones mediating copper insertion into the CuA site of the aa_3-type cytochrome *c* oxidase of *Paracoccus denitrificans*. *Biochimica et Biophysica Acta*, *1847*, 202–211.

Davies, K. M., Anselmi, C., Wittig, I., Faraldo-Gómez, J. D., & Kühlbrandt, W. (2012). Structure of the yeast F1Fo-ATP synthase dimer and its role in shaping the mitochondrial cristae. *Proceedings of the National Academy of Sciences of the United States of America*, *109*, 13602–13607.

de Hoffmann, E., & Stroobant, V. (2007). *Mass spectrometry: Principles and applications*. Hoboken, NJ: Wiley.

De Rasmo, D., Palmisano, G., Scacco, S., Technikova-Dobrova, Z., Panelli, D., Cocco, T., et al. (2010). Phosphorylation pattern of the NDUFS4 subunit of complex I of the mammalian respiratory chain. *Mitochondrion*, *10*, 464–471.

De Rasmo, D., Signorile, A., Santeramo, A., Larizza, M., Lattanzio, P., Capitanio, G., et al. (2015). Intramitochondrial adenylyl cyclase controls the turnover of nuclear-encoded subunits and activity of mammalian complex I of the respiratory chain. *Biochimica et Biophysica Acta*, *1853*, 183–191.

de Souza, G. A., & Wiker, H. G. (2011). A proteomic view of mycobacteria. *Proteomics*, *11*, 3118–3127.

Dengjel, J., Kratchmarova, I., & Blagoev, B. (2010). Mapping protein-protein interactions by quantitative proteomics. *Methods in Molecular Biology*, *658*, 267–278.

Desmurs, M., Foti, M., Raemy, E., Vaz, F. M., Martinou, J. C., Bairoch, A., et al. (2015). C11orf83, a mitochondrial cardiolipin-binding protein involved in bc_1 complex assembly and supercomplex stabilization. *Molecular and Cellular Biology*, *35*, 1139–1156.

Douglas, D. J., Frank, A. J., & Mao, D. (2005). Linear ion traps in mass spectrometry. *Mass Spectrometry Reviews*, *24*, 1–29.

Dresler, J., Klimentova, J., & Stulik, J. (2011). *Francisella tularensis* membrane complexome by blue native/SDS-PAGE. *Journal of Proteomics*, *75*, 257–269.

Dröse, S., Brandt, U., & Wittig, I. (2014). Mitochondrial respiratory chain complexes as sources and targets of thiol-based redox-regulation. *Biochimica et Biophysica Acta*, *1844*, 1344–1354.

Dröse, S., Krack, S., Sokolova, L., Zwicker, K., Barth, H. D., Morgner, N., et al. (2011). Functional dissection of the proton pumping modules of mitochondrial complex I. *PLoS Biology*, *9*, e1001128.

du Plessis, D. J., Nouwen, N., & Driessen, A. J. (2011). The Sec translocase. *Biochimica et Biophysica Acta*, *1808*, 851–865.

Duarte, M., & Videira, A. (2009). Effects of mitochondrial complex III disruption in the respiratory chain of *Neurospora crassa*. *Molecular Microbiology*, *72*, 246–258.

Dudkina, N. V., Eubel, H., Keegstra, W., Boekema, E. J., & Braun, H. P. (2005). Structure of a mitochondrial supercomplex formed by respiratory-chain complexes I and III. *Proceedings of the National Academy of Sciences of the United States of America*, *A102*, 3225–3229.

Dudkina, N. V., Heinemeyer, J., Keegstra, W., Boekema, E. J., & Braun, H. P. (2005). Structure of dimeric ATP synthase from mitochondria: An angular association of monomers induces the strong curvature of the inner membrane. *FEBS Letters*, *579*, 5769–5772.

Dudkina, N. V., Kouril, R., Peters, K., Braun, H. P., & Boekema, E. J. (2010). Structure and function of mitochondrial supercomplexes. *Biochimica et Biophysica Acta*, *1797*, 664–670.

Eichacker, L. A., Granvogl, B., Mirus, O., Müller, B. C., Miess, C., & Schleiff, E. (2004). Hiding behind hydrophobicity. Transmembrane segments in mass spectrometry. *The Journal of Biological Chemistry*, *279*, 50915–50922.

Eichacker, L. A., Weber, G., Sukop-Köppel, U., & Wildgruber, R. (2015). Free flow electrophoresis for separation of native membrane protein complexes. *Methods in Molecular Biology*, *1295*, 415–425.

Ekici, S., Pawlik, G., Lohmeyer, E., Koch, H. G., & Daldal, F. (2012). Biogenesis of *cbb*$_3$-type cytochrome *c* oxidase in *Rhodobacter capsulatus*. *Biochimica et Biophysica Acta*, *1817*, 898–8910.

Elias, D. A., Monroe, M. E., Marshall, M. J., Romine, M. F., Belieav, A. S., Fredrickson, J. K., et al. (2005). Global detection and characterization of hypothetical proteins in *Shewanella oneidensis* MR-1 using LC-MS based proteomics. *Proteomics*, *5*, 3120–3130.

Eng, J. K., McCormack, A. L., & Yates, J. R. (1994). An approach to correlate tandem mass spectral data of peptides with amino acid sequences in a protein database. *Journal of the American Society for Mass Spectrometry*, *5*, 976–989.

Erhardt, H., Dempwolff, F., Pfreundschuh, M., Riehle, M., Schäfer, C., Pohl, T., et al. (2014). Organization of the *Escherichia coli* aerobic enzyme complexes of oxidative phosphorylation in dynamic domains within the cytoplasmic membrane. *MicrobiologyOpen*, *3*, 316–326.

Erhardt, H., Steimle, S., Muders, V., Pohl, T., Walter, J., & Friedrich, T. (2012). Disruption of individual *nuo*-genes leads to the formation of partially assembled NADH: Ubiquinone oxidoreductase (complex I) in *Escherichia coli*. *Biochimica et Biophysica Acta*, *1817*, 863–871.

Ermler, U., Grabarse, W., Shima, S., Goubeaud, M., & Thauer, R. K. (1997). Crystal structure of methyl-coenzyme M reductase: The key enzyme of biological methane formation. *Science*, *278*, 1457–1462.

Eubel, H., Heinemeyer, J., Sunderhaus, S., & Braun, H. P. (2004). Respiratory chain supercomplexes in plant mitochondria. *Plant Physiology and Biochemistry*, *42*, 937–942.

Everberg, H., Gustavasson, N., & Tjerned, F. (2008). Enrichment of membrane proteins by partitioning in detergent/polymer aqueous two-phase systems. *Methods in Molecular Biology, 424*, 403–412.

Everberg, H., Leiding, T., Schiöth, A., Tjerneld, F., & Gustavsson, N. (2006). Efficient and non-denaturing membrane solubilization combined with enrichment of membrane protein complexes by detergent/polymer aqueous two-phase partitioning for proteome analysis. *Journal of Chromatography, A1122*, 35–46.

Eymann, C., Becher, D., Bernhardt, J., Gronau, K., Klutzny, A., & Hecker, M. (2007). Dynamics of protein phosphorylation on Ser/Thr/Tyr in *Bacillus subtilis*. *Proteomics, 7*, 3509–3526.

Eymann, C., Dreisbach, A., Albrecht, D., Bernhardt, J., Becher, D., Gentner, S., et al. (2004). A comprehensive proteome map of growing *Bacillus subtilis* cells. *Proteomics, 4*, 2849–2876.

Fahey, R. C. (2013). Glutathione analogs in prokaryotes. *Biochimica et Biophysica Acta, 1830*, 3182–3198.

Fang, R., Elias, D. A., Monroe, M. E., Shen, Y., McIntosh, M., Wang, P., et al. (2006). Differential label-free quantitative proteomic analysis of *Shewanella oneidensis* cultured under aerobic and suboxic conditions by accurate mass and time tag approach. *Molecular & Cellular Proteomics, 5*, 714–725.

Farhoud, M. H. (2011). *Disease biology of mitochondrial complex-I: Proteomic insights*. PhD thesis. Radboud University Nijmegen.

Farhoud, M. H., Wessels, H. J., Steenbakkers, P. J., Mattijssen, S., Wevers, R. A., van Engelen, B. G., et al. (2005). Protein complexes in the archaeon *Methanothermobacter thermautotrophicus* analyzed by blue native/SDS-PAGE and mass spectrometry. *Molecular & Cellular Proteomics, 4*, 1653–1663.

Faust, K., Lahti, L., Gonze, D., de Vos, W. M., & Raes, J. (2015). Metagenomics meets time series analysis: Unraveling microbial community dynamics. *Current Opinion in Microbiology, 25*, 56–66.

Fearnley, I. M., Carroll, J., & Walker, J. E. (2007). Proteomic analysis of the subunit composition of complex I (NADH:ubiquinone oxidoreductase) from bovine heart mitochondria. *Methods in Molecular Biology, 357*, 103–125.

Feng, Y., De Franceschi, G., Kahraman, A., Soste, M., Melnik, A., Boersema, P. J., et al. (2014). Global analysis of protein structural changes in complex proteomes. *Nature Biotechnology, 32*, 1036–1044.

Feng, S., Powell, S. M., Wilson, R., & Bowman, J. P. (2015). Proteomic insight into functional changes of proteorhodopsin-containing bacterial species *Psychroflexus torquis* under different illumination and salinity levels. *Journal of Proteome Research, 14*, 3848–3858.

Ferrer, M., Golyshina, O. V., Beloqui, A., Golyshin, P. N., & Timmis, K. N. (2007). The cellular machinery of *Ferroplasma acidiphilum* is iron-protein-dominated. *Nature, 445*, 91–94.

Fischer, F., & Poetsch, A. (2006). Protein cleavage strategies for an improved analysis of the membrane proteome. *Proteome Science, 4*, 2.

Fischer, F., Wolters, D., Rögner, M., & Poetsch, A. (2005). Toward the complete membrane proteome: High coverage of integral membrane proteins through transmembrane peptide detection. *Molecular & Cellular Proteomics, 5*, 444–453.

Fishov, I., & Norris, V. (2012). Membrane heterogeneity created by transertion is a global regulator in bacteria. *Current Opinion in Microbiology, 15*, 724–730.

Fontana, A., de Laureto, P. P., Spolaore, B., & Frare, E. (2012). Identifying disordered regions in proteins by limited proteolysis. *Methods in Molecular Biology, 896*, 297–318.

Fontana, A., de Laureto, P. P., Spolaore, B., Frare, E., Picotti, P., & Zambonin, M. (2004). Probing protein structure by limited proteolysis. *Acta Biochimica Polonica, 51*, 299–321.

Formosa, L. E., Mimaki, M., Frazier, A. E., McKenzie, M., Stait, T. L., Thorburn, D. R., et al. (2015). Characterization of mitochondrial FOXRED1 in the assembly of respiratory chain complex I. *Human Molecular Genetics, 24*, 2952–2965.

Frank, A., & Pevzner, P. (2005). PepNovo: *De novo* peptide sequencing via probabilistic network modeling. *Analytical Chemistry*, *77*, 964–973.

Fränzel, B., Poetsch, A., Trötschel, C., Persicke, M., Kalinowski, J., & Wolters, D. A. (2010). Quantitative proteomic overview on the *Corynebacterium glutamicum* L-lysine producing strain DM1730. *Journal of Proteomics*, *73*, 2336–2353.

Fränzel, B., & Wolters, D. A. (2011). Advanced MudPIT as a next step toward high proteome coverage. *Proteomics*, *11*, 3651–3656.

Freeman, E., & Ivanov, A. R. (2011). Proteomics under pressure: Development of essential sample preparation techniques in proteomics using ultrahigh hydrostatic pressure. *Journal of Proteome Research*, *10*, 5536–5546.

Frezza, C., Cipolat, S., Martins de Brito, O., Micaroni, M., Beznoussenko, G. V., Rudka, T., et al. (2006). OPA1 controls apoptotic cristae remodeling independently from mitochondrial fusion. *Cell*, *126*, 177–189.

Frielingsdorf, S., Schubert, T., Pohlmann, A., Lenz, O., & Friedrich, B. (2011). A trimeric supercomplex of the oxygen-tolerant membrane-bound [NiFe]-hydrogenase from *Ralstonia eutropha* H16. *Biochemistry*, *50*, 10836–10843.

Fröbel, J., Rose, P., & Müller, M. (2012). Twin-arginine-dependent translocation of folded proteins. *Philosophical Transactions of the Royal Society of London Series B, Biological Sciences*, *367*, 1029–1046.

Fu, H., Yuan, J., & Gao, H. (2015). Microbial oxidative stress response: Novel insights from environmental facultative anaerobic bacteria. *Archives of Biochemistry and Biophysics*, *584*, 28–35.

Fujiki, Y., Hubbard, A. L., Fowler, S., & Lazarow, P. B. (1982). Isolation of intracellular membranes by means of sodium carbonate treatment: Application to endoplasmic reticulum. *The Journal of Cell Biology*, *93*, 97–102.

Fujita, Y. (2009). Carbon catabolite control of the metabolic network in *Bacillus subtilis*. *Bioscience, Biotechnology, and Biochemistry*, *73*, 245–259.

Fukuyama, Y., Iwamoto, S., & Tanaka, K. (2006). Rapid sequencing and disulfide mapping of peptides containing disulfide bonds by using 1,5-diaminonaphthalene as a reductive matrix. *Journal of Mass Spectrometry*, *41*, 191–201.

Galkin, A., Meyer, B., Wittig, I., Karas, M., Schägger, H., Vinogradov, A., et al. (2008). Identification of the mitochondrial ND3 subunit as a structural component involved in the active/deactive enzyme transition of respiratory complex I. *The Journal of Biological Chemistry*, *283*, 20907–20913.

Gallien, S., Duriez, E., & Domon, B. (2011). Selected reaction monitoring applied to proteomics. *Journal of Mass Spectrometry*, *46*, 298–312.

Gama, M. R., Collins, C. H., & Bottoli, C. B. (2013). Nano-liquid chromatography in pharmaceutical and biomedical research. *Journal of Chromatographic Science*, *51*, 694–703.

García Montes de Oca, L. Y., Chagolla-López, A., González de la Vara, L., Cabellos-Avelar, T., Gómez-Lojero, C., & Gutiérrez Cirlos, E. B. (2012). The composition of the *Bacillus subtilis* aerobic respiratory chain supercomplexes. *Journal of Bioenergetics and Biomembranes*, *44*, 473–486.

Ge, R., Sun, X., Xiao, C., Yin, X., Shan, W., Chen, Z., et al. (2011). Phosphoproteome analysis of the pathogenic bacterium *Helicobacter pylori* reveals over-representation of tyrosine phosphorylation and multiply phosphorylated proteins. *Proteomics*, *11*, 1449–1461.

Geer, L. Y., Markey, S. P., Kowalak, J. A., Wagner, L., Xu, M., Maynard, D. M., et al. (2004). Open mass spectrometry search algorithm. *Journal of Proteome Research*, *3*, 958–964.

Genova, M. L., & Lenaz, G. (2013). A critical appraisal of the role of respiratory supercomplexes in mitochondria. *Biological Chemistry*, *394*, 631–639.

Genova, M. L., & Lenaz, G. (2014). Functional role of mitochondrial respiratory supercomplexes. *Biochimica et Biophysica Acta*, *1837*, 427–443.

Gerber, G. K. (2014). The dynamic microbiome. *FEBS Letters*, *588*, 4131–4139.

Gerber, S. A., Rush, J., Stemman, O., Kirschner, M. W., & Gygi, S. P. (2003). Absolute quantification of proteins and phosphoproteins from cell lysates by tandem MS. *Proceedings of the National Academy of Sciences of the United States of America*, *100*, 6940–6945.

Giese, H., Ackermann, J., Heide, H., Bleier, L., Dröse, S., Wittig, I., et al. (2015). NOVA: A software to analyze complexome profiling data. *Bioinformatics*, *31*, 440–441.

Gil Borlado, M. C., Moreno Lastres, D., Gonzalez Hoyuela, M., Moran, M., Blazquez, A., Pello, R., et al. (2010). Impact of the mitochondrial genetic background in complex III deficiency. *PLoS One*, *5*, e12801.

Gillette, M., & Carr, S. A. (2013). Quantitative analysis of peptides and proteins in biomedicine by targeted mass spectrometry. *Nature Methods*, *10*, 28–34.

Gilmore, J. M., & Washburn, M. P. (2010). Advances in shotgun proteomics and the analysis of membrane proteomes. *Journal of Proteomics*, *73*, 2078–2091.

Gingras, A. C., Gstaiger, M., Raught, B., & Aebersold, R. (2007). Analysis of protein complexes using mass spectrometry. *Nature Reviews Molecular Cell Biology*, *8*, 645–654.

Glew, M. D., Veith, P. D., Chen, D., Seers, C. A., Chen, Y. Y., & Reynolds, E. C. (2014). Blue native-PAGE analysis of membrane protein complexes in *Porphyromonas gingivalis*. *Journal of Proteomics*, *110*, 72–92.

Godzik, A. (2011). Metagenomics and the protein universe. *Current Opinion in Structural Biology*, *21*, 398–403.

Gómez, L. A., Monette, J. S., Chavez, J. D., Maier, C. S., & Hagen, T. M. (2009). Supercomplexes of the mitochondrial electron transport chain decline in the aging rat heart. *Archives of Biochemistry and Biophysics*, *490*, 30–35.

Goosens, V. J., Monteferrante, C. G., & van Dijl, J. M. (2014a). Co-factor insertion and disulfide bond requirements for twin-arginine translocase-dependent export of the *Bacillus subtilis* Rieske protein QcrA. *The Journal of Biological Chemistry*, *289*, 13124–13131.

Goosens, V. J., Monteferrante, C. G., & van Dijl, J. M. (2014b). The Tat system of Gram-positive bacteria. *Biochimica et Biophysica Acta*, *1843*, 1698–1706.

Goosens, V. J., Otto, A., Glasner, C., Monteferrante, C. C., van der Ploeg, R., Hecker, M., et al. (2013). Novel twin-arginine translocation pathway-dependent phenotypes of *Bacillus subtilis* unveiled by quantitative proteomics. *Journal of Proteome Research*, *12*, 796–807.

Görg, A., Weiss, W., & Dunn, M. J. (2004). Current two-dimensional electrophoresis technology for proteomics. *Proteomics*, *4*, 3665–3685.

Gorman, J. J., Ferguson, B. L., & Nguyen, T. B. (1996). Use of 2,6-dihydroxyacetophenone for analysis of fragile peptides, disulphide bonding and small proteins by matrix-assisted laser desorption/ionization. *Rapid Communications in Mass Spectrometry*, *10*, 529–536.

Götzke, H., Muheim, C., Altelaar, A. F., Heck, A. J., Maddalo, G., & Daley, D. O. (2015). Identification of putative substrates for the periplasmic chaperone YfgM in *Escherichia coli* using quantitative proteomics. *Molecular & Cellular Proteomics*, *14*, 216–226.

Gouw, J. W., & Krijgsveld, J. (2012). MSQuant: A platform for stable isotope-based quantitative proteomics. *Methods in Molecular Biology*, *893*, 511–522.

Grabarse, W., Mahlert, F., Duin, E. C., Goubeaud, M., Shima, S., Thauer, R. K., et al. (2001). On the mechanism of biological methane formation: Structural evidence for conformational changes in methyl-coenzyme M reductase upon substrate binding. *Journal of Molecular Biology*, *309*, 315–330.

Grangeasse, C., Cozzone, A. J., Deutscher, J., & Mijakovic, I. (2007). Tyrosine phosphorylation: An emerging regulatory device of bacterial physiology. *Trends in Biochemical Sciences*, *32*, 86–94.

Grangeasse, C., Stülke, J., & Mijakovic, I. (2015). Regulatory potential of post-translational modifications in bacteria. *Frontiers in Microbiology*, *6*, 500.

Grein, F., Ramos, A. R., Venceslau, S. S., & Pereira, I. A. (2013). Unifying concepts in anaerobic respiration: Insights from dissimilatory sulfur metabolism. *Biochimica et Biophysica Acta*, *1827*, 145–160.

Grimaldi, S., Schoepp-Cothenet, B., Ceccaldi, P., Guigliarelli, B., & Magalon, A. (2013). The prokaryotic Mo/W-bisPGD enzymes family: A catalytic workhorse in bioenergetic. *Biochimica et Biophysica Acta*, *1827*, 1048–1085.

Grossmann, J., Roos, F. F., Cieliebak, M., Liptak, Z., Mathis, L. K., Muller, M., et al. (2005). AUDENS: A tool for automated peptide *de novo* sequencing. *Journal of Proteome Research*, *4*, 1768–1774.

Gu, L., Chen, Y., Wang, Q. T., Li, X., Mi, K., & Deng, H. (2015). Functional characterization of sirtuin-like protein in *Mycobacterium smegmatis*. *Journal of Proteome Research*, *14*, 4441–4449.

Guan, K. L., & Xiong, Y. (2011). Regulation of intermediary metabolism by protein acetylation. *Trends in Biochemical Sciences*, *36*, 108–116.

Guan, K. L., Yu, W., Lin, Y., Xiong, Y., & Zhao, S. (2010). Generation of acetyllysine antibodies and affinity enrichment of acetylated peptides. *Nature Protocols*, *5*, 1583–1595.

Gurumoorthy, P., & Ludwig, B. (2015). Deciphering protein-protein interactions during the biogenesis of cytochrome *c* oxidase from *Paracoccus denitrificans*. *The FEBS Journal*, *282*, 537–549.

Guthals, A., & Bandeira, N. (2012). Peptide identification by tandem mass spectrometry with alternate fragmentation modes. *Molecular & Cellular Proteomics*, *11*, 550–557.

Gygi, S. P., Rist, B., Gerber, S. A., Turecek, F., Gelb, M. H., & Aebersold, R. (1999). Quantitative analysis of complex protein mixtures using isotope-coded affinity tags. *Nature Biotechnology*, *17*, 994–999.

Habersetzer, J., Larrieu, I., Priault, M., Salin, B., Rossignol, R., Brèthes, D., et al. (2013). Human F_1F_0 ATP synthase, mitochondrial ultrastructure and OXPHOS impairment: A (super-)complex matter? *PLoS One*, *8*, e75429.

Hakhverdyan, Z., Domanski, M., Hough, L. E., Oroskar, A. A., Oroskar, A. R., Keegan, S., et al. (2015). Rapid, optimized interactomic screening. *Nature Methods*, *12*, 553–560.

Han, H., Hemp, J., Pace, L. A., Ouyang, H., Ganesan, K., Roh, J. H., et al. (2011). Adaptation of aerobic respiration to low O_2 environments. *Proceedings of the National Academy of Sciences of the United States of America*, *A108*, 14109–14114.

Han, M. J., & Lee, S. Y. (2006). The *Escherichia coli* proteome: Past, present, and future prospects. *Microbiology and Molecular Biology Reviews*, *70*, 362–439.

Hannappel, A., Bundschuh, F. A., & Ludwig, B. (2012). Role of Surf1 in heme recruitment for bacterial COX biogenesis. *Biochimica et Biophysica Acta*, *1817*, 928–937.

Hartmann, T., Schwanhold, N., & Leimkühler, S. (2015). Assembly and catalysis of molybdenum or tungsten-containing formate dehydrogenases from bacteria. *Biochimica et Biophysica Acta*, *1854*, 1090–1100.

Hasan, S. S., Proctor, E. A., Yamashita, E., Dokholyan, N. V., & Cramer, W. A. (2014). Traffic within the cytochrome b_6f lipoprotein complex: Gating of the quinone portal. *Biophysical Journal*, *107*, 1620–1628.

Hasan, S. S., Yamashita, E., & Cramer, W. A. (2013). Transmembrane signaling and assembly of the cytochrome b_6f-lipidic charge transfer complex. *Biochimica et Biophysica Acta*, *1827*, 1295–1308.

Hasan, S. S., Yamashita, E., Ryan, C. M., Whitelegge, J. P., & Cramer, W. A. (2011). Conservation of lipid functions in cytochrome *bc* complexes. *Journal of Molecular Biology*, *414*, 145–162.

Haselberg, R., de Jong, G. J., & Somsen, G. W. (2013). CE-MS for the analysis of intact proteins 2010-2012. *Electrophoresis*, *34*, 99–112.

Hatzixanthis, K., Richardson, D. J., & Sargent, F. (2005). Chaperones involved in assembly and export of N-oxide reductases. *Biochemical Society Transactions*, *33*, 124–126.

Heide, H., Bleier, L., Steger, M., Ackermann, J., Dröse, S., Schwamb, B., et al. (2012). Complexome profiling identifies TMEM126B as a component of the mitochondrial complex I assembly complex. *Cell Metabolism*, *16*, 538–549.

Heide, H., & Wittig, I. (2013). Methods to analyse composition and dynamics of macromolecular complexes. *Biochemical Society Transactions*, *41*, 1235–1241.

Heimann, A. C., Blodau, C., Postma, D., Larsen, F., Viet, P. H., Nhan, P. Q., et al. (2007). Hydrogen thresholds and steady-state concentrations associated with microbial arsenate respiration. *Environmental Science & Technology*, *41*, 2311–2317.

Heintz, D., Gallien, S., Wischgoll, S., Ullmann, A. K., Schaeffer, C., Kretzschmar, A. K., et al. (2009). Differential membrane proteome analysis reveals novel proteins involved in the degradation of aromatic compounds in *Geobacter metallireducens*. *Molecular & Cellular Proteomics*, *8*, 2159–2169.

Helbig, A. O., de Groot, M. J., van Gestel, R. A., Mohammed, S., de Hulster, E. A., Luttik, M. A., et al. (2009). A three-way proteomics strategy allows differential analysis of yeast mitochondrial membrane protein complexes under anaerobic and aerobic conditions. *Proteomics*, *9*, 4787–4798.

Helbig, A. O., Heck, A. J., & Slijper, M. (2010). Exploring the membrane proteome—Challenges and analytical strategies. *Journal of Proteomics*, *73*, 868–878.

Helling, S., Hüttemann, M., Kadenbach, B., Ramzan, R., Vogt, S., & Marcus, K. (2012a). Discovering the phosphoproteome of the hydrophobic cytochrome *c* oxidase membrane protein complex. *Methods in Molecular Biology*, *893*, 345–358.

Helling, S., Hüttemann, M., Ramzan, R., Kim, S. H., Lee, I., Müller, T., et al. (2012b). Multiple phosphorylations of cytochrome *c* oxidase and their functions. *Proteomics*, *12*, 950–959.

Helling, S., Schmitt, E., Joppich, C., Schulenborg, T., Müllner, S., Felske-Müller, S., et al. (2006). 2-D differential membrane proteome analysis of scarce protein samples. *Proteomics*, *6*, 4506–4513.

Helling, S., Vogt, S., Rhiel, A., Ramzan, R., Wen, L., Marcus, K., et al. (2008). Phosphorylation and kinetics of mammalian cytochrome c oxidase. *Molecular & Cellular Proteomics*, 7, 1714–1724.

Hemp, J., & Gennis, R. B. (2008). Diversity of the heme-copper superfamily in archaea: Insights from genomics and structural modeling. *Results and Problems in Cell Differentiation*, *45*, 1–31.

Hendrickson, C. L., Quinn, J. P., Kaiser, N. K., Smith, D. F., Blakney, G. T., Chen, T., et al. (2015). 21 Tesla Fourier transform ion cyclotron resonance mass spectrometer: A national resource for ultrahigh resolution mass analysis. *Journal of the American Society for Mass Spectrometry*, *26*, 1626–1632.

Hendrickx, S., Adams, E., & Cabooter, D. (2015). Recent advances in the application of hydrophilic interaction chromatography for the analysis of biological matrices. *Bioanalysis*, 7, 2927–2945.

Hentchel, K. L., & Escalante-Semerena, J. C. (2015). Acylation of biomolecules in prokaryotes: A widespread strategy for the control of biological function and metabolic stress. *Microbiology and Molecular Biology Reviews*, *79*, 321–346.

Henzel, W. J., Billeci, T. M., Stults, J. T., Wong, S. C., Grimley, C., & Watanabe, C. (1993). Identifying proteins from two-dimensional gels by molecular mass searching of peptide fragments in protein sequence databases. *Proceedings of the National Academy of Sciences of the United States of America*, *90*, 5011–5015.

Henzel, W. J., Watanabe, C., & Stults, J. T. (2003). Protein identification: The origins of peptide mass fingerprinting. *Journal of the American Society for Mass Spectrometry*, *14*, 931–942.

Hillenkamp, F., & Karas, M. (1990). Mass spectrometry of peptides and proteins by matrix-assisted ultraviolet laser desorption/ionization. *Methods in Enzymology, 193*, 280–295.

Hillenkamp, F., Karas, M., Beavis, R. C., & Chait, B. T. (1991). Matrix-assisted laser desorption/ionization mass spectrometry of biopolymers. *Analytical Chemistry, 63*, 1193A–1203A.

Hillion, M., & Antelmann, H. (2015). Thiol-based redox switches in prokaryotes. *Biological Chemistry, 396*, 415–444.

Hitchcock, A., Hall, S. J., Myers, J. D., Mulholland, F., Jones, M. A., & Kelly, D. J. (2010). Roles of the twin-arginine translocase and associated chaperones in the biogenesis of the electron transport chains of the human pathogen *Campylobacter jejuni*. *Microbiology, 156*, 2994–3010.

Hochgräfe, F., Mostertz, J., Pöther, D. C., Becher, D., Helmann, J. D., & Hecker, M. (2007). *S*-Cysteinylation is a general mechanism for thiol protection of *Bacillus subtilis* proteins after oxidative stress. *The Journal of Biological Chemistry, 282*, 25981–25985.

Hoffmann, J., Sokolova, L., Preiss, L., Hicks, D. B., Krulwich, T. A., Morgner, N., et al. (2010). ATP synthases: Cellular nanomotors characterized by LILBID mass spectrometry. *Physical Chemistry Chemical Physics, 12*, 13375–13382.

Hogan, J. M., Pitteri, S. J., Chrisman, P. A., & McLuckey, S. A. (2005). Complementary structural information from a tryptic N-linked glycopeptide via electron transfer ion/ion reactions and collision-induced dissociation. *Journal of Proteome Research, 4*, 628–632.

Hoke, D. E., Zhang, K., Egan, S., Hatfaludi, T., Buckle, A. M., & Adler, B. (2011). Membrane proteins of *Pseudoalteromonas tunicata* during the transition from planktonic to extracellular matrix-adherent state. *Environmental Microbiology Reports, 3*, 405–413.

Hopkins, A., Buchanan, G., & Palmer, T. (2014). Role of the twin arginine protein transport pathway in the assembly of the *Streptomyces coelicolor* cytochrome bc_1 complex. *Journal of Bacteriology, 196*, 50–59.

Hopper, J. T., & Robinson, C. V. (2014). Mass spectrometry quantifies protein interactions—From molecular chaperones to membrane porins. *Angewandte Chemie (International Ed in English), 53*, 14002–14015.

Hopper, J. T., Yu, Y. T., Li, D., Raymond, A., Bostock, M., Liko, I., et al. (2013). Detergent-free mass spectrometry of membrane protein complexes. *Nature Methods, 10*, 1206–1208.

Huang, T. Y., & McLuckey, S. A. (2010). Gas-phase chemistry of multiply charged bioions in analytical mass spectrometry. *Annual Review of Analytical Chemistry (Palo Alto, California), 3*, 365–385.

Hui, S., Silverman, J. M., Chen, S. S., Erickson, D. W., Basan, M., Wang, J., et al. (2015). Quantitative proteomic analysis reveals a simple strategy of global resource allocation in bacteria. *Molecular Systems Biology, 11*, 784.

Hutchings, M. I., Palmer, T., Harrington, D. J., & Sutcliffe, I. C. (2009). Lipoprotein biogenesis in Gram-positive bacteria: Knowing when to hold 'em, knowing when to fold 'em. *Trends in Microbiology, 17*, 13–21.

Hüttemann, M., Helling, S., Sanderson, T. H., Sinkler, C., Samavati, L., Mahapatra, G., et al. (2012). Regulation of mitochondrial respiration and apoptosis through cell signaling: Cytochrome *c* oxidase and cytochrome *c* in ischemia/reperfusion injury and inflammation. *Biochimica et Biophysica Acta, 1817*, 598–609.

Huynen, M. A., Mühlmeister, M., Gotthardt, K., Guerrero-Castillo, S., & Brandt, U. (2015). Evolution and structural organization of the mitochondrial contact site (MICOS) complex and the mitochondrial intermembrane space bridging (MIB) complex. *Biochimica et Biophysica Acta, 1863*, 91–101.

Hyung, S. J., & Ruotolo, B. T. (2012). Integrating mass spectrometry of intact protein complexes into structural proteomics. *Proteomics, 12*, 1547–1564.

Iacob, R. E., & Engen, J. R. (2012). Hydrogen exchange mass spectrometry: Are we out of the quicksand? *Journal of the American Society for Mass Spectrometry*, *23*, 1003–1010.

Imlay, J. A. (2013). The molecular mechanisms and physiological consequences of oxidative stress: Lessons from a model bacterium. *Nature Reviews. Microbiology*, *11*, 443–454.

Ishihama, Y., Oda, Y., Tabata, T., Sato, T., Nagasu, T., Rappsilber, J., et al. (2005). Exponentially modified protein abundance index (emPAI) for estimation of absolute protein amount in proteomics by the number of sequenced peptides per protein. *Molecular & Cellular Proteomics*, *4*, 1265–1272.

Issaq, H., & Veenstra, T. (2008). Two-dimensional polyacrylamide gel electrophoresis (2D-PAGE): Advances and perspectives. *Biotechniques*, *44*, 697–700.

James, P., Quadroni, M., Carafoli, E., & Gonnet, G. (1993). Protein identification by mass profile fingerprinting. *Biochemical and Biophysical Research Communications*, *195*, 58–64.

Jeong, K., Kim, S., & Pevzner, P. A. (2013). UniNovo: A universal tool for de novo peptide sequencing. *Bioinformatics*, *29*, 1953–1962.

Jers, C., Kobir, A., Søndergaard, E. O., Jensen, P. R., & Mijakovic, I. (2011). *Bacillus subtilis* two-component system sensory kinase DegS is regulated by serine phosphorylation in its input domain. *PLoS One*, *6*, e14653.

Jiang, C. Y., Liu, L. J., Guo, X., You, X. Y., Liu, S. J., & Poetsch, A. (2014). Resolution of carbon metabolism and sulfur-oxidation pathways of *Metallosphaera cuprina* Ar-4 via comparative proteomics. *Journal of Proteomics*, *109*, 276–289.

Jo, I., Chung, I. Y., Bae, H. W., Kim, J. S., Song, S., Cho, Y. H., et al. (2015). Structural details of the OxyR peroxide-sensing mechanism. *Proceedings of the National Academy of Sciences of the United States of America*, *A112*, 6443–6448.

Johnson, T. L., Sikora, A. E., Zielke, R. A., & Sandkvist, M. (2013). Fluorescence microscopy and proteomics to investigate subcellular localization, assembly, and function of the type II secretion system. *Methods in Molecular Biology*, *966*, 157–172.

Johnson, A. S., van Horck, S., & Lewis, P. J. (2004). Dynamic localization of membrane proteins in *Bacillus subtilis*. *Microbiology*, *150*, 2815–2824.

Jones, J. D., & O'Connor, C. D. (2011). Protein acetylation in prokaryotes. *Proteomics*, *11*, 3012–3022.

Juhasz, P., Costello, C. E., & Biemann, K. (1993). Matrix-assisted laser desorption ionization mass spectrometry with 2-(4-hydroxyphenylazo)benzoic acid matrix. *Journal of the American Society for Mass Spectrometry*, *4*, 399–409.

Kane, L. A., & Van Eyk, J. E. (2009). Post-translational modifications of ATP synthase in the heart: Biology and function. *Journal of Bioenergetics and Biomembranes*, *41*, 145–150.

Kane, L. A., Youngman, M. J., Jensen, R. E., & Van Eyk, J. E. (2010). Phosphorylation of the F_1F_o ATP synthase beta subunit: Functional and structural consequences assessed in a model system. *Circulation Research*, *106*, 504–513.

Karabacak, N. M., Li, L., Tiwari, A., Hayward, L. J., Hong, P., Easterling, M. L., et al. (2009). Sensitive and specific identification of wild type and variant proteins from 8 to 669 kDa using top-down mass spectrometry. *Molecular & Cellular Proteomics*, *8*, 846–856.

Karas, M., Bachmann, D., & Hillenkamp, F. (1985). Influence of the wavelength in high-irradiance ultraviolet laser desorption mass spectrometry of organic molecules. *Analytical Chemistry*, *57*, 2935–2939.

Kartal, B., de Almeida, N. M., Maalcke, W. J., Op den Camp, H. J., Jetten, M. S., & Keltjens, J. T. (2013). How to make a living from anaerobic ammonium oxidation. *FEMS Microbiology Reviews*, *37*, 428–461.

Kashino, Y., Harayama, T., Pakrasi, H. B., & Satoh, K. (2007). Preparation of membrane proteins for analysis by two-dimensional gel electrophoresis. *Journal of Chromatography, B: Analytical Technologies in the Biomedical and Life Sciences*, *849*, 282–292.

Kaster, A. K., Moll, J., Parey, K., & Thauer, R. K. (2011). Coupling of ferredoxin and heterodisulfide reduction via electron bifurcation in hydrogenotrophic methanogenic archaea. *Proceedings of the National Academy of Sciences of the United States of America, 108*, 2981–2986.

Kero, F. A., Pedder, R. E., & Yost, R. A. (2004). *Encyclopedia of genetics, genomics, proteomics and bioinformatics*. John Wiley and Sons Ltd.

Kettenbach, A. N., Rush, J., & Gerber, S. A. (2011). Absolute quantification of protein and post-translational modification abundance with stable isotope-labeled synthetic peptides. *Nature Protocols, 6*, 175–186.

Kim, J. H., Bothe, J. R., Alderson, T. R., & Markley, J. L. (2015). Tangled web of interactions among proteins involved in iron–sulfur cluster assembly as unraveled by NMR, SAXS, chemical crosslinking, and functional studies. *Biochimica et Biophysica Acta, 1853*, 1416–1428.

Kim, M. S., Jang, J., Ab Rahman, N. B., Pethe, K., Berry, E. A., & Huang, L. S. (2015). Isolation and characterization of a hybrid respiratory supercomplex consisting of *Mycobacterium tuberculosis* cytochrome *bcc* and *Mycobacterium smegmatis* cytochrome aa_3. *The Journal of Biological Chemistry, 290*, 14350–14360.

Kim, S. H., Kim, S. K., Jung, K. H., Kim, Y. K., Hwang, H. C., Ryu, S. G., et al. (2013). Proteomic analysis of the oxidative stress response induced by low-dose hydrogen peroxide in *Bacillus anthracis*. *Journal of Microbiology and Biotechnology, 23*, 750–758.

Kim, G. W., & Yang, X. J. (2011). Comprehensive lysine acetylomes emerging from bacteria to humans. *Trends in Biochemical Sciences, 36*, 211–220.

Kim, D., Yu, B. J., Kim, J. A., Lee, Y. J., Choi, S. G., Kang, S., et al. (2013). The acetylproteome of Gram-positive model bacterium *Bacillus subtilis*. *Proteomics, 13*, 1726–1736.

Kislinger, T., Gramolini, A. O., MacLennan, D. H., & Emili, A. (2005). Multidimensional protein identification technology (MudPIT): Technical overview of a profiling method optimized for the comprehensive proteomic investigation of normal and diseased heart tissue. *Journal of the American Society for Mass Spectrometry, 16*, 1207–1220.

Kliman, M., May, J. C., & McLean, J. A. (2011). Lipid analysis and lipidomics by structurally selective ion mobility-mass spectrometry. *Biochimica et Biophysica Acta, 1811*, 935–945.

Klodmann, J., & Braun, H. P. (2011). Proteomic approach to characterize mitochondrial complex I from plants. *Phytochemistry, 72*, 1071–1080.

Klodmann, J., Lewejohann, D., & Braun, H. P. (2011). Low-SDS Blue native PAGE. *Proteomics, 11*, 1834–1839.

Klodmann, J., Sunderhaus, S., Nimtz, M., Jänsch, L., & Braun, H. P. (2010). Internal architecture of mitochondrial complex I from *Arabidopsis thaliana*. *Plant Cell, 22*, 797–810.

Kobayashi, K., Fujikawa, M., & Kozawa, T. (2014). Oxidative stress sensing by the iron-sulfur cluster in the transcription factor, SoxR. *Journal of Inorganic Biochemistry, 133*, 87–91.

Kohlmann, Y., Pohlmann, A., Otto, A., Becher, D., Cramm, R., Lütte, S., et al. (2011). Analyses of soluble and membrane proteomes of *Ralstonia eutropha* H16 reveal major changes in the protein complement in adaptation to lithoautotrophy. *Journal of Proteome Research, 10*, 2767–2776.

Kohlmann, Y., Pohlmann, A., Schwartz, E., Zühlke, D., Otto, A., Albrecht, D., et al. (2014). Coping with anoxia: A comprehensive proteomic and transcriptomic survey of denitrification. *Journal of Proteome Research, 13*, 4325–4338.

Konermann, L., & Pan, Y. (2012). Exploring membrane protein structural features by oxidative labeling and mass spectrometry. *Expert Review of Proteomics, 9*, 497–504.

Konermann, L., Pan, J., & Liu, Y. H. (2011). Hydrogen exchange mass spectrometry for studying protein structure and dynamics. *Chemical Society Reviews, 40*, 1224–1234.

Konermann, L., Tong, X., & Pan, Y. (2008). Protein structure and dynamics studied by mass spectrometry: H/D exchange, hydroxyl radical labeling, and related approaches. *Journal of Mass Spectrometry, 43*, 1021–1036.

Konijnenberg, A., Bannwarth, L., Yilmaz, D., Koçer, A., Venien-Bryan, C., & Sobott, F. (2015). Top-down mass spectrometry of intact membrane protein complexes reveals oligomeric state and sequence information in a single experiment. *Protein Science, 24*, 1292–1300.

Kosono, S., Tamura, M., Suzuki, S., Kawamura, Y., Yoshida, A., Nishiyama, M., et al. (2015). Changes in the acetylome and succinylome of *Bacillus subtilis* in response to carbon source. *PLoS One, 10*, e0131169.

Kouyianou, K., Aivaliotis, M., Gevaert, K., Karas, M., & Tsiotis, G. (2010). Membrane proteome of the green sulfur bacterium *Chlorobium tepidum* (syn. *Chlorobaculum tepidum*) analyzed by gel-based and gel-free methods. *Photosynthesis Research, 104*, 153–162.

Kraft, B., Strous, M., & Tegetmeyer, H. E. (2011). Microbial nitrate respiration—Genes, enzymes and environmental distribution. *Journal of Biotechnology, 155*, 104–117.

Kranz, R. G., Richard-Fogal, C., Taylor, J. S., & Frawley, E. R. (2009). Cytochrome *c* biogenesis: Mechanisms for covalent modifications and trafficking of heme and for heme-iron redox control. *Microbiology and Molecular Biology Reviews, 73*, 510–528.

Krause, F. (2006). Detection and analysis of protein-protein interactions in organellar and prokaryotic proteomes by native gel electrophoresis: (Membrane) protein complexes and supercomplexes. *Electrophoresis, 27*, 2759–2781.

Krause, F., Reifschneider, N. H., Vocke, D., Seelert, H., Rexroth, S., & Dencher, N. A. (2004). "Respirasome"-like supercomplexes in green leaf mitochondria of spinach. *The Journal of Biological Chemistry, 279*, 48369–48375.

Krijgsveld, J., Ketting, R. F., Mahmoudi, T., Johansen, J., Artal-Sanz, M., Verrijzer, C. P., et al. (2003). Metabolic labeling of *C. elegans* and *D. melanogaster* for quantitative proteomics. *Nature Biotechnology, 21*, 927–931.

Krug, K., Carpy, A., Behrends, G., Matic, K., Soares, N. C., & Macek, B. (2013). Deep coverage of the *Escherichia coli* proteome enables the assessment of false discovery rates in simple proteogenomic experiments. *Molecular & Cellular Proteomics, 12*, 3420–3430.

Kuhn, M. L., Zemaitaitis, B., Hu, L. I., Sahu, A., Sorensen, D., Minasov, G., et al. (2014). Structural, kinetic and proteomic characterization of acetyl phosphate-dependent bacterial protein acetylation. *PLoS One, 9*, e94816.

Kyte, J., & Doolittle, R. F. (1982). A simple method for displaying the hydropathic character of a protein. *Journal of Molecular Biology, 157*, 105–132.

Laganowsky, A., Reading, E., Allison, T. M., Ulmschneider, M. B., Degiacomi, M. T., Baldwin, A. J., et al. (2014). Membrane proteins bind lipids selectively to modulate their structure and function. *Nature, 510*, 172–175.

Laganowsky, A., Reading, E., Hopper, J. T., & Robinson, C. V. (2013). Mass spectrometry of intact membrane protein complexes. *Nature Protocols, 8*, 639–651.

Lai, E. M., Nair, U., Phadke, N. D., & Maddock, J. R. (2004). Proteomic screening and identification of differentially distributed membrane proteins in *Escherichia coli*. *Molecular Microbiology, 52*, 1029–1044.

Lam, H. T., Antonioli, P., Righetti, P. G., Citterio, A., & Girault, H. (2007). Gel-free IEF in a membrane-sealed multicompartment cell for proteome prefractionation. *Electrophoresis, 28*, 1860–1866.

Lancaster, W. A., Praissman, J. L., Poole, F. L., 2nd., Cvetkovic, A., Menon, A. L., Scott, J. W., et al. (2011). A computational framework for proteome-wide pursuit and prediction of metalloproteins using ICP-MS and MS/MS data. *BMC Bioinformatics, 12*, 64.

Lanciano, P., Vergnes, A., Grimaldi, S., Guigliarelli, B., & Magalon, A. (2007). Biogenesis of a respiratory complex is orchestrated by a single accessory protein. *The Journal of Biological Chemistry, 282*, 17468–17474.

Land, M., Hauser, L., Jun, S. R., Nookaew, I., Leuze, M. R., Ahn, T. H., et al. (2015). Insights from 20 years of bacterial genome sequencing. *Functional & Integrative Genomics, 15*, 141–161.

Landreh, M., & Robinson, C. V. (2015). A new window into the molecular physiology of membrane proteins. *The Journal of Physiology, 93*, 355–362.

Lanouette, S., Mongeon, V., Figeys, D., & Couture, J. F. (2014). The functional diversity of protein lysine methylation. *Molecular Systems Biology, 10*, 724.

Lanucara, F., Holman, S. W., Gray, C. J., & Eyers, C. E. (2014). The power of ion mobility-mass spectrometry for structural characterization and the study of conformational dynamics. *Nature Chemistry, 6*, 281–294.

Lapuente-Brun, E., Moreno-Loshuertos, R., Acín-Pérez, R., Latorre-Pellicer, A., Colás, C., Balsa, E., et al. (2013). Supercomplex assembly determines electron flux in the mitochondrial electron transport chain. *Science, 340*, 1567–1570.

Larbi, N. B., & Jefferies, C. (2009). 2D-DIGE: Comparative proteomics of cellular signalling pathways. *Methods in Molecular Biology, 517*, 105–132.

Lasker, K., Förster, F., Bohn, S., Walzthoeni, T., Villa, E., Unverdorben, P., et al. (2012). Molecular architecture of the 26S proteasome holocomplex determined by an integrative approach. *Proceedings of the National Academy of Sciences of the United States of America, 109*, 1380–1387.

Lasserre, J. P., Beyne, E., Pyndiah, S., Lapaillerie, D., Claverol, S., & Bonneu, M. (2006). A complexomic study of *Escherichia coli* using two-dimensional blue native/SDS polyacrylamide gel electrophoresis. *Electrophoresis, 27*, 3306–3321.

Leary, S. C. (2012). Blue native polyacrylamide gel electrophoresis: A powerful diagnostic tool for the detection of assembly defects in the enzyme complexes of oxidative phosphorylation. *Methods in Molecular Biology, 837*, 195–206.

LeDuc, R. D., Taylor, G. K., Kim, Y. B., Januszyk, T. E., Bynum, L. H., Sola, J. V., et al. (2004). ProSight PTM: An integrated environment for protein identification and characterization by top-down mass spectrometry. *Nucleic Acids Research, 32*, W340–W345.

Lee, A. G. (2003). Lipid-protein interactions in biological membranes: A structural perspective. *Biochimica et Biophysica Acta, 1612*, 1–40.

Lee, D. W., Kim, D., Lee, Y. J., Kim, J. A., Choi, J. Y., Kang, S., et al. (2013). Proteomic analysis of acetylation in thermophilic *Geobacillus kaustophilus*. *Proteomics, 13*, 2278–2282.

Lee, A. M., Sevinsky, J. R., Bundy, J. L., Grunden, A. M., & Stephenson, J. L., Jr. (2009). Proteomics of *Pyrococcus furiosus*, a hyperthermophilic archaeon refractory to traditional methods. *Journal of Proteome Research, 8*, 3844–3851.

Leichert, L. I., & Jakob, U. (2004). Protein thiol modifications visualized in vivo. *PLoS Biology, 2*, e333.

Leichert, L. I., & Jakob, U. (2006). Global methods to monitor the thiol-disulfide state of proteins in vivo. *Antioxidants & Redox Signaling, 8*, 763–772.

Leitner, A., Joachimiak, L. A., Unverdorben, P., Walzthoeni, T., Frydman, J., Förster, F., et al. (2014). Chemical cross-linking/mass spectrometry targeting acidic residues in proteins and protein complexes. *Proceedings of the National Academy of Sciences of the United States of America, A111*, 9455–9460.

Leitner, A., Walzthoeni, T., Kahraman, A., Herzog, F., Rinner, O., Beck, M., et al. (2010). Probing native protein structures by chemical cross-linking, mass spectrometry, and bioinformatics. *Molecular & Cellular Proteomics, 9*, 1634–1649.

Lemaire, C., & Dujardin, G. (2008). Preparation of respiratory chain complexes from *Saccharomyces cerevisiae* wild-type and mutant mitochondria: Activity measurement and subunit composition analysis. *Methods in Molecular Biology, 432*, 65–81.

Lenaz, G., & Genova, M. L. (2009). Mobility and function of coenzyme Q (ubiquinone) in the mitochondrial respiratory chain. *Biochimica et Biophysica Acta, 1787*, 563–573.

Lenn, T., & Leake, M. C. (2016). Single-molecule studies of the dynamics and interactions of bacterial OXPHOS complexes. *Biochimica et Biophysica Acta, 1857*(3), 224–231.

Lenn, T., Leake, M. C., & Mullineaux, C. W. (2008). Clustering and dynamics of cytochrome *bd*-I complexes in the *Escherichia coli* plasma membrane in vivo. *Molecular Microbiology, 70*, 1397–1407.

Leversen, N. A., de Souza, G. A., Målen, H., Prasad, S., Jonassen, I., & Wiker, H. G. (2009). Evaluation of signal peptide prediction algorithms for identification of mycobacterial signal peptides using sequence data from proteomic methods. *Microbiology, 155*, 2375–2383.

Liao, G., Xie, L., Li, X., Cheng, Z., & Xie, J. (2014). Unexpected extensive lysine acetylation in the trump-card antibiotic producer *Streptomyces roseosporus* revealed by proteome-wide profiling. *Journal of Proteomics, 106*, 260–269.

Lin, M. H., Hsu, T. L., Lin, S. Y., Pan, Y. J., Jan, J. T., Wang, J. T., et al. (2009). Phosphoproteomics of *Klebsiella pneumoniae* NTUH-K2044 reveals a tight link between tyrosine phosphorylation and virulence. *Molecular & Cellular Proteomics, 8*, 2613–2623.

Lin, M. H., Sugiyama, N., & Ishihama, Y. (2015). Systematic profiling of the bacterial phosphoproteome reveals bacterium-specific features of phosphorylation. *Science Signaling, 8*, rs10.

Lindahl, M., Mata-Cabana, A., & Kieselbach, T. (2011). The disulfide proteome and other reactive cysteine proteomes: Analysis and functional significance. *Antioxidants & Redox Signaling, 14*, 2581–2642.

Lindemann, C., & Leichert, L. I. (2012). Quantitative redox proteomics: The NOxICAT method. *Methods in Molecular Biology, 893*, 387–403.

Lindemann, C., Lupilova, N., Müller, A., Warscheid, B., Meyer, H. E., Kuhlmann, K., et al. (2013). Redox proteomics uncovers peroxynitrite-sensitive proteins that help *Escherichia coli* to overcome nitrosative stress. *The Journal of Biological Chemistry, 288*, 19698–19714.

Listinsky, J. J., Siegal, G. P., & Listinsky, C. M. (2013). Glycoengineering in cancer therapeutics: A review with fucose-depleted trastuzumab as the model. *Anti-Cancer Drugs, 24*, 219–227.

Liu, X., Gong, X., Hicks, D. B., Krulwich, T. A., Yu, L., & Yu, C. A. (2007). Interaction between cytochrome *caa*$_3$ and F_1F_0-ATP synthase of alkaliphilic *Bacillus pseudofirmus* OF4 is demonstrated by saturation transfer electron paramagnetic resonance and differential scanning calorimetry assays. *Biochemistry, 46*, 306–313.

Liu, X., Sirotkin, Y., Shen, Y., Anderson, G., Tsai, Y. S., Ting, Y. S., et al. (2012). Protein identification using top-down. *Molecular & Cellular Proteomics, 11*, M111008524.

Liu, T., Tian, C. F., & Chen, W. X. (2015). Site-specific Ser/Thr/Tyr phosphoproteome of *Sinorhizobium meliloti* at stationary phase. *PLoS One, 10*, e0139143.

Liu, F., Yang, M., Wang, X., Yang, S., Gu, J., Zhou, J., et al. (2014). Acetylome analysis reveals diverse functions of lysine acetylation in *Mycobacterium tuberculosis*. *Molecular & Cellular Proteomics, 13*, 3352–3366.

Llorente-Garcia, I., Lenn, T., Erhardt, H., Harriman, O. L., Liu, L. N., Robson, A., et al. (2014). Single-molecule in vivo imaging of bacterial respiratory complexes indicates delocalized oxidative phosphorylation. *Biochimica et Biophysica Acta, 1837*, 811–824.

Loi, V. V., Rossius, M., & Antelmann, H. (2015). Redox regulation by reversible protein *S*-thiolation in bacteria. *Frontiers in Microbiology, 6*, 187.

Loll, P. J. (2014). Membrane proteins, detergents and crystals: What is the state of the art? *Acta Crystallographica. Section F, Structural Biology Communications, 70*, 1576–1583.

Lopez-Campistrous, A., Semchuk, P., Burke, L., Palmer-Stone, T., Brokx, S. J., Broderick, G., et al. (2005). Localization, annotation, and comparison of the *Escherichia coli* K-12 proteome under two states of growth. *Molecular & Cellular Proteomics*, *4*, 1205–1209.

Lorenzi, M., Sylvi, L., Gerbaud, G., Mileo, E., Halgand, F., Walburger, A., et al. (2012). Conformational selection underlies recognition of a molybdoenzyme by its dedicated chaperone. *PLoS One*, 7, e49523.

Lu, B., McClatchy, D. B., Kim, J. Y., & Yates, J. R., 3rd. (2008). Strategies for shotgun identification of integral membrane proteins by tandem mass spectrometry. *Proteomics*, *8*, 3947–3955.

Luchini, A., Espina, V., & Liotta, L. A. (2014). Protein painting reveals solvent-excluded drug targets hidden within native protein-protein interfaces. *Nature Communications*, *5*, 4413.

Luirink, J., Yu, Z., Wagner, S., & de Gier, J. W. (2012). Biogenesis of inner membrane proteins in *Escherichia coli*. *Biochimica et Biophysica Acta*, *1817*, 965–976.

Lundgren, D. H., Hwang, S. I., Wu, L., & Han, D. K. (2010). Role of spectral counting in quantitative proteomics. *Expert Review of Proteomics*, 7, 39–53.

Lutter, P., Meyer, H. E., Langer, M., Witthohn, K., Dormeyer, W., Sickmann, A., et al. (2001). Investigation of charge variants of rViscumin by two-dimensional gel electrophoresis and mass spectrometry. *Electrophoresis*, *22*, 2888–2897.

Ma, B. (2015). Novor: Real-time peptide *de Novo* sequencing software. *Journal of the American Society for Mass Spectrometry*, *26*, 1885–1894.

Maas, M. F., Krause, F., Dencher, N. A., & Sainsard-Chanet, A. (2009). Respiratory complexes III and IV are not essential for the assembly/stability of complex I in fungi. *Journal of Molecular Biology*, *387*, 259–269.

Macek, B., Gnad, F., Soufi, B., Kumar, C., Olsen, J. V., Mijakovic, I., et al. (2008). Phosphoproteome analysis of *E. coli* reveals evolutionary conservation of bacterial Ser/Thr/Tyr phosphorylation. *Molecular & Cellular Proteomics*, 7, 299–307.

Macek, B., Mann, M., & Olsen, J. V. (2009). Global and site-specific quantitative phosphoproteomics: Principles and applications. *Annual Review of Pharmacology and Toxicology*, *49*, 199–221.

Macek, B., & Mijakovic, I. (2011). Site-specific analysis of bacterial phosphoproteomes. *Proteomics*, *11*, 3002–3011.

Macek, B., Mijakovic, I., Olsen, J. V., Gnad, F., Kumar, C., Jensen, P. R., et al. (2007). The serine/threonine/tyrosine phosphoproteome of the model bacterium *Bacillus subtilis*. *Molecular & Cellular Proteomics*, *6*, 697–707.

MacLean, B., Tomazela, D. M., Abbatiello, S. E., Zhang, S., Whiteaker, J. R., Paulovich, A. G., et al. (2010a). Effect of collision energy optimization on the measurement of peptides by selected reaction monitoring (SRM) mass spectrometry. *Analytical Chemistry*, *82*, 10116–10124.

MacLean, B., Tomazela, D. M., Shulman, N., Chambers, M., Finney, G. L., Frewen, B., et al. (2010b). Skyline: An open source document editor for creating and analyzing targeted proteomics experiments. *Bioinformatics*, *26*, 966–968.

Maddalo, G., Chovanec, P., Stenberg-Bruzell, F., Nielsen, H. V., Jensen-Seaman, M. I., Ilag, L. L., et al. (2011). A reference map of the membrane proteome of *Enterococcus faecalis*. *Proteomics*, *11*, 3935–3941.

Magalon, A., & Alberge, F. (2015). Distribution and dynamics of OXPHOS complexes in the bacterial cytoplasmic membrane. *Biochimica et Biophysica Acta*, *1857*(3), 198–213. http://dx.doi.org/10.1016/j.bbabio.2015.10.015. pii: S0005-2728(15)00222-4.

Magalon, A., Arias-Cartin, R., & Walburger, A. (2012). Supramolecular organization in prokaryotic respiratory systems. *Advances in Microbial Physiology*, *61*, 217–266.

Magalon, A., & Mendel, R. R. (2015). Biosynthesis and insertion of the molybdenum cofactor. *EcoSal Plus*, *6*(2). http://dx.doi.org/10.1128/ecosalplus.ESP-0006-2013.

Makarov, A. (2000). Electrostatic axially harmonic orbital trapping: A high-performance technique of mass analysis. *Analytical Chemistry*, *72*, 1156–1162.

Målen, H., Berven, F. S., Søfteland, T., Arntzen, M. Ø., D'Santos, C. S., De Souza, G. A., et al. (2008). Membrane and membrane-associated proteins in Triton X-114 extracts of *Mycobacterium bovis* BCG identified using a combination of gel-based and gel-free fractionation strategies. *Proteomics*, *8*, 1859–1870.

Mamyrin, B. A. (2001). Time-of-flight mass spectrometry (concepts, achievements, and prospects). *International Journal of Mass Spectrometry*, *206*, 251–266.

Mandell, J. G., Baerga-Ortiz, A., Croy, C. H., Falick, A. M., & Komives, E. A. (2005a). Application of amide proton exchange mass spectrometry for the study of protein-protein interactions. *Current Protocols in Protein Science*. Chapter 20:Unit 20.9.

Mandell, J. G., Baerga-Ortiz, A., Falick, A. M., & Komives, E. A. (2005b). Measurement of solvent accessibility at protein-protein interfaces. *Methods in Molecular Biology*, *305*, 65–80.

Maranzana, E., Barbero, G., Falasca, A. I., Lenaz, G., & Genova, M. L. (2013). Mitochondrial respiratory supercomplex association limits production of reactive oxygen species from complex I. *Antioxidants & Redox Signaling*, *19*, 1469–1480.

March, R. E. (1997). An introduction to quadrupole ion trap mass spectrometry. *Journal of Mass Spectrometry*, *32*, 351–369.

Marcoux, J., & Cianférani, S. (2015). Towards integrative structural mass spectrometry: Benefits from hybrid approaches. *Methods*, *89*, 4–12.

Maret, W. (2010). Metalloproteomics, metalloproteomes, and the annotation of metalloproteins. *Metallomics*, *2*, 117–125.

Marín-Buera, L., García-Bartolomé, A., Morán, M., López-Bernardo, E., Cadenas, S., Hidalgo, B., et al. (2015). Differential proteomic profiling unveils new molecular mechanisms associated with mitochondrial complex III deficiency. *Journal of Proteomics*, *113*, 38–56.

Marozava, S., Röling, W. F., Seifert, J., Küffner, R., von Bergen, M., & Meckenstock, R. U. (2014). Physiology of *Geobacter metallireducens* under excess and limitation of electron donors. Part I. Batch cultivation with excess of carbon sources. *Systematic and Applied Microbiology*, *37*, 277–286.

Marques, I., Dencher, N. A., Videira, A., & Krause, F. (2007). Supramolecular organization of the respiratory chain in *Neurospora crassa* mitochondria. *Eukaryotic Cell*, *6*, 2391–2405.

Marshall, A. G., Hendrickson, C. L., & Jackson, G. S. (1998). Fourier transform ion cyclotron resonance mass spectrometry: A primer. *Mass Spectrometry Reviews*, *17*, 1–35.

Martfeld, A. N., Rajagopalan, V., Greathouse, D. V., & Koeppe, R. E., 2nd. (2015). Dynamic regulation of lipid-protein interactions. *Biochimica et Biophysica Acta*, *1848*, 1849–1859.

Masuda, T., Saito, N., Tomita, M., & Ishihama, Y. (2009). Unbiased quantitation of *Escherichia coli* membrane proteome using phase transfer surfactants. *Molecular & Cellular Proteomics*, *8*, 2770–2777.

Mathy, G., & Sluse, F. E. (2008). Mitochondrial comparative proteomics: Strengths and pitfalls. *Biochimica et Biophysica Acta*, *1777*, 1072–1077.

Matias, P. M., Pereira, I. A., Soares, C. M., & Carrondo, M. A. (2005). Sulphate respiration from hydrogen in *Desulfovibrio* bacteria: A structural biology overview. *Progress in Biophysics and Molecular Biology*, *89*, 292–329.

Matsumoto, K., Hara, H., Fishov, I., Mileykovskaya, E., & Norris, V. (2015). The membrane: Transertion as an organizing principle in membrane heterogeneity. *Frontiers in Microbiology*, *6*, 572.

Matus-Ortega, M. G., Cárdenas-Monroy, C. A., Flores-Herrera, O., Mendoza-Hernández, G., Miranda, M., González-Pedrajo, B., et al. (2015). New complexes containing the internal alternative NADH dehydrogenase (Ndi1) in mitochondria of *Saccharomyces cerevisiae*. *Yeast*, *32*, 629–641.

Mawuenyega, K. G., Forst, C. V., Dobos, K. M., Belisle, J. T., Chen, J., Bradbury, E. M., et al. (2005). *Mycobacterium tuberculosis* functional network analysis by global subcellular protein profiling. *Molecular Biology of the Cell, 16*, 396–404.

May, J. C., & McLean, J. A. (2015). Ion mobility-mass spectrometry: Time-dispersive instrumentation. *Analytical Chemistry, 87*, 1422–1436.

McAlister, G. C., Huttlin, E. L., Haas, W., Ting, L., Jedrychowski, M. P., Rogers, J. C., et al. (2012). Increasing the multiplexing capacity of TMTs using reporter ion isotopologues with isobaric masses. *Analytical Chemistry, 84*, 7469–7478.

McDonald, C. A., Yang, J. Y., Marathe, V., Yen, T. Y., & Macher, B. A. (2009). Combining results from lectin affinity chromatography and glycocapture approaches substantially improves the coverage of the glycoproteome. *Molecular & Cellular Proteomics, 8*, 287–301.

McKenzie, M., Lazarou, M., Thorburn, D. R., & Ryan, M. T. (2007). Analysis of mitochondrial subunit assembly into respiratory chain complexes using Blue Native polyacrylamide gel electrophoresis. *Analytical Biochemistry, 364*, 128–137.

McLuckey, S. A., Van Berkel, G. J., Goeringer, D. E., & Glish, G. L. (1994). Ion trap mass spectrometry. Using high-pressure ionization. *Analytical Chemistry, 66*, 737A–743A.

Melo, A. M., & Teixeira, M. (2016). Supramolecular organization of bacterial aerobic respiratory chains: From cells and back. *Biochimica et Biophysica Acta, 1857*(3), 190–197.

Méndez-García, C., Peláez, A. I., Mesa, V., Sánchez, J., Golyshina, O. V., & Ferrer, M. (2015). Microbial diversity and metabolic networks in acid mine drainage habitats. *Frontiers in Microbiology, 6*, 475.

Menon, A. L., Poole, F. L., 2nd., Cvetkovic, A., Trauger, S. A., Kalisiak, E., Scott, J. W., et al. (2009). Novel multiprotein complexes identified in the hyperthermophilic archaeon *Pyrococcus furiosus* by non-denaturing fractionation of the native proteome. *Molecular & Cellular Proteomics, 8*, 735–751.

Meyer, K., & Selbach, M. (2015). Quantitative affinity purification mass spectrometry: A versatile technology to study protein-protein interactions. *Frontiers in Genetics, 6*, 237.

Mijakovic, I., & Deutscher, J. (2015). Protein-tyrosine phosphorylation in *Bacillus subtilis*: A 10-year retrospective. *Frontiers in Microbiology, 6*, 18.

Mijakovic, I., & Macek, B. (2012). Impact of phosphoproteomics on studies of bacterial physiology. *FEMS Microbiology Reviews, 36*, 877–892.

Mikkat, S., Fulda, S., & Hagemann, M. (2014). A 2D gel electrophoresis-based snapshot of the phosphoproteome in the cyanobacterium *Synechocystis* sp. strain PCC 6803. *Microbiology, 160*, 296–306.

Mileykovskaya, E., & Dowhan, W. (2014). Cardiolipin-dependent formation of mitochondria respiratory supercomplexes. *Chemistry and Physics of Lipids, 179*, 42–48.

Millea, K. M., Krull, I. S., Cohen, S. A., Gebler, J. C., & Berger, S. J. (2006). Integration of multidimensional chromatographic protein separations with a combined "top-down" and "bottom-up" proteomic strategy. *Journal of Proteome Research, 5*, 135–146.

Mimaki, M., Wang, X., McKenzie, M., Thorburn, D. R., & Ryan, M. T. (2012). Understanding mitochondrial complex I assembly in health and disease. *Biochimica et Biophysica Acta, 1817*, 851–862.

Minden, J. S. (2012). Two-dimensional difference gel electrophoresis. *Methods in Molecular Biology, 869*, 287–304.

Mitsopoulos, P., Chang, Y. H., Wai, T., König, T., Dunn, S. D., Langer, T., et al. (2015). Stomatin-like protein 2 is required for in vivo mitochondrial respiratory chain supercomplex formation and optimal cell function. *Molecular and Cellular Biology, 35*, 1838–1847.

Molloy, M. P. (2008). Isolation of bacterial cell membranes proteins using carbonate extraction. *Methods in Molecular Biology, 424*, 397–401.

Moore, K. E., Carlson, S. M., Camp, N. D., Cheung, P., James, R. G., Chua, K. F., et al. (2013). A general molecular affinity strategy for global detection and proteomic analysis of lysine methylation. *Molecular Cell, 50*, 444–456.

Moraes, I., Evans, G., Sanchez-Weatherby, J., Newstead, S., & Stewart, P. D. (2014). Membrane protein structure determination—The next generation. *Biochimica et Biophysica Acta, 1838*, 78–87.

Moreno-Beltrán, B., Díaz-Moreno, I., González-Arzola, K., Guerra-Castellano, A., Velázquez-Campoy, A., De la Rosa, M. A., et al. (2015). Respiratory complexes III and IV can each bind two molecules of cytochrome *c* at low ionic strength. *FEBS Letters, 589*, 476–483.

Morgner, N., Montenegro, F., Barrera, N. P., & Robinson, C. V. (2012). Mass spectrometry—From peripheral proteins to membrane motors. *Journal of Molecular Biology, 423*, 1–13.

Muster, B., Kohl, W., Wittig, I., Strecker, V., Joos, F., Haase, W., et al. (2010). Respiratory chain complexes in dynamic mitochondria display a patchy distribution in life cells. *PLoS One, 5*, e11910.

Nag, B., Arimilli, S., Koukis, B., Rhodes, E., Baichwal, V., & Sharma, S. D. (1994). Intramolecular charge heterogeneity in purified major histocompatibility class II alpha and beta polypeptide chains. *The Journal of Biological Chemistry, 269*, 10061–10070.

Nambi, S., Gupta, K., Bhattacharyya, M., Ramakrishnan, P., Ravikumar, V., Siddiqui, N., et al. (2013). Cyclic AMP-dependent protein lysine acylation in mycobacteria regulates fatty acid and propionate metabolism. *The Journal of Biological Chemistry, 288*, 14114–14124.

Neidhardt, F. C. (2011). How microbial proteomics got started. *Proteomics, 11*, 2943–2946.

Neumann, S., Wessels, H. J., Rijpstra, W. I., Sinninghe Damsté, J. S., Kartal, B., Jetten, M. S., et al. (2014). Isolation and characterization of a prokaryotic cell organelle from the anammox bacterium *Kuenenia stuttgartiensis*. *Molecular Microbiology, 94*, 794–802.

Nicke, T., Schnitzer, T., Münch, K., Adamczack, J., Haufschildt, K., Buchmeier, S., et al. (2013). Maturation of the cytochrome cd_1 nitrite reductase NirS from *Pseudomonas aeruginosa* requires transient interactions between the three proteins NirS, NirN and NirF. *Bioscience Reports, 33*, e00048.

Niebisch, A., & Bott, M. (2003). Purification of a cytochrome bc-aa_3 supercomplex with quinol oxidase activity from *Corynebacterium glutamicum*. Identification of a fourth subunity of cytochrome aa_3 oxidase and mutational analysis of diheme cytochrome c_1. *The Journal of Biological Chemistry, 278*, 4339–4346.

Niederman, R. A. (2013). Membrane development in purple photosynthetic bacteria in response to alterations in light intensity and oxygen tension. *Photosynthesis Research, 116*, 333–348.

Nijtmans, L. G., Huynen, M. A., & Thorburn, D. R. (2013). Mutations in the UQCC1-interacting protein, UQCC2, cause human complex III deficiency associated with perturbed cytochrome *b* protein expression. *PLoS Genetics, 9*, e1004034.

Niu, S., & Ruotolo, B. T. (2015). Collisional unfolding of multiprotein complexes reveals cooperative stabilization upon ligand binding. *Protein Science, 24*, 1272–1281.

Nothwang, H. G., & Schindler, J. (2009). Two-dimensional separation of membrane proteins by 16-BAC-SDS-PAGE. *Methods in Molecular Biology, 528*, 269–277.

Nouws, J., Nijtmans, L., Houten, S. M., van den Brand, M., Huynen, M., Venselaar, H., et al. (2010). Acyl-CoA dehydrogenase 9 is required for the biogenesis of oxidative phosphorylation complex I. *Cell Metabolism, 12*, 283–294.

Nowak, G., & Bakajsova, D. (2015). Protein kinase C-α interaction with F_0F_1-ATPase promotes F_0F_1-ATPase activity and reduces energy deficits in injured renal cells. *The Journal of Biological Chemistry, 290*, 7054–7066.

Nübel, E., Wittig, I., Kerscher, S., Brandt, U., & Schägger, H. (2009). Two-dimensional native electrophoretic analysis of respiratory supercomplexes from *Yarrowia lipolytica*. *Proteomics, 9*, 2408–2418.

Nyon, M. P., Prentice, T., Day, J., Kirkpatrick, J., Sivalingam, G. N., Levy, G., et al. (2015). An integrative approach combining ion mobility mass spectrometry, X-ray crystallography, and nuclear magnetic resonance spectroscopy to study the conformational dynamics of α1-antitrypsin upon ligand binding. *Protein Science, 24*, 1301–1312.

O'Farrell, P. H. (1975). High resolution two-dimensional electrophoresis of proteins. *The Journal of Biological Chemistry, 250*, 4007–4021.

Okanishi, H., Kim, K., Masui, R., & Kuramitsu, S. (2013). Acetylome with structural mapping reveals the significance of lysine acetylation in *Thermus thermophilus*. *Journal of Proteome Research, 12*, 3952–3968.

Oliveira, B. M., Coorssen, J. R., & Martins-de-Souza, D. (2014). 2DE: The phoenix of proteomics. *Journal of Proteomics, 104*, 140–150.

Oliveira, A. P., & Picotti, P. (2014). Global analysis of protein structural changes in complex proteomes. *Nature Biotechnology, 32*, 1036–1044.

Oliver, P. M., Crooks, J. A., Leidl, M., Yoon, E. J., Saghatelian, A., & Weibel, D. B. (2014). Localization of anionic phospholipids in *Escherichia coli* cells. *Journal of Bacteriology, 196*, 3386–3398.

Olsen, J. V., & Macek, B. (2009). High accuracy mass spectrometry in large-scale analysis of protein phosphorylation. *Methods in Molecular Biology, 492*, 131–142.

Olsen, J. V., & Mann, M. (2013). Status of large-scale analysis of post-translational modifications by mass spectrometry. *Molecular & Cellular Proteomics, 12*, 3444–3452.

Ong, S. E., Blagoev, B., Kratchmarova, I., Kristensen, D. B., Steen, H., Pandey, A., et al. (2002). Stable isotope labeling by amino acids in cell culture, SILAC, as a simple and accurate approach to expression proteomics. *Molecular & Cellular Proteomics, 1*, 376–386.

Orellana, R., Hixson, K. K., Murphy, S., Mester, T., Sharma, M. L., Lipton, M. S., et al. (2014). Proteome of *Geobacter sulfurreducens* in the presence of U(VI). *Microbiology, 160*, 2607–2617.

Oren, A., & Garrity, G. M. (2014). Then and now: A systematic review of the systematics of prokaryotes in the last 80 years. *Antonie Van Leeuwenhoek, 106*, 43–56.

Osborne, B., Cooney, G. J., & Turner, N. (2014). Are sirtuin deacylase enzymes important modulators of mitochondrial energy metabolism? *Biochimica et Biophysica Acta, 1840*, 1295–1302.

Owens, R. (2011). Methods in structural proteomics. *Methods, 55*, 1–2.

Paetzel, M. (2014). Structure and mechanism of *Escherichia coli* type I signal peptidase. *Biochimica et Biophysica Acta, 1843*, 1497–1508.

Pagliarini, D. J., & Dixon, J. E. (2006). Mitochondrial modulation: Reversible phosphorylation takes center stage? *Trends in Biochemical Sciences, 31*, 26–34.

Palmisano, G., Sardanelli, A. M., Signorile, A., Papa, S., & Larsen, M. R. (2007). The phosphorylation pattern of bovine heart complex I subunits. *Proteomics*, 7, 1575–1583.

Pan, J., Ye, Z., Cheng, Z., Peng, X., Wen, L., & Zhao, F. (2014). Systematic analysis of the lysine acetylome in *Vibrio parahemolyticus*. *Journal of Proteome Research, 13*, 3294–3302.

Pang, C. N., Gasteiger, E., & Wilkins, M. R. (2010). Identification of arginine- and lysine-methylation in the proteome of *Saccharomyces cerevisiae* and its functional implications. *BMC Genomics, 11*, 92.

Papanicolaou, K. N., O'Rourke, B., & Foster, D. B. (2014). Metabolism leaves its mark on the powerhouse: Recent progress in post-translational modifications of lysine in mitochondria. *Frontiers in Physiology, 5*, 301.

Pappin, D. J., Hojrup, P., & Bleasby, A. J. (1993). Rapid identification of proteins by peptide-mass fingerprinting. *Current Biology, 3*, 327–332.

Paradies, G., Paradies, V., De Benedictis, V., Ruggiero, F. M., & Petrosillo, G. (2014). Functional role of cardiolipin in mitochondrial bioenergetics. *Biochimica et Biophysica Acta, 1837*, 408–417.

Parker, J. L., Jones, A. M., Serazetdinova, L., Saalbach, G., Bibb, M. J., & Naldrett, M. J. (2010). Analysis of the phosphoproteome of the multicellular bacterium *Streptomyces coelicolor* A3(2) by protein/peptide fractionation, phosphopeptide enrichment and high-accuracy mass spectrometry. *Proteomics, 10*, 2486–2497.

Paton, L. N., Gerrard, J. A., & Bryson, W. G. (2008). Investigations into charge heterogeneity of wool intermediate filament proteins. *Journal of Proteomics, 71*, 513–529.

Pedrioli, P. G. (2010). Trans-proteomic pipeline: A pipeline for proteomic analysis. *Methods in Molecular Biology, 604*, 213–238.

Pereira, S. F., Goss, L., & Dworkin, J. (2011). Eukaryote-like serine/threonine kinases and phosphatases in bacteria. *Microbiology and Molecular Biology Reviews, 75*, 192–212.

Perkins, D. N., Pappin, D. J., Creasy, D. M., & Cottrell, J. S. (1999). Probability-based protein identification by searching sequence databases using mass spectrometry data. *Electrophoresis, 20*, 3551–3567.

Perry, R. H., Cooks, R. G., & Noll, R. J. (2008). Orbitrap mass spectrometry: Instrumentation, ion motion and applications. *Mass Spectrometry Reviews, 27*, 661–699.

Peterman, S. M., Dufresne, C. P., & Horning, S. (2005). The use of a hybrid linear trap/FT-ICR mass spectrometer for on-line high resolution/high mass accuracy bottom-up sequencing. *Journal of Biomolecular Techniques, 16*, 112–124.

Peters, J. W., Schut, G. J., Boyd, E. S., Mulder, D. W., Shepard, E. M., Broderick, J. B., et al. (2015). [FeFe]- and [NiFe]-hydrogenase diversity, mechanism, and maturation. *Biochimica et Biophysica Acta, 1853*, 1350–1369.

Petrotchenko, E. V., & Borchers, C. H. (2010). Crosslinking combined with mass spectrometry for structural proteomics. *Mass Spectrometry Reviews, 29*, 862–876.

Petrotchenko, E. V., & Borchers, C. H. (2014). Modern mass spectrometry-based structural proteomics. *Advances in Protein Chemistry and Structural Biology, 95*, 193–213.

Pevtsov, S., Fedulova, I., Mirzaei, H., Buck, C., & Zhang, X. (2006). Performance evaluation of existing de novo sequencing algorithms. *Journal of Proteome Research, 5*, 3018–3028.

Phillips, N. J., Steichen, C. T., Schilling, B., Post, D. M., Niles, R. K., Bair, T. B., et al. (2012). Proteomic analysis of *Neisseria gonorrhoeae* biofilms shows shift to anaerobic respiration and changes in nutrient transport and outermembrane proteins. *PLoS One, 7*, e38303.

Planas-Iglesias, J., Dwarakanath, H., Mohammadyani, D., Yanamala, N., Kagan, V. E., & Klein-Seetharaman, J. (2015). Cardiolipin interactions with proteins. *Biophysical Journal, 109*, 1282–1294.

Pocsfalvi, G., Cuccurullo, M., Schlosser, G., Cacace, G., Siciliano, R. A., Mazzeo, M. F., et al. (2006). Shotgun proteomics for the characterization of subunit composition of mitochondrial complex I. *Biochimica et Biophysica Acta, 1757*, 1438–1450.

Poetsch, A., & Wolters, D. (2008). Bacterial membrane proteomics. *Proteomics, 8*, 4100–4122.

Pohl, S., Tu, W. Y., Aldridge, P. D., Gillespie, C., Hahne, H., Mäder, U., et al. (2011). Combined proteomic and transcriptomic analysis of the response of *Bacillus anthracis* to oxidative stress. *Proteomics, 11*, 3036–3055.

Poland, T., Rabilloud, T., & Sihna, P. (2005). Silver staining of 2D gels. In J. M. Walker (Ed.), *The proteomics protocols handbook* (pp. 177–184). Totowa, NJ: Human Press.

Ponomarova, O., & Patil, K. R. (2015). Metabolic interactions in microbial communities: Untangling the Gordian knot. *Current Opinion in Microbiology, 27*, 37–44.

Price, C. E., & Driessen, A. J. (2010). Biogenesis of membrane bound respiratory complexes in *Escherichia coli*. *Biochimica et Biophysica Acta, 1803*, 748–766.

Prime, T. A., Blaikie, F. H., Evans, C., Nadtochiy, S. M., James, A. M., Dahm, C. C., et al. (2009). A mitochondria-targeted *S*-nitrosothiol modulates respiration, nitrosates thiols, and protects against ischemia-reperfusion injury. *Proceedings of the National Academy of Sciences of the United States of America, 106*, 10764–10769.

Prisic, S., Dankwa, S., Schwartz, D., Chou, M. F., Locasale, J. W., Kang, C. M., et al. (2010). Extensive phosphorylation with overlapping specificity by *Mycobacterium tuberculosis* serine/threonine protein kinases. *Proceedings of the National Academy of Sciences of the United States of America*, *107*, 7521–7526.

Prunetti, L., Infossi, P., Brugna, M., Ebel, C., Giudici-Orticoni, M. T., & Guiral, M. (2010). New functional sulfide oxidase-oxygen reductase supercomplex in the membrane of the hyperthermophilic bacterium *Aquifex aeolicus*. *The Journal of Biological Chemistry*, *285*, 41815–41826.

Puig, O., Caspary, F., Rigaut, G., Rutz, B., Bouveret, E., Bragado-Nilsson, E., et al. (2001). The tandem affinity purification (TAP) method: A general procedure of protein complex purification. *Methods*, *24*, 218–229.

Puts, C. F., Lenoir, G., Krijgsveld, J., Williamson, P., & Holthuis, J. C. (2010). A P4-ATPase protein interaction network reveals a link between aminophospholipid transport and phosphoinositide metabolism. *Journal of Proteome Research*, *9*, 833–842.

Pyndiah, S., Lasserre, J. P., Ménard, A., Claverol, S., Prouzet-Mauléon, V., Mégraud, F., et al. (2007). Two-dimensional blue native/SDS gel electrophoresis of multiprotein complexes from *Helicobacter pylori*. *Molecular & Cellular Proteomics*, *6*, 193–206.

Rabilloud, T. (2009). Membrane proteins and proteomics: Love is possible, but so difficult. *Electrophoresis*, *30*(Suppl. 1), S174–S180.

Rabilloud, T., Chevallet, M., Luche, S., & Lelong, C. (2008a). Fully denaturing two-dimensional electrophoresis of membrane proteins: A critical update. *Proteomics*, *8*, 3965–3973.

Rabilloud, T., Chevallet, M., Luche, S., & Lelong, C. (2008b). Two-dimensional gel electrophoresis in proteomics: Past, present and future. *Journal of Proteomics*, *73*, 2064–2077.

Rabuck, J. N., Hyung, S. J., Ko, K. S., Fox, C. C., Soellner, M. B., & Ruotolo, B. T. (2013). Activation state-selective kinase inhibitor assay based on ion mobility-mass spectrometry. *Analytical Chemistry*, *85*, 6995–7002.

Ramírez-Aguilar, S. J., Keuthe, M., Rocha, M., Fedyaev, V. V., Kramp, K., Gupta, K. J., et al. (2011). The composition of plant mitochondrial supercomplexes changes with oxygen availability. *The Journal of Biological Chemistry*, *286*, 43045–43053.

Ramisetty, S. R., & Washburn, M. P. (2011). Unraveling the dynamics of protein interactions with quantitative mass spectrometry. *Critical Reviews in Biochemistry and Molecular Biology*, *46*, 216–228.

Rand, K. D., Zehl, M., & Jørgensen, T. J. (2014). Measuring the hydrogen/deuterium exchange of proteins at high spatial resolution by mass spectrometry: Overcoming gas-phase hydrogen/deuterium scrambling. *Accounts of Chemical Research*, *47*, 3018–3027.

Rappsilber, J. (2011). The beginning of a beautiful friendship: Cross-linking/mass spectrometry and modelling of proteins and multi-protein complexes. *Journal of Structural Biology*, *173*, 530–540.

Rappsilber, J., & Mann, M. (2002). What does it mean to identify a protein in proteomics? *Trends in Biochemical Sciences*, *27*, 74–78.

Rappsilber, J., Siniossoglou, S., Hurt, E. C., & Mann, M. (2000). A generic strategy to analyze the spatial organization of multi-protein complexes by cross-linking and mass spectrometry. *Analytical Chemistry*, *72*, 267–275.

Ravichandran, A., Sugiyama, N., Tomita, M., Swarup, S., & Ishihama, Y. (2009). Ser/Thr/Tyr phosphoproteome analysis of pathogenic and non-pathogenic *Pseudomonas* species. *Proteomics*, *9*, 2764–2775.

Ravikumar, V., Shi, L., Krug, K., Derouiche, A., Jers, C., Cousin, C., et al. (2014). Quantitative phosphoproteome analysis of *Bacillus subtilis* reveals novel substrates of the kinase PrkC and phosphatase PrpC. *Molecular & Cellular Proteomics*, *13*, 1965–1978.

Reiland, S., Messerli, G., Baerenfaller, K., Gerrits, B., Endler, A., Grossmann, J., et al. (2009). Large-scale *Arabidopsis* phosphoproteome profiling reveals novel chloroplast kinase substrates and phosphorylation networks. *Plant Physiology*, *150*, 889–903.

Reinders, J., Wagner, K., Zahedi, R. P., Stojanovski, D., Eyrich, B., van der Laan, M., et al. (2007). Profiling phosphoproteins of yeast mitochondria reveals a role of phosphorylation in assembly of the ATP synthase. *Molecular & Cellular Proteomics*, *6*, 1896–1906.

Remmerie, N., De Vijlder, T., Valkenborg, D., Laukens, K., Smets, K., Vreeken, J., et al. (2011). Unraveling tobacco BY-2 protein complexes with BN PAGE/LC-MS/MS and clustering methods. *Journal of Proteomics*, *74*, 1201–1217.

Renier, S., Chafsey, I., Chambon, C., Caccia, N., Charbit, A., Hébraud, M., et al. (2015). Contribution of the multiple Type I signal peptidases to the secretome of *Listeria monocytogenes*: Deciphering their specificity for secreted exoproteins by exoproteomic analysis. *Journal of Proteomics*, *117*, 95–105.

Rhein, V. F., Carroll, J., Ding, S., Fearnley, I. M., & Walker, J. E. (2013). NDUFAF7 methylates arginine 85 in the NDUFS2 subunit of human complex I. *The Journal of Biological Chemistry*, *288*, 33016–33026.

Richter, K., Schicklberger, M., & Gescher, J. (2012). Dissimilatory reduction of extracellular electron acceptors in anaerobic respiration. *Applied and Environmental Microbiology*, *78*, 913–921.

Rigaut, G., Shevchenko, A., Rutz, B., Wilm, M., Mann, M., & Séraphin, B. (1999). A generic protein purification method for protein complex characterization and proteome exploration. *Nature Biotechnology*, *17*, 1030–1032.

Robinson, C., Matos, C. F., Beck, D., Ren, C., Lawrence, J., Vasisht, N., et al. (2011). Transport and proofreading of proteins by the twin-arginine translocation (Tat) system in bacteria. *Biochimica et Biophysica Acta*, *1808*, 876–884.

Robinson, N. J., & Winge, D. R. (2010). Copper metallochaperones. *Annual Review of Biochemistry*, *79*, 537–562.

Robledo, V. R., & Smyth, W. F. (2014). Review of the CE-MS platform as a powerful alternative to conventional couplings in bio-omics and target-based applications. *Electrophoresis*, *35*, 2292–2308.

Roepstorff, P., & Fohlman, J. (1984). Proposal for a common nomenclature for sequence ions in mass spectra of peptides. *Biomedical Mass Spectrometry*, *11*, 601.

Rosca, M. G., Vazquez, E. J., Kerner, J., Parland, W., Chandler, M. P., Stanley, W., et al. (2008). Cardiac mitochondria in heart failure: Decrease in respirasomes and oxidative phosphorylation. *Cardiovascular Research*, *80*, 30–39.

Ross, P. L., Huang, Y. N., Marchese, J. N., Williamson, B., Parker, K., Hattan, S., et al. (2004). Multiplexed protein quantitation in *Saccharomyces cerevisiae* using amine-reactive isobaric tagging reagents. *Molecular & Cellular Proteomics*, *3*, 1154–1169.

Rybak, J. N., Scheurer, S. B., Neri, D., & Elia, G. (2004). Purification of biotinylated proteins on streptavidin resin: A protocol for quantitative elution. *Proteomics*, *4*, 2296–2299.

Ryšlavá, H., Doubnerová, V., Kavan, D., & Vaněk, O. (2013). Effect of posttranslational modifications on enzyme function and assembly. *Journal of Proteomics*, *92*, 80–109.

Saada, A., Vogel, R. O., Hoefs, S. J., van den Brand, M. A., Wessels, H. J., Willems, P. H., et al. (2009). Mutations in NDUFAF3 (C3ORF60), encoding an NDUFAF4 (C6ORF66)-interacting complex I assembly protein, cause fatal neonatal mitochondrial disease. *American Journal of Human Genetics*, *84*, 718–727.

Saddar, S., Dienhart, M. K., & Stuart, R. A. (2008). The F_1F_0-ATP synthase complex influences the assembly state of the cytochrome bc_1-cytochrome oxidase supercomplex and its association with the TIM23 machinery. *The Journal of Biological Chemistry*, *283*, 6677–6686.

Sakakura, M., & Takayama, M. (2010). In-source decay and fragmentation characteristics of peptides using 5-aminosalicylic acid as a matrix in matrix-assisted laser desorption/ionization mass spectrometry. *Journal of the American Society for Mass Spectrometry*, *21*, 979–988.

Sala, A., Bordes, P., & Genevaux, P. (2014). Multitasking SecB chaperones in bacteria. *Frontiers in Microbiology*, *5*, 666.

Santoni, V., Molloy, M., & Rabilloud, T. (2000). Membrane proteins and proteomics: Un amour impossible? *Electrophoresis*, *21*, 1054–1070.

Sapra, R. (2009). The use of difference in-gel electrophoresis for quantitation of protein expression. *Methods in Molecular Biology*, *492*, 93–112.

Saraogi, I., & Shan, S. O. (2014). Co-translational protein targeting to the bacterial membrane. *Biochimica et Biophysica Acta*, *1843*, 1433–1441.

Sargent, F. (2007). Constructing the wonders of the bacterial world: Biosynthesis of complex enzymes. *Microbiology*, *153*, 633–651.

Savas, J. N., Stein, B. D., Wu, C. C., & Yates, J. R., 3rd. (2011). Mass spectrometry accelerates membrane protein analysis. *Trends in Biochemical Sciences*, *36*, 388–396.

Schäfer, E., Dencher, N. A., Vonck, J., & Parcej, D. N. (2007). Three-dimensional structure of the respiratory chain supercomplex I1III2IV1 from bovine heart mitochondria. *Biochemistry*, *46*, 12579–12585.

Schägger, H. (2002). Respiratory chain supercomplexes of mitochondria and bacteria. *Biochimica et Biophysica Acta*, *1555*, 154–159.

Schägger, H., Aquila, H., & Von Jagow, G. (1988). Coomassie blue-sodium dodecyl sulfate-polyacrylamide gel electrophoresis for direct visualization of polypeptides during electrophoresis. *Analytical Biochemistry*, *173*, 201–205.

Schägger, H., & Pfeiffer, K. (2000). Supercomplexes in the respiratory chains of yeast and mammalian mitochondria. *The EMBO Journal*, *19*, 1777–1783.

Schägger, H., & von Jagow, G. (1991). Blue native electrophoresis for isolation of membrane protein complexes in enzymatically active form. *Analytical Biochemistry*, *199*, 223–231.

Scherl, A., François, P., Bento, M., Deshusses, J. M., Charbonnier, Y., Converset, V., et al. (2005). Correlation of proteomic and transcriptomic profiles of *Staphylococcus aureus* during the post-exponential phase of growth. *Journal of Microbiological Methods*, *60*, 247–257.

Scheurer, S. B., Rybak, J. N., Roesli, C., Brunisholz, R. A., Potthast, F., Schlapbach, R., et al. (2005). Identification and relative quantification of membrane proteins by surface biotinylation and two-dimensional peptide mapping. *Proteomics*, *5*, 2718–2728.

Schiapparelli, L. M., McClatchy, D. B., Liu, H. H., Sharma, P., Yates, J. R., 3rd., & Cline, H. T. (2014). Direct detection of biotinylated proteins by mass spectrometry. *Journal of Proteome Research*, *13*, 3966–3978.

Schluesener, D., Fischer, F., Kruip, J., Rögner, M., & Poetsch, A. (2005). Mapping the membrane proteome of *Corynebacterium glutamicum*. *Proteomics*, *5*, 1317–1330.

Schmidl, S. R., Gronau, K., Pietack, N., Hecker, M., Becher, D., & Stülke, J. (2010). The phosphoproteome of the minimal bacterium *Mycoplasma pneumoniae*: Analysis of the complete known Ser/Thr kinome suggests the existence of novel kinases. *Molecular & Cellular Proteomics*, *9*, 1228–1242.

Schmidt, A., Müller, N., Schink, B., & Schleheck, D. (2013). A proteomic view at the biochemistry of syntrophic butyrate oxidation in *Syntrophomonas wolfei*. *PLoS One*, *8*, e56905.

Schmidt, C., Zhou, M., Marriott, H., Morgner, N., Politis, A., & Robinson, C. V. (2013). Comparative cross-linking and mass spectrometry of an intact F-type ATPase suggest a role for phosphorylation. *Nature Communications*, *4*, 1985.

Schmitt, E., Naveau, M., & Mechulam, Y. (2010). Eukaryotic and archaeal translation initiation factor 2: A heterotrimeric tRNA carrier. *FEBS Letters*, *584*, 405–412.

Schroeter, R., Voigt, B., Jürgen, B., Methling, K., Pöther, D. C., Schäfer, H., et al. (2011). The peroxide stress response of *Bacillus licheniformis*. *Proteomics*, *11*, 2851–2866.
Schubert, H. L., Blumenthal, R. M., & Cheng, X. (2003). Many paths to methyltransfer: A chronicle of convergence. *Trends in Biochemical Sciences*, *28*, 329–335.
Seidler, J., Zinn, N., Boehm, M. E., & Lehmann, W. D. (2010). *De novo* sequencing of peptides by MS/MS. *Proteomics*, *10*, 634–649.
Sestak, J., Moravcova, D., & Kahle, V. (2015). Instrument platforms for nano liquid chromatography. *Journal of Chromatography*, *A1421*, 2–17.
Sevcenco, A. M., Hagen, W. R., & Hagedoorn, P. L. (2012). Microbial metalloproteomes explored using MIRAGE. *Chemistry & Biodiversity*, *9*, 1967–1980.
Shan, Y., Lai, Y., & Yan, A. (2012). Metabolic reprogramming under microaerobic and anaerobic conditions in bacteria. *Sub-Cellular Biochemistry*, *64*, 159–179.
Shi, W., & Chance, M. R. (2011). Metalloproteomics: Forward and reverse approaches in metalloprotein structural and functional characterization. *Current Opinion in Chemical Biology*, *15*, 144–148.
Shi, L., Pigeonneau, N., Ravikumar, V., Dobrinic, P., Macek, B., Franjevic, D., et al. (2014). Cross-phosphorylation of bacterial serine/threonine and tyrosine protein kinases on key regulatory residues. *Frontiers in Microbiology*, *5*, 495.
Shi, L., & Tu, B. P. (2015). Acetyl-CoA and the regulation of metabolism: Mechanisms and consequences. *Current Opinion in Cell Biology*, *33*, 125–131.
Shoubridge, E. A. (2012). Supersizing the mitochondrial respiratory chain. *Cell Metabolism*, *15*, 271–272.
Sieber, J. R., Crable, B. R., Sheik, C. S., Hurst, G. B., Rohlin, L., Gunsalus, R. P., et al. (2015). Proteomic analysis reveals metabolic and regulatory systems involved in the syntrophic and axenic lifestyle of *Syntrophomonas wolfei*. *Frontiers in Microbiology*, *6*, 115.
Simon, J., & Kern, M. (2008). Quinone-reactive proteins devoid of haem *b* form widespread membrane-bound electron transport modules in bacterial respiration. *Biochemical Society Transactions*, *36*, 1011–1016.
Sinha, S., Kosalai, K., Arora, S., Namane, A., Sharma, P., Gaikwad, A. N., et al. (2005). Immunogenic membrane-associated proteins of *Mycobacterium tuberculosis* revealed by proteomics. *Microbiology*, *151*, 2411–2419.
Sinz, A., Arlt, C., Chorev, D., & Sharon, M. (2015). Chemical cross-linking and native mass spectrometry: A fruitful combination for structural biology. *Protein Science*, *24*, 1193–1209.
Smith, D. R., Doucette-Stamm, L. A., Deloughery, C., Lee, H., Dubois, J., Aldredge, T., et al. (1997). Complete genome sequence of *Methanobacterium thermoautotrophicum* deltaH: Functional analysis and comparative genomics. *Journal of Bacteriology*, *179*, 7135–7155.
Smith, P. M., Fox, J. L., & Winge, D. R. (2012). Biogenesis of the cytochrome bc_1 complex and role of assembly factors. *Biochimica et Biophysica Acta*, *1817*, 276–286.
Smith, L. M., Kelleher, N. L., & Consortium for Top Down Proteomics (2013). Proteoform: A single term describing protein complexity. *Nature Methods*, *10*, 186–187.
Snijders, A. P., Hung, M. L., Wilson, S. A., & Dickman, M. J. (2010). Analysis of arginine and lysine methylation utilizing peptide separations at neutral pH and electron transfer dissociation mass spectrometry. *Journal of the American Society for Mass Spectrometry*, *21*, 88–96.
Soares, N. C., Spät, P., Krug, K., & Macek, B. (2013). Global dynamics of the *Escherichia coli* proteome and phosphoproteome during growth in minimal medium. *Journal of Proteome Research*, *12*, 2611–2621.
Sokolova, L., Wittig, I., Barth, H. D., Schägger, H., Brutschy, B., & Brandt, U. (2010). Laser-induced liquid bead ion desorption-MS of protein complexes from blue-native gels, a sensitive top-down proteomic approach. *Proteomics*, *10*, 1401–1407.

Soppa, J. (2010). Protein acetylation in archaea, bacteria, and eukaryotes. *Archaea, 2010*. Article ID 820681.

Soufi, B., Gnad, F., Jensen, P. R., Petranovic, D., Mann, M., Mijakovic, I., et al. (2008). The Ser/Thr/Tyr phosphoproteome of *Lactococcus lactis* IL1403 reveals multiply phosphorylated proteins. *Proteomics*, *8*, 3486–3493.

Soufi, B., Krug, K., Harst, A., & Macek, B. (2015). Characterization of the *E. coli* proteome and its modifications during growth and ethanol stress. *Frontiers in Microbiology*, *6*, 103.

Soufi, B., & Macek, B. (2015). Global analysis of bacterial membrane proteins and their modifications. *International Journal of Medical Microbiology*, *305*, 203–208.

Soufi, B., Soares, N. C., Ravikumar, V., & Macek, B. (2012). Proteomics reveals evidence of cross-talk between protein modifications in bacteria: Focus on acetylation and phosphorylation. *Current Opinion in Microbiology*, *15*, 357–363.

Soung, G. Y., Miller, J. L., Koc, H., & Koc, E. C. (2009). Comprehensive analysis of phosphorylated proteins of *Escherichia coli* ribosomes. *Journal of Proteome Research*, *8*, 3390–3402.

Sousa, P. M., Silva, S. T., Hood, B. L., Charro, N., Carita, J. N., Vaz, F., et al. (2011). Supramolecular organizations in the aerobic respiratory chain of *Escherichia coli*. *Biochimie*, *93*, 418–425.

Sousa, P. M., Videira, M. A., Bohn, A., Hood, B. L., Conrads, T. P., Goulao, L. F., et al. (2012). The aerobic respiratory chain of *Escherichia coli*: From genes to supercomplexes. *Microbiology*, *158*, 2408–2418.

Sousa, P. M., Videira, M. A., & Melo, A. M. (2013a). The formate:oxygen oxidoreductase supercomplex of *Escherichia coli* aerobic respiratory chain. *FEBS Letters*, *587*, 2559–2564.

Sousa, P. M., Videira, M. A., Santos, F. A., Hood, B. L., Conrads, T. P., & Melo, A. M. (2013b). The *bc*:caa$_3$ supercomplexes from the Gram positive bacterium *Bacillus subtilis* respiratory chain: A megacomplex organization? *Archives of Biochemistry and Biophysics*, *537*, 153–160.

Spät, P., Maček, B., & Forchhammer, K. (2015). Phosphoproteome of the cyanobacterium *Synechocystis* sp. PCC 6803 and its dynamics during nitrogen starvation. *Frontiers in Microbiology*, *6*, 248.

Spiro, S. (2012). Nitrous oxide production and consumption: Regulation of gene expression by gas-sensitive transcription factors. *Philosophical Transactions of the Royal Society of London. Series B, Biological Sciences*, *367*, 1213–1225.

Srinivasa, S., Ding, X., & Kast, J. (2015). Formaldehyde cross-linking and structural proteomics: Bridging the gap. *Methods*, *89*, 91–98.

Stangl, M., & Schneider, D. (2015). Functional competition within a membrane: Lipid recognition vs. transmembrane helix oligomerization. *Biochimica et Biophysica Acta*, *1848*, 1886–1896.

Stemmann, O., Zou, H., Gerber, S. A., Gygi, S. P., & Kirschner, M. W. (2001). Dual inhibition of sister chromatid separation at metaphase. *Cell*, *107*, 715–726.

Stenberg, F., Chovanec, P., Maslen, S. L., Robinson, C. V., Ilag, L. L., Heijne, G., et al. (2005). Protein complexes of the *Escherichia coli* cell envelope. *The Journal of Biological Chemistry*, *280*, 34409–34419.

Stojko, J., Fieulaine, S., Petiot-Bécard, S., Van Dorsselaer, A., Meinnel, T., Giglione, C., et al. (2015). Ion mobility coupled to native mass spectrometry as a relevant tool to investigate extremely small ligand-induced conformational changes. *Analyst*, *140*, 7234–7245.

Strader, M. B., Verberkmoes, N. C., Tabb, D. L., Connelly, H. M., Barton, J. W., Bruce, B. D., et al. (2004). Characterization of the 70S Ribosome from *Rhodopseudomonas palustris* using an integrated "top-down" and "bottom-up" mass spectrometric approach. *Journal of Proteome Research*, *3*, 965–978.

Strecker, V., Wumaier, Z., Wittig, I., & Schägger, H. (2010). Large pore gels to separate mega protein complexes larger than 10 MDa by blue native electrophoresis: Isolation of putative respiratory strings or patches. *Proteomics*, *10*, 3379–3387.

Stroh, A., Anderka, O., Pfeiffer, K., Yagi, T., Finel, M., Ludwig, B., et al. (2004). Assembly of respiratory complexes I, III, and IV into NADH oxidase supercomplex stabilizes complex I in *Paracoccus denitrificans*. *The Journal of Biological Chemistry*, *279*, 5000–5007.

Stülke, J., & Hillen, W. (1998). Coupling physiology and gene regulation in bacteria: The phosphotransferase sugar uptake system delivers the signals. *Naturwissenschaften*, *85*, 583–592.

Suenaga, H. (2015). Targeted metagenomics unveils the molecular basis for adaptive evolution of enzymes to their environment. *Frontiers in Microbiology*, *6*, 1018.

Suh, M. J., Kuntumalla, S., Yu, Y., & Pieper, R. (2014). Proteomes of pathogenic *Escherichia coli/Shigella* group surveyed in their host environments. *Expert Review of Proteomics*, *11*, 593–609.

Sun, X., Chiu, J. F., & He, Q. Y. (2008). Fractionation of proteins by immobilized metal affinity chromatography. *Methods in Molecular Biology*, *424*, 205–212.

Sun, X., Ge, F., Xiao, C. L., Yin, X. F., Ge, R., Zhang, L. H., et al. (2010). Phosphoproteomic analysis reveals the multiple roles of phosphorylation in pathogenic bacterium *Streptococcus pneumoniae*. *Journal of Proteome Research*, *9*, 275–282.

Sun, X., Jia, H. L., Xiao, C. L., Yin, X. F., Yang, X. Y., Lu, J., et al. (2011). Bacterial proteome of *Streptococcus pneumoniae* through multidimensional separations coupled with LC-MS/MS. *OMICS*, *15*, 477–482.

Syka, J. E., Marto, J. A., Bai, D. L., Horning, S., Senko, M. W., Schwartz, J. C., et al. (2004). Novel linear quadrupole ion trap/FT mass spectrometer: Performance characterization and use in the comparative analysis of histone H3 post-translational modifications. *Journal of Proteome Research*, *3*, 621–626.

Tahara, M., Ohsawa, A., Saito, S., & Kimura, M. (2004). In vitro phosphorylation of initiation factor 2 alpha (aIF2 alpha) from hyperthermophilic archaeon *Pyrococcus horikoshii* OT3. *Journal of Biochemistry*, *135*, 479–485.

Tainer, J. A., & Adams, M. W. (2010). Microbial metalloproteomes are largely uncharacterized. *Nature*, *466*, 779–782.

Tang, S., Guo, Y., Xiong, C., Liu, S., Liu, X., & Jiang, S. (2014). Nanoparticle-based monoliths for chromatographic separations. *Analyst*, *139*, 4103–4117.

Taylor, J. A., & Johnson, R. S. (1997). Sequence database searches via *de novo* peptide sequencing by tandem mass spectrometry. *Rapid Communications in Mass Spectrometry*, *11*, 1067–1075.

Temperton, B., & Giovannoni, S. J. (2012). Metagenomics: Microbial diversity through a scratched lens. *Current Opinion in Microbiology*, *15*, 605–612.

Teramoto, H., Inui, M., & Yukawa, H. (2013). OxyR acts as a transcriptional repressor of hydrogen peroxide-inducible antioxidant genes in *Corynebacterium glutamicum* R. *The FEBS Journal*, *280*, 3298–3312.

Terpe, K. (2003). Overview of tag protein fusions: From molecular and biochemical fundamentals to commercial systems. *Applied Microbiology and Biotechnology*, *60*, 523–533.

Thauer, R. K., Kaster, A. K., Seedorf, H., Buckel, W., & Hedderich, R. (2008). Methanogenic archaea: Ecologically relevant differences in energy conservation. *Nature Reviews Microbiology*, *6*, 579–591.

Thompson, A., Schafer, J., Kuhn, K., Kienle, S., Schwarz, J., Schmidt, G., et al. (2003). Tandem mass tags: A novel quantification strategy for comparative analysis of complex protein mixtures by MS/MS. *Analytical Chemistry*, *75*, 1895–1904.

Tipton, J. D., Tran, J. C., Catherman, A. D., Ahlf, D. R., Durbin, K. R., & Kelleher, N. L. (2011). Analysis of intact protein isoforms by mass spectrometry. *The Journal of Biological Chemistry*, *286*, 25451–25458.

Tosha, T., & Shiro, Y. (2013). Crystal structures of nitric oxide reductases provide key insights into functional conversion of respiratory enzymes. *IUBMB Life*, *65*, 2217–2226.

Tran, B. Q., Goodlett, D. R., & Goo, Y. A. (2015). Advances in protein complex analysis by chemical cross-linking coupled with mass spectrometry (CXMS) and bioinformatics. *Biochimica et Biophysica Acta, 1864*, 123–129.

Trinkle-Mulcahy, L. (2012). Resolving protein interactions and complexes by affinity purification followed by label-based quantitative mass spectrometry. *Proteomics, 12*, 1623–1638.

Tripodi, F., Nicastro, R., Reghellin, V., & Coccetti, P. (2015). Post-translational modifications on yeast carbon metabolism: Regulatory mechanisms beyond transcriptional control. *Biochimica et Biophysica Acta, 1850*, 620–627.

Trouillard, M., Meunier, B., & Rappaport, F. (2011). Questioning the functional relevance of mitochondrial supercomplexes by time-resolved analysis of the respiratory chain. *Proceedings of the National Academy of Sciences of the United States of America, 108*, E1027–E1034.

Tsou, C. C., Tsai, C. F., Tsui, Y. H., Sudhir, P. R., Wang, Y. T., Chen, Y. J., et al. (2010). IDEAL-Q, an automated tool for label-free quantitation analysis using an efficient peptide alignment approach and spectral data validation. *Molecular & Cellular Proteomics, 9*, 131–144.

Tsutsui, Y., & Wintrode, P. L. (2007). Hydrogen/deuterium exchange-mass spectrometry: A powerful tool for probing protein structure, dynamics and interactions. *Current Medicinal Chemistry, 14*, 2344–2358.

Tucker, E. J., Wanschers, B. F., Szklarczyk, R., Mountford, H. S., Wijeyeratne, X. W., van den Brand, M. A., et al. (2013). Mutations in the UQCC1-interacting protein, UQCC2, cause human complex III deficiency associated with perturbed cytochrome *b* protein expression. *PLoS Genetics, 9*, e1004034.

Valsecchi, F., Konrad, C., & Manfredi, G. (2014). Role of soluble adenylyl cyclase in mitochondria. *Biochimica et Biophysica Acta, 1842*, 2555–2560.

Valsecchi, F., Ramos-Espiritu, L. S., Buck, J., Levin, L. R., & Manfredi, G. (2013). cAMP and mitochondria. *Physiology (Bethesda), 28*, 199–209.

van den Ecker, D., van den Brand, M. A., Bossinger, O., Mayatepek, E., Nijtmans, L. G., & Distelmaier, F. (2010). Blue native electrophoresis to study mitochondrial complex I in *C. elegans*. *Analytical Biochemistry, 407*, 287–289.

van Gestel, R. A., Rijken, P. J., Surinova, S., O'Flaherty, M., Heck, A. J., Killian, J. A., et al. (2010). The influence of the acyl chain composition of cardiolipin on the stability of mitochondrial complexes; an unexpected effect of cardiolipin in alpha-ketoglutarate dehydrogenase and prohibitin complexes. *Journal of Proteomics, 73*, 806–814.

van Noort, V., Seebacher, J., Bader, S., Mohammed, S., Vonkova, I., Betts, M. J., et al. (2012). Cross-talk between phosphorylation and lysine acetylation in a genome-reduced bacterium. *Molecular Systems Biology, 8*, 571.

van Schaik, W. (2015). The human gut resistome. *Philosophical Transactions of the Royal Society of London. Series B, Biological Sciences, 370*, 20140087. http://dx.doi.org/10.1098/rstb.2014.0087.

VanBogelen, R. A., Olson, E. R., Wanner, B. L., & Neidhardt, F. C. (1996). Global analysis of proteins synthesized during phosphorus restriction in *Escherichia coli*. *Journal of Bacteriology, 178*, 4344–4366.

Vartak, R., Porras, C. A., & Bai, Y. (2013). Respiratory supercomplexes: Structure, function and assembly. *Protein & Cell, 4*, 582–590.

Vartak, R. S., Semwal, M. K., & Bai, Y. (2014). An update on complex I assembly: The assembly of players. *Journal of Bioenergetics and Biomembranes, 46*, 323–328.

Vassilopoulos, A., Pennington, J. D., Andresson, T., Rees, D. M., Bosley, A. D., Fearnley, I. M., et al. (2014). SIRT3 deacetylates ATP synthase F1 complex proteins in response to nutrient- and exercise-induced stress. *Antioxidants & Redox Signaling, 21*, 551–564.

Venable, J. D., Steckler, C., Ou, W., Grünewald, J., Agarwalla, S., & Brock, A. (2015). Isotope-coded labeling for accelerated protein interaction profiling using MS. *Analytical Chemistry*, *87*, 7540–7544.

Venne, A. S., Kollipara, L., & Zahedi, R. P. (2014). The next level of complexity: Crosstalk of posttranslational modifications. *Proteomics*, *14*, 513–524.

Verissimo, A. F., & Daldal, F. (2014). Cytochrome *c* biogenesis system I: An intricate process catalyzed by a maturase supercomplex? *Biochimica et Biophysica Acta*, *1837*, 989–998.

Vermeulen, M., Hubner, N. C., & Mann, M. (2008). High confidence determination of specific protein–protein interactions using quantitative mass spectrometry. *Current Opinion in Biotechnology*, *19*, 331–337.

Villén, J., & Gygi, S. P. (2008). The SCX/IMAC enrichment approach for global phosphorylation analysis by mass spectrometry. *Nature Protocols*, *3*, 1630–1638.

Viswanathan, S., Unlü, M., & Minden, J. S. (2006). Two-dimensional difference gel electrophoresis. *Nature Protocols*, *1*, 1351–1358.

Vogel, R. O., Dieteren, C. E., van den Heuvel, L. P., Willems, P. H., Smeitink, J. A., Koopman, W. J., et al. (2007). Identification of mitochondrial complex I assembly intermediates by tracing tagged NDUFS3 demonstrates the entry point of mitochondrial subunits. *The Journal of Biological Chemistry*, *282*, 7582–7590.

Vogel, R. O., Janssen, R. J., Ugalde, C., Grovenstein, M., Huijbens, R. J., Visch, H. J., et al. (2005). Human mitochondrial complex I assembly is mediated by NDUFAF1. *The FEBS Journal*, *272*, 5317–5326.

Vogel, R. O., Janssen, R. J., van den Brand, M. A., Dieteren, C. E., Verkaart, S., Koopman, W. J., et al. (2007). Cytosolic signaling protein Ecsit also localizes to mitochondria where it interacts with chaperone NDUFAF1 and functions in complex I assembly. *Genes & Development*, *21*, 615–624.

Vogel, R. O., Smeitink, J. A., & Nijtmans, L. G. (2007). Human mitochondrial complex I assembly: A dynamic and versatile process. *Biochimica et Biophysica Acta*, *1767*, 1215–1227.

Voisin, S., Watson, D. C., Tessier, L., Ding, W., Foote, S., Bhatia, S., et al. (2007). The cytoplasmic phosphoproteome of the Gram-negative bacterium *Campylobacter jejuni*: Evidence for modification by unidentified protein kinases. *Proteomics*, 7, 4338–4348.

Völkel, P., Le Faou, P., & Angrand, P. O. (2010). Interaction proteomics: Characterization of protein complexes using tandem affinity purification-mass spectrometry. *Biochemical Society Transactions*, *38*, 883–887.

Völker, U., & Hecker, M. (2005). From genomics via proteomics to cellular physiology of the Gram-positive model organism *Bacillus subtilis*. *Cellular Microbiology*, 7, 1077–1085.

Vonck, J., & Schäfer, E. (2009). Supramolecular organization of protein complexes in the mitochondrial inner membrane. *Biochimica et Biophysica Acta*, *1793*, 117–124.

Vuckovic, D., Dagley, L. F., Purcell, A. W., & Emili, A. (2013). Membrane proteomics by high performance liquid chromatography-tandem mass spectrometry: Analytical approaches and challenges. *Proteomics*, *13*, 404–423.

Wagner, G. R., & Hirschey, M. D. (2014). Nonenzymatic protein acylation as a carbon stress regulated by sirtuin deacylases. *Molecular Cell*, *54*, 5–16.

Walpole, T. B., Palmer, D. N., Jiang, H., Ding, S., Fearnley, I. M., & Walker, J. E. (2015). Conservation of complete trimethylation of lysine-43 in the rotor ring of c-subunits of metazoan adenosine triphosphate (ATP) synthases. *Molecular & Cellular Proteomics*, *14*, 828–840.

Walzthoeni, T., Leitner, A., Stengel, F., & Aebersold, R. (2013). Mass spectrometry supported determination of protein complex structure. *Current Opinion in Structural Biology*, *23*, 252–260.

Wang, Y. K., Ma, Z., Quinn, D. F., & Fu, E. W. (2002). Inverse 15N-metabolic labeling/mass spectrometry for comparative proteomics and rapid identification of protein markers/targets. *Rapid Communications in Mass Spectrometry*, *16*, 1389–1397.

Wang, L., Pan, L., & Tao, W. A. (2014). Specific visualization and identification of phosphoproteome in gels. *Analytical Chemistry*, *86*, 6741–6747.

Wang, H., Straubinger, R. M., Aletta, J. M., Cao, J., Duan, X., Yu, H., et al. (2009). Accurate localization and relative quantification of arginine methylation using nanoflow liquid chromatography coupled to electron transfer dissociation and orbitrap mass spectrometry. *Journal of the American Society for Mass Spectrometry*, *20*, 507–519.

Wang, B., Xu, M., & Sun, G. (2010). Comparative analysis of membranous proteomics of *Shewanella decolorationis* S12 grown with azo compound or Fe (III) citrate as sole terminal electron acceptor. *Applied Microbiology and Biotechnology*, *86*, 1513–1523.

Wang, Q., Zhang, Y., Yang, C., Xiong, H., Lin, Y., Yao, J., et al. (2010). Acetylation of metabolic enzymes coordinates carbon source utilization and metabolic flux. *Science*, *327*, 1004–1007.

Washburn, M. P. (2004). Utilisation of proteomics datasets generated via multidimensional protein identification technology (MudPIT). *Briefings in Functional Genomics & Proteomics*, *3*, 280–286.

Washburn, M. P., Wolters, D., & Yates, J. R., 3rd. (2001). Large-scale analysis of the yeast proteome by multidimensional protein identification technology. *Nature Biotechnology*, *19*, 242–247.

Watanabe, M., Kubota, H., Wada, H., Narikawa, R., & Ikeuchi, M. (2011). Novel supercomplex organization of photosystem I in *Anabaena* and *Cyanophora paradoxa*. *Plant & Cell Physiology*, *52*, 162–168.

Watanabe, M., Semchonok, D. A., Webber-Birungi, M. T., Ehira, S., Kondo, K., Narikawa, R., et al. (2014). Attachment of phycobilisomes in an antenna-photosystem I supercomplex of cyanobacteria. *Proceedings of the National Academy of Sciences of the United States of America*, *111*, 2512–2517.

Wayne Outten, F. (2015). Recent advances in the Suf Fe-S cluster biogenesis pathway: Beyond the Proteobacteria. *Biochimica et Biophysica Acta*, *1853*, 1464–1469.

Weatherly, D. B., Atwood, J. A., 3rd, Minning, T. A., Cavola, C., Tarleton, R. L., & Orlando, R. (2005). A Heuristic method for assigning a false-discovery rate for protein identifications from Mascot database search results. *Molecular & Cellular Proteomics*, *4*, 762–772.

Weber, T. A., Koob, S., Heide, H., Wittig, I., Head, B., van der Bliek, A., et al. (2013). APOOL is a cardiolipin-binding constituent of the Mitofilin/MINOS protein complex determining cristae morphology in mammalian mitochondria. *PLoS One*, *8*, e63683.

Wegener, K. M., Singh, A. K., Jacobs, J. M., Elvitigala, T., Welsh, E. A., Keren, N., et al. (2010). Global proteomics reveal an atypical strategy for carbon/nitrogen assimilation by a cyanobacterium under diverse environmental perturbations. *Molecular & Cellular Proteomics*, *9*, 2678–2689.

Wei, J., Sun, J., Yu, W., Jones, A., Oeller, P., Keller, M., et al. (2005). Global proteome discovery using an online three-dimensional LC-MS/MS. *Journal of Proteome Research*, *4*, 801–808.

Weiner, J. H., & Li, L. (2008). Proteome of the *Escherichia coli* envelope and technological challenges in membrane proteome analysis. *Biochimica et Biophysica Acta*, *1778*, 1698–1713.

Weinert, B. T., Schölz, C., Wagner, S. A., Iesmantavicius, V., Su, D., Daniel, J. A., et al. (2013). Lysine succinylation is a frequently occurring modification in prokaryotes and eukaryotes and extensively overlaps with acetylation. *Cell Reports*, *4*, 842–851.

Wenz, T., Hielscher, R., Hellwig, P., Schägger, H., Richers, S., & Hunte, C. (2009). Role of phospholipids in respiratory cytochrome bc_1 complex catalysis and supercomplex formation. *Biochimica et Biophysica Acta*, *1787*, 609–616.

Wepf, A., Glatter, T., Schmidt, A., Aebersold, R., & Gstaiger, M. (2009). Quantitative interaction proteomics using mass spectrometry. *Nature Methods*, *6*, 203–205.

Werner, T., Sweetman, G., Savitski, M. F., Mathieson, T., Bantscheff, M., & Savitski, M. M. (2014). Ion coalescence of neutron encoded TMT 10-plex reporter ions. *Analytical Chemistry*, *86*, 3594–3601.

Wessels, H. J., Vogel, R. O., Lightowlers, R. N., Spelbrink, J. N., Rodenburg, R. J., van den Heuvel, L. P., et al. (2013). Analysis of 953 human proteins from a mitochondrial HEK293 fraction by complexome profiling. *PLoS One*, *8*, e68340.

Wessels, H. J., Vogel, R. O., van den Heuvel, L., Smeitink, J. A., Rodenburg, R. J., Nijtmans, L. G., et al. (2009). LC-MS/MS as an alternative for SDS-PAGE in blue native analysis of protein complexes. *Proteomics*, *9*, 4221–4228.

Whitelegge, J. P. (2002). Plant proteomics: BLASTing out of a MudPIT. *Proceedings of the National Academy of Sciences of the United States of America*, *99*, 11564–11566.

Wickström, D., Wagner, S., Baars, L., Ytterberg, A. J., Klepsch, M., van Wijk, K. J., et al. (2011a). Consequences of depletion of the signal recognition particle in *Escherichia coli*. *The Journal of Biological Chemistry*, *286*, 4598–4609.

Wickström, D., Wagner, S., Simonsson, P., Pop, O., Baars, L., Ytterberg, A. J., et al. (2011b). Characterization of the consequences of YidC depletion on the inner membrane proteome of *E. coli* using 2D blue native/SDS-PAGE. *Journal of Molecular Biology*, *409*, 124–135.

Wilmes, P., & Bond, P. L. (2004). The application of two-dimensional polyacrylamide gel electrophoresis and downstream analyses to a mixed community of prokaryotic microorganisms. *Environmental Microbiology*, *6*, 911–920.

Wilson, M. M., Anderson, D. E., & Bernstein, H. D. (2015). Analysis of the outer membrane proteome and secretome of *Bacteroides fragilis* reveals a multiplicity of secretion mechanisms. *PLoS One*, *10*, e0117732.

Wittig, I., Beckhaus, T., Wumaier, Z., Karas, M., & Schägger, H. (2010). Mass estimation of native proteins by blue native electrophoresis: Principles and practical hints. *Molecular & Cellular Proteomics*, *9*, 2149–2161.

Wittig, I., Braun, H. P., & Schägger, H. (2006). Blue native PAGE. *Nature Protocols*, *1*, 418–428.

Wittig, I., Carrozzo, R., Santorelli, F. M., & Schägger, H. (2006). Supercomplexes and subcomplexes of mitochondrial oxidative phosphorylation. *Biochimica et Biophysica Acta*, *1757*, 1066–1072.

Wittig, I., Karas, M., & Schägger, H. (2007). High resolution clear native electrophoresis for in-gel functional assays and fluorescence studies of membrane protein complexes. *Molecular & Cellular Proteomics*, *6*, 1215–1225.

Wittig, I., & Schägger, H. (2005). Advantages and limitations of clear-native PAGE. *Proteomics*, *5*, 4338–4346.

Wittig, I., & Schägger, H. (2008a). Features and applications of blue-native and clear-native electrophoresis. *Proteomics*, *8*, 3974–3990.

Wittig, I., & Schägger, H. (2008b). Structural organization of mitochondrial ATP synthase. *Biochimica et Biophysica Acta*, *1777*, 592–598.

Wittig, I., & Schägger, H. (2009a). Native electrophoretic techniques to identify protein-protein interactions. *Proteomics*, *9*, 5214–5223.

Wittig, I., & Schägger, H. (2009b). Supramolecular organization of ATP synthase and respiratory chain in mitochondrial membranes. *Biochimica et Biophysica Acta*, *1787*, 672–680.

Wittig, I., Velours, J., Stuart, R., & Schägger, H. (2008). Characterization of domain interfaces in monomeric and dimeric ATP synthase. *Molecular & Cellular Proteomics*, *7*, 995–1004.

Wolff, S., Antelmann, H., Albrecht, D., Becher, D., Bernhardt, J., Bron, S., et al. (2007). Towards the entire proteome of the model bacterium *Bacillus subtilis* by gel-based and gel-free approaches. *Journal of Chromatography B, Analytical Technologies in the Biomedical and Life Sciences*, *849*, 129–140.

Wolters, D. A., Washburn, M. P., & Yates, J. R., 3rd. (2001). An automated multidimensional protein identification technology for shotgun proteomics. *Analytical Chemistry, 73*, 5683–5690.

Wright, J. C., Collins, M. O., Yu, L., Kall, L., Brosch, M., & Choudhary, J. S. (2012). Enhanced peptide identification by electron transfer dissociation using an improved Mascot Percolator. *Molecular & Cellular Proteomics, 11*, 478–491.

Wu, Y. T., Lee, H. C., Liao, C. C., & Wei, Y. H. (2013). Regulation of mitochondrial F_oF_1ATPase activity by Sirt3-catalyzed deacetylation and its deficiency in human cells harboring 4977 bp deletion of mitochondrial DNA. *Biochimica et Biophysica Acta, 1832*, 216–227.

Wu, S., Lourette, N. M., Tolić, N., Zhao, R., Robinson, E. W., Tolmachev, A. V., et al. (2009). An integrated top-down and bottom-up strategy for broadly characterizing protein isoforms and modifications. *Journal of Proteome Research, 8*, 1347–1357.

Wu, C. C., MacCoss, M. J., Howell, K. E., & Yates, J. R., 3rd. (2003). A method for the comprehensive proteomic analysis of membrane proteins. *Nature Biotechnology, 21*, 532–538.

Wu, X., Vellaichamy, A., Wang, D., Zamdborg, L., Kelleher, N. L., Huber, S. C., et al. (2013). Differential lysine acetylation profiles of *Erwinia amylovora* strains revealed by proteomics. *Journal of Proteomics, 79*, 60–71.

Xie, L., Wang, X., Zeng, J., Zhou, M., Duan, X., Li, Q., et al. (2015). Proteome-wide lysine acetylation profiling of the human pathogen *Mycobacterium tuberculosis*. *The International Journal of Biochemistry & Cell Biology, 59*, 193–202.

Xu, H., Hegde, S. S., & Blanchard, J. S. (2011). Reversible acetylation and inactivation of *Mycobacterium tuberculosis* acetyl-CoA synthetase is dependent on cAMP. *Biochemistry, 50*, 5883–5892.

Yamashita, M., & Fenn, J. B. (1984). Electrospray ion source. Another variation on the free-jet theme. *The Journal of Physical Chemistry, 88*, 4451–4459.

Yates, J. R., Ruse, C. I., & Nakorchevsky, A. (2009). Proteomics by mass spectrometry: Approaches, advances, and applications. *Annual Review of Biomedical Engineering, 11*, 49–79.

Yates, J. R., 3rd., Speicher, S., Griffin, P. R., & Hunkapiller, T. (1993). Peptide mass maps: A highly informative approach to protein identification. *Analytical Biochemistry, 214*, 397–408.

Yeagle, P. L. (2014). Non-covalent binding of membrane lipids to membrane proteins. *Biochimica et Biophysica Acta, 1838*, 1548–1559.

Yokoyama, K., & Leimkühler, S. (2015). The role of FeS clusters for molybdenum cofactor biosynthesis and molybdoenzymes in bacteria. *Biochimica et Biophysica Acta, 1853*, 1335–1349.

Young, N. L., Plazas-Mayorca, M. D., & Garcia, B. A. (2010). Systems-wide proteomic characterization of combinatorial post-translational modification patterns. *Expert Review of Proteomics, 7*, 79–92.

Young, M. M., Tang, N., Hempel, J. C., Oshiro, C. M., Taylor, E. W., Kuntz, I. D., et al. (2000). High throughput protein fold identification by using experimental constraints derived from intramolecular cross-links and mass spectrometry. *Proceedings of the National Academy of Sciences of the United States of America, 97*, 5802–5806.

Yu, B. J., Kim, J. A., Moon, J. H., Ryu, S. E., & Pan, J. G. (2008). The diversity of lysine-acetylated proteins in *Escherichia coli*. *Journal of Microbiology and Biotechnology, 18*, 1529–1536.

Zahedi, R. P., Meisinger, C., & Sickmann, A. (2005). Two-dimensional benzyldimethyl-n-hexadecylammonium chloride/SDS-PAGE for membrane proteomics. *Proteomics, 5*, 3581–3588.

Zahedi, R. P., Moebius, J., & Sickmann, A. (2007). Two-dimensional BAC/SDS-PAGE for membrane proteomics. *Sub-Cellular Biochemistry, 43*, 13–20.

Zamdborg, L., LeDuc, R. D., Glowacz, K. J., Kim, Y. B., Viswanathan, V., Spaulding, I. T., et al. (2007). ProSight PTM 2.0: Improved protein identification and characterization for top down mass spectrometry. *Nucleic Acids Research*, *35*, W701–W706.

Zhang, Y., Fonslow, B. R., Shan, B., Baek, M. C., & Yates, J. R., 3rd. (2013). Protein analysis by shotgun/bottom-up proteomics. *Chemical Reviews*, *113*, 2343–2394.

Zhang, M., Mileykovskaya, E., & Dowhan, W. (2002). Gluing the respiratory chain together. Cardiolipin is required for supercomplex formation in the inner mitochondrial membrane. *The Journal of Biological Chemistry*, *277*, 43553–43556.

Zhang, M., Mileykovskaya, E., & Dowhan, W. (2005). Cardiolipin is essential for organization of complexes III and IV into a supercomplex in intact yeast mitochondria. *The Journal of Biological Chemistry*, *280*, 29403–29408.

Zhang, Y., Wu, Z. X., Wan, X. L., Liu, P., Zhang, J. B., Ye, Y., et al. (2014). Comprehensive profiling of lysine acetylome in *Staphylococcus aureus*. *Science China: Chemistry*, *57*, 732–738.

Zhang, J., Xin, L., Shan, B., Chen, W., Xie, M., Yuen, D., et al. (2012). PEAKS DB: *De novo* sequencing assisted database search for sensitive and accurate peptide identification. *Molecular & Cellular Proteomics*, *11*, M111010587.

Zhang, K., Zheng, S., Yang, J. S., Chen, Y., & Cheng, Z. (2013). Comprehensive profiling of protein lysine acetylation in *Escherichia coli*. *Journal of Proteome Research*, *12*, 844–851.

Zhao, Y., & Jensen, O. N. (2009). Modification-specific proteomics: Strategies for characterization of post-translational modifications using enrichment techniques. *Proteomics*, *9*, 4632–4641.

Zhao, S., Xu, W., Jiang, W., Yu, W., Lin, Y., Zhang, T., et al. (2010). Regulation of cellular metabolism by protein lysine acetylation. *Science*, *327*, 1000–1004.

Zheng, J., & Jia, Z. (2010). Structure of the bifunctional isocitrate dehydrogenase kinase/phosphatase. *Nature*, *465*, 961–965.

Zheng, J., Yates, S. P., & Jia, Z. (2012). Structural and mechanistic insights into the bifunctional enzyme isocitrate dehydrogenase kinase/phosphatase AceK. *Philosophical Transactions of the Royal Society of London. Series B, Biological Sciences*, *367*, 2656–2668.

Zhou, Y., Aebersold, R., & Zhang, H. (2007). Isolation of N-linked glycopeptides from plasma. *Analytical Chemistry*, *79*, 5826–5837.

Zhou, Y., Dong, J., & Vachet, R. W. (2011). Electron transfer dissociation of modified peptides and proteins. *Current Pharmaceutical Biotechnology*, *12*, 1558–1567.

Zhou, M., Morgner, N., Barrera, N. P., Politis, A., Isaacson, S. C., Matak-Vinković, D., et al. (2011). Mass spectrometry of intact V-type ATPases reveals bound lipids and the effects of nucleotide binding. *Science*, *334*, 380–385.

Zhou, M., & Robinson, C. V. (2010). When proteomics meets structural biology. *Trends in Biochemical Sciences*, *35*, 522–529.

Zhou, M., & Robinson, C. V. (2014). Flexible membrane proteins: Functional dynamics captured by mass spectrometry. *Current Opinion in Structural Biology*, *28*, 122–130.

Zijnge, V., Kieselbach, T., & Oscarsson, J. (2012). Proteomics of protein secretion by *Aggregatibacter actinomycetemcomitans*. *PLoS One*, 7, e41662.

Zischka, H., Braun, R. J., Marantidis, E. P., Büringer, D., Bornhövd, C., Hauck, S. M., et al. (2006). Differential analysis of *Saccharomyces cerevisiae* mitochondria by free flow electrophoresis. *Molecular & Cellular Proteomics*, *5*, 2185–2200.

Zumft, W. G., & Kroneck, P. M. (2007). Respiratory transformation of nitrous oxide (N_2O) to dinitrogen by Bacteria and Archaea. *Advances in Microbial Physiology*, *52*, 107–227.

CHAPTER SEVEN

Nitrous Oxide Metabolism in Nitrate-Reducing Bacteria: Physiology and Regulatory Mechanisms

M.J. Torres*, J. Simon†, G. Rowley‡, E.J. Bedmar*, D.J. Richardson‡,§, A.J. Gates‡,§, M.J. Delgado*,1

*Estación Experimental del Zaidín, CSIC, Granada, Spain
†Technische Universität Darmstadt, Darmstadt, Germany
‡School of Biological Sciences, University of East Anglia, Norwich Research Park, Norwich, United Kingdom
§Centre for Molecular and Structural Biochemistry, University of East Anglia, Norwich, United Kingdom
[1]Corresponding author: e-mail address: mariajesus.delgado@eez.csic.es

Contents

Abstract

Nitrous oxide (N_2O) is an important greenhouse gas (GHG) with substantial global warming potential and also contributes to ozone depletion through photochemical nitric oxide (NO) production in the stratosphere. The negative effects of N_2O on climate

Advances in Microbial Physiology, Volume 68
ISSN 0065-2911
http://dx.doi.org/10.1016/bs.ampbs.2016.02.007

and stratospheric ozone make N_2O mitigation an international challenge. More than 60% of global N_2O emissions are emitted from agricultural soils mainly due to the application of synthetic nitrogen-containing fertilizers. Thus, mitigation strategies must be developed which increase (or at least do not negatively impact) on agricultural efficiency whilst decrease the levels of N_2O released. This aim is particularly important in the context of the ever expanding population and subsequent increased burden on the food chain. More than two-thirds of N_2O emissions from soils can be attributed to bacterial and fungal denitrification and nitrification processes. In ammonia-oxidizing bacteria, N_2O is formed through the oxidation of hydroxylamine to nitrite. In denitrifiers, nitrate is reduced to N_2 via nitrite, NO and N_2O production. In addition to denitrification, respiratory nitrate ammonification (also termed dissimilatory nitrate reduction to ammonium) is another important nitrate-reducing mechanism in soil, responsible for the loss of nitrate and production of N_2O from reduction of NO that is formed as a by-product of the reduction process. This review will synthesize our current understanding of the environmental, regulatory and biochemical control of N_2O emissions by nitrate-reducing bacteria and point to new solutions for agricultural GHG mitigation.

ABBREVIATIONS

AOB ammonia-oxidizing bacteria
Crp cAMP receptor protein
Cys cysteine
Cyt cytochrome
DNRA dissimilatory nitrate reduction to ammonium
ETC electron transport chain
FMN flavin mononucleotide
FNR fumarate and nitrate reductase regulatory protein
HCO haem–copper oxidase
His histidine
Hmp flavohaemoglobin
H-T-H helix-turn-helix motif
LbNO nitrosyl-leghaemoglobin
MK menaquinone
MKH_2 menaquinol
$Mo[MGD]_2$ molybdenum *bis* molybdopterin guanine dinucleotide
N_2OR nitrous oxide reductase
Nap periplasmic nitrate reductase
Nar membrane-bound nitrate reductase
NDH NADH dehydrogenase
Nir nitrite reductase
NirBD NADH-dependent assimilatory sirohaem-containing nitrite reductase
NirK Cu-type nitrite reductase
NirS cd_1-type nitrite reductase
NnrR nitrite and nitric oxide reductase regulator
Nor nitric oxide reductase
NorVW flavorubredoxin

NOS nitric oxide synthase
Nrf nitrite reduction by formate
NrfA cytochrome *c* nitrite reductase
NssR nitrosative stress-sensing regulator
Pmf proton motive force
RNAP RNA polymerase
Tat twin arginine translocation
UQ ubiquinone
UQH_2 ubiquinol

1. INTRODUCTION

Nitrous oxide (N_2O) is a powerful greenhouse gas (GHG) and a major cause of ozone layer depletion with an atmospheric lifetime of 114 years. Although N_2O only accounts for around 0.03% of total GHG emissions, it has an estimated 300-fold greater potential for global warming compared with that of carbon dioxide (CO_2), based on its radiative capacity. Hence, when the impact of individual GHGs on global warming was expressed in terms of the Intergovernmental Panel on Climate Change (IPCC, 2014) approved unit of CO_2 equivalents, N_2O accounts for approximately 10% of total emissions (IPCC, 2014). Human activities are currently considered to emit 6.7 Tg N–N_2O per year mainly from agriculture, which accounts for about 60% of N_2O emissions (IPCC, 2014; Smith et al., 2008; Smith, Mosier, Crutzen, & Winiwarter, 2012). This contribution has been amplified through the intensification of agriculture, the so-called green revolution, which has increased the presence of nitrogen (N) in soil through the application, since the early 1900s, of synthetic nitrogen-based fertilizers whose production steadily increased after the invention of the Haber–Bosch process. Since 1997, many of the nonbiological emissions of N_2O, for example, those associated with the transport industry, have been systematically lowered, whereas emissions from agriculture are essentially unchanged (IPCC, 2014). Given the clear evidence that now highlights the damaging effect on climate of atmospheric N_2O, strategies to mitigate N_2O emissions arising from intensive agricultural practices have to be developed in order to increase agricultural efficiency and decrease current levels of N_2O production in particular in the context of the continuing population growth (Richardson, Felgate, Watmough, Thomson, & Baggs, 2009; Thomson, Giannopoulos, Pretty, Baggs, & Richardson, 2012). Strategies that might be adopted include: (i) management of soil chemistry and microbiology to ensure that bacterial denitrification proceeds to completion, to form N_2 instead of N_2O; (ii) reducing the dependence on fertilizers through engineering crops to fix dinitrogen themselves or, alternatively through the application of nitrogen-fixing bacteria; (iii) promotion of sustainable agriculture, ie, producing more output from the same area of land while reducing negative environmental impacts and (iv) an increased understanding of the environmental and molecular factors that contribute to the biological generation and consumption of N_2O. Pathways for biological N_2O production include dissimilatory nitrate/nitrite reduction to N_2 (denitrification) (Zumft, 1997), dissimilatory nitrate reduction to ammonia (DNRA) (Bleakley & Tiedje, 1982), nitrifier denitrification, hydroxylamine oxidation by ammonia-oxidizing bacteria (AOB) and NO detoxification (also known as nitrosative stress defence). N_2O is also produced by methane-oxidizing bacteria (Campbell et al., 2011) and ammonia-oxidizing archaea (Liu, Wu, Ding, Shi, & Chen, 2010; Stieglmeier et al., 2014). N_2O production by nitrite-oxidizing bacteria, anaerobic methane (N-AOM) and anaerobic AOB (anammox bacteria) has also been reported (for reviews, see Butterbach-Bahl, Baggs, Dannenmann, Kiese, & Zechmeister-Boltenstern, 2014; Schreiber, Wunderlin, Udert, & Wells, 2012; Stein, 2011). Among these processes, denitrification and DNRA are two important nitrate-reducing mechanisms in soil (Fig. 1). During denitrification, nitrate is reduced to the gaseous products, N_2O and dinitrogen gas (N_2), in a step-wise manner via nitrite and nitric oxide (NO) as intermediates (Zumft, 1997). N_2O and N_2 release to the atmosphere causes N loss from terrestrial and aquatic environments, and N_2O is an ozone-depleting greenhouse gas. DNRA shares the nitrate to nitrite reduction step with denitrification but reduces nitrite to ammonium (Bleakley & Tiedje, 1982; Simon & Klotz, 2013). In contrast to nitrate and nitrite, ammonium is retained in soils and sediments and has a higher tendency for incorporation into microbial or plant biomass. Hence, the relative contributions of denitrification vs respiratory ammonification activities have important consequences for N retention, plant growth and climate. In addition to denitrification that produces N_2O when abiotic conditions or the lack of an N_2O reductase (N_2OR) encoding gene prevent its reduction to N_2, DNRA seemingly releases N_2O as a by-product of the nitrate/nitrite reduction process (Fig. 1). In denitrifiers, it has been well established the role of the Cu-containing (NirK) and cd_1-type (NirS) nitrite reductases as well as the membrane-bound respiratory NO reductases (cNor and qNor enzymes)

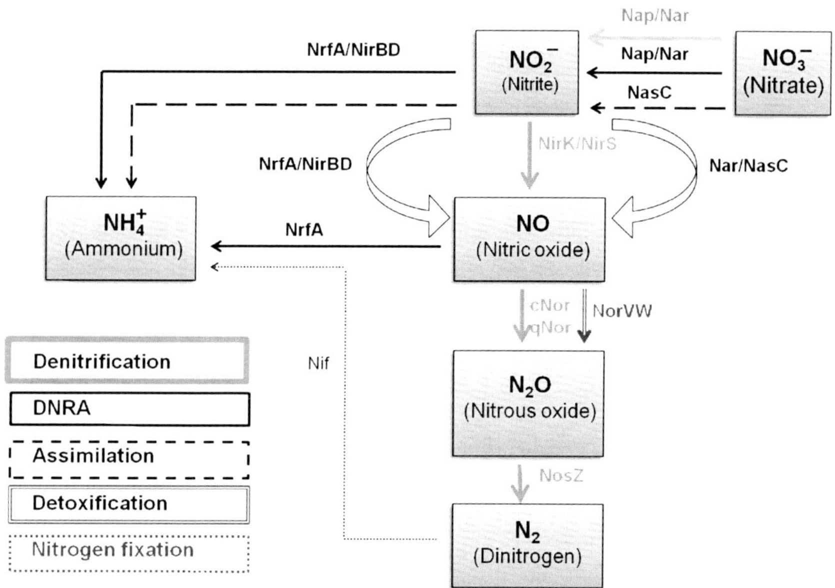

Fig. 1 Biological pathways of N_2O metabolism in nitrate-reducing bacteria. The major processes involved in nitrate transformation to N_2O are denitrification, dissimilatory nitrate reduction to ammonium (DNRA), assimilation and NO detoxification. The main enzymes involved are *Nar*, membrane-bound dissimilatory nitrate reductase; *Nap*, periplasmic dissimilatory nitrate reductase; *NasC*, assimilatory nitrate reductase; *NirBD*, sirohaem containing nitrite reductase; *NrfA*, cytochrome *c* nitrite reductase; *NirK/NirS*, Cu-containing/cd_1-type nitrite reductase, nitric oxide reductases (*cNor* and *qNor*); *NorVW*, anaerobic nitric oxide reducing flavorubredoxin; *NosZ*, nitrous oxide reductase; *Nif*, nitrogenase.

in NO and N_2O formation (Fig. 1). In nitrate ammonifiers, the ammonium-generating respiratory cytochrome *c* nitrite reductase (NrfA), the assimilatory sirohaem-containing nitrite reductase (NirBD) and the NO-detoxifying flavorubredoxin (NorVW) are the main candidates to be involved in NO and N_2O production (Fig. 1). Recent findings have proposed the involvement of the *Salmonella enterica* serovar Typhimurium membrane-bound respiratory nitrate reductase (Nar) (Gilberthorpe & Poole, 2008; Rowley et al., 2012) and the *Bradyrhizobium japonicum* assimilatory nitrate reductase (NasC) (Cabrera et al., 2016) in NO and N_2O metabolism (Fig. 1).

While there are several enzymatic and microbial routes to N_2O production, the bacterial N_2OR is the only known enzyme capable of reducing N_2O to N_2 (Fig. 1). The typical N_2OR enzyme, NosZ, from denitrifiers has been considered for long time the only enzyme involved in N_2O mitigation. Recently, however, a closely related enzyme variant named atypical NosZ has been identified in diverse microbial taxa forming a distinct clade of N_2OR (Jones, Graf, Bru, Philippot, & Hallin, 2013; Sanford et al., 2012). Organisms containing atypical NosZ enzymes also possess divergent *nos* clusters with genes that are evolutionarily distinct from the typical *nos* genes of denitrifiers (Table 1). Interestingly, nitrate-ammonifying bacteria such as *Wolinella succinogenes* as well as some other nondenitrifiers contain this atypical N_2OR that probably acts on the N_2O produced by detoxifying activities that remove the NO formed as a by-product of the DNRA process (Jones et al., 2013; Sanford et al., 2012; Simon, Einsle, Kroneck, & Zumft, 2004). However, another group of nitrate ammonifiers including enterobacteria such as *Escherichia coli* or *S.* Typhimurium that also can produce N_2O do not have an enzyme that can consume it. Thus, these bacteria might contribute significantly to global N_2O emissions. A greater understanding of the key enzymes and environmental and regulatory factors involved in N_2O metabolism in denitrifiers and organisms performing DNRA may allow the development of more effective N_2O mitigation strategies in soil nitrate-reducing communities. The goal of this review is to present an overview of the enzymatic mechanisms of N_2O production and consumption by nitrate-reducing bacteria, as well as the environmental signals and the regulatory networks involved.

Table 1 Properties of Typical and Atypical N_2O Reductases and *nos* Gene Clusters

	Typical	**Atypical**
Metabolic type of the host cell	Denitrifier	Nondenitrifier[a]
NosZ signal peptide	Tat dependent	Sec dependent
Characteristic motifs of Cu_Z centre ligands	DXHHXH, EPHD	DXHH, EPH
Haem *c* domain (*c*NosZ-type enzymes)	Absent	Often present
nosB, -*G*, -*C1*, -*C2*, -*H* genes	Absent	Present
nosR and -*X* genes	Present	Absent
Representative model organisms	*Paracoccus denitrificans*	*Wolinella succinogenes*
	Bradyrhizobium japonicum	*Anaeromyxobacter dehalogenans*

[a]Often lacking an NO-generating nitrite reductase of the NirS- or NirK-type.

2. N_2O METABOLISM IN NITRATE-AMMONIFYING BACTERIA

The metabolism of N_2O in organisms that grow by respiratory nitrate or nitrite ammonification is poorly understood. The respective organisms reduce nitrate to nitrite using a membrane-bound nitrate reductase (Nar) and/or a periplasmic nitrate reductase (Nap) (Kern & Simon, 2009a; Richardson, Berks, Russell, Spiro, & Taylor, 2001; Simon & Klotz, 2013). Subsequently, nitrite is reduced to ammonium by a NrfA, which obtains electrons from the quinone/quinol pool through one of the several different electron transport enzyme systems, depending on the organism (Kern & Simon, 2009a; Simon, 2002; Simon & Klotz, 2013). Prominent examples of respiratory nitrate ammonifiers are Gamma-, Delta- and Epsilonproteobacteria such as *E. coli*, *S.* Typhimurium, *Shewanella oneidensis*, *Shewanella loihica*, *Anaeromyxobacter dehalogenans*, *Campylobacter jejuni* and *W. succinogenes* but also some less well-known members of the genus *Bacillus* (phylum Firmicutes), for example, *Bacillus vireti, Bacillus azotoformans* or *Bacillus bataviensis* (Heylen & Keltjens, 2012; Mania, Heylen, van Spanning, & Frostegard, 2014; Simon, 2002; Simon & Klotz, 2013). With the exception of *Sh. loihica,* nitrate-ammonifying bacteria usually lack both NirK and NirS as well as the typical membrane-bound respiratory NO reductases (cNor and qNor) found in denitrifiers. Apparently, however, the catalysis of respiratory nitrate/nitrite ammonification is also a source of N_2O. In a first step leading to N_2O production, NO is generated either chemically and/or enzymatically from nitrite. The detailed mechanisms of these conversions, however, are yet to be elucidated. Since NO is a highly toxic compound that exerts nitrosative stress on cells and organisms, it needs to be detoxified (Poole, 2005). It is therefore not surprising that N_2O generation from NO has been described for numerous nonrespiratory enzymes, including flavodiiron proteins (Fdp), flavorubredoxin (NorVW), cytochrome c_{554} (CycA; present in nitrifiers), cytochrome c'-beta (CytS) and cytochrome c'-alpha (CytP) (Simon & Klotz, 2013; references therein). In these cases, NO reduction to N_2O is thought to serve predominantly in NO detoxification. In the light of such an N_2O-producing capacity, it is not surprising that some nitrate ammonifiers such as *W. succinogenes*, *A. dehalogenans* and *B. vireti* have been reported to grow by anaerobic N_2O respiration using N_2O as sole electron acceptor (Kern & Simon, 2016; Mania, Heylen, van Spanning, & Frostegard, 2016; Sanford et al., 2012;

Yoshinari, 1980). Moreover, the cells of some other species have been reported to reduce N_2O and many genomes of ammonifiers indeed contain a *nos* gene cluster (see Section 2.2.2). These *nos* clusters comprise a *nosZ* gene encoding the 'atypical' nitrous oxide reductase (N_2OR) and some of them even a cytochrome *c* N_2OR (*c*NosZ) (Table 1; Jones et al., 2013; Kern & Simon, 2009a; Sanford et al., 2012; Simon et al., 2004; Simon & Klotz, 2013; Zumft & Kroneck, 2007). The *c*NosZ enzyme is a variant of the canonical NosZ found in denitrifiers that contains a C-terminal monohaem cytochrome *c* domain, which is thought to donate electrons to the active copper site (Simon et al., 2004). Export of *c*NosZ to the periplasm is accomplished by the Sec secretion pathway rather than by the twin arginine translocation (Tat) pathway used by the canonical NosZ.

2.1 Gammaproteobacteria

N_2O metabolism by Gammaproteobacteria that perform dissimilatory nitrate/nitrite reduction to ammonia (DNRA) has been mainly investigated in *E. coli* and *S.* Typhimurium. These bacteria belong to the Enterobacteriaceae family which have their natural habitats in soil, water (fresh and marine) environments or the intestines of both warm- and cold-blooded animals. In humans, while *Salmonella* species are pathogenic and can result in an inflamed intestine and gastroenteritits, nonpathogenic *E. coli* strains can form part of the normal flora with beneficial traits for humans.

In many species of Enterobacteriaceae, there are two biochemically distinct nitrate reductases: Nar and Nap. Nar enzymes have been most studied in *E. coli* and *Paracoccus* sp. (reviewed by Gonzalez, Correia, Moura, Brondino, & Moura, 2006; Potter, Angove, Richardson, & Cole, 2001; Richardson, 2011; Richardson et al., 2001; Richardson, van Spanning, & Ferguson, 2007). Nar is common to both ammonification and denitrification and has been crystallographically resolved from *E. coli* (Bertero et al., 2003; Jormakka, Richardson, Byrne, & Iwata, 2004). It is a three-subunit enzyme composed of NarGHI, where NarG is the catalytic subunit of about 140 kDa that contains a molybdenum *bis* molybdopterin guanine dinucleotide (Mo[MGD]$_2$), and a [4Fe–4S] cluster. NarH, of about 60 kDa, contains one [3Fe–4S] and three [4Fe–4S] clusters. NarG and NarH are located in the cytoplasm and associate with NarI, an integral membrane dihaem cytochrome *b* quinol oxidase about 25 kDa (Fig. 2A). Nar proteins are encoded by the *narGHJI* operon. Whereas *narGHI* encodes the structural subunits, *narJ* codes for a cognate chaperone required for the proper

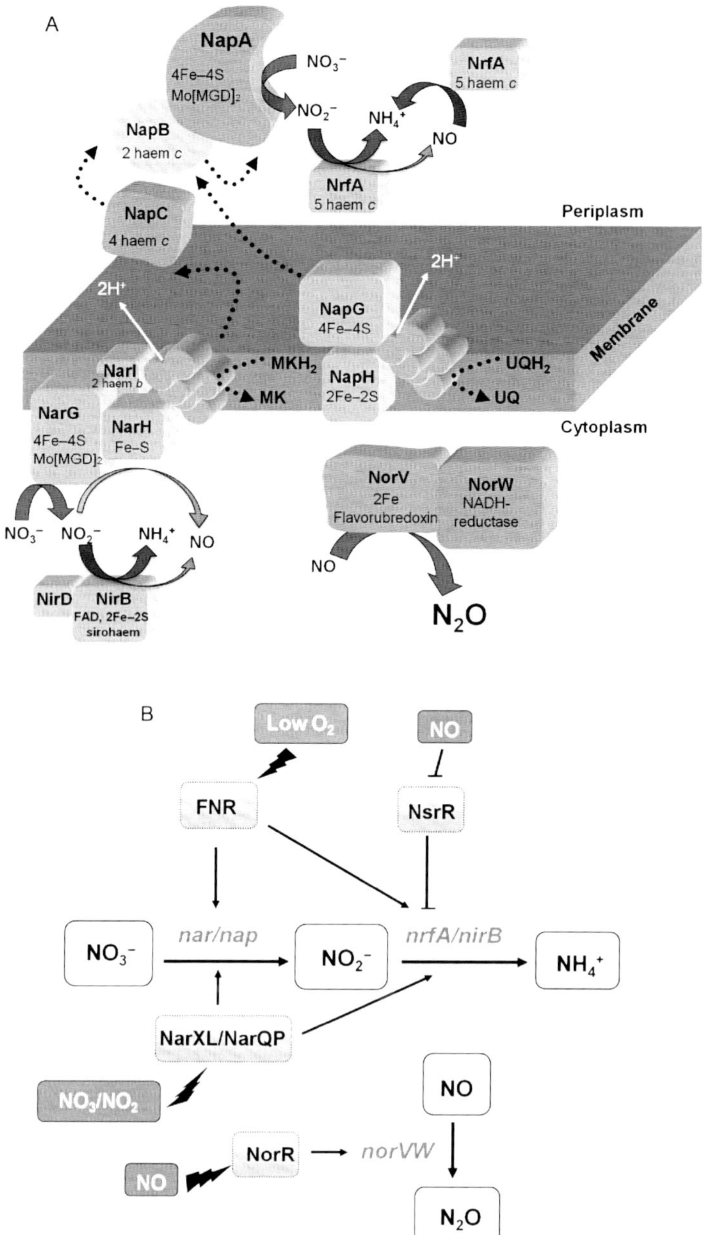

Fig. 2 Enzymes and regulators involved in NO and N_2O metabolism in *E. coli* and *Salmonella* Typhimurium. (A) Enzymes involved in nitrate reduction (NapABC, NapGH and NarGHI), nitrite reduction (NrfA, NirBD), NO production (NrfA/NirBD, Nar) and N_2O production (NorVW) are shown. For the Nrf system (cf. Fig. 3A), only the catalytic subunit NrfA is shown. (B) Regulators involved in NO production (NarXL/NarQP, FNR, NsrR) and N_2O production (NorR). Positive regulation is denoted by *arrows*, and negative regulation is indicated by *perpendicular lines*. See text for details.

maturation and membrane insertion of Nar. The organization of this operon is conserved in most species that express Nar. *E. coli* and *S.* Typhimurium have a functional duplicate of the *narGHJI* operon named *narZYWV*, which has a central role in the physiology of starved and stressed cells, rather than anaerobic respiration per se (Blasco, Iobbi, Ratouchniak, Bonnefoy, & Chippaux, 1990; Spector et al., 1999). In the cytoplasm, a NADH-dependent assimilatory sirohaem-containing nitrite reductase (NirBD) reduces nitrite to ammonia as rapidly as it is formed from nitrate by Nar (Fig. 2A). The *nir* operon includes *nirB* and *nirD* as structural genes for the two enzyme subunits; a third gene, *nirC*, probably encodes a nitrite transport protein; and finally *cysG*, the product of which is required for the synthesis of the novel haem group, sirohaem (Peakman, Busby, & Cole, 1990).

Enteric bacteria such as *E. coli* and *S.* Typhimurium have evolved a second respiratory pathway to survive in electron acceptor-limited anaerobic conditions. Under anoxic and microoxic conditions in the presence of low levels of nitrate, Nap and the periplasmic nitrite reduction by formate (Nrf) system are expressed (Figs 2A and 3A). NapA is the catalytic subunit responsible for the two-electron reduction of nitrate to nitrite, while NrfA reduces nitrite to ammonium through a six-electron reduction proposed to involve bound intermediates of NO and hydroxylamine (Einsle,

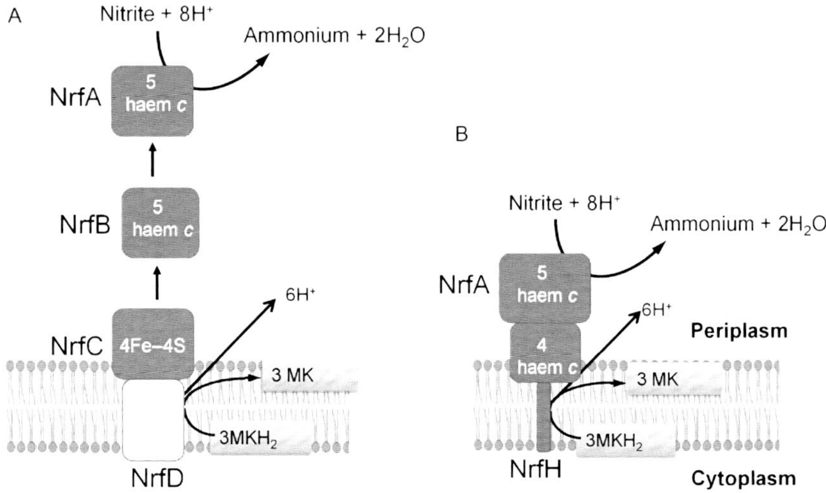

Fig. 3 Model of respiratory Nrf systems. (A) Nrf system of *E. coli*. (B) Nrf system of *W. succinogenes*. See text for details. For simplicity, only monomeric enzyme forms are shown. *MK*, menaquinone; *MKH₂*, menaquinol.

Messerschmidt, Huber, Kroneck, & Neese, 2002). In *E. coli*, the reduction of nitrate to ammonium can be coupled to energy-conserving electron transport pathways with formate as an electron donor (Potter et al., 2001). The Nap system is found in many different Gram-negative bacteria (reviewed by Gonzalez et al., 2006; Potter et al., 2001; Richardson, 2011; Richardson et al., 2007; Simon & Klotz, 2013). The best-studied Nap enzymes were isolated from *Paracoccus pantotrophus, E. coli, Rhodobacter sphaeroides* and *Desulfovibrio desulfuricans*. The crystal structure of *E. coli* NapA has been solved (Jepson et al., 2007). Similar to NarG, NapA binds Mo $[MGD]_2$ and a [4Fe–4S] cluster. In the majority of known cases, NapA forms a complex with the dihaem cytochrome *c* NapB. Generally, mature NapA is transported across the membrane by the Tat apparatus, and this process requires the cytoplasmic chaperone NapD, which is encoded in all known *nap* gene clusters (Grahl, Maillard, Spronk, Vuister, & Sargent, 2012). In the majority of Nap systems, electron transfer from quinol to NapAB complex requires a tetrahaem cytochrome *c* NapC, a member of the NapC/NrfH family (Fig. 2A). However, in *E. coli* a second quinol-oxidizing system has been identified, the NapGH complex which consists of two proposed Fe/S proteins. NapH is a membrane-bound quinol dehydrogenase, while NapG is a periplasmic electron transfer adapter protein (Fig. 2A). The structure and detailed function of the NapGH proteins, however, remain unclear as these have not been purified. In addition to *napDAGHBC* genes directly involved in nitrate reduction, *E. coli napFDAGHBC* operon also contains *napF* encoding an accessory protein. NapF is a cytoplasmic Fe/S protein that is thought to have a role in the posttranslational modification of NapA prior to the export of folded NapA into the periplasm (Nilavongse et al., 2006).

The best-known periplasmic ammonium-generating nitrite reductase is NrfA (Figs 2A and 3A; reviewed by Clarke et al., 2008; Einsle, 2011; Simon & Klotz, 2013). This enzyme reduces nitrite produced by Nap to ammonium by using six electrons that are commonly obtained through the oxidation of formate. This allows nitrite to be used as a terminal electron acceptor, facilitating anaerobic respiration while allowing nitrogen to remain in a biologically available form. NrfA, first described in *E. coli*, is expressed within the periplasm of a wide range of Gamma-, Delta- and Epsilonproteobacteria. In *E. coli*, *nrfABCDEFG* genes are involved in the synthesis and activity of NrfA with *nrfA* coding for the actual enzyme, *nrfB* coding for a small, pentahaem electron transfer protein, *nrfC* and *nrfD* for a membrane-integral quinol dehydrogenase (Fig. 3A) and *nrfE*, *nrfF* and *nrfG* for components of a dedicated assembly machinery required for attachment of the active site haem group. The electron transfer between NrfCD and NrfA in *E. coli* is mediated by NrfB (Clarke, Cole, Richardson, & Hemmings, 2007). Crystal structures of NrfA from *E. coli* are currently available (Bamford et al., 2002; Clarke et al., 2008). NrfA contains four His/His-ligated *c*-haems for electron transfer and a structurally differentiated haem that provides the catalytic centre for nitrite reduction. The catalytic haem has proximal ligation from lysine, or histidine, and an exchangeable distal ligand bound within a pocket that includes a conserved His. Recent experiments where electrochemical, structural and spectroscopic analyses were combined revealed that the distal His is proposed to play a key role in orienting the nitrite for N–O bond cleavage (Lockwood et al., 2015).

2.1.1 Enzymes Involved in NO and N_2O Metabolism

The cytotoxin NO is the major precursor of N_2O in many biological pathways, and the accumulation of N_2O in bacteria that lack NosZ can be used as a direct reporter of intracellular NO production (Rowley et al., 2012). In prokaryotes, NO formation was considered to occur only in denitrification, anaerobic ammonium oxidation and other related respiratory pathways (Bothe & Beyer, 2007; Jetten, 2008; Maia & Moura, 2014; Schreiber et al., 2012; Zumft, 1997). NO formation from nitrite constitutes the first committed step in denitrification and is an essential step in anaerobic ammonium oxidation and other respiratory pathways where nitrogen compounds are used to derive energy. For those respiratory functions, prokaryotes developed NirS or NirK to reduce nitrite to NO. Several studies have suggested that NO is also generated in prokaryotes by nonrespiratory pathways via NO synthase (NOS) enzymes, homologous to the oxygenase domain of the mammalian NOS. NOS catalyses aerobic NO formation from arginine, using cellular redox equivalents that are not normally committed to NO production (reviewed by Maia & Moura, 2014; Spiro, 2011). *Salmonella* species and *E. coli* lack the typical respiratory NirS or NirK enzymes, as well as NOS; however, they do produce NO as a side-product of nitrate or nitrite metabolism. Studies with *E. coli* mutants suggested that nitrite-dependent NO formation was assumed to arise from the 'side' activity of NirBD, as well as from NrfA that both catalyse nitrite reduction to ammonium (Corker & Poole, 2003; Weiss, 2006; Fig. 2A). However, NO formation from nitrite in *S.* Typhimurium does not involve NirBD or NrfA. Recently, reduction of nitrite by Nar has been proposed as one major source of NO in *E. coli* and *S. enterica* serovar Typhimurium (Fig. 2A). By contrast, a small contribution (<3%) from the periplasmatic Nap to NO formation has been reported in both bacteria (Gilberthorpe & Poole, 2008; Rowley et al., 2012; Vine & Cole, 2011).

In addition to catalysing the six-electron reduction of nitrite to ammonium, *E. coli* NrfA also has the ability to act as an NO reductase. Kinetic, spectroscopic, voltammetric and crystallization studies with purified NrfA have demonstrated the capacity of this enzyme to reduce NO (Clarke et al., 2008; Einsle, 2011; van Wonderen, Burlat, Richardson, Cheesman, & Butt, 2008). This capacity has also been reported in whole cells studies using wild-type and *nrf* mutant strains of *E. coli and S.* Typhimurium where a contribution by NrfA to NO stress tolerance has been demonstrated (Mills, Rowley, Spiro, Hinton, & Richardson, 2008; Poock, Leach, Moir, Cole, & Richardson, 2002; Poole, 2005). *E. coli* and *S.* Typhimurium are known to possess other NO-consuming systems to overcome NO produced by the immune system as well as to defend themselves against their own toxic metabolites. They comprise the soluble flavohaemoglobin Hmp, and the diiron-centred flavorubredoxin NorV with its NADH-dependent oxidoreductase NorW. Hmp is phylogenetically widespread, being found in denitrifying bacteria and nondenitrifiers (Vinogradov, Tinajero-Trejo, Poole, & Hoogewijs, 2013). This enzyme has a globin-like domain, and an FAD-containing domain that binds NAD(P)H. In the presence of oxygen, Hmp oxidizes NO to nitrate, an activity that has been described as an NO dioxygenase or NO denitrosylase. A detailed description of Hmp enzymatic and structural properties has been published in several reviews (Forrester & Foster, 2012; Gardner, 2005; Poole, 2005; Spiro, 2011). Aside from NO dioxygenation, Hmp has also been shown to execute NO reduction to N_2O under anoxic conditions (Kim, Orii, Lloyd, Hughes, & Poole, 1999), which operates at approximately 1% of the rate of the aerobic dioxygenation reaction (Mills, Sedelnikova, Soballe, Hughes, & Poole, 2001). Although this Hmp-based NO reduction may operate under anaerobic conditions, it remains somewhat unclear whether it provides physiologically relevant protection from nitrosative stress. Consequently, Hmp may not be a significant source of N_2O. The main candidate to reduce NO to N_2O in nondenitrifying bacteria is NorVW (Fig. 2A). The physiological role of this enzyme seems to be NO detoxification under anaerobic or microoxic conditions. This reaction may be particularly important in organisms (such as *E. coli* or *S.* Typhimurium) which make low concentrations of NO as a by-product of the reduction of nitrite to ammonium but lack the respiratory Nor enzymes typical from denitrifiers (reviewed by Poole, 2005; Spiro, 2011). In NorVW, NO is reduced by a flavodiiron protein, which receives electrons from a rubredoxin domain or protein. The rubredoxin is itself reduced by an NADH-dependent flavoenzyme. The flavodiiron protein of *E. coli* and *S.* Typhimurium has a fused rubredoxin domain, and so is called flavorubredoxin (also called NorV). In complex with the NADH-dependent oxidoreductase (NorW), this enzyme functions as an NO reductase in vitro (Gomes et al., 2002). Consistently, in *E. coli* and *S.* Typhimurium it has been reported that protection against NO stress during anaerobic respiratory conditions was mainly attributed to the action of NorVW (Gardner & Gardner, 2002; Gardner, Helmick, & Gardner, 2002; Mills et al., 2008; Mühlig, Kabisch, Pichner, Scherer, & Muller-Herbst, 2014). However, it should be noted that *S.* Typhimurium mutant strains lacking functional copies of *hmpA*, *norV* and *nrfA* are still able to resist anaerobic NO stress, albeit very poorly, indicating a role for other NO detoxification mechanisms in this bacterium (Mills et al., 2008). As observed in *S.* Typhimurium, *E. coli* single mutants defective in NirB, NrfA, NorV or Hmp and even the mutant defective in all four proteins reduced NO at the same rate as the parent. Clearly, therefore, there are mechanisms of NO reduction by enteric bacteria that remain to be characterized (Vine & Cole, 2011).

Although it has been proposed in nitrate ammonifiers that N_2O is the product of NO reduction, however, studies about the contribution of this bacterial group to N_2O emissions from agricultural soils as well as the mechanisms behind this are poorly understood. In this context, there have been a few reports of N_2O release by pure cultures of Enterobacteriaceae, including *E. coli*, *Klebsiella pneumoniae* and *S.* Typhimurium during nitrate metabolism that presumably reflects NO being converted into N_2O (Bleakley & Tiedje, 1982; Smith, 1983). In complex medium nutrient-sufficient batch culture experiments, the rate of N_2O production during nitrate ammonification was around 5% of nitrate (Bleakley & Tiedje, 1982). Thus, it has been suggested that enteric nitrate-ammonifying bacteria could be a significant source of N_2O in soil (Bleakley & Tiedje, 1982). Recently, it has been demonstrated the potential for N_2O production by soil-isolated nitrate-ammonifying bacteria under different C and N availabilities. By performing chemostat cultures, it has been shown that maximum N_2O production was correlated with high nitrite production under C-limitation/nitrate sufficiency conditions (Streminska, Felgate, Rowley, Richardson, & Baggs, 2012).

As mentioned earlier, one major source of N_2O in *S.* Typhimurium is the reduction of NO produced by Nar (Gilberthorpe & Poole, 2008; Fig. 2A). This process has been studied in detail using chemostat cultures where kinetics analyses of nitrate consumption, nitrite accumulation and N_2O production by *S.* Thyphimurium *nap* or *nar* mutants have confirmed that Nar is the major enzymatic route for nitrate catabolism associated with N_2O production (Rowley et al., 2012). Under nitrate-sufficient conditions, a *narG* mutant produced ~30-fold lower N_2O than wild-type *S.* Thyphimurium, while under nitrate-limited conditions, *nap*, but not *nar*, was upregulated and very little N_2O production was observed. Accordingly, Rowley and coworkers conclude that a combination of nitrate sufficiency, nitrite accumulation and an active Nar-type nitrate reductase leads to NO and thence N_2O production, and this can account for up to 20% of the nitrate catabolized (Rowley et al., 2012).

2.1.2 Regulatory Proteins

The main regulators that mediate expression of NO detoxification systems and consequently N_2O formation in *S.* Typhimurium and *E. coli* include NorR, NsrR and FNR (reviewed by Arkenberg, Runkel, Richardson, & Rowley, 2011; Mettert & Kiley, 2015; Spiro, 2007, 2011, 2012; Tucker, Ghosh, Bush, Zhang, & Dixon, 2010; Tucker, Le Brun, Dixon, & Hutchings, 2010; Fig. 2B). NorR is a member of the σ^{54}-dependent enhancer-binding protein (EBP) family of transcriptional activators that has a three-domain structure that is typical of EBPs, including a C-terminal DNA-binding domain, a central AAA^+ family domain that has ATPase activity and interacts with RNA polymerase (RNAP) (Bush, Ghosh, Tucker, Zhang, & Dixon, 2010), and an N-terminal signal transduction domain. The N-terminal regulatory GAF domain of NorR contains a mononuclear nonhaem iron centre, which reversibly binds NO. Binding of NO stimulates the ATPase activity of NorR, which drives the transcriptional activation by RNAP. The mechanism of NorR reveals an unprecedented biological role for a nonhaem mononitrosyl–iron complex in NO perception in bacteria (D'Autreaux, Tucker, Dixon, & Spiro, 2005; Tucker et al., 2008). When activated by NO, NorR is a key transcriptional activator of *E. coli norVW* gene expression (Gardner, Gessner, & Gardner, 2003; Hutchings, Mandhana, & Spiro, 2002; Fig. 2B).

NsrR is a NO-sensitive transcriptional repressor that contains an [Fe–S] cluster. Nitrosylation of this cluster leads to a loss of DNA-binding activity and, hence, derepression of genes controlled by NsrR

(Bodenmiller & Spiro, 2006; Crack et al., 2015; Tucker et al., 2008; Yukl, Elbaz, Nakano, & Moenne-Loccoz, 2008). The NsrR-binding site is an 11-1-11 bp inverted repeat of the following A/T-rich consensus recognition sequence: AAGATGCYTTT (Bodenmiller & Spiro, 2006), although chromatin immunoprecipitation (ChIP-chip) analysis has suggested that a single 11 bp consensus sequence (AANATGCATTT) can function as an NsrR-binding site in vivo (Partridge, Bodenmiller, Humphrys, & Spiro, 2009). Very recently, it has been demonstrated that although *nsrR* is expressed from a strong promoter, the translational efficiency of this regulator is extremely inefficient, leading to low cellular NsrR protein levels. Therefore, it has been proposed that promoters with low-affinity NsrR-binding sites may partially escape NsrR-mediated repression (Chhabra & Spiro, 2015). Microarray analyses revealed that NsrR represses 9 operons encoding 20 genes in *E. coli*, including *hmp*, and the well-studied *nrfA* promoter (Filenko et al., 2007). Notably, regulation of the *nrf* operon by NsrR is consistent with the ability of the periplasmic nitrite reductase to reduce NO and thus protect the cell against reactive nitrogen species (Fig. 2B).

FNR (fumarate–nitrate reduction regulator) belongs to the subgroup of the cyclic-AMP receptor protein family of bacterial transcription regulators. FNR is a O_2-sensitive protein involved in gene expression to coordinate the switch from aerobic to anaerobic metabolism, when facultative anaerobes like *E. coli* are starved of O_2 (Constantinidou et al., 2006; Myers et al., 2013; Partridge et al., 2007; Rolfe et al., 2012). The N-terminal region of FNR contains four essential cysteine (Cys) residues that coordinate an O_2-sensitive [4Fe–4S] cluster (Crack, Green, Hutchings, Thomson, & Le Brun, 2012; Zhang et al., 2012). In the absence of O_2, the [4Fe–4S] cluster is stable, and FNR exists as a homodimer that is capable of high affinity, site-specific DNA binding to an FNR box (TTGATNNNNATCAA). When bound to target DNA, FNR activates the expression of genes encoding proteins required for anaerobic metabolism and represses those utilized under aerobic conditions. In addition to its primary function in mediating an adaptive response to O_2-limitation, FNR also plays a role in sensing and responding to NO. It has been demonstrated that NO damages the *E. coli* FNR [4Fe–4S] cluster in vitro, resulting in decreased DNA-binding activity by the FNR protein (Crack, Stapleton, Green, Thomson, & Le Brun, 2013). In the absence of nitrogen oxides *hmp* is repressed by FNR; however, the addition of either nitrite or nitrate causes a derepression of *hmp* gene expression (Cruz-Ramos et al., 2002). Conversely, in *E. coli*, transcription of the *nrf* operon is activated by FNR in the absence of oxygen and induced further by

NarL and NarP in response to low concentrations of nitrate and/or nitrite (Tyson, Cole, & Busby, 1994; Fig. 2B). Consistent with the additional NO-detoxifying function of Nrf, recent studies have suggested that the *nrf* promoter (*pnrf*) is also regulated by the global transcription repressor NsrR (Filenko et al., 2007; Partridge et al., 2009). Supporting those findings, it has been demonstrated that FNR-dependent activation of the *E. coli pnrf* is downregulated by NsrR together with the nucleoid-associated protein integration host factor (IHF), which bind to overlapping targets adjacent to the DNA site for FNR binding (Browning, Lee, Spiro, & Busby, 2010). Interestingly, alignment of the *pnrf* sequence from *S.* Typhimurium with that of *E. coli* revealed a base difference in the DNA site for NsrR that would be expected to decrease NsrR binding. In fact, anaerobic expression of *pnrf* in *S.* Typhimurium is unaffected by the disruption of *nsrR* (Browning et al., 2010), implying that in contrast to *E. coli,* the *S.* Typhimurium *pnrf* appears to have become 'blind' to repression by NsrR, though it remains to be seen if this experimental observation has any biological significance.

2.1.3 Nitrate Ammonification and Denitrification in Sh. loihica

Until recently, it was widely envisaged that pathways for denitrification and respiratory nitrate ammonification did not coexist within a single organism. However, recent genome analyses have revealed that at least three different bacterial species, including *Opitutus terrae* strain PB90-1, *Marivirga tractuosa* strain DSM 4126 and the Gammaproteobacterium *Sh. loihica* strain PV-4, possess the complete sets of genes encoding the pathways for denitrification and respiratory ammonification (Sanford et al., 2012). In fact, *Sh. loihica* strain PV-4 possesses two copies of *nrfA*, as well as the complete suite of genes encoding denitrification enzymes (*nirK*, *norB* and *nosZ*) (Sanford et al., 2012; Yoon, Sanford, & Loffler, 2013). The functionality of both the denitrification and the respiratory ammonification pathways has been recently confirmed (Yoon, Cruz-Garcia, Sanford, Ritalahti, & Loffler, 2015). Batch and continuous culture experiments using *Sh. loihica* strain PV-4 revealed that denitrification dominated at low C/N ratios (ie, electron donor-limiting growth conditions), whereas ammonium was the predominant product at high C/N ratios (ie, electron acceptor-limiting growth conditions) (Yoon, Cruz-Garcia, et al., 2015). In addtion to C/N ratio, pH and temperature also affected the fate of nitrate/nitrite with ammonium formation being favoured by incubation above pH 7.0 and temperatures of 30°C (Yoon, Cruz-Garcia, et al., 2015). Recent findings revealed that the nitrite/nitrate ratio also affected the distribution of reduced products, and respiratory ammonification dominated at high nitrite/nitrate ratios, whereas low nitrite/nitrate ratios favoured denitrification (Yoon, Sanford, & Loffler, 2015). These findings implicate nitrite as a relevant modulator of nitrate fate in *Sh. loihica* strain PV-4 and, by extension, suggest that nitrite is a relevant determinant for N retention (by ammonification) vs N loss and greenhouse gas emission (by denitrification).

2.2 Epsilonproteobacteria

2.2.1 Respiratory Reduction of Nitrate and Nitrite, Detoxification of NO and the Concomitant Generation of N_2O

Epsilonproteobacteria comprise host-associated heterotrophic species (exemplary genera include *Campylobacter, Helicobacter* and *Wolinella*) as well as free-living species that have been isolated mostly from sulphidic terrestrial and marine habitats (*Sulfurospirillum, Sulfurimonas, Nautilia*) (Campbell, Engel, Porter, & Takai, 2006). Epsilonproteobacterial cells usually grow at the expense of microaerobic or anaerobic respiration and many species use hydrogen, formate or reduced sulphur compounds, such as sulphide or thiosulfate as electron donor substrates. Nitrate is a prominent electron acceptor in Epsilonproteobacteria and is initially reduced to nitrite by Nap. The nonfermentative rumen bacterium *W. succinogenes* has been used for a long time as an epsilonproteobacterial model organism to investigate the multitude of electron transport chains (ETCs) that couple anaerobic respiration to ATP generation. *W. succinogenes* cells may use formate, hydrogen gas or sulphide as electron donors and either fumarate, nitrate, nitrite, N_2O, dimethyl sulfoxide, polysulphide or sulphite as electron acceptors (Hermann, Kern, La Pietra, Simon, & Einsle, 2015; Kern, Klotz, & Simon, 2011; Kern & Simon, 2009a, 2016; Klimmek et al., 2004; Kröger et al., 2002; Simon, 2002; Simon & Klotz, 2013; Simon & Kroneck, 2013, 2014; references therein). The cells are also capable of microaerobic respiration, and the complete genome sequence suggests the existence of further electron acceptors such as arsenate or tetrathionate (Baar et al., 2003). With respect to the physiology and enzymology of respiratory nitrate ammonification, *W. succinogenes* is arguably the best-characterized member of the Epsilonproteobacteria (reviewed by Kern & Simon, 2009a; Simon, 2002; Simon & Klotz, 2013). Like many other Epsilonproteobacteria, the cells employ Nap for nitrate reduction to nitrite, and the latter is subsequently reduced to ammonium by NrfA.

Epsilonproteobacterial *nap* gene clusters generally lack a *napC* gene but, instead, NapG and NapH proteins are encoded (Kern & Simon, 2008).

The NapGH complex is proposed to constitute a menaquinol (MKH_2)-oxidizing complex, in which NapH may act as a membrane-bound quinol dehydrogenase, while NapG is a periplasmic Fe–S protein that is thought to deliver electrons to NapB (Fig. 2A). In *W. succinogenes*, Nap encoded by the *napAGHBFLD* gene cluster. The role of individual *nap* genes in *W. succinogenes* has been assessed by characterizing nonpolar gene inactivation mutants (Kern, Mager, & Simon, 2007; Kern & Simon, 2008, 2009b). NapB and NapD were shown to be essential for growth by nitrate respiration, with NapD being required for the production of mature NapA. The inactivation of either *napH* or *napG* almost abolished growth without affecting the formation and activity of NapA. The cytoplasmic Fe/S protein NapF was shown to interact with NapH. NapF could be involved in electron transfer to immature NapA. Inactivation of *napL* did only slightly affect the growth behaviour of mutant cells, although the NapA-dependent nitrate reductase activity was clearly reduced. The function of NapL, however, is not known.

In contrast to *E. coli* and other Gammaproteobacteria, the epsilonbacterial NrfA forms a subunit of a membrane-bound menaquinol-reactive complex that also contains a tetrahaem cytochrome *c* of the NapC-type called NrfH (Einsle, 2011; Kern & Simon, 2008; Rodrigues, Oliveira, Pereira, & Archer, 2006; Simon et al., 2000; Simon & Kroneck, 2014; Fig. 3B). Such NrfHA complexes form a membrane-associated respiratory complex on the extracellular side of the cytoplasmic membrane that catalyses electroneutral menaquinol oxidation by nitrite. In *W. succinogenes* the structural genes *nrfA* and *nrfH* are part of an *nrfHAIJ* gene cluster. The product of the *nrfI* gene is a membrane-bound cytochrome *c* synthase of the CcsBA-type, which belongs to the so-called system II of cytochrome *c* biogenesis (Simon & Hederstedt, 2011). *W. succinogenes* NrfI was shown to play a crucial role in NrfA biogenesis as it is required for the covalent attachment of the active site cytochrome *c* moeity (haem 1) to NrfA via the novel CX_2CK motif, which includes the axial haem iron ligand lysine (Kern, Eisel, Scheithauer, Kranz, & Simon, 2010; Pisa, Stein, Eichler, Gross & Simon, 2002). By contrast, no function in nitrite respiration could be assigned to NrfJ as concluded from the characterization of a corresponding gene deletion mutant (Simon et al., 2000).

W. succinogenes NrfA has a remarkable substrate range since it catalyses the reduction of nitrite, NO and hydroxylamine to ammonium (Simon & Hederstedt, 2011; Simon & Kroneck, 2014; Stach, Einsle, Schumacher, Kurun, & Kroneck, 2000). NrfA was also reported to produce N_2O as a product of NO reduction under suitable conditions (Costa et al., 1990) and to react with N_2O to a so far unidentified product (Stach et al., 2000). Furthermore, NrfA catalyses the decomposition of hydrogen peroxide and the reduction of sulphite to hydrogen sulphide, which is an isoelectronic reaction to ammonium production from nitrite (Kern, Volz, & Simon, 2011; Lukat et al., 2008). The reactive promiscuity of NrfA has been shown to mediate the stress response to nitrite, NO, hydroxylamine and hydrogen peroxide in *W. succinogenes* cells indicating that NrfA has an important detoxifying function in cell physiology (Kern, Volz, et al., 2011) Aside from NrfA, a cytoplasmic Fdp has been proposed to be involved in nitrosative stress defence in *W. succinogenes* (Kern, Volz, et al., 2011). As proposed previously for these type of Fdps (Saraiva, Vicente, & Teixeira, 2004), *W. succinogenes* Fdp is assumed to reduce NO to N_2O. However, this reaction has not been demonstrated for *W. succinogenes* Fdp since the protein has not been purified. Further possible candidates for enzymatic NO reduction in *W. succinogenes* include the hybrid cluster protein (Hcp) and a homolog of *Helicobacter pylori* NorH (Ws1903) (Justino, Ecobichon, Fernandes, Boneca, & Saraiva, 2012; Kern, Volz, et al., 2011; Luckmann et al., 2014). However, the contribution of these proteins to N_2O production has yet to be clarified.

The capacity of *W. succinogenes* to produce N_2O during growth by nitrate ammonification has been recently examined using nitrate-sufficient or nitrate-limited medium containing formate as electron donor (Luckmann et al., 2014). It was found that cells growing in nitrate-sufficient medium (80 mM formate and 50 mM nitrate) produced small amounts of N_2O (about 0.15% of nitrate-N), which derived from accumulated nitrite and, most likely, from the presence of NO. In contrast, nitrite is only transiently formed during growth in nitrate-limited medium (80 mM formate and 10 mM nitrate), and both NO and N_2O could not be detected under these conditions (Luckmann et al., 2014). However, questions remain concerning how NO is generated from nitrite by *W. succinogenes* since NapA and NrfA are unlikely to release NO as a by-product (as opposed to Nar; see Section 2.1). In the experiments described by Luckmann et al. (2014), conceivably NO may have been generated by chemical reactions between components of the medium and nitrite. Taken together, there is clear evidence that *W. succinogenes* cells are able to produce N_2O when faced with NO by subsequent detoxification reactions. It is quite likely that these features do also hold true for other Epsilonproteobacteria that contain similar *nap*, *nrf* and *nos* gene clusters, for example, free-living species of the genus *Sulfurospirillum* (Kern & Simon, 2009a) as well as host-associated

Campylobacter species (Payne, Grant, Shapleigh, & Hoffman, 1982; Schumacher & Kroneck, 1992). Interestingly, Kaspar and Tiedje (1981) reported that the nitrate-ammonifying rumen microbiota accumulated up to 0.3% of the added nitrate-N as N_2O.

2.2.2 Growth by N_2O Respiration and Reduction of N_2O by the Atypical Cytochrome c N_2OR System

More than three decades ago, *W. succinogenes* and *Campylobacter fetus* cells were reported to grow by N_2O respiration using formate as electron donor (Payne et al., 1982; Yoshinari, 1980). However, only recently a corresponding growth curve for *W. succinogenes* has been provided that allowed determination of a doubling time of 1.2 h and an estimate for growth yield of about 10 g dry cells per mole formate (Kern & Simon, 2016). Interestingly, this value is higher than the reported maximal cell yield of fumarate respiration (8.5 g of dry cells per mole formate; Bronder, Mell, Stupperich, & Kroger, 1982) as well as of nitrate and nitrite respiration (5.6 and 5.3 g of dry cells per mole formate, respectively) (Bokranz, Katz, Schröder, Robertson, & Kröger, 1983). In the latter three modes for anaerobic respiration, the proton motive force (*pmf*) is formed up by the redox loop mechanism of the membrane-bound formate dehydrogenase complex (FdhABC) (Richardson & Sawers, 2002; Simon, van Spanning, & Richardson, 2008). Furthermore, it has been shown that menaquinol oxidation by fumarate or nitrite are electroneutral processes (Kröger et al., 2002; Lancaster et al., 2005; Simon et al., 2000) and, originally, the same was expected for menaquinol oxidation by nitrate or N_2O given the postulated architecture of the corresponding ETCs that are envisaged to comprise homologous menaquinol dehydrogenases (NapGH or NosGH) (Figs 2A and 5; Kern & Simon, 2008; Simon et al., 2004; Simon & Klotz, 2013; references therein). In the light of the cell yield of N_2O respiration, it remains to be seen whether menaquinol oxidation by N_2O might involve a hitherto undiscovered *pmf*-generating process that is absent in classical nitrate respiratory pathways. Conceivable scenarios include involvement of the cytochrome bc_1 complex (electrogenic menaquinol oxidation through the Q cycle mechanism) and/or the as-yet uncharacterized polytopic membrane protein, NosB, that might work as a menaquinol-reactive proton pump (Fig. 5). The presence of the corresponding gene is conserved in epsilonproteobacterial *nos* gene clusters with *nosB* flanked by the *nosZ* and *nosD* genes in most cases (Fig. 4; Sanford et al., 2012).

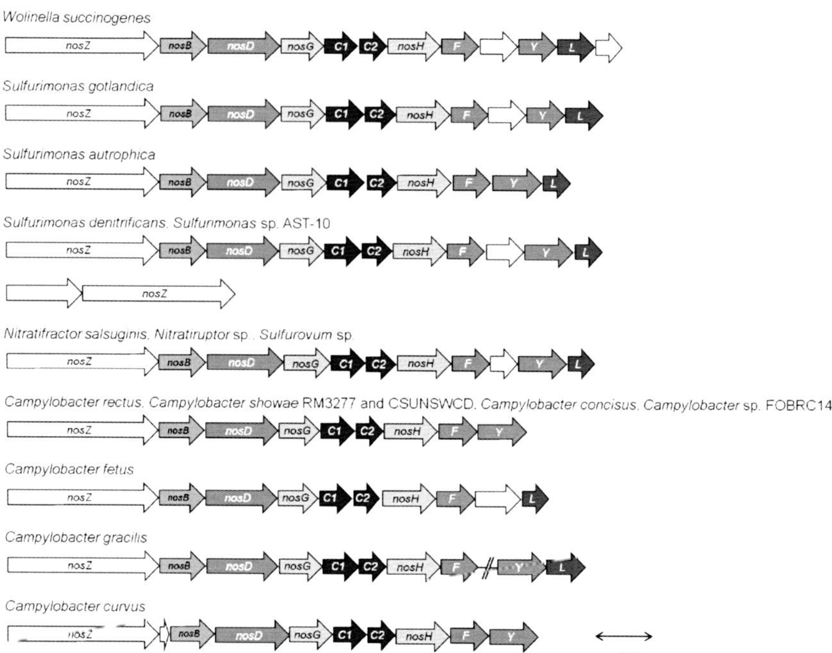

Fig. 4 Compilation of selected *nos* gene clusters in epsilonproteobacterial genomes. The *nosZ* genes encode cytochrome *c* nitrous oxide reductases (*c*NosZ enzymes) that belong to the so-called atypical N_2ORs. The presence of *nosB, -G, -H, -C1* and *-C2* genes as well as the absence of *nosR* and *-X* genes is indicative for atypical *nos* gene clusters. Undesignated genes shown in white encode hypothetical proteins.

The *W. succinogenes nos* gene cluster belongs to the atypical clusters and contains *nosZ, -B, -D, -G, -C1, -C2, -H, -F, -Y* and *-L* genes (Sanford et al., 2012; Simon et al., 2004; Fig. 4). The NosG, -C1, -C2 and -H proteins have been postulated to encode a putative electron transport pathway from menaquinol to *c*NosZ (Fig. 5). This pathway comprises a NosGH menaquinol dehydrogenase complex and two cytochromes *c* (NosC1 and NosC2). NosG and NosH are highly similar to NapG and NapH and therefore expected to form a NosGH complex that is functionally equivalent to NapGH. NosC1 and NosC2 are monohaem cytochromes *c* located either in the periplasm, or monotopic membrane proteins anchored to the outer face of the cytoplasmic membrane via an N-terminal helix

Ultimately, electrons are thought to be transferred via the cytochrome *c* domain of *c*NosZ to the copper-containing catalytic site of N_2O reduction. The *nosF, -Y* and *-D* genes are likely to encode a membrane-bound ABC transporter, and the *nosL* gene is predicted to encode a copper chaperone

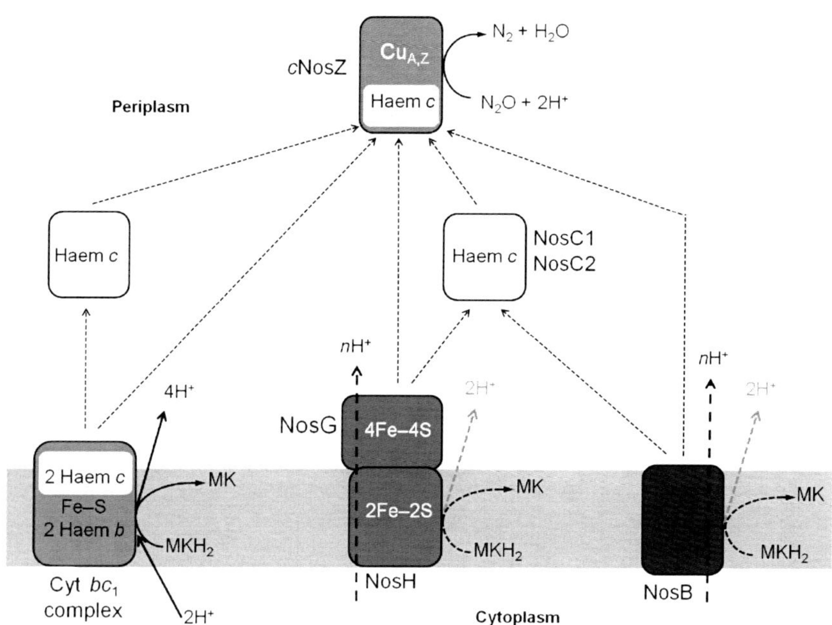

Fig. 5 Putative electron transport pathways connecting the membranous menaquinone/menaquinol pool with periplasmic *c*NosZ in Epsilonproteobacteria. *Dashed arrows* indicate speculative reactions, interactions or proton pathways. See Fig. 4 for the organization of genes encoding NosGH, NosC1, NosC2 and NosB. Protons shown in *grey* or *black* contribute to electroneutral or electrogenic reactions, respectively. Note that epsilonproteobacterial cytochrome bc_1 complexes are predicted to contain a dihaem cytochrome *c* that is thought to interact with the cytochrome *c* domain of *c*NosZ or another cytochrome *c*. In *W. succinogenes*, a suitable candidate for such a small soluble cytochrome *c* is Ws0700. *MK*, menaquinone; *MKH*$_2$, menaquinol.

involved in metallocentre assembly (Zumft, 2005; Zumft & Kroneck, 2007). In analogy to that which has been proposed for denitrifiers (see Fig. 8), the NosF, -Y, -D- and -L proteins might be involved in the maturation of atypical Nos systems. Many other Epsilonproteobacteria also possess atypical *nos* gene clusters resembling that of *W. succinogenes*, and it is conspicuous that the presence and arrangement of the *nosB*, -*G*, -*H*, -*C1* and -*C2* genes seem to be strictly conserved (Fig. 4).

2.2.3 Transcriptional Regulation of the W. succinogenes nos Gene Cluster

In *W. succinogenes*, the respiratory Nap, Nrf and *c*Nos enzymes involved in N_2O metabolism are upregulated in response to the presence of either nitrate, each of the NO-releasing compounds sodium nitroprusside

(SNP), *S*-nitrosoglutathione (GSNO) and spermine NONOate or N_2O, but not to nitrite or hydroxylamine (Kern & Simon, 2016; Kern, Winkler, & Simon, 2011). However, nitrate-responsive two-component systems homologous to NarXL/NarQP from *E. coli* and other enteric bacteria are not encoded in the *W. succinogenes* genome. Furthermore, well-characterized NO-responsive proteins such as NsrR and NorR as well as NO-reactive transcription regulators of the (cAMP receptor protein) Crp/FNR superfamily, for example, the FNR, NNR/NnrR, Dnr and NarR proteins, are also absent in *W. succinogenes*. Instead, this bacterium employs three transcription regulators of the Crp/FNR superfamily (homologs of *C. jejuni* NssR (nitrosative stress-sensing regulator) Elvers et al., 2005), designated NssA, NssB and NssC, to mediate upregulation of Nap, Nrf and *c*Nos via dedicated signal transduction routes (Fig. 6; Kern & Simon, 2016). Analysis of single *nss* mutants revealed that NssA controls production of the Nap and Nrf systems in fumarate-grown cells, while NssB was required to induce the Nap, Nrf and *c*Nos systems specifically in response to NO-generators (Fig. 6). NssC was indispensable for *c*Nos production under any tested condition. Moreover, N_2O apparently induced the Nap and Nrf systems independently of any Nss protein. The data implied the presence of an N_2O-sensing mechanism since upregulation of Nap, Nrf and *c*Nos was found in N_2O-gassed formate/fumarate medium, ie, in the absence of notable amounts of nitrate or NO.

Nss proteins contain an N-terminal effector domain and a C-terminal DNA-binding domain. In *C. jejuni*, which lacks *nos* genes altogether, NssR was found to be involved in the expression of genes encoding a single domain haemoglobin (Cgb) and a truncated haemoglobin (Ctb) in response to NO/nitrosative stress conditions (Elvers et al., 2005; Monk, Pearson, Mulholland, Smith, & Poole, 2008). An *nssR* disruption mutant was found to be hypersensitive to NO-related stress conditions (Elvers et al., 2005). The *C. jejuni* NssR protein was purified and shown to bind specifically to the *ctb* promoter by electrophoretic mobility shift assays (Smith, Shepherd, Monk, Green, & Poole, 2011). Most likely, this binding was accomplished via an FNR-like binding site with a TTAAC-N_4-GTTAA consensus sequence (Elvers et al., 2005) is absent upstream of the *C. jejuni nap* and *nrf* gene clusters. Interestingly, DNA regions upstream of the *W. succinogenes nap*, *nrf* and *nos* gene clusters contain potential Nss-binding sites (consensus sequence TTGA-N_6-TCAA) within appropriate distances to the respective transcriptional start sites. In the future, it will be most interesting to characterize the different N-terminal effector domains

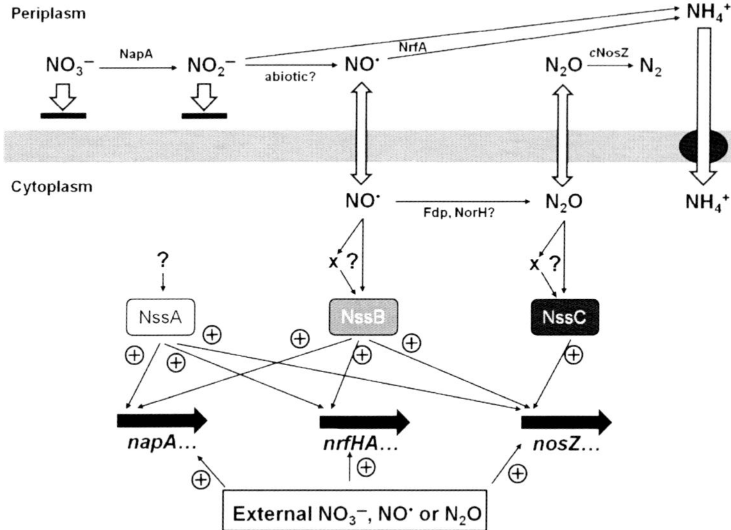

Fig. 6 Working model depicting the dissimilatory/detoxifying metabolism of nitrogen compounds in *W. succinogenes* cells and the predicted roles of the NssA, NssB and NssC proteins. NO and N_2O are thought to passively cross the cell membrane, whereas ammonium is probably taken up by an Amt-type transporter. There are no obvious candidates for nitrate or nitrite uptake systems encoded in the genome. Externally supplied nitrate, NO or N_2O were found to be capable of inducing each of the three respiratory systems (bottom). The assumed interaction of NssA, NssB and NssC with regulatory elements of the *nap*, *nrf* and *nos* gene clusters is shown and the *encircled* + denotes that an Nss protein is required to upregulate the corresponding enzyme system. *Question marks* denote that the signal transduction pathways for NssA, NssB (responsive to NO) and NssC (responsive to N_2O) are not known. It cannot be excluded, however, that NO and N_2O directly interact with NssB and NssC, respectively. *Adapted from Kern, M., & Simon, J. (2016). Three transcription regulators of the Nss family mediate the adaptive response induced by nitrate, nitric oxide or nitrous oxide in Wolinella succinogenes.* Environmental Microbiology. *http://dx.doi.org/10.1111/1462-2920.13060.*

of NssR, NssA, NssB and NssC and to assess whether these are directly or indirectly involved in cytoplasmic signal sensing. To date, it cannot be excluded that such domains are reactive with nitrogen compounds such as nitrate, NO or even N_2O.

2.3 Nitrate-Ammonifying *Bacillus* Species

Streminska et al. (2012) demonstrated that nitrate-ammonifying soil isolates of the genus *Bacillus* formed N_2O (up to 2.7% of nitrate was found to be reduced to N_2O) under nitrate-sufficient conditions (low C/nitrate ratio). Furthermore, the genomes of several other *Bacillus* species including

B. vireti, *B. azotoformans* and *B. bataviensis* were reported to encode a cytochrome *c* nitrite reductase complex (NrfHA) in addition to the presence of one or more atypical *nos* gene clusters (Heylen & Keltjens, 2012; Mania et al., 2014). In fact, the *B. azotoformans* genome encodes three atypical N_2OR (lacking the monohaem cytochrome *c* domain found in Epsilonproteobacteria) in different genetic contexts (Heylen & Keltjens, 2012). Each of the gene clusters includes a copy of *nosB* but lacks *nosG*, *-H*, *-R* and *-X* genes. Cells of *B. vireti* have been described to grow as nitrate ammonifiers in the presence of 5 mM nitrate, although their N_2OR was also found to be active in generating N_2 under these conditions (Mania et al., 2014). More recently, evidence was provided that N_2O reduction is coupled to growth of *B. vireti* cells (Mania et al., 2016). On the other hand, the *B. vireti* genome does not encode any obvious gene for an NO-generating nitrite reductase (NirS or NirK) and thus the cells do not qualify to be termed a classical denitrifier (Liu et al., 2015; Mania et al., 2014). It seems reasonable to assume that the mentioned *Bacillus* species are respiratory nitrate ammonifiers that are also capable to reduce N_2O formed as a product of NO detoxification. Surely, it is desirable to explore these environmentally important organisms using suitable gene deletion mutants; however, corresponding genetic systems remain to be established in most cases. Recently, mutants of *B. vireti* lacking either *narG* or *nrfA* have been successfully constructed, and their physiology will be investigated in the future (M. Nothofer, T. Heß, & J. Simon, unpublished data).

3. N_2O METABOLISM IN DENITRIFYING BACTERIA

Despite various sources for N_2O emission in soils (see Section 1), it has been estimated that over 65% of atmospheric N_2O is derived from microbial nitrification and denitrification (Thomson et al., 2012). Of these processes, denitrification is currently considered to be the largest source of N_2O. Denitrification commonly proceeds with respiratory reduction of the water-soluble nitrogen (N)-oxyanion nitrate, which is readily bioavailable and abundant in many terrestrial and aquatic ecosystems. The nitrite formed from dissimilatory nitrate reduction is subsequently converted to gaseous N-oxide intermediates, including the highly reactive cytotoxic free radical and ozone-depleting agent NO, and the potent and long-lived GHG N_2O, which can be further reduced to N_2. Here, each of the N-oxyanions and N-oxides described may act as an individual terminal electron acceptor. Therefore, the reactions of denitrification underpin alternative and elaborate respiratory chains that function in the absence of the terminal oxidant, oxygen (O_2) to enable facultative aerobic microorganisms to survive and multiply under anaerobic conditions.

When faced with a shortage of O_2, although many bacterial species may have the potential to tailor their respiratory pathways, the identity (ie, genetic complement coding for active denitrification enzymes) and environmental conditions largely determine whether a denitrifier serves as a source or sink for N_2O (Thomson et al., 2012). Denitrification is widespread within the domain of *Bacteria* and appears to be dominant within *Proteobacteria* (Shapleigh, 2006). However, there is evidence that some fungi (Prendergast-Miller, Baggs, & Johnson, 2011; Takaya, 2002) and archaea (Treusch et al., 2005) may also denitrify. Most of the studies about denitrification have been focused on Gram-negative bacteria that occupy terrestrial niches, using the alpha-proteobacterium *Paracoccus* (*Pa.*) *denitrificans* as well as the gammaproteobacteria *Pseudomonas* (*Ps.*) *stutzeri* and *Ps. aeruginosa* as model organisms (Zumft, 1997). The reactions of denitrification are catalysed by Nap or Nar, nitrite reductases (NirK/NirS), nitric oxide reductases (cNor, qNor or Cu_ANor) and N_2OR encoded by *nap*/*nar*, *nirK*/*nirS*, *nor* and *nos* genes, respectively (Fig. 7). Reviews covering the physiology, biochemistry and molecular genetics of denitrification have been published elsewhere (Bueno, Mesa, Bedmar, Richardson, & Delgado, 2012; Kraft, Strous, & Tegetmeyer, 2011; Richardson, 2011; van Spanning, Delgado, & Richardson, 2005; van Spanning, Richardson, & Ferguson, 2007; Zumft, 1997).

Most denitrifiers have Nap and Nar enzymes, and depending on the species, Nap is employed for anaerobic nitrate respiration as a part of bacterial ammonification (see Sections 2.1 and 2.2), to promote denitrification (see Section 4) or as electron sink during aerobic (photo) organoheterotrophic growth on reduced carbon sources to ensure redox homeostasis to dissipate excess reductant. This is the case of *Pa. denitrificans* where in addition to Nar, it also synthesizes Nap. In this bacterium, Nar reduces nitrate as the first step of growth-linked anaerobic denitrification, however, Nap, which is non-electrogenic and serves to dissipate excess reducing equivalents formed during aerobic growth. These enzymes have been studied at the biochemical level and derive electrons from the ubiquinol pool (reviewed by Gonzalez et al., 2006; Potter et al., 2001; Richardson, 2011; Richardson et al., 2007; Simon & Klotz, 2013; for detailed information, see Section 2.1). With the exception of some archaeal and bacterial examples of Nar-type nitrate reductases with an active site on the outside of the

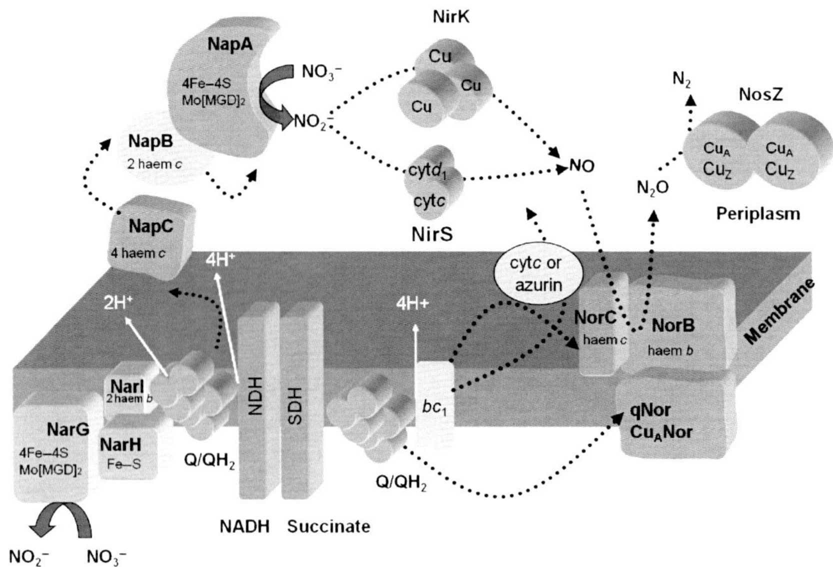

Fig. 7 Topological organization of denitrification enzymes. The membrane-bound (NarGHI), and periplasmic (NapABC) nitrate reductases as well as the nitrite reductases (NirK and NirS), nitric oxide reductases (*c*Nor, qNor and Cu_ANor) and nitrous oxide reductase (NosZ) are shown. *Adapted from Bueno, E., Mesa, S., Bedmar, E. J., Richardson, D. J., & Delgado, M. J. (2012). Bacterial adaptation of respiration from oxic to microoxic and anoxic conditions: Redox control.* Antioxidants & Redox Signaling, 16, *819–852.*

cytoplasmic membrane (Martinez-Espinosa et al., 2007), most Nar enzymes are oriented such that the active site for nitrate reduction is exposed to the cytoplasm and is dependent on a nitrate transport system situated in the cytoplasmic membrane for function. In *Pa. denitrificans*, NarK has been identified as a nitrate importer that moves nitrate into the cytoplasm and also exports nitrite, the product of nitrate reduction, to the periplasm to support respiratory denitrification. NarK is a fusion protein of two transmembrane domains NarK1 and NarK2, NarK1 is a proposed proton-linked nitrate importer, and NarK2 is a putative nitrate/nitrite antiporter (Goddard, Moir, Richardson, & Ferguson, 2008; Wood, Alizadeh, Richardson, Ferguson, & Moir, 2002).

As described earlier, two types of respiratory nitrite reductases have been described in denitrifying bacteria, NirS and NirK (reviewed by Rinaldo et al., 2008; Rinaldo & Cutruzzolá, 2007; van Spanning, 2011). These enzymes catalyse the single electron reduction of nitrite to nitric oxide, however, neither of the enzymes is electrogenic. Both are located in the periplasmic space and receive electrons from cytochrome *c* and/or a blue copper protein, pseudoazurin, via the cytochrome bc_1 complex (Fig. 7). NirS is a homodimeric enzyme with haems *c* and d_1. Electrons are transferred via the haem *c* of NirS to haem d_1, where nitrite binds and is reduced to NO (Rinaldo et al., 2008). The best-characterized *nirS* gene clusters are those from *Ps. aeruginosa* (*nirSMCFDLGHJEN*) and *Pa. denitrificans* (*nirXISECFDLGHJN*). In the model denitrifier *Ps. stutzeri,* there are two *nir* clusters (*nirSTBMCFDLGH* and *nirJEN*) which are separated by one part of *nor* gene cluster encoding nitric oxide reductase. The *nirS* gene encodes the functional subunits of the dimeric NirS. All other genes are required for proper synthesis and assembly and/or insertion of the active site haem d_1 cofactor required for NirS function. Biosynthesis of the haem d_1 cofactor has been a subject of much research (for a review, see Bali, Palmer, Schroeder, Ferguson, & Warren, 2014).

NirK enzymes are homotrimeric complexes harbouring three type I, and three type II copper centres, which form the active site (Fig. 7). Nitrite binds to the type II site where it is reduced to NO by electrons transferred from the type I copper site. In contrast to the complex organization of the genes encoding the NirS proteins, NirK is encoded by the *nirK* gene (Rinaldo & Cutruzzolá, 2007; van Spanning, 2011). Here, it must be noted that expression of NirK requires only a single gene, sometimes accompanied with a second one expressing a protein called NirV. The latter enzyme is related to desulfurase and may well be required for proper insertion of the copper reaction centre. As yet, no organism has been found that contains both types of nitrite reductase, so apparently the presence of one class of nitrite reductase excludes the option of gaining the second type.

The major contributor to the biological production of N_2O in many environments is the respiratory Nor found in denitrifying bacteria and in some ammonia-oxidizing organisms. Subsequently, the N_2O produced is consumed through respiratory reduction to N_2, which is catalysed by the N_2OR. Therefore, N_2OR completes the final reduction step in the denitrification pathway (Zumft & Kroneck, 2007) and is generally considered the sole enzyme able to turnover N_2O. However, various authors have suggested the existence of an alternative N_2O consumption pathway in which N_2O is reduced to ammonium by nitrogenase, the enzyme involved in N_2 fixation (Burgess & Lowe, 1996; Jensen & Burris, 1986; Yamazaki, Yoshida, Wada, & Matsuo, 1987). In fact, both N_2OR and nitrogenase are found in many denitrifiers (Shapleigh, 2006). Recent isotope tracing experiments by using *Ps. stutzeri* showed that consumption of N_2O via assimilatory reduction to ammonium did not occur (Desloover et al.,

2014). However, the latter studies showed that respiratory N_2O reduction can be coupled to N_2 fixation as N_2O is first reduced to N_2 before is further reduced to ammonium and incorporated into cellular biomass. This mechanism may play a significant role as an additional sink for N_2O and thus may be relevant to strategies to mitigate N_2O-driven climate change.

Given the importance of Nor and N_2OR enzymes for N_2O formation during denitrification, it appear to be essential to expand our current knowledge concerning these enzymes considered natural targets in the search for options to mitigate N_2O emission from agricultural soils.

3.1 Nitric Oxide Reductases

Nitric oxide reductase enzymes catalyse NO reduction at the outer face of the cytoplasmic membrane, and most of them have been characterized in denitrifying *Proteobacteria* (reviewed by de Vries, Suharti, & Pouvreau, 2007; Hendriks et al., 2000; Richardson, 2011; Zumft, 2005). The best-known NO reductases are cNor and qNor that use either cytochrome *c*/cupredoxins or quinones as immediate redox partners and both belong to the superfamily of haem–copper oxidases (HCOs) (Fig. 7). The catalytic site of NO reduction harbours a dinuclear haem b_3::FeB active site that is reduced by another haem *b* group bound by the same protein (NorB). In cNor enzymes, NorB receives electrons from the monohaem cytochrome *c* subunit NorC, while qNor enzymes are quinol-reactive single-subunit enzymes that resemble NorB (Fig. 7). In *Pa. denitrificans*, pseudoazurin or cytochrome c_{550} was found to donate electrons to the NorC subunits of a heterotetrameric $(NorBC)_2$ complex (Hendriks et al., 1998; Fig. 7). The best-characterized cNors are those from *Pa. denitrificans*, *Ps. stutzeri* and *Ps. aeruginosa*. The structure of the NorBC complex from *Ps. aeruginosa* (Hino et al., 2010) confirmed the predicted presence of 12 membrane-spanning α-helices in NorB, while NorC is anchored to the membrane by a single membrane-spanning segment. Biochemical experiments indicated that the protons required for NO reduction are taken from the periplasmic side of the membrane and that NorB does not function as a proton pump (Bell, Richardson, & Ferguson, 1992). The latter is confirmed in the structure by the absence of transmembrane proton channels in NorB analogous to those found in the proton-translocating HCOs (Hino et al., 2010). Based on the crystal structure of cNor from *Ps. aeruginosa* and molecular dynamics simulations, three different proton transfer pathways were proposed, all leading from the periplasmic side of the membrane (Hino et al.,

2010; Pisliakov, Hino, Shiro, & Sugita, 2012; Shiro, Sugimoto, Tosha, Nagano, & Hino, 2012). It has been demonstrated by site-directed mutation that *Pa. denitrificans* cNor is sensitive to mutations along the previously suggested proton transfer pathway 1 but not the others. Thus, although no energy is conserved, proton transfer still occurs through a specific pathway in *Pa. denitrificans* cNor (ter Beek, Krause, Reimann, Lachmann, & Adelroth, 2013). Furthermore, the formation of the hyponitrite (HO—N=N—O—) species in the haem b_3 Fe–FeB dinuclear centre of cNor from *Pa. denitrificans* has been recently demonstrated (Daskalakis, Ohta, Kitagawa, & Varotsis, 2015).

In contrast to cNor, qNor enzymes are reactive with ubiquinol and/or menaquinol and contain an N-terminal extension that is absent from NorB in the cNor complex. While this N-terminal extension shows similarity to NorC, a haem *c*-binding motif is lacking. The crystal structure of *Geobacillus stearothermophilus* qNor revealed a water channel from the cytoplasm that might serve in proton delivery (Matsumoto et al., 2012). Thus, the possibility that qNor might catalyse electrogenic quinol oxidation coupled to nitric oxide reduction cannot be excluded. An unusual qNor subgroup (qCu_ANor), exemplified by the enzyme from *B. azotoformans*, contains NorB in a complex with a subunit harbouring a Cu_A site (typically found in oxygen-reducing HCOs), which makes this enzyme competent in receiving electrons from membrane-bound cytochrome c_{551} in addition to the menaquinol pool (de Vries et al., 2007). However, it has been recently reported that the *Bacillus* enzyme lacks menaquinol activity and has changed its name from qCu_ANor to Cu_ANor (Al-Attar & de Vries, 2015).

NorCB structural subunits of cNor are encoded by *norCB* genes, respectively, which are usually cotranscribed with accessory genes designed *norD*, *norE*, *norF* and *norQ*. The gene order *norEFCBQD* is not universal; *norQ* and *norD* are always linked to *norCB*, however, *norE* and *norF* may be distantly located or absent in some genomes (Zumft, 2005). Intriguingly, in ancient thermophilic bacteria belonging to the Thermales and Aquificales phylogenetic groups, the *norC* and *norB* genes are always followed by a third gene (*norH*) encoding a small membrane protein that is required for efficient denitrification in vivo, likely allowing more efficient electron transport to cNor (Bricio et al., 2014). The functions of the accessory genes and their protein products are not well understood. It has been shown that NorD and NorE are integral membrane proteins required for successful heterologous assembly of the NorCB complex (Butland, Spiro, Watmough, & Richardson, 2001). NorE is a member of the subunit III of the cytochrome *c* oxidases family. Inactivation of *norEF* genes has been shown to slow NO reduction in both *Pa. denitrificans* and *R. sphaeroides* 2.4.3 (de Boer et al., 1996; Hartsock & Shapleigh, 2010). Recent physiological experiments have shown that *norEF* are not essential for Nor activity; however, their absence does affect activity under conditions where endogenous Nir activity generates prolonged exposure to NO (Bergaust, Hartsock, Liu, Bakken, & Shapleigh, 2014).

3.2 Nitrous Oxide Reductase

N_2OR is the terminal enzyme of bacterial denitrification and reduces N_2O by two electrons, breaking the N–O bond to release N_2 and H_2O (reviewed by Solomon et al., 2014; Spiro, 2012; van Spanning, 2011; Zumft & Kroneck, 2007; Fig. 7). Since this discovery, N_2OR had been purified and biochemically characterized from 11 denitrifying bacteria including *Ps. stutzeri* (Coyle, Zumft, Kroneck, Korner, & Jakob, 1985) and *Pa. denitrificans* (Snyder & Hollocher, 1987) among others. The crystal structure of the *Pa. denitrificans* N_2OR enzyme at 1.6 Å resolution has been revealed (Haltia et al., 2003). N_2ORs are homodimers with molecular weights of 120–160 kDa, a copper content of ~12 Cu atoms per dimer and a sulphide content of ~2 S^{2-} ions per dimer (Rasmussen et al., 2000). N_2OR contains two copper sites: Cu_A, a binuclear copper site with two Cys residues, two His residues, one Met residues, and the backbone carbonyl of a Trp residue as ligands, which acts as an electron transfer site (as in the HCOs), and Cu_Z, a tetranuclear μ4-sulphide-bridged cluster liganded by seven His residues, which is thought to be the site of N_2O binding and reduction. The ligands of the Cu_A site were identified from mutagenesis studies, and its structure was determined by analogy to the structurally characterized Cu_A site in the HCOs, which has close to identical properties to Cu_A in NosZ. By contrast, the structure of Cu_Z was determined by X-ray crystallography and is still a matter of active study (Einsle, 2011; Kroneck, Antholine, Riester, & Zumft, 1989; Zumft et al., 1992). There is a high degree of similarity between NosZ isolated from different sources, with the exception of *Thiobacillus denitrificans* which is a membrane-bound protein (Hole et al., 1996), most are periplasmic. The sequence of NosZ is conserved, showing a distinct two-domain architecture with an N-terminal, seven-bladed β-propeller domain and a smaller, C-terminal domain that adopts a conserved cupredoxin fold typical for copper-binding proteins (for a detail description of NosZ structural properties, see Wüst et al., 2012). Each domain harbours one of the copper-based metal centres of the enzyme, the binuclear Cu_A site in the cupredoxin domain and the tetranuclear copper-sulphide centre Cu_Z in the centre of the β-propeller (Johnston, Dell'Acqua, Pauleta, Moura, & Solomon, 2015; Johnston et al., 2014).

A gene cluster has been identified that is required for N_2O reduction, which encodes the NosZ protein and several ancillary proteins required for its expression, maturation and maintenance (Zumft, 2005). The core of this cluster, which is the minimum required for N_2O reduction, contains six genes (*nosRZDFYL*) and is sometimes associated with a further gene, *nosX* (Fig. 8). The cluster *nosZDFYL* is found in every N_2O-reducing prokaryote, whereas *nosR*, *nosX* and other *nos* genes such as *nosC*, *nosG* and

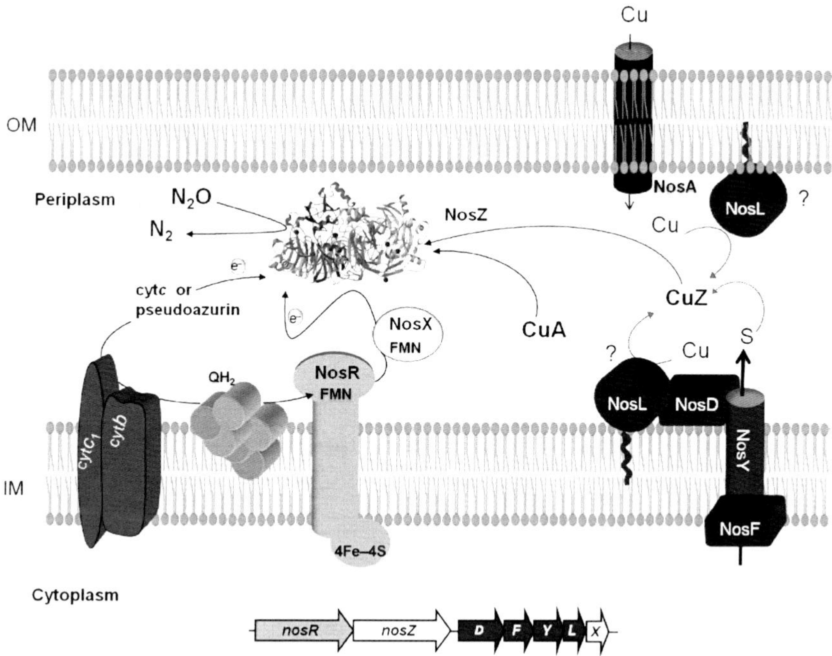

Fig. 8 Model of the membrane topology of components for N_2O respiration and for the typical NosZ biogenesis. The *nosRZDFYLX* gene cluster from denitrifiers, *nos* gene products and proposed operating electron transfer pathways from quinol (QH_2) to NosZ via the cytochrome bc_1-complex (cytbc_1) and cytochrome *c* (cyt*c*) or pseudoazurin, the other providing electrons to Cu_Z via NosR and NosX FMN-proteins. NosDFYL required for Cu_Z assemblage in NosZ is also shown. *Question marks* denote that the location of NosL in the OM or IM is still unclear. *IM,* inner membrane; *OM,* outer membrane. *Adapted from van Spanning, R. J. (2011). Structure, function, regulation and evolution of the nitrite and nitrous oxide reductase: denitrification enzymes with a b-propeller fold. In J. W. B. Moir (Ed.),* Nitrogen Cycling in Bacteria *(pp. 135–161). Norkfolk, UK: Caister Academic Press.*

nosH are distributed mostly according to taxonomic patterns and are not ubiquitous (Table 1; Zumft & Kroneck, 2007). In this context, it is worth to mention the case of the atypical N_2OR of *W. succinogenes* that is encoded in a gene cluster that also contain *nosG*, *-C1*, *-C2* and *-H* genes which were postulated to encode a putative menaquinol dehydrogenase pathway to cNosZ alternative to the conventional cytochrome bc_1 complex (see Section 2.2.2; Figs 4 and 5). These gene clusters lack *nosR* or *nosX* that in α-, β- and γ-Proteobacteria encode two flavin mononucleotide (FMN)-binding flavoproteins (NosR and NosX) that might constitute yet another electron transport pathway from the quinone pool to NosZ (Fig. 8; Table 1). In fact, NosR resembles NosH but contains an additional periplasmic FMN-binding domain (Wunsch & Zumft, 2005). In contrast to the majority of *nosRZDFYLX* gene clusters present in denitrifiers, in *Pa. denitrificans* biosynthesis of N_2OR requires the expression of *nosCRZDFYLX* genes where a *nosC* gene initiates the *nos* cluster. The gene's product, NosC, is a hypothetical protein with unknown function and close (>50% identical) homologs appear to be only distributed among other *Paracoccus* sp. Notably, all known homologs of NosC contain a CXXCXXC motif that may bind a redox-active cofactor, the significance of which is unknown.

The Tat system is responsible for transporting the NosZ apoprotein into the periplasm, where its maturation is completed. In addition to NosZ, NosX is another component of N_2O respiration system exported by the Tat system (Wunsch, Herb, Wieland, Schiek, & Zumft, 2003; Zumft & Kroneck, 2007). An increasing list of NosZ proteins (besides the NosZ of *W. succinogenes*) have Sec-type signal peptides and, in contrast to the usual Tat export pathway, seem to be exported by the Sec system (Table 1; Simon et al., 2004). NosR and NosY are integral membrane proteins and have Sec-specific signal peptides. Thus, it is clear that both the Tat and the Sec translocation system have to cooperate to assemble a functional N_2O respiratory system (Zumft & Kroneck, 2007). NosZ, despite to be targeted to the Tat system, makes an exception to the concept that cofactor acquisition occurs prior to translocation, since Cu_A and Cu_Z are assemblaged in the periplasm (Zumft, 2005; Fig. 8).

Mutation analyses demonstrated that NosDFY or NosL are involved in the maturation of the NosZ Cu_Z, but not in the biogenesis of the Cu_A site (reviewed by van Spanning, 2011; Zumft & Kroneck, 2007). Cu_A is thought to be loaded in vivo by the same route used for the loading of Cu_A in the HCOs. The sequence similarity between Cu_A centre and the subunit II of cytochrome *c* oxidase led to the issue of a putative evolutionary relationship of the two enzymes (Zumft, 2005). Thus, maturation of the NosZ Cu_A site may be well mediated via SenC-like proteins, which are homologous of the family Sco proteins. By contrast to Cu_A site, the biogenesis of the Cu_Z site and its maintenance in vivo depends on the *nosDFY* or *nosL* ancillary genes. NosDFY encodes an ABC-type transporter where NosY is a membrane-spanning protein, NosF is a cytoplasmic ATPase and NosD is a periplasmic protein from the carbohydrate-binding and sugar hydrolase protein family (Fig. 8). Mutant strains lacking NosDFY express Cu_Z-deficient N_2OR, indicating that NosDFY is essential for Cu_Z biogenesis. The exact role of this transporter system is not known, but it is proposed to be the sulphur transporter that supplies the sulphide required for Cu_Z biogenesis (Fig. 8). The *nosL* product was predicted as an outer membrane lipoprotein that was proposed to be the Cu transporter associated with Cu_Z assembly (see Zumft & Kroneck, 2007). The affinity of NosL for Cu is markedly higher for Cu(I) than Cu(II), supporting the role as a chaperone. However, active N_2OR containing both Cu sites can be obtained in the absence of NosL, so an alternative Cu chaperone must exist (Dreusch, Burgisser, Heizmann & Zumft, 1997). In archaea, the *nosD* gene is fused to the *nosL* gene, which may then suggest synthesis of a fusion protein in which the NosL domain delivers Cu to the NosD domain (see van Spanning, 2011). These observations suggest that, in addition to transport a sulphur compound via the NosFY proteins, NosD may gathers Cu ions from the NosL protein. In fact, in most N_2O-respiring bacteria *nosL* location downstream of *nosDFY* is strongly conserved (Figs 4 and 8). In spite of the experimental data that propose a scenario for Cu_Z assemblage, however, the mechanism for Cu_Z biogenesis remains still unclear.

Once assembled the NosZ copper centres, it would be expected the existence of mechanisms that preserve and maintain catalytically active the protein and the proper state of the reaction centre even in the case of changes in the cellular environment. For example, if oxygen enters a denitrifying cell, it may react with the Cu_Z reaction centre rendering as a redox inactive Cu_Z★ state (Rasmussen et al., 2000; Wüst et al., 2012). This Cu_Z★ also appears when there is insufficient supply of the natural electron donors. These differences in redox properties lead several studies to propose that Cu_Z and Cu_Z★ are structurally different. In fact, the recent X-ray crystal structure of anaerobic NosZ indicates a significant structural difference, with Cu_Z containing two bridging sulphide ligands, while the previously described structure of Cu_Z★ contains only one (Pomowski, Zumft, Kroneck & Einsle, 2011). In order to rescue an already assembled NosZ enzyme it would make sense to mobilize an electron transfer machinery that is able to maintain Cu_Z or to reactivate the Cu_Z* reaction centre. In addition to low potential electron donors as cytochrome *c* or pseudoazurin, NosR and NosX proteins have also been proposed as candidates to make up such an electron-donating mechanism (Wunsch & Zumft, 2005; Fig. 8). NosR encodes a transmembrane portion with six transmembrane helices, a flavin-binding site in the N-terminal (periplasmic) domain and two [4Fe–4S] ferrodoxin-type iron–sulphur clusters in the C-terminal (cytoplasmic) domain (Wunsch & Zumft, 2005). In the presence of modified forms of NosR where the flavin-binding domain is deleted or the ferrodoxin sites are modified, NosZ is obtained that contains both Cu_A and Cu_Z, but the spectroscopic and redox properties of Cu_Z are modified (Wunsch & Zumft, 2005). A similar phenotype is obtained in the absence of the NosX gene product for organisms that contain NosX, which codes for another periplasmic flavoprotein (Wunsch et al., 2005). This suggests that NosR and NosX are not involved in Cu_Z biogenesis but play a role in N_2O reduction in vivo altering the state of the Cu_Z site during turnover and sustaining the catalytic activity of NosZ. Taken together, these results propose the existence of an electron donation pathway via NosR as a quinol-NosX oxidoreductase. This route may be paralleled by one involving cytochrome bc_1, cytochrome c_{550} and pseudoazurin (Fig. 8).

In addition to its proposed role as electron donor to NosZ, it has also been suggested a regulatory role for NosR since it was showed to be required for the transcription of *nosZ* and *nosD* operons in *Ps. stutzeri* (Honisch & Zumft, 2003). However, the membrane location and domain organization of NosR, as well as the absence of a predicted DNA-binding-domain indicate an indirect control of NosR on its target genes. Moreover, deletion analyses of NosR showed that only the periplasmic flavin-containing domain is required for *nosZ* expression (Wunsch & Zumft, 2005).

3.3 Regulators

In general, the environmental requirements for expression of the denitrification pathway are: (a) restricted O_2 availability, (b) the presence of a nitrogen oxide (NOx) as terminal electron acceptor and (c) suitable electron donors such as organic carbon compounds. Thus, the key molecules that act as signals for the regulation of denitrification genes are oxygen, a NOx (nitrate, nitrite or NO) and the redox state of the cell. These environmental signals are perceived by a diversed number of transcriptional

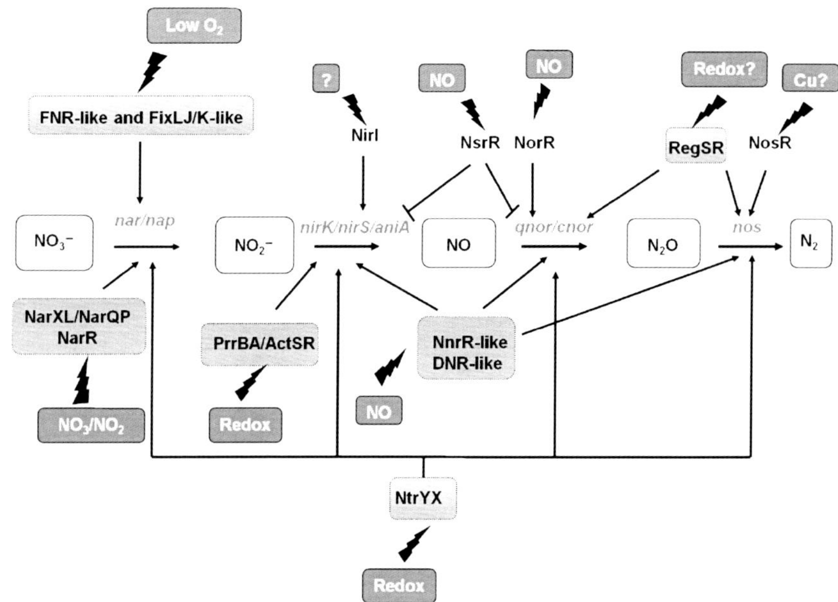

Fig. 9 Regulatory network of denitrification in response to O_2 concentration, nitrate/nitrite (NO_3^-/NO_2^-), nitric oxide (NO) and redox conditions. Positive regulation is denoted by *arrows*, and negative regulation is indicated by *perpendicular lines*. *Question marks* denote that the redox signal involved in RegSR control as well as the mechanism implicated in Cu response of *nos* genes mediated by NosR are still unknown. *Adapted from Bueno, E., Mesa, S., Bedmar, E. J., Richardson, D. J., & Delgado, M. J. (2012). Bacterial adaptation of respiration from oxic to microoxic and anoxic conditions: Redox control.* Antioxidants & Redox Signaling, 16, *819–852.*

regulators that integrate them into regulatory networks (Fig. 9; for reviews, see Bueno et al., 2012; Shapleigh, 2011; Spiro, 2011, 2012; van Spanning, 2011).

Oxygen strongly influences the growth and physiology of bacteria, as well as the expression of denitrification genes. Generally, denitrification is regarded as an anoxic or microoxic process. Since denitrifiers are facultative aerobes, this means that they must choose between oxygen and nitrate if both are available. Due to the organization and structural features of the denitrification enzymes, the maximum efficiency of free energy transduction during denitrification is only 60% of that during aerobic respiration (Richardson, 2000; Simon et al., 2008). Thus, oxygen is preferred as terminal electron acceptor than nitrate, and hence the regulation of expression of either type of respiration occurs according to an energetic hierarchy. In all species, the onset of denitrification is triggered by oxygen depletion and nitrate availability. Expression of *nar, nir, nor* and *nos* genes in most denitrifiers is tightly controlled, only occurring under microoxic conditions and in the presence of a NOx. By contrary, *nap* expression is quite variable, with this enzyme being maximally expressed under oxic conditions in some bacteria, but under microoxic conditions in others, adjusting to fit the physiological role it plays (Bueno et al., 2012; Shapleigh, 2011). It has been reported that NosZ has a greater sensitivity to O_2 compared to other denitrification enzymes, with important implications for N_2O emissions from habitats where O_2 fluctuates (Morley, Baggs, Dorsch, & Bakken, 2008). However, it has been recently demonstrated the capacity of *Ps. stutzeri* species to consume N_2O under oxic conditions (Desloover et al., 2014), supporting previous observations showing that the *nosZ* gene can also be expressed at high O_2 concentrations (Miyahara et al., 2010). Supporting these findings, it has been recently reported in *Pa. denitrificans* the reduction of N_2O at high O_2 partial pressure (Qu, Bakken, Molstad, Frostegard, & Bergaust, 2016).

In addition to O_2, nitrate/nitrite and NO have been proposed as signal molecules that are required for induction of denitrification. NO is a potent cytotoxin, and consequently both NO-generating (Nir) and NO-consuming (Nor) enzymes of denitrification are very tightly controlled by this molecule in order to avoid NO accumulation. With respect to N_2O, there is an absence of regulation of denitrification genes by this molecule presumably because it is nontoxic gas, so the denitrifying populations do not apparently respond to N_2O accumulation by making more of the N_2OR.

3.3.1 Oxygen Response

The most important types of O_2 sensors involved in regulation of denitrification are FixL and FNR (Fig. 9). FixL is a membrane-bound O_2 sensor found in rhizobial species which together with its cognate response regulator FixJ, belong to the group of two-component regulatory systems. In *B. japonicum*, phosphorylated FixJ activates transcription of *fixK$_2$*. In turn, $FixK_2$ activates expression of genes involved in denitrification, among others (for detailed information, see Section 4.1). FNR is an oxygen responsive regulator that belongs to the Crp/FNR superfamily of transcription factors that has been extensively described in Section 2.1.2. Orthologous of FNR from other organisms (such as FnrP, ANR and FnrN) are presumed to work in a similar way. For example, the *nar* and *nap* operons in *E. coli* and *B. subtilis* are activated by FNR under anoxic conditions (Reents, Munch, Dammeyer, Jahn, & Hartig, 2006; Stewart & Bledsoe, 2005; Tolla &

Savageau, 2011). *Pa. denitrificans* FnrP controls expression of the *nar* gene cluster and the *cco*-gene cluster encoding the *cbb*$_3$-type oxidase (Bouchal et al., 2010; Veldman, Reijnders, & van Spanning, 2006). Oxygen tension is sensed in *Ps. aeruginosa* by the Anr regulator, which activates transcription of the *narK1K2GHJI* operon encoding nitrate reductase and two transporters in response to oxygen limitation (Schreiber et al., 2007).

3.3.2 Nitrate/Nitrite Response

Denitrifying bacteria as well as those that reduce anaerobically nitrate to ammonium (DNRA, see Section 2.1.2) respond to nitrate/nitrite through three types of regulatory systems: NarXL, NarQP and NarR (Fig. 9). NarXL and NarQP are two-component regulatory systems being the NarX and NarQ proteins the signal sensors, and NarL and NarP proteins their cognate response regulators, respectively (Stewart, 2003). The sensing mechanism of the kinase NarX has been established (Cheung & Hendrickson, 2009; Stewart & Chen, 2010). In *E. coli* NarL and NarP bind DNA to control induction of the *nar* and *nap* operons (Darwin, Ziegelhoffer, Kiley, & Stewart, 1998; Stewart, 2003; Stewart & Bledsoe, 2005). The effects of nitrate and nitrite on the *E. coli* transcriptome during anaerobic growth have been investigated, revealing in a novel group of operons that are regulated by all FNR, NarL and NarP (Constantinidou et al., 2006). To date, *narXL* and *narQP* genes are confined to species classified in the γ and β subdivisions of the proteobacteria such as *Escherichia, Salmonella, Klebsiella, Yersinia, Burkholderia, Ralstonia, Neisseria* and *Pseudomonas* species among others. In *Ps. aeruginosa*, NarL in concert with the regulators Anr and Dnr and an IHF activate transcription of the *narK1K2GHJI* operon encoding nitrate reductase and two transporters in response to oxygen limitation, nitrate and N-oxides (Schreiber et al., 2007). Recently, it has been shown that during anaerobic growth of *Ps. aeruginosa* PAO1, NarL directly represses expression of Nap, while induces maximal expression of Nar (van Alst, Sherrill, Iglewski, & Haidaris, 2009).

NarR is a member of the CRP/FNR family of transcription activators, but it lacks a [4Fe–4S] cluster. Genes encoding NarR are found in the α-proteobacteria *Brucella suis, Brucella melitensis, Pa. denitrificans* and *Pa. pantotrophus*. In *Pa. pantotrophus* NarR controls expression of the *narKGHJI* genes encoding the respiratory nitrate reductase, NarGHI and the nitrate transport system, NarK, in response to nitrate and/or nitrite (Wood et al., 2001). The mechanism of the response is not clear, but since NarR can also be activated by azide, which normally binds to metal centres, it might be possible that NarR is a metalloprotein. There are no indications that they have counterparts of *narXL*. It therefore seems that NarR substitutes the NarXL system in the α-proteobacteria (for reviews, see Bueno et al., 2012; van Spanning et al., 2007).

3.3.3 NO Response

In addition to low oxygen conditions and nitrate/nitrite, expression of denitrification genes also requires a fine-tuned regulation in order to keep the free concentrations of nitrite and NO below cytotoxic levels. In this context, NO has been proposed as an additional key molecule that is involved in denitrification genes regulation (reviewed by Rodionov, Dubchak, Arkin, Alm, & Gelfand, 2005; Spiro, 2011; Stein, 2011; Stern & Zhu, 2014). As yet, several NO response transcription factors have been proposed to be involved in denitrification; NorR, NnrR, NsrR and DNR (Fig. 9). Among them, NorR and NsrR have been already described in Section 2.1.2 as regulators of NO-detoxifying enzymes such as Hmp or NorVW. NorR was first identified in *Ralstonia eutropha* (Pohlmann, Cramm, Schmelz, & Friedrich, 2000). This bacterium has two copies of the *norR* gene, both of which are located upstream of their *norAB* gene clusters where *norB* encodes a single-subunit NorB of the qNor type. In response to anaerobiosis and the presence of NO, NorR specifically activates transcription of the σ^{54}-dependent *norAB* promoters (Büsch, Strube, Friedrich, & Cramm, 2005; Fig. 9). NsrR has also a regulatory role in denitrifying bacteria coordinating production of Nir and Nor to prevent the build up of NO (reviewed by Tucker, Ghosh, et al., 2010). Intriguingly, the same role is performed by Nnr homologs in denitrifying bacteria that do not contain NsrR. In the denitrifying pathogenic organisms *Neisseria meningitidis* and *Neisseria gonorrhoeae*, NsrR represses both the membrane-bound Nir (AniA) and the respiratory NorB expression in the absence of NO (Heurlier, Thomson, Aziz, & Moir, 2008; Isabella, Lapek, Kennedy, & Clark, 2009; Overton et al., 2006; Fig. 9). Exposure to NO inactivates this repressor by a NO-mediated modification of the protein-bound [Fe–S cluster] (for details, see Section 2.1.2).

NnrR (nitrite and nitric oxide reductase regulator) and DNR are members of the Crp/FNR family of transcription factors, but NnrR, just like NarR, lacks the cysteines to incorporate a [4Fe–4S] cluster. NnrR and DNR orthologs, sometimes named as Nnr, or DnrR have been described in denitrifying bacteria including *Pa. denitrificans*, *Ps. stutzeri*, *Ps. aeruginosa*, *B. japonicum*, *Ensifer meliloti* and *Rhizobium etli*, and they orchestrate the expression of the *nir* and *nor* gene clusters (Fig. 9; reviewed by Rodionov et al., 2005; Spiro, 2011; Stern & Zhu, 2014). The promoters of these operons contain NnrR-binding sites that resemble the consensus FNR box to a large extent. The mechanism of NO sensing by NnrR and DNR is less well defined than NorR and NsrR. The crystal structures of DNR have only been obtained without prosthetic groups, but reveal a hydrophobic pocket that might be a haem-binding site, and purified apo-DNR can bind haem (Giardina et al., 2008). The current model proposes that DNA-binding activity of DNR in vitro requires haem and NO, and perturbation of the haem synthesis capabilities of the cell reduced the capacity of DNR to activate transcription of the *nor* promoter (Castiglione, Rinaldo, Giardina, & Cutruzzola, 2009). Molecular details of the NO-sensing mechanism employed by DNR have been recently reported (Cutruzzola et al., 2014; Rinaldo et al., 2012). In the case of NnrR, it has been proposed that NNR is activated in vivo by physiological (eg, nitrate and nitrite) and nonphysiological (eg, nitroprusside) sources of NO (Hutchings & Spiro, 2000; van Spanning et al., 1999). Heterologous expression of the *Pa. denitrificans nnr* gene in *E. coli* indicated that activation of NNR by NO does not require de novo synthesis of the NNR polypeptide. In anaerobic cultures, NNR is inactivated slowly following removal of the source of NO. In contrast, exposure of anaerobically grown cultures to oxygen causes rapid inactivation of NNR, suggesting that the protein is inactivated directly by oxygen (Lee, Shearer, & Spiro, 2006). NNR site-directed mutagenesis and structural modelling suggested that an Arg-80 closed to the C-helix that forms the monomer–monomer interface in other members of the Crp/FNR family might play an important role in transducing the activating signal between the regulatory and DNA-binding domains (Lee et al., 2006). Furthermore, assays of NNR activity in a haem-deficient mutant of *E. coli* provided preliminary evidence to indicate that NNR activity is haem dependent (Lee et al., 2006). However, the mechanism of NO or O_2 sensing by NNR has not been demonstrated in vitro.

In *Pa. denitrificans*, the global role of FnrP, NNR and NarR during the transition from aerobic to anaerobic respiration has been confirmed using proteomics, with data validation at the transcript and genome levels (Bouchal et al., 2010). Interestingly, these studies demonstrated that a mutation in the *fnrP* gene resulted in a significant decrease of the N_2OR level under semiaerobic conditions. The involvement of FnrP is also consistent with the presence of two FNR-binding sites TTGAGAATTGTCAA and TTGACCTAAGTCAA in the *nos* promoter encoding N_2OR.

Another group of proteins controlled by FnrP, NNR and NarR included SSU ribosomal protein S305/σ^{54} modulation protein (Bouchal et al., 2010). Thus, in addition to transcription regulators, sigma (σ) factors may play an important role in the FNR-mediated regulatory network as well. In this context, it has been proposed that specific classes of σ-factor binding to promoter sites downstream of the FNR box may be essential for the observed specificity of any of the 3 FNR-type transcription activators *in Pa. denitrificans* (Veldman et al., 2006). Denitrification phenotypes of the *Pa. denitrificans* FnrP, NNR and NarR transcriptional regulators have been analysed by using a robotized incubation system that monitor changes in concentrations of oxygen and nitrogen gases produced during the transition from oxic to anoxic respiration. These experiments have completed the current understanding about the involvement of these regulators in transcriptional activation of *nar, nir* and *nor* genes involved in N_2O production (Bergaust, van Spanning, Frostegard, & Bakken, 2012). With regard to the regulation of N_2O reduction, results from these studies indicate that N_2OR is subjected to a robust regulation being FnrP and NNR alternative and equally effective inducers in response to oxygen depletion (via FnrP) or an NO signal (via NNR) (Bergaust et al., 2012).

3.3.4 Redox Response

Redox changes can regulate the expression of genes involved in denitrification (for reviews, see Bueno et al., 2012; van Spanning, 2011). Redox-responsive two-component regulatory systems are present in a large number of Proteobacteria. These proteins are named RegBA in *Rhodobacter capsulatus, Rhodovulum sulfidophilum*, and *Roseobacter denitrificans*, PrrBA in *R. sphaeroides*, ActSR in *E. meliloti* and *Agrobacterium tumefaciens*, RegSR in *B. japonicum* and RoxSR in *Ps. aeruginosa*. In *Rhodobacter* species, the RegBA/PrrBA regulon encodes proteins involved in numerous energy-generating and energy-utilizing processes such as photosynthesis, carbon fixation, nitrogen fixation, hydrogen utilization, aerobic respiration and denitrification, among others (reviewed by Bueno et al., 2012; Elsen, Swem, Swem, & Bauer, 2004; Wu & Bauer, 2008). The RegBA/PrrBA two-component systems comprise the membrane-associated RegB/PrrB histidine protein kinase, which senses changes in redox state, and its cognate PrrA/RegA response regulator. Under conditions where the redox state of the cell is altered due to generation of an excess of reducing potential, produced by either an increase in the input of reductants into the system (eg, presence of reduced carbon source) or a shortage of the terminal respiratory electron acceptor (eg, oxygen deprivation), the kinase activity of RegB/PrrB is stimulated relative to its phosphatase activity. This increases phosphorylation of the partner response regulators RegA/PrrA, which are transcription factors that bind DNA and activate or repress gene expression. The membrane-bound sensor kinase proteins RegB/PrrB contain an H-box site of autophosphorylation (His^{225}), a highly conserved quinone-binding site (the heptapeptide consensus sequence GGXXNPF, which is totally conserved among all known RegB homologues), and a conserved redox-active Cys^{265} located in a 'redox box'. The mechanism by which RegB controls kinase activity in response to redox changes has been an active area of investigation. A previous study demonstrated that RegB Cys^{265} is partially responsible for redox control of kinase activity. Under oxidizing growth conditions, Cys^{265} can form an intermolecular disulphide bond to convert active RegB dimers into inactive tetramers (Swem et al., 2003). The highly conserved sequence, GGXXNPF, located in a short periplasmic loop of the RegB transmembrane domain has also being implicated in redox sensing by interacting with the ubiquinone pool (Swem, Gong, Yu, & Bauer, 2006).

RegA/PrrA contains conserved domains that are typical in two-component response regulators such as a phosphate-accepting aspartate, an 'acid box' containing two highly conserved aspartate residues and a helix-turn-helix (H-T-H) DNA-binding motif. The phosphorylated form of RegA/PrrA has increased DNA-binding capacity (Laguri, Stenzel, Donohue, Phillips-Jones, & Williamson, 2006; Ranson-Olson, Jones, Donohue, & Zeilstra-Ryalls, 2006). Under oxidizing conditions, RegB/PrrB shifts the relative equilibrium from the kinase to the phosphatase mode resulting in a dephosphorylated inactive RegA/PrrA form. Despite this evidence, it has been reported that inactivation of the *regA* gene affects expression of many different genes under oxidizing (aerobic) conditions, suggesting that both, phosphorylated and unphosphorylated RegA/PrrA, may be active transcriptional regulators (Swem et al., 2001). In this context, it has been shown that both phosphorylated and unphosphorylated forms of RegA/PrrA are capable of binding DNA in vitro and activating transcription (Ranson-Olson et al., 2006).

The PrrBA from *R. sphaeroides* (Laratta, Choi, Tosques, & Shapleigh, 2002), ActSR from *A. tumefaciens* (Baek, Hartsock, & Shapleigh, 2008) and RegSR from *B. japonicum* control denitrification (Torres, Argandona, et al., 2014; for detailed information, see Section 4.1). In *R. sphaeroides* 2.4.3, inactivation of *prrA* impaired ability to grow both photosynthetically and anaerobically in the dark on nitrite-amended medium (Laratta et al., 2002). The PrrA-deficient strain exhibited a severe decrease in both nitrite reductase activity and expression of a *nirK–lacZ* fusion when environmental oxygen tension was limited (Fig. 9). This regulation is not mediated by NnrR, since *nnrR* is fully expressed in a PrrA mutant background. Instead, Laratta et al. (2002) proposed a model where, under low oxygen tension, the kinase activity of PrrB is increased relative to its phosphatase activity, resulting in an increased concentration of PrrA-P. Thus, under microoxic conditions in the presence of NO, PrrA-P activates transcription of *nirK* in collaboration with NnrR. Insertional inactivation of the response regulator ActR in *A. tumefaciens* significantly reduced *nirK* expression and Nir activity but not *nnrR* expression (Fig. 9). In *A. tumefaciens*, a putative ActR-binding site was identified in the *nirK* promoter region using mutational analysis and an in vitro-binding assay (Baek et al., 2008). These studies also showed that purified ActR bound to the *nirK* promoter but not to the *nor* or *nnrR* promoter.

In addition to PrrBA, ActSR and RegSR (Fig. 9), it has been recently reported that the NtrYX two component system of *Brucella* spp. acts as a redox sensor and regulates the expression of *nar*, *nir*, *nor* and *nos* operons in response to microoxic conditions (Fig. 9; Roop & Caswell, 2012) and that PrrBA and NtrYX coordinately regulate the expression of denitrification (Carrica, Fernandez, Sieira, Paris, & Goldbaum, 2013). NtrYX two-component system is also involved in the expression of respiratory nitrite reductase (AniA) and nitric oxide reductase (NorB) in the human pathogen *N. gonorrhoeae* (Atack et al., 2013).

3.3.5 Copper and pH as Emerging Regulatory Factors

The enzymes of denitrification are complex metalloenzymes that require a suite of redox-active cofactors including molybdenum, iron and/or copper for their respective activities. In particular, the reduction of N_2O by denitrifying bacteria is heavily reliant on the availability of copper, a key constituent of N_2OR. This phenomenon has been explored in detail in *Pa. denitrificans*, where bacterial cultures lacking the trace element copper accumulate significant amounts of N_2O (Felgate et al., 2012). Furthermore, mathematical models have been developed that quantitatively predict the levels of N_2O emitted by bacterial denitrification in response to copper availability (Woolfenden et al., 2013). A recent global transcriptomic study by Sullivan and coworkers has revealed that copper deficiency not only affects functional maturation of N_2OR, but it has an important impact on gene expression in *Pa. denitrificans*, including expression of *nosZ* that is downregulated during copper-limited growth (Sullivan, Gates, Appia-Ayme, Rowley, & Richardson, 2013). In addition, *nosZ* transcript levels in both *Pa. denitrificans nosC* or *nosR* mutants were found to be similar in copper-limited or copper-sufficient growth conditions indicating that repression of *nosZ* during copper-limited growth was deregulated in response to metal availability. Therefore, these results strongly suggest a role of NosC and NosR in copper regulation of *nosZ* expression, although the mechanism involved in this control remains to be established. Interestingly, these transcriptomics studies also revealed that the high levels of N_2O produced as a consequence of decreased NosZ activity lead to *Pa. denitrificans* switching from vitamin B_{12}-dependent to vitamin B_{12}-independent biosynthetic pathways through the transcriptional modulation of genes controlled by vitamin B_{12} riboswitches (Sullivan et al., 2013).

In addition to copper availability, pH is another key factor that has been demonstrated to significantly influence microbial N_2O emissions. Soil pH is known to be a major driver of denitrifier N_2O:N_2 ratios, and numerous studies have shown that the reduction of N_2O to N_2 is impaired by low soil pH, suggesting that liming of acidic soils may be an effective strategy to lower N_2O emissions (Liu, Morkved, Frostegard, & Bakken, 2010; Van den Heuvel, Bakker, Jetten, & Hefting, 2011). A series of experiments involving *Pa. denitrificans* have shown that modulating pH has little effect on the transcription of the *nosZ* gene (Bergaust, Mao, Bakken, & Frostegard, 2010). Instead, the enzymatic rate of N_2O reduction was significantly attenuated at low pH levels, implying that environmental pH may have a direct posttranslational effect on the assembly and/or activity of the N_2OR holoenzyme. Consistent with these findings, spectroscopic and steady-state kinetics studies in N_2OR from *Achromobacter cycloclastes* suggest that $[H^+]$ has multiple effects on both the activation and the catalytic reactions (Fujita & Dooley, 2007). One plausible explanation for these observations is that low pH may influence the assembly of the enzyme, which takes place in the periplasm. That said, a link between metal availability and pH has yet to be explored. Recent analyses of growth-linked NO, N_2O and N_2 profiles alongside relevant denitrification gene transcript levels (ie, for *nirS*, *nirK* and *nosZ*), using cells extracted from soils with different pH values, suggests that low pH may interfere with the manufacture of N_2OR rather than the function of the enzyme once properly assembled (Liu, Frostegård, & Bakken, 2014).

4. *B. JAPONICUM* AS A MODEL OF LEGUME-ASSOCIATED RHIZOBIAL DENITRIFIERS

Legume plants, which includes lentils, peas, beans, peanuts and soya, are hugely important as a source of food due to their high protein content. They are second only to cereals in agriculture importance, and many species as alfalfa are also used for forage, hay, silage and green manure, and it constitutes an important component for fodder animal feeding. Moreover, legume family has the unique ability to establish a N_2-fixing symbiotic association with soil bacteria collectively referred as rhizobia (Sprent, 2009). During this process, an exchange of molecular signals occurs between the two partners, leading to the formation of root nodules, where biological nitrogen fixation takes place by rhizobia (for a recent review, see Udvardi & Poole, 2013). Legumes can safe huge amounts of environment polluting nitrogen fertilizers protecting ground water from toxicity while increasing soil fertility and contribute to the improvement of soil structure with a turn-over effects on the subsequent crops (Sprent, 2009). Thus, inoculation of legumes with rhizobia is an economical and environmental friendly recommended worldwide agricultural practice to increase crop yield and to improve soil fertility without adding N fertilizers. More than 60% of N_2O emissions globally are emitted from agricultural soils due to the synthetic N addition into them. Thus, one strategy for N_2O mitigation is reducing the dependence on chemical fertilizers in agriculture enhancing biological nitrogen fixation. However, legume crops also contribute to N_2O emissions by several ways: (i) biologically fixed N may be nitrified and denitrified, thus providing a source of N_2O (Inaba et al., 2012; Saggar et al., 2013); (ii) by providing N-rich residues for decomposition (Baggs, Rees, Smith, & Vinten, 2000) and (iii) directly by some rhizobia that are able to denitrify under free-living conditions or under symbiotic association with legume plants (Bedmar et al., 2013; Bedmar, Robles, & Delgado, 2005; Hirayama, Eda, Mitsui, & Minamisawa, 2011; Inaba et al., 2009, 2012).

Although denitrification among rhizobia is rare, several of the most agronomical interesting species contain denitrification genes in their genomes (Table 2). So, *Pseudomonas* sp. G-179 (actually *Rhizobium galegae*) (Bedzyk et al., 1999) has been shown to contain Nap, Nor and NirK. *Rhizobium sullae* (formerly *R. hedysari*) only expresses NirK (Toffanin et al., 1996). The genetic determinants for expression of NirK and cNor are present in *R. etli* CFN42 (Bueno, Gomez-Hernandez, Girard,

Table 2 Denitrification Genes in Rhizobia

Species and Strain	Denitrification Genes: *nap*	*nirK*	*nor*	*nos*	References
Rhizobium galegae (formerly *Pseudomonas* sp. G-179)	*EFDABC*	*nirK*	*EFCBQD*	–	Bedzyk, Wang, and Ye (1999)
Rhizobium sullae (formerly *R. hedysari*)	–	*niK*	–	–	Toffanin et al. (1996)
Rizobium etli CFN42	–	*nirK*	*ECBQD*	–	Gomez-Hernandez et al. (2011)
Ensifer meliloti 1021 (formerly *Sinorhizobium meliloti*)	*EFDABC*	*nirK*	*ECBQD*	*nosRZDFYLX*	Torres, Rubia, Bedmar, and Delgado (2011)
Bradyrhizobium japonicum USDA110	*EDABC*	*nirK*	*CBQD*	*nosRZDFYLX*	Bedmar et al. (2005)
Rhizobium sp. NGR234	*EFDABC*	*nirK*	*CBQD*	–	http://genome.microbedb.jp/rhizobase/

Bedmar, & Delgado, 2005; Gomez-Hernandez et al., 2011). *E. meliloti* (formerly *Sinorhizobium meliloti*) (Galibert et al., 2001; Holloway, McCormick, Watson, & Chan, 1996; Torres et al., 2011) and *B. japonicum* (recently reclassified as *Bradyrhizobium diazoefficiens* USDA 110, Delamuta et al., 2013; Bedmar et al., 2005; Kaneko et al., 2002) contain *nap*, *nirK*, *nor* and *nos* genes (http://www.kazusa.or.jp/rhizobase). Among them, *B. japonicum* is the only rhizobial species that has the ability to grow under anoxic conditions with nitrate through denitrification pathway and where this process has been extensively investigated not only under free-living but also under symbiotic conditions (for reviews, see Bedmar et al., 2013; Bedmar et al., 2005; Delgado, Casella, & Bedmar, 2007; Sanchez, Bedmar, & Delgado, 2011; Sanchez, Cabrera, et al., 2011).

B. japonicum occupies two distinct niches: free-living in the soil and establishing symbiotic associations with soybean (*Glycine max*), siratro

(*Macroptilium atropurpureum*), mung bean (*Vigna radiata*) and other *Vigna* species. Soybeans are unique in legumes with contents of 40% protein and 21% oil as well as isoflavones. Thus, soybean crops represent 50% of the total legume crop area and 68% of global production, able to fix 16.4 Tg N annually, representing 77% of the N fixed by legume crops (Herridge, Peoples, & Boddey, 2008). Soybean has an industrial and economical interest for oil, food and protein, pharmaceuticals for protective coating or biodiesel production that represents the largest individual element of international oilseed production (59%), with United States (34%), Brazil (30%) and Argentina (18%) being the main contributers to world soybean production. Soybean is the first legume species with a complete genome sequence (Schmutz et al., 2010). It is, therefore, a key reference for the more than 20,000 legume species, and for the remarkable evolutionary innovation of nitrogen-fixing symbiosis. The genome sequence is also an essential framework for vast new experimental information such as tissue-specific expression and whole-genome association data. The genome sequence opens the door to crop improvements that are needed for sustainable human and animal food production, energy production and environmental balance in agriculture worldwide. *B. japonicum* strain USDA110 was originally isolated from soybean nodules in Florida, USA in 1957, and has been widely used for the purpose of molecular genetics, physiology and ecology. Taken in consideration this background, *B. japonicum* USDA110 is considered a model rhizobial species for studying denitrification in legume-associated bacteria under both free-living and symbiotic conditions.

4.1 Regulation of *B. japonicum* Denitrification

In *B. japonicum*, denitrification is dependent on the *napEDABC* (Delgado, Bonnard, Tresierra-Ayala, Bedmar, & Muller, 2003), *nirK* (Velasco, Mesa, Delgado, & Bedmar, 2001), *norCBQD* (Mesa, Velasco, Manzanera, Delgado, & Bedmar, 2002) and *nosRZDYFLX* genes (Velasco, Mesa, Xu, Delgado, & Bedmar, 2004; Table 2). In addition, accessory cytochromes such as cytochrome c_{550}, encoded by *cycA*, are necessary to support electron transport during denitrification being essential for the electron delivery to NirK (Bueno, Bedmar, Richardson, & Delgado, 2008). Neither azurin- nor pseudoazurin-like copper proteins have been annotated in the genome sequence of *B. japonicum* (http://www.kazusa.jp/rhizobase/).

Similarly to many other denitrifiers, expression of denitrification genes in *B. japonicum* requires both oxygen limitation and the presence of nitrate or a derived NOx (Bedmar et al., 2005). In this bacterium, perception and transduction of the 'low oxygen signal' are mediated by two interlinked oxygen responsive regulatory cascades, the FixLJ–FixK$_2$–NnrR and the RegSR–NifA (reviewed by Bueno et al., 2012; Torres et al., 2011; Fig. 10). A moderate decrease in the oxygen concentration in the gas phase ($\leq$5%) is sufficient to activate expression of FixLJ–FixK$_2$-dependent targets (Sciotti, Chanfon, Hennecke, & Fischer, 2003). The haem-based sensory kinase FixL senses this 'low oxygen' signal and autophosphorylates and transfers the phosphoryl group to the FixJ response regulator which then activates transcription of *fixK$_2$* gene. In turn, the Crp/FNR-like transcriptional regulator FixK$_2$ induces expression of *nap, nirK* and *nor* denitrification genes involved in N$_2$O production (Mesa et al., 2002; Robles, Sanchez, Bonnard, Delgado, & Bedmar, 2006; Velasco et al., 2001) as well as regulatory genes such as *rpoN$_1$*, *fixK$_1$* and *nnrR* (Mesa, Bedmar, Chanfon, Hennecke, & Fischer, 2003; Mesa et al., 2008; Nellen-Anthamatten et al., 1998). Thus, *B. japonicum* NnrR expands the FixLJ–FixK$_2$ regulatory cascade probably by an additional control level that integrates the N-oxide signal required for maximal induction of denitrification genes (Fig. 10). The NO-sensing mechanism by *B. japonicum* NnrR is still unknown. It has been recently found that *nap, nirK* or *nor* promoters exhibit differences with regard to their dependence on FixK$_2$ and NnrR. In fact, purified FixK$_2$

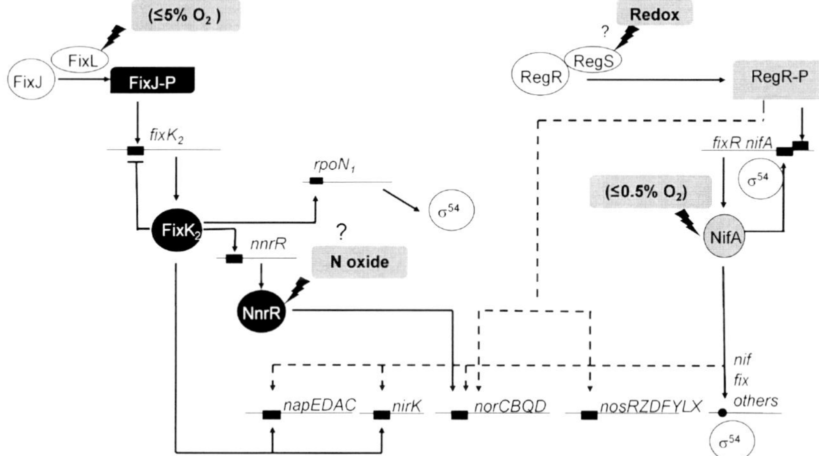

Fig. 10 Regulatory network of *B. japonicum* denitrification. Positive regulation is denoted by *arrows*, negative regulation is indicated by *perpendicular lines* and unknown control mechanisms are indicated by *dashed lines*. *Question marks* denote that the N-oxide sensing mechanism by NnrR as well as the redox signal involved in RegSR control are still unknown. *Adapted from Bueno, E., Mesa, S., Bedmar, E. J., Richardson, D. J., & Delgado, M. J. (2012). Bacterial adaptation of respiration from oxic to microoxic and anoxic conditions: Redox control.* Antioxidants & Redox Signaling, 16, *819–852.*

activates transcription from *nap*- or *nirK*-dependent promoters but not from *nor*-dependent promoter. By contrast, NnrR bound to a specific DNA fragment from the promoter region of the *nor* genes, but not to those from the *nap* and *nirK* genes (Fig. 10; E. Bueno, unpublished work).

In addition to FixLJ–FixK_2–NnrR, the second oxygen responsive regulatory cascade, RegSR–NifA, that respond to very low oxygen concentrations ($\leq 0.5\%$), has been reported to be involved in the maximal induction of *B. japonicum* denitrification genes. In the RegSR–NifA cascade, the response regulator RegR of the RegSR two-component regulatory system induces expression of the *fixR–nifA* operon (Barrios, Fischer, Hennecke, & Morett, 1995; Barrios, Grande, Olvera, & Morett, 1998; Bauer, Kaspar, Fischer, & Hennecke, 1998) under all oxygen conditions. Moreover, upon a switch to low oxygen or anoxic conditions, the redox-responsive NifA protein in concert with RNAP-containing RpoN (σ^{54}) enhances its own synthesis. In *B. japonicum*, RpoN is encoded by the two highly similar and functionally equivalent genes ($rpoN_1$ and $rpoN_2$) (Kullik et al., 1991). Since $rpoN_1$ is under the control of FixK_2, this gene represents the link between the two regulatory cascades. Targets of NifA include *nif* and *fix* genes, which are directly or indirectly involved in nitrogen fixation (Hauser et al., 2007; Nienaber, Huber, Gottfert, Hennecke, & Fischer, 2000). Recent results from our group showed that NifA is also required for maximal expression of *nap*, *nirK* and *nor* genes (Fig. 10; Bueno, Mesa, Sanchez, Bedmar, & Delgado, 2010). Whether or not these genes are direct or indirect targets of NifA is under investigation. In addition to NifA, it has been recently demonstrated the involvement of RegR in the control of denitrification genes in *B. japonicum* (Torres, Argandona, et al., 2014). In this context, comparative transcriptomic analyses of wild-type and *regR* strains revealed that almost 620 genes induced in the wild type under denitrifying conditions were regulated (directly or indirectly) by RegR, pointing out the important role of this protein as a global regulator of denitrification. Genes controlled by RegR included *nor* and *nos* structural genes (Fig. 10), as well as genes encoding electron transport proteins such as *cycA* or *cy2*, among others. It has also been demonstrated the capacity of purified RegR to interact with the promoters of *norC* and *nosR* (Torres, Argandona, et al., 2014). Expression studies with a *norC–lacZ* fusion and haem *c*-staining analyses revealed that anoxia and nitrate are required for RegR-dependent induction of *nor* genes and that this control is independent of the sensor protein RegS (Torres, Argandona, et al., 2014).

Taken together, these results suggest the existence of a complex regulatory network of the *B. japonicum* denitrification process (Fig. 10) and therefore, of N_2O emissions by soybean root nodules. While a progress on the knowledge about the regulation of *nap*, *nir* and *nor* genes involved in N_2O synthesis has been made in *B. japonicum*, much remains to be discovered regarding the regulatory mechanisms and networks involved in the control of *nosRZDYFLX* genes involved in N_2O reduction to N_2, the key step to N_2O mitigation.

4.2 NO and N_2O Metabolism in Soybean Nodules

Several studies have reported the evolution of N_2O from sliced or detached soybean nodules (Inaba et al., 2012; Mesa, Alche, Bedmar, & Delgado, 2004; Sameshima-Saito et al., 2006). It has been recently demonstrated that nitrate is essential for N_2O emissions from nodules of plants inoculated with *B. japonicum* USDA110, and its concentration enhanced N_2O fluxes showing a statistical linear correlation. In addition to nitrate, N_2O emission from soybean nodules is significantly induced when plants were subjected to flooding, especially during long (7 days)-term flooding (Tortosa et al., 2015).

Flooding and nitrate also induce the formation in detached nodules of the precursor of N_2O, the cytotoxic and ozone-depleting gas NO (Meakin et al., 2007; Sanchez et al., 2010). This molecule contributes to the formation of nitrosyl-leghaemoglobin (LbNO) complexes in soybean nodules (Fig. 11; Sanchez et al., 2010) and is an inhibitor of nitrogenase

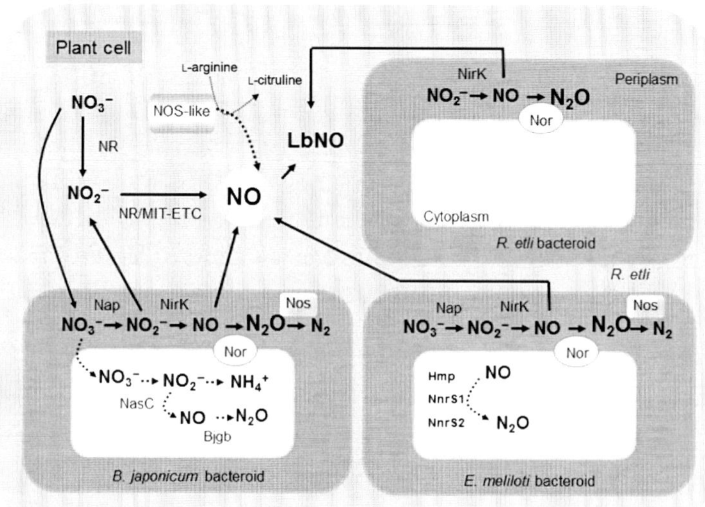

Fig. 11 Schematic representation of NO and N_2O metabolism in root nodules from *G. max–B. japonicum*, *M. truncatula–E. meliloti* and *P. vulgaris–R. etli* symbiosis. The large *grey square* represents the plant cell, and the small *grey squares* represent the bacteroids where the periplasm (in *grey*) and the cytoplasm (in *white*) are shown.

activity (Kato, Kanahama, & Kanayama, 2010) and expression of the *nifH* and *nifD* genes (Sanchez et al., 2010).

The main process involved in NO and N_2O production in soybean nodules is *B. japonicum* denitrification (Fig. 11; Inaba et al., 2012; Meakin et al., 2007; Sanchez et al., 2010). Thus, the main candidate for N_2O synthesis in nodules is Nor that reduces NO to N_2O. It has also been demonstrated that the *B. japonicum* N_2OR is a key enzyme to mitigate N_2O emissions from soybean nodules (Horchani et al., 2011; Inaba et al., 2012; Tortosa et al., 2015). Based on this, Itakura et al. (2013) hypothesized and proved that N_2O emission from soil could be reduced by inoculating soybean plants with a *nosZ*-overexpressing strain of *B. japonicum*. Thus, inoculation with *nosZ*$^+$ *B. japonicum* strains can be used as a strategy to mitigate N_2O emissions from increasing soybean fields.

4.3 A New System Involved in NO and N_2O Metabolism in *B. japonicum*

It is well established that *B. japonicum* denitrification is the main process involved in NO and N_2O production in soybean nodules. Nevertheless, basal levels of NO and N_2O were recorded in nodules from soybean plants subjected to nitrate and flooding conditions and inoculated with a *napA* mutant where denitrification is blocked (Sanchez et al., 2010; Tortosa et al., 2015). These observations suggest that other mechanisms different to denitrification pathway could be involved in NO and N_2O production in nodules. In this context, it has been recently identified in *B. japonicum* a putative haemoglobin, Bjgb, implicated in NO detoxification (Cabrera et al., 2011; Sanchez, Bedmar et al., 2011; Sanchez, Cabrera, et al., 2011). Similarly to other bacterial haemoglobins, Bjgb might reduce NO to N_2O under anoxic free-living conditions or inside the nodules. In *B. japonicum*, the Bjgb is encoded in a gene cluster that also codes for a number of proteins with important roles in nitrate assimilation (Cabrera et al., 2016) including the large catalytic subunit of the assimilatory nitrate reductase (NasC), a major-facilitator superfamily-type nitrate/nitrite transporter, an FAD-dependent NAD(P)H oxidoreductase (Fig. 12). A ferredoxin-dependent assimilatory nitrite reductase (NirA) is present a distinct locus on the chromosome. This *nirA* gene lies immediately downstream of genes recently reported to code for a nitrate/nitrite responsive regulatory system (NasS–NasT) in *B. japonicum* (Sanchez et al., 2014). This integrated system for NO detoxification and nitrate assimilation has been demonstrated to be another source of NO and probably to N_2O. In fact, the importance of NasC not only in nitrate assimilation but also in NO production has been demonstrated (Cabrera et al., 2016). Although, the biochemical basis for NO-formation during anaerobic bacterial respiration has been shown to result from NR-catalysed reduction of nitrite by Nar (Gilberthorpe & Poole, 2008; Rowley et al., 2012; Vine & Cole, 2011), to our knowledge, this is the first time where a combined nitrate assimilation/NO-detoxification system represents a novel method by which bacteria protect against cytoplasmic NO produced by NasC during anaerobic nitrate-dependent growth, where pathways for both respiratory denitrification and nitrate assimilation are active (Fig. 12; Cabrera et al., 2016).

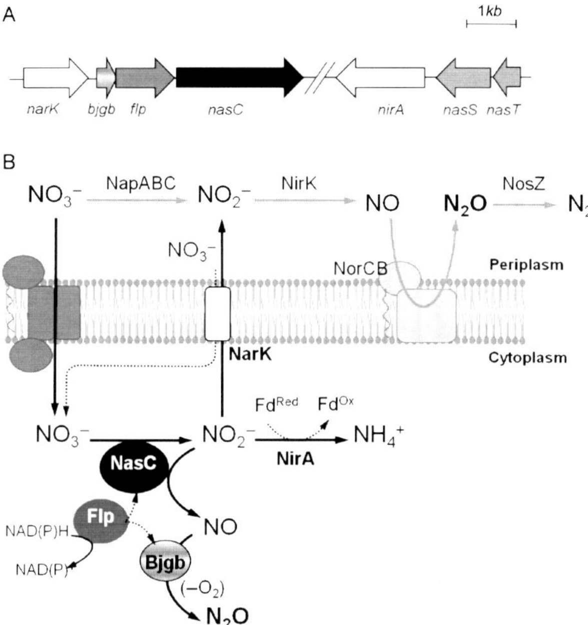

Fig. 12 (A) Organization of regulatory and structural genes for the assimilatory nitrate pathway in *B. japonicum*. (B) Proposed biochemical pathway for nitrate assimilation and NO detoxification system, alongside well-characterized denitrification pathway in *B. japonicum*. Assimilatory reduction of nitrate to ammonium is performed by sequential action of the nitrate reductase NasC and ferredoxin (Fd)-dependent nitrite reductase NirA. Electrons from NAD(P)H are supplied to NasC and also Bjgb by Flp. During assimilatory nitrate reduction, cytoplasmic nitrite may accumulate and be further reduced, by NasC, to generate cytotoxic NO. NarK can counteract accumulation of nitrite by exporting it to the periplasm. Bjgb might detoxify NO to N_2O in the absence of O_2. *Adapted from Cabrera, J. J., Salas, A., Torres, M. J., Bedmar, E. J., Richardson, D. J., Gates, A. J., et al. (2016). An integrated biochemical system for nitrate assimilation and nitric oxide detoxification in Bradyrhizobium japonicum.* The Biochemical Journal, 473, *297–309.*

These observations strongly suggest that in addition to denitrification, rhizobial nitrate assimilation might be another important source of NO and N_2O in nodules. Further investigations are being carried out to establish the role of this nitrate assimilation/NO-detoxification system in NO and N_2O metabolism in soybean nodules.

5. NO AND N_2O METABOLISM IN OTHER RHIZOBIA-LEGUME SYMBIOSIS

5.1 *E. meliloti–Medicago truncatula*

E. meliloti is an aerobic soil bacterium which establishes symbiotic N_2-fixing associations with plants of the genera *Medicago, Melilotus* and *Trigonella. Medicago sativa* (also known as alfalfa or lucerne) is one of the most widely forage legume crops in the world. In addition to the traditional uses as an animal feed, alfalfa has a great potential as a bioenergy crop and different studies considered alfalfa (especially stems) as a good sustainable crop for second-generation bioethanol production. These plants also possess therapeutic virtues that have been used in veterinary and medicine. Among *Medicago* species, *Medicago truncatula* plays a prominent role in fundamental research on legume biology and symbiotic nitrogen fixation due to favourable characteristics including diploid genetics, small genome ($\sim$500 Mbp), ease of transformation, short life cycle and high levels of natural diversity (Cook, 1999). The genome of this model legume was sequenced in the first decade of the 21st century (Young et al., 2011). *E. meliloti* 1021 is a model rhizobial strain that has been extensively used to better understand the interaction between *E. meliloti* and *M. truncatula* that has been the subject of extensive biochemical, molecular and genetic investigation (Jones, Kobayashi, Davies, Taga, & Walker, 2007; Young et al., 2011). Inspection of the *E. meliloti* 1021 genome sequence shows a composite architecture, consisting of three replicons with distinctive structural and functional: a 3.65 Mb chromosome and two megaplasmids, pSymA (1.35 Mb) and pSymB (1.68 Mb) (Galibert et al., 2001). pSymA contains a large fraction of the genes known to be specifically involved in symbiosis and genes likely to be involved in nitrogen and carbon metabolism, transport, stress and resistance responses that give *E. meliloti* an advantage in its specialized niche (Barnett et al., 2001). A 53 kb segment of pSymA is particularly rich in genes encoding proteins related to nitrogen metabolism, including *napEFDABC*, *nirK*, *norECBQD* and *nosRZDFYLX* denitrification genes (Table 2). Transcriptomic analyses have shown that *E. meliloti* denitrification genes are induced in response to microoxic and symbiotic conditions (Becker et al., 2004). Under free-living microoxic conditions, the expression of denitrification genes is coordinated via the two-component regulatory system, FixLJ, and via the transcriptional regulator, FixK (Bobik, Meilhoc, & Batut, 2006). Furthermore, transcriptomic studies demonstrated that denitrification genes (*nirK* and *norC*) and other genes related to denitrification (*azu*1, *hemN*, *nnrU* and *nnrS*) are also induced in response to NO and that the regulatory protein NnrR is involved in the control of this process (Meilhoc, Cam, Skapski, & Bruand, 2010). However, and despite possessing and expressing the complete set of denitrification genes, *E. meliloti* has been considered a partial denitrifier due to its inability to grow under anaerobic conditions with nitrate or nitrite as terminal electron acceptors. Despite the inability of *E. meliloti* to grow under denitrifying conditions, *napA*, *nirK*, *norC* and *nosZ* structural genes are functional since they are involved in the expression of denitrification enzymes under specific growth conditions (initial oxygen concentrations of 2%) (Torres, Rubia, et al., 2014). By using a robotized incubation system it has been recently confirmed the incapacity of *E. meliloti* to respire nitrate and reduce it to N_2O or N_2 under anoxic conditions (Bueno et al., 2015). By contrast, in the latter studies the capacity of *E. meliloti* to grow through anaerobic respiration of N_2O to N_2 was demonstrated. N_2OR activity was not dependent on the presence of nitrogen oxyanions or NO, thus the expression could be induced by oxygen depletion alone. When incubated at pH 6, the capacity of *E. meliloti* to reduce N_2O was severely impair, corroborating previous observations found in both, extracted soil bacteria and *Pa. denitrificans* pure cultures, where expression of functional N_2OR is difficult at low pH (Bergaust et al., 2010; Liu et al., 2014). Furthermore, the presence in the medium of highly reduced C-substrates, such as butyrate, negatively affected N_2OR activity. The emission of N_2O from soils can be lowered if legumes plants are inoculated with rhizobial strains overexpressing N_2OR. This study demonstrates that strains like *E. meliloti* 1021, which do not produce N_2O from nitrate respiration but are able to reduce the N_2O emitted by other organisms, could act as potential N_2O sinks. These results could be expanded to competitive and efficient N_2-fixers *E. meliloti* strains in order to develop strategies to reduce N_2O emissions from alfalfa crops.

5.1.1 NO in M. truncatula *Nodules*

It is well known that NO is produced at various stages of *E. meliloti*–*M. truncatula* symbiosis, and this molecule has a beneficial role during infection, nodule development and mature nodule functioning (for a recent review, see Hichri et al., 2015). On the other hand NO was also shown to have inhibitory effects on nitrogenase, induces senescence, and it has been recently reported to contribute to the plant glutamine synthetase posttranslational modification in nitrogen-fixing nodules (Blanquet et al., 2015). In the nodules, both the plant and the bacterial partners should be considered as potential sources of NO. In plants, beside a nonenzymatic conversion of nitrite to NO in the apoplast (Bethke, Badger, & Jones, 2004), seven enzymatic pathways for NO production have been described (Gupta, Bauwe, & Mur, 2011). In the reductive pathways, nitrite can be reduced to NO through the action of either nitrate reductase (NR), plasma membrane-bound nitrite:NO reductase, xanthine oxidoreductase or the mitochondrial ETC, particularly in a low O_2 environment (Gupta et al., 2011; Mur et al., 2013; Fig. 11). Oxidative pathways that lead to NO production depend on arginine, polyamines or hydroxylamine as primary substrates. This oxidative NO production, mediated by still uncharacterized enzymes [NOS-like, polyamine oxidase (PAOx)], occurs under normoxic conditions (Gupta et al., 2011; Mur et al., 2013). In addition to plant sources, *E. meliloti napA* and *nirK* denitrification genes were shown to participate significantly in NO synthesis, at least in mature nodules (Fig. 11; Horchani et al., 2011). Given the clear evidences of NO production in *M. truncatula* nodules, NO-detoxification systems in nodules are essential in maintaining a balanced NO concentration and an efficient symbiosis. In this context, plant haemoglobins (nonsymbiotic haemoglobins but also leghaemoglobins or truncated haemoglobins) have been shown to be involved in NO degradation (Gupta et al., 2011). From the bacterial side, two *E. meliloti* proteins, Hmp and Nor are the major NO-detoxifying enzymes essential in maintaining a balanced NO concentration and an efficient symbiosis (Cam et al., 2012; Meilhoc, Blanquet, Cam, & Bruand, 2013; Fig. 11). Furthermore, it has been recently demonstrated the involvement of *E. meliloti* $nnrS_1$ and $nnrS_2$ in NO degradation under both in free-living and symbiotic conditions (Blanquet et al., 2015; Fig. 11). $NnrS_1$ and $NnrS_2$ are haem- and copper-containing membrane proteins whose homologues in *Vibrio cholerae* and *R. sphaeroides* 2.4.1 have been shown to be important in resisting to nitrosative stress in culture (Arai, Roh, Eraso, & Kaplan, 2013; Stern et al., 2012) Hence, *E. meliloti* possesses at least four systems (Hmp, Nor, $NnrS_1$ and $NnrS_2$) to detoxify NO, which belong to the NO stimulon (Meilhoc et al., 2010), and their expression is dependent upon the NO-specific regulator NnrR. These proteins might not have the same role and/or not function in the same conditions inside nodules. Indeed they have different localization in the bacterial cell, and on the other hand, they display a different expression pattern within the different zones of the nodules (Meilhoc et al., 2013; Roux et al., 2014). Although the involvement of Hmp, Nor, $NnrS_1$ and $NnrS_2$ in NO detoxification has been demonstrated, the potential impact of those NO-consuming proteins on the emission of the GHG N_2O by alfalfa nodules is poorly investigated.

5.2 *R. etli–Phaseolus vulgaris*

R. etli fixes nitrogen in association with *Phaseolus vulgaris* L., or common bean which is the most important legume for human consumption. This crop is the principal source of protein for hundreds of millions of people and more than 18 million tonnes of dry common bean are produced annually (Broughton et al., 2003). *P. vulgaris* is also a model species for the study of symbiosis in association with nitrogen-fixing bacteria from the genus *Rhizobium*. The genome sequence of *P. vulgaris* has been recently released (Schmutz et al., 2014). *R. etli* is the natural microsymbiont of *P. vulgaris* that has been isolated from diverse geographical regions across Latin America given the strong integration of beans into the diet of this continent. *R. etli* CFN42 was originally isolated from bean nodules in Mexico, and since its sequence is known (Gonzalez et al., 2006), this strain has been widely used for molecular genetics, physiology and ecology studies. *R. etli* CFN42 contains a chromosome and six large plasmids (pCFN42a to pCFN42f) whose sizes range from 184.4 to 642.5 kb (Gonzalez et al., 2006). In *R. etli* CFN42, genes encoding denitrification enzymes were identified on plasmid pCFN42f. Genes located in this region include those encoding proteins with significant similarity to *nirK* and *norCBQD* (Table 2). Neither genes encoding for a respiratory nitrate reductase (*nap* or *nar* genes) nor for the respiratory N_2OR (*nos* genes) were found in the *R. etli* genome. Plasmid pCFN42f also includes regulatory genes such as *fixK* and *fixL*. In contrast to *E. meliloti* or *B. japonicum*, the transcriptional activator with functional homology with FixJ is absent in *R. etli*. Instead, it has been recently identified FxkR as the missing regulator that allows the transduction of the microaerobic signal for the activation of the FixKf regulon (Zamorano-Sanchez et al., 2012). In the *nirK–norC* region of pCFN42f is also located the *nnrR* gene which encodes NnrR, the FNR-type transcriptional regulator of denitrification genes. Although *R. etli* is unable to respire nitrate and to perform a complete denitrification pathway, the presence of

NirK- and Nor-coding regions in this bacterium suggests an NO-detoxifying role for these enzymes, preventing accumulation of NO inside the free-living cells or in the nodules. In fact, in vivo experiments demonstrated that NirK is required for nitrite reduction to NO and that Nor is required to detoxify NO under free-living conditions (Bueno et al., 2005; Gomez-Hernandez et al., 2011). In *R. etli*, microaerobic expression of *nirK* and *norC* promoters requires a functional FixKf, whereas the response to NO is mediated by NnrR. As reported in *B. japonicum*, microaerobic expression of *R. etli nnrR* is controlled by FixKf. By contrary, in *E. meliloti* NnrR and FixK are part of two different regulatory pathways (for a review, see Cabrera et al., 2011). Additionally, the N_2-fixation regulator NifA has a negative effect on the transcription of the *nirK* operon (Gomez-Hernandez et al., 2011). This finding contradicts those reported in *B. japonicum* where NifA is involved in maximal expression of *nap*, *nirK* and *norC* denitrification genes (Bueno et al., 2010).

R. etli nirK and *norC* denitrification genes are also functional in common bean nodules. NirK is an important contributor to the formation of NO in response to nitrate, since levels of LbNO complexes in nodules exposed to nitrate increased in those produced by the *norC* mutant, but decreased in *nirK* nodules compared with LbNO levels detected in wild-type nodules (Gomez-Hernandez et al., 2011; Fig. 11). Interestingly, the presence of nitrate in the plant nutrient solution declined nitrogenase-specific activity in both the wild-type and the *norC* nodules. However, the inhibition of nitrogenase activity by nitrate was not detected in *nirK* nodules (Gomez-Hernandez et al., 2011). Taken together, these results clearly demonstrate the capacity of common bean nodules to produce NO from nitrate present in the nutrient solution. *R. etli* lacks genes encoding Nap or Nar but have a gene (RHE_CH01780) that encodes a putative assimilatory nitrate reductase (Nas) (http://genome.microbedb.jp/rhizobase/). In addition to the bacterial Nas, nitrate can be reduced to nitrite in the nodule through the action of the plant nitrate reductase (NR) that has been reported to be a source of NO in nodules (see Section 5.1.1). Thus, plant NR or *R. etli* Nas are candidates to reduce nitrate to nitrite inside the nodules. Thus, both enzymes should be considered as potential sources of nitrate-dependent NO production. However, the contribution of these enzymes to NO formation in *P. vulgaris* nodules is unknown. While a progress has been made on the study of NO metabolism in *R. etli* free-living cells as well as in common bean nodules, very little is known about N_2O metabolism in the *R. etli–P. vulgaris* symbiosis.

6. CONCLUSIONS

The negative impact of N_2O on climate change and stratospheric ozone has been clearly reported. It is currently believed that microbial denitrification and nitrification are the most important biological pathways for N_2O emission from soils mainly due to the application of synthetic nitrogen-based fertilizers as part of the agricultural practices. One important strategy to ameliorate N_2O emission would be an increased understanding of the environmental and molecular factors, which contribute to the biological generation and consumption of N_2O. Denitrification and DNRA are the major microbial processes in soil that are capable of removing nitrate through its reduction to N_2 or ammonium, respectively. Both energy-conserving processes compete for nitrate since they share nitrate reduction to nitrite. While denitrification causes N loss from terrestrial and aquatic environments and releases N_2O and N_2 to the atmosphere, DNRA retains ammonium in soils and sediments and has a higher tendency for incorporation into microbial or plant biomass. Hence, the relative contributions of denitrification vs respiratory ammonification activities have important consequences for N retention, plant growth and climate. In addition to denitrifiers, recent studies in *E. coli* and *S.* Typhimurium propose the involvement of nitrate-ammonifying bacteria in N_2O emissions; however, the metabolism of N_2O in these organisms is poorly understood. Nitrate-ammonifying bacteria usually lack the respiratory nitrite reductases (NirK/NirS) as well as the typical membrane-bound respiratory NO reductases (cNor and qNor enzymes) found in denitrifiers. Instead, *E. coli* produces NO during nitrate/nitrite reduction to ammonium catalysed by the periplasmic Nap/Nrf and the cytosolic Nar/Nir nitrate reductase and nitrite reductase complexes (Fig. 2A). By contrast to *E. coli*, NO formation from nitrite reduction by Nrf or Nir does not occur in *S.* Typhimurium. Interestingly, the respiratory Nar has been proposed as one major source of NO in *E. coli* and *S.* Typhimurium (Fig. 2A). Given the high toxicity of NO, this molecule has to be removed in order to avoid nitrosative stress conditions. Since nitrate ammonifiers do not have the typical NO reductases found in denitrifiers, other enzymes need to overcome the NO-detoxification role. In this context, NrfA and NorVW are considered the main candidates to function as NO reductases in vivo and in vitro. While NrfA reduces NO to ammonium, NorVW reduces NO to N_2O (Fig. 2A). The key molecules that act as signals for the regulation of NO-production and NO-detoxification proteins are oxygen, and a NOx (nitrate, nitrite or NO). These environmental signals are perceived by a diversed number of transcriptional regulators (NarXL/QP, FNR, NorR and NsrR) that integrate them into regulatory networks in order to allow the cells to respire nitrate/nitrite and avoid NO accumulation as by-product of the reduction process (Fig. 2B).

It was believed for long time that respiratory nitrate ammonification is typical from Gamma-, Delta- and Epsilonproteobacteria and denitrification from Alpha-, Beta- and Gammaproteobacteria, and both pathways do not coexist within a single organism. However, the functionality of both the denitrification and the respiratory ammonification pathways has been recently demonstrated in the Gammaproteobacterium *Sh. loihica* strain PV-4.

Epsilonproteobacteria represent another interesting group of ammonifiers, in which cells employ Nap for nitrate reduction to nitrite and the latter is subsequently reduced to ammonium by NrfA. The capacity of the Epsilonproteobacterium *W. succinogenes* to produce N_2O during growth by nitrate ammonification has been recently demonstrated. However, the question remains how NO is generated from nitrite by *W. succinogenes* since NapA and NrfA are unlikely to release NO as a by-product (as opposed to the *E. coli* NrfA and Nar enzymes). In addition to respire nitrite, *W. succinogenes* NrfA has a detoxifying function in cell physiology given its demonstrated capacity to mediate the stress response to nitrite, NO, hydroxylamine and hydrogen peroxide. In contrast to *E. coli* or *S.* Typhimurim, *W. succinogenes* lacks NorVW; however, a cytoplasmic Fdp, a Hcp and a protein homologous to *H. pylori* NorH have been proposed to be involved in nitrosative stress defence in *W. succinogenes*. The contribution of these proteins to N_2O production, however, has to be clarified in the future.

Given the capacity of nitrate-ammonifying bacteria to produce N_2O during growth by nitrate respiration, it seems reasonable to assume that these bacteria are also capable to reduce N_2O formed as a product of NO detoxification. However, the capacity to reduce N_2O is restricted to Epsilonproteobacteria and some nitrate-ammonifying *Bacillus* species. In fact, the capacity to grow by anaerobic N_2O respiration using N_2O as sole electron acceptor has been recently reported for *W. succinogenes*, *A. dehalogenans* and *B. vireti*. These ammonifiers as well as some other nondenitrifiers contain a *nos* gene cluster encoding the 'atypical' N_2OR NosZ and some of them even a cytochrome *c* nitrous oxide reductase (*c*NosZ) (Table 1; Fig. 5). By contrast, other nitrate-ammonifying bacteria including enterobacteria such as *E. coli* or *S.* Typhimurium that also can produce N_2O do not have an enzyme that can consume it. Thus, these bacteria might contribute significantly to global N_2O emissions.

In the model Epsilonproteobacterium *W. succinogenes*, the respiratory Nap, Nrf and *c*NosZ enzymes are upregulated by the presence of nitrate, NO and N_2O. In contrast to *E. coli* and other nitrate-ammonifying bacteria, *W. succinogenes* lacks the typical nitrate- or NO-responsive proteins such as NarXL/NarQP, NsrR and NorR. Instead, *W. succinogenes* cells employ three transcription regulators of the Crp/FNR superfamily designated NssA, NssB and NssC to mediate upregulation of Nap, Nrf and *c*Nos via dedicated signal transduction routes (Fig. 6).

Denitrification is currently considered to be the largest source of N_2O in soils. In addition to free-living soil bacteria, legume-associated endosymbiotic denitrifiers also contribute to N_2O emissions in free-living conditions as well as inside the root nodules. The environmental signals as well as the regulatory networks involved in the control of denitrification are well known. In addition to oxygen, a NOx (nitrate, nitrite or NO), and the redox state of the cell, new factors such as pH and Cu have been identified recently to be involved in the control of denitrification and more precisely in the regulation of the *nos* genes. In contrast to the atypical *c*NosZ from *W. succinogenes* that responds to N_2O in the absence of nitrate or NO, regulation of the typical NosZ by N_2O seems to be absent. The well-established regulatory mechanisms and networks involved in the control of denitrification (see Fig. 9) become more complex in rhizobial denitrifiers where denitrification and nitrogen fixation processes share common regulators (FixK, NifA, RegR; see Fig. 10).

In denitrifiers, the role of NirK, NirS as well as the NO reductases (cNor and qNor enzymes) in NO and N_2O formation has been well established. However, new enzymes are emerging as candidates to be involved in NO and N_2O metabolism. In particular, it has been recently demonstrated that the assimilatory nitrate reductase (NasC) from *B. japonicum* is important not only in nitrate assimilation but also in NO production. In this context, it has been recently identified in *B. japonicum* a putative haemoglobin, Bjgb, implicated in NO detoxification. Similar to other bacterial haemoglobins, Bjgb might reduce NO to N_2O under anoxic free-living conditions or inside the nodules. Furthermore, *E. meliloti* possesses, in addition to Nor, at least three systems (Hmp, $NnrS_1$ and $NnrS_2$) to detoxify NO under free-living conditions which are also essential in maintaining a balanced NO concentration in nodules and an efficient symbiosis. However, the potential impact of those new NO-consuming proteins on the emission of the GHG N_2O by root nodules has to be demonstrated.

ACKNOWLEDGEMENTS

Work in M.J. Delgado's laboratory was supported by European Regional Development Fund (ERDF) cofinanced Grant AGL2013-45087-R from Ministerio de Economía y Competitividad (Spain) and PE2012-AGR1968 from Junta de Andalucía. Continuous support from Junta de Andalucía to group BIO275 is also acknowledged. Work in J. Simon's laboratory was supported by the Deutsche Forschungsgemeinschaft. Work performed at the University of East Anglia was funded by the Biotechnology and Biological Sciences Research Council [Grant numbers: BB/M00256X/1 (to A.J.G.), BB/L022796/1 (to G.R., A.J.G. and D.J.R) and BB/H012796/1 (to D.J.R.)] and the Royal Society [Grant number IE140222 (to A.J.G. and M.J.D.)]. D.J.R. is a Royal Society Wolfson Foundation Merit Award holder. The authors are grateful to Monique Luckmann for help with figure preparation.

REFERENCES

Al-Attar, S., & de Vries, S. (2015). An electrogenic nitric oxide reductase. *FEBS Letters*, *589*, 2050–2057.

Arai, H., Roh, J. H., Eraso, J. M., & Kaplan, S. (2013). Transcriptome response to nitrosative stress in *Rhodobacter sphaeroides* 2.4.1. *Bioscience, Biotechnology, and Biochemistry*, 77, 111–118.

Arkenberg, A., Runkel, S., Richardson, D. J., & Rowley, G. (2011). The production and detoxification of a potent cytotoxin, nitric oxide, by pathogenic enteric bacteria. *Biochemical Society Transactions*, *39*, 1876–1879.

Atack, J. M., Srikhanta, Y. N., Djoko, K. Y., Welch, J. P., Hasri, N. H., Steichen, C. T., et al. (2013). Characterization of an *ntrX* mutant of *Neisseria gonorrhoeae* reveals a response regulator that controls expression of respiratory enzymes in oxidase-positive proteobacteria. *Journal of Bacteriology*, *195*, 2632–2641.

Baar, C., Eppinger, M., Raddatz, G., Simon, J., Lanz, C., Klimmek, O., et al. (2003). Complete genome sequence and analysis of *Wolinella succinogenes*. *Proceedings of the National Academy of Sciences of the United States of America*, *100*, 11690–11695.

Baek, S. H., Hartsock, A., & Shapleigh, J. P. (2008). *Agrobacterium tumefaciens* C58 uses ActR and FnrN to control *nirK* and *nor* expression. *Journal of Bacteriology*, *190*, 78–86.

Baggs, E. M., Rees, R. M., Smith, K. A., & Vinten, A. J. A. (2000). Nitrous oxide emission from soils after incorporation of crop residues. *Soil Use and Management*, *16*, 82–87.

Bali, S., Palmer, D. J., Schroeder, S., Ferguson, S. J., & Warren, M. J. (2014). Recent advances in the biosynthesis of modified tetrapyrroles: The discovery of an alternative pathway for the formation of heme and heme d 1. *Cellular and Molecular Life Sciences*, *71*, 2837–2863.

Bamford, V. A., Angove, H. C., Seward, H. E., Thomson, A. J., Cole, J. A., Butt, J. N., et al. (2002). Structure and spectroscopy of the periplasmic cytochrome *c* nitrite reductase from *Escherichia coli*. *Biochemistry*, *41*, 2921–2931.

Barnett, M. J., Fisher, R. F., Jones, T., Komp, C., Abola, A. P., Barloy-Hubler, F., et al. (2001). Nucleotide sequence and predicted functions of the entire *Sinorhizobium meliloti* pSymA megaplasmid. *Proceedings of the National Academy of Sciences of the United States of America*, *98*, 9883–9888.

Barrios, H., Fischer, H. M., Hennecke, H., & Morett, E. (1995). Overlapping promoters for two different RNA polymerase holoenzymes control *Bradyrhizobium japonicum nifA* expression. *Journal of Bacteriology*, *177*, 1760–1765.

Barrios, H., Grande, R., Olvera, L., & Morett, E. (1998). In vivo genomic footprinting analysis reveals that the complex *Bradyrhizobium japonicum fixRnifA* promoter region is differently occupied by two distinct RNA polymerase holoenzymes. *Proceedings of the National Academy of Sciences of the United States of America*, *95*, 1014–1019.

Bauer, E., Kaspar, T., Fischer, H. M., & Hennecke, H. (1998). Expression of the *fixR-nifA* operon in *Bradyrhizobium japonicum* depends on a new response regulator, RegR. *Journal of Bacteriology*, *180*, 3853–3863.

Becker, A., Berges, H., Krol, E., Bruand, C., Ruberg, S., Capela, D., et al. (2004). Global changes in gene expression in *Sinorhizobium meliloti* 1021 under microoxic and symbiotic conditions. *Molecular Plant-Microbe Interactions*, *17*, 292–303.

Bedmar, E. J., Bueno, E., Correa, D., Torres, M. J., Delgado, M. J., & Mesa, S. (2013). Ecology of denitrification in soils and plant-associated. In B. Rodelas & J. Gonzalez-López (Eds.), *Beneficial plant-microbial interactions: Ecology and applications* (pp. 164–182). Boca Ratón, Florida, USA: CRC Press.

Bedmar, E. J., Robles, E. F., & Delgado, M. J. (2005). The complete denitrification pathway of the symbiotic, nitrogen-fixing bacterium *Bradyrhizobium japonicum*. *Biochemical Society Transactions*, *33*, 141–144.

Bedzyk, L., Wang, T., & Ye, R. W. (1999). The periplasmic nitrate reductase in *Pseudomonas* sp. strain G-179 catalyzes the first step of denitrification. *Journal of Bacteriology*, *181*, 2802–2806.

Bell, L. C., Richardson, D. J., & Ferguson, S. J. (1992). Identification of nitric oxide reductase activity in *Rhodobacter capsulatus*: The electron transport pathway can either use or bypass both cytochrome c_2 and the cytochrome bc_1 complex. *Journal of General Microbiology*, *138*, 437–443.

Bergaust, L., Mao, Y., Bakken, L. R., & Frostegard, A. (2010). Denitrification response patterns during the transition to anoxic respiration and posttranscriptional effects of suboptimal pH on nitrous oxide reductase in *Paracoccus denitrificans*. *Applied and Environmental Microbiology*, *76*, 6387–6396.

Bergaust, L., van Spanning, R. J., Frostegard, A., & Bakken, L. R. (2012). Expression of nitrous oxide reductase in *Paracoccus denitrificans* is regulated by oxygen and nitric oxide through FnrP and NNR. *Microbiology*, *158*, 826–834.

Bergaust, L. L., Hartsock, A., Liu, B., Bakken, L. R., & Shapleigh, J. P. (2014). Role of *norEF* in denitrification, elucidated by physiological experiments with *Rhodobacter sphaeroides*. *Journal of Bacteriology*, *196*, 2190–2200.

Bertero, M. G., Rothery, R. A., Palak, M., Hou, C., Lim, D., Blasco, F., et al. (2003). Insights into the respiratory electron transfer pathway from the structure of nitrate reductase A. *Nature Structural Biology*, *10*, 681–687.

Bethke, P. C., Badger, M. R., & Jones, R. L. (2004). Apoplastic synthesis of nitric oxide by plant tissues. *Plant Cell*, *16*, 332–341.

Blanquet, P., Silva, L., Catrice, O., Bruand, C., Carvalho, H., & Meilhoc, E. (2015). *Sinorhizobium meliloti* controls nitric oxide-mediated post-translational modification of a *Medicago truncatula* nodule protein. *Molecular Plant-Microbe Interactions*, *28*, 1353–1363.

Blasco, F., Iobbi, C., Ratouchniak, J., Bonnefoy, V., & Chippaux, M. (1990). Nitrate reductases of *Escherichia coli*: Sequence of the second nitrate reductase and comparison with that encoded by the *narGHJI* operon. *Molecular & General Genetics*, *222*, 104–111.

Bleakley, B. H., & Tiedje, J. M. (1982). Nitrous oxide production by organisms other than nitrifiers or denitrifiers. *Applied and Environmental Microbiology*, *44*, 1342–1348.

Bobik, C., Meilhoc, E., & Batut, J. (2006). FixJ: A major regulator of the oxygen limitation response and late symbiotic functions of *Sinorhizobium meliloti*. *Journal of Bacteriology*, *188*, 4890–4902.

Bodenmiller, D. M., & Spiro, S. (2006). The *yjeB* (*nsrR*) gene of *Escherichia coli* encodes a nitric oxide-sensitive transcriptional regulator. *Journal of Bacteriology, 188*, 874–881.

Bokranz, M., Katz, J., Schröder, I., Robertson, A. M., & Kröger, A. (1983). Energy metabolism and biosynthesis of *Vibrio succinogenes* growing with nitrate or nitrite as terminal electron acceptor. *Archives of Microbiology, 135*, 36–41.

Bothe, J. R., & Beyer, K. D. (2007). Experimental determination of the NH_4NO_3/$(NH_4)_2SO_4$/H_2O phase diagram. *The Journal of Physical Chemistry A, 111*, 12106–12117.

Bouchal, P., Struharova, I., Budinska, E., Sedo, O., Vyhlidalova, T., Zdrahal, Z., et al. (2010). Unraveling an FNR based regulatory circuit in *Paracoccus denitrificans* using a proteomics-based approach. *Biochimica et Biophysica Acta, 1804*, 1350–1358.

Bricio, C., Alvarez, L., San Martin, M., Schurig-Briccio, L. A., Gennis, R. B., & Berenguer, J. (2014). A third subunit in ancestral cytochrome *c*-dependent nitric oxide reductases. *Applied and Environmental Microbiology, 80*, 4871–4878.

Bronder, M., Mell, H., Stupperich, E., & Kroger, A. (1982). Biosynthetic pathways of *Vibrio succinogenes* growing with fumarate as terminal electron acceptor and sole carbon source. *Archives of Microbiology, 131*, 216–223.

Broughton, W. J., Hernández, G., Blair, M., Beebe, S., Gepts, P., & Vanderleyden, J. (2003). Beans (*Phaseolus* spp.)—Model food legumes. *Plant and Soil, 252*, 55–128.

Browning, D. F., Lee, D. J., Spiro, S., & Busby, S. J. (2010). Down-regulation of the *Escherichia coli* K-12 *nrf* promoter by binding of the NsrR nitric oxide-sensing transcription repressor to an upstream site. *Journal of Bacteriology, 192*, 3824–3828.

Bueno, E., Bedmar, E. J., Richardson, D. J., & Delgado, M. J. (2008). Role of *Bradyrhizobium japonicum* cytochrome c_{550} in nitrite and nitrate respiration. *FEMS Microbiology Letters, 279*, 188–194.

Bueno, E., Gomez-Hernandez, N., Girard, L., Bedmar, E. J., & Delgado, M. J. (2005). Function of the *Rhizobium etli* CFN42 *nirK* gene in nitrite metabolism. *Biochemical Society Transactions, 33*, 162–163.

Bueno, E., Mania, D., Frostegard, A., Bedmar, E. J., Bakken, L. R., & Delgado, M. J. (2015). Anoxic growth of Ensifer meliloti 1021 by N2O-reduction, a potential mitigation strategy. *Frontiers in Microbiology, 6*, 537.

Bueno, E., Mesa, S., Bedmar, E. J., Richardson, D. J., & Delgado, M. J. (2012). Bacterial adaptation of respiration from oxic to microoxic and anoxic conditions: Redox control. *Antioxidants & Redox Signaling, 16*, 819–852.

Bueno, E., Mesa, S., Sanchez, C., Bedmar, E. J., & Delgado, M. J. (2010). NifA is required for maximal expression of denitrification genes in *Bradyrhizobium japonicum*. *Environmental Microbiology, 12*, 393–400.

Burgess, B. K., & Lowe, D. J. (1996). Mechanism of molybdenum nitrogenase. *Chemical Reviews, 96*, 2983–3012.

Büsch, A., Strube, K., Friedrich, B., & Cramm, R. (2005). Transcriptional regulation of nitric oxide reduction in *Ralstonia eutropha* H16. *Biochemical Society Transactions, 33*, 193–194.

Bush, M., Ghosh, T., Tucker, N., Zhang, X., & Dixon, R. (2010). Nitric oxide-responsive interdomain regulation targets the σ54-interaction surface in the enhancer binding protein NorR. *Molecular Microbiology, 77*, 1278–1288.

Butland, G., Spiro, S., Watmough, N. J., & Richardson, D. J. (2001). Two conserved glutamates in the bacterial nitric oxide reductase are essential for activity but not assembly of the enzyme. *Journal of Bacteriology, 183*, 189–199.

Butterbach-Bahl, K., Baggs, E. M., Dannenmann, M., Kiese, R., & Zechmeister-Boltenstern, S. (2014). Nitrous oxide emissions from soils: How well do we understand the processes and their controls? *Philosophical Transactions of the Royal Society B, 368*, 20130122.

Cabrera, J. J., Salas, A., Torres, M. J., Bedmar, E. J., Richardson, D. J., Gates, A. J., et al. (2016). An integrated biochemical system for nitrate assimilation and nitric oxide detoxification in *Bradyrhizobium japonicum*. *The Biochemical Journal*, *473*, 297–309.

Cabrera, J. J., Sanchez, C., Gates, A. J., Bedmar, E. J., Mesa, S., Richardson, D. J., et al. (2011). The nitric oxide response in plant-associated endosymbiotic bacteria. *Biochemical Society Transactions*, *39*, 1880–1885.

Cam, Y., Pierre, O., Boncompagni, E., Herouart, D., Meilhoc, E., & Bruand, C. (2012). Nitric oxide (NO): A key player in the senescence of *Medicago truncatula* root nodules. *The New Phytologist*, *196*, 548–560.

Campbell, B. J., Engel, A. S., Porter, M. L., & Takai, K. (2006). The versatile epsilon-proteobacteria: Key players in sulphidic habitats. *Nature Reviews. Microbiology*, *4*, 458–468.

Campbell, M. A., Nyerges, G., Kozlowski, J. A., Poret-Peterson, A. T., Stein, L. Y., & Klotz, M. G. (2011). Model of the molecular basis for hydroxylamine oxidation and nitrous oxide production in methanotrophic bacteria. *FEMS Microbiology Letters*, *322*, 82–89.

Carrica, M. D., Fernandez, I., Sieira, R., Paris, G., & Goldbaum, F. A. (2013). The two-component systems PrrBA and NtrYX co-ordinately regulate the adaptation of *Brucella abortus* to an oxygen-limited environment. *Molecular Microbiology*, *88*, 222–233.

Castiglione, N., Rinaldo, S., Giardina, G., & Cutruzzola, F. (2009). The transcription factor DNR from *Pseudomonas aeruginosa* specifically requires nitric oxide and haem for the activation of a target promoter in *Escherichia coli*. *Microbiology*, *155*, 2838–2844.

Clarke, T. A., Cole, J. A., Richardson, D. J., & Hemmings, A. M. (2007). The crystal structure of the pentahaem *c*-type cytochrome NrfB and characterization of its solution-state interaction with the pentahaem nitrite reductase NrfA. *The Biochemical Journal*, *406*, 19–30.

Clarke, T. A., Mills, P. C., Poock, S. R., Butt, J. N., Cheesman, M. R., Cole, J. A., et al. (2008). *Escherichia coli* cytochrome *c* nitrite reductase NrfA. *Methods in Enzymology*, *437*, 63–77.

Constantinidou, C., Hobman, J. L., Griffiths, L., Patel, M. D., Penn, C. W., Cole, J. A., et al. (2006). A reassessment of the FNR regulon and transcriptomic analysis of the effects of nitrate, nitrite, NarXL, and NarQP as *Escherichia coli* K12 adapts from aerobic to anaerobic growth. *The Journal of Biological Chemistry*, *281*, 4802–4815.

Cook, D. R. (1999). *Medicago truncatula*—A model in the making! *Current Opinion in Plant Biology*, *2*, 301–304.

Corker, H., & Poole, R. K. (2003). Nitric oxide formation by *Escherichia coli*. Dependence on nitrite reductase, the NO-sensing regulator Fnr, and flavohemoglobin Hmp. *The Journal of Biological Chemistry*, *278*, 31584–31592.

Costa, C., Macedo, A., Moura, I., Moura, J. J., Le Gall, J., Berlier, Y., et al. (1990). Regulation of the hexaheme nitrite/nitric oxide reductase of *Desulfovibrio desulfuricans*, *Wolinella succinogenes* and *Escherichia coli*. A mass spectrometric study. *FEBS Letters*, *276*, 67–70.

Coyle, C. L., Zumft, W. G., Kroneck, P. M., Korner, H., & Jakob, W. (1985). Nitrous oxide reductase from denitrifying *Pseudomonas perfectomarina*. Purification and properties of a novel multicopper enzyme. *European Journal of Biochemistry*, *153*, 459–467.

Crack, J. C., Green, J., Hutchings, M. I., Thomson, A. J., & Le Brun, N. E. (2012). Bacterial iron-sulfur regulatory proteins as biological sensor-switches. *Antioxidants & Redox Signaling*, *17*, 1215–1231.

Crack, J. C., Munnoch, J., Dodd, E. L., Knowles, F., Al Bassam, M. M., Kamali, S., et al. (2015). NsrR from *Streptomyces coelicolor* is a nitric oxide-sensing [4Fe-4S] cluster protein with a specialized regulatory function. *The Journal of Biological Chemistry*, *290*, 12689–12704.

Crack, J. C., Stapleton, M. R., Green, J., Thomson, A. J., & Le Brun, N. E. (2013). Mechanism of [4Fe-4S](Cys)4 cluster nitrosylation is conserved among NO-responsive regulators. *The Journal of Biological Chemistry*, *288*, 11492–11502.

Cruz-Ramos, H., Crack, J., Wu, G., Hughes, M. N., Scott, C., Thomson, A. J., et al. (2002). NO sensing by FNR: Regulation of the *Escherichia coli* NO-detoxifying flavohaemoglobin, Hmp. *The EMBO Journal*, *21*, 3235–3244.

Cutruzzola, F., Arcovito, A., Giardina, G., della Longa, S., D'Angelo, P., & Rinaldo, S. (2014). Distal-proximal crosstalk in the heme binding pocket of the NO sensor DNR. *Biometals*, *27*, 763–773.

Cheung, J., & Hendrickson, W. A. (2009). Structural analysis of ligand stimulation of the histidine kinase NarX. *Structure*, *17*, 190–201.

Chhabra, S., & Spiro, S. (2015). Inefficient translation of *nsrR* constrains behaviour of the NsrR regulon in *Escherichia coli*. *Microbiology*, *161*, 2029–2038.

D'Autreaux, B., Tucker, N. P., Dixon, R., & Spiro, S. (2005). A non-haem iron centre in the transcription factor NorR senses nitric oxide. *Nature*, *437*, 769–772.

Darwin, A. J., Ziegelhoffer, E. C., Kiley, P. J., & Stewart, V. (1998). Fnr, NarP, and NarL regulation of *Escherichia coli* K-12 *napF* (periplasmic nitrate reductase) operon transcription in vitro. *Journal of Bacteriology*, *180*, 4192–4198.

Daskalakis, V., Ohta, T., Kitagawa, T., & Varotsis, C. (2015). Structure and properties of the catalytic site of nitric oxide reductase at ambient temperature. *Biochimica et Biophysica Acta*, *1847*, 1240–1244.

de Boer, A. P., van der Oost, J., Reijnders, W. N., Westerhoff, H. V., Stouthamer, A. H., & van Spanning, R. J. (1996). Mutational analysis of the *nor* gene cluster which encodes nitric-oxide reductase from *Paracoccus denitrificans*. *European Journal of Biochemistry*, *242*, 592–600.

de Vries, S., Suharti, & Pouvreau, L. A. M. (2007). Nitric oxide reductase: Structural variations and catalytic mechanism. In H. Bothe, S. J. Ferguson, & W. E. Newton (Eds.), *Biology of the nitrogen cycle* (pp. 57–67). Amsterdam: Elservier Science.

Delamuta, J. R., Ribeiro, R. A., Ormeno-Orrillo, E., Melo, I. S., Martinez-Romero, E., & Hungria, M. (2013). Polyphasic evidence supporting the reclassification of *Bradyrhizobium japonicum* group Ia strains as *Bradyrhizobium diazoefficiens* sp. nov. *International Journal of Systematic and Evolutionary Microbiology*, *63*, 3342–3351.

Delgado, M. J., Bonnard, N., Tresierra-Ayala, A., Bedmar, E. J., & Muller, P. (2003). The *Bradyrhizobium japonicum napEDABC* genes encoding the periplasmic nitrate reductase are essential for nitrate respiration. *Microbiology*, *149*, 3395–3403.

Delgado, M. J., Casella, S., & Bedmar, E. J. (2007). Denitrification in rhizobia-legume symbiosis. In H. Bothe, S. J. Ferguson, & W. E. Newton (Eds.), *Biology of the nitrogen cycle* (pp. 83–93). Amsterdam: Elsevier Science.

Desloover, J., Roobroeck, D., Heylen, K., Puig, S., Boeckx, P., Verstraete, W., et al. (2014). Pathway of nitrous oxide consumption in isolated *Pseudomonas stutzeri* strains under anoxic and oxic conditions. *Environmental Microbiology*, *16*, 3143–3152.

Dreusch, A., Burgisser, D. M., Heizmann, C. W., & Zumft, W. G. (1997). Lack of copper insertion into unprocessed cytoplasmic nitrous oxide reductase generated by an R20D substitution in the arginine consensus motif of the signal peptide. *Biochimica et Biophysica Acta*, *1319*, 311–318.

Einsle, O. (2011). Structure and function of formate-dependent cytochrome *c* nitrite reductase, NrfA. *Methods in Enzymology*, *496*, 399–422.

Einsle, O., Messerschmidt, A., Huber, R., Kroneck, P. M., & Neese, F. (2002). Mechanism of the six-electron reduction of nitrite to ammonia by cytochrome *c* nitrite reductase. *Journal of the American Chemical Society*, *124*, 11737–11745.

Elsen, S., Swem, L. R., Swem, D. L., & Bauer, C. E. (2004). RegB/RegA, a highly conserved redox-responding global two-component regulatory system. *Microbiology and Molecular Biology Reviews*, *68*, 263–279.

Elvers, K. T., Turner, S. M., Wainwright, L. M., Marsden, G., Hinds, J., Cole, J. A., et al. (2005). NssR, a member of the Crp-Fnr superfamily from *Campylobacter jejuni*, regulates a nitrosative stress-responsive regulon that includes both a single-domain and a truncated haemoglobin. *Molecular Microbiology, 57*, 735–750.

Felgate, H., Giannopoulos, G., Sullivan, M. J., Gates, A. J., Clarke, T. A., Baggs, E., et al. (2012). The impact of copper, nitrate and carbon status on the emission of nitrous oxide by two species of bacteria with biochemically distinct denitrification pathways. *Environmental Microbiology, 14*, 1788–1800.

Filenko, N., Spiro, S., Browning, D. F., Squire, D., Overton, T. W., Cole, J., et al. (2007). The NsrR regulon of *Escherichia coli* K-12 includes genes encoding the hybrid cluster protein and the periplasmic, respiratory nitrite reductase. *Journal of Bacteriology, 189*, 4410–4417.

Forrester, M. T., & Foster, M. W. (2012). Protection from nitrosative stress: A central role for microbial flavohemoglobin. *Free Radical Biology & Medicine, 52*, 1620–1633.

Fujita, K., & Dooley, D. M. (2007). Insights into the mechanism of N_2O reduction by reductively activated N_2O reductase from kinetics and spectroscopic studies of pH effects. *Inorganic Chemistry, 46*, 613–615.

Galibert, F., Finan, T. M., Long, S. R., Puhler, A., Abola, P., Ampe, F., et al. (2001). The composite genome of the legume symbiont *Sinorhizobium meliloti*. *Science, 293*, 668–672.

Gardner, A. M., & Gardner, P. R. (2002). Flavohemoglobin detoxifies nitric oxide in aerobic, but not anaerobic, Escherichia coli. Evidence for a novel inducible anaerobic nitric oxide-scavenging activity. *The Journal of Biological Chemistry, 277*, 8166–8171.

Gardner, A. M., Gessner, C. R., & Gardner, P. R. (2003). Regulation of the nitric oxide reduction operon (*norRVW*) in *Escherichia coli*. Role of NorR and σ^{54} in the nitric oxide stress response. *The Journal of Biological Chemistry, 278*, 10081–10086.

Gardner, A. M., Helmick, R. A., & Gardner, P. R. (2002). Flavorubredoxin, an inducible catalyst for nitric oxide reduction and detoxification in *Escherichia coli*. *The Journal of Biological Chemistry, 277*, 8172–8177.

Gardner, P. R. (2005). Nitric oxide dioxygenase function and mechanism of flavohemoglobin, hemoglobin, myoglobin and their associated reductases. *Journal of Inorganic Biochemistry, 99*, 247–266.

Giardina, G., Rinaldo, S., Johnson, K. A., Di Matteo, A., Brunori, M., & Cutruzzola, F. (2008). NO sensing in *Pseudomonas aeruginosa*: Structure of the transcriptional regulator DNR. *Journal of Molecular Biology, 378*, 1002–1015.

Gilberthorpe, N. J., & Poole, R. K. (2008). Nitric oxide homeostasis in *Salmonella typhimurium*: Roles of respiratory nitrate reductase and flavohemoglobin. *The Journal of Biological Chemistry, 283*, 11146–11154.

Goddard, A. D., Moir, J. W., Richardson, D. J., & Ferguson, S. J. (2008). Interdependence of two NarK domains in a fused nitrate/nitrite transporter. *Molecular Microbiology, 70*, 667–681.

Gomes, C. M., Giuffre, A., Forte, E., Vicente, J. B., Saraiva, L. M., Brunori, M., et al. (2002). A novel type of nitric-oxide reductase. Escherichia coli flavorubredoxin. *The Journal of Biological Chemistry, 277*, 25273–25276.

Gomez-Hernandez, N., Reyes-Gonzalez, A., Sanchez, C., Mora, Y., Delgado, M. J., & Girard, L. (2011). Regulation and symbiotic role of *nirK* and *norC* expression in *Rhizobium etli*. *Molecular Plant-Microbe Interactions, 24*, 233–245.

Gonzalez, P. J., Correia, C., Moura, I., Brondino, C. D., & Moura, J. J. (2006). Bacterial nitrate reductases: Molecular and biological aspects of nitrate reduction. *Journal of Inorganic Biochemistry, 100*, 1015–1023.

Grahl, S., Maillard, J., Spronk, C. A., Vuister, G. W., & Sargent, F. (2012). Overlapping transport and chaperone-binding functions within a bacterial twin-arginine signal peptide. *Molecular Microbiology, 83*, 1254–1267.

Gupta, K. J., Bauwe, H., & Mur, L. A. (2011). Nitric oxide, nitrate reductase and UV-B tolerance. *Tree Physiology, 31*, 795–797.

Haltia, T., Brown, K., Tegoni, M., Cambillau, C., Saraste, M., Mattila, K., et al. (2003). Crystal structure of nitrous oxide reductase from *Paracoccus denitrificans* at 1.6 A resolution. *The Biochemical Journal, 369*, 77–88.

Hartsock, A., & Shapleigh, J. P. (2010). Identification, functional studies, and genomic comparisons of new members of the NnrR regulon in *Rhodobacter sphaeroides*. *Journal of Bacteriology, 192*, 903–911.

Hauser, F., Pessi, G., Friberg, M., Weber, C., Rusca, N., Lindemann, A., et al. (2007). Dissection of the *Bradyrhizobium japonicum* NifA+σ^{54} regulon, and identification of a ferredoxin gene (*fdxN*) for symbiotic nitrogen fixation. *Molecular Genetics and Genomics, 278*, 255–271.

Hendriks, J., Oubrie, A., Castresana, J., Urbani, A., Gemeinhardt, S., & Saraste, M. (2000). Nitric oxide reductases in bacteria. *Biochimica et Biophysica Acta, 1459*, 266–273.

Hendriks, J., Warne, A., Gohlke, U., Haltia, T., Ludovici, C., Lubben, M., et al. (1998). The active site of the bacterial nitric oxide reductase is a dinuclear iron center. *Biochemistry, 37*, 13102–13109.

Hermann, B., Kern, M., La Pietra, L., Simon, J., & Einsle, O. (2015). The octahaem MccA is a haem *c*-copper sulfite reductase. *Nature, 520*, 706–709.

Herridge, D. F., Peoples, M. B., & Boddey, R. M. (2008). Global inputs of biological nitrogen fixation in agricultural systems. *Plant and Soil, 311*, 1–18.

Heurlier, K., Thomson, M. J., Aziz, N., & Moir, J. W. (2008). The nitric oxide (NO)-sensing repressor NsrR of *Neisseria meningitidis* has a compact regulon of genes involved in NO synthesis and detoxification. *Journal of Bacteriology, 190*, 2488–2495.

Heylen, K., & Keltjens, J. (2012). Redundancy and modularity in membrane-associated dissimilatory nitrate reduction in *Bacillus*. *Frontiers in Microbiology, 3*, 371.

Hichri, I., Boscari, A., Castella, C., Rovere, M., Puppo, A., & Brouquisse, R. (2015). Nitric oxide: A multifaceted regulator of the nitrogen-fixing symbiosis. *Journal of Experimental Botany, 66*, 2877–2887.

Hino, T., Matsumoto, Y., Nagano, S., Sugimoto, H., Fukumori, Y., Murata, T., et al. (2010). Structural basis of biological N_2O generation by bacterial nitric oxide reductase. *Science, 330*, 1666–1670.

Hirayama, J., Eda, S., Mitsui, H., & Minamisawa, K. (2011). Nitrate-dependent N_2O emission from intact soybean nodules via denitrification by *Bradyrhizobium japonicum* bacteroids. *Applied and Environmental Microbiology, 77*, 8787–8790.

Hole, U. H., Vollack, K. U., Zumft, W. G., Eisenmann, E., Siddiqui, R. A., Friedrich, B., et al. (1996). Characterization of the membranous denitrification enzymes nitrite reductase (cytochrome cd_1) and copper-containing nitrous oxide reductase from *Thiobacillus denitrificans*. *Archives of Microbiology, 165*, 55–61.

Holloway, P., McCormick, W., Watson, R. J., & Chan, Y. K. (1996). Identification and analysis of the dissimilatory nitrous oxide reduction genes, *nosRZDFY*, of *Rhizobium meliloti*. *Journal of Bacteriology, 178*, 1505–1514.

Honisch, U., & Zumft, W. G. (2003). Operon structure and regulation of the nos gene region of *Pseudomonas stutzeri*, encoding an ABC-Type ATPase for maturation of nitrous oxide reductase. *Journal of Bacteriology, 185*, 1895–1902.

Horchani, F., Prevot, M., Boscari, A., Evangelisti, E., Meilhoc, E., Bruand, C., et al. (2011). Both plant and bacterial nitrate reductases contribute to nitric oxide production in *Medicago truncatula* nitrogen-fixing nodules. *Plant Physiology, 155*, 1023–1036.

Hutchings, M. I., Mandhana, N., & Spiro, S. (2002). The NorR protein of *Escherichia coli* activates expression of the flavorubredoxin gene *norV* in response to reactive nitrogen species. *Journal of Bacteriology, 184*, 4640–4643.

Hutchings, M. I., & Spiro, S. (2000). The nitric oxide regulated nor promoter of *Paracoccus denitrificans*. *Microbiology, 146*, 2635–2641.

Inaba, S., Ikenishi, F., Itakura, M., Kikuchi, M., Eda, S., Chiba, N., et al. (2012). N_2O emission from degraded soybean nodules depends on denitrification by *Bradyrhizobium japonicum* and other microbes in the rhizosphere. *Microbes and Environments, 27*, 470–476.

Inaba, S., Tanabe, K., Eda, S., Ikeda, S., Higashitani, A., Mitsui, H., et al. (2009). Nitrous oxide emission and microbial community in the rhizosphere of nodulated soybeans during the late growth period. *Microbes and Environments, 24*, 64–67.

IPCC. (2014). Climate change 2014: Synthesis report. In *Contribution of working groups I, II and III to the fifth assessment report of the intergovernmental panel on climate change*. Geneva, Switzerland: IPCC.

Isabella, V. M., Lapek, J. D., Jr., Kennedy, E. M., & Clark, V. L. (2009). Functional analysis of NsrR, a nitric oxide-sensing Rrf2 repressor in Neisseria gonorrhoeae. *Molecular Microbiology, 71*, 227–239.

Itakura, M., Uchida, Y., Akiyama, H., Hoshino, Y. T., Shimomura, Y., Morimoto, S., et al. (2013). Mitigation of nitrous oxide emissions from soils by *Bradyrhizobium japonicum* inoculation. *Nature Climate Change, 3*, 208–212.

Jensen, B. B., & Burris, R. H. (1986). N_2O as a substrate and as a competitive inhibitor of nitrogenase. *Biochemistry, 25*, 1083–1088.

Jepson, B. J., Mohan, S., Clarke, T. A., Gates, A. J., Cole, J. A., Butler, C. S., et al. (2007). Spectropotentiometric and structural analysis of the periplasmic nitrate reductase from *Escherichia coli*. *The Journal of Biological Chemistry, 282*, 6425–6437.

Jetten, M. S. (2008). The microbial nitrogen cycle. *Environmental Microbiology, 10*, 2903–2909.

Johnston, E. M., Dell'Acqua, S., Pauleta, S. R., Moura, I., & Solomon, E. I. (2015). Protonation state of the CuS Cu site in nitrous oxide reductase: Redox dependence and insight into reactivity. *Chemical Science, 6*, 5670–5679.

Johnston, E. M., Dell'Acqua, S., Ramos, S., Pauleta, S. R., Moura, I., & Solomon, E. I. (2014). Determination of the active form of the tetranuclear copper sulfur cluster in nitrous oxide reductase. *Journal of the American Chemical Society, 136*, 614–617.

Jones, C. M., Graf, D. R., Bru, D., Philippot, L., & Hallin, S. (2013). The unaccounted yet abundant nitrous oxide-reducing microbial community: A potential nitrous oxide sink. *The ISME Journal, 7*, 417–426.

Jones, K. M., Kobayashi, H., Davies, B. W., Taga, M. E., & Walker, G. C. (2007). How rhizobial symbionts invade plants: The *Sinorhizobium-Medicago* model. *Nature Reviews. Microbiology, 5*, 619–633.

Jormakka, M., Richardson, D., Byrne, B., & Iwata, S. (2004). Architecture of NarGH reveals a structural classification of Mo-bisMGD enzymes. *Structure, 12*, 95–104.

Justino, M. C., Ecobichon, C., Fernandes, A. F., Boneca, I. G., & Saraiva, L. M. (2012). *Helicobacter pylori* has an unprecedented nitric oxide detoxifying system. *Antioxidants & Redox Signaling, 17*, 1190–1200.

Kaneko, T., Nakamura, Y., Sato, S., Minamisawa, K., Uchiumi, T., Sasamoto, S., et al. (2002). Complete genomic sequence of nitrogen-fixing symbiotic bacterium *Bradyrhizobium japonicum* USDA110. *DNA Research, 9*, 189–197.

Kaspar, H. F., & Tiedje, J. M. (1981). Dissimilatory reduction of nitrate and nitrite in the bovine rumen: Nitrous oxide production and effect of acetylene. *Applied and Environmental Microbiology, 41*, 705–709.

Kato, K., Kanahama, K., & Kanayama, Y. (2010). Involvement of nitric oxide in the inhibition of nitrogenase activity by nitrate in *Lotus* root nodules. *Journal of Plant Physiology, 167*, 238–241.

Kern, M., Eisel, F., Scheithauer, J., Kranz, R. G., & Simon, J. (2010). Substrate specificity of three cytochrome *c* haem lyase isoenzymes from *Wolinella succinogenes*: Unconventional haem *c* binding motifs are not sufficient for haem *c* attachment by NrfI and CcsA1. *Molecular Microbiology, 75*, 122–137.

Kern, M., Klotz, M. G., & Simon, J. (2011). The *Wolinella succinogenes mcc* gene cluster encodes an unconventional respiratory sulphite reduction system. *Molecular Microbiology, 82*, 1515–1530.

Kern, M., Mager, A. M., & Simon, J. (2007). Role of individual *nap* gene cluster products in NapC-independent nitrate respiration of *Wolinella succinogenes*. *Microbiology, 153*, 3739–3747.

Kern, M., & Simon, J. (2008). Characterization of the NapGH quinol dehydrogenase complex involved in *Wolinella succinogenes* nitrate respiration. *Molecular Microbiology, 69*, 1137–1152.

Kern, M., & Simon, J. (2009a). Electron transport chains and bioenergetics of respiratory nitrogen metabolism in Wolinella succinogenes and other Epsilonproteobacteria. *Biochimica et Biophysica Acta, 1787*, 646–656.

Kern, M., & Simon, J. (2009b). Periplasmic nitrate reduction in *Wolinella succinogenes*: Cytoplasmic NapF facilitates NapA maturation and requires the menaquinol dehydrogenase NapH for membrane attachment. *Microbiology, 155*, 2784–2794.

Kern, M., & Simon, J. (2016). Three transcription regulators of the Nss family mediate the adaptive response induced by nitrate, nitric oxide or nitrous oxide in *Wolinella succinogenes*. *Environmental Microbiology*. http://dx.doi.org/10.1111/1462-2920.13060.

Kern, M., Volz, J., & Simon, J. (2011). The oxidative and nitrosative stress defence network of *Wolinella succinogenes*: Cytochrome *c* nitrite reductase mediates the stress response to nitrite, nitric oxide, hydroxylamine and hydrogen peroxide. *Environmental Microbiology, 13*, 2478–2494.

Kern, M., Winkler, C., & Simon, J. (2011). Respiratory nitrogen metabolism and nitrosative stress defence in ε-proteobacteria: The role of NssR-type transcription regulators. *Biochemical Society Transactions, 39*, 299–302.

Kim, S. O., Orii, Y., Lloyd, D., Hughes, M. N., & Poole, R. K. (1999). Anoxic function for the *Escherichia coli* flavohaemoglobin (Hmp): Reversible binding of nitric oxide and reduction to nitrous oxide. *FEBS Letters, 445*, 389–394.

Klimmek, O., Dietrich, W., Dancea, F., Lin, Y. L., Pfeiffer, S., Löhr, F., et al. (2004). Sulfur respiration. In D. Zannoni (Ed.), *Respiration in archaea and bacteria* (pp. 217–232). Dordrecht: Kluwer Scientific.

Kraft, B., Strous, M., & Tegetmeyer, H. E. (2011). Microbial nitrate respiration—Genes, enzymes and environmental distribution. *Journal of Biotechnology, 155*, 104–117.

Kröger, A., Biel, S., Simon, J., Gross, R., Unden, G., & Lancaster, C. R. (2002). Fumarate respiration of *Wolinella succinogenes*: Enzymology, energetics and coupling mechanism. *Biochimica et Biophysica Acta, 1553*, 23–38.

Kroneck, P. M., Antholine, W. A., Riester, J., & Zumft, W. G. (1989). The nature of the cupric site in nitrous oxide reductase and of Cu_A in cytochrome *c* oxidase. *FEBS Letters, 248*, 212–213.

Kullik, I., Fritsche, S., Knobel, H., Sanjuan, J., Hennecke, H., & Fischer, H. M. (1991). *Bradyrhizobium japonicum* has two differentially regulated, functional homologs of the σ_{54} gene (*rpoN*). *Journal of Bacteriology, 173*, 1125–1138.

Laguri, C., Stenzel, R. A., Donohue, T. J., Phillips-Jones, M. K., & Williamson, M. P. (2006). Activation of the global gene regulator PrrA (RegA) from *Rhodobacter sphaeroides*. *Biochemistry, 45*, 7872–7881.

Lancaster, C. R., Sauer, U. S., Gross, R., Haas, A. H., Graf, J., Schwalbe, H., et al. (2005). Experimental support for the "E pathway hypothesis" of coupled transmembrane e^- and H^+ transfer in dihemic quinol:fumarate reductase. *Proceedings of the National Academy of Sciences of the United States of America, 102*, 18860–18865.

Laratta, W. P., Choi, P. S., Tosques, I. E., & Shapleigh, J. P. (2002). Involvement of the PrrB/PrrA two-component system in nitrite respiration in *Rhodobacter sphaeroides* 2.4.3: Evidence for transcriptional regulation. *Journal of Bacteriology, 184*, 3521–3529.

Lee, Y. Y., Shearer, N., & Spiro, S. (2006). Transcription factor NNR from *Paracoccus denitrificans* is a sensor of both nitric oxide and oxygen: Isolation of *nnr** alleles encoding effector-independent proteins and evidence for a haem-based sensing mechanism. *Microbiology, 152*, 1461–1470.

Liu, B., Frostegård, Å., & Bakken, L. R. (2014). Impaired reduction of N_2O to N_2 in acid soil is due to a posttranscriptional interference with the expression of *nosZ*. *mBio. 5*. http://dx.doi.org/10.1128/mBio.01383-14. e01383-14.

Liu, B., Morkved, P. T., Frostegard, A., & Bakken, L. R. (2010). Denitrification gene pools, transcription and kinetics of NO, N_2O and N_2 production as affected by soil pH. *FEMS Microbiology Ecology, 72*, 407–417.

Liu, G. H., Liu, B., Wang, J. P., Che, J. M., Chen, Q. Q., & Chen, Z. (2015). High-quality genome sequence of *Bacillus vireti* DSM 15602 T for setting up phylogenomics for the genomic taxonomy of *Bacillus*-like bacteria. *Genome Announcements. 3*(4). http://dx.doi.org/10.1128/genomeA.00864-15.3. pii: e00864-15.

Liu, J. J., Wu, W. X., Ding, Y., Shi, D. Z., & Chen, Y. X. (2010). Ammonia-oxidizing archaea and their important roles in nitrogen biogeochemical cycling: A review. *Ying Yong Sheng Tai Xue Bao, 21*, 2154–2160.

Lockwood, C. W., Burlat, B., Cheesman, M. R., Kern, M., Simon, J., Clarke, T. A., et al. (2015). Resolution of key roles for the distal pocket histidine in cytochrome *c* nitrite reductases. *Journal of the American Chemical Society, 137*, 3059–3068.

Luckmann, M., Mania, D., Kern, M., Bakken, L. R., Frostegard, A., & Simon, J. (2014). Production and consumption of nitrous oxide in nitrate-ammonifying *Wolinella succinogenes* cells. *Microbiology, 160*, 1749–1759.

Lukat, P., Rudolf, M., Stach, P., Messerschmidt, A., Kroneck, P. M., Simon, J., et al. (2008). Binding and reduction of sulfite by cytochrome *c* nitrite reductase. *Biochemistry, 47*, 2080–2086.

Maia, L. B., & Moura, J. J. (2014). How biology handles nitrite. *Chemical Reviews, 114*, 5273–5357.

Mania, D., Heylen, K., van Spanning, R. J., & Frostegard, A. (2014). The nitrate-ammonifying and *nosZ*-carrying bacterium *Bacillus vireti* is a potent source and sink for nitric and nitrous oxide under high nitrate conditions. *Environmental Microbiology, 16*, 3196–3210.

Mania, D., Heylen, K., van Spanning, R. J., & Frostegard, A. (2016). Regulation of nitrogen metabolism in the nitrate-ammonifying soil bacterium *Bacillus vireti* and evidence for its ability to grow using NO as electron acceptor. *Environmental Microbiology*. http://dx.doi.org/10.1111/1462-2920.13124.

Martinez-Espinosa, R. M., Dridge, E. J., Bonete, M. J., Butt, J. N., Butler, C. S., Sargent, F., et al. (2007). Look on the positive side! The orientation, identification and bioenergetics of 'Archaeal' membrane-bound nitrate reductases. *FEMS Microbiology Letters, 276*, 129–139.

Matsumoto, Y., Tosha, T., Pisliakov, A. V., Hino, T., Sugimoto, H., Nagano, S., et al. (2012). Crystal structure of quinol-dependent nitric oxide reductase from *Geobacillus stearothermophilus*. *Nature Structural & Molecular Biology, 19*, 238–245.

Meakin, G. E., Bueno, E., Jepson, B., Bedmar, E. J., Richardson, D. J., & Delgado, M. J. (2007). The contribution of bacteroidal nitrate and nitrite reduction to the formation of nitrosylleghaemoglobin complexes in soybean root nodules. *Microbiology, 153*, 411–419.

Meilhoc, E., Blanquet, P., Cam, Y., & Bruand, C. (2013). Control of NO level in rhizobium-legume root nodules: Not only a plant globin story. *Plant Signaling & Behavior, 8*(10), e25923. http://dx.doi.org/10.4161/psb.25923.

Meilhoc, E., Cam, Y., Skapski, A., & Bruand, C. (2010). The response to nitric oxide of the nitrogen-fixing symbiont *Sinorhizobium meliloti*. *Molecular Plant-Microbe Interactions, 23*, 748–759.

Mesa, S., Alche, J. D., Bedmar, E., & Delgado, M. J. (2004). Expression of *nir, nor* and *nos* denitrification genes from *Bradyrhizobium japonicum* in soybean root nodules. *Physiologia Plantarum, 120*, 205–211.

Mesa, S., Bedmar, E. J., Chanfon, A., Hennecke, H., & Fischer, H. M. (2003). *Bradyrhizobium japonicum* NnrR, a denitrification regulator, expands the FixLJ-FixK_2 regulatory cascade. *Journal of Bacteriology, 185*, 3978–3982.

Mesa, S., Hauser, F., Friberg, M., Malaguti, E., Fischer, H. M., & Hennecke, H. (2008). Comprehensive assessment of the regulons controlled by the FixLJ-FixK_2-FixK_1 cascade in *Bradyrhizobium japonicum*. *Journal of Bacteriology, 190*, 6568–6579.

Mesa, S., Velasco, L., Manzanera, M. E., Delgado, M. J., & Bedmar, E. J. (2002). Characterization of the *norCBQD* genes, encoding nitric oxide reductase, in the nitrogen fixing bacterium *Bradyrhizobium japonicum*. *Microbiology, 148*, 3553–3560.

Mettert, E. L., & Kiley, P. J. (2015). Fe-S proteins that regulate gene expression. *Biochimica et Biophysica Acta, 1853*, 1284–1293.

Mills, C. E., Sedelnikova, S., Soballe, B., Hughes, M. N., & Poole, R. K. (2001). *Escherichia coli* flavohaemoglobin (Hmp) with equistoichiometric FAD and haem contents has a low affinity for dioxygen in the absence or presence of nitric oxide. *The Biochemical Journal, 353*, 207–213.

Mills, P. C., Rowley, G., Spiro, S., Hinton, J. C., & Richardson, D. J. (2008). A combination of cytochrome *c* nitrite reductase (NrfA) and flavorubredoxin (NorV) protects *Salmonella enterica* serovar *Typhimurium* against killing by NO in anoxic environments. *Microbiology, 154*, 1218–1228.

Miyahara, M., Kim, S. W., Fushinobu, S., Takaki, K., Yamada, T., Watanabe, A., et al. (2010). Potential of aerobic denitrification by *Pseudomonas stutzeri* TR2 to reduce nitrous oxide emissions from wastewater treatment plants. *Applied and Environmental Microbiology, 76*, 4619–4625.

Monk, C. E., Pearson, B. M., Mulholland, F., Smith, H. K., & Poole, R. K. (2008). Oxygen- and NssR-dependent globin expression and enhanced iron acquisition in the response of campylobacter to nitrosative stress. *The Journal of Biological Chemistry, 283*, 28413–28425.

Morley, N., Baggs, E. M., Dorsch, P., & Bakken, L. (2008). Production of NO, N_2O and N_2 by extracted soil bacteria, regulation by NO_2^- and O_2 concentrations. *FEMS Microbiology Ecology, 65*, 102–112.

Mühlig, A., Kabisch, J., Pichner, R., Scherer, S., & Muller-Herbst, S. (2014). Contribution of the NO-detoxifying enzymes HmpA, NorV and NrfA to nitrosative stress protection of *Salmonella Typhimurium* in raw sausages. *Food Microbiology, 42*, 26–33.

Mur, L. A., Mandon, J., Persijn, S., Cristescu, S. M., Moshkov, I. E., Novikova, G. V., et al. (2013). Nitric oxide in plants: An assessment of the current state of knowledge. *AoB PLANTS, 5*, pls052. http://dx.doi.org/10.1093/aobpla/pls052.

Myers, K. S., Yan, H., Ong, I. M., Chung, D., Liang, K., Tran, F., et al. (2013). Genome-scale analysis of *Escherichia coli* FNR reveals complex features of transcription factor binding. *PLoS Genetics, 9*, e1003565.

Nellen-Anthamatten, D., Rossi, P., Preisig, O., Kullik, I., Babst, M., Fischer, H. M., et al. (1998). *Bradyrhizobium japonicum* FixK_2, a crucial distributor in the FixLJ-dependent regulatory cascade for control of genes inducible by low oxygen levels. *Journal of Bacteriology, 180*, 5251–5255.

Nienaber, A., Huber, A., Gottfert, M., Hennecke, H., & Fischer, H. M. (2000). Three new NifA-regulated genes in the *Bradyrhizobium japonicum* symbiotic gene region discovered by competitive DNA-RNA hybridization. *Journal of Bacteriology, 182*, 1472–1480.

Nilavongse, A., Brondijk, T. H., Overton, T. W., Richardson, D. J., Leach, E. R., & Cole, J. A. (2006). The NapF protein of the *Escherichia coli* periplasmic nitrate reductase system: Demonstration of a cytoplasmic location and interaction with the catalytic subunit, NapA. *Microbiology, 152*, 3227–3237.

Overton, T. W., Whitehead, R., Li, Y., Snyder, L. A., Saunders, N. J., Smith, H., et al. (2006). Coordinated regulation of the *Neisseria gonorrhoeae*-truncated denitrification pathway by the nitric oxide-sensitive repressor, NsrR, and nitrite-insensitive NarQ-NarP. *The Journal of Biological Chemistry, 281*, 33115–33126.

Partridge, J. D., Bodenmiller, D. M., Humphrys, M. S., & Spiro, S. (2009). NsrR targets in the *Escherichia coli* genome: New insights into DNA sequence requirements for binding and a role for NsrR in the regulation of motility. *Molecular Microbiology, 73*, 680–694.

Partridge, J. D., Sanguinetti, G., Dibden, D. P., Roberts, R. E., Poole, R. K., & Green, J. (2007). Transition of *Escherichia coli* from aerobic to micro-aerobic conditions involves fast and slow reacting regulatory components. *The Journal of Biological Chemistry, 282*, 11230–11237.

Payne, W. J., Grant, M. A., Shapleigh, J., & Hoffman, P. (1982). Nitrogen oxide reduction in *Wolinella succinogenes* and *Campylobacter* species. *Journal of Bacteriology, 152*, 915–918.

Peakman, T., Busby, S., & Cole, J. (1990). Transcriptional control of the *cysG* gene of *Escherichia coli* K-12 during aerobic and anaerobic growth. *European Journal of Biochemistry, 191*, 325–331.

Pisa, R., Stein, T., Eichler, R., Gross, R., & Simon, J. (2002). The *nrfI* gene is essential for the attachment of the active site haem group of *Wolinella succinogenes* cytochrome *c* nitrite reductase. *Molecular Microbiology, 43*, 763–770.

Pisliakov, A. V., Hino, T., Shiro, Y., & Sugita, Y. (2012). Molecular dynamics simulations reveal proton transfer pathways in cytochrome *c*-dependent nitric oxide reductase. *PLoS Computational Biology, 8,* e1002674.

Pohlmann, A., Cramm, R., Schmelz, K., & Friedrich, B. (2000). A novel NO-responding regulator controls the reduction of nitric oxide in *Ralstonia eutropha*. *Molecular Microbiology, 38*, 626–638.

Pomowski, A., Zumft, W. G., Kroneck, P. M., & Einsle, O. (2011). N_2O binding at a [4Cu:2S] copper-sulphur cluster in nitrous oxide reductase. *Nature, 477*, 234–237.

Poock, S. R., Leach, E. R., Moir, J. W., Cole, J. A., & Richardson, D. J. (2002). Respiratory detoxification of nitric oxide by the cytochrome *c* nitrite reductase of *Escherichia coli*. *The Journal of Biological Chemistry, 277*, 23664–23669.

Poole, R. K. (2005). Nitric oxide and nitrosative stress tolerance in bacteria. *Biochemical Society Transactions, 33*, 176–180.

Potter, L., Angove, H., Richardson, D., & Cole, J. (2001). Nitrate reduction in the periplasm of gram-negative bacteria. *Advances in Microbial Physiology, 45*, 51–112.

Prendergast-Miller, M. T., Baggs, E. M., & Johnson, D. (2011). Nitrous oxide production by the ectomycorrhizal fungi *Paxillus involutus* and *Tylospora fibrillosa*. *FEMS Microbiology Letters, 316*, 31–35.

Qu, Z., Bakken, L. R., Molstad, L., Frostegard, A., & Bergaust, L. (2016). Transcriptional and metabolic regulation of denitrification in *Paracoccus denitrificans* allows low but significant activity of nitrous oxide reductase under oxic conditions. *Environmental Microbiology*. http://dx.doi.org/10.111/1462-2902.13128.

Ranson-Olson, B., Jones, D. F., Donohue, T. J., & Zeilstra-Ryalls, J. H. (2006). In vitro and in vivo analysis of the role of PrrA in *Rhodobacter sphaeroides* 2.4.1 *hemA* gene expression. *Journal of Bacteriology, 188*, 3208–3218.

Rasmussen, T., Berks, B. C., Sanders-Loehr, J., Dooley, D. M., Zumft, W. G., & Thomson, A. J. (2000). The catalytic center in nitrous oxide reductase, Cu_Z, is a copper-sulfide cluster. *Biochemistry, 39*, 12753–12756.

Reents, H., Munch, R., Dammeyer, T., Jahn, D., & Hartig, E. (2006). The Fnr regulon of *Bacillus subtilis*. *Journal of Bacteriology, 188*, 1103–1112.

Richardson, D., Felgate, H., Watmough, N., Thomson, A., & Baggs, E. (2009). Mitigating release of the potent greenhouse gas N_2O from the nitrogen cycle could enzymic regulation hold the key? *Trends in Biotechnology, 27*, 388–397.

Richardson, D., & Sawers, G. (2002). Structural biology. PMF through the redox loop. *Science*, *295*, 1842–1843.

Richardson, D. J. (2000). Bacterial respiration: A flexible process for a changing environment. *Microbiology*, *146*, 551–571.

Richardson, D. J. (2011). Redox complexes of the nitrogen cycle. In J. W. B. Moir (Ed.), *Nitrogen cycling in bacteria* (pp. 23–39). Norkfolk, UK: Caister Academic Press.

Richardson, D. J., Berks, B. C., Russell, D. A., Spiro, S., & Taylor, C. J. (2001). Functional, biochemical and genetic diversity of prokaryotic nitrate reductases. *Cellular and Molecular Life Sciences*, *58*, 165–178.

Richardson, D. J., van Spanning, R. J., & Ferguson, S. J. (2007). The prokaryotic nitrate reductases. In H. Bothe, S. J. Ferguson, & W. E. Newton (Eds.), *Biology of the nitrogen cycle* (pp. 21–35). The Nerthelands: Elsevier.

Rinaldo, S., Arcovito, A., Giardina, G., Castiglione, N., Brunori, M., & Cutruzzola, F. (2008). New insights into the activity of *Pseudomonas aeruginosa* cd_1 nitrite reductase. *Biochemical Society Transactions*, *36*, 1155–1159.

Rinaldo, S., Castiglione, N., Giardina, G., Caruso, M., Arcovito, A., Longa, S. D., et al. (2012). Unusual heme binding properties of the dissimilative nitrate respiration regulator, a bacterial nitric oxide sensor. *Antioxidants & Redox Signaling*, *17*, 1178–1189.

Rinaldo, S., & Cutruzzolá, F. (2007). Nitrite reductases in denitrification. In H. Bothe, S. J. Ferguson, & W. E. Newton (Eds.), *Biology of the nitrogen cycle* (pp. 37–56). The Nerthelands: Elsevier.

Robles, E. F., Sanchez, C., Bonnard, N., Delgado, M. J., & Bedmar, E. J. (2006). The *Bradyrhizobium japonicum napEDABC* genes are controlled by the FixLJ-FixK_2-NnrR regulatory cascade. *Biochemical Society Transactions*, *34*, 108–110.

Rodionov, D. A., Dubchak, I. L., Arkin, A. P., Alm, E. J., & Gelfand, M. S. (2005). Dissimilatory metabolism of nitrogen oxides in bacteria: Comparative reconstruction of transcriptional networks. *PLoS Computational Biology*, *1*, e55.

Rodrigues, M. L., Oliveira, T. F., Pereira, I. A., & Archer, M. (2006). X-ray structure of the membrane-bound cytochrome *c* quinol dehydrogenase NrfH reveals novel haem coordination. *The EMBO Journal*, *25*, 5951–5960.

Rolfe, M. D., Ocone, A., Stapleton, M. R., Hall, S., Trotter, E. W., Poole, R. K., et al. (2012). Systems analysis of transcription factor activities in environments with stable and dynamic oxygen concentrations. *Open Biology*, *2*, 120091.

Roop, R. M., 2nd, & Caswell, C. C. (2012). Redox-responsive regulation of denitrification genes in *Brucella*. *Molecular Microbiology*, *85*, 5–7.

Roux, B., Rodde, N., Jardinaud, M. F., Timmers, T., Sauviac, L., Cottret, L., et al. (2014). An integrated analysis of plant and bacterial gene expression in symbiotic root nodules using laser-capture microdissection coupled to RNA sequencing. *The Plant Journal*, *77*, 817–837.

Rowley, G., Hensen, D., Felgate, H., Arkenberg, A., Appia-Ayme, C., Prior, K., et al. (2012). Resolving the contributions of the membrane-bound and periplasmic nitrate reductase systems to nitric oxide and nitrous oxide production in *Salmonella enterica* serovar *Typhimurium*. *The Biochemical Journal*, *441*, 755–762.

Saggar, S., Jha, N., Deslippe, J., Bolan, N. S., Luo, J., Giltrap, D. L., et al. (2013). Denitrification and N_2O:N_2 production in temperate grasslands: Processes, measurements, modelling and mitigating negative impacts. *The Science of the Total Environment*, *465*, 173–195.

Sameshima-Saito, R., Chiba, K., Hirayama, J., Itakura, M., Mitsui, H., Eda, S., et al. (2006). Symbiotic *Bradyrhizobium japonicum* reduces N_2O surrounding the soybean root system via nitrous oxide reductase. *Applied and Environmental Microbiology*, *72*, 2526–2532.

Sanchez, C., Bedmar, E. J., & Delgado, M. J. (2011). Denitrification in Legume-associated endosymbiotic *bacteria*. In J. W. B. Moir (Ed.), *Nitrogen cycling in bacteria* (pp. 197–210). Norfolk, UK: Caister Academic Press.

Sanchez, C., Cabrera, J. J., Gates, A. J., Bedmar, E. J., Richardson, D. J., & Delgado, M. J. (2011). Nitric oxide detoxification in the rhizobia-legume symbiosis. *Biochemical Society Transactions*, *39*, 184–188.

Sanchez, C., Gates, A. J., Meakin, G. E., Uchiumi, T., Girard, L., Richardson, D. J., et al. (2010). Production of nitric oxide and nitrosylleghemoglobin complexes in soybean nodules in response to flooding. *Molecular Plant-Microbe Interactions*, *23*, 702–711.

Sanchez, C., Itakura, M., Okubo, T., Matsumoto, T., Yoshikawa, H., Gotoh, A., et al. (2014). The nitrate-sensing NasST system regulates nitrous oxide reductase and periplasmic nitrate reductase in *Bradyrhizobium japonicum*. *Environmental Microbiology*, *16*, 3263–3274.

Sanford, R. A., Wagner, D. D., Wu, Q., Chee-Sanford, J. C., Thomas, S. H., Cruz-Garcia, C., et al. (2012). Unexpected nondenitrifier nitrous oxide reductase gene diversity and abundance in soils. *Proceedings of the National Academy of Sciences of the United States of America*, *109*, 19709–19714.

Saraiva, L. M., Vicente, J. B., & Teixeira, M. (2004). The role of the flavodiiron proteins in microbial nitric oxide detoxification. *Advances in Microbial Physiology*, *49*, 77–129.

Sciotti, M. A., Chanfon, A., Hennecke, H., & Fischer, H. M. (2003). Disparate oxygen responsiveness of two regulatory cascades that control expression of symbiotic genes in *Bradyrhizobium japonicum*. *Journal of Bacteriology*, *185*, 5639–5642.

Schmutz, J., Cannon, S. B., Schlueter, J., Ma, J., Mitros, T., Nelson, W., et al. (2010). Genome sequence of the palaeopolyploid soybean. *Nature*, *463*, 178–183.

Schmutz, J., McClean, P. E., Mamidi, S., Wu, G. A., Cannon, S. B., Grimwood, J., et al. (2014). A reference genome for common bean and genome-wide analysis of dual domestications. *Nature Genetics*, *46*, 707–713.

Schreiber, F., Wunderlin, P., Udert, K. M., & Wells, G. F. (2012). Nitric oxide and nitrous oxide turnover in natural and engineered microbial communities: Biological pathways, chemical reactions, and novel technologies. *Frontiers in Microbiology*, *3*, 372.

Schreiber, K., Krieger, R., Benkert, B., Eschbach, M., Arai, H., Schobert, M., et al. (2007). The anaerobic regulatory network required for *Pseudomonas aeruginosa* nitrate respiration. *Journal of Bacteriology*, *189*, 4310–4314.

Schumacher, W., & Kroneck, P. M. H. (1992). Anaerobic energy metabolism of the sulfur-reducing bacterium "*Spirillum*" 5175 during dissimilatory nitrate reduction to ammonia. *Archives of Microbiology*, *157*, 464–470.

Shapleigh, J. P. (2006). The denitrifying prokaryotes. In M. Dworkin, S. Falkow, E. Rosenberg, K. H. Schleifer, & E. Stackebrandt (Eds.), *The prokaryotes: A handbook on the biology of bacteria* (3rd ed., pp. 769–792). New York: Springer Science + Business Media.

Shapleigh, J. P. (2011). Oxygen control of nitrogen oxide respiration, focusing on *alpha*-proteobacteria. *Biochemical Society Transactions*, *39*, 179–183.

Shiro, Y., Sugimoto, H., Tosha, T., Nagano, S., & Hino, T. (2012). Structural basis for nitrous oxide generation by bacterial nitric oxide reductases. *Philosophical Transactions of the Royal Society of London Series B, Biological Sciences*, *367*, 1195–1203.

Simon, J. (2002). Enzymology and bioenergetics of respiratory nitrite ammonification. *FEMS Microbiology Reviews*, *26*, 285–309.

Simon, J., Einsle, O., Kroneck, P. M., & Zumft, W. G. (2004). The unprecedented *nos* gene cluster of *Wolinella succinogenes* encodes a novel respiratory electron transfer pathway to cytochrome *c* nitrous oxide reductase. *FEBS Letters*, *569*, 7–12.

Simon, J., Gross, R., Einsle, O., Kroneck, P. M., Kroger, A., & Klimmek, O. (2000). A NapC/NirT-type cytochrome *c* (NrfH) is the mediator between the quinone pool and the cytochrome *c* nitrite reductase of *Wolinella succinogenes*. *Molecular Microbiology*, *35*, 686–696.

Simon, J., & Hederstedt, L. (2011). Composition and function of cytochrome *c* biogenesis System II. *The FEBS Journal*, *278*, 4179–4188.

Simon, J., & Klotz, M. G. (2013). Diversity and evolution of bioenergetic systems involved in microbial nitrogen compound transformations. *Biochimica et Biophysica Acta, 1827*, 114–135.

Simon, J., & Kroneck, P. M. (2013). Microbial sulfite respiration. *Advances in Microbial Physiology, 62*, 45–117.

Simon, J., & Kroneck, P. M. (2014). The production of ammonia by multiheme cytochromes *c*. *Metal Ions in Life Sciences, 14*, 211–236.

Simon, J., van Spanning, R. J., & Richardson, D. J. (2008). The organisation of proton motive and non-proton motive redox loops in prokaryotic respiratory systems. *Biochimica et Biophysica Acta, 1777*, 1480–1490.

Smith, H. K., Shepherd, M., Monk, C., Green, J., & Poole, R. K. (2011). The NO-responsive hemoglobins of *Campylobacter jejuni*: Concerted responses of two globins to NO and evidence in vitro for globin regulation by the transcription factor NssR. *Nitric Oxide, 25*, 234–241.

Smith, K. A., Mosier, A. R., Crutzen, P. J., & Winiwarter, W. (2012). The role of N_2O derived from crop-based biofuels, and from agriculture in general, in Earth's climate. *Philosophical Transactions of the Royal Society of London. Series B, Biological Sciences, 367*, 1169–1174.

Smith, M. S. (1983). Nitrous oxide production by *Escherichia coli* is correlated with nitrate reductase activity. *Applied and Environmental Microbiology, 45*, 1545–1547.

Smith, P., Martino, D., Cai, Z., Gwary, D., Janzen, H., Kumar, P., et al. (2008). Greenhouse gas mitigation in agriculture. *Philosophical Transactions of the Royal Society of London. Series B, Biological Sciences, 363*, 789–813.

Snyder, S. W., & Hollocher, T. C. (1987). Purification and some characteristics of nitrous oxide reductase from *Paracoccus denitrificans*. *The Journal of Biological Chemistry, 262*, 6515–6525.

Solomon, E. I., Heppner, D. E., Johnston, E. M., Ginsbach, J. W., Cirera, J., Qayyum, M., et al. (2014). Copper active sites in biology. *Chemical Reviews, 114*, 3659–3853.

Spector, M. P., Garcia del Portillo, F., Bearson, S. M., Mahmud, A., Magut, M., Finlay, B. B., et al. (1999). The *rpoS*-dependent starvation-stress response locus *stiA* encodes a nitrate reductase (*narZYWV*) required for carbon-starvation-inducible thermotolerance and acid tolerance in *Salmonella typhimurium*. *Microbiology, 145*, 3035–3045.

Spiro, S. (2007). Regulators of bacterial responses to nitric oxide. *FEMS Microbiology Reviews, 31*, 193–211.

Spiro, S. (2011). Nitric oxide metabolism: Physiology and regulatory mechanisms. In J. W. Moir (Ed.), *Nitrogen cycling in bacteria* (pp. 177–197). Norfolk, UK: Caister Academic Press.

Spiro, S. (2012). Nitrous oxide production and consumption: Regulation of gene expression by gas-sensitive transcription factors. *Philosophical Transactions of the Royal Society of London. Series B, Biological Sciences, 367*, 1213–1225.

Sprent, J. I. (2009). An interdisciplinary look at legumes and their bacterial symbionts: Some thoughts from Big Sky. *The New Phytologist, 184*, 15–17.

Stach, P., Einsle, O., Schumacher, W., Kurun, E., & Kroneck, P. M. (2000). Bacterial cytochrome *c* nitrite reductase: New structural and functional aspects. *Journal of Inorganic Biochemistry, 79*, 381–385.

Stein, L. Y. (2011). Surveying N_2O-producing pathways in bacteria. *Methods in Enzymology, 486*, 131–152.

Stern, A. M., Hay, A. J., Liu, Z., Desland, F. A., Zhang, J., Zhong, Z., et al. (2012). The NorR regulon is critical for *Vibrio cholerae* resistance to nitric oxide and sustained colonization of the intestines. *mBio, 3*. e00013-00012.

Stern, A. M., & Zhu, J. (2014). An introduction to nitric oxide sensing and response in bacteria. *Advances in Applied Microbiology, 87*, 187–220.

Stewart, V. (2003). Biochemical Society Special Lecture. Nitrate- and nitrite-responsive sensors NarX and NarQ of proteobacteria. *Biochemical Society Transactions, 31*, 1–10.

Stewart, V., & Bledsoe, P. J. (2005). Fnr-, NarP- and NarL-dependent regulation of transcription initiation from the *Haemophilus influenzae* Rd *napF* (periplasmic nitrate reductase) promoter in *Escherichia coli* K-12. *Journal of Bacteriology, 187*, 6928–6935.

Stewart, V., & Chen, L. L. (2010). The S helix mediates signal transmission as a HAMP domain coiled-coil extension in the NarX nitrate sensor from *Escherichia coli* K-12. *Journal of Bacteriology, 192*, 734–745.

Stieglmeier, M., Mooshammer, M., Kitzler, B., Wanek, W., Zechmeister-Boltenstern, S., Richter, A., et al. (2014). Aerobic nitrous oxide production through N-nitrosating hybrid formation in ammonia-oxidizing *archaea*. *The ISME Journal, 8*, 1135–1146.

Streminska, M. A., Felgate, H., Rowley, G., Richardson, D. J., & Baggs, E. M. (2012). Nitrous oxide production in soil isolates of nitrate-ammonifying bacteria. *Environmental Microbiology Reports, 4*, 66–71.

Sullivan, M. J., Gates, A. J., Appia-Ayme, C., Rowley, G., & Richardson, D. J. (2013). Copper control of bacterial nitrous oxide emission and its impact on vitamin B_{12}-dependent metabolism. *Proceedings of the National Academy of Sciences of the United States of America, 110*, 19926–19931.

Swem, L. R., Elsen, S., Bird, T. H., Swem, D. L., Koch, H. G., Myllykallio, H., et al. (2001). The RegB/RegA two-component regulatory system controls synthesis of photosynthesis and respiratory electron transfer components in *Rhodobacter capsulatus*. *Journal of Molecular Biology, 309*, 121–138.

Swem, L. R., Gong, X., Yu, C. A., & Bauer, C. E. (2006). Identification of a ubiquinone-binding site that affects autophosphorylation of the sensor kinase RegB. *The Journal of Biological Chemistry, 281*, 6768–6775.

Swem, L. R., Kraft, B. J., Swem, D. L., Setterdahl, A. T., Masuda, S., Knaff, D. B., et al. (2003). Signal transduction by the global regulator RegB is mediated by a redox-active cysteine. *The EMBO Journal, 22*, 4699–4708.

Takaya, N. (2002). Dissimilatory nitrate reduction metabolisms and their control in *fungi*. *Journal of Bioscience and Bioengineering, 94*, 506–510.

ter Beek, J., Krause, N., Reimann, J., Lachmann, P., & Adelroth, P. (2013). The nitric-oxide reductase from *Paracoccus denitrificans* uses a single specific proton pathway. *The Journal of Biological Chemistry, 288*, 30626–30635.

Thomson, A. J., Giannopoulos, G., Pretty, J., Baggs, E. M., & Richardson, D. J. (2012). Biological sources and sinks of nitrous oxide and strategies to mitigate emissions. *Philosophical Transactions of the Royal Society of London. Series B, Biological Sciences, 367*, 1157–1168.

Toffanin, A., Wu, Q., Maskus, M., Caselia, S., Abruna, H. D., & Shapleigh, J. P. (1996). Characterization of the gene encoding nitrite reductase and the physiological consequences of its expression in the nondenitrifying *Rhizobium "hedysari"* strain HCNT1. *Applied and Environmental Microbiology, 62*, 4019–4025.

Tolla, D. A., & Savageau, M. A. (2011). Phenotypic repertoire of the FNR regulatory network in *Escherichia coli*. *Molecular Microbiology, 79*, 149–165.

Torres, M. J., Argandona, M., Vargas, C., Bedmar, E. J., Fischer, H. M., Mesa, S., et al. (2014). The global response regulator RegR controls expression of denitrification genes in *Bradyrhizobium japonicum*. *PLoS One, 9*, e99011.

Torres, M. J., Rubia, M. I., Bedmar, E. J., & Delgado, M. J. (2011). Denitrification in *Sinorhizobium meliloti*. *Biochemical Society Transactions, 39*, 1886–1889.

Torres, M. J., Rubia, M. I., de la Pena, T. C., Pueyo, J. J., Bedmar, E. J., & Delgado, M. J. (2014). Genetic basis for denitrification in *Ensifer meliloti*. *BMC Microbiology, 14*, 142.

Tortosa, G., Hidalgo, A., Salas, A., Bedmar, E., Mesa, S., & Delgado, M. (2015). Nitrate and flooding induce N2O emissions from soybean nodules. *Symbiosis, 67*, 125–133.

Treusch, A. H., Leininger, S., Kletzin, A., Schuster, S. C., Klenk, H. P., & Schleper, C. (2005). Novel genes for nitrite reductase and Amo-related proteins indicate a role of uncultivated mesophilic crenarchaeota in nitrogen cycling. *Environmental Microbiology*, 7, 1985–1995.

Tucker, N. P., Ghosh, T., Bush, M., Zhang, X., & Dixon, R. (2010). Essential roles of three enhancer sites in σ^{54}-dependent transcription by the nitric oxide sensing regulatory protein NorR. *Nucleic Acids Research*, *38*, 1182–1194.

Tucker, N. P., Hicks, M. G., Clarke, T. A., Crack, J. C., Chandra, G., Le Brun, N. E., et al. (2008). The transcriptional repressor protein NsrR senses nitric oxide directly via a [2Fe-2S] cluster. *PLoS One*, *3*, e3623.

Tucker, N. P., Le Brun, N. E., Dixon, R., & Hutchings, M. I. (2010). There's NO stopping NsrR, a global regulator of the bacterial NO stress response. *Trends in Microbiology*, *18*, 149–156.

Tyson, K. L., Cole, J. A., & Busby, S. J. (1994). Nitrite and nitrate regulation at the promoters of two *Escherichia coli* operons encoding nitrite reductase: Identification of common target heptamers for both NarP- and NarL-dependent regulation. *Molecular Microbiology*, *13*, 1045–1055.

Udvardi, M., & Poole, P. S. (2013). Transport and metabolism in legume-rhizobia symbioses. *Annual Review of Plant Biology*, *64*, 781–805.

van Alst, N. E., Sherrill, L. A., Iglewski, B. H., & Haidaris, C. G. (2009). Compensatory periplasmic nitrate reductase activity supports anaerobic growth of *Pseudomonas aeruginosa* PAO1 in the absence of membrane nitrate reductase. *Canadian Journal of Microbiology*, *55*, 1133–1144.

Van den Heuvel, R. N., Bakker, S. E., Jetten, M. S., & Hefting, M. M. (2011). Decreased N_2O reduction by low soil pH causes high N_2O emissions in a riparian ecosystem. *Geobiology*, *9*, 294–300.

van Spanning, R. J. (2011). Structure, function, regulation and evolution of the nitrite and nitrous oxide reductase: Denitrification enzymes with a *b*-propeller fold. In J. W. B. Moir (Ed.), *Nitrogen cycling in bacteria* (pp. 135–161). Norkfolk, UK: Caister Academic Press.

van Spanning, R. J., Delgado, M. J., & Richardson, D. J. (2005). The nitrogen cycle: Denitrification and its relationship to N_2 fixation. In D. Werner & W. E. Newton (Eds.), *Nitrogen fixation in agriculture, forestry, ecology and the environment* (pp. 277–342). Netherlands: Springer.

van Spanning, R. J., Houben, E., Reijnders, W. N., Spiro, S., Westerhoff, H. V., & Saunders, N. (1999). Nitric oxide is a signal for NNR-mediated transcription activation in *Paracoccus denitrificans*. *Journal of Bacteriology*, *181*, 4129–4132.

van Spanning, R. J., Richardson, D. J., & Ferguson, S. J. (2007). Introduction to the biochemistry and molecular biology of denitrification. In H. Bothe, S. J. Ferguson, & W. E. Newton (Eds.), *Biology of the nitrogen cycle: Vol. 3–20*. Amsterdam: Elsevier Science.

van Wonderen, J. H., Burlat, B., Richardson, D. J., Cheesman, M. R., & Butt, J. N. (2008). The nitric oxide reductase activity of cytochrome c nitrite reductase from Escherichia coli. *The Journal of Biological Chemistry*, *283*, 9587–9594.

Velasco, L., Mesa, S., Delgado, M. J., & Bedmar, E. J. (2001). Characterization of the *nirK* gene encoding the respiratory, Cu-containing nitrite reductase of *Bradyrhizobium japonicum*. *Biochimica et Biophysica Acta*, *1521*, 130–134.

Velasco, L., Mesa, S., Xu, C. A., Delgado, M. J., & Bedmar, E. J. (2004). Molecular characterization of *nosRZDFYLX* genes coding for denitrifying nitrous oxide reductase of *Bradyrhizobium japonicum*. *Antonie Van Leeuwenhoek*, *85*, 229–235.

Veldman, R., Reijnders, W. N., & van Spanning, R. J. (2006). Specificity of FNR-type regulators in *Paracoccus denitrificans*. *Biochemical Society Transactions*, *34*, 94–96.

Vine, C. E., & Cole, J. A. (2011). Nitrosative stress in *Escherichia coli:* Reduction of nitric oxide. *Biochemical Society Transactions*, *39*, 213–215.

Vinogradov, S. N., Tinajero-Trejo, M., Poole, R. K., & Hoogewijs, D. (2013). Bacterial and archaeal globins—A revised perspective. *Biochimica et Biophysica Acta, 1834*, 1789–1800.

Weiss, B. (2006). Evidence for mutagenesis by nitric oxide during nitrate metabolism in *Escherichia coli*. *Journal of Bacteriology, 188*, 829–833.

Wood, N. J., Alizadeh, T., Bennett, S., Pearce, J., Ferguson, S. J., Richardson, D. J., et al. (2001). Maximal expression of membrane-bound nitrate reductase in *Paracoccus* is induced by nitrate via a third FNR-like regulator named NarR. *Journal of Bacteriology, 183*, 3606–3613.

Wood, N. J., Alizadeh, T., Richardson, D. J., Ferguson, S. J., & Moir, J. W. (2002). Two domains of a dual-function NarK protein are required for nitrate uptake, the first step of denitrification in *Paracoccus pantotrophus*. *Molecular Microbiology, 44*, 157–170.

Woolfenden, H. C., Gates, A. J., Bocking, C., Blyth, M. G., Richardson, D. J., & Moulton, V. (2013). Modeling the effect of copper availability on bacterial denitrification. *MicrobiologyOpen, 2*, 756–765.

Wu, J., & Bauer, C. E. (2008). RegB/RegA, a global redox-responding two-component system. *Advances in Experimental Medicine and Biology, 631*, 131–148.

Wunsch, P., Herb, M., Wieland, H., Schiek, U. M., & Zumft, W. G. (2003). Requirements for Cu_A and Cu-S center assembly of nitrous oxide reductase deduced from complete periplasmic enzyme maturation in the nondenitrifier *Pseudomonas putida*. *Journal of Bacteriology, 185*, 887–896.

Wunsch, P., Korner, H., Neese, F., van Spanning, R. J., Kroneck, P. M., & Zumft, W. G. (2005). NosX function connects to nitrous oxide (N_2O) reduction by affecting the Cu(Z) center of NosZ and its activity in vivo. *FEBS Letters, 579*, 4605–4609.

Wunsch, P., & Zumft, W. G. (2005). Functional domains of NosR, a novel transmembrane iron-sulfur flavoprotein necessary for nitrous oxide respiration. *Journal of Bacteriology, 187*, 1992–2001.

Wüst, A., Schneider, L., Pomowski, A., Zumft, W. G., Kroneck, P. M., & Einsle, O. (2012). Nature's way of handling a greenhouse gas: The copper-sulfur cluster of purple nitrous oxide reductase. *Biological Chemistry, 393*, 1067–1077.

Yamazaki, T., Yoshida, N., Wada, E., & Matsuo, S. (1987). N_2O reduction by Azotobacter vinelandii with emphasis on kinetic nitrogen isotope effects. *Plant & Cell Physiology, 28*, 263–271.

Yoon, S., Cruz-Garcia, C., Sanford, R., Ritalahti, K. M., & Loffler, F. E. (2015). Denitrification versus respiratory ammonification: Environmental controls of two competing dissimilatory NO_3^-/NO_2^- reduction pathways in *Shewanella loihica* strain PV-4. *The ISME Journal, 9*, 1093–1104.

Yoon, S., Sanford, R. A., & Loffler, F. E. (2013). *Shewanella* spp. Use acetate as an electron donor for denitrification but not ferric iron or fumarate reduction. *Applied and Environmental Microbiology, 79*, 2818–2822.

Yoon, S., Sanford, R. A., & Loffler, F. E. (2015). Nitrite control over dissimilatory nitrate/nitrite reduction pathways in *Shewanella loihica* Strain PV-4. *Applied and Environmental Microbiology, 81*, 3510–3517.

Yoshinari, T. (1980). N_2O reduction by *Vibrio succinogenes*. *Applied and Environmental Microbiology, 39*, 81–84.

Young, N. D., Debelle, F., Oldroyd, G. E., Geurts, R., Cannon, S. B., Udvardi, M. K., et al. (2011). The *Medicago* genome provides insight into the evolution of rhizobial symbioses. *Nature, 480*, 520–524.

Yukl, E. T., Elbaz, M. A., Nakano, M. M., & Moenne-Loccoz, P. (2008). Transcription factor NsrR from *Bacillus subtilis* senses nitric oxide with a 4Fe-4S cluster (dagger). *Biochemistry, 47*, 13084–13092.

Zamorano-Sanchez, D., Reyes-Gonzalez, A., Gomez-Hernandez, N., Rivera, P., Georgellis, D., & Girard, L. (2012). FxkR provides the missing link in the *fixL-fixK* signal transduction cascade in *Rhizobium etli* CFN42. *Molecular Plant-Microbe Interactions, 25*, 1506–1517.

Zhang, B., Crack, J. C., Subramanian, S., Green, J., Thomson, A. J., Le Brun, N. E., et al. (2012). Reversible cycling between cysteine persulfide-ligated [2Fe-2S] and cysteine-ligated [4Fe-4S] clusters in the FNR regulatory protein. *Proceedings of the National Academy of Sciences of the United States of America, 109*, 15734–15739.

Zumft, W. G. (1997). Cell biology and molecular basis of denitrification. *Microbiology and Molecular Biology Reviews, 61*, 533–616.

Zumft, W. G. (2005). Biogenesis of the bacterial respiratory Cu_A, Cu-S enzyme nitrous oxide reductase. *Journal of Molecular Microbiology and Biotechnology, 10*, 154–166.

Zumft, W. G., Dreusch, A., Lochelt, S., Cuypers, H., Friedrich, B., & Schneider, B. (1992). Derived amino acid sequences of the *nosZ* gene (respiratory N_2O reductase) from *Alcaligenes eutrophus, Pseudomonas aeruginosa* and *Pseudomonas stutzeri* reveal potential copper-binding residues. Implications for the Cu_A site of N_2O reductase and cytochrome-*c* oxidase. *European Journal of Biochemistry, 208*, 31–40.

Zumft, W. G., & Kroneck, P. M. (2007). Respiratory transformation of nitrous oxide (N_2O) to dinitrogen by *bacteria* and *archaea*. *Advances in Microbial Physiology, 52*, 107–227.

CHAPTER EIGHT

The Model [NiFe]-Hydrogenases of *Escherichia coli*

F. Sargent[1]

School of Life Sciences, University of Dundee, Dundee, Scotland, United Kingdom

[1]Corresponding author: e-mail address: f.sargent@dundee.ac.uk

Contents

Abstract

In *Escherichia coli*, hydrogen metabolism plays a prominent role in anaerobic physiology. The genome contains the capability to produce and assemble up to four [NiFe]-hydrogenases, each of which are known, or predicted, to contribute to different aspects of cellular metabolism. In recent years, there have been major advances in the understanding of the structure, function, and roles of the *E. coli* [NiFe]-hydrogenases. The

Advances in Microbial Physiology, Volume 68
ISSN 0065-2911
http://dx.doi.org/10.1016/bs.ampbs.2016.02.008

membrane-bound, periplasmically oriented, respiratory Hyd-1 isoenzyme has become one of the most important paradigm systems for understanding an important class of oxygen-tolerant enzymes, as well as providing key information on the mechanism of hydrogen activation per se. The membrane-bound, periplasmically oriented, Hyd-2 isoenzyme has emerged as an unusual, bidirectional redox valve able to link hydrogen oxidation to quinone reduction during anaerobic respiration, or to allow disposal of excess reducing equivalents as hydrogen gas. The membrane-bound, cytoplasmically oriented, Hyd-3 isoenzyme is part of the formate hydrogenlyase complex, which acts to detoxify excess formic acid under anaerobic fermentative conditions and is geared towards hydrogen production under those conditions. Sequence identity between some Hyd-3 subunits and those of the respiratory NADH dehydrogenases has led to hypotheses that the activity of this isoenzyme may be tightly coupled to the formation of transmembrane ion gradients. Finally, the *E. coli* genome encodes a homologue of Hyd-3, termed Hyd-4, however strong evidence for a physiological role for *E. coli* Hyd-4 remains elusive. In this review, the versatile hydrogen metabolism of *E. coli* will be discussed and the roles and potential applications of the spectrum of different types of [NiFe]-hydrogenases available will be explored.

1. INTRODUCTION

Hydrogenases are enzymes that catalyse the oxidation of molecular hydrogen (H_2) into protons an electrons, and of course catalyse the reverse reaction of reduction of protons to molecular hydrogen. Such enzymes are widespread in microbial systems and at least three broad classes of hydrogenases have coevolved: [NiFe]-hydrogenases, [FeFe]-hydrogenases, and [Fe]-hydrogenases. Numerous comprehensive reviews have been written on the different classes of microbial hydrogenases, including detailed structural and biochemical analyses (Fontecilla-Camps, Volbeda, Cavazza, & Nicolet, 2007; Lubitz, Ogata, Rudiger, & Reijerse, 2014; Peters et al., 2014), phylogenetic analyses (Greening et al., 2016; Vignais & Billoud, 2007; Vignais, Billoud, & Meyer, 2001), studies of cofactor biosynthesis and enzyme assembly (Böck, King, Blokesch, & Posewitz, 2006; Pinske & Sawers, 2014; Watanabe, Sasaki, Tominaga, & Miki, 2012; Wu, Chanal, & Rodrigue, 2000), as well as more focused studies on specific biological systems (Burgdorf et al., 2005). This review will concentrate on the four nickel-dependent hydrogenases produced by *Escherichia coli*, which play distinct but central roles in the versatile metabolism of that model organism. Indeed, in recent years, the hydrogenases of *E. coli* have become model systems in their own right, and so have served as test beds and paradigms for many aspects of hydrogenase biochemistry.

E. coli has the capability to produce four distinct [NiFe]-hydrogenase isoenzymes and does not contain any other classes of hydrogenase (Andrews et al., 1997; Sawers, 1994). In general, the central 'core' module of a [NiFe]-hydrogenase comprises two subunits: a large (or α-) subunit, typically of ~60 kDa, that harbours the Ni–Fe–CO–2CN$^-$ active site and a small (or β-) subunit, typically of ~30 kDa, that contains at least one, but most commonly three, [Fe–S] clusters. The small subunit [Fe–S] cluster closest to the [NiFe] active site in the large subunit is termed the 'proximal cluster', while the small subunit cluster furthest from the active site at the surface of the protein in termed the 'distal cluster', and the [Fe–S] cluster between these two is called the 'medial cluster'. Such 'core' [NiFe]-hydrogenase modules can either be associated with periplasmic or membrane-bound cytochrome subunits, alternative electron transfer subunits such as additional [Fe–S] cluster proteins, or be parts of much larger multisubunit redox enzymes involved in, for example, carbon monoxide or formate metabolism. The first crystal structure of a core [NiFe]-hydrogenase was reported from *Desulfovibrio gigas* in 1995 (Volbeda et al., 1995) and additional structures have been regularly described ever since, including a stunning recent structure of a [NiFe]-hydrogenase from *Desulfovibrio vulgaris* where protons and hydrides could be resolved (Ogata, Nishikawa, & Lubitz, 2015). The *D. gigas* enzyme remains one of the paradigm [NiFe]-hydrogenase structures representing a classical 'standard' (O_2 sensitive) core hydrogenase module. The small subunit contains a [4Fe–4S] proximal cluster, a [3Fe–4S] medial cluster, and a special [4Fe–4S] distal cluster where one proteinaceous cluster ligand is provided by a histidine side chain (Volbeda et al., 1995).

The *E. coli* [NiFe]-hydrogenases are termed Hyd-1, Hyd-2, Hyd-3, and Hyd-4 and all four are known or predicted to be membrane-bound proteins. Hyd-1 is expressed under anaerobic conditions at stationary phase (Atlung, Knudsen, Heerfordt, & Brondsted, 1997) and is predominantly a respiratory, hydrogen-oxidising enzyme (Volbeda et al., 2013). Hyd-1 is tolerant to attack by oxygen (Lukey et al., 2010) and is coexpressed with a cytochrome oxidase (Brondsted & Atlung, 1996), which suggests its physiological role is most likely to operate at the anaerobic/aerobic switch when *E. coli* is transitioning from an anaerobic to an aerobic environment. Hyd-2 is a standard oxygen-sensitive enzyme (Lukey et al., 2010) expressed under anaerobic respiratory conditions where it couples hydrogen oxidation to the generation of a transmembrane electrochemical gradient (Pinske et al., 2015). Under some conditions, Hyd-2 can operate as a redox pressure release valve, allowing the oxidation of reduced quinol to be coupled to hydrogen production (Pinske et al., 2015; Trchounian, Soboh, Sawers, & Trchounian, 2013). Hyd-3 is expressed under anaerobic fermentative conditions (ie, in the absence of all exogenous respiratory electron acceptors) at low pH and high formic acid concentrations. It is a component of the formate hydrogenlyase complex (Bohm, Sauter, & Böck, 1990), which connects formate oxidation to proton reduction and is responsible for the majority of sustained H_2 production by *E. coli*. Finally, the *E. coli* genome contains an operon that encodes a Hyd-4 isoenzyme that would be closely related to Hyd-3 (Andrews et al., 1997). The physiological or biochemical role of Hyd-4 is not fully understood, although genetic analysis suggests this isoenzyme may be able to transduce free energy released during the formate hydrogenlyase reaction to the generation of a transmembrane ion gradient (Batista, Marreiros, & Pereira, 2013; Marreiros, Batista, Duarte, & Pereira, 2013).

2. [NiFe]-HYDROGENASE-1: AN O_2-TOLERANT PARADIGM

The *E. coli* Hyd-1 isoenzyme is encoded within a six-gene operon located at 22′ on the chromosome (Fig. 1). The sequence of the *hyaABCDEF* operon was reported in 1990 (Menon et al., 1990) and was shown to encode a [NiFe]-hydrogenase small subunit (HyaA), a [NiFe]-hydrogenase large subunit (HyaB), an integral membrane *b*-type cytochrome (HyaC), a system-specific protease (HyaD), and two other gene products of unknown function (HyaE and HyaF). Transcription of the operon is controlled by anaerobiosis (Brondsted & Atlung, 1994; Richard, Sawers, Sargent, McWalter, & Boxer, 1999), which is in line with earlier work focused on following active Hyd-1 levels by immunological techniques (Ballantine & Boxer, 1985; Sawers, Ballantine, & Boxer, 1985) and is further increased in stationary phase (Atlung et al., 1997).

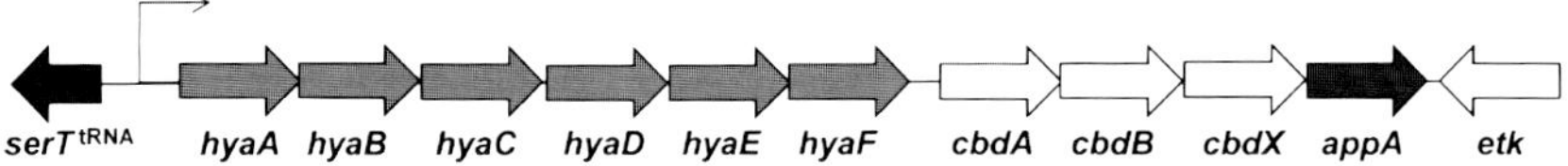

Fig. 1 Organisation of the *E. coli hya* locus. Depiction of the genetic organisation around the *hyaABCDEF* operon. The *hyaABC* genes encode the structural components of Hyd-1 and *hyaDEF* are accessory genes involved in enzyme assembly. The *cbdABX* genes encode a cytochrome *bd*-II oxidase and *appA* encodes an acid phosphatase. The locus is flanked by a gene for a serine tRNA and *etk*, a gene encoding a kinase involved in O-antigen biosynthesis. Not to scale.

Interestingly, although coupling H_2 oxidation (standard redox potential $E_0 = -414$ mV) to nitrate reduction to nitrite (standard redox potential $E_0 = +433$ mV) within an anaerobic respiratory electron transport chain would potentially release a large amount of free energy that could be coupled to ATP generation or other processes (Thauer, Jungermann, & Decker, 1977), *E. coli* chooses not to take this option and instead completely represses transcription of *hyaABCDEF* in the presence of excess nitrate (Richard et al., 1999). The reason for this is not altogether clear, but is probably linked to a preference for water-soluble formate as a respiratory electron donor rather than the more volatile H_2. In addition, because formate is an important (sometimes solitary) electron donor for H_2 production by enteric bacteria, by coupling all formate oxidation to nitrate reduction using coexpressed enzymes (Enoch & Lester, 1975) endogenous H_2 production would be kept to absolute minimum and therefore expending energy on [NiFe]-hydrogenase biosynthesis would be pointless under these growth conditions.

Transcription of the *hyaABCDEF* operon is repressed by IscR, a master regulator of [Fe–S] cluster biosynthesis in *E. coli* (Giel, Rodionov, Liu, Blattner, & Kiley, 2006; Nesbit, Giel, Rose, & Kiley, 2009). In addition, the AppY protein was found to coregulate *hyaABCDEF* with the gene encoding a periplasmic acid phosphatase (*appA*) and the genes for a high-affinity cytochrome oxidase (*cbdAB*) (Atlung et al., 1997; Brondsted & Atlung, 1996), which are all located immediately downstream of *hyaABCDEF* on the *E. coli* chromosome. This not only implicated Hyd-1 activity in a role related to pH homeostasis (King & Przybyla, 1999), but also gave an initial indication that Hyd-1 activity could possibly be linked to oxygen respiration.

Sequence analysis of the *hyaABCDEF* operon suggested that the core [NiFe]-hydrogenase component of Hyd-1 would be a membrane-bound heterodimeric complex of HyaA and HyaB (Menon et al., 1990), and this corroborated work on the purification of *E. coli* Hyd-1 from detergent-dispersed membranes (Sawers & Boxer, 1986). The HyaAB subunits are held at the periplasmic side of the cytoplasmic membrane (Fig. 2) by a single transmembrane domain located at the extreme C-terminus of the small subunit, HyaA (Hatzixanthis, Palmer, & Sargent, 2003). The purified heterodimeric form of Hyd-1 was found to be active with benzyl viologen as artificial electron acceptor; however, this form of the enzyme was unable to reduce quinones (Sawers & Boxer, 1986). This was most likely because the preparation lacked the HyaC cytochrome *b* subunit that, by analogy with

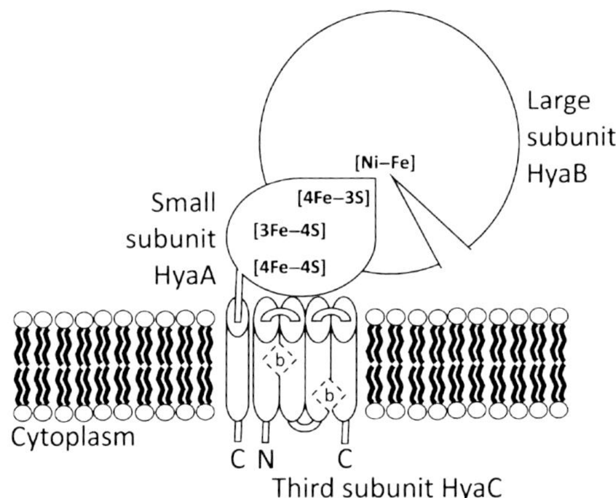

Fig. 2 Architecture of [NiFe]-hydrogenase-1. A cartoon of *E. coli* Hyd-1. The protein comprises a large subunit (HyaA), containing a [NiFe] cofactor, tightly bound to a small subunit (HyaA), containing three [Fe–S] clusters, at the periplasmic side of the membrane. HyaAB is anchored to the lipid bilayer by a single transmembrane domain located at the C-terminus of HyaA. The HyaAB core hydrogenase is associated with an integral membrane *b*-type cytochrome encoded by the *hyaC* gene. HyaC has four transmembrane domains. This is the basic functional unit of Hyd-1. Structural analysis suggests the final physiological form is most likely $(HyaABC)_2$.

formate dehydrogenase-N and nitrate reductase-A (Jormakka, Tornroth, Byrne, & Iwata, 2002; Rothery, Blasco, Magalon, & Weiner, 2001), would be expected to be the primary point of contact for quinones. The presence of a HyaC subunit with juxtaposed haem groups allows the prediction to be made that Hyd-1 activity should generate a proton electrochemical gradient by a redox loop mechanism (Simon, van Spanning, & Richardson, 2008).

The purification strategy that lead to crystallisation and structure determination of the Hyd-1 enzyme (Fig. 3) depended on a recombineering approach, where an affinity tag encoding sequence was engineered onto the native *hyaA* gene on the *E. coli* chromosome (Lukey et al., 2010). This allowed the native regulation and biosynthetic pathways for Hyd-1 to remain naturally balanced and coordinated, while also allowing rapid purification of the enzyme (Lukey et al., 2010). Using this approach, an enzymatically active HyaAB version of Hyd-1 could be isolated (Lukey et al., 2010). This preparation of Hyd-1 could be readily assayed in electrochemical experiments and was shown to have a K_m for H_2 of 9 μM (Lukey et al., 2010), which was in good agreement with earlier studies using alternative methodology reporting a K_m of 2 μM (Sawers & Boxer, 1986).

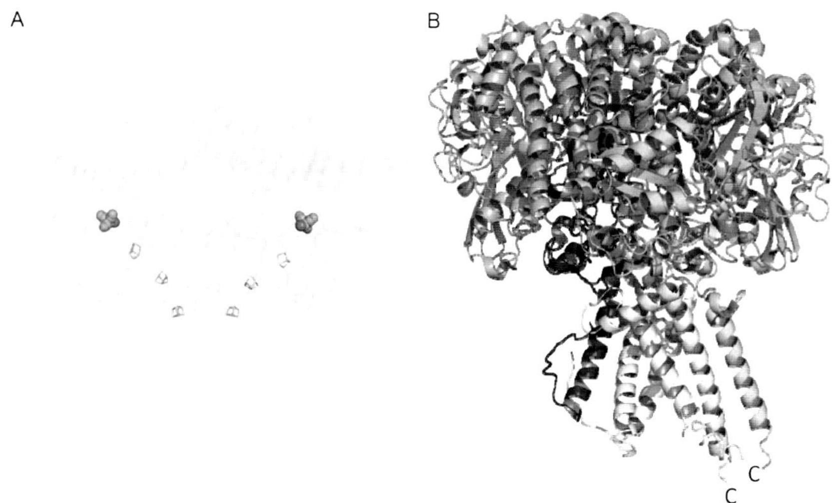

Fig. 3 The crystal structure of [NiFe]-hydrogenase-1. (A) The crystal structure of native *E. coli* Hyd-1 showing the protein in its $(HyaAB)_2$ conformation. This enzyme was isolated in the presence of detergent via an affinity tag located at the end of a C-terminal transmembrane helix on the small subunit. Data for this part of the protein were not obtained (Volbeda et al., 2012). The positions of the metallocofactors have been *highlighted*. (B) The structure of a small subunit P242C variant of Hyd-1 (4GD3, Volbeda et al., 2013). The protein is in a $(HyaAB)_2$ conformation with a single HyaC subunit associated. The large subunits (coloured *orange* and *pink*) are tightly associated with small subunits (*blue* and *cyan*) and in this case the C-terminal transmembrane helices can be modelled. The HyaC subunit (*yellow*) comprises four transmembrane helices and contains a single *b*-type haem (highlighted in *red*). (See the colour plate.)

The X-ray crystal structure of native *E. coli* Hyd-1 revealed a $(HyaAB)_2$ dimeric complex (Fig. 3) containing two large subunits and two small subunits (Volbeda et al., 2012). Despite the affinity tags themselves being placed at the extreme C-termini of the small subunits, the transmembrane domains of Hyd-1 were not detectable in the X-ray diffraction data and the HyaC cytochrome subunit was not present either (Volbeda et al., 2012). However, in other regards, the structure of Hyd-1 corroborated perfectly earlier biochemical work since the $(HyaAB)_2$ arrangement probably reflects the 200,000 Da complex characterised by size exclusion chromatography (Sawers & Boxer, 1986). Moreover, the crystal structure revealed four different metallocofactors within the enzyme: a Ni–Fe–CO–2CN$^-$ cofactor within the large subunits and three [Fe–S] clusters within the small subunits, including a [4Fe–3S] proximal cluster, a [3Fe–4S] medial cluster, and a [4Fe–4S] distal cluster (Volbeda et al., 2012). One striking aspect of [NiFe]-hydrogenase structures, which was already noted in the original work in this area (Volbeda et al., 1995), is the extensive contact between the two key subunits and the positioning of the [NiFe] active site and the proximal [Fe–S] cluster very close to the interface between the subunits. Indeed, there is an invariant histidine side chain in [NiFe]-hydrogenase large subunits (His229 in *E. coli* HyaB) that points directly across the subunit interface towards the proximal [Fe–S] cluster (Fig. 4), and there is some evidence that this may contribute to enzyme activity (Bowman et al., 2014; Dance, 2015). Indeed, it is a very attractive proposition to consider that the [NiFe] and proximal [Fe–S] clusters, and their surrounding proteinaceous canopy, operate in unison as a single active site for hydrogenase (Pandelia, Lubitz, & Nitschke, 2012) (Fig. 5).

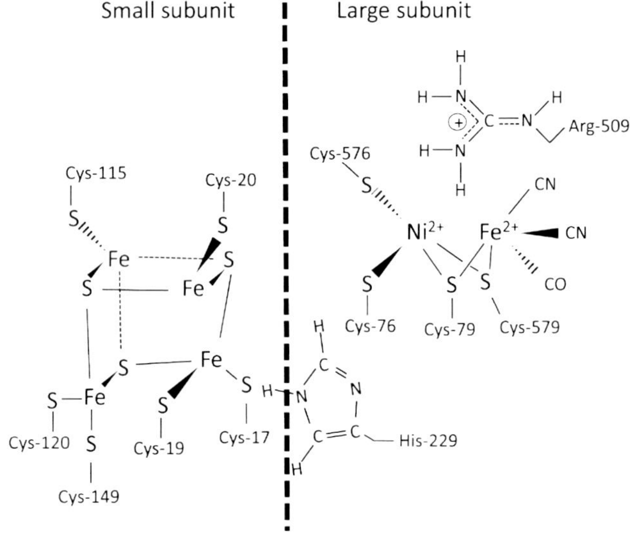

Fig. 4 A shared active site between two subunits. A depiction of the Hyd-1 active site showing some of the key cofactor coordinating and catalytic residues from the large and small subunits. The large subunit [NiFe] cofactor cysteine ligands are shown with the 'pendant arginine' that is critical for catalysis. His-229 from the large subunit is shown close to the subunit interface and pointing directly at the proximal cluster. The six cysteine ligands to the [4Fe–3S] proximal cluster are shown with numbering from the crystal structure, so missing residues for the entire twin-arginine signal peptide.

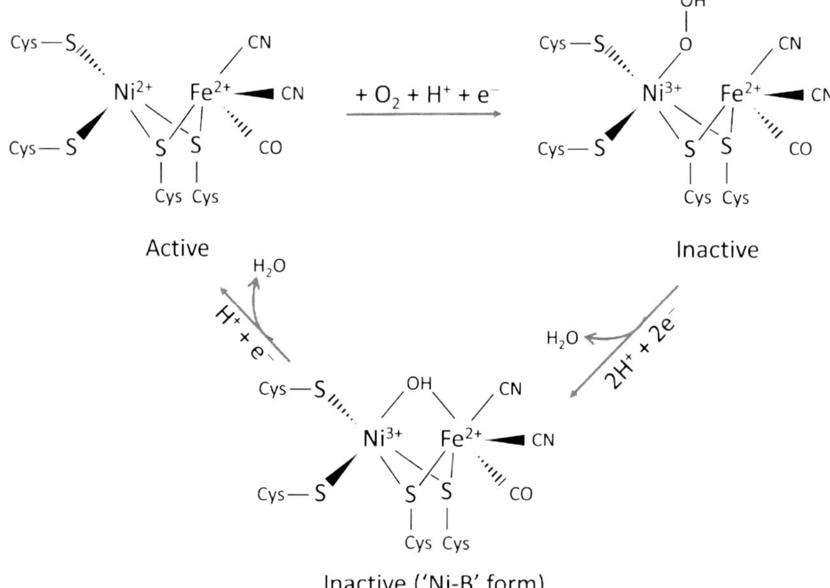

Fig. 5 A mechanism to survive oxygen attack. A summarised mechanism, based on many studies on several related O_2-tolerant [NiFe]-hydrogenases including *E. coli* Hyd-1, for the intrinsic, rapid self-rescue mechanism for surviving O_2 attack displayed by this type of enzyme. The scheme begins with a fully reduced, active [NiFe] cofactor (*top left*). When O_2 attacks, Ni(II) is oxidised to Ni(III) and a one electron reduction of water gives an intermediate bound partially reduced dioxygen species (*top right*). Next, the special [4Fe–3S] proximal cluster sequentially releases two electrons to the active site, releasing one H_2O molecule and leaving OH^- as a bridging ligand on the metal (*bottom centre*). The OH^- bound form of the cofactor gives a characteristic spectrum when observed by electron paramagnetic resonance spectroscopy known as 'Ni-B'. The Ni-B inactive form can be rapidly reduced by a fourth electron (provided by the medial [3Fe–4S] cluster), releasing a second H_2O molecule and regenerating the fully reduced active site.

Further recombineering allowed isolation of a variant form of Hyd-1 carrying a P242C substitution in the small subunit (Volbeda et al., 2013). This was originally designed to help decipher the role of the medial [3Fe–4S] cluster in Hyd-1 function (Evans et al., 2013), but had the good fortune to result in a purified enzyme with a more stably attached HyaC subunit (Volbeda et al., 2013). In this case, the $(HyaAB)_2$ core hydrogenase retained its dimer conformation but the HyaA transmembrane helices could now be modelled together with a single HyaC subunit bearing one haem (Fig. 3). Given both HyaC subunits are normally lost upon detergent treatment, this was a step forward in understanding the biochemistry of *E. coli*

Hyd-1. Further stabilisation of the complete Hyd-1 isoenzyme may be possible if an appropriate alternative detergent is employed. For example, switching from Triton X-114 (Fritsch, Scheerer, et al., 2011; Goris et al., 2011) to digitonin (Frielingsdorf, Schubert, Pohlmann, Lenz, & Friedrich, 2011) as detergent allowed the isolation of a complete trimeric hydrogenase enzyme from *Rastonia eutropha*.

2.1 The Mechanism of O_2-Tolerance Displayed by Hyd-1

E. coli Hyd-1 falls into the Group 1d class of O_2-tolerant, membrane-bound, respiratory enzymes (Greening et al., 2016). The phenomenon of 'O_2-tolerance' in [NiFe]-hydrogenases has long interested science since early biochemical work focused on 'standard' hydrogenases from strict anaerobes such as *Desulfovibrio* sp., which were often irreversibly inactivated by exposure to oxygen (De Lacey, Fernandez, Rousset, & Cammack, 2007). Pioneering work on the frequently renamed knallgas bacterium *Alcaligenes eutrophus/R. eutropha/Cupriavidus necator* (referred to here as *R. eutropha*), which couples H_2 oxidation directly to O_2 reduction to generate a transmembrane electrochemical gradient, identified several [NiFe]-hydrogenases that retained significant activity in the presence of air (Burgdorf et al., 2005). Electrochemical characterisation of purified versions of those enzymes established a robust assay for the ability of a [NiFe]-hydrogenase to tolerate attack by O_2 during turnover (Cracknell, Wait, Lenz, Friedrich, & Armstrong, 2009), and this technique has proved insightful in delineating the mechanism (or mechanisms, as several may have coevolved) of O_2-tolerance utilised by some [NiFe]-hydrogenases.

The identification of *E. coli* Hyd-1 as an O_2-tolerant enzyme was slightly surprising, not least since it was already well established that Hyd-1 is maximally expressed under anaerobic conditions (Sawers, 1994). Following some genetic engineering to place affinity tag sequences within the native *hya* operon, Hyd-1 could be isolated in a single chromatographic step (Lukey et al., 2010). The enzyme was found to bind efficiently to a graphite electrode and produced some of the highest currents (ie, highest H_2 oxidation activity) ever recorded by this technique (Lukey et al., 2010). This electrochemical approach showed graphically that Hyd-1 could recover all of its activity following transient exposure to O_2 and was a bona fide O_2-tolerant enzyme (Lukey et al., 2010). The O_2-tolerant properties of Hyd-1 were quickly exploited to generate a biobattery in conjunction with a fungal multicopper oxidase, which when placed in an H_2/O_2 atmosphere was able to produce enough current to power a digital timepiece (Wait, Parkin, Morley, dos Santos, & Armstrong, 2010).

Progress has been made in understanding the molecular basis of O_2-tolerance by *E. coli* Hyd-1 and related enzymes. Initially, bioinformatics and sequence analysis were employed to search for obvious differences between O_2-tolerant and standard [NiFe]-hydrogenases (Goris et al., 2011; Pandelia et al., 2012). This identified an unusual sequence motif in the small subunit of O_2-tolerant hydrogenases, including the *E. coli* HyaA subunit. Normally, the binding motifs for a classical [4Fe–4S] cluster are relatively straightforward to identify in a protein sequence, since they are typically a single Cys-Xxx-Xxx-Cys metal-binding motif together with two other conserved cysteines. Sequences like these are present coordinating the proximal clusters of all [NiFe]-hydrogenases. However, in the small subunits of enzymes known or predicted to be O_2-tolerant there is a consistent conservation of a Cys-Thr-Cys-Cys motif (which is Cys-Thr-Gly-Cys in standard hydrogenases) together with and a sixth conserved cysteine elsewhere in the sequence (Pandelia et al., 2012). Structure predictions suggested all six cysteine side chains would surround the proximal [Fe–S] cluster in an O_2-tolerant hydrogenase and site-directed mutagenesis experiments with *R. eutropha* membrane-bound hydrogenase (MBH) (Goris et al., 2011) and *E. coli* Hyd-1 (Lukey et al., 2011) demonstrated that the extra cysteines were not required for enzyme assembly or hydrogen oxidation, but were critically required for O_2-tolerance. Subsequent structural analysis of the *R. eutropha* MBH that is related to Hyd-1 (Fritsch, Scheerer, et al., 2011), the O_2-tolerant [NiFe]-hydrogenase from *Hydrogenovibrio marinus* (Shomura, Yoon, Nishihara, & Higuchi, 2011), and *E. coli* Hyd-1 itself (Volbeda et al., 2012) revealed that the proximal cluster was in fact a unique [4Fe–3S] cluster that was stabilised by six conserved cysteine side chains.

The probable molecular mechanism that *E. coli* Hyd-1, and its homologues, has evolved to protect itself from O_2 attack is now based on a remarkable body of extensive spectroscopic and biochemical evidence collected over a relatively short time period (Evans et al., 2013; Frielingsdorf et al., 2014; Goris et al., 2011; Lukey et al., 2010, 2011; Murphy, Sargent, & Armstrong, 2014; Radu, Frielingsdorf, Evans, Lenz, & Jeuken, 2014; Roessler, Evans, Davies, Harmer, & Armstrong, 2012; Siebert et al., 2015; Wulff, Day, Sargent, & Armstrong, 2014). The strategy hypothesised to be used has the hydrogenase operating as an oxidase, reducing O_2 to water in a $4e^-$ reaction, as soon as the active site is attacked by O_2. Hyd-1, and Hyd-1-like, enzymes can do this thanks primarily to the special chemistry exhibited by the proximal [4Fe–3S] cluster and the medial [3Fe–4S] cluster. Under 'normal' conditions, the enzyme will oxidise H_2 and pass electrons to the quinone pool via its chain of electron-transferring cofactors. This effectively results in a fully reduced enzyme loaded with electrons. When O_2 hits the [NiFe] active site the strategy of the enzyme is to rapidly carry out an initial three-electron reduction of O_2 to generate an hydroxide-bound version of the [NiFe] active site that can then be easily further reduced (Wulff et al., 2014). To do this, the [4Fe–3S] proximal cluster is able to change its conformation and, very unusually, quickly and sequentially release two electrons towards the active site (Wulff et al., 2014). This would result in a uniquely reversible 'superoxidised' $[4Fe–3S]^{5+}$ version of the proximal cluster stabilised by the surrounding proteinaceous shell (Dance, 2015; Wulff et al., 2014). The third electron would be supplied by the very high potential [3Fe–4S] medial cluster (Evans et al., 2013; Roessler et al., 2012), and the evidence for this is strong given that a P242C variant of *E. coli* Hyd-1 with a deliberately engineered [4Fe–4S] cluster at the medial position (Volbeda et al., 2013) is no longer O_2-tolerant (Evans et al., 2013). The ultimate result is the reduction of O_2 to H_2O by the enzyme, this regenerating the active site ready for H_2 oxidation, and this has been clearly shown for *E. coli* Hyd-1 and a *R. eutropha* hydrogenase using oxygen isotopes and mass spectrometry (Lauterbach & Lenz, 2013; Wulff et al., 2014).

The crystal structure of *E. coli* Hyd-1 revealed the stable association of two core [NiFe]-hydrogenase units (Volbeda et al., 2012), consistent with the $(\alpha\beta)_2$/$(HyaAB)_2$ stoichiometry predicted from the original biochemical analysis of the purified enzyme (Sawers & Boxer, 1986). One striking feature from the Hyd-1 crystal structure was the proximity of the two distal [4Fe–4S] clusters within the small subunits. In fact, these clusters are only 10 Å apart, which is close enough for rapid electron transfer. One attractive hypothesis then, related to the possible mechanism of O_2-tolerance, is that the $(HyaAB)_2$ stoichiometry could have functional significance, with electron flow from one large subunit being able to rescue the other from O_2 attack. To address this, Wulff, Bielak, Sargent, and Armstrong (2016) were able to use limited detergent treatment in order to isolate HyaAB 'monomers' and $(HyaAB)_2$ dimers from the *E. coli* inner membrane. The dimer form was found to be better equipped to withstand exposure to oxygen during catalysis, suggesting that inter-subunit electron transfer may be both possible and important (Wulff et al., 2016).

2.2 A General Mechanism for Hydrogen Oxidation by [NiFe] Hydrogenases Based on Analysis of Hyd-1 Function

The ability of *E. coli* Hyd-1 to withstand a program of site-directed mutagenesis of its small subunit (Evans et al., 2013; Lukey et al., 2011) was initially surprising. This was because the danger of incorrect or failed cofactor insertion into a [NiFe]-hydrogenase would most likely lead to complete loss of assembled enzyme (Wu et al., 2000). Indeed, attempts to engineer 'O_2-tolerant' [4Fe–3S] proximal clusters into *E. coli* Hyd-2 and Hyd-3 have proven unsuccessful (Lukey et al., 2011; McDowall, Hjersing, Palmer, & Sargent, 2015). Nevertheless, using *E. coli* Hyd-1 as a model, attention has switched to dissecting the mechanism of hydrogen activation per se using site-directed mutagenesis and crystallographic approaches (Evans et al., 2016). Bioinformatic analysis of [NiFe]-hydrogenase large subunits had identified an invariant arginine side chain that, in available crystal structures, is located immediately adjacent to the Ni–Fe–CO–2CN$^-$ cofactor (Fig. 6). This side chain had been targeted for substitution previously, but the required gene mutation led to unstable protein product that could not be purified (DeLacey, Fernandez, Rousset, Cavazza, & Hatchikian, 2003).

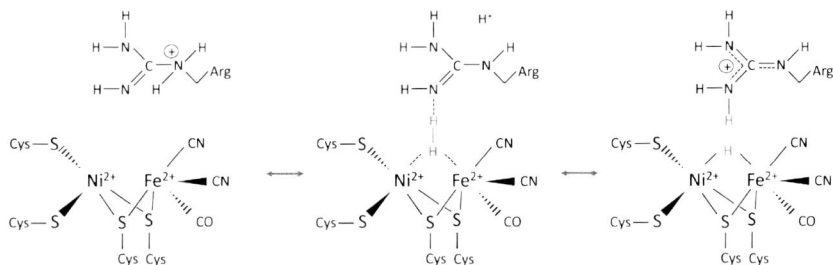

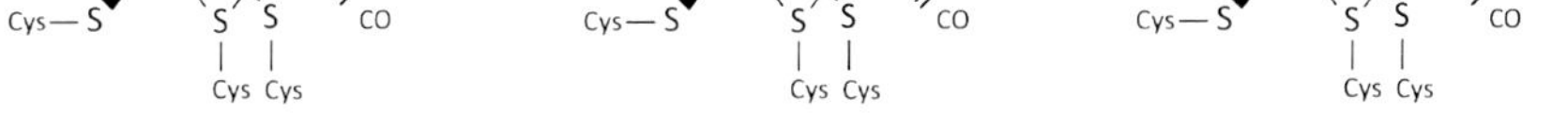

Fig. 6 A key arginine residue for the activation of hydrogen. A scheme, based on evidence provided by Evans et al. (2016), for the initial oxidation of H_2 by [NiFe]-hydrogenases (proceeding from the left panel to the right). It is proposed that the arginine side chain operates as a general base in a frustrated Lewis pair mechanism. Initially, when protonated the positive charge is delocalised throughout the guanidinium head group of the arginine. In an attempt to clarify the proposed mechanism, the initial state is represented with the charge resting on the δ nitrogen of the arginine side chain. H_2 is bound between the metal ions (which together operate as an acid) and the arginine base (*middle panel*), which becomes deprotonated. H_2 is then split into a proton, which reprotonates the guanidinium group, and a hydride is formed as a bridging ligand between the metals (*right panel*). The second half of the reaction (not shown) would be the oxidation of the hydride to release the second proton and two electrons towards the proximal cluster.

Fortuitously, however, in the case of *E. coli* Hyd-1, a HyaB R509K variant could be isolated (Evans et al., 2016). Indeed, whereas the R509K variant could even be crystallised and its structure solved, the modified enzyme was found to be completely devoid of any enzymatic activity (Evans et al., 2016). The current hypothesis is that the Ni–Fe–CO–2CN$^-$ cofactor acts on conjunction with the arginine side chain as a frustrated Lewis pair, which allows the H_2 molecule to be split into a proton and hydride (Fig. 6). This model relies on the guanidine group on the arginine side chain operating as a general base, which would be unusual but not unique in biological systems (Evans et al., 2016; Guillen Schlippe & Hedstrom, 2005).

3. [NiFe]-HYDROGENASE-2: A BIDIRECTIONAL REDOX VALVE

Immunochemical studies originally identified Hyd-2 as the major respiratory 'uptake' hydrogenase in *E. coli* (Ballantine & Boxer, 1985). Hyd-2 production was found to be maximal under anaerobic respiratory conditions, especially in exponential growth phase with glycerol (a non-fermentable carbon source) as electron donor and fumarate as terminal electron acceptor (Ballantine & Boxer, 1985, 1986). Hyd-2 is encoded by the *hybOABCDEFG* operon (Fig. 7) located at 67′ on the *E. coli* chromosome (Menon et al., 1994; Sargent, Ballantine, Rugman, Palmer, & Boxer, 1998). The operon sequence was initially reported as a truncated *hybABCDEFG* version (Menon et al., 1994), however this was subsequently updated when protein chemistry coupled with analysis of the now complete *E. coli* genome sequence identified the missing *hybO* gene (originally dubbed '*hyb0*'; zero) upstream of *hybA* (Sargent, Ballantine, et al., 1998). The transcription of the

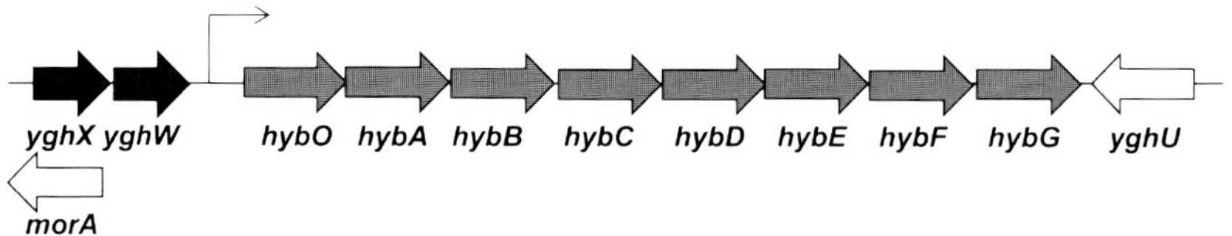

Fig. 7 Organisation of the *E. coli hyb* locus. Depiction of the genetic organisation around the *hybOABCDEFG* operon. The *hybOABC* genes encode the structural components of Hyd-2 and *hybDEFG* are accessory genes involved in enzyme assembly. The flanking *yghW*, *yghX*, and *yghU* genes are not well characterised although there is a crystal structure available for YghU (Stourman et al., 2011). An alternative hypothetical gene, *morA*, has been identified transcribed antisense to the *yghWX* genes (Kurata et al., 2013). Not to scale.

hyb operon is regulated by IscR and NarL and, like the *hyaABCDEF* operon encoding Hyd-1, is completely repressed by exogenous nitrate (Giel et al., 2006; Nesbit et al., 2009; Richard et al., 1999; Sawers et al., 1985). A mutant strain expressing only Hyd-2 (ie, carrying complete deletions in the *hya* and *hyc* operons), following anaerobic growth in minimal media with glucose (Laurinavichene & Tsygankov, 2001), was able to catalyse H_2-dependent reduction of nitrate, as well as dimethyl sulphoxide and fumarate (Laurinavichene & Tsygankov, 2001). This at least demonstrates that Hyd-2 is capable of reducing menaquinone, since *E. coli* fumarate reductase and dimethyl sulphoxide reductase preferentially use menaquinol as electron donor (Cecchini, Schroder, Gunsalus, & Maklashina, 2002). Under these growth conditions the NarZVY nitrate reductase, a homologue to the nitrate-inducible NarGHI (nitrate reductase-A) system, as well as the periplasmic NapAB enzyme, would be expressed and these enzymes can use both ubiquinol and menaquinol as electron donors (Potter, Angove, Richardson, & Cole, 2001; Rothery et al., 2001). Similar whole-cell experiments suggested Hyd-2 activity was best suited to low redox potential (−80 mV) environments (ie, strictly anaerobic) (Laurinavichene, Zorin, & Tsygankov, 2002).

The core catalytic hydrogenase of Hyd-2 is comprised of HybO and HybC and the enzyme falls into class 1c of [NiFe]-hydrogenases (Greening et al., 2016). HybO is the small subunit that is synthesised with a twin-arginine signal peptide at the N-terminus and contains a single transmembrane domain at the C-terminus (Dubini, Pye, Jack, Palmer, & Sargent, 2002; Hatzixanthis et al., 2003; Sargent, Ballantine, et al., 1998). The small subunit is predicted to contain three Fe–S clusters and is a standard O_2-sensitive enzyme as assayed by protein film electrochemistry (Lukey et al., 2010), and as such it carries a [4Fe–4S] cluster at the proximal position. HybC is the large subunit and contains the [NiFe] catalytic centre. The active site base in the HybC large subunit is predicted to be Arg-509, and the cross-subunit histidine ligand from the large subunit to the proximal cluster in the small subunit is His-214.

The core Hyd-2 enzyme was first purified using an ingenious approach of protease treatment of sphaeroplasts (Ballantine & Boxer, 1986). This was important in determining the topology of Hyd-2, which although tightly membrane bound in its native state was obviously exposed in its entirety at the periplasmic side of the membrane (Fig. 8). The trypsin treatment, coupled with the relatively high activity shown by Hyd-2 with benzyl viologen as electron acceptor, meant the enzyme could be purified using

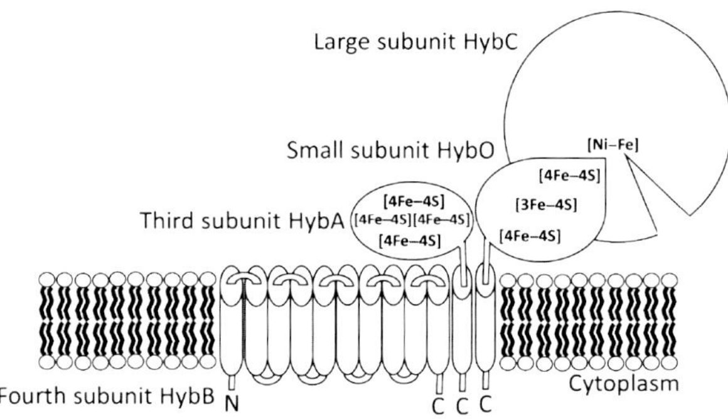

Fig. 8 The architecture of [NiFe]-hydrogenase-2. A cartoon of *E. coli* Hyd-2. The protein comprises a large subunit (HybC), containing a [NiFe] cofactor, together with a small subunit (HybO), containing three [Fe–S] clusters, at the periplasmic side of the membrane. HybOC is anchored to the lipid bilayer by a single transmembrane domain located at the C-terminus of HybO. The HybOC core hydrogenase is associated with HybA, which is predicted to bind four [4Fe–4S] clusters and anchor itself to the periplasmic side of the membrane via a single C-terminal transmembrane domain. The fourth subunit is HybB, which is a cofactor less member of the NrfD/DmsC/PsrC family of proteins. HybB is predicted to contain 10 transmembrane domains. This is the basic functional unit of Hyd-2. Structural and biochemical analysis of Hyd-2 and related proteins suggests the final physiological form is most likely $(HybOC\text{-}AB)_2$.

conventional means and characterised. An initial K_m for H_2 was reported as 4 μM, and in vitro activity was completely inhibited by Cu(II) (Ballantine & Boxer, 1986). This core HybOC active fragment was studied by analytical ultracentrifugation and found to behave with a relative molecular mass of $180{,}000 \pm 8000$ Da (Ballantine & Boxer, 1986), which matches perfectly with a $(HybOC)_2$ stoichiometry, much like the crystal structures of other [NiFe]-hydrogenases from enteric bacteria (Bowman et al., 2014; Volbeda et al., 2012). Once the gene sequence was known the enzyme was purified again using the same trypsin treatment protocol (Sargent, Ballantine, et al., 1998). The small subunit was found to be processed at the N-terminus, thus delineating a 37 amino acid twin-arginine signal peptide, and the predicted sequence allowed possible trypsin cleavage sites to be identified at the N-side of a single transmembrane domain (Sargent, Ballantine, et al., 1998). Interestingly, the HybO TM and its tryspin cleavage sites are transposable onto other enzymes (Hatzixanthis et al., 2003). Fusion of the HybO TM to the C-terminus of the *E. coli* Tat-dependent periplasmic TMAO reductase (TorA) resulted in a membrane-bound chimaera that could also be released by trypsinolysis (Hatzixanthis et al., 2003).

The entire *E. coli hybOABCDEFG* operon clearly encodes many more polypeptides than the $(HybOC)_2$ tryptic fragment that had been biochemically characterised. The *hybA* gene is predicted to encode a Tat-targeted [Fe–S] protein and the *hybB* gene is predicted to encode a large integral membrane protein (Dubini et al., 2002; Menon et al., 1994). Together, the HybOC-AB group is proposed to make up the complete, membrane-bound Hyd-2 system (Fig. 8). The remaining *hybDEFG* genes are involved with [NiFe]-cofactor biosynthesis and Tat proofreading functions. This genetic arrangement and predicted enzyme structure bears absolutely no resemblance to that of the O_2-tolerant Hyd-1 isoenzyme. Instead, the four-subunit HybOC-AB arrangement looks like the Hmc electron transport chains displayed by strict anaerobes such as *D. vulgaris* that grow a low redox potentials (Dolla, Pohorelic, Voordouw, & Voordouw, 2000; Rossi et al., 1993).

Genetic engineering of the *hyb* operon resulted in a natively produced version of Hyd-2 containing a full length $HybO^{His}$ protein (Lukey et al., 2010). Isolation of this species from detergent-solubilised extracts resulted in the copurification of HybO with both HybC, the large subunit, and also HybA (Lukey et al., 2010). The HybA protein is predicted to harbour four [4Fe–4S] clusters and be synthesised with both an N-terminal twin-arginine signal peptide at the N-terminus and a short, single, transmembrane domain at the C-terminus (Dubini et al., 2002; Menon et al., 1994). Since this HybA protein is never found copurifying with trypsin-treated Hyd-2 (Ballantine & Boxer, 1986), it appears likely that HybO and HybA interact via their C-terminal transmembrane domains. The HybB protein was not identified in these preparations, and it must be concluded that it is not stably attached to HybOCA when the lipid bilayer is dispersed with the detergent so far tested (Lukey et al., 2010).

Isolation of affinity-tagged, active, Hyd-2 allowed its spectroscopic and electrochemical characterisation for the first time (Lukey et al., 2010). Using electrode-bound enzyme, the Hyd-2 K_m for H_2 (with the electrode poised at −175 mV) was calculated as 17 μM (Lukey et al., 2010). Moreover, purified Hyd-2 was found to be a standard, O_2-sensitive hydrogenase that was active at relatively low potentials, being inactivated above −80 mV (Lukey et al., 2010). These data are in good agreement with those obtained in whole cells producing only Hyd-2 (Laurinavichene & Tsygankov, 2001) and solidify the proposed role of Hyd-2 as strictly anaerobically expressed respiratory enzyme.

One of the most striking discoveries of the electrochemical experiments was that *E. coli* Hyd-2 was a fully reversible enzyme in vitro (Lukey et al., 2010). This was perhaps not surprising in itself, but given the reluctance of Hyd-1 to perform proton reduction when attached to an electrode (Lukey et al., 2010; Murphy et al., 2014), and the intrinsic lability of Hyd-3 under the same conditions (McDowall et al., 2014), the high hydrogen evolution activity displayed by Hyd-2 made this an interesting isoenzyme for further study. Hyd-2 was subsequently exploited as a component of a biological device designed to replicate the water–gas shift reaction at bench scale. A graphite platelet was incubated with purified Hyd-2 and purified carbon monoxide dehydrogenase from *Carboxydothermus hydrogenoformans* and then incubated in an anaerobic atmosphere containing CO (Lazarus et al., 2009). With the graphite acting as an electron transfer material, the combined enzymes were able to generate H_2 and CO_2 from the carbon monoxide substrate, thus replicating the water–gas shift reaction (Lazarus et al., 2009).

3.1 Generation of a Transmembrane Electrochemical Gradient by Hyd-2

The in vitro studies of Hyd-2 (Lazarus et al., 2009; Lukey et al., 2010), combined with a revised interest in Hyd-2 as a hydrogen producer in vivo (Trchounian, Poladyan, Vassilian, & Trchounian, 2012; Trchounian, Soboh, et al., 2013), led to a mini-renaissance in the study of the physiological role of Hyd-2 in anaerobic *E. coli.* As previously described, hydrogen oxidation in whole cells by Hyd-2 could be coupled to menaquinone-dependent enzymes such as fumarate reductase. However, a strain producing only Hyd-2 could be seen to produce H_2 gas when incubated with the nonfermentable carbon source/respiratory electron donor glycerol (Pinske et al., 2015). This activity was dependent on a functional *hybA* gene, while activity of HybOC with redox dyes such as benzyl viologen were already known to be independent of HybA (Dubini et al., 2002; Jack et al., 2004), thus showing that HybA was part of the physiological electron transfer chain to-and-from HybOC (Pinske et al., 2015). This experiment in fact demonstrated reverse electron transfer through Hyd-2 and suggested that the isoenzyme operates as a 'redox pressure release valve' in vivo: oxidising H_2 at low redox potentials in the presence of suitable electron acceptors, and reducing protons to molecular hydrogen if the quinone pool becomes over-reduced when electron acceptors become scarce. Importantly, however, the 'reverse reaction' where Hyd-2 was forced to perform H_2 production was completely inhibited by the addition of 100 μM CCCP (a proton ionophore) to the cells (Pinske et al., 2015). The simplest interpretation of these data is that the 'forward reaction' (H_2 oxidation linked to menaquinone reduction) carried out by Hyd-2 is tightly coupled to the generation of proton motive force, and thus the 'reverse reaction' must be *driven by* proton motive force.

The mechanism by which Hyd-2 couples H_2 oxidation to proton translocation must be a novel one. The core HybOC hydrogenase is located at the periplasmic side of the inner membrane together with HybA. The key to proton translocation, then, lies with the HybB protein that, presumably is also the site of menaquinone reduction. HybB is a member of the NrfD/PsrC/DmsC family of integral membrane proteins (Rothery, Workun, & Weiner, 2008; Simon & Kern, 2008). Although this type of membrane protein is often annotated as 'cytochromes *b*' in genomic studies, sequence analysis fails to identify conserved classical haem ligands in these proteins (Pandelia et al., 2012; Rothery et al., 2008; Simon & Kern, 2008). Thus, rather than generating a transmembrane proton gradient by a redox loop mechanism, *ala* Hyd-1, it seems more likely that NrfD/PsrC/DmsC-type proteins are true conformational ion pumps, perhaps more akin to the cofactor-less NuoH/L/M/N proteins found in respiratory Complex I. A crystal structure of PsrC from the *Thermus thermophilus* polysulphide reductase has been solved and this shows that the protein does not contain metallocofactors and also binds menaquinone at the periplasmic side of the membrane (Jormakka et al., 2008). There are also structural evidence and biochemical arguments to suggest that PsrC is a conformational proton pump (Jormakka et al., 2008).

E. coli HybB is predicted to comprise 10 transmembrane helices with an N-in/C-in topology. Thus, taken together with HybO and HybA, the complete Hyd-2 membrane-bound complex could display 12 transmembrane domains, a fact that was once considered important in classifying membrane proteins (Kilty & Amara, 1992). The development of facile assays to follow hydrogen production from *E. coli* Hyd-2 (Pinske et al., 2015) could open up new avenues of research to understand further the different molecular mechanisms that have evolved for proton motive force generation using HybB-like proteins. Given that uncouplers can prevent H_2 production, but not hydrogen oxidation, by *E. coli* Hyd-2 (Pinske et al., 2015) it should be possible to adequately control new experiments and then begin to understand the relationship between quinone reduction and proton pumping in these systems.

4. [NiFe]-HYDROGENASE-3: A CENTRAL COMPONENT OF FORMATE HYDROGENLYASE

The ability of the bacterium eventually named *E. coli* to produce H_2 under fermentative conditions has been long studied. In the 1930s, seminal work by Marjory Stephenson and coworkers demonstrated H_2 production by the disproportionation of formic acid in a reaction carried out by a notional 'hydrogenlyase' (Stephenson & Stickland, 1933). Immediate follow-up work by D.D. Woods demonstrated the reverse reaction in whole cells; that is the ability of *E. coli* to produce formic acid using just H_2 and CO_2 as substrates (Woods, 1936). However, characterisation of the enzyme responsible for this activity was to remain many years off. The advent of immunochemistry and molecular genetics during the 1970s and 1980s, and the fortuitous focus on *E. coli* as a model organism during the establishment of those fields, reignited the study of formate hydrogenlyase.

By first raising antisera to membranes isolated from anaerobically grown *E. coli*, and then by raising specific antisera to purified Hyd-1 and Hyd-2 and performing immunoprecipitations, it became clear that there was a third, immunologically distinct, nickel-dependent hydrogenase activity that was maximally expressed during fermentative growth (Ballantine & Boxer, 1985; Sawers et al., 1985). This activity, although initially relatively high and accessible to dye-linked assays in crude cell extracts, was found to be very labile and short-lived but clearly correlated with formate hydrogenlyase activity (Sawers et al., 1985). The genes encoding the third hydrogenase, Hyd-3, were cloned and sequenced and found to be located in an operon designated *hycABCDEFGH* at 58′ on the *E. coli* chromosome (Bohm et al., 1990), later revised to *hycABCDEFGHI* (Fig. 9) (Rossmann, Maier, Lottspeich, & Böck, 1995). Sequence and mutagenic analysis of these

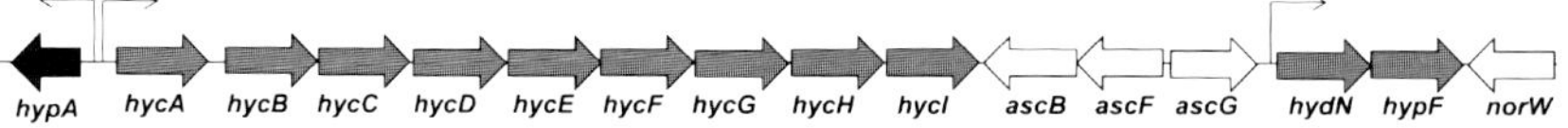

Fig. 9 Organisation of the *E. coli hyc* locus. Depiction of the genetic organisation around the *hycABCDEFGHI* operon. The *hycBEFG* genes encode the structural components of Hyd-3 soluble domain, the *hycCD* genes encode the Hyd-3 membrane domain, and *hycAHI* are accessory genes involved in gene transcription and enzyme assembly. The *hyc* operon is flanked by the *hypA* operon and genes encoding a putative β-glucosidase (*ascB*). Downstream of the *hyc* operon lies the FhlA-regulated *hydN-hypF* bicistronic operon. Not to scale.

nine genes uncovered a truly fascinating enzyme system that has, arguably, its roots in the very dawning of life on Earth (Nitschke & Russell, 2009).

Expression of the *hyc* operon is regulated by FhlA, which is a formate-responsive transcriptional regulator (Schlensog & Böck, 1990; Schlensog, Lutz, & Böck, 1994). Formate is produced endogenously by *E. coli* under generally anaerobic (not only fermentative) conditions, which is when pyruvate formatelyase (PFL) becomes active (Sawers & Watson, 1998). PFL uses radical-based biochemistry to generate formate and acetyl CoA from pyruvate and coenzyme A. PFL is coexpressed, and physically interacts, with an integral membrane formate channel (FocA) that instantly secretes formate from the cytoplasm upon its biosynthesis (Doberenz et al., 2014). Under respiratory conditions, this secreted formate will be oxidised by the electrogenic respiratory formate dehydrogenases in the periplasm, but under fermentative conditions the formate instead builds up in the extracellular melieu. The pK_a of formate is 3.75, thus as the concentration increases the external pH begins to drop to a critical level that triggers the FocA channel to act as an import, rather than export, channel. Thus FocA has the perhaps unusual ability to allow selective movement of single molecule in both directions across the energy-transducing membrane. The crystal structure of the *E. coli* FocA protein has been reported (Wang et al., 2009). It shows a complex of five protomers, each with six transmembrane domains, with an overall architecture reminiscent of an aquaporin (Falke et al., 2010; Wang et al., 2009). The mechanism of FocA operation is not fully understood, but is an area of intense research. Electrophysiological analysis of the purified *S. enterica* FocA homologue suggest this protein can allow transport of several different organic acids, and chloride ions, as well as formate (Lu et al., 2012). Computational modelling and molecular dynamic simulations have been employed to assess the activity, and apparent pH-dependent directional switching (Lu et al., 2011), of FocA (Feng, Hou, & Li, 2012; Lv, Liu, Ke, & Gong, 2013). Combined structural, computational, and biochemical work have identified an invariant histidine residue, located at a constriction point in the substrate channel buried within the lipid bilayer, as being critical for channel function (Hunger, Doberenz, & Sawers, 2014). The bioenergetic implications of formate import are not well enough understood (for example, whether formate is cotransported with a proton counter-ion), and this will be a critical issue to address in order to obtain a complete picture of formate-dependent energy metabolism in *E. coli*.

Transcription of *hyc* is also regulated by ModE (Self, Grunden, Hasona, & Shanmugam, 1999) and internal FhlA promoters lie within the *hycA* gene itself. Overall, the genetic regulation is perfectly poised, therefore, for environmental conditions that favour the production of H_2 by FHL, that being a relatively low pH and a relatively high formate concentration, as well as favourable conditions to assemble an active moldenum-dependent formate dehydrogenase.

The protein products of the *hyc* operon (Fig. 9) comprise various regulatory, biosynthetic, and structural proteins. The HycA protein is a small 153 amino acid polypeptide that was quickly identified as repressor of *hyc* transcription (Bohm et al., 1990; Sauter, Bohm, & Böck, 1992), however its mode of function has not been extensively explored. Genes encoding HycA-like proteins are common in operons encoding Hyd-3-like enzymes in most enteric bacteria. Likewise, the *E. coli* HycH and HycI proteins are not components of Hyd-3 but instead involved in the biosynthesis of active Hyd-3, where HycI is most likely a protease required for C-terminal proteolytic processing of the catalytic large subunit.

The *hycBCDEFG* gene products are all key components of the formate hydrogenlyase enzyme itself (McDowall et al., 2014). Based on sequence identity with the respiratory NADH dehydrogenase Complex I it can be inferred that FHL comprises a membrane domain, made up of HycC and HycD, and a soluble cytoplasmic domain, made up of HycBEFG and the formate dehydrogenase FdhF (Fig. 10). The soluble cytoplasmic domain would be expected to interact extensively with HycCD to generate the mature membrane-bound enzyme (Fig. 10).

The soluble cytoplasmic domain harbours the catalytic subunits of *E. coli* Hyd-3, a [NiFe]-hydrogenase of the Group 4a family (Greening et al., 2016) and so is somewhat different in primary structure from Hyd-1 and Hyd-2. The catalytic large subunit of Hyd-3, encoded by *hycE*, comprises of two quite distinct domains. The C-terminal domain, some 368 amino acid residues, contains a paired-down version of a periplasmic [NiFe]-hydrogenase large subunits (which are typically ~550 amino acids) with clearly conserved motifs for binding of the [NiFe] cofactor and performing hydrogen chemistry. This type of hydrogenase domain is closely related to NuoD (or Nqo4 or 49 kDa protein family) from Complex I (Efremov & Sazanov, 2012), which has evolved a quinone binding site to replace the [NiFe] cofactor binding site. In fact, the predicted structural similarity between the whole of Hyd-3 and the respiratory Complex I has been a point of discussion since the outset (Bohm et al., 1990). Most of the components of *E. coli* Hyd-3 have equivalents in Complex I (Efremov & Sazanov, 2012; Marreiros et al., 2013). The N-terminal domain of *E. coli* HycE (166 amino acid

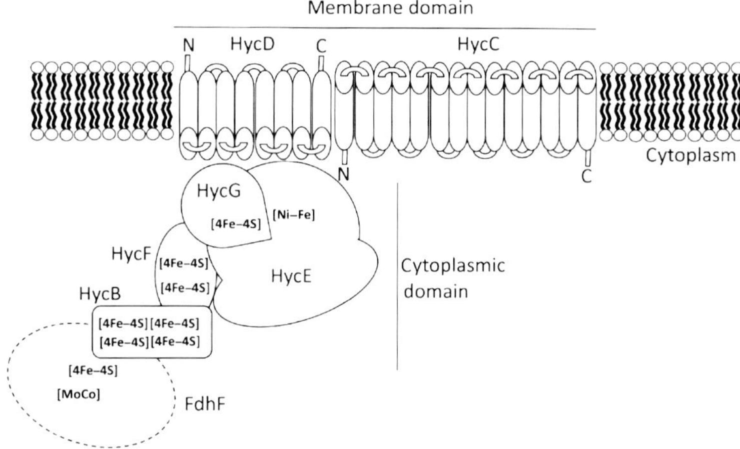

Fig. 10 The architecture of [NiFe]-hydrogenase-3 and formate hydrogenlyase-1. A cartoon of *E. coli* Hyd-3, which when in complex with formate dehydrogenase-H forms the compete formate hydrogenlyase-1 enzyme. The complex comprises two large domains: a membrane domain and a cytoplasmic domain. The membrane domain contains two subunit: the HycC and HycD proteins. These proteins are not expected to bind any cofactors and HycD is predicted to be the principle point of contact with the cytoplasmic domain. The cytoplasmic domain contains a hydrogenase large subunit (HycE), containing a [NiFe] cofactor, together with a small subunit (HycG), containing a single [Fe–S] cluster, at the cytoplasmic side of the membrane. The HycEG core hydrogenase is associated with HycF, which is predicted to bind two [4Fe–4S] clusters, and another Fe–S protein, HycB, which is often referred to as the small subunit of formate dehydrogenase-H (FdhF). This is the basic structural unit of Hyd-3. Formate dehydrogenase-H (FdhF) is a selenoprotein that contains molybdenum cofactor and a [4Fe–4S] cluster. This protein is normally only transiently attached to Hyd-3.

residues) is not known to bind any cofactors and shares limited sequence identity with the NuoC (Nqo5/30 kDa protein family) family of Complex I proteins. The small subunit of *E. coli* Hyd-3 is predicted to be the HycG protein, since this protein contains a single [Fe–S] cluster similar to the proximal [4Fe–4S] cluster of standard hydrogenases (McDowall et al., 2015). HycG is a member of the NuoB (Nqo6/PSST family) family of Complex I subunits. The core HycEG [NiFe]-hydrogenase subunits associate with a further two [Fe–S] proteins, HycB and HycF, to complete the soluble cytoplasmic domain of *E. coli* Hyd-3, and must further interact with the molybdenum-dependent formate dehydrogenase-H (FdhF) to generate a formate hydrogenlyase system.

The HycF protein (the equivalent of NuoI/Nqo9/TYKY family in Complex I) is predicted to bind two [4Fe–4S] clusters and its proposed role is to link HycG to the formate dehydrogenase small subunit HycB (Fig. 10). The HycB protein contains four [4Fe–4S] clusters and is the final component of Hyd-3 (Fig. 10). The active Hyd-3 portion of FHL is assayable, and reversible, with redox dyes both in vitro (Sawers et al., 1985) and in vivo (Maeda, Sanchez-Torres, & Wood, 2007b). However, to function as a formate hydrogenlyase Hyd-3 must associate with an integral membrane domain and a formate dehydrogenase subunit.

The Hyd-3 membrane domain is also evolutionarily related to subunits of the respiratory Complex I (Efremov & Sazanov, 2012; Marreiros et al., 2013) but comprises just two gene products, HycC and HycD. The HycC protein is a large polytopic integral membrane protein comprising 608 amino acid residues. *E. coli* HycC shares 29% sequence identity with *E. coli* NuoL over a 100–400 amino acid central portion of sequence; slightly less identity with *E. coli* NuoM, sharing 28% overall sequence identity over a short section spanning residues 203–394; and similar 28% overall sequence identity with *E. coli* NuoN across the central region of 202–394 HycC residues. The Complex I NuoL/M/N proteins belong to an enigmatic family of predicted ion transporters that contain between 12 and 16 transmembrane domains (N- and C-termini at the cytoplasmic side of the membrane) and often contain amphipathic helices at their extreme C-termini (Batista et al., 2013). Indeed, when the crystal structures of prokaryotic Complex I membrane domains were solved this amphipathic helix was immediately implied as being central to proton pumping capability (Efremov & Sazanov, 2011). Sequence analysis suggests *E. coli* HycC may contain such an amphipathic helix. What is clear is that HycC shares a conserved 'belt', or perhaps 'nest' is more appropriate for a three-dimensional structure, of charged and polar side chains with NuoL/M/N that is located within the plane of the lipid bilayer. The positioning of this nest-of-charges midway between the cytoplasmic and periplasmic faces of the membrane has also attracted hypotheses on its role in ion transport (Marreiros et al., 2013). Indeed, most of the key conserved amino acid residues found in NuoL/M/N are also present in *E. coli* HycC, save for the equivalent of NuoM Lys-234, which was reportedly essential for Complex I activity (Euro, Belevich, Verkhovsky, Wikstrom, & Verkhovskaya, 2008). It is, of course, tempting to speculate that *E. coli* Hyd-3/formate hydrogenlyase could be an energy-conserving system.

The HycD protein comprises 307 amino acid residues arranged in eight transmembrane helices with an N-out/C-out topology. HycD shares 26% overall sequence identity (41% similarity) with the *E. coli* NuoH/Nqo8 protein, which is the primary point of contact between the Complex I membrane and cytoplasmic domains. HycD also contains conserved charged and polar residues with two glutamate side chains (Glu-138 and Glu-189) predicted to lie close together in the middle of the lipid bilayer, possibly equivalent to Glu-157 from *E. coli* NuoH (Hirst, 2013).

Although it is natural to predict that HycC and HycD should form a membrane subcomplex together (Fig. 10), the current evidence for this is not particularly strong. In *E. coli* Complex I, for example, NuoH is separated from the NuoL/M/N trimeric complex by the NuoA, NuoJ, and NuoK membrane proteins. A partial homologue of NuoK (NrfE) is encoded within the *hyf* operon encoding *E. coli* Hyd-4, but this has never been reported as copurifying with Hyd-3. Thus, if HycC and HycD do form a stable interaction, it must be through contacts not present in the Complex I equivalents.

E. coli Hyd-3 therefore comprises six proteins arranged as a membrane domain (HycCD) and a cytoplasmic domain (HycBEFG). The addition of formate dehydrogenase-H (encoded by the *fdhF* gene, which is not part of the *hyc* locus) generates the final formate hydrogenlyase-1 (FHL-1) complex (Fig. 10). The five-subunit FdhF-HycBEFG cytoplasmic domain is predicted to operate as a closed electron transport system connecting formate oxidation directly to proton reduction. Given the evolutionary links with Complex I, it is thought that connecting this thermodynamically favourable activity to the membrane could result in proton (or other ion) translocation and thus energy conservation (Batista et al., 2013; Nitschke & Russell, 2009). This is certainly true for some biological systems (Lim, Mayer, Kang, & Muller, 2014), but has never been convincingly shown for *E. coli*. Interestingly, although all of the cofactors and substrate binding sites are located on the cytoplasmic domain, genetic deletion of the membrane domain components (Δ*hycC* or Δ*hycD*) results in a completely inactive FHL-1 (Sauter et al., 1992). Moreover, while the FdhF component could be assayed with redox dyes in the Δ*hycC* or Δ*hycD* backgrounds, the dye-linked Hyd-3 activity was almost totally abolished (Sauter et al., 1992). This is surprising since active Hyd-3 can be purified from both in detergent-solubilised membranes, and also in a way that strips the membrane domain off the cytoplasmic domain, without loss of activity (McDowall et al., 2014). One possibility is that final attachment of Hyd-3 to the membrane domain is required for ultimate activation of the enzyme.

Similar to Hyd-1 and Hyd-2, a recombineering strategy was employed to facilitate purification of FHL-1 from *E. coli* (McDowall et al., 2014). The native *hyc* locus was engineered to encode affinity-tagged versions of the catalytic large subunit HycE, the small subunit HycG, as well as the HycB and HycF proteins. Tagging of the HycE protein was particularly effective, with an internal stretch of 10 histidine residues incorporated into the cofactor-free N-terminal domain of the protein (McDowall et al., 2014). This approach meant that natural expression levels and biosynthesis pathways of FHL-1 were not impaired and allowed the isolation of all seven detergent-solubilised components, including HycC and HycD, in a single chromatographic step (McDowall et al., 2014). No other proteins consistently purified with FHL-1 using this methodology and the protein was able to generate H_2 from formate in vitro (McDowall et al., 2014), so it must be concluded that this complex is sufficient for the reaction.

4.1 Hydrogen-Linked Formate Dehydrogenase

The formate dehydrogenase component of FHL was historically named FDH-H, for hydrogen-linked formate dehydrogenase (Zinoni, Birkmann, Stadtman, & Böck, 1986). FDH-H is encoded by a single gene, *fdhF*, which is located at 93 min on the *E. coli* genome, so its product is more properly termed FdhF. Thus, despite the close functional relationship between the *fdhF* and *hycBCDEFG* gene products, these genes are found at quite different loci on the *E. coli* chromosome. The initial cloning of the *fdhF* gene led to some great surprises and opened up a whole new research field in bacterial physiology (Böck et al., 1991). The predicted transcribed gene sequence contained a UGA nonsense codon, perfectly in-frame, and located exactly where a UGU or UGC cysteine codon might be expected for other formate dehydrogenases of this type. In this case, UGA was found to code for selenocysteine (Zinoni, Birkmann, Leinfelder, & Böck, 1987) and this special side chain is essential for full activity of formate dehydrogenase in *E. coli* (Gladyshev, Khangulov, Axley, & Stadtman, 1994). The biosynthesis of a selenoprotein, like FdhF, is therefore not straightforward and is constrained by the availability and activity of a number of biosynthetic processes, including specialist tRNA and accessory factors (Labunskyy, Hatfield, & Gladyshev, 2014).

As well as selenocysteine, FdhF also contains a single [4Fe–4S] cluster and a *bis*-molybdopterin guanine dinucleotide cofactor, which is complex between a redox-active molybdenum atom and two cyclic pyranopterin organic moieties (Mendel & Leimkuhler, 2015). The crystal structure of *E. coli* FdhF (Fig. 11) confirmed clearly that selenocysteine was a direct

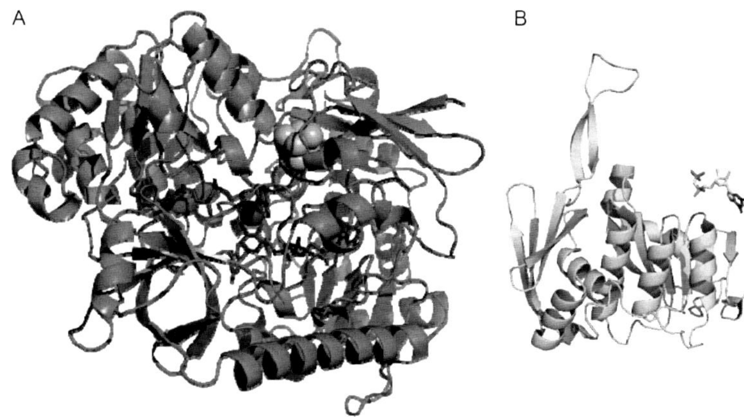

Fig. 11 The crystal structure of formate dehydrogenase-H (FdhF) and FdhD. (A) The crystal structure of *E. coli* FdhF. This protein binds bis-molybdopterin guanine dinucleotide (shown in *blue*) and an 4Fe–4S cluster (shown in *orange*) as cofactors. This structure was solved in 1997 and remains the only component of formate hydrogenlyase that has been structurally characterised (Boyington, Gladyshev, Khangulov, Stadtman, & Sun, 1997). (B) The crystal structure of *E. coli* FdhD in complex with GDP. This accessory protein is essential for efficient activation or biosynthesis of FdhF. PDB file 4PDE is shown (Arnoux et al., 2015). (See the colour plate.)

ligand to the molybdenum atom at the active site (Boyington et al., 1997). Mononuclear molybdenum enzymes are well known for performing two-electron oxo-transferase reactions, since Mo can readily cycle between Mo^{4+} and Mo^{6+} valence states (Hille, Hall, & Basu, 2014). However, by performing experiments with ^{13}C-labelled formate in ^{18}O-labelled water, it was shown that the mechanism of the *E. coli* FdhF enzyme directed abstraction of a proton directly from formate to result in a CO_2 product (Khangulov, Gladyshev, Dismukes, & Stadtman, 1998). Indeed, it has been reported that most formate dehydrogenases produce CO_2 as a product, rather than HCO_3^- (Thauer et al., 1977). FdhF also requires extra steps in the biosynthesis of its molybdenum-containing active site. The product of the *fdhD* gene (Fig. 11) is required for activation or assembly of FdhF (Arnoux et al., 2015; Thome et al., 2012).

The FdhF protein is only loosely attached to the greater FHL complex (McDowall et al., 2014; Sauter et al., 1992) and is unusual when compared with the other formate dehydrogenases of *E. coli* in that it readily reacts with BV as an artificial electron acceptor. This facile assay, together with its relative stability and water solubility, meant that biochemical characterisation of FdhF has far outpaced that of FHL. Indeed, FdhF remains the only component of FHL for which there exists an X-ray structure (Boyington et al., 1997). The kinetics of the FdhF reaction are also well understood, especially as the enzyme interacts with electrodes and so allows electrochemical characterisation (Bassegoda, Madden, Wakerley, Reisner, & Hirst, 2014). In fact, the electrochemical characterisation of FdhF was a breakthrough in understanding, and potentially exploiting, the enzyme. While FdhF readily reduces BV with formate as electron donor, it cannot, most likely for thermodynamic but possibly also for biochemical reasons, operate in reverse with either BV or MV as electron donor (Bassegoda et al., 2014). Attachment to a tuneable graphite electrode overcame the limitations of using redox dyes and demonstrated clearly that FdhF was reversible and capable of 'fixing' dissolved CO_2 to formate (Bassegoda et al., 2014). This enzyme, and FHL as a whole, therefore has potential to be harnessed for biotechnological purposes, perhaps in the recycling of industrially produced CO_2 off-gas.

As long ago as 1936, intact whole cells of *E. coli* were shown to be able to produce formate when incubated with CO_2 and H_2 (Woods, 1936). Of course, in those early experiments it was not yet known that *E. coli* contained three formate dehydrogenases and four hydrogenases, and so definitive evidence that the observed reaction was catalysed solely by FHL is yet to be reported. Nevertheless, taken together with new insights into FdhF (Bassegoda et al., 2014), the opportunity remains to engineer and optimise this activity. There are perhaps three challenges to overcome in order to engineer *E. coli* to process CO_2 into formate. The first challenge is that FdhF is only loosely attached to Hyd-3 and so electron transfer between the two halves of the enzyme may be limited by the rate of protein–protein interactions. It should be noted that it is not unusual for redox enzymes to only transiently interact with their cognate electron transfer partners. The *E. coli* trimethylamine *N*-oxide (TMAO) reductase (TorA), for example, is a highly active periplasmic enzyme that supports anaerobic growth of *E. coli* with TMAO as sole respiratory electron acceptor (Mejean et al., 1994). Upon cellular fractionation, TorA appears completely water soluble and not tightly associated with any other proteins, however it receives electrons only from a membrane-bound *c*-type cytochrome termed TorC, with which TorA must interact during catalysis (Mejean et al., 1994). Similar transient protein–protein interactions are seen with periplasmic nitrate reductases (Potter et al., 2001). In order to improve the efficiency of interaction between FdhF and Hyd-3, a synthetic biology approach has been taken where FdhF was physically attached to HycB using a peptide linker

(McDowall et al., 2015). This strategy led to a functional fused enzyme (McDowall et al., 2015).

The second challenge in using *E. coli* FHL activity to fix CO_2 to formate is that expression of the genes encoding the enzyme is biased towards conditions that favour the opposite reaction, ie, relatively low pH and relatively high formate conditions. That means cells will have to be pregrown under permissive conditions, or the regulation will have to be artificially controlled (McDowall et al., 2015). The third challenge is that *E. coli* cannot grow with formate or CO_2 as the sole carbon source and this is a problem that is unlikely to be overcome. However, given imaginative genetic and metabolic engineering approaches in the future it may be possible to couple industrial-scale CO_2 capture and recycling using harnessed FHL activity to other microbial processes.

4.2 On the Role of HydN

The *E. coli* HydN protein has long enjoyed an enigmatic role in hydrogen metabolism. The *E. coli hydN* gene is located in a bicistronic operon with *hypF* (Fig. 9), the gene encoding the carbamoyl phosphate phosphatase required for [NiFe] cofactor biosynthesis (Karube, Tomiyama, & Kikuchi, 1984; Maier, Binder, & Böck, 1996). The *hydNF* operon is regulated by FhlA, which initially implicated HydN as being a component of the FHL complex (Maier et al., 1996). However, deletion of the *hydN* gene had no obvious effect on cellular FHL activity nor total hydrogenase activity (Maier et al., 1996). Interestingly, however, a Δ*hydN* strain displayed less than half of the normal level of BV-dependent formate dehydrogenase activity, which can be attributed solely to FdhF, suggesting some connection with formate dehydrogenase function or assembly (Maier et al., 1996). This is important, since *E. coli* HydN shares 38% overall sequence identity (and 52% overall sequence similarity) with the *E. coli* HycB protein. HycB is a confirmed component of the FHL complex (a His-tagged version copurifies the entire FHL system; McDowall et al., 2014) and is most likely the direct point-of-contact between FdhF and Hyd-3 (McDowall et al., 2015, 2014). This could suggest HydN might substitute for HycB under some circumstances, however a single *hycB* mutation is sufficient to prevent gas production by *E. coli* and prevent FdhF association with the remainder of Hyd-3 (McDowall et al., 2015; Sauter et al., 1992). In addition, HydN was not identified as a component of purified *E. coli* FHL (McDowall et al., 2015, 2014).

The physiological function of HydN remains to be determined. One possibility is that there are two pools of FdhF in the cell, one fraction associated with HycB and the known FHL-1 complex, and a second fraction associated with HydN. In both cases the interactions must be transient, as overproduced FdhF can be recovered as a soluble monomer from the cytoplasm. Moreover, given the sequence similarity between HycB and HydN, it is difficult to tease out genetic evidence for coevolution or coexpression between *fdhF* and either of *hycB* or *hydN* homologues, for example, using the STRING algorithm (Szklarczyk et al., 2015). However, it is notable that in plant pathogens related to *E. coli*, such as the γ-proteobacteria *Pectobacterium atrosepticum* and *Dickya dadantii*, genes related to *hydN* and *fdhF* are colocalised together in the genome and probably also coexpressed (Babujee et al., 2012).

4.3 Engineering Hyd-3 for Biohydrogen Production

The ability of *E. coli* to produce H_2, especially under fermentative conditions via FHL, has long since attracted the attention of biotechnologists interested in developing renewable sources of biohydrogen. The theoretical amount of H_2 available from hydrolysis of a single mole of glucose is 12 mol of H_2. Normally, during mixed acid fermentation *E. coli* typically generates 0.5–2.0 mol H_2 per mol glucose (Davila-Vazquez et al., 2008; Maeda, Sanchez-Torres, & Wood, 2007a) and it has been proposed that yields of must reach 6 mol before it becomes economically viable to market microbially produced biohydrogen (Benemann, 1996). The feasibility of reaching this level while maintaining redox balance during fermentation is a challenge, however various approaches have been taken to improve the yield of H_2 from *E. coli*. One obvious route is to channel as much carbon as possible from glucose through to formate and therefore FHL, which is the major route of H_2 production under fermentative conditions. This has been attempted by removing all respiratory hydrogenases and formate dehydrogenases, either directly or indirectly by including *tatABC* or *fdhE* mutations, and blocking off potential formate secretion routes by removing FocA and related proteins. Although results can be rather contradictory (Orozco et al., 2010), on balance this type of metabolic engineering of *E. coli* approach was successful in improving H_2 yields slightly (Bisaillon, Turcot, & Hallenbeck, 2006; Maeda et al., 2007a; Penfold, Sargent, & Macaskie, 2006; Yoshida, Nishimura, Kawaguchi, Inui, & Yukawa, 2006). Further work led to a focus on overproduction of FhlA, the formate-responsive transcriptional regulator of the formate regulon (Schlensog et al., 1994). First, overproduction of native (Yoshida, Nishimura, Kawaguchi, Inui, & Yukawa, 2005), then a constitutively active truncated form (Maeda, Sanchez-Torres, & Wood, 2008a; Turcot, Bisaillon, & Hallenbeck, 2008), then mutagenesis (Sanchez-Torres, Maeda, & Wood, 2009), of FhlA was able to further enhance H_2 production from formate in *E. coli* by maximising production of active FHL in the cell. Indeed, combining both metabolic engineering with also modified expression levels of FHL was also found to be a partly successful strategy (Yoshida, Nishimura, Kawaguchi, Inui, & Yukawa, 2007).

Bioengineering of the *E. coli* Hyd-3 catalytic subunit (HycE) itself has been attempted. In order to carry out directed evolution or high-throughput genetic experiments, a facile screen for in vivo FHL activity would be required. One approach would be to adopt the MacConkey formate/fumarate plate test (Wu & Mandrand-Berthelot, 1986), which can select for FHL-positive strains. However, an alternative approach took advantage of a hydrogen-responsive chromogenic material, which resulted in the isolation of a number of interesting HycE variants with altered hydrogen production activities (Maeda, Sanchez-Torres, & Wood, 2008b). One of the most surprising outcomes of this study was the identification of C-terminally truncated versions of HycE that retained hydrogen production activity in vivo (Maeda et al., 2008b). Despite the absence of the C-terminal assembly peptide, correct processing of which is important for Hyd-3 activity (Theodoratou, Paschos, Mintz, & Böck, 2000), and more than half of the active site residues (Evans et al., 2016), truncated versions of HycE retained the ability to generate H_2 in vivo (Maeda et al., 2008b). The biochemical reasons for this enhanced activity are not clear, though it is worth remembering that *E. coli* HycE is a fusion between a potentially cofactor-less NuoC-like N-terminal domain (residues 1–166) and a minimalist [NiFe]-hydrogenase domain (residues 172–539), and so has the potential to display an alternative modus operandi than standard [NiFe]-hydrogenases (Maeda et al., 2008b). However, similar mutagenic work has been carried out on the closely related HycE protein from *Klebsiella oxytoca* HP1 (Huang et al., 2015), a γ-proteobacterium similar to *E. coli* but with a potentially wider metabolic capability (Minnan et al., 2005). There, in silico modelling was used to predict the tertiary structure of HycE and amino acid residues located at the extreme C-terminus were indeed implicated in active-site coordination similar to other [NiFe]-hydrogenases (Huang et al., 2015).

In recent years, there has been an increasing interest in using glycerol as a carbon source for biohydrogen production by *E. coli*. This is chiefly because glycerol is a plentiful by-product of biodiesel (fatty acid methyl/ethyl esters) production by the transesterification of plant-derived oils (triacylglycerides) (Du & Liu, 2012). One challenge to overcome is that *E. coli* does not perform fermentation of glycerol as some other bacteria can, but instead glycerol is the classic 'nonfermentable carbon source' where the oxidation of glycerol 3-phosphate to glyceraldehyde 3-phosphate (which can then enter glycolysis) is catalysed by quinone-dependent respiratory enzymes. Thus, in a stringent minimal medium *E. coli* cannot grow anaerobically on glycerol alone. However, modification of the culture medium slightly will allow *E. coli* to grow anaerobically and, once growing, it will metabolise exogenous glycerol (Gonzalez, Murarka, Dharmadi, & Yazdani, 2008). Traditionally, glycerol would be added to the growth medium at 0.5% (v/v), which is the equivalent to 0.63% (w/v) and a final molar concentration of 68.5 mM (Sambrook & Russell, 2001). To induce anaerobic growth and metabolism of glycerol it is important to supplement the media with tryptone (trypsin-digested milk proteins); to ensure relatively low potassium and phosphate levels; and to increase the initial glycerol concentration to 110 mM (0.8%, v/v or 1%, w/v) (Gonzalez et al., 2008). Under these revised conditions hydrogen is produced (Dharmadi, Murarka, & Gonzalez, 2006; Gonzalez et al., 2008) and the formate dehydrogenase component of FHL (FdhF) was found to be essential for this type of glycerol fermentation, possibly because a high level of dissolved CO_2 is required to drive the process (Dharmadi et al., 2006; Gonzalez et al., 2008). Metabolic- and genetic-engineering approaches have been taken to enhance hydrogen production from glycerol by *E. coli* (Hu & Wood, 2010; Tran, Maeda, & Wood, 2014; Valle, Cabrera, Cantero, & Bolivar, 2015).

There is also a contribution from *E. coli* Hyd-2 to hydrogen production under glycerol fermentation conditions (Blbulyan & Trchounian, 2015; Trchounian, Blbulyan, & Trchounian, 2013; Trchounian, Pinske, Sawers, & Trchounian, 2011; Trchounian, Soboh, et al., 2013). This makes sense, given that Hyd-2 has been shown to a reversible [NiFe]-hydrogenase that can be used as a redox pressure release valve and so reduce protons to H_2 by reverse electron transport (Pinske et al., 2015). Indeed, under such conditions of very high relative glycerol concentrations and very low terminal electron acceptor availability, the quinone-linked glycerol 3-phosphate dehydrogenases will likely over-reduce the quinone pool resulting in the ideal environment for reverse electron flow through Hyd-2.

An alternative approach to enhancing biohydrogen production by *E. coli* is heterologous expression of other hydrogenases from other biological systems. There have been several approaches taken to this, however many of them have been attempted in the *E. coli* laboratory strain BL21 (DE3), presumably to take advantage of the integrated T7 polymerase that this strain harbours (Studier & Moffatt, 1986). Unfortunately, *E. coli* BL21 (DE3) is not particularly proficient at either expression or assembly of anaerobic metalloenzymes, presumably due to aerobic conditioning and accumulated mutations under laboratory conditions (Pinske, Bonn, Kruger, Lindenstrauss, & Sawers, 2011). Nevertheless, some data have been accumulated on using *E. coli* BL21(DE3) as a host for H_2 production. A plasmid system encoding only the core subunits of the Tat-dependent *E. coli* Hyd-1 isoenzyme was prepared and used to transform BL21(DE3) (Kim, Jo, & Cha, 2010). This was sufficient to induce hydrogen production by the normally gas-minus strain (Kim et al., 2010). Heterologous overproduction of an analagous Tat-dependent O_2-tolerant [NiFe]-hydrogenase from *H. marinus* was also able to induce H_2 production in the *E. coli* BL21 (DE3) strain (Kim, Jo, & Cha, 2011), as was overexpression of two Hyd-3-like genes from *K. oxytoca* (Bai, Wu, Jiang, Liu, & Long, 2012). An interesting twist to these tales suggests that much of the hydrogen detected may have been resulting from the normally silent endogenous FHL complex in BL21(DE3) (Jo & Cha, 2015). In this work, the core genes for the Tat-dependent [NiFe]-hydrogenases from *H. marinus*, *Rhodobacter sphaeroides*, and *E. coli* Hyd-1 were all expressed in *E. coli* BL21(DE3) and, as before, induced H_2 production. Interestingly, however, when BL21(DE3) was modified to carry clean deletions in either *fdhF* or *hycE*, all hydrogen production ceased (Jo & Cha, 2015). Moreover, activation of BL21(DE3) FHL by the *H. marinus* [NiFe]-hydrogenase was found to be dependent upon the small subunit only, and specifically a version of the small subunit containing its Tat-signal peptide (Jo & Cha, 2015). This suggests the possibility of membrane or envelope stress being exerted on the system that could be inducing expression or assembly of FHL, as the isolated, overproduced *H. marinus* small subunit may not be an efficient Tat substrate (DeLisa, Lee, Palmer, & Georgiou, 2004).

5. [NiFe]-HYDROGENASE-4: FOSSIL OR FUNCTIONAL?

The discovery of genes encoding a fourth [NiFe]-hydrogenase in *E. coli* came as a surprise to the field (Andrews et al., 1997). Until then, biochemical and genetic analysis had been able to identify only three active [NiFe]-hydrogenases in *E. coli* (Sawers, 1994). The *hyfABCDEFGHIJRfocB* operon is located at 56′ on the *E. coli* chromosome (Fig. 12) and encodes Hyd-4 (Fig. 13), an enzyme that is closely related to Hyd-3 (Fig. 10) and as such is also predicted to comprise a cytoplasmic domain and a membrane domain, both with evolutionary links to Complex I (Andrews et al., 1997). In *E. coli* K-12, the *hyf* operon is not normally transcribed and it is therefore not possible to measure-related hydrogenase activity with redox dyes (Andrews et al., 1997; Self, Hasona, & Shanmugam, 2004; Skibinski et al., 2002). Overproduction of *hyfR in trans* will induce expression from the *hyfA* promoter (Self et al., 2004; Skibinski et al., 2002), but detection of Hyd-4 activity remained a challenge (Skibinski et al., 2002).

The *E. coli* HyfG protein shares 70% overall sequence identity (and 82% similarity) with *E. coli* HycE, the [NiFe] cofactor containing subunit of Hyd-3. HyfG is undoubtedly a hydrogenase large subunit, containing all of the conserved active site residues and a 32 amino acid C-terminal assembly peptide required for its biosynthesis. HyfG also contains the cofactor less NuoC-like N-terminal domain as displayed by HycE. The Hyd-4 small subunit is probably HyfI, which shares 62% overall sequence identity (73% similarity) with HycG, which is most likely the small subunit of Hyd-3. The remainder of the Hyd-4 cytoplasmic domain comprises HyfH, which is the equivalent of HycF sharing 45% overall sequence identity (57% similarity), and HyfA, which is the equivalent of HycB, the predicted small subunit of formate dehydrogenase (50% overall sequence identity, 63% similarity). Incidentally, HyfA is no more closely related to HydN as HycB is, sharing 35% overall sequence identity and 47% similarity. Thus, the soluble domain of Hyd-4 contains four subunits (HyfAGHI) in a similar arrangement to Hyd-3 (HycBEFG).

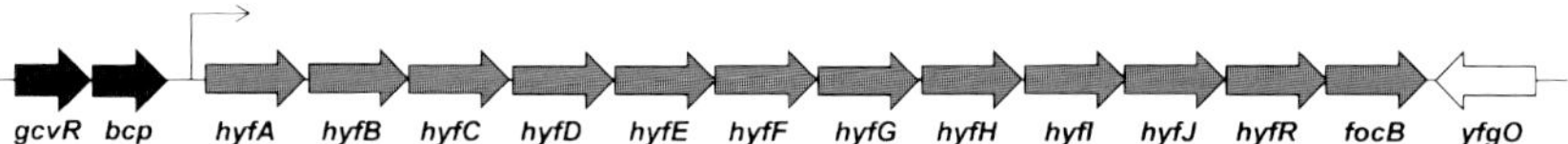

Fig. 12 Organisation of the *E. coli hyf* locus. Depiction of the genetic organisation around the *hyfABCDEFGHIJRfocB* operon. The *hyfAGHI* genes encode the structural components of Hyd-4 soluble domain, the *hyfBCDEF* genes encode the Hyd-4 membrane domain, and *hyfJR* are accessory genes involved in gene regulation and enzyme assembly. The *focB* gene encodes a putative formate channel related to FocA. The *hyf* operon is flanked by the *yfgO*, encoding a putative inner membrane protein and *bcp* encoding peroxiredoxin. Not to scale.

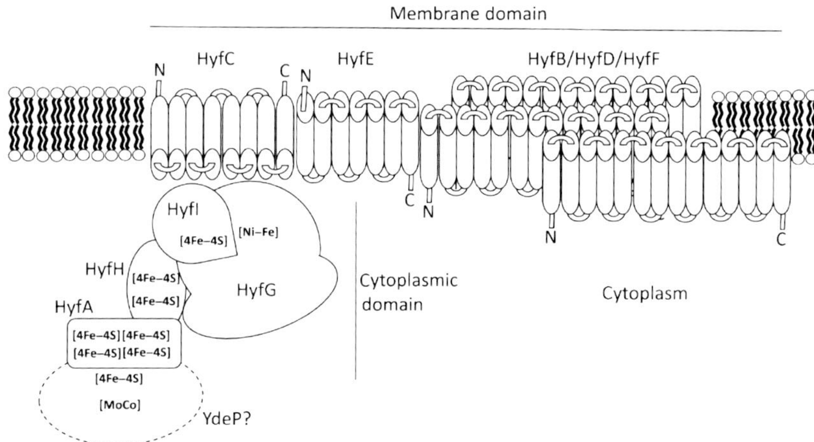

Fig. 13 The architecture of [NiFe]-hydrogenase-4 and formate hydrogenlyase-2. A cartoon of *E. coli* Hyd-4, which when in complex with a formate dehydrogenase would be expected form a formate hydrogenlyase-2 enzyme. The complex is predicted to comprise two domains: a large membrane domain and a cytoplasmic domain similar to Hyd-3. The membrane domain is predicted to contains at least five subunits, the related HyfB/D/F proteins, a HyfC protein and a HyfE protein. These polypeptides are not expected to bind any cofactors and HyfC is predicted to be the main point-of-contact with the cytoplasmic domain. The cytoplasmic domain contains a hydrogenase large subunit (HyfG), containing a [NiFe] cofactor, together with a small subunit (HyfI), containing a single [Fe–S] cluster, at the cytoplasmic side of the membrane. The HycGI core hydrogenase is associated with HycH, which is predicted to bind two [4Fe–4S] clusters, and another Fe–S protein, HycA, which shares identity with HycB and HydN, so could be a partner subunit for a formate dehydrogenase. Together, these nine subunits make up Hyd-4. To generate a formate dehydrogenase-2 enzyme, Hyd-4 must either share FdhF with Hyd-3 or, as depicted here, engage an alternative enzyme such as YdeP.

As with the operon encoding *E. coli* Hyd-3 (Fig. 9), there is no gene at the *hyf* locus (Fig. 12) that would encode a formate dehydrogenase and so directly implicate this enzyme as a second formate hydrogenlyase complex, but the presence of a gene encoding the HycB homologue HyfA; a gene encoding a putative formate channel (FocB); a gene encoding an putative formate-responsive regulatory protein (*hyfR*); as well as some disparate experimental evidence under defined growth regimens (reviewed by) (Trchounian et al., 2012), suggest that this is probably the case. It is thus acceptable to label Hyd-4 as a component of an 'FHL-2' complex. Whether FHL-2 shares FdhF with the Hyd-3-dependent 'FHL-1' has never been proven, however, since a fully assembled or active version of FHL-2 has not been isolated or characterised. The *E. coli* genome does encode alternatives to FdhF and one homologue is termed YdeP. In this case, YdeP is not a selenoprotein but instead a cytoplasmic molybdoenzyme (no N-terminal signal peptide) and otherwise most closely related to formate dehydrogenases and with a role in acid stress resistance in *E. coli* (Masuda & Church, 2003), which could fit with a physiological function as a component of an FHL complex (Vivijs et al., 2015).

The membrane domain of Hyd-4 is significantly different to that of Hyd-3. The *hyf* operon encodes five integral membrane proteins—HyfB, HyfC, HyfD, HyfE, and HyfF (Fig. 13). The HyfB protein comprises 672 amino acids and shares 37% overall sequence identity (57% similarity) with the *E. coli* HycC protein. Moreover, HyfD and HyfF can also be classified as members of the HycC-like group of proteins. HyfD is 479 amino acids and shares 24% sequence identity with HycC across the initial 400 amino acids. HyfF is slightly larger at 526 amino acids, but only shares sequence identity with HycC across a very specific central region spanning HycC residues 226–408. Nevertheless, like HycC, HyfB/D/F are clearly members of the NuoL/M/N proteins of Complex I (Marreiros et al., 2013), however in this case it is striking that the *hyf* operon encodes three of these types of proteins. Furthermore, while the *E. coli* HyfC protein is a homologue of *E. coli* HycD (52% overall sequence identity), there is an extra membrane protein encoded by the *hyf* operon that is not present in FHL-1. That extra protein is HyfE, and this polypeptide is essentially unique when compared to the remainder of the *E. coli* genome. HyfE is only 216 amino acids in length but contains seven-transmembrane helices with an N-out, C-in topology. Its closest homologue in Complex I is the 100 amino acid, 3 TM, NuoK subunit. Indeed, NuoK and be used to construct a 3D model of the C-terminal helices of HyfE. Hyd-4 therefore has a considerably richer membrane domain, much more closely identifiable with Complex I, than that of Hyd-3 (Marreiros et al., 2013; Moparthi & Hagerhall, 2011). Indeed, the presence of a HyfE-encoding gene, plus additional *hycC*-like genes, could potentially be used as bioinformatic markers to differentiate FHL-2 from FHL-1 enzymes (Marreiros et al., 2013).

By comparison with Complex I (Marreiros et al., 2013), the soluble domain of Hyd-4 is expected to contact directly with HyfC, which would then be linked to HyfB/D/F trimer via the HyfE protein (Fig. 13). This structural arrangement in the membrane domain perhaps suggests that Hyd-4 activity is more likely than Hyd-3 to be coupled to the generation of a transmembrane electrochemical gradient. However, it should be noted that the *Thermococcus onnurineus* NA1 genome, and organism that does have a proton-pumping FHL (Lim et al., 2014), does not encode a HyfE homologue. Alternative evidence has been put forward that Hyd-4 hydrogen production activity may be driven by, rather than generate, a transmembrane ion gradient (Kuniyoshi, Balan, Schenberg, Severino, & Hallenbeck, 2015). In order to experimentally generate a proton motive force in *E. coli*, one approach is to heterologously produce an active proteorhodopsin in the inner membrane, which can be used to generate a transmembrane proton gradient in the presence of light (Tipping, Steel, Delalez, Berry, & Armitage, 2013). When this is done in the presence of excess HyfR, but notably also in the presence of active Hyd-3, a slight increase in hydrogen production was observed (Kuniyoshi et al., 2015).

The ability to unequivocally demonstrate Hyd-4 activity in *E. coli* has proven to be a challenge. Whole cell assays measuring extracellular redox changes hint that, under some growth conditions, Hyd-4 may be active (Bagramyan, Vassilian, Mnatsakanyan, & Trchounian, 2001), however the overall picture remains opaque (Trchounian et al., 2012). One sticking point with understanding *E. coli* Hyd-4 is a potentially serious issue with biosynthesis of the [NiFe] active site. While the Hyd-4 catalytic subunit HyfG is produced as a precursor with a C-terminal assembly peptide, the *hyf* operon does not encode a specific protease for the removal of this peptide and this would be expected to preclude activation of the enzyme (Andrews et al., 1997). This, together with the poor levels of transcription normally observed, has led to the suggestion that this is a silent, cryptic operon, with no physiological or biochemical activity. For once, in order to understand the function of a Hyd-4-type [NiFe]-hydrogenase it may be necessary to establish an alternative model organism.

6. BIOSYNTHESIS OF HYDROGENASES

6.1 Biosynthesis of the [NiFe] Cofactor

Students of modern bacterial genomics would be underwhelmed at the news that individual bacterial genomes contain several related genes encoding several closely related isoenzymes that perform identical or similar chemical reactions. However, early bacterial physiologists would not have suspected this at all and so set-up robust genetic screens, based on dye-based assays, to identify mutants defective in hydrogenase activity expecting, perhaps, to identify the structural genes for the enzyme. Instead, these genetic screens uncovered numerous loci (for example, the originally named *hydA*, *hydB*, and *hydC* loci) that had pleiotropic effects on total cellular hydrogenase activity, and all were eventually found to be involved in [NiFe] cofactor biosynthesis. Although numerous gene names have been used over the years, the field finally settled on the terminology '*hyp*-' (where 'p' represents 'pleiotropic') to represent genes that were important in the biosynthesis of the [NiFe] cofactor.

The first steps in the biosynthesis of the Ni–Fe–CO–2CN$^-$ cofactor are the synthesis of the nonproteinaceous ligands. The CN$^-$ ligand is made by from a carbamoyl phosphate substrate molecule (Reissmann et al., 2003). The original *E. coli hydA* locus was found to encode an important protein (HypF) involved in the initial stages of carbamoyl phosphate modification, which explains why mutations in this gene rendered strains completely devoid of all hydrogenase activity. The *hypF* gene is located at 59′ on the *E. coli* chromosome nestled between *hydN* and *norW*, close to the *hycABCDEFGHI* operon encoding FHL (Fig. 9). The HypF protein consists of 750 amino acids separated into two domains. The N-terminal 90 amino acids contain an acylphosphatase domain with fascinating properties, being able to form amyloid fibrils under various conditions. Indeed, this 'HypF-N' domain has become a well-studied model protein for exploring the biophysics of fibrillation (Bhavsar, Prasad, & Roy, 2013; Campioni et al., 2012; Magherini et al., 2009; Relini et al., 2004; Tatini et al., 2013). The C-terminal 92–750 residues of HypF contain a zinc finger-like domain and two nucleotide-binding domains, and these domains are critically important in the binding and remodelling of carbamoyl phosphate (Petkun et al., 2011; Shomura & Higuchi, 2012). Structural and biochemical analyses suggest that HypF first hydrolyses carbamoyl phosphate at the acylphosphatase domain thus releasing a reactive carbamate, which, given the proximity of the different domains in HypF, is never released from the protein but instead quickly reacts with bound ATP to give a carbamoyl adenylate moiety and pyrophosphate (Shomura & Higuchi, 2012).

The *E. coli* HypF protein interacts with the *E. coli* HypE protein (Rangarajan et al., 2008). HypE is a 336 amino acid protein encoded by the *hypABCDE* operon, which located at 58′ on the *E. coli* chromosome and transcribed divergently from the *hycABCDEFGHI* operon (Fig. 14). HypE has a special C-terminal cysteine residue and the close interaction between HypF and HypE allows the product of the HypF reaction, carbamoyl adenylate, to rapidly react with the HypE cysteine to form an S-carbamoyl modified version of HypE (Shomura & Higuchi, 2012; Tominaga et al., 2013). The HypE protein has carbamoyl dehydratase activity and next uses ATP hydrolysis to modify the bound carbamoyl moiety to

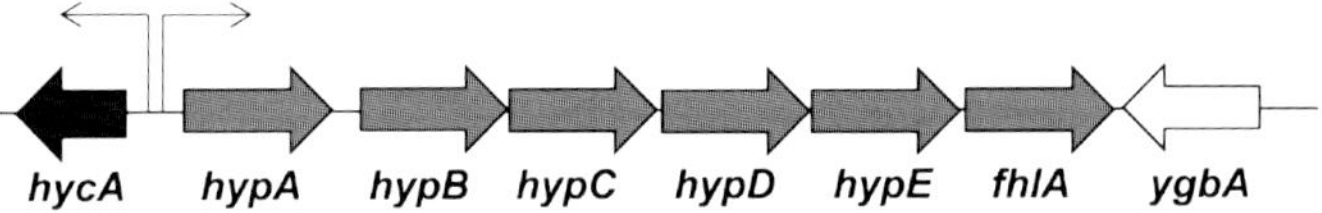

Fig. 14 Organisation of the *E. coli hyp* locus. Depiction of the genetic organisation around the *hypABCDE-fhlA* operon. The *hypAB* genes encode proteins dedicated to nickel processing, the *hypCDE* genes encode proteins dedicated to assembling the Fe–CO–2CN$^-$ half of the [NiFe] cofactor, and *fhlA* encodes a formate-responsive regulator. The *hyp* operon is flanked by the *hyc* operon encoding Hyd-3 and the *ygbA* gene, which is part of the NsrR regulon (Filenko et al., 2007). Not to scale.

either a bound thiocyanate (Tominaga et al., 2013) or perhaps more likely a bound isothiocyanate (Stripp, Lindenstrauss, Sawers, & Soboh, 2015). The isothiocyanate-loaded version of HypE is now ready to donate CN$^-$ to Fe(II) in the biosynthesis of the [NiFe] cofactor.

The source of the second nonproteinaceous ligand in the active site, CO, is less well understood than that of CN$^-$. In other biological systems, CO is often enzymatically derived from tyrosine (Kuchenreuther et al., 2014), however for the *E. coli* [NiFe]-hydrogenases the source of CO is still an area requiring further research. Latest results suggest that the source of CO may come from CO_2 that can bind naturally to an Fe(II) ion held by the HypCD complex (Soboh et al., 2013), which serves a scaffold for building the Fe–CO–2CN$^-$ half of the [NiFe] cofactor (Blokesch, Albracht, et al., 2004; Watanabe et al., 2007; Watanabe, Matsumi, Atomi, Imanaka, & Miki, 2012). *E. coli* HypD is a 373 amino acid protein that contains a [4Fe–4S] cluster as a cofactor (Blokesch, Albracht, et al., 2004; Blokesch & Böck, 2006). The HypD protein also binds at least one additional iron ion that serves as the basis of the [NiFe] cofactor itself (Soboh et al., 2012; Watanabe, Matsumi, et al., 2012), and this Fe(II) may be coligated by the HypC protein within the complex (Soboh et al., 2012; Stripp et al., 2013; Watanabe, Matsumi, et al., 2012). The initial starting point for building an [NiFe] cofactor, then, is a HypCD complex holding an Fe–CO_2 species that presumably is reduced to an Fe–CO moiety (Soboh et al., 2013). This order of events is disputed somewhat, with other biochemical evidence pointing to loaded of CN$^-$ first onto a HypCD-Fe scaffold (Burstel et al., 2012). Next, the isothiocyanate-carrying HypE protein can interact with the HypCD complex (Blokesch, Albracht, et al., 2004; Watanabe et al., 2007; Watanabe, Matsumi, et al., 2012). HypE-bound cyanate is repeatedly deposited onto the Fe–CO until the Fe–CO–2CN$^-$ species is formed (Stripp et al., 2013).

Having now formed half of the [NiFe] cofactor on the HypCD scaffold complex (Fig. 15), the next step is to insert this into an empty apoenzyme (Fig. 16). Here, the HypC protein acts as a link between the cofactor biosynthesis pathway and the empty apoenzymes since HypC, and its close homologue HybG, can also interact directly with the immature catalytic subunits of all three [NiFe]-hydrogenases in *E. coli* (Blokesch, Magalon, & Böck, 2001; Drapal & Böck, 1998). Presumably, the interaction between HypC/HybG and the large subunit apoenzymes brings the HypD, HypE and HypF proteins into the closest possible proximity at allows efficient assembly and insertion of the Fe–CO–2CN$^-$ half of the cofactor (Fig. 16). Having assembled half the cofactor into the large subunit, the second half of the cofactor biosynthesis process is set to occur in situ. The remaining steps involve nickel processing and assembly (Fig. 16).

Nickel itself is scarce in the environment and under anaerobic conditions *E. coli* produces a high affinity nickel transporter. The system was initially discovered as the *hydC* locus that, when mutated, resulted in strains that were devoid of hydrogenase activity but rescuable by adding 10–500 μM Ni(II) salts to the growth medium (Wu & Mandrand-Berthelot, 1986; Wu et al., 1989). Under these conditions, Ni(II) is smuggled through a

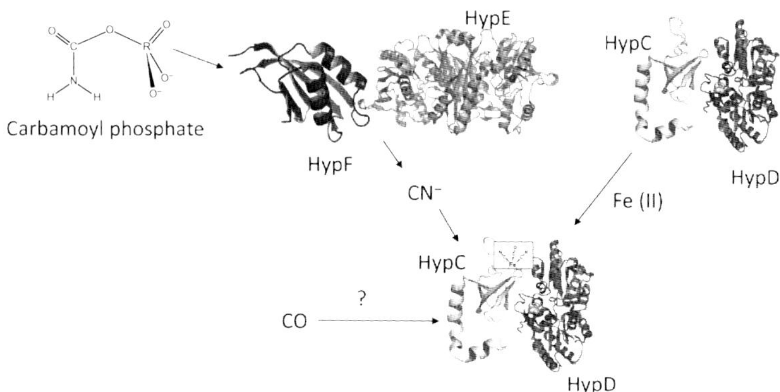

Fig. 15 Biosynthesis of the [NiFe] cofactor: the first half of the process takes place on a common scaffold. A depiction of the initial biochemical processes required to begin the biosynthesis of the Ni–Fe–CO–2CN$^-$ cofactor. First, carbamoyl phosphate is remodelled by HypF (PDB file 1GXT is shown) and HypE (PDB file 2RB9) to provide a source of CN$^-$. In the meantime, the HypC-HybD complex (illustrated here with PDB files 2OT2 and 1CFZ) is loaded with Fe(II) ready to receive both CN$^-$ and CO. The source of CO could be from tyrosine, as in the biosynthesis of [FeFe]-hydrogenase cofactors, or from reduction of an initially bound CO_2 molecule. Finally, the Fe–CO–2CN$^-$ 'half' of the cofactor is assembled on a HypC–HypD scaffold. (See the colour plate.)

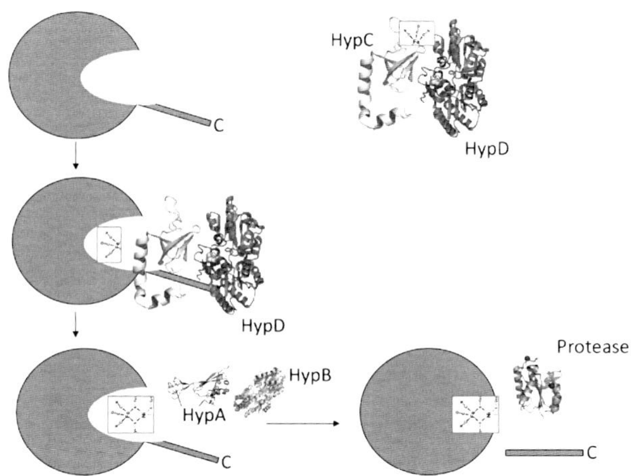

Fig. 16 Biosynthesis of the [NiFe] cofactor: the second half of the process is completed in situ on the large subunit. A depiction of the insertion, and completion of biosynthesis, of the Ni–Fe–CO–CN^- cofactor in the large subunit of a [NiFe]-hydrogenase. The HypC–HypD complex (illustrated here with PDB files 2OT2 and 1CFZ) already preloaded with Fe–CO–2CN^- (Fig. 15) interacts with the empty apoenzyme of a hydrogenase, which is held in an 'open' conformation due to the presence of a C-terminal assembly peptide. After insertion of the Fe–CO–CN^- 'half' of the cofactor, the HypC–HypD complex is replaced by a HypA–HypB nickel-binding system (illustrated here with PDB files 3A43 and 2HF9). Finally, the whole process is rendered irreversible by the proteolytic cleavage of the [NiFe]-hydrogenase precursor by a specific protease (PDB file 2E85). (See the colour plate.)

magnesium transport system instead (Wu et al., 1989). The fully characterised *nikABCDER* operon encodes an ATP-binding cassette (ABC) transporter for nickel (Wu, Navarro, & Mandrand-Berthelot, 1991). NikA is the periplasmic-binding protein of this uptake system; NikB and NikC make-up the transmembrane section; NikD and NikE are the ABCs; while NikR is a nickel-reponsive transcriptional regulator. Thus, ATP hydrolysis is required to drive nickel uptake from the environment to facilitate [NiFe]-hydrogenase biosynthesis.

At 56 kDa, the NikA protein is a relatively large periplasmic-binding protein that binds nickel(II) with high affinity (de Pina et al., 1995). The protein has been studied by solution NMR (Rajesh et al., 2005) and crystallographic approaches (Cherrier, Cavazza, Bochot, Lemaire, & Fontecilla-Camps, 2008; Cherrier et al., 2005; Heddle, Scott, Unzai, Park, & Tame, 2003; Lebrette et al., 2014; Lebrette, Iannello, Fontecilla-Camps, & Cavazza, 2013), and the mechanism of metal binding has proven to be rich and complex. The *E. coli* NikA protein prefers to bind Ni(II) together with a coligand or 'nickelophore' (Lebrette et al., 2014). The nature of the nickelophore is the subject of some debate and much research. Under some circumstances, such as when haem export is upregulated, NikA can bind to free haem in vivo as well as in vitro (Shepherd, Heath, & Poole, 2007). In this case, however, there appears to be no clear link between haem- or porphyrin-binding activity and nickel uptake. NikA will also bind Fe–EDTA complexes (Cherrier et al., 2008) and efforts to obtain a metal-free apostructure led to the suggestion, based on crystallographic evidence, that butane-1,2,4-tricarboxylate may be the small organic molecule assisting with nickel binding in *E. coli* NikA (Cherrier et al., 2008). Careful physiological studies revealed that nickel uptake by *E. coli* was significantly enhanced by the addition of L-histidine to the medium (Chivers, Benanti, Heil-Chapdelaine, Iwig, & Rowe, 2012). Indeed, purified NikA was found to bind to a Ni-(L-His)$_2$ complex in vitro (Chivers et al., 2012; Lebrette et al., 2013). This is perhaps a mechanism to prevent mismetallation of NikA in the *E. coli* periplasm and so ensure only nickel is presented to the NikBCDE membrane-embedded transporter module.

The *E. coli nikABCDER* operon is regulated by FNR (thus high-affinity nickel transport is only available under anaerobic conditions) and NikR, the last gene in the operon (Wu et al., 1989). General anaerobiosis therefore activates nickel uptake in *E. coli*, in order to coincide with the more elaborate anaerobic regulation of the genes encoding three [NiFe]-hydrogenases, and the nickel-responsive NikR repressor protein tempers transcription from the *nikA* promoter to maintain a balance between nickel uptake and hydrogenase biosynthesis. The NikR protein has been structurally characterised (Chivers & Sauer, 1999; Phillips, Schreiter, Stultz, & Drennan, 2010; Schreiter et al., 2003; Schreiter, Wang, Zamble, & Drennan, 2006). NikR is a small protein (133 amino acid residues) that comprises an N-terminal DNA-binding domain and a C-terminal Ni(II)-binding domain, and it is a member of the ribbon–helix–helix superfamily of transcription regulators (Schreiter & Drennan, 2007). NikR forms a homotetramer and binds to Ni(II) at both high affinity and low affinity sites, which in turn activate NikR into a tight DNA-binding mode (Chivers & Sauer, 2002).

Once taken into the cytoplasm, nickel must be trafficked towards the hydrogenase apoenzymes, ready loaded with their 'half active centre' comprising Fe–CO–2CN$^-$. This is achieved primarily by the HypA and HypB proteins, while the histidine-rich SlyD protein can function as a nickel-storage-and-supply system (Kaluarachchi, Zhang, & Zamble, 2011). In short, HypA and HypB are both nickel-binding proteins that work in tandem to deliver Ni(II) to the hydrogenase apoenzymes and under 'normal' physiological conditions (ie, very low nickel availability in the environment) nucleotide hydrolysis is required to offload the metal from the HypAB proteins (Watanabe, Sasaki, et al., 2012).

Mutations within the *hypB* gene were among the first pleiotropic hydrogenase mutations described in *E. coli* (Waugh & Boxer, 1986). The originally named *hydB* locus was mapped to 58′ on the *E. coli* chromosome (Waugh & Boxer, 1986) and eventually pin-pointed to the *hypB* gene (Maier, Jacobi, Sauter, & Böck, 1993). The most remarkable characteristic of *hypB* mutants was the ability to suppress the hydrogenase-null phenotype by culturing in 600 μM Ni(II) salts (Waugh & Boxer, 1986). Characterisation of the *E. coli hypB* gene and its protein product revealed a GTPase and, remarkably, inactivating amino acid substitutions in the GTP-binding motifs gave rise to the nickel-suppressible phenotype (Maier et al., 1993; Maier, Lottspeich, & Böck, 1995). The molecular basis of the suppression of the *hypB* phenotype with Ni(II) has never been fully explained, especially as any surplus nickel taken into the *E. coli* cytoplasm is rapidly excreted again via the RcnABR system (Rodrigue, Effantin, & Mandrand-Berthelot, 2005). However, biochemical characterisation of *E. coli* HypB (Leach, Sandal, Sun, & Zamble, 2005), and structural studies of homologues from other biological systems (Watanabe et al., 2015), are beginning to shed light on the process.

The *E. coli* HypB protein is a homodimer that binds GTP and contains two different Ni(II) binding sites (Leach et al., 2005; Maier et al., 1993). Both metal binding sites operate in tandem to facilitate metalation of the hydrogenase apoenzymes (Chung et al., 2008; Dias et al., 2008). The HypB protein seems to operate as a secondary Ni(II)-buffering protein the cell cytoplasm without a direct role in Ni(II) loading into hydrogenase. Instead, HypB interacts with a second nickel-binding protein—HypA—that again has a function that can be overcome by flooding the system with Ni(II) (Lutz et al., 1991)—and the protein–protein interaction induces formation of a new Ni(II) binding site within the HypA protein (Douglas, Ngu, Kaluarachchi, & Zamble, 2013; Watanabe et al., 2015). Nucleotide hydrolysis (*E. coli* HypB has a preference for GTP although other biological systems utilise ATP) then drives metal transfer from HypB to HypA and then on to the waiting hydrogenase (Douglas et al., 2013; Watanabe et al., 2015). Indeed, HypA-like proteins can be considered the link between nickel trafficking and the hydrogenase apoenzymes. In *E. coli*, a homologue of HypA is encoded in the Hyd-2 operon and the HybF protein can substitute for HypA, particularly under anaerobic respiration when Hyd-1 and Hyd-2 levels exceed that of Hyd-3 (Blokesch, Rohrmoser, Rode, & Böck, 2004; Hube, Blokesch, & Böck, 2002). As an interesting aside, a Δ*hypA*, Δ*hypB*, Δ*hybF* triple-mutant-strain is devoid of hydrogenase activity but can be phentotypically suppressed by adding excess Ni(II) salts to the growth medium (Blokesch, Rohrmoser, et al., 2004).

In summary, the Fe–CO–2CN$^-$ 'half' of the [NiFe] cofactor is built on the HypCD scaffold by the HypEF enzymes (Fig. 15) and transferred first into the empty hydrogenase apoenzyme (Fig. 16). Next, nickel is added to the waiting cofactor in situ, normally by the action of nickel processing enzymes HypAB (Fig. 16) and the nonessential general accessory protein SlyD. Finally, the cofactor is 'locked' in position by C-terminal proteolytic processing of the hydrogenase catalytic subunit by a specific protease (Fig. 16). Each of *E. coli* Hyd-1, Hyd-2, and Hyd-3 has a dedicated private protease for this purpose encoded by the *hyaD*, *hybD*, and *hycI* genes, respectively.

The whole system can be reconstituted in vitro (Soboh et al., 2014), as can the final nickel insertion process (Maier & Böck, 1996). The *hypABCDE* operon is regulated by FhlA, while an FNR binding site is located within *hypA* itself and it is used to boost anaerobic levels of *hypBCDE* where *hybF* takes over from *hypA* (Hube et al., 2002; Lutz et al., 1991; Messenger & Green, 2003).

6.2 Biosynthesis of [Fe–S] Clusters

Iron and iron–sulphur clusters are critical for the biosynthesis and activity of [NiFe]-hydrogenases (Pinske & Sawers, 2014). In *E. coli*, Fe–S cluster biosynthesis proceeds via two separate pathways—the 'ISC' and sulphur formation ('SUF') pathways (Roche et al., 2013). The ISC pathway is encoded by the *iscRSUA* operon and *hscAB* genes, which encode a transcriptional regulator (IscR); a cysteine desulphurylase (IscS); a scaffold protein for Fe–S cluster assembly (IscU); an 'A-type' Fe–S cluster carrier (IscA), which is a protein with a debatable function but has been proposed to act as a conduit between IscU and empty apoenzymes in the cell; and the HscAB ATPase that aides transfer of completed clusters (Blanc, Gerez, & Ollagnier de Choudens, 2015). The SUF pathway, which is proposed to be ancient in origin since it is found in Archaea and chloroplasts rather than aerobic bacteria and mitochondria, is encoded by the *sufABCDSE* operon (Outten, 2015). In this case, SufS is the cysteine desulphurylase that forms a complex with SufE and hands liberated sulphur onto a scaffolding complex comprising SufBCD, which builds clusters in a flavin- and ATP-dependent manner. Like IscA, SufA is also an 'A-type' Fe–S cluster carrier protein and a third member of this family, ErpA, is also found in *E. coli* (Roche et al., 2013).

The roles of these systems in *E. coli* hydrogenase biosynthesis has been explored, however given cofactor biosynthetic proteins such as HypD and MoaA contain Fe–S clusters it can be challenging to adequately control experiments with such systems. Nevertheless, the roles of the 'A-type' cluster carriers IscA, SufA, and ErpA were addressed (Pinske & Sawers, 2012). In this study, the *suf* system was found to be not involved, since a Δ*sufA* strain retained Hyd-1 and Hyd-2 activity. Interestingly, both Δ*iscA* and Δ*erpA* strains were devoid of hydrogenase activity, suggesting these two gene products are working in series to assemble Hyd-1 and Hyd-2, rather than in parallel. The loss of hydrogenase activity in the mutant strains was probably not due to a defect in [NiFe] cofactor biosynthesis, since the Hyd-1 and Hyd-2 large subunits were correctly C-terminally processed, but instead the small subunits appeared destabilised (Pinske & Sawers, 2012).

The functional relationship between the ISC biosynthetic machinery and the [NiFe]-hydrogenase is stronger than with most other metalloenzymes in *E. coli*. That is because the master regulator, IscR, not only controls the *iscRSUA* and *sufABCDSE* operons, but also represses *hya* and *hyb* under aerobic conditions (Giel et al., 2006; Nesbit et al., 2009).

6.3 Biosynthesis of Cytochromes *b*

Hyd-1 contains a membrane-integral *b*-type cytochrome that most likely operates as a quinone reductase. The protein, HyaC, contains four transmembrane domains with an N-in/C-in orientation and, although the available crystal structure contains only one *b*-type haem (Volbeda et al., 2013), the protein sequence has all the signature motifs required to correctly bind two haems, as seen with similar subunits of the Tat-dependent formate dehydrogenase-N from *E. coli* (Jormakka et al., 2002) or the *E. coli* Tat-independent nitrate reductase-A (Rothery et al., 2001). The haem in HyaC is not covalently bound and the biosynthesis of such cytochromes is not fully understood. Current wisdom suggests the protein is integrated into the inner membrane in a cotranslational manner (Elvekrog & Walter, 2015) and then picks up available haem from the cytoplasm or periplasm without the aid of specific accessory proteins or chaperones. The pathway of haem b (protohaem IX or ferrohaem b) biosynthesis itself is well understood (Heinemann, Jahn, & Jahn, 2008; Layer, Reichelt, Jahn, & Heinz, 2010). The committed steps involve the HemE (uroporphyrinogen decarboxylase), HemF (coproporphyrinogen III oxidase), and HemG (protoporphyrinogen oxidase) proteins to result in protoporphyrin IX. Finally, a ferrochetalase enzyme (HemH) is used to load iron into the organic moiety resulting in ferrohaem b. The *hemG* and *hemH* genes are considered essential in *E. coli* (Baba et al., 2006; Kato & Hashimoto, 2007).

6.4 The Twin-Arginine Translocation Pathway

Arguably, the final major biosynthetic hurdle for Hyd-1 and Hyd-2 is the transmembrane translocation of the enzymes. This is achieved by a remarkable molecular machine that has evolved to transport fully folded proteins across ionically sealed membranes—the twin-arginine translocation (Tat) system (Fig. 17) (Palmer & Berks, 2012). Studies of [NiFe]-hydrogenases played central roles in the discovery of the Tat pathway. Early gene sequencing revealed that the small subunits were synthesised as precursors with relatively long N-terminal signal peptides and, while these shared the physicochemical properties of better-known Sec signal peptides, it became clear that the [NiFe]-hydrogenase signal peptides contained regions of conserved amino acid sequence (Niviere, Wong, & Voordouw, 1992; Prickril, Czechowski, Przybyla, Peck, & LeGall, 1986; Wu & Mandrand, 1993). Tat signal peptides contain a conserved SRRxFLK 'twin-arginine' amino acid motif (Berks, 1996; Wu & Mandrand, 1993), and directed mutations in this motif of [NiFe]-hydrogenase signal peptides were found to prevent transport activity (Bernhard, Schwartz, Rietdorf, & Friedrich, 1996; Niviere et al., 1992).

In *E. coli*, the genes require for export of proteins bearing Tat signal peptides are *tatABC* (Fig. 17) and the *tatA* homologue *tatE* (Bogsch et al., 1998; Sargent, Bogsch, et al., 1998; Sargent, Stanley, Berks, & Palmer, 1999; Weiner et al., 1998). An *E. coli* strain bearing a double deletion in *tatA* and *tatE* was unable to transport Hyd-1 and Hyd-2 to the periplasmic side of the cytoplasmic membrane (Sargent, Bogsch, et al., 1998), likewise single deletion mutants in either *tatB* (Chanal, Santini, & Wu, 1998; Sargent et al., 1999) or *tatC* (Bogsch et al., 1998) were blocked in hydrogenase translocation. Mislocalisation of Hyd-1 and Hyd-2 meant that the enzymes were

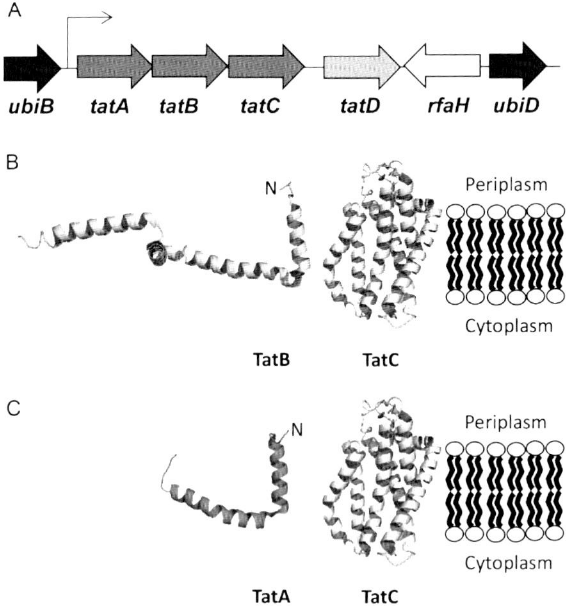

Fig. 17 The twin-arginine transport system. (A) Depiction of the genetic organisation around the *E. coli tatABC* operon, encoding the key structural components of the *E. coli* twin-arginine translocase. Not to scale. (B) The crystal structure of monomeric TatC from *Aquifex aeolicus* (PDB file 4B4A) compared to the solution NMR structure of *E. coli* TatB (PDB file 2MI2). Together, the TatBC complex makes up the initial signal recognition complex of the twin-arginine translocase. (C) The solution NMR structure of a monomeric variant (T22P) of *E. coli* TatA (PDB file 2LZR) compared to the crystal structure of TatC. (See the colour plate.)

physiologically inactive and the cells displayed no hydrogen uptake activity (Sargent et al., 1999). In other respects, however, the mislocalised enzymes were fully assembled and could be readily assayed for hydrogenase activity using redox dyes (Bogsch et al., 1998; Chanal et al., 1998; Sargent, Bogsch, et al., 1998; Sargent et al., 1999). This was significant since, together with preceding work on cofactor insertion pathways (Rodrigue, Boxer, Mandrand-Berthelot, & Wu, 1996), it helped establish that the hydrogenase heterodimers were fully folded, assembled, and active before the transmembrane translocation step, thus adding weight to the argument that Tat was dedicated to the export of folded proteins. Given the large subunit carries no signal peptide of its own, this phenomenon of two different subunits being targeted by a single signal peptide was termed 'hitch-hiking' (Rodrigue, Chanal, Beck, Muller, & Wu, 1999). Importantly, the *tat* mutants were completely unaffected in their ability to generate hydrogen via formate hydrogenlyase (Sargent et al., 1999), thus demonstrating no role for the Tat pathway in the biosynthesis of FHL.

In terms of structure and function, advances have been made in the understanding of the mechanism of the Tat translocase (Berks, Lea, & Stansfeld, 2014; Palmer & Berks, 2012). The Tat signal peptide is recognised by the TatC component of the transport system and crystal structures of a TatC homologue from *Aquifex aeolicus* have been solved (Ramasamy, Abrol, Suloway, & Clemons, 2013; Rollauer et al., 2012). *E. coli* TatC comprises six transmembrane helices (Fig. 17) with a cytoplasmic loop between helices 2 and 3 being the initial contact point for the twin-arginine motif of the Tat signal peptide. In *E. coli*, multiple copies of TatC associate with equimolar amounts of the TatB protein to form a large oligomeric complex (Tarry et al., 2009). The NMR solution structure of detergent-solubilised *E. coli* TatB (Fig. 17) shows a small integral membrane with a single, N-terminal transmembrane domain followed at the cytoplasmic side of the membrane by an amphipathic helix and a short soluble domain (Zhang, Wang, Hu, & Jin, 2014). The precise role of TatB remains unclear, since it can be dispensed with in some bacteria (Benoit & Maier, 2014), but in *E. coli* TatB was found to be essential for the transport of all substrates tested (Palmer & Berks, 2012). It is thought that TatB could cooperate with TatC to modulate signal peptide binding as well as regulate the recruitment of TatA (and possibly TatE) to the final translocase complex (Blummel, Haag, Eimer, Muller, & Frobel, 2015; Frobel et al., 2012; Lausberg et al., 2012). The TatA protein, and its homologue TatE, are evolutionarily related to TatB, sharing some of its molecular features (Berks et al., 2014). Solution NMR of detergent-solubilised *E. coli* TatA (Fig. 17) has been used to model its structure (Rodriguez et al., 2013; Zhang, Hu, Li, & Jin, 2014). Each *E. coli* TatA monomer comprises only 89 amino acids, but is thought to oligomerise into the primary channel structure required to accommodate very large Tat substrate proteins. Like TatB, the TatA protomers consist of an N-terminal transmembrane helix and a short, cytoplasmic, amphipathic helix linked via a hinge region. Upon recruitment of a Tat substrate to TatBC, it is thought that TatA oligomerises until a transport channel is formed (Alcock et al., 2013; Rose, Frobel, Graumann, & Muller, 2013). The *E. coli* TatE protein is similar to TatA and can functionally replace it (Sargent, Bogsch, et al., 1998) and there is some evidence that TatA and TatE operate in tandem while generating a transport channel (Eimer, Frobel, Blummel, & Muller, 2015).

Translocation of [NiFe]-hydrogenases through the Tat translocase poses several challenges to the cell. One challenge regards the coordination of the different biosynthetic and translocation pathways. This is important because the minimum transported unit is a large subunit/small subunit heterodimer where only one signal peptide is present on the small subunit. As with other Tat-dependent metalloenzymes (Sargent, 2007), it is possible that mechanisms exist to suppress signal peptide activity until all other assembly processes are complete. Indeed, for both Hyd-1 and Hyd-2, regulation of signal peptide activity is crucial for correct assembly of the isoenzymes—however, it is notable different solutions have been found, probably reflecting different evolutionary origins of these isoenzymes.

For *E. coli* Hyd-1, the Tat signal peptide itself seems to contain an intrinsic regulatory activity (Bowman, Palmer, & Sargent, 2013). The Hyd-1 Tat signal peptide has an extended *n*-region preceding the twin-arginine motif and genetic removal of this region of sequence completely disrupts assembly of the enzyme (Bowman et al., 2013). Moreover, when the Hyd-1 signal peptide is isolated and placed on a reporter protein, its transport activity is relatively slow and only when the extended *n*-region is removed does the signal peptide operate efficiently (Bowman et al., 2013). It is possible that the Hyd-1 signal peptide *n*-region regulates transport activity and allows the enzyme to be fully assembled before transport. Interestingly, the sequence of the Hyd-1 signal peptide, including the *n*-region, are conserved in other O_2-tolerant [NiFe]-hydrogenases, including the MBH from *R. eutropha*. Genetic deletion of the *n*-region of the MBH Tat signal peptide also disrupted assembly of the enzyme, however in this case there is evidence that the Tat signal peptide *n*-region may be a binding site for specific chaperones (Schubert, Lenz, Krause, Volkmer, & Friedrich, 2007).

The *E. coli* Hyd-2, small subunit HybO has a Tat signal peptide that does not share extensive sequence identity with those of Hyd-1 or other O_2-tolerant [NiFe]-hydrogenases. Regulation of Hyd-2 signal peptide activity is therefore by a different route. In this case, a soluble, cytoplasmic, specific chaperone protein (HybE) has been identified. Using an in vivo bacterial two-hybrid assay, the HybE protein was found to bind directly to the full-length HybO precursor protein and apparently did not interact with the HybO mature protein (Dubini & Sargent, 2003). The phenotype of a Δ*hybE* mutant is particularly revealing. In this mutant, the Hyd-2 small subunit HybO was translocated to the membrane in the absence of the large subunit HybC (Jack et al., 2004). In other words, the coordination of cofactor insertion and protein transport pathways is lost in the Δ*hybE* background and this led to the hypothesis that some chaperones act as 'Tat proofreading' proteins, in a sense double-checking the assembly state of metalloenzymes before Tat transport occurs (Jack et al., 2004). Signal-peptide-swapping experiments with Hyd-2 helped to identify other Tat proofreading chaperones. Replacement of the natural HybO Tat signal peptide with that from the *E. coli* TMAO reductase TorA inhibited Hyd-2 assembly, presumably because HybE could no longer carry out its function efficiently (Jack et al., 2004). However, coexpression with TorD, a protein that interacts with the TorA Tat signal peptide, was able to rescue Hyd-2 biosynthesis (Jack et al., 2004). This was an important observation because it demonstrated that Tat proofreading was independent of cofactor type—the TorD chaperone normally helps to maturate a protein that contains a molybdenum cofactor and clearly this is absent from Hyd-2.

Biosynthesis of Hyd-2 is further complicated because the *hybOABCDEFG* operon encodes two proteins with Tat signal peptides—the HybOC core hydrogenase and the HybA protein. Affinity purification studies have shown that the three proteins interact (Lukey et al., 2010), and this raises questions around whether the complex will form before or after Tat transport. It would seem logical to suggest that HybOC would be transported separately from HybA and the final complex would assembled at the periplasmic side of the membrane. However, work on the *Salmonella enterica* tetrathionate reductase complex suggests an alternative hypothesis. The *S. enterica* tetrathionate reductase is encoded by the *ttrBCA* operon and although TtrB and TtrA form a complex both contain Tat signal peptides (James, Coulthurst, Palmer, & Sargent, 2013). Moreover, when one tetrathionate reductase Tat signal peptide is inactivated the other can compensate (James et al., 2013). This at least raises the possibility for *E. coli* Hyd-2 that a HybAOC trimer, containing two Tat different signal peptides, could form prior to translocation.

The question of whether multisignal complexes are common Tat substrates is an intriguing one and even has implications for the biosynthesis of *E. coli* Hyd-1. The crystal structures (Volbeda et al., 2012, 2013), and original biochemical characterisation for that matter (Sawers & Boxer, 1986), of *E. coli* Hyd-1 show the enzyme as a 'dimer-of dimers', that is the core hydrogenase of HyaAB is duplicated to give a stable $(HyaAB)_2$ structure. The two core hydrogenase units interact extensively across all four subunits in the structure and the orientation places the distal [Fe–S] clusters closer than 12 Å apart and the small subunit N-termini (signal peptide cleavage sites) at opposite sides of the complex. It is conceivable that a single Tat signal peptide in this tetramer will interact with the Tat translocase during transport, however, given both signals are cleanly removed by the LepB signal peptidase after transport, it is more likely that both signal peptides are active during translocation.

Following translocation on the Tat pathway, the biosynthesis of Hyd-1 and Hyd-2 is still not yet complete. These [NiFe]-hydrogenases are somewhat unusual for Tat substrates in that they are bona fide integral membrane proteins. The C-termini of HyaA, HybA, and HybO all contain transmembrane domains that serve to anchor the enzymes close to the periplasmic side of the inner membrane, but also act as interaction interfaces, either with HyaC in the case of Hyd-1 or between HybO and HybA in Hyd-2 and presumably between HybAOC and HybB as well. Exactly, how the Tat translocase handles transmembrane domains is not fully understood. Certainly, proteins that are both Sec- and Tat-dependent because they contain transmembrane domains and extracytoplasmic cofactor-containing domains have been characterised (Keller, de Keyzer, Driessen, & Palmer, 2012), but Hyd-1 and Hyd-2 do not fall into that category. Instead, integration of the C-terminal transmembrane domains of HyaA, HybO, and HybA is completely Tat-dependent, so much so that the transmembrane domains can be attached to other, normally soluble, Tat substrates and so render them membrane bound (Hatzixanthis et al., 2003). Likewise, genetic removal of the transmembrane domain from either Hyd-1 (Evans et al., 2016) or Hyd-2 (Hatzixanthis et al., 2003) does not affect assembly of the core hydrogenase but instead results in periplasmic water-soluble enzymes.

Overall, it is clear that the Tat translocase plays a critical role in the biosynthesis of Hyd-1 and Hyd-2 and would also pose an obvious bottleneck if these, or similar, proteins were to be grossly overproduced in *E. coli*. Indeed, part of the historical difficulty in heterologous expression of nonnative [NiFe]-hydrogenases in *E. coli* may stem from problems in the *E. coli* Tat translocase recognising alien twin-arginine signal peptides.

6.5 Accessory Proteins Specific to Hyd-1

The *hyaABCDEF* operon encodes three proteins that are not part of the final Hyd-1 enzyme—HyaD, HyaE, and HyaF. The *hyaD* gene product is closely related to the structurally defined HybD protein (Fritsche, Paschos, Beisel, Böck, & Huber, 1999), and as such is predicted to be a metal-dependent endopeptidase required for the final step in [NiFe] cofactor insertion (Böck et al., 2006). Although the phenotype of a clean Δ*hyaD* strain has not been studied in detail, a plasmid carrying a Δ*hyaD* version of the *hya* operon has been described (Menon, Robbins, Wendt,

Shanmugam, & Przybyla, 1991). In that work, the Hyd-1 large subunit HyaB remained largely unprocessed in the absence of HyaD (Menon et al., 1991).

HyaE-like proteins are commonly encoded within operons that encode O_2-tolerant, Tat-dependent enzymes that follow the *E. coli* Hyd-1/*R. eutropha* MBH paradigm. *E. coli* HyaE is a soluble, cytoplasmic protein that has had its structure examined by NMR techniques (Parish et al., 2008). The protein clearly adopts a classical thioredoxin fold (Fig. 18), but this was a surprise because *E. coli* HyaE contains precisely zero cysteine residues and so cannot be functioning as a bona fide thioredoxin (Parish et al., 2008). Instead, the surface loop where a functional thioredoxin would display a redox-active CxxC motif has instead a conserved negatively charged sequence. However, defining a phenotype for an *E. coli* Δ*hyaE* mutant, which would be a first step in allowing dissection of HyaE function, has not been straightforward. When grown anaerobically in rich media, at least, Hyd-1 appears to be assembled correctly in the membrane and the core hydrogenase retains enzymatic activity with redox dyes (Dubini et al., 2002). This at least provides evidence that HyaE is not strictly required for [NiFe] active site assembly under anaerobic conditions, and genetic two-hybrid studies suggest that HyaE interacts with the small subunit, HyaA (Dubini & Sargent, 2003). This points to a role for HyaE in small subunit maturation. Indeed, this hypothesis is corroborated by studies of the HyaE homologues from *R. eutropha* and *Rhizobium leguminosarum*. The *Ralsonia*

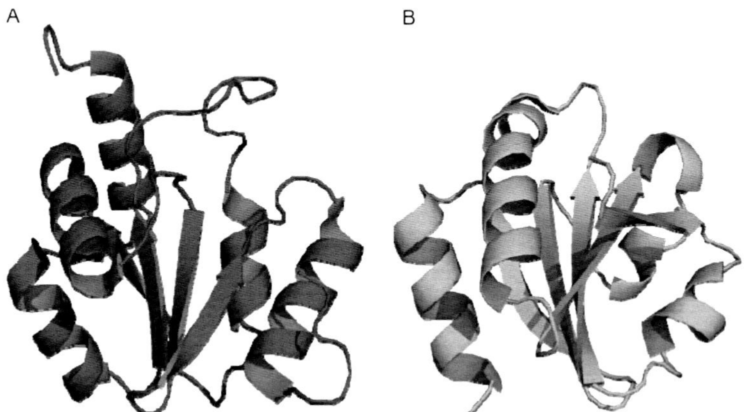

Fig. 18 Accessory proteins for Hyd-1 biosynthesis. (A) The solution NMR structure of the *E. coli* HyaE protein (PDB file 2HFD). (B) For comparison, the crystal structure of the *E. coli* thioredoxin-1 protein is shown (PDB file 2TRX) (Katti, LeMaster, & Eklund, 1990). (See the colour plate.)

gene, *hoxO*, is essential for MBH assembly and HoxO interacts exclusively with the small subunit, most likely via the signal peptide (Schubert et al., 2007). Moreover, the *Rh. leguminosarum* gene, *hupG*, is essential for maturation of the Tat-dependent hydrogenase in that organism, but only under aerobic conditions (Manyani, Rey, Palacios, Imperial, & Ruiz-Argueso, 2005). Although there may be other reasons for this observation, this may point to HoxQ/HyaE as having a role in assembly of the special [4Fe–3S] proximal cluster of the small subunit under aerobic conditions. However, this is a hypothesis that would be a challenge to test in *E. coli*, since the *hya* operon is maximally expressed only under anaerobic conditions. It should be considered, however, that as *E. coli* transitions from anaerobic to microaerobic, or fully aerobic, conditions that the activity of O_2-tolerant Hyd-1 coupled to the *cbdABX* high affinity terminal oxidase electron transport chain may be an important feature of energy metabolism. Under these conditions, any residual de novo Hyd-1 synthesis that occurs before *hya* transcription is completely shut down would have to take place in the presence of O_2, and so perhaps here the function of HyaE becomes critical. A role for Hyd-1 at this anaerobic/aerobic interface, where anaerobically conditioned *E. coli* would encounter some transient and variable amounts of O_2 before perhaps switching back again to fully anaerobic, perhaps accounts for some features of the *hya* operon that are *not* present. For example, the *E. coli hyaABCDEF* operon lacks genes that would encode equivalents of either *Ralstonia* HoxR or HoxV. HoxR is a rubredoxin-type protein and HoxV is a scaffold that allows assembly of the [NiFe] cofactor in an environment protected from O_2 attack, and both of these are essential for MBH assembly under aerobic conditions (Fritsch, Lenz, & Friedrich, 2011; Ludwig et al., 2009). Homologues are encoded in organisms that possess aerobically expressed [NiFe]-hydrogenases, such as *Rh. leguminosarum* (Albareda et al., 2014) and *S. enterica* (Parkin et al., 2012), although in *S. enterica* HoxR is found as a fusion with a HyaF-like protein. This combined genetic evidence further points towards Hyd-1 as a predominantly anaerobically assembled enzyme designed to withstand transient exposure to oxygen.

The *E. coli* HyaF protein, although conserved in many biological systems and sometimes fused to HoxR rubredoxin domains, is also expendable for Hyd-1 biosynthesis in an anaerobic laboratory setting (Dubini et al., 2002; Menon et al., 1991). A crystal structure of a homologue of HyaF has been solved and this sheds absolutely no light on its physiological function (Parkin & Sargent, 2012). The *Ralsonia* homologue of HyaF is HoxQ, and this behaves in a similar way to HoxO in that organism, forming a complex with the MBH small subunit via the Tat signal peptide (Schubert et al., 2007).

6.6 Accessory Proteins Specific to Hyd-2

The *hybOABCDEFG* operon for Hyd-2 encodes four structural proteins (HybOABC), two proteins with homologues encoded in the *hypABCDE* operon—HybF, which is a homologue of HypA (Blokesch, Rohrmoser, et al., 2004; Hube et al., 2002) and HybG, which is a homologue of HypC (Blokesch et al., 2001)—and two proteins, HybD and HybE, that are specific for Hyd-2 maturation. The presence of the HybD protein is essential for C-terminal proteolytic processing of the HybC large subunit and is most probably a metal-dependent endopeptidase (Fritsche et al., 1999). The crystal structure of *E. coli* HybD revealed a Glu-Asp-His metal-binding triad that was ligated to cadmium (Fig. 19), presumably derived from the crystallisation liquor (Fritsche et al., 1999). Although the position of the cleavage site at the C-terminus of the Hyd-2 large subunit HybC is known, it is not clear how HybD recognises its substrate or determines if the [NiFe] cofactor is loaded. Although it could be assumed that the HybC C-terminal extension itself would be the primary recognition site, it has been shown that substitution of the *E. coli* Hyd-1 HyaA C-terminal peptide onto mature

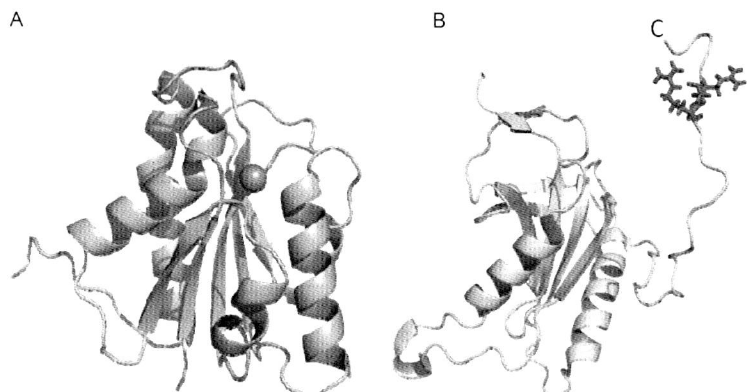

Fig. 19 Accessory proteins for Hyd-2 biosynthesis. (A) The crystal structure of the *E. coli* HybD protein with a cadmium ion locating the metal binding site (PDB file 1CFZ). This protein is a putative endopeptidase specifically required for the C-terminal proteolytic processing of HybC, the Hyd-2 large subunit, after [NiFe] cofactor biosynthesis. (B) The solution NMR structure of the *E. coli* HybE protein (PDB file 2KC5). The protein has a flexible C-terminus with a conserved Arg–Arg motif highlighted in *red*. (See the colour plate.)

HybC had no deleterious effects on Hyd-2 assembly (Thomas, Muhr, & Sawers, 2015). The protease must, therefore, recognise a conformation of the mature domain that is induced by the presence of a C-terminal assembly peptide (Thomas et al., 2015).

While the role of HybD is dedicated to the assembly of *E. coli* HybC large subunit, the *E. coli* HybE protein appears to have a role primarily focused on maturation of the small subunit, HybO. Genetic two-hybrid analysis suggested that *E. coli* HybE interacts with the full-length precursor form of HybO (Dubini & Sargent, 2003) and analysis of Hyd-2 assembly in a Δ*hybE* strain suggests the chaperone is required for coordination of large and small subunit maturation (Jack et al., 2004). Indeed, the phenotype of an *E. coli* Δ*hybE* mutant is quite distinctive and so deserves further comment here. Analysis of the Hyd-2 large subunit revealed that this protein had been successfully C-terminally processed in the Δ*hybE* strain, which helped establish that HybE was not involved in cofactor insertion or maturation of the large subunit (Jack et al., 2004). It was notable, however, that the Hyd-2 small subunit was disconnected from the large subunit in the Δ*hybE* background, being targeted to the membrane alone leaving the otherwise processed large subunit bereft in the cell cytoplasm (Jack et al., 2004). One interpretation of this data was that Tat transport of the small subunit had been 'decoupled' from the normal assembly processes of the Hyd-2 heterodimer, since normally only a fully formed HybOC complex would be allowed to engage with the Tat translocon. This led to the hypothesis that HybE was a 'Tat proofreading' chaperone that exerted a type of quality control check on the Hyd-2 assembly process, and when this protein was absent the Hyd-2 subunits were assembled as separate units without coordination (Jack et al., 2004).

Although the *E. coli* HybE protein appears to be not required for assembly of *E. coli* Hyd-1, since the entire *hyb* operon can be deleted without obvious effects on Hyd-1 activity (Laurinavichene & Tsygankov, 2001; Laurinavichene et al., 2002), homologues of HybE are present in other bacteria—and in bacteria that produce Hyd-1-like O_2-tolerant [NiFe]-hydrogenases rather than Hyd-2-like isoenzymes. HoxT from *R. eutropha*, for example, is involved in the biosynthesis of the Hyd-1-like O_2-tolerant MBH from that organism and interacts with the enzyme at a very late stage of biosynthesis, particularly under high O_2 levels (Fritsch, Lenz, et al., 2011). In Rhizobia, the HybE/HoxT homologue is called HupJ. Studies in *Rh. leguminosarum* corroborate, in part, the other research in this area in that it is clear that HupJ is not required for maturation of the

O_2-tolerant hydrogenase large subunit but instead dedicated to biosynthesis of the Tat-dependent small subunit (Manyani et al., 2005).

The biochemical and physiological functions of HybE remain to be described in detail. Its role is probably not dedicated to assembly of O_2-tolerant hydrogenases, since it is needed for Hyd-2 maturation in *E. coli* but not Hyd-1 maturation. Moreover, its role cannot overlap with that of HyaE- or HyaF-like proteins since homologues of all three are required for biosynthesis of a fully active MBH in *Ralstonia*, for example (Burgdorf et al., 2005). A solution NMR structure model of *E. coli* HybE is available (Fig. 19) and revealed a protein fold that was unlike other any other structurally characterised Tat proofreading chaperones such as those of the TorD/DmsD or NapD families (Shao, Lu, Hu, Xia, & Jin, 2009; Shao, Lu, Xia, & Jin, 2009). Although the HybE protein can bind to the full-length precursor of HybO—and not to the mature form (Dubini & Sargent, 2003)—direct biophysical or structural evidence of signal peptide binding has never been reported. However, the *E. coli* HybE protein does have a hydrophobic cleft on its surface that may accommodate a signal peptide (Shao, Lu, Hu, et al., 2009), and a conserved Arg-Arg motif in its flexible C-terminus that could be hypothesised to operate as a Tat-translocase 'decoy' to prevent premature export of the immature complex.

6.7 Accessory Proteins Specific to Hyd-3

The *hycABCDEFGHI* operon encodes three proteins that are not part of the purified, active, enzyme. These are a transcriptional repressor (HycA), the poorly characterised HycH protein, and the maturation protease, HycI.

The discovery of *E. coli* HycI was a pivotal moment in the understanding of [NiFe]-hydrogenase biosynthesis. Initially, because of significant overlap between *hycH* and *hycI* genes in the operon, HycI was missed and the sequence encoding Hyd-3 was reported as *hycABCDEFGH* (Bohm et al., 1990). In the original work, a deletion in *hycH* (unbeknownst at the time actually a *hycHI* double mutant) was devoid of Hyd-3 activity and produced an unprocessed HycE precursor (Sauter et al., 1992). Classical biochemical and molecular genetic approaches identified that a protease responsible for processing and activation of Hyd-3 was encoded at the extreme 3′ end of the *hyc* operon (Rossmann et al., 1995; Rossmann, Sauter, Lottspeich, & Böck, 1994). Revised clean deletions in *hycH* and *hycI* established that HycI was essential for Hyd-3 activity while HycH was not (Binder, Maier, & Böck, 1996).

In the absence of HycI, the Hyd-3 large subunit (HycE) is already fully loaded with its [NiFe] cofactor but is not proteolytically processed (Theodoratou, Paschos, Magalon, et al., 2000), thus clearly demonstrating a role for HycI at the very conclusion of the assembly of the catalytic subunit. By analogy with Hyd-2, the fact that this form of the Hyd-3 large subunit remains inactive probably reflects the inability of this extended form of the protein to interact with the small subunit, HycG (Thomas et al., 2015). Structures of the *E. coli* HycI protein reveal a metal-dependent endopeptidase with an apparent catalytic metallocentre akin to the HybD protein (Kumarevel, Tanaka, Bessho, Shinkai, & Yokoyama, 2009; Yang et al., 2007). Unlike the crystal structure of HybD, which contains cadmium (Fritsche et al., 1999), the crystal form of HycI (Fig. 20) contains calcium in the putative active site pocket (Kumarevel et al., 2009). It should be noted, however, that the true catalytic metal has yet to be unequivocally identified. Although perhaps most likely zinc for such a cytoplasmic enzyme (Fritsche et al., 1999), the apo-form of purified HycI will bind Ni(II) in vitro (Yang et al., 2007), and it is tempting to speculate that the requirement for a nickel-dependent protease for [NiFe]-hydrogenase assembly would add some poetic synergy to this biological system.

Establishing the physiological role of HycH has proven to be a challenge, simply because a strain carrying a clean deletion in this gene retains a high proportion of Hyd-3 activity (Rossmann et al., 1995). However, conditions

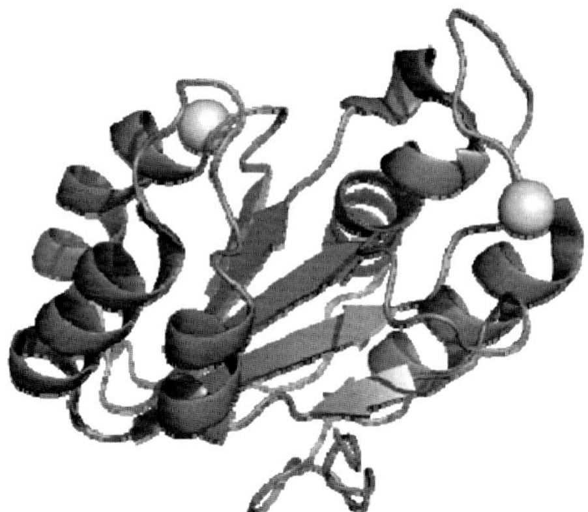

Fig. 20 Accessory protein for Hyd-3 biosynthesis. The crystal structure of the *E. coli* HycI protein (PDB file 2E85). This protein is a putative metal-dependent endopeptidase specifically required for the C-terminal proteolytic processing of HycE, the Hyd-3 large subunit, after [NiFe] cofactor biosynthesis. This structure contains two calcium atoms per protomer and these are coloured in *wheat*. (See the colour plate.)

must exist, most likely in the polymicrobial natural environment rather than a laboratory setting, where the *hycH* gene product becomes critically important, otherwise it would not have been so highly conserved in FHL-encoding gene clusters. Indeed, HycH proteins are not normally found encoded within operons for Tat-dependent [NiFe]-hydrogenases. The *E. coli* HycH protein comprises only 136 amino acid residues and contains intriguing, conserved, twin-Tyr and twin-His motifs, which may point to some role in metal coordination. Sequence analysis of this type of protein provides very little clues to function, however, and structural modelling is similarly unable to shed light on HycH. However, a recent possible hint to the function of HycH was uncovered when the *E. coli* HycE protein was purified from a Δ*hycG* strain, which would lack the small subunit of Hyd-3 (McDowall et al., 2015). In this experiment, HycH was identified copurifying with mature HycE in the absence of HycG (McDowall et al., 2015). This gives the first indication that HycH operates after HycI-dependent processing of HycE and suggests HycH may shield the fully formed [NiFe] active site until the small subunit, HycG, successfully docks.

6.8 Accessory Proteins Missing for Hyd-4

Sequence analysis of the *hyfABCDEFGHIJRfocB* operon encoding the putative Hyd-4 isoenzyme suggests that it encodes only one accessory protein that would be predicted to be involved in Hyd-4 maturation. This protein is HyfJ, which shares sequence identity with the HycH protein that has a subtle and nonessential role in Hyd-3 maturation (Rossmann et al., 1994). A stark omission from the *E. coli hyf* operon is a gene encoding a specific processing protease for the C-terminal assembly peptide of the HyfG subunit. The absence of such a gene could be considered a major clue that the *E. coli hyf* operon is not only transcriptionally silent (Self et al., 2004) but may also incapable of producing an active [NiFe]-hydrogenase. It should be considered, however, that the Hyd-3 protease HycI, or indeed HyaD or HybD, could be responsible for processing of HyfG. Sequence analysis suggests this may be likely. For example, *P. atrosepticum* contains an expressed operon that encodes a putative Hyd-4 (Babujee et al., 2012). The *P. atrosepticum* operon in question carries all the attributes of a bona fide Hyd-4—namely two extra genes encoding NuoL/M/N homologues, a gene encoding a HyfE homolog, and a HyfG protein that fits all the subtle sequence criteria, including an Arg-Val cleavage site at its C-terminus, rather than the Arg-Met motif seen in Hyd-3. Indeed, the *P. atrosepticum* HyfG shares 78% overall sequence identity (87% similarity) with *E. coli* HyfG. Significantly, however, the *P. atrosepticum hyf* operon contains a gene (*hyfK*) that would encode a maturation protease. Clearly a [NiFe]-hydrogenase protease, the putative protein (HyfK) shares 71% overall sequence identity (81% similarity) with the *E. coli* HycI protein, which is known to be essential for processing of the Hyd-3 catalytic subunit, HycE. Thus, if *E. coli* HyfG requires a HyfK-like protein for processing and activation, then it seems highly likely that *E. coli* HycI could fulfil that role.

7. CONCLUDING REMARKS

E. coli dedicates a large amount of genetic material to anaerobic hydrogen metabolism. Furthermore, the chemical energy required to build and activate [NiFe]-hydrogenases is extensive, with nickel uptake and cofactor biosynthesis alone requiring ATP and GTP. For this level of investment, the ultimate payback must be considerable and it is now clear that Hyd-1 and Hyd-2 activity will generate a transmembrane electrochemical gradient that will be transduced for ATP synthesis. The possibility also remains that Hyd-3 and Hyd-4 could similarly contribute directly to bacterial energy metabolism. Even though hydrogen metabolism would appear to be a non-essential activity under laboratory conditions, the animal gut is a rich and diverse microbial environment where molecular H_2 would be competed for along with other energy sources, including formate.

Through the great advances made in understanding the roles, structure, mechanism, and biosynthesis of [NiFe]-hydrogenases, it is becoming increasingly clear that the distinctly different [NiFe]-hydrogenase isoenzymes in *E. coli* contribute in distinctly different ways to the anaerobic physiology of the organism. However, there remains much more to uncover: the physiological role of Hyd-1, for example, is still not satisfactorily understood; the structure and function of Hyd-2, especially how it couples H_2 oxidation to proton pumping, deserves closer attention; and understanding the molecular structures and mechanisms of FHL-1 and FHL-2 promises to reveal insights into energy metabolism, life on Earth, and inspire ideas on how to design new fuel sources. New advances in structural biology, especially cryoelectron microscopy, will be critical in advancing knowledge for all these areas. Furthermore, *E. coli* will remain one of the paradigm organisms for understanding all aspects of [NiFe]-hydrogenase biosynthesis, and be the first-choice 'chassis' for testing innovative synthetic biology approaches to enhancing microbial hydrogen metabolism.

ACKNOWLEDGEMENTS

Hydrogenase research in Dundee is sponsored by the Biotechnology & Biological Sciences Research Council through responsive-mode research awards, the EASTBIO Doctoral Training Partnership, and the C1net Network in Industrial Biotechnology & Bioenergy, as well as Tenovus Scotland and the Marie-Sklodowska-Curie Individual Fellowships scheme.

REFERENCES

Albareda, M., Pacios, L. F., Manyani, H., Rey, L., Brito, B., Imperial, J., et al. (2014). Maturation of *Rhizobium leguminosarum* hydrogenase in the presence of oxygen requires the interaction of the chaperone HypC and the scaffolding protein HupK. *The Journal of Biological Chemistry*, *289*, 21217–21229.

Alcock, F., Baker, M. A., Greene, N. P., Palmer, T., Wallace, M. I., & Berks, B. C. (2013). Live cell imaging shows reversible assembly of the TatA component of the twin-arginine protein transport system. *Proceedings of the National Academy of Sciences of the United States of America*, *110*, E3650–E3659.

Andrews, S. C., Berks, B. C., McClay, J., Ambler, A., Quail, M. A., Golby, P., et al. (1997). A 12-cistron *Escherichia coli* operon (*hyf*) encoding a putative proton-translocating formate hydrogenlyase system. *Microbiology*, *143*, 3633–3647.

Arnoux, P., Ruppelt, C., Oudouhou, F., Lavergne, J., Siponen, M. I., Toci, R., et al. (2015). Sulphur shuttling across a chaperone during molybdenum cofactor maturation. *Nature Communications*, *6*, 6148.

Atlung, T., Knudsen, K., Heerfordt, L., & Brondsted, L. (1997). Effects of sigma^{S} and the transcriptional activator AppY on induction of the *Escherichia coli hya* and *cbdAB-appA* operons in response to carbon and phosphate starvation. *Journal of Bacteriology*, *179*, 2141–2146.

Baba, T., Ara, T., Hasegawa, M., Takai, Y., Okumura, Y., Baba, M., et al. (2006). Construction of *Escherichia coli* K-12 in-frame, single-gene knockout mutants: The Keio collection. *Molecular Systems Biology*, *2*(2006), 0008.

Babujee, L., Apodaca, J., Balakrishnan, V., Liss, P., Kiley, P. J., Charkowski, A. O., et al. (2012). Evolution of the metabolic and regulatory networks associated with oxygen availability in two phytopathogenic enterobacteria. *BMC Genomics*, *13*, 110.

Bagramyan, K., Vassilian, A., Mnatsakanyan, N., & Trchounian, A. (2001). Participation of *hyf*-encoded hydrogenase 4 in molecular hydrogen release coupled with proton-potassium exchange in *Escherichia coli*. *Membrane & Cell Biology*, *14*, 749–763.

Bai, L., Wu, X., Jiang, L., Liu, J., & Long, M. (2012). Hydrogen production by over-expression of hydrogenase subunit in oxygen-tolerant *Klebsiella oxytoca* HP1. *International Journal of Hydrogen Energy*, *37*, 13227–13233.

Ballantine, S. P., & Boxer, D. H. (1985). Nickel-containing hydrogenase isoenzymes from anaerobically grown *Escherichia coli* K-12. *Journal of Bacteriology*, *163*, 454–459.

Ballantine, S. P., & Boxer, D. H. (1986). Isolation and characterisation of a soluble active fragment of hydrogenase isoenzyme 2 from the membranes of anaerobically grown *Escherichia coli*. *European Journal of Biochemistry*, *156*, 277–284.

Bassegoda, A., Madden, C., Wakerley, D. W., Reisner, E., & Hirst, J. (2014). Reversible interconversion of CO_2 and formate by a molybdenum-containing formate dehydrogenase. *Journal of the American Chemical Society*, *136*, 15473–15476.

Batista, A. P., Marreiros, B. C., & Pereira, M. M. (2013). The antiporter-like subunit constituent of the universal adaptor of complex I, group 4 membrane-bound [NiFe]-hydrogenases and related complexes. *Biological Chemistry*, *394*, 659–666.

Benemann, J. (1996). Hydrogen biotechnology: Progress and prospects. *Nature Biotechnology, 14*, 1101–1103.

Benoit, S. L., & Maier, R. J. (2014). Twin-arginine translocation system in *Helicobacter pylori*: TatC, but not TatB, is essential for viability. *mBio, 5*, e01016-13.

Berks, B. C. (1996). A common export pathway for proteins binding complex redox cofactors? *Molecular Microbiology, 22*, 393–404.

Berks, B. C., Lea, S. M., & Stansfeld, P. J. (2014). Structural biology of Tat protein transport. *Current Opinion in Structural Biology, 27*, 32–37.

Bernhard, M., Schwartz, E., Rietdorf, J., & Friedrich, B. (1996). The *Alcaligenes eutrophus* membrane-bound hydrogenase gene locus encodes functions involved in maturation and electron transport coupling. *Journal of Bacteriology, 178*, 4522–4529.

Bhavsar, R. D., Prasad, S., & Roy, I. (2013). Effect of osmolytes on the fibrillation of HypF-N. *Biochimie, 95*, 2190–2193.

Binder, U., Maier, T., & Böck, A. (1996). Nickel incorporation into hydrogenase 3 from *Escherichia coli* requires the precursor form of the large subunit. *Archives of Microbiology, 165*, 69–72.

Bisaillon, A., Turcot, J., & Hallenbeck, P. C. (2006). The effect of nutrient limitation on hydrogen production by batch cultures of *Escherichia coli*. *International Journal of Hydrogen Energy, 31*, 1504–1508.

Blanc, B., Gerez, C., & Ollagnier de Choudens, S. (2015). Assembly of Fe/S proteins in bacterial systems: Biochemistry of the bacterial ISC system. *Biochimica et Biophysica Acta, 1853*, 1436–1447.

Blbulyan, S., & Trchounian, A. (2015). Impact of membrane-associated hydrogenases on the F_1F_o-ATPase in *Escherichia coli* during glycerol and mixed carbon fermentation: ATPase activity and its inhibition by N,N′-dicyclohexylcarbodiimide in the mutants lacking hydrogenases. *Archives of Biochemistry and Biophysics, 579*, 67–72.

Blokesch, M., Albracht, S. P., Matzanke, B. F., Drapal, N. M., Jacobi, A., & Böck, A. (2004). The complex between hydrogenase-maturation proteins HypC and HypD is an intermediate in the supply of cyanide to the active site iron of [NiFe]-hydrogenases. *Journal of Molecular Biology, 344*, 155–167.

Blokesch, M., & Böck, A. (2006). Properties of the [NiFe]-hydrogenase maturation protein HypD. *FEBS Letters, 580*, 4065–4068.

Blokesch, M., Magalon, A., & Böck, A. (2001). Interplay between the specific chaperone-like proteins HybG and HypC in maturation of hydrogenases 1, 2, and 3 from *Escherichia coli*. *Journal of Bacteriology, 183*, 2817–2822.

Blokesch, M., Rohrmoser, M., Rode, S., & Böck, A. (2004). HybF, a zinc-containing protein involved in NiFe hydrogenase maturation. *Journal of Bacteriology, 186*, 2603–2611.

Blummel, A. S., Haag, L. A., Eimer, E., Muller, M., & Frobel, J. (2015). Initial assembly steps of a translocase for folded proteins. *Nature Communications, 6*, 7234.

Böck, A., Forchhammer, K., Heider, J., Leinfelder, W., Sawers, G., Veprek, B., et al. (1991). Selenocysteine: The 21st amino acid. *Molecular Microbiology, 5*, 515–520.

Böck, A., King, P. W., Blokesch, M., & Posewitz, M. C. (2006). Maturation of hydrogenases. *Advances in Microbial Physiology, 51*, 1–71.

Bogsch, E. G., Sargent, F., Stanley, N. R., Berks, B. C., Robinson, C., & Palmer, T. (1998). An essential component of a novel bacterial protein export system with homologues in plastids and mitochondria. *The Journal of Biological Chemistry, 273*, 18003–18006.

Bohm, R., Sauter, M., & Böck, A. (1990). Nucleotide sequence and expression of an operon in *Escherichia coli* coding for formate hydrogenlyase components. *Molecular Microbiology, 4*, 231–243.

Bowman, L., Flanagan, L., Fyfe, P. K., Parkin, A., Hunter, W. N., & Sargent, F. (2014). How the structure of the large subunit controls function in an oxygen-tolerant [NiFe]-hydrogenase. *The Biochemical Journal, 458*, 449–458.

Bowman, L., Palmer, T., & Sargent, F. (2013). A regulatory domain controls the transport activity of a twin-arginine signal peptide. *FEBS Letters*, *587*, 3365–3370.

Boyington, J. C., Gladyshev, V. N., Khangulov, S. V., Stadtman, T. C., & Sun, P. D. (1997). Crystal structure of formate dehydrogenase H: Catalysis involving Mo, molybdopterin, selenocysteine, and an Fe_4S_4 cluster. *Science*, *275*, 1305–1308.

Brondsted, L., & Atlung, T. (1994). Anaerobic regulation of the hydrogenase 1 (*hya*) operon of *Escherichia coli*. *Journal of Bacteriology*, *176*, 5423–5428.

Brondsted, L., & Atlung, T. (1996). Effect of growth conditions on expression of the acid phosphatase (*cyx-appA*) operon and the *appY* gene, which encodes a transcriptional activator of *Escherichia coli*. *Journal of Bacteriology*, *178*, 1556–1564.

Burgdorf, T., Lenz, O., Buhrke, T., van der Linden, E., Jones, A. K., Albracht, S. P., et al. (2005). [NiFe]-hydrogenases of *Ralstonia eutropha* H16: Modular enzymes for oxygen-tolerant biological hydrogen oxidation. *Journal of Molecular Microbiology and Biotechnology*, *10*, 181–196.

Burstel, I., Siebert, E., Winter, G., Hummel, P., Zebger, I., Friedrich, B., et al. (2012). A universal scaffold for synthesis of the $Fe(CN)_2(CO)$ moiety of [NiFe] hydrogenase. *The Journal of Biological Chemistry*, *287*, 38845–38853.

Campioni, S., Mannini, B., Lopez-Alonso, J. P., Shalova, I. N., Penco, A., Mulvihill, E., et al. (2012). Salt anions promote the conversion of HypF-N into amyloid-like oligomers and modulate the structure of the oligomers and the monomeric precursor state. *Journal of Molecular Biology*, *424*, 132–149.

Cecchini, G., Schroder, I., Gunsalus, R. P., & Maklashina, E. (2002). Succinate dehydrogenase and fumarate reductase from *Escherichia coli*. *Biochimica et Biophysica Acta*, *1553*, 140–157.

Chanal, A., Santini, C., & Wu, L. (1998). Potential receptor function of three homologous components, TatA, TatB and TatE, of the twin-arginine signal sequence-dependent metalloenzyme translocation pathway in *Escherichia coli*. *Molecular Microbiology*, *30*, 674–676.

Cherrier, M. V., Cavazza, C., Bochot, C., Lemaire, D., & Fontecilla-Camps, J. C. (2008). Structural characterization of a putative endogenous metal chelator in the periplasmic nickel transporter NikA. *Biochemistry*, *47*, 9937–9943.

Cherrier, M. V., Martin, L., Cavazza, C., Jacquamet, L., Lemaire, D., Gaillard, J., et al. (2005). Crystallographic and spectroscopic evidence for high affinity binding of FeEDTA(H_2O)- to the periplasmic nickel transporter NikA. *Journal of the American Chemical Society*, *127*, 10075–10082.

Chivers, P. T., Benanti, E. L., Heil-Chapdelaine, V., Iwig, J. S., & Rowe, J. L. (2012). Identification of Ni-$(L\text{-}His)_2$ as a substrate for NikABCDE-dependent nickel uptake in *Escherichia coli*. *Metallomics*, *4*, 1043–1050.

Chivers, P. T., & Sauer, R. T. (1999). NikR is a ribbon–helix–helix DNA-binding protein. *Protein Science*, *8*, 2494–2500.

Chivers, P. T., & Sauer, R. T. (2002). NikR repressor: High-affinity nickel binding to the C-terminal domain regulates binding to operator DNA. *Chemistry & Biology*, *9*, 1141–1148.

Chung, K. C., Cao, L., Dias, A. V., Pickering, I. J., George, G. N., & Zamble, D. B. (2008). A high-affinity metal-binding peptide from *Escherichia coli* HypB. *Journal of the American Chemical Society*, *130*, 14056–14057.

Cracknell, J. A., Wait, A. F., Lenz, O., Friedrich, B., & Armstrong, F. A. (2009). A kinetic and thermodynamic understanding of O_2 tolerance in [NiFe]-hydrogenases. *Proceedings of the National Academy of Sciences of the United States of America*, *106*, 20681–20686.

Dance, I. (2015). What is the trigger mechanism for the reversal of electron flow in oxygen-tolerant [NiFe] hydrogenases? *Chemical Science*, *6*, 1433–1443.

Davila-Vazquez, G., Arriaga, S., Alatriste-Mondragón, F., León-Rodríguez, A., Rosales-Colunga, L., & Razo-Flores, E. (2008). Fermentative biohydrogen production: Trends and perspectives. *Reviews in Environmental Science and Biotechnology*, *7*, 27–45.

De Lacey, A. L., Fernandez, V. M., Rousset, M., & Cammack, R. (2007). Activation and inactivation of hydrogenase function and the catalytic cycle: Spectroelectrochemical studies. *Chemical Reviews*, *107*, 4304–4330.

DeLacey, A. L., Fernandez, V. M., Rousset, M., Cavazza, C., & Hatchikian, E. C. (2003). Spectroscopic and kinetic characterization of active site mutants of *Desulfovibrio fructosovorans* Ni-Fe hydrogenase. *Journal of Biological Inorganic Chemistry*, *8*, 129–134.

DeLisa, M. P., Lee, P., Palmer, T., & Georgiou, G. (2004). Phage shock protein PspA of *Escherichia coli* relieves saturation of protein export via the Tat pathway. *Journal of Bacteriology*, *186*, 366–373.

de Pina, K., Navarro, C., McWalter, L., Boxer, D. H., Price, N. C., Kelly, S. M., et al. (1995). Purification and characterization of the periplasmic nickel-binding protein NikA of *Escherichia coli* K12. *European Journal of Biochemistry*, *227*, 857–865.

Dharmadi, Y., Murarka, A., & Gonzalez, R. (2006). Anaerobic fermentation of glycerol by *Escherichia coli*: A new platform for metabolic engineering. *Biotechnology and Bioengineering*, *94*, 821–829.

Dias, A. V., Mulvihill, C. M., Leach, M. R., Pickering, I. J., George, G. N., & Zamble, D. B. (2008). Structural and biological analysis of the metal sites of *Escherichia coli* hydrogenase accessory protein HypB. *Biochemistry*, *47*, 11981–11991.

Doberenz, C., Zorn, M., Falke, D., Nannemann, D., Hunger, D., Beyer, L., et al. (2014). Pyruvate formate-lyase interacts directly with the formate channel FocA to regulate formate translocation. *Journal of Molecular Biology*, *426*, 2827–2839.

Dolla, A., Pohorelic, B. K., Voordouw, J. K., & Voordouw, G. (2000). Deletion of the *hmc* operon of *Desulfovibrio vulgaris* subsp. Hildenborough hampers hydrogen metabolism and low-redox-potential niche establishment. *Archives of Microbiology*, *174*, 143–151.

Douglas, C. D., Ngu, T. T., Kaluarachchi, H., & Zamble, D. B. (2013). Metal transfer within the *Escherichia coli* HypB-HypA complex of hydrogenase accessory proteins. *Biochemistry*, *52*, 6030–6039.

Drapal, N., & Böck, A. (1998). Interaction of the hydrogenase accessory protein HypC with HycE, the large subunit of *Escherichia coli* hydrogenase 3 during enzyme maturation. *Biochemistry*, *37*, 2941–2948.

Du, W., & Liu, D. H. (2012). Biodiesel from conventional feedstocks. *Advances in Biochemical Engineering/Biotechnology*, *128*, 53–68.

Dubini, A., Pye, R. L., Jack, R. L., Palmer, T., & Sargent, F. (2002). How bacteria get energy from hydrogen: A genetic analysis of periplasmic hydrogen oxidation in *Escherichia coli*. *International Journal of Hydrogen Energy*, *27*, 1413–1420.

Dubini, A., & Sargent, F. (2003). Assembly of Tat-dependent [NiFe] hydrogenases: Identification of precursor-binding accessory proteins. *FEBS Letters*, *549*, 141–146.

Efremov, R. G., & Sazanov, L. A. (2011). Respiratory complex I: 'steam engine' of the cell? *Current Opinion in Structural Biology*, *21*, 532–540.

Efremov, R. G., & Sazanov, L. A. (2012). The coupling mechanism of respiratory complex I—A structural and evolutionary perspective. *Biochimica et Biophysica Acta*, *1817*, 1785–1795.

Eimer, E., Frobel, J., Blummel, A. S., & Muller, M. (2015). TatE as a regular constituent of bacterial twin-arginine protein translocases. *The Journal of Biological Chemistry*, *290*(49), 29281–29289.

Elvekrog, M. M., & Walter, P. (2015). Dynamics of co-translational protein targeting. *Current Opinion in Chemical Biology*, *29*, 79–86.

Enoch, H. G., & Lester, R. L. (1975). The purification and properties of formate dehydrogenase and nitrate reductase from *Escherichia coli*. *The Journal of Biological Chemistry*, *250*, 6693–6705.

Euro, L., Belevich, G., Verkhovsky, M. I., Wikstrom, M., & Verkhovskaya, M. (2008). Conserved lysine residues of the membrane subunit NuoM are involved in energy conversion by the proton-pumping NADH: Ubiquinone oxidoreductase (Complex I). *Biochimica et Biophysica Acta*, *1777*, 1166–1172.

Evans, R. M., Brooke, E. J., Wehlin, S. A., Nomerotskaia, E., Sargent, F., Carr, S. B., et al. (2016). Mechanism of hydrogen activation by [NiFe] hydrogenases. *Nature Chemical Biology*, *12*(1), 46–50.

Evans, R. M., Parkin, A., Roessler, M. M., Murphy, B. J., Adamson, H., Lukey, M. J., et al. (2013). Principles of sustained enzymatic hydrogen oxidation in the presence of oxygen—The crucial influence of high potential Fe–S clusters in the electron relay of [NiFe]-hydrogenases. *Journal of the American Chemical Society*, *135*, 2694–2707.

Falke, D., Schulz, K., Doberenz, C., Beyer, L., Lilie, H., Thiemer, B., et al. (2010). Unexpected oligomeric structure of the FocA formate channel of *Escherichia coli*: A paradigm for the formate-nitrite transporter family of integral membrane proteins. *FEMS Microbiology Letters*, *303*, 69–75.

Feng, Z., Hou, T., & Li, Y. (2012). Concerted movement in pH-dependent gating of FocA from molecular dynamics simulations. *Journal of Chemical Information and Modeling*, *52*, 2119–2131.

Filenko, N., Spiro, S., Browning, D. F., Squire, D., Overton, T. W., Cole, J., et al. (2007). The NsrR regulon of *Escherichia coli* K-12 includes genes encoding the hybrid cluster protein and the periplasmic, respiratory nitrite reductase. *Journal of Bacteriology*, *189*, 4410–4417.

Fontecilla-Camps, J. C., Volbeda, A., Cavazza, C., & Nicolet, Y. (2007). Structure/function relationships of [NiFe]- and [FeFe]-hydrogenases. *Chemical Reviews*, *107*, 4273–4303.

Frielingsdorf, S., Fritsch, J., Schmidt, A., Hammer, M., Lowenstein, J., Siebert, E., et al. (2014). Reversible [4Fe–3S] cluster morphing in an O_2-tolerant [NiFe] hydrogenase. *Nature Chemical Biology*, *10*, 378–385.

Frielingsdorf, S., Schubert, T., Pohlmann, A., Lenz, O., & Friedrich, B. (2011). A trimeric supercomplex of the oxygen-tolerant membrane-bound [NiFe]-hydrogenase from *Ralstonia eutropha* H16. *Biochemistry*, *50*, 10836–10843.

Fritsch, J., Lenz, O., & Friedrich, B. (2011). The maturation factors HoxR and HoxT contribute to oxygen tolerance of membrane-bound [NiFe] hydrogenase in *Ralstonia eutropha* H16. *Journal of Bacteriology*, *193*, 2487–2497.

Fritsch, J., Scheerer, P., Frielingsdorf, S., Kroschinsky, S., Friedrich, B., Lenz, O., et al. (2011). The crystal structure of an oxygen-tolerant hydrogenase uncovers a novel iron-sulphur centre. *Nature*, *479*, 249–252.

Fritsche, E., Paschos, A., Beisel, H. G., Böck, A., & Huber, R. (1999). Crystal structure of the hydrogenase maturating endopeptidase HYBD from *Escherichia coli*. *Journal of Molecular Biology*, *288*, 989–998.

Frobel, J., Rose, P., Lausberg, F., Blummel, A. S., Freudl, R., & Muller, M. (2012). Transmembrane insertion of twin-arginine signal peptides is driven by TatC and regulated by TatB. *Nature Communications*, *3*, 1311.

Giel, J. L., Rodionov, D., Liu, M., Blattner, F. R., & Kiley, P. J. (2006). IscR-dependent gene expression links iron–sulphur cluster assembly to the control of O_2-regulated genes in *Escherichia coli*. *Molecular Microbiology*, *60*, 1058–1075.

Gladyshev, V. N., Khangulov, S. V., Axley, M. J., & Stadtman, T. C. (1994). Coordination of selenium to molybdenum in formate dehydrogenase H from *Escherichia coli*. *Proceedings of the National Academy of Sciences of the United States of America*, *91*, 7708–7711.

Gonzalez, R., Murarka, A., Dharmadi, Y., & Yazdani, S. S. (2008). A new model for the anaerobic fermentation of glycerol in enteric bacteria: Trunk and auxiliary pathways in *Escherichia coli*. *Metabolic Engineering*, *10*, 234–245.

Goris, T., Wait, A. F., Saggu, M., Fritsch, J., Heidary, N., Stein, M., et al. (2011). A unique iron–sulfur cluster is crucial for oxygen tolerance of a [NiFe]-hydrogenase. *Nature Chemical Biology*, *7*, 310–318.

Greening, C., Biswas, A., Carere, C. R., Jackson, C. J., Taylor, M. C., Stott, M. B., et al. (2016). Genomic and metagenomic surveys of hydrogenase distribution indicate H_2 is a widely utilised energy source for microbial growth and survival. *The ISME Journal*, *10*, 761–777.

Guillen Schlippe, Y. V., & Hedstrom, L. (2005). A twisted base? The role of arginine in enzyme-catalyzed proton abstractions. *Archives of Biochemistry and Biophysics*, *433*, 266–278.

Hatzixanthis, K., Palmer, T., & Sargent, F. (2003). A subset of bacterial inner membrane proteins integrated by the twin-arginine translocase. *Molecular Microbiology*, *49*, 1377–1390.

Heddle, J., Scott, D. J., Unzai, S., Park, S. Y., & Tame, J. R. (2003). Crystal structures of the liganded and unliganded nickel-binding protein NikA from *Escherichia coli*. *The Journal of Biological Chemistry*, *278*, 50322–50329.

Heinemann, I. U., Jahn, M., & Jahn, D. (2008). The biochemistry of heme biosynthesis. *Archives of Biochemistry and Biophysics*, *474*, 238–251.

Hille, R., Hall, J., & Basu, P. (2014). The mononuclear molybdenum enzymes. *Chemical Reviews*, *114*, 3963–4038.

Hirst, J. (2013). Mitochondrial complex I. *Annual Review of Biochemistry*, *82*, 551–575.

Hu, H., & Wood, T. K. (2010). An evolved *Escherichia coli* strain for producing hydrogen and ethanol from glycerol. *Biochemical and Biophysical Research Communications*, *391*, 1033–1038.

Huang, G. F., Wu, X. B., Bai, L. P., Liu, K., Jiang, L. J., Long, M. N., et al. (2015). Improved O_2-tolerance in variants of a H_2-evolving [NiFe]-hydrogenase from *Klebsiella oxytoca* HP1. *FEBS Letters*, *589*, 910–918.

Hube, M., Blokesch, M., & Böck, A. (2002). Network of hydrogenase maturation in *Escherichia coli*: Role of accessory proteins HypA and HybF. *Journal of Bacteriology*, *184*, 3879–3885.

Hunger, D., Doberenz, C., & Sawers, R. G. (2014). Identification of key residues in the formate channel FocA that control import and export of formate. *Biological Chemistry*, *395*, 813–825.

Jack, R. L., Buchanan, G., Dubini, A., Hatzixanthis, K., Palmer, T., & Sargent, F. (2004). Coordinating assembly and export of complex bacterial proteins. *The EMBO Journal*, *23*, 3962–3972.

James, M. J., Coulthurst, S. J., Palmer, T., & Sargent, F. (2013). Signal peptide etiquette during assembly of a complex respiratory enzyme. *Molecular Microbiology*, *90*, 400–414.

Jo, B. H., & Cha, H. J. (2015). Activation of formate hydrogen-lyase via expression of uptake [NiFe]-hydrogenase in *Escherichia coli* BL21(DE3). *Microbial Cell Factories*, *14*, 151.

Jormakka, M., Tornroth, S., Byrne, B., & Iwata, S. (2002). Molecular basis of proton motive force generation: Structure of formate dehydrogenase-N. *Science*, *295*, 1863–1868.

Jormakka, M., Yokoyama, K., Yano, T., Tamakoshi, M., Akimoto, S., Shimamura, T., et al. (2008). Molecular mechanism of energy conservation in polysulfide respiration. *Nature Structural & Molecular Biology*, *15*, 730–737.

Kaluarachchi, H., Zhang, J. W., & Zamble, D. B. (2011). *Escherichia coli* SlyD, more than a Ni(II) reservoir. *Biochemistry*, *50*, 10761–10763.

Karube, I., Tomiyama, M., & Kikuchi, A. (1984). Molecular-cloning and physical mapping of the *hyd* gene of *Escherichia coli* K-12. *FEMS Microbiology Letters*, *25*, 165–168.

Kato, J., & Hashimoto, M. (2007). Construction of consecutive deletions of the *Escherichia coli* chromosome. *Molecular Systems Biology, 3*, 132.

Katti, S. K., LeMaster, D. M., & Eklund, H. (1990). Crystal structure of thioredoxin from *Escherichia coli* at 1.68 A resolution. *Journal of Molecular Biology, 212*, 167–184.

Keller, R., de Keyzer, J., Driessen, A. J., & Palmer, T. (2012). Co-operation between different targeting pathways during integration of a membrane protein. *The Journal of Cell Biology, 199*, 303–315.

Khangulov, S. V., Gladyshev, V. N., Dismukes, G. C., & Stadtman, T. C. (1998). Selenium-containing formate dehydrogenase H from *Escherichia coli*: A molybdopterin enzyme that catalyzes formate oxidation without oxygen transfer. *Biochemistry, 37*, 3518–3528.

Kilty, J. E., & Amara, S. G. (1992). Families of twelve transmembrane domain transporters. *Current Opinion in Biotechnology, 3*, 675–682.

Kim, J. Y., Jo, B. H., & Cha, H. J. (2010). Production of biohydrogen by recombinant expression of [NiFe]-hydrogenase 1 in *Escherichia coli*. *Microbial Cell Factories, 9*, 54.

Kim, J. Y., Jo, B. H., & Cha, H. J. (2011). Production of biohydrogen by heterologous expression of oxygen-tolerant *Hydrogenovibrio marinus* [NiFe]-hydrogenase in *Escherichia coli*. *Journal of Biotechnology, 155*, 312–319.

King, P. W., & Przybyla, A. E. (1999). Response of *hya* expression to external pH in *Escherichia coli*. *Journal of Bacteriology, 181*, 5250–5256.

Kuchenreuther, J. M., Myers, W. K., Suess, D. L., Stich, T. A., Pelmenschikov, V., Shiigi, S. A., et al. (2014). The HydG enzyme generates an $Fe(CO)_2(CN)$ synthon in assembly of the FeFe hydrogenase H-cluster. *Science, 343*, 424–427.

Kumarevel, T., Tanaka, T., Bessho, Y., Shinkai, A., & Yokoyama, S. (2009). Crystal structure of hydrogenase maturating endopeptidase HycI from *Escherichia coli*. *Biochemical and Biophysical Research Communications, 389*, 310–314.

Kuniyoshi, T. M., Balan, A., Schenberg, A. C., Severino, D., & Hallenbeck, P. C. (2015). Heterologous expression of proteorhodopsin enhances H_2 production in *Escherichia coli* when endogenous Hyd-4 is overexpressed. *Journal of Biotechnology, 206*, 52–57.

Kurata, T., Katayama, A., Hiramatsu, M., Kiguchi, Y., Takeuchi, M., Watanabe, T., et al. (2013). Identification of the set of genes, including nonannotated *morA*, under the direct control of ModE in *Escherichia coli*. *Journal of Bacteriology, 195*, 4496–4505.

Labunskyy, V. M., Hatfield, D. L., & Gladyshev, V. N. (2014). Selenoproteins: Molecular pathways and physiological roles. *Physics Review, 94*, 739–777.

Laurinavichene, T. V., & Tsygankov, A. A. (2001). H_2 consumption by *Escherichia coli* coupled via hydrogenase 1 or hydrogenase 2 to different terminal electron acceptors. *FEMS Microbiology Letters, 202*, 121–124.

Laurinavichene, T. V., Zorin, N. A., & Tsygankov, A. A. (2002). Effect of redox potential on activity of hydrogenase 1 and hydrogenase 2 in *Escherichia coli*. *Archives of Microbiology, 178*, 437–442.

Lausberg, F., Fleckenstein, S., Kreutzenbeck, P., Frobel, J., Rose, P., Muller, M., et al. (2012). Genetic evidence for a tight cooperation of TatB and TatC during productive recognition of twin-arginine (Tat) signal peptides in *Escherichia coli*. *PLoS ONE, 7*, e39867.

Lauterbach, L., & Lenz, O. (2013). Catalytic production of hydrogen peroxide and water by oxygen-tolerant [NiFe]-hydrogenase during H_2 cycling in the presence of O_2. *Journal of the American Chemical Society, 135*, 17897–17905.

Layer, G., Reichelt, J., Jahn, D., & Heinz, D. W. (2010). Structure and function of enzymes in heme biosynthesis. *Protein Science, 19*, 1137–1161.

Lazarus, O., Woolerton, T. W., Parkin, A., Lukey, M. J., Reisner, E., Seravalli, J., et al. (2009). Water-gas shift reaction catalyzed by redox enzymes on conducting graphite platelets. *Journal of the American Chemical Society, 131*, 14154–14155.

Leach, M. R., Sandal, S., Sun, H., & Zamble, D. B. (2005). Metal binding activity of the *Escherichia coli* hydrogenase maturation factor HypB. *Biochemistry, 44*, 12229–12238.

Lebrette, H., Brochier-Armanet, C., Zambelli, B., de Reuse, H., Borezee-Durant, E., Ciurli, S., et al. (2014). Promiscuous nickel import in human pathogens: Structure, thermodynamics, and evolution of extracytoplasmic nickel-binding proteins. *Structure*, *22*, 1421–1432.

Lebrette, H., Iannello, M., Fontecilla-Camps, J. C., & Cavazza, C. (2013). The binding mode of Ni-(L-His)$_2$ in NikA revealed by X-ray crystallography. *Journal of Inorganic Biochemistry*, *121*, 16–18.

Lim, J. K., Mayer, F., Kang, S. G., & Muller, V. (2014). Energy conservation by oxidation of formate to carbon dioxide and hydrogen via a sodium ion current in a hyperthermophilic archaeon. *Proceedings of the National Academy of Sciences of the United States of America*, *111*, 11497–11502.

Lu, W., Du, J., Schwarzer, N. J., Gerbig-Smentek, E., Einsle, O., & Andrade, S. L. (2012). The formate channel FocA exports the products of mixed-acid fermentation. *Proceedings of the National Academy of Sciences of the United States of America*, *109*, 13254–13259.

Lu, W., Du, J., Wacker, T., Gerbig-Smentek, E., Andrade, S. L., & Einsle, O. (2011). pH-dependent gating in a FocA formate channel. *Science*, *332*, 352–354.

Lubitz, W., Ogata, H., Rudiger, O., & Reijerse, E. (2014). Hydrogenases. *Chemical Reviews*, *114*, 4081–4148.

Ludwig, M., Schubert, T., Zebger, I., Wisitruangsakul, N., Saggu, M., Strack, A., et al. (2009). Concerted action of two novel auxiliary proteins in assembly of the active site in a membrane-bound [NiFe] hydrogenase. *The Journal of Biological Chemistry*, *284*, 2159–2168.

Lukey, M. J., Parkin, A., Roessler, M. M., Murphy, B. J., Harmer, J., Palmer, T., et al. (2010). How *Escherichia coli* is equipped to oxidize hydrogen under different redox conditions. *The Journal of Biological Chemistry*, *285*, 3928–3938.

Lukey, M. J., Roessler, M. M., Parkin, A., Evans, R. M., Davies, R. A., Lenz, O., et al. (2011). Oxygen-tolerant [NiFe]-hydrogenases: The individual and collective importance of supernumerary cysteines at the proximal Fe–S cluster. *Journal of the American Chemical Society*, *133*, 16881–16892.

Lutz, S., Jacobi, A., Schlensog, V., Bohm, R., Sawers, G., & Böck, A. (1991). Molecular characterization of an operon (*hyp*) necessary for the activity of the three hydrogenase isoenzymes in *Escherichia coli*. *Molecular Microbiology*, *5*, 123–135.

Lv, X., Liu, H., Ke, M., & Gong, H. (2013). Exploring the pH-dependent substrate transport mechanism of FocA using molecular dynamics simulation. *Biophysical Journal*, *105*, 2714–2723.

Maeda, T., Sanchez-Torres, V., & Wood, T. K. (2007a). Enhanced hydrogen production from glucose by metabolically engineered *Escherichia coli*. *Applied Microbiology and Biotechnology*, *77*, 879–890.

Maeda, T., Sanchez-Torres, V., & Wood, T. K. (2007b). *Escherichia coli* hydrogenase 3 is a reversible enzyme possessing hydrogen uptake and synthesis activities. *Applied Microbiology and Biotechnology*, *76*, 1035–1042.

Maeda, T., Sanchez-Torres, V., & Wood, T. K. (2008a). Metabolic engineering to enhance bacterial hydrogen production. *Microbial Biotechnology*, *1*, 30–39.

Maeda, T., Sanchez-Torres, V., & Wood, T. K. (2008b). Protein engineering of hydrogenase 3 to enhance hydrogen production. *Applied Microbiology and Biotechnology*, *79*, 77–86.

Magherini, F., Pieri, L., Guidi, F., Giangrande, C., Amoresano, A., Bucciantini, M., et al. (2009). Proteomic analysis of cells exposed to prefibrillar aggregates of HypF-N. *Biochimica et Biophysica Acta*, *1794*, 1243–1250.

Maier, T., Binder, U., & Böck, A. (1996). Analysis of the *hydA* locus of *Escherichia coli*: Two genes (*hydN* and *hypF*) involved in formate and hydrogen metabolism. *Archives of Microbiology*, *165*, 333–341.

Maier, T., & Böck, A. (1996). Generation of active [NiFe] hydrogenase in vitro from a nickel-free precursor form. *Biochemistry*, *35*, 10089–10093.

Maier, T., Jacobi, A., Sauter, M., & Böck, A. (1993). The product of the *hypB* gene, which is required for nickel incorporation into hydrogenases, is a novel guanine nucleotide-binding protein. *Journal of Bacteriology*, *175*, 630–635.

Maier, T., Lottspeich, F., & Böck, A. (1995). GTP hydrolysis by HypB is essential for nickel insertion into hydrogenases of *Escherichia coli*. *European Journal of Biochemistry*, *230*, 133–138.

Manyani, H., Rey, L., Palacios, J. M., Imperial, J., & Ruiz-Argueso, T. (2005). Gene products of the *hupGHIJ* operon are involved in maturation of the iron-sulfur subunit of the [NiFe] hydrogenase from *Rhizobium leguminosarum* bv. viciae. *Journal of Bacteriology*, *187*, 7018–7026.

Marreiros, B. C., Batista, A. P., Duarte, A. M., & Pereira, M. M. (2013). A missing link between complex I and group 4 membrane-bound [NiFe] hydrogenases. *Biochimica et Biophysica Acta*, *1827*, 198–209.

Masuda, N., & Church, G. M. (2003). Regulatory network of acid resistance genes in *Escherichia coli*. *Molecular Microbiology*, *48*, 699–712.

McDowall, J. S., Hjersing, M. C., Palmer, T., & Sargent, F. (2015). Dissection and engineering of the *Escherichia coli* formate hydrogenlyase complex. *FEBS Letters*, *589*, 3141–3147.

McDowall, J. S., Murphy, B. J., Haumann, M., Palmer, T., Armstrong, F. A., & Sargent, F. (2014). Bacterial formate hydrogenlyase complex. *Proceedings of the National Academy Sciences of the United States of America*, *111*, E3948–E3956.

Mejean, V., Iobbi-Nivol, C., Lepelletier, M., Giordano, G., Chippaux, M., & Pascal, M. C. (1994). TMAO anaerobic respiration in *Escherichia coli*: Involvement of the *tor* operon. *Molecular Microbiology*, *11*, 1169–1179.

Mendel, R. R., & Leimkuhler, S. (2015). The biosynthesis of the molybdenum cofactors. *Journal of Biological Inorganic Chemistry*, *20*, 337–347.

Menon, N. K., Chatelus, C. Y., Dervartanian, M., Wendt, J. C., Shanmugam, K. T., Peck, H. D., Jr., et al. (1994). Cloning, sequencing, and mutational analysis of the *hyb* operon encoding *Escherichia coli* hydrogenase 2. *Journal of Bacteriology*, *176*, 4416–4423.

Menon, N. K., Robbins, J., Peck, H. D., Jr., Chatelus, C. Y., Choi, E. S., & Przybyla, A. E. (1990). Cloning and sequencing of a putative *Escherichia coli* [NiFe] hydrogenase-1 operon containing six open reading frames. *Journal of Bacteriology*, *172*, 1969–1977.

Menon, N. K., Robbins, J., Wendt, J. C., Shanmugam, K. T., & Przybyla, A. E. (1991). Mutational analysis and characterization of the *Escherichia coli hya* operon, which encodes [NiFe] hydrogenase 1. *Journal of Bacteriology*, *173*, 4851–4861.

Messenger, S. L., & Green, J. (2003). FNR-mediated regulation of *hyp* expression in *Escherichia coli*. *FEMS Microbiology Letters*, *228*, 81–86.

Minnan, L., Jinli, H., Xiaobin, W., Huijuan, X., Jinzao, C., Chuannan, L., et al. (2005). Isolation and characterization of a high H_2-producing strain *Klebsiella oxytoca* HP1 from a hot spring. *Research in Microbiology*, *156*, 76–81.

Moparthi, V. K., & Hagerhall, C. (2011). The evolution of respiratory chain complex I from a smaller last common ancestor consisting of 11 protein subunits. *Journal of Molecular Evolution*, *72*, 484–497.

Murphy, B. J., Sargent, F., & Armstrong, F. A. (2014). Transforming an oxygen-tolerant [NiFe] uptake hydrogenase into a proficient, reversible hydrogen producer. *Energy & Environmental Science*, *7*, 1426–1433.

Nesbit, A. D., Giel, J. L., Rose, J. C., & Kiley, P. J. (2009). Sequence-specific binding to a subset of IscR-regulated promoters does not require IscR Fe–S cluster ligation. *Journal of Molecular Biology*, *387*, 28–41.

Nitschke, W., & Russell, M. J. (2009). Hydrothermal focusing of chemical and chemiosmotic energy, supported by delivery of catalytic Fe, Ni, Mo/W, Co, S and Se, forced life to emerge. *Journal of Molecular Evolution*, *69*, 481–496.

Niviere, V., Wong, S. L., & Voordouw, G. (1992). Site-directed mutagenesis of the hydrogenase signal peptide consensus box prevents export of a beta-lactamase fusion protein. *Journal of General Microbiology*, *138*, 2173–2183.

Ogata, H., Nishikawa, K., & Lubitz, W. (2015). Hydrogens detected by subatomic resolution protein crystallography in a [NiFe] hydrogenase. *Nature*, *520*, 571–574.

Orozco, R. L., Redwood, M. D., Yong, P., Caldelari, I., Sargent, F., & Macaskie, L. E. (2010). Towards an integrated system for bio-energy: Hydrogen production by *Escherichia coli* and use of palladium-coated waste cells for electricity generation in a fuel cell. *Biotechnology Letters*, *32*, 1837–1845.

Outten, W. F. (2015). Recent advances in the Suf Fe–S cluster biogenesis pathway: Beyond the proteobacteria. *Biochimica et Biophysica Acta*, *1853*, 1464–1469.

Palmer, T., & Berks, B. C. (2012). The twin-arginine translocation (Tat) protein export pathway. *Nature Reviews Microbiology*, *10*, 483–496.

Pandelia, M. E., Lubitz, W., & Nitschke, W. (2012). Evolution and diversification of Group 1 [NiFe] hydrogenases. Is there a phylogenetic marker for O_2-tolerance? *Biochimica et Biophysica Acta*, *1817*, 1565–1575.

Parish, D., Benach, J., Liu, G., Singarapu, K. K., Xiao, R., Acton, T., et al. (2008). Protein chaperones Q8ZP25_SALTY from *Salmonella typhimurium* and HYAE_ECOLI from *Escherichia coli* exhibit thioredoxin-like structures despite lack of canonical thioredoxin active site sequence motif. *Journal of Structural and Functional Genomics*, *9*, 41–49.

Parkin, A., Bowman, L., Roessler, M. M., Davies, R. A., Palmer, T., Armstrong, F. A., et al. (2012). How *Salmonella* oxidises H_2 under aerobic conditions. *FEBS Letters*, *586*, 536–544.

Parkin, A., & Sargent, F. (2012). The hows and whys of aerobic H_2 metabolism. *Current Opinion in Chemical Biology*, *16*, 26–34.

Penfold, D. W., Sargent, F., & Macaskie, L. E. (2006). Inactivation of the *Escherichia coli* K-12 twin-arginine translocation system promotes increased hydrogen production. *FEMS Microbiology Letters*, *262*, 135–137.

Peters, J. W., Schut, G. J., Boyd, E. S., Mulder, D. W., Shepard, E. M., Broderick, J. B., et al. (2014). [FeFe]- and [NiFe]-hydrogenase diversity, mechanism, and maturation. *Biochimica et Biophysica Acta*, *1853*(6), 1350–1369.

Petkun, S., Shi, R., Li, Y., Asinas, A., Munger, C., Zhang, L., et al. (2011). Structure of hydrogenase maturation protein HypF with reaction intermediates shows two active sites. *Structure*, *19*, 1773–1783.

Phillips, C. M., Schreiter, E. R., Stultz, C. M., & Drennan, C. L. (2010). Structural basis of low-affinity nickel binding to the nickel-responsive transcription factor NikR from *Escherichia coli*. *Biochemistry*, *49*, 7830–7838.

Pinske, C., Bonn, M., Kruger, S., Lindenstrauss, U., & Sawers, R. G. (2011). Metabolic deficiences revealed in the biotechnologically important model bacterium *Escherichia coli* BL21(DE3). *PLoS ONE*, *6*, e22830.

Pinske, C., Jaroschinsky, M., Linek, S., Kelly, C. L., Sargent, F., & Sawers, R. G. (2015). Physiology and bioenergetics of [NiFe]-hydrogenase 2-catalyzed H_2-consuming and H_2-producing reactions in *Escherichia coli*. *Journal of Bacteriology*, *197*, 296–306.

Pinske, C., & Sawers, R. G. (2012). Delivery of iron–sulfur clusters to the hydrogen-oxidizing [NiFe]-hydrogenases in *Escherichia coli* requires the A-type carrier proteins ErpA and IscA. *PLoS ONE*, 7, e31755.

Pinske, C., & Sawers, R. G. (2014). The importance of iron in the biosynthesis and assembly of [NiFe]-hydrogenases. *Biomolecular Concepts*, *5*, 55–70.

Potter, L., Angove, H., Richardson, D., & Cole, J. (2001). Nitrate reduction in the periplasm of gram-negative bacteria. *Advances in Microbial Physiology*, *45*, 51–112.

Prickril, B. C., Czechowski, M. H., Przybyla, A. E., Peck, H. D., Jr., & LeGall, J. (1986). Putative signal peptide on the small subunit of the periplasmic hydrogenase from *Desulfovibrio vulgaris*. *Journal of Bacteriology*, *167*, 722–725.

Radu, V., Frielingsdorf, S., Evans, S. D., Lenz, O., & Jeuken, L. J. (2014). Enhanced oxygen-tolerance of the full heterotrimeric membrane-bound [NiFe]-hydrogenase of *Ralstonia eutropha*. *Journal of the American Chemical Society*, *136*, 8512–8515.

Rajesh, S., Heddle, J. G., Kurashima-Ito, K., Nietlispach, D., Shirakawa, M., Tame, J. R., et al. (2005). Backbone ^{1}H, ^{13}C, and ^{15}N assignments of a 56 kDa *E. coli* nickel binding protein NikA. *Journal of Biomolecular NMR*, *32*, 177.

Ramasamy, S., Abrol, R., Suloway, C. J., & Clemons, W. M., Jr. (2013). The glove-like structure of the conserved membrane protein TatC provides insight into signal sequence recognition in twin-arginine translocation. *Structure*, *21*, 777–788.

Rangarajan, E. S., Asinas, A., Proteau, A., Munger, C., Baardsnes, J., Iannuzzi, P., et al. (2008). Structure of [NiFe] hydrogenase maturation protein HypE from *Escherichia coli* and its interaction with HypF. *Journal of Bacteriology*, *190*, 1447–1458.

Reissmann, S., Hochleitner, E., Wang, H., Paschos, A., Lottspeich, F., Glass, R. S., et al. (2003). Taming of a poison: Biosynthesis of the NiFe-hydrogenase cyanide ligands. *Science*, *299*, 1067–1070.

Relini, A., Torrassa, S., Rolandi, R., Gliozzi, A., Rosano, C., Canale, C., et al. (2004). Monitoring the process of HypF fibrillization and liposome permeabilization by protofibrils. *Journal of Molecular Biology*, *338*, 943–957.

Richard, D. J., Sawers, G., Sargent, F., McWalter, L., & Boxer, D. H. (1999). Transcriptional regulation in response to oxygen and nitrate of the operons encoding the [NiFe] hydrogenases 1 and 2 of *Escherichia coli*. *Microbiology*, *145*, 2903–2912.

Roche, B., Aussel, L., Ezraty, B., Mandin, P., Py, B., & Barras, F. (2013). Iron/sulfur proteins biogenesis in prokaryotes: Formation, regulation and diversity. *Biochimica et Biophysica Acta*, *1827*, 455–469.

Rodrigue, A., Boxer, D. H., Mandrand-Berthelot, M. A., & Wu, L. F. (1996). Requirement for nickel of the transmembrane translocation of NiFe-hydrogenase 2 in *Escherichia coli*. *FEBS Letters*, *392*, 81–86.

Rodrigue, A., Chanal, A., Beck, K., Muller, M., & Wu, L. F. (1999). Co-translocation of a periplasmic enzyme complex by a hitchhiker mechanism through the bacterial tat pathway. *The Journal of Biological Chemistry*, *274*, 13223–13228.

Rodrigue, A., Effantin, G., & Mandrand-Berthelot, M. A. (2005). Identification of *rcnA* (*yohM*), a nickel and cobalt resistance gene in *Escherichia coli*. *Journal of Bacteriology*, *187*, 2912–2916.

Rodriguez, F., Rouse, S. L., Tait, C. E., Harmer, J., De Riso, A., Timmel, C. R., et al. (2013). Structural model for the protein-translocating element of the twin-arginine transport system. *Proceedings of the National Academy of Sciences of the United States of America*, *110*, E1092–E1101.

Roessler, M. M., Evans, R. M., Davies, R. A., Harmer, J., & Armstrong, F. A. (2012). EPR spectroscopic studies of the Fe–S clusters in the O_2-tolerant [NiFe]-hydrogenase Hyd-1 from *Escherichia coli* and characterization of the unique [4Fe–3S] cluster by HYSCORE. *Journal of the American Chemical Society*, *134*, 15581–15594.

Rollauer, S. E., Tarry, M. J., Graham, J. E., Jaaskelainen, M., Jager, F., Johnson, S., et al. (2012). Structure of the TatC core of the twin-arginine protein transport system. *Nature*, *492*, 210–214.

Rose, P., Frobel, J., Graumann, P. L., & Muller, M. (2013). Substrate-dependent assembly of the Tat translocase as observed in live *Escherichia coli* cells. *PLoS ONE*, *8*, e69488.

Rossi, M., Pollock, W. B., Reij, M. W., Keon, R. G., Fu, R., & Voordouw, G. (1993). The *hmc* operon of *Desulfovibrio vulgaris* subsp. Hildenborough encodes a potential transmembrane redox protein complex. *Journal of Bacteriology, 175*, 4699–4711.

Rossmann, R., Maier, T., Lottspeich, F., & Böck, A. (1995). Characterisation of a protease from *Escherichia coli* involved in hydrogenase maturation. *European Journal of Biochemistry, 227*, 545–550.

Rossmann, R., Sauter, M., Lottspeich, F., & Böck, A. (1994). Maturation of the large subunit (HYCE) of *Escherichia coli* hydrogenase 3 requires nickel incorporation followed by C-terminal processing at Arg537. *European Journal of Biochemistry, 220*, 377–384.

Rothery, R. A., Blasco, F., Magalon, A., & Weiner, J. H. (2001). The diheme cytochrome b subunit (NarI) of *Escherichia coli* nitrate reductase A (NarGHI): Structure, function, and interaction with quinols. *Journal of Molecular Microbiology and Biotechnology, 3*, 273–283.

Rothery, R. A., Workun, G. J., & Weiner, J. H. (2008). The prokaryotic complex iron-sulfur molybdoenzyme family. *Biochimica et Biophysica Acta, 1778*, 1897–1929.

Sambrook, J., & Russell, D. W. (2001). *Molecular cloning: A laboratory manual*. Cold Spring Harbor, NY: Cold Spring Harbor Laboratory Press.

Sanchez-Torres, V., Maeda, T., & Wood, T. K. (2009). Protein engineering of the transcriptional activator FhlA To enhance hydrogen production in *Escherichia coli*. *Applied and Environmental Microbiology, 75*, 5639–5646.

Sargent, F. (2007). Constructing the wonders of the bacterial world: Biosynthesis of complex enzymes. *Microbiology, 153*, 633–651.

Sargent, F., Ballantine, S. P., Rugman, P. A., Palmer, T., & Boxer, D. H. (1998). Reassignment of the gene encoding the *Escherichia coli* hydrogenase 2 small subunit—identification of a soluble precursor of the small subunit in a *hypB* mutant. *European Journal of Biochemistry, 255*, 746–754.

Sargent, F., Bogsch, E. G., Stanley, N. R., Wexler, M., Robinson, C., Berks, B. C., et al. (1998). Overlapping functions of components of a bacterial Sec-independent protein export pathway. *The EMBO Journal, 17*, 3640–3650.

Sargent, F., Stanley, N. R., Berks, B. C., & Palmer, T. (1999). Sec-independent protein translocation in *Escherichia coli*. A distinct and pivotal role for the TatB protein. *The Journal of Biological Chemistry, 274*, 36073–36082.

Sauter, M., Bohm, R., & Böck, A. (1992). Mutational analysis of the operon (*hyc*) determining hydrogenase 3 formation in *Escherichia coli*. *Molecular Microbiology, 6*, 1523–1532.

Sawers, G. (1994). The hydrogenases and formate dehydrogenases of *Escherichia coli*. *Antonie Van Leeuwenhoek, 66*, 57–88.

Sawers, R. G., Ballantine, S. P., & Boxer, D. H. (1985). Differential expression of hydrogenase isoenzymes in *Escherichia coli* K-12: Evidence for a third isoenzyme. *Journal of Bacteriology, 164*, 1324–1331.

Sawers, R. G., & Boxer, D. H. (1986). Purification and properties of membrane-bound hydrogenase isoenzyme 1 from anaerobically grown *Escherichia coli* K12. *European Journal of Biochemistry, 156*, 265–275.

Sawers, G., & Watson, G. (1998). A glycyl radical solution: Oxygen-dependent interconversion of pyruvate formate-lyase. *Molecular Microbiology, 29*, 945–954.

Schlensog, V., & Böck, A. (1990). Identification and sequence analysis of the gene encoding the transcriptional activator of the formate hydrogenlyase system of *Escherichia coli*. *Molecular Microbiology, 4*, 1319–1327.

Schlensog, V., Lutz, S., & Böck, A. (1994). Purification and DNA-binding properties of FHLA, the transcriptional activator of the formate hydrogenlyase system from *Escherichia coli*. *The Journal of Biological Chemistry, 269*, 19590–19596.

Schreiter, E. R., & Drennan, C. L. (2007). Ribbon-helix-helix transcription factors: Variations on a theme. *Nature Reviews Microbiology, 5*, 710–720.

Schreiter, E. R., Sintchak, M. D., Guo, Y., Chivers, P. T., Sauer, R. T., & Drennan, C. L. (2003). Crystal structure of the nickel-responsive transcription factor NikR. *Nature Structural Biology*, *10*, 794–799.

Schreiter, E. R., Wang, S. C., Zamble, D. B., & Drennan, C. L. (2006). NikR-operator complex structure and the mechanism of repressor activation by metal ions. *Proceedings of the National Academy of Sciences of the United States of America*, *103*, 13676–13681.

Schubert, T., Lenz, O., Krause, E., Volkmer, R., & Friedrich, B. (2007). Chaperones specific for the membrane-bound [NiFe]-hydrogenase interact with the Tat signal peptide of the small subunit precursor in *Ralstonia eutropha* H16. *Molecular Microbiology*, *66*, 453–467.

Self, W. T., Grunden, A. M., Hasona, A., & Shanmugam, K. T. (1999). Transcriptional regulation of molybdoenzyme synthesis in *Escherichia coli* in response to molybdenum: ModE-molybdate, a repressor of the *modABCD* (molybdate transport) operon is a secondary transcriptional activator for the *hyc* and *nar* operons. *Microbiology*, *145*, 41–55.

Self, W. T., Hasona, A., & Shanmugam, K. T. (2004). Expression and regulation of a silent operon, *hyf*, coding for hydrogenase 4 isoenzyme in *Escherichia coli*. *Journal of Bacteriology*, *186*, 580–587.

Shao, X., Lu, J., Hu, Y., Xia, B., & Jin, C. (2009a). Solution structure of the *Escherichia coli* HybE reveals a novel fold. *Proteins*, *75*, 1051–1056.

Shao, X., Lu, J., Xia, B., & Jin, C. (2009b). ^{1}H, ^{13}C and ^{15}N resonance assignments of the chaperone HybE of hydrogenase-2 from *Escherichia coli*. *Biomolecular NMR Assignments*, *3*, 129–131.

Shepherd, M., Heath, M. D., & Poole, R. K. (2007). NikA binds heme: A new role for an *Escherichia coli* periplasmic nickel-binding protein. *Biochemistry*, *46*, 5030–5037.

Shomura, Y., & Higuchi, Y. (2012). Structural basis for the reaction mechanism of S-carbamoylation of HypE by HypF in the maturation of [NiFe]-hydrogenases. *The Journal of Biological Chemistry*, *287*, 28409–28419.

Shomura, Y., Yoon, K. S., Nishihara, H., & Higuchi, Y. (2011). Structural basis for a [4Fe–3S] cluster in the oxygen-tolerant membrane-bound [NiFe]-hydrogenase. *Nature*, *479*, 253–256.

Siebert, E., Rippers, Y., Frielingsdorf, S., Fritsch, J., Schmidt, A., Kalms, J., et al. (2015). Resonance Raman spectroscopic analysis of the [NiFe] active site and the proximal [4Fe–3S] cluster of an O_2-tolerant membrane-bound hydrogenase in the crystalline state. *The Journal of Physical Chemistry B*, *119*, 13785–13796.

Simon, J., & Kern, M. (2008). Quinone-reactive proteins devoid of haem *b* form widespread membrane-bound electron transport modules in bacterial respiration. *Biochemical Society Transactions*, *36*, 1011–1016.

Simon, J., van Spanning, R. J., & Richardson, D. J. (2008). The organisation of proton motive and non-proton motive redox loops in prokaryotic respiratory systems. *Biochimica et Biophysica Acta*, *1777*, 1480–1490.

Skibinski, D. A., Golby, P., Chang, Y. S., Sargent, F., Hoffman, R., Harper, R., et al. (2002). Regulation of the hydrogenase-4 operon of *Escherichia coli* by the sigma(54)-dependent transcriptional activators FhlA and HyfR. *Journal of Bacteriology*, *184*, 6642–6653.

Soboh, B., Lindenstrauss, U., Granich, C., Javed, M., Herzberg, M., Thomas, C., et al. (2014). [NiFe]-hydrogenase maturation in vitro: Analysis of the roles of the HybG and HypD accessory proteins1. *The Biochemical Journal*, *464*, 169–177.

Soboh, B., Stripp, S. T., Bielak, C., Lindenstrauss, U., Braussemann, M., Javaid, M., et al. (2013). The [NiFe]-hydrogenase accessory chaperones HypC and HybG of *Escherichia coli* are iron- and carbon dioxide-binding proteins. *FEBS Letters*, *587*, 2512–2516.

Soboh, B., Stripp, S. T., Muhr, E., Granich, C., Braussemann, M., Herzberg, M., et al. (2012). [NiFe]-hydrogenase maturation: Isolation of a HypC-HypD complex carrying diatomic CO and CN^{-} ligands. *FEBS Letters*, *586*, 3882–3887.

Stephenson, M., & Stickland, L. H. (1933). Hydrogenlyases: Further experiments on the formation of formic hydrogenlyase by *Bact. coli*. *The Biochemical Journal*, *27*, 1528–1532.

Stourman, N. V., Branch, M. C., Schaab, M. R., Harp, J. M., Ladner, J. E., & Armstrong, R. N. (2011). Structure and function of YghU, a nu-class glutathione transferase related to YfcG from *Escherichia coli*. *Biochemistry*, *50*, 1274–1281.

Stripp, S. T., Lindenstrauss, U., Sawers, R. G., & Soboh, B. (2015). Identification of an isothiocyanate on the HypEF complex suggests a route for efficient cyanyl-group channeling during [NiFe]-hydrogenase cofactor generation. *PLoS ONE*, *10*, e0133118.

Stripp, S. T., Soboh, B., Lindenstrauss, U., Braussemann, M., Herzberg, M., Nies, D. H., et al. (2013). HypD is the scaffold protein for Fe-(CN)2CO cofactor assembly in [NiFe]-hydrogenase maturation. *Biochemistry*, *52*, 3289–3296.

Studier, F. W., & Moffatt, B. A. (1986). Use of bacteriophage T7 RNA polymerase to direct selective high-level expression of cloned genes. *Journal of Molecular Biology*, *189*, 113–130.

Szklarczyk, D., Franceschini, A., Wyder, S., Forslund, K., Heller, D., Huerta-Cepas, J., et al. (2015). STRING v10: Protein–protein interaction networks, integrated over the tree of life. *Nucleic Acids Research*, *43*, D447–D452.

Tarry, M. J., Schafer, E., Chen, S., Buchanan, G., Greene, N. P., Lea, S. M., et al. (2009). Structural analysis of substrate binding by the TatBC component of the twin-arginine protein transport system. *Proceedings of the National Academy of Sciences of the United States of America*, *106*, 13284–13289.

Tatini, F., Pugliese, A. M., Traini, C., Niccoli, S., Maraula, G., Ed Dami, T., et al. (2013). Amyloid-beta oligomer synaptotoxicity is mimicked by oligomers of the model protein HypF-N. *Neurobiology of Aging*, *34*, 2100–2109.

Thauer, R. K., Jungermann, K., & Decker, K. (1977). Energy conservation in chemotrophic anaerobic bacteria. *Bacteriological Reviews*, *41*, 100–180.

Theodoratou, E., Paschos, A., Magalon, A., Fritsche, E., Huber, R., & Böck, A. (2000a). Nickel serves as a substrate recognition motif for the endopeptidase involved in hydrogenase maturation. *European Journal of Biochemistry*, *267*, 1995–1999.

Theodoratou, E., Paschos, A., Mintz, W., & Böck, A. (2000b). Analysis of the cleavage site specificity of the endopeptidase involved in the maturation of the large subunit of hydrogenase 3 from *Escherichia coli*. *Archives of Microbiology*, *173*, 110–116.

Thomas, C., Muhr, E., & Sawers, R. G. (2015). Coordination of synthesis and assembly of a modular membrane-associated [NiFe]-hydrogenase is determined by cleavage of the C-terminal peptide. *Journal of Bacteriology*, *197*, 2989–2998.

Thome, R., Gust, A., Toci, R., Mendel, R., Bittner, F., Magalon, A., et al. (2012). A sulfurtransferase is essential for activity of formate dehydrogenases in *Escherichia coli*. *The Journal of Biological Chemistry*, *287*, 4671–4678.

Tipping, M. J., Steel, B. C., Delalez, N. J., Berry, R. M., & Armitage, J. P. (2013). Quantification of flagellar motor stator dynamics through in vivo proton-motive force control. *Molecular Microbiology*, *87*, 338–347.

Tominaga, T., Watanabe, S., Matsumi, R., Atomi, H., Imanaka, T., & Miki, K. (2013). Crystal structures of the carbamoylated and cyanated forms of HypE for [NiFe] hydrogenase maturation. *Proceedings of the National Academy of Sciences of the United States of America*, *110*, 20485–20490.

Tran, K. T., Maeda, T., & Wood, T. K. (2014). Metabolic engineering of *Escherichia coli* to enhance hydrogen production from glycerol. *Applied Microbiology and Biotechnology*, *98*, 4757–4770.

Trchounian, K., Blbulyan, S., & Trchounian, A. (2013). Hydrogenase activity and proton-motive force generation by *Escherichia coli* during glycerol fermentation. *Journal of Bioenergetics and Biomembranes*, *45*, 253–260.

Trchounian, K., Pinske, C., Sawers, R. G., & Trchounian, A. (2011). Dependence on the F_0F_1-ATP synthase for the activities of the hydrogen-oxidizing hydrogenases 1 and 2 during glucose and glycerol fermentation at high and low pH in *Escherichia coli*. *Journal of Bioenergetics and Biomembranes*, *43*, 645–650.

Trchounian, K., Poladyan, A., Vassilian, A., & Trchounian, A. (2012). Multiple and reversible hydrogenases for hydrogen production by *Escherichia coli*: Dependence on fermentation substrate, pH and the F_0F_1-ATPase. *Critical Reviews in Biochemistry and Molecular Biology*, *47*, 236–249.

Trchounian, K., Soboh, B., Sawers, R. G., & Trchounian, A. (2013). Contribution of hydrogenase 2 to stationary phase H_2 production by *Escherichia coli* during fermentation of glycerol. *Cell Biochemistry and Biophysics*, *66*, 103–108.

Turcot, J., Bisaillon, A., & Hallenbeck, P. C. (2008). Hydrogen production by continuous cultures of *Escherchia coli* under different nutrient regimes. *International Journal of Hydrogen Energy*, *33*, 1465–1470.

Valle, A., Cabrera, G., Cantero, D., & Bolivar, J. (2015). Identification of enhanced hydrogen and ethanol *Escherichia coli* producer strains in a glycerol-based medium by screening in single-knock out mutant collections. *Microbial Cell Factories*, *14*, 93.

Vignais, P. M., & Billoud, B. (2007). Occurrence, classification, and biological function of hydrogenases: An overview. *Chemical Reviews*, *107*, 4206–4272.

Vignais, P. M., Billoud, B., & Meyer, J. (2001). Classification and phylogeny of hydrogenases. *FEMS Microbiology Reviews*, *25*, 455–501.

Vivijs, B., Haberbeck, L. U., Baiye Mfortaw Mbong, V., Bernaerts, K., Geeraerd, A. H., Aertsen, A., et al. (2015). Formate hydrogen lyase mediates stationary-phase deacidification and increases survival during sugar fermentation in acetoin-producing enterobacteria. *Frontiers in Microbiology*, *6*, 150.

Volbeda, A., Amara, P., Darnault, C., Mouesca, J. M., Parkin, A., Roessler, M. M., et al. (2012). X-ray crystallographic and computational studies of the O_2-tolerant [NiFe]-hydrogenase 1 from *Escherichia coli*. *Proceedings of the National Academy of Sciences of the United States of America*, *109*, 5305–5310.

Volbeda, A., Charon, M. H., Piras, C., Hatchikian, E. C., Frey, M., & Fontecilla-Camps, J. C. (1995). Crystal structure of the nickel-iron hydrogenase from *Desulfovibrio gigas*. *Nature*, *373*, 580–587.

Volbeda, A., Darnault, C., Parkin, A., Sargent, F., Armstrong, F. A., & Fontecilla-Camps, J. C. (2013). Crystal structure of the O_2-tolerant membrane-bound hydrogenase 1 from *Escherichia coli* in complex with its cognate cytochrome *b*. *Structure*, *21*, 184–190.

Wait, A. F., Parkin, A., Morley, G. M., dos Santos, L., & Armstrong, F. A. (2010). Characteristics of enzyme-based hydrogen fuel cells using an oxygen-tolerant hydrogenase as the anodic catalyst. *Journal of Physical Chemistry C*, *114*, 12003–12009.

Wang, Y., Huang, Y., Wang, J., Cheng, C., Huang, W., Lu, P., et al. (2009). Structure of the formate transporter FocA reveals a pentameric aquaporin-like channel. *Nature*, *462*, 467–472.

Watanabe, S., Kawashima, T., Nishitani, Y., Kanai, T., Wada, T., Inaba, K., et al. (2015). Structural basis of a Ni acquisition cycle for [NiFe] hydrogenase by Ni-metallochaperone HypA and its enhancer. *Proceedings of the National Academy of Sciences of the United States of America*, *112*, 7701–7706.

Watanabe, S., Matsumi, R., Arai, T., Atomi, H., Imanaka, T., & Miki, K. (2007). Crystal structures of [NiFe] hydrogenase maturation proteins HypC, HypD, and HypE: Insights into cyanation reaction by thiol redox signaling. *Molecular Cell*, *27*, 29–40.

Watanabe, S., Matsumi, R., Atomi, H., Imanaka, T., & Miki, K. (2012). Crystal structures of the HypCD complex and the HypCDE ternary complex: Transient intermediate complexes during [NiFe] hydrogenase maturation. *Structure*, *20*, 2124–2137.

Watanabe, S., Sasaki, D., Tominaga, T., & Miki, K. (2012). Structural basis of [NiFe] hydrogenase maturation by Hyp proteins. *Biological Chemistry*, *393*, 1089–1100.

Waugh, R., & Boxer, D. H. (1986). Pleiotropic hydrogenase mutants of *Escherichia coli* K12: Growth in the presence of nickel can restore hydrogenase activity. *Biochimie, 68*, 157–166.

Weiner, J. H., Bilous, P. T., Shaw, G. M., Lubitz, S. P., Frost, L., Thomas, G. H., et al. (1998). A novel and ubiquitous system for membrane targeting and secretion of cofactor-containing proteins. *Cell, 93*, 93–101.

Woods, D. D. (1936). Hydrogenlyases: The synthesis of formic acid by bacteria. *The Biochemical Journal, 30*, 515–527.

Wu, L. F., Chanal, A., & Rodrigue, A. (2000). Membrane targeting and translocation of bacterial hydrogenases. *Archives of Microbiology, 173*, 319–324.

Wu, L. F., & Mandrand, M. A. (1993). Microbial hydrogenases: Primary structure, classification, signatures and phylogeny. *FEMS Microbiology Reviews, 10*, 243–269.

Wu, L. F., & Mandrand-Berthelot, M. A. (1986). Genetic and physiological characterization of new *Escherichia coli* mutants impaired in hydrogenase activity. *Biochimie, 68*, 167–179.

Wu, L. F., Mandrand-Berthelot, M. A., Waugh, R., Edmonds, C. J., Holt, S. E., & Boxer, D. H. (1989). Nickel deficiency gives rise to the defective hydrogenase phenotype of *hydC* and *fnr* mutants in *Escherichia coli*. *Molecular Microbiology, 3*, 1709–1718.

Wu, L. F., Navarro, C., & Mandrand-Berthelot, M. A. (1991). The *hydC* region contains a multi-cistronic operon (*nik*) involved in nickel transport in *Escherichia coli*. *Gene, 107*, 37–42.

Wulff, P., Bielak, C., Sargent, F., & Armstrong, F. A. (2016). How the oxygen tolerance of a [NiFe]-hydrogenase depends on quaternary structure. *Journal of Biological Inorganic Chemistry*, (in press).

Wulff, P., Day, C. C., Sargent, F., & Armstrong, F. A. (2014). How oxygen reacts with oxygen-tolerant respiratory [NiFe]-hydrogenases. *Proceedings of the National Academy of Sciences of the United States of America, 111*(18), 6606–6611.

Yang, F., Hu, W., Xu, H., Li, C., Xia, B., & Jin, C. (2007). Solution structure and backbone dynamics of an endopeptidase HycI from *Escherichia coli*: Implications for mechanism of the [NiFe] hydrogenase maturation. *The Journal of Biological Chemistry, 282*, 3856–3863.

Yoshida, A., Nishimura, T., Kawaguchi, H., Inui, M., & Yukawa, H. (2005). Enhanced hydrogen production from formic acid by formate hydrogen lyase-overexpressing *Escherichia coli* strains. *Applied and Environmental Microbiology, 71*, 6762–6768.

Yoshida, A., Nishimura, T., Kawaguchi, H., Inui, M., & Yukawa, H. (2006). Enhanced hydrogen production from glucose using *ldh*- and *frd*-inactivated *Escherichia coli* strains. *Applied Microbiology and Biotechnology, 73*, 67–72.

Yoshida, A., Nishimura, T., Kawaguchi, H., Inui, M., & Yukawa, H. (2007). Efficient induction of formate hydrogen lyase of aerobically grown *Escherichia coli* in a three-step biohydrogen production process. *Applied Microbiology and Biotechnology, 74*, 754–760.

Zhang, Y., Hu, Y., Li, H., & Jin, C. (2014). Structural basis for TatA oligomerization: An NMR study of *Escherichia coli* TatA dimeric structure. *PLoS ONE, 9*, e103157.

Zhang, Y., Wang, L., Hu, Y., & Jin, C. (2014). Solution structure of the TatB component of the twin-arginine translocation system. *Biochimica et Biophysica Acta, 1838*, 1881–1888.

Zinoni, F., Birkmann, A., Leinfelder, W., & Böck, A. (1987). Cotranslational insertion of selenocysteine into formate dehydrogenase from *Escherichia coli* directed by a UGA codon. *Proceedings of the National Academy of Sciences of the United States of America, 84*, 3156–3160.

Zinoni, F., Birkmann, A., Stadtman, T. C., & Böck, A. (1986). Nucleotide sequence and expression of the selenocysteine-containing polypeptide of formate dehydrogenase (formate-hydrogen-lyase-linked) from *Escherichia coli*. *Proceedings of the National Academy of Sciences of the United States of America, 83*, 4650–4654.

AUTHOR INDEX

Note: Page numbers followed by "*f*" indicate figures, and "*t*" indicate tables.

A

B

C

E

F

L

M

N

O

P

Q

R

S

T

X

Y

Z

SUBJECT INDEX

Note: Page numbers followed by "*f*" indicate figures and "*t*" indicate tables.

C

D

E

F

G

H

I

K

L

M

N

O

P

T

U

W

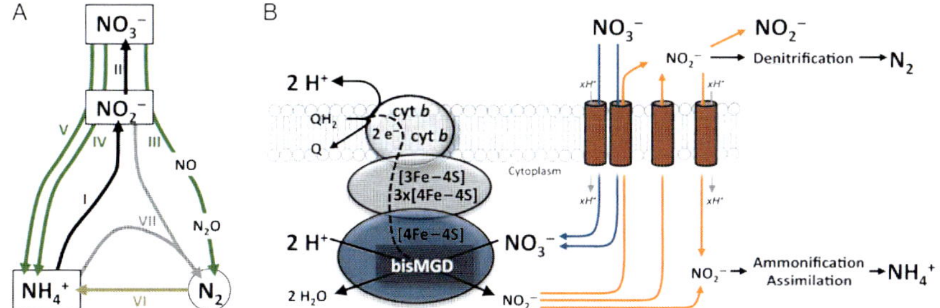

R.G. Sawers *et al.*, Fig. 5 Nitrate reduction and the nitrogen cycle. (A) A simplified scheme of the bacterial nitrogen cycle shows nitrification (I and II), denitrification (III), nitrate/nitrite ammonification (IV), assimilatory nitrate/nitrite reduction (V), nitrogen fixation (VI) and anaerobic ammonia oxidation (anammox; VII). Note that scheme I also includes ammonia-oxidising archaea (AOA—see text for details). (B) Schematic representation of respiratory nitrate reductase including the bis-MGD (bis-molybdenum guanine dinucleotide) cofactor in the catalytic subunit. Nitrate reduction requires nitrate uptake into the cytoplasm, which can occur by nitrite-driven antiport (Zheng, Wisedchaisri, & Gonen, 2013) or by proton-symport (Moir & Wood, 2001). Nitrite extrusion can theoretically be achieved without being coupled to nitrate import. Toxic nitrite can be reduced by ammonification in the cytoplasm or on the periplasmic side of the membrane by denitrification.

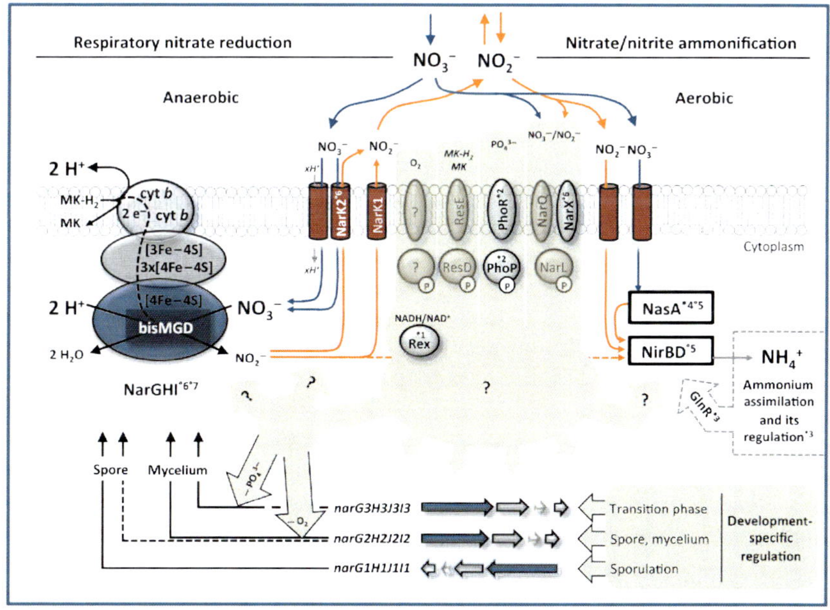

R.G. Sawers *et al.*, Fig. 8 Model of the regulation and function of respiratory and assimilatory nitrate reduction in *S. coelicolor* A3(2). The regulatory features or functions of particular proteins are based on the following papers, which are included within the diagram: *1, Brekasis & Paget, 2003; *2, Sola-Landa, Moura, & Martín, 2003; *3, Fink, Weissschuh, Reuther, Wohlleben, & Engels, 2002; *4, Wang & Zhao, 2009. Those papers, which are based on our work are as follows: °5, Fischer et al., 2012; °6, Fischer et al., 2014; °7, Fischer et al., 2010 and Fischer et al., 2013. Hypothetical regulatory control nodes are indicated by *question marks*.

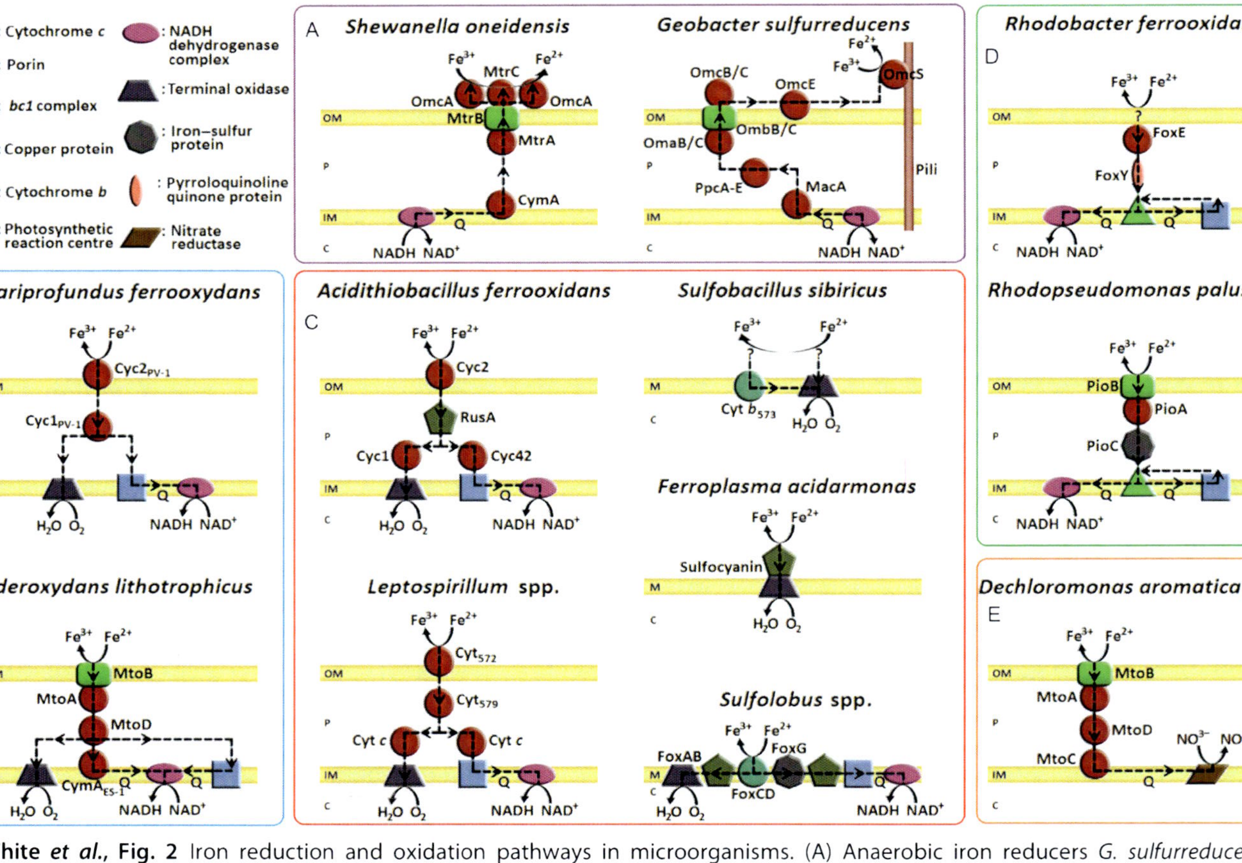

G.F. White *et al.*, Fig. 2 Iron reduction and oxidation pathways in microorganisms. (A) Anaerobic iron reducers *G. sulfurreducens* and *S. oneidensis* (Liu et al., 2014; Lovley, 2006; Shi, Rosso, Zachara, & Fredrickson, 2012). (B) Neutrophilic aerobic iron oxidisers *M. ferrooxydans* and *S. lithotrophicus* (Barco et al., 2015; Beckwith et al., 2015). (C) Acidophilic aerobic iron oxidisers *A. ferrooxidans*, *Leptospirillum* spp., *S. sibiricus*, *F. acidarmonas*, *Sulfolobus* spp. (Ilbert & Bonnefoy, 2013). (D) Neutrophilic photosynthetic iron oxidisers *R. ferrooxidans* and *R. palustris* (Jiao & Newman, 2007; Saraiva, Newman, & Louro, 2012). (E) Neutrophilic anaerobic iron oxidisers dependent

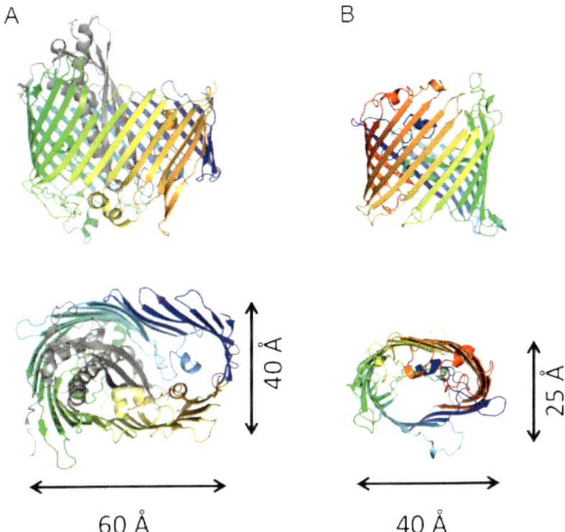

G.F. White *et al.*, Fig. 4 Crystal structures of porins similar in size to Cyc2 and MtrB. (A) Lipopolysaccharide translocon consisting of a 26 β-strand porin LptD and α-helical LptE plug (*light grey*) (PDB ID: 4N4R). (B) 18 β-strand monomer of the trimeric sucrose transporter ScrY (PDB ID: 1A0T).

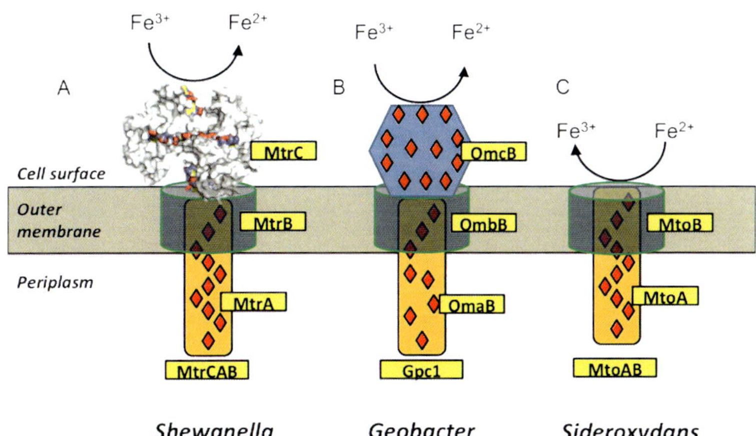

G.F. White *et al.*, Fig. 5 Theoretical structures of known outer membrane porin–cytochromes complexes. (A) MtrCAB complex of *Shewanella oneidensis* consisting of the decahaem MtrA, the 28 β-strand porin MtrB and the decahaem cytochrome MtrC (PDB ID: 4LMB). (B) Gpc1/Gpc2 complex of *Geobacter sulfurreducens* consisting of the octahaem OmaB, 20 β-strand porin OmbB and dodecahaem OmcB. (C) MtoAB complex from *Sideroxydans lithotrophicus ES-1* consisting of the decahaem MtoA and 28 β-strand porin MtoB. *Red diamonds* represent c-type haems.

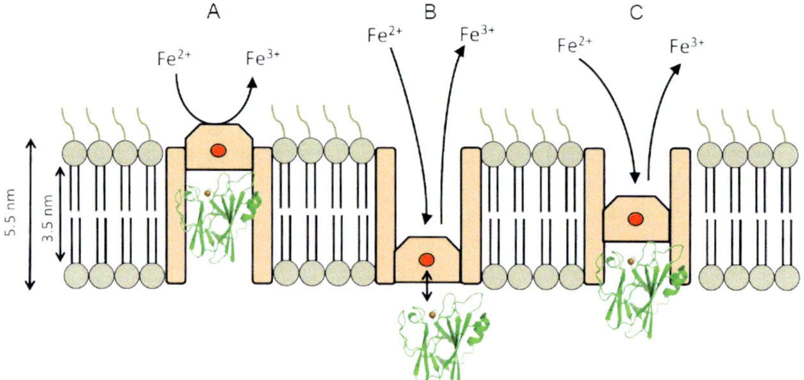

G.F. White *et al.*, Fig. 6 Possible configurations of a Cyc2 fused porin–cytochrome. (A) The haem domain is located on the cell surface, allowing access to Fe^{2+} and insertion of rustacyanin into the barrel. (B) The haem domain located on the periplasmic face of the membrane, allowing reversible association with rustacyanin and Fe^{2+} diffusion through the extracellular-facing barrel entrance. (C) The haem domain is in the centre of the barrel with Fe^{2+} diffusing into the extracellular side and electrons being transferred to a bound rustacyanin at the periplasmic face of the membrane.

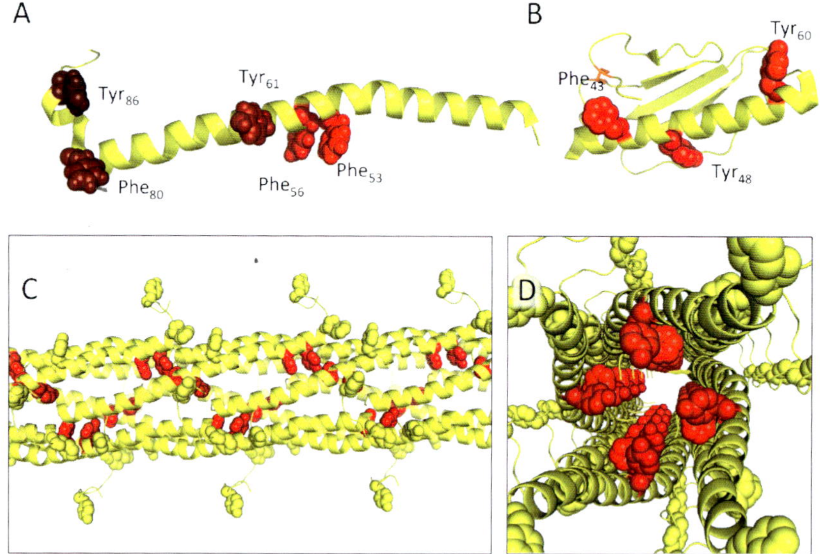

G.F. White *et al.*, Fig. 8 The structures and configuration of pilin associated with mineral-reducing bacteria. (A) The *G. sulfurreducens* PilA structure (Reardon & Mueller, 2013, PDB ID: 2M7G) with aromatic residues implicated in electron transfer shown as *spheres*. (B) PilA structure from *S. oneidensis* (Gorgel et al., 2015, PDB ID: 4D40) with aromatic residues associated with the pilin centre shown as *spheres*. (C) Side view of proposed assembly of conductive pilin based on homology modelling using *P. aeruginosa* pilin assembly as a template. Aromatic residues shown as *spheres* with proposed stacking residues highlighted in *red*. (D) End-on view of *G. pilin* assembly. *Structural coordinates for pilin assembly obtained from Malvankar et al. (2015).*